Databasing the Brain

DATABASING THE BRAIN

From Data to Knowledge (Neuroinformatics)

Edited by

STEPHEN H. KOSLOW, PH.D.

SHANKAR SUBRAMANIAM, PH.D.

WILEY-LISS

A John Wiley & Sons, Inc., Publication

Cover Illustration: Local and global coordinate systems for describing different levels of localization in the rat brain. (See Chapter 21)

Contents

Preface

In the Fourth Century B.C., Aristotle considered the brain to be a secondary organ that served as a cooling agent for the heart and a place in which spirit circulated freely. By the First Century A.D., Alexandrian anatomists, such as Rufus of Ephesus, had provided a physical description of the brain. Basic structures such as the *pia mater* and *dura mater* (the soft and hard layers encasing the brain) were identified in addition to the basic divisions of the brain itself. Building upon this research in the next century, the Roman physician Galen concluded that mental activity occurred in the brain rather than the heart, as Aristotle had suggested. His observations of the effect of brain injuries on mental activity formed an important practical basis for his conclusions. In the Middle Ages, the anatomy of the brain had consolidated around three principle divisions, or "cells," which were eventually called ventricles. Each cell localized the site of different mental activity. Traditionally imagination was located in the anterior ventricle, memory in the posterior ventricle, and reason located in between. Rene Descartes (1596–1650) paraphrased the intense interest in the human brain and its connection to the human condition by saying, "cogito, ergo sum" (I think, therefore I am). Following Descartes, in 1669 Nicolaus Steno hailed, "The brain, the masterpiece of creation, is almost unknown to us."

During the Renaissance period, Leonardo da Vinci drew and dissected the brain and his images were considerably more anatomical. He began to examine the relationship between the brain and the olfactory and optical nerves through experimenting with wax injections that helped him to model the ventricles. He sketched the brain from many different perspectives, looking closely at the ventricles and the origins of the nerves in the medulla.

In more recent times, one of the first great neuroscientists, Santiago Ramon y Cajal said in his book, *Recollections of My Life*, (1937), **"To know the brain ... is equivalent to ascertaining the material course of thought and will, to discovering the intimate history of life in its perpetual duel with external forces."** Francis H.C. Crick wrote that **there is no scientific study more vital to man than the study of his own brain. Our entire view of the universe depends on it**. (*Scientific American*, September, 1979)

As we begin the Twenty-First century, we realize that we have witnessed a dramatic change in our understanding of the intricacies of the brain through the advent of innovative technologies. The challenge now is to organize and integrate this knowledge in a manner which enhances our understanding of the intact brain. Fortuitously, in this same recent time period, we have had a revolution in the field of Information Technology (IT). Using IT, it will be possible to preserve, share, and integrate the wealth of nervous system data being generated from the most granular levels of the gene and cellular signaling to the most integrative behavioral data. The field of *neuroinformatics*, which is creating databases and the associated tools, is just in its infancy and will usher in the ability of neuroscientists to take a "discovery" approach to understanding the nervous system, similar to that which has occurred in the field of genomics. Neuroscientists will now have the ability to do both hypothese-driven research and discovery research. Each will accelerate our understanding of brain structure and function.

The Human Brain Project, initiated more than a decade ago, paralleled the human genome project to enable going from genome to cognition with a clear understanding of the daunting challenges faced in organizing the information at various levels and achieving interoperability among them. From the genome, we obtain, for the first time, an ability to map genotype to phenotype in the brain, where specific diseases are associated with changes in gene structure, expression, or regulation. From advances in imaging methods, we are able to obtain detailed maps of brain regions and in some cases to associate the regions with neuronal function. Considerable advances have also been made in the computation of neuronal processes at the molecular and cellular levels using myriad experimental measurements.

The largest challenge in our understanding and computation of the brain has, however, been the large chasm between neuroscience and computational science. Neuroscience was founded on very detailed and descriptive physiology. Computer Science has evolved from mathematics, physics, engineering, and information science and its applications have been restricted until a few decades ago to hard quantitative sciences. Structuring the vast amounts of information in neuroscience, relating seemingly diverse types of data and information, transforming data to knowledge in neuroscience, and obtaining functional understanding of the brain are some of the most important tasks at the interface of neuroscience and computer science. In this book, authored by eminent scientists of our century who have pioneered new developments in the interface areas, we present the synthesis of neuroscience and computer science and lay the foundations for the future of brain research. The book is divided conceptually into three parts which reflect this thinking. In the Part I, Computer Science Meets Neuroscience, we present the foundation for the application of computer science for building databases, tools and workflow environments in neuroscience. In Part II, Systems Approaches, we present an analysis of neuronal systems in a quantitative context and in Part III, we provide specific applications to neuroscience that reflect the present and future information infrastructures for the researcher.

Neuroscientists measure and collect vast amounts of data that range from molecular to behavioral. As a first step, this data needs to be organized and stored. In order to organize the data, we need unequivocal ontologies that describe the stored objects and their interrelationships. In neuroscience the anatomical context is essential to frame the relationship of diverse data. Most of the measured data are contextual to regions in the brain; much like climate data needs to be mapped to a geographical region. It is essential to have detailed atlases of brains of each animal based on high-resolution, histological images. In Part I, the first two chapters describe the underlying structure of objects and provide the basis for their relationship to brain anatomy. Bloom and Young analyze neuroscience data entities in a hierarchical format, listing molecule, cellular, circuitry, systems, behavioral and cognitive data as types that require structured terminology in order to be organized into four levels: raw data, interpretations of the raw data, consensus understanding, and encyclopedic information that introduces the contextual understanding of the knowledge. They discuss their prototype database, "the brainbrowser" in detail emphasizing schema and atlases for the above data types. Bowden and Duback provide a systematic nomenclature based on their pioneering work on primate brains. They argue that brain atlases are essential tools of neuroinformatics since every point in the brain is identified unambiguously within a structure and only one structure at the finest level of segmentation. Further, the anatomical atlas provides the template for interoperability between different types of data and relates the data to physiological function. Bowden and Duback, in their conclusions, argue the need for robust ontologies that will become the *lingua franca* of neuroscience Web portals to knowledge.

Ontologies provide the scaffold on which databases are built. However, databases have the content that need to be appropriately connected, stored, and accessed. What are the desirable features for neuroscience databases? This question is explored in detail by Gardner and colleagues in the chapter on the role, design, and use of databases. Gardner argues for complementing publications in more standard forms such as journal articles with structured and portable databases that contain both data and meta data to aid analyses by the research community. Accessing complex heterogeneous data in multiple formats has been a great challenge for the research community and Gardner argues for standards for exporting data in formats like XML. This is the only method to achieve interoperability and they demonstrate the concepts through a description of BrainML developed by them for exporting data and metadata. Another problem with research databases is their transience. It is common for data and databases and Websites to disappear either through negligence or due to the excessive cost of maintenance. Moore in his chapter describes the computer science issues associated with persistent collections. How can persistent collections of data and databases be appraised, accessed, organized, described, preserved, and maintained? Moore provides the framework for persistence of collections and demonstrates this through numerous examples from the sciences. Storage Resource Broker, developed by Moore and his collaborators, serves as the community standard for collections of databases and implements the persistent collection capabilities that are necessary for long sustenance of databases.

Scientific databases can be free format (known as flatfile databases), relational, or object-oriented. In some cases they can be combinations of the above types of databases. What is good for one type of data may not be the best format for another. In the following chapter, Miller and his colleagues argue for a new strategy for databasing neuroscience data. They argue that while flat files have the scope and ease for describing data in linguistically rich and nuanced ways scientists are accustomed, to which relational databases are highly structured and tend to be limiting when it comes to describing evolving data. They propose an intermediate solution that has been implemented in their SenseLab databases. The-Entity-Attribute-Value (EAV) approach provides enormous flexibility with evolving data and allows easy querying of complex heterogeneous data. Miller and colleagues also describe the difficulty in building interfaces to query neuroscience databases.

The chapter by Foster and another by Martone and colleagues, describe the emerging paradigm of Grid computing. Foster, one of the pioneers of Grid computing, describes data grids, their accession and relationship to compute grids. Foster argues the inevitability of heterogeneous data repository locations and the need to have cross-access for querying and computing. Arguably, the best methods for accessing and computing such distributed data is to use the emerging Grid technology. Globus was developed by Foster and his colleagues for grid computing and is extensible to data grids; Foster provides a detailed discussion of this extensibility. A large scale neuroscience project utilizing the Grid, called BIRN (Bioinformatics Research Network) is the subject of the chapter by Martone and her collaborators. They introduce the telescience portal, a revolutionary grid-based resource for electron tomography. The cell-centered database and the mediator enable seamless querying of data across the database. Martone provides the rationale for such efforts that promote interoperability in neuroscience through data federation.

Humans relate to the world through sight. Visualization is the first cognitive entrée into our processing these information about the world. While this is often taken for granted, the computer science aspects of visualization that allow transformation of large dimensional data into something a brain can process more readily merits detailed understanding. Cox straddles the world between art and the science of visualization and provides a very detailed perspective on visualizing multidimensional data; she describes how to transition between different dimensional data, dimension reduction, as well as reconstruction into higher dimensions, and provides an array of applications demonstrating these the concepts. Part I concludes with a detailed analysis of the construction and deployment of workbenches in biology. Maer and colleagues describe detailed architecture of workbenches, how to combine databases, interfaces, and tools into a single environment, and how Web portals serve as entry points into a very complex world of databases and computation. They illustrate workbenches from different life-science disciplines and argue for the need for improved construction of such infrastructures for neuroscience.

Part II deals with systems approaches to neuroscience brings forth the quantitative aspects of neuroscience. Genomes provide the framework for understanding cellular function. However study of neuronal cells is enormously complicated since the brain functions as a highly connected system. This further highlights the difficulties in processing data between several levels of function only some of which is hierarchical. The expert authors in this part describe genomic, cellular, and morphological systems of brain function.

Each cell type in an animal expresses a distinct repertoire of genes. Specialized and distinct function of cells and tissues is a consequence of the changed gene structure. While a major part of the cellular and tissue function involves post-transcriptional processes, the expressed genes and changes in their expression under given input conditions serve as a metric to the state of the cell and its underlying function. In the first chapter of this section, Hovatta and Barlow study complex neurological diseases and behavior through gene expression profiling in murine models and human brain tissue. They have been pioneers in using new technological advances in gene expression for studying brain function and diseases and highlight the importance of these methods for modern neuroscience. Besides providing a comprehensive review of the technology and data analysis methods, Hovatta and Barlow link traits from gene expression profiling changes to complex neuronal diseases in Alzheimer's and Schizophrenia patients. In the following chapter, McClung and Nestler, explore synaptic activity dependent changes in gene expression responsible for long lasting neuronal adaptations. They relate the changes to cell growth and differentiation for studying learning and memory and complex neuronal diseases. These two chapters bridge genome science and neuroscience and emphasize the link provided by informatics and data analysis.

This is followed by two chapters reinforcing the links among genome science, molecular biology, and neuroscience. Neuronal processes are mediated by and are a consequence of myriad signaling phenomena. Iyengar and colleagues describe the role of cellular signaling networks in long-term potentiation and memory. This chapter serves as a state-of-the-art review of the organizing principles of the post-synaptic long-term potentiation (LTP) signaling networks, which include signal transduction, modules, and cross-talk in signaling networks, and the role of feedback in modulating the flow of signal in the circuitry of the cell. The rationale for systemic and quantitative modeling of the signaling networks is also provided. The chapter by Kandel and colleagues introduces the use of transgenic methods for investigation of LTP and systematically dissects the connectivity of the network through designed knockouts. Using the phenotypic response, they connect behavioral properties such as learned fear to the amygdala and visual cortex circuitry and their signaling systems. Kandel and colleagues convey the promise of these methods for future research in neuroscience.

The following two chapters describe the software engineering aspects associated with systemic description of cellular circuitry and morphology. Representation of cellular circuits, organizing the information in the pathways represented by the circuitry and functional analysis of the networks is a complex task. Rzhetsky and colleagues describe an integrated software system for extracting, visualizing, analyzing, and integrating molecular pathway data. Rzhetsky's "GeneWays" system provides insights into the type of infrastructure neuroscience will require for cellular data analysis and visualization. In their chapter, Turner and collaborators describe the challenges associated with describing and presenting cellular brain morphology. For instance, they pose: What is the relationship of neuron appearance and structure to dendritic functions? Current approaches to mapping such relationships require sophisticated mathematical modeling

when vast amount of measured data are available. This chapter describes the type of data sets needed for describing morphology and defines the guidelines for organizing these data sets.

The last three chapters in Part II serve as seminal reviews on systems level modeling of cellular and neuronal entities. Winslow provides a thorough and lucid review of strategies for modeling excitable cells using the myocyte as an exemplar. The cellular networks are parsed through interacting components, which are then cast into a mathematical formulation and analyzed for emergent properties. This integrative modeling approach serves as a paradigm for cellular computation in excitable cells. Nelson's chapter provides a historical development perspective of mathematical models for electrophysiology. Beginning with the well-known Hodgkin–Huxley model, Nelson provides a perspective on individual channel, synaptic, multicompartment, and network models for studying electrophysiology. The Genesis software, of which Nelson is a developer, provides the neuroscientist easy access to modeling electrophysiology. The last chapter by Donohue and Ascoli provides a survey of computational models of neurite growth. This has the potential to provide insights into developmental biology associated with the brain and aid in studying brain morphology.

Part III of this book deals with applications of databases and models to exemplar problems in neuroscience. The experts who have written these chapters have been leaders in bridging computational and neuroscience approaches, and serve as the harbingers of the future of neuroscience research. The applications are in the critical neuroscience research areas, the methods employed are sophisticated by computer science standards and resulting synthesis paves the way for future applications.

Markram and his colleagues describe, in the first chapter of Part III, the Neocortical Microcircuit Database (NMDB), its structure and organization, and user access to the database. They also point out the need for community tools and interfaces. A neuroscientist will be able to submit data to or download data from the NMDB and work interactively with the data to edit and analyze it. In the following chapter, Shepherd and coworkers describe the design and challenges associated with the construction of SenseLab, a collection of multilevel neuronal databases and tools.

Bjaalie and Leergaard's chapter provides an intimate integration between very complex neuroscience and computer science. The process of traversing the knowledge path from animal brain images to wiring patterns is arguably the most challenging task for a practicing neuroscientist and this chapter provides the details of this process. Starting from tracing methods and the data obtained from tracing, Bjaalie and Leergaard describe the reconstruction of 3D models, methods for traversing local and global coordinate systems, and integrating data, models, and tools into a single integrated infrastructure. Next, Van Essen and colleagues provide a very detailed exposition of linking surface-based atlases to actual data from individual brains and provide a thorough discussion of data management and modeling issues associated with surface-based analysis. These two chapters serve as excellent introductions into the world of data and analysis for experimental neuroscientists who are engaged in detailed imaging of brain structure and function.

The chapters by Mazziota, and Toga and colleagues provide mapping between real data on individual human subjects and brain atlases and also provide mapping for normal and diseased populations taking into account individual variations in structure and morphology through probabilistic descriptions. Mazziota describes projects that create databases from raw data available to the community, use targeted data obtained for probabilistic atlas descriptions and databases and atlases constructed from only derived data. Toga and his colleagues go into details on brain asymmetry, gender, and other factors in describing the probabilistic atlas of the brain in normal and diseased populations. This chapter also illustrates the role of databases and atlases in studying brain pathology and provides the rationale for systematic mapping of real data to neuroanatomy.

Neuroimaging has served as the most important experimental window into the brain. In the chapters by Smith, and Van Horn and Gazzaniga, the interface between neuroimaging data and neuroscience is explored in detail. Smith emphasizes the great importance of metadata in sharing neuroimage data, the need for data assurance through metadata, and the risks associated with incorrect implementation modeling of the data. Van Horn and Gazzaniga also emphasize the importance of metadata and describe in detail the role of sharing and archiving data in maximizing information for studying human cognition.

For several years, we, as editors of this volume, have been engaged in discussions on the importance of the interface among computer science, biocomputation, and neuroscience.

Our communities are slowly embracing the realization of importance of this integration with the largest barrier being the "cultures" problem. Nevertheless, we are aware of the leading-edge neuroscientists who have pioneered the integration of these disciplines and have the ability to educate the community incisively. This was the *leit motif* for this volume. The authors have been gracious and we have learned immensely from reading the content of the contributions. We believe that this volume will be of great value to practitioners of neuroscience research and we hope it will motivate a large number of students to embrace interdisciplinary paradigms. The systems research approach will serve as the final frontier in our exploration of the human brain.

We are grateful to all the authors who have been exceptionally helpful and patient and the outstanding nature of this volume is due to their excellent contributions and hard work. We feel fortunate to serve as editors of this volume.

STEPHEN H. KOSLOW
SHANKAR SUBRAMANIAM

Bethesda, Maryland
San Diego, California
December 2004

Contributors

Michael Abato   Laboratory of Neuroinformatics, Department of Physiology and Biophysics, Weill Medical College of Cornell University, New York, New York

Giorgio A. Ascoli   Krasnow Institute and Psychology Department, George Mason University, Fairfax, Virginia

Carrolee Barlow   The Salk Institute for Biological Studies, Laboratory of Genetics and Merck Research Laboratories, San Diego, California

Christopher J. Beaver   Departments of Surgery (Neurosurgery) and Neurobiology, Duke University Medical Center, Durham, North Carolina

Jan G. Bjaalie   Neural Systems and Graphics Computing Laboratory, Center for Molecular Biology and Neuroscience, and Department of Anatomy, University of Oslo, Oslo, Norway

Robert D. Blitzer   Departments of Pharmacology, Biological Chemistry, and Psychiatry, Mount Sinai School of Medicine, New York, New York, and Psychiatry Service, Bronx VA Medical Center, Bronx, New York

Floyd E. Bloom   Neurome, Inc., La Jolla, California

Douglas M. Bowden   National Primate Research Center, University of Washington, Seattle, Washington

Robert C. Cannon   Institute of Adaptive and Neural Computation Division of Informatics, Edinburgh, United Kingdom

Donna J. Cox   National Center for Supercomputing Applications, School of Art and Design, University of Illinois at Urbana-Champaign, Champaign, Illinois

Chiquito J. Crasto   Center for Medical Informatics, Yale University School of Medicine, New Haven, Connecticut

James Dickson   Department of Anatomy and Neurobiology, Washington University School of Medicine, St. Louis, Missouri

Duncan E. Donohue   Krasnow Institute and Psychology Department, George Mason University, Fairfax, Virginia

Mark Dubach   National Primate Research Center, University of Washington, Seattle, Washington

Pablo Ariel Duboue   Department of Computer Science, Columbia University, New York, New York

Mark H. Ellisman   National Center for Microscopy and Imaging Research, Department of Neurosciences, University of California, San Diego, San Diego, California

Amit Etkin   Center for Neurobiology, Columbia University, New York, New York

Ian Foster   Department of Mathematics and Computer Science, Argonne National Laboratory, Argonne, Illinois

Carol Friedman   Department of Biomedical Informatics, Columbia University, New York, New York

Daniel Gardner   Laboratory of Neuroinformatics, Department of Physiology and Biophysics, Weill Medical College of Cornell University, New York, New York

Michael S. Gazzaniga   The fMRI Data Center, Dartmouth Brain Imaging Center, Center for Cognitive Neuroscience, Dartmouth College, Hanover, New Hampshire

Elizabeth A. Grace   Departments of Pharmacology and Biological Chemistry, Mount Sinai School of Medicine, New York, New York

Anirudh Gupta   Brain Mind Institute, Lausanne, Switzerland

Donna Hanlon   Department of Anatomy and Neurobiology, Washington University School of Medicine, St. Louis, Missouri

John Harwell   Department of Anatomy and Neurobiology, Washington University School of Medicine, St. Louis, Missouri

Vasileios Hatzivassiloglou   Department of Computer Science, Columbia University, New York, New York

Michael L. Hines   Section on Neurobiology, Yale University School of Medicine, New Haven, Connecticut

Iiris Hovatta   The Salk Institute for Biological Studies, Laboratory of Genetics and Merck Research Laboratories, San Diego, California

Ivan Iossifov   Columbia Genome Center, Department of Biomedical Informatics, Columbia University, New York, New York

Ravi Iyengar   Departments of Pharmacology and Biological Chemistry, Mount Sinai School of Medicine, New York, New York

Eric R. Kandel   Center for Neurobiology and Behavior and Howard Hughes Medical Institute, Columbia University, College of Physicians and Surgeons, New York, New York

Kevin H. Knuth   Laboratory of Neuroinformatics, Department of Physiology and Biophysics, Weill Medical College of Cornell University, New York, New York

Tomohiro Koike   Hitachi Software Engineering Co., Ltd., Yokohama, Japan

Stephen H. Koslow   National Institute of Mental Health, National Institutes of Health, Bethesda Maryland (formerly) and Allen Institute for Brain Science, Seattle, Washington

Pauline Kra   Department of Biomedical Informatics, Columbia University, New York, New York

Michael Krauthammer   Columbia Genome Center, Columbia University, New York, New York

Emmanuel M. Landau   Departments of Pharmacology, Biological Chemistry, and Psychiatry, Mount Sinai School of Medicine, New York, New York, and Psychiatry Service, Bronx VA Medical Center, Bronx, New York

Trygve B. Leergaard   Neural Systems and Graphics Computing Laboratory, Center for Molecular Biology and Neuroscience, and Department of Anatomy, University of Oslo, Oslo, Norway

Jian Liu   Section on Neurobiology, Yale University School of Medicine, New Haven, Connecticut

Nian Liu   Section on Neurobiology, Yale University School of Medicine, New Haven, Connecticut

Xiaozhong Luo   Brain Mind Institute, Lausanne, Switzerland

Andreia Maer   San Diego Supercomputer Center and the Departments of Bioengineering and Chemistry and Biochemistry, University of California San Diego, La Jolla, California

Luis Marenco   Center for Medical Informatics, Yale University School of Medicine, New Haven, Connecticut

Henry Markram   Brain Mind Institute, Lausanne, Switzerland

Maryann E. Martone   National Center for Microscopy and Imaging Research, Department of Neurosciences, University of California, San Diego, San Diego, California

John Mazziotta   Department of Neurology and Brain Mapping Center, University of California Los Angeles, Los Angeles, California

Colleen A. McClung   Department of Psychiatry and Center for Basic Neuroscience, University of Texas Southwestern Medical Center, Dallas, Texas

Michele Migliore   Section on Neurobiology, Yale University School of Medicine, New Haven, Connecticut

Perry L. Miller   Center for Medical Informatics, Yale University School of Medicine, New Haven, Connecticut

Reagan W. Moore   San Diego Supercomputer Center, University of California San Diego, La Jolla, California

Mitzi Morris   Columbia Genome Center, Columbia University, New York, New York

Thomas M. Morse   Section on Neurobiology, Yale University School of Medicine, New Haven, Connecticut

Prakash M. Nadkarni   Center for Medical Informatics, Yale University School of Medicine, New Haven, Connecticut

Katherine L. Narr   Laboratory of Neuro Imaging, Department of Neurology, University of California Los Angeles School of Medicine, Los Angeles, California

Mark E. Nelson   Department of Molecular and Integrative Physiology, University of Illinois at Urbana-Champaign, Urbana, Illinois

Eric J. Nestler   Department of Psychiatry and Center for Basic Neuroscience, University of Texas Southwestern Medical Center, Dallas, Texas

Steven T. Peltier   National Center for Microscopy and Imaging Research, Department of Neurosciences, University of California, San Diego, San Diego, California

Christopher J. Pittenger   Center for Neurobiology, Columbia University, New York, New York

Adrian Robert   Laboratory of Neuroinformatics, Department of Physiology and Biophysics, Weill Medical College of Cornell University, New York, New York

Andrey Rzhetsky   Columbia Genome Center, Department of Biomedical Informatics, and Center for Computational Biology and Bioinformatics, Columbia University, New York, New York

Brian Saunders   San Diego Supercomputer Center and the Departments of Bioengineering and Chemistry and Biochemistry, University of California San Diego, La Jolla, California

Gordon M. Shepherd   Center for Neurobiology, Yale University School of Medicine, New Haven, Connecticut

Gleb Shumyatsky   Center for Neurobiology, Columbia University, New York, New York

Gilad Silberberg Brain Mind Institute, Lausanne, Switzerland

Kenneth Smith MITRE Corporation, McLean, Virginia

Elizabeth R. Sowell Laboratory of Neuro Imaging, Department of Neurology, University of California Los Angeles School of Medicine, Los Angeles, California

Shankar Subramaniam San Diego Supercomputer Center and the Departments of Bioengineering and Chemistry and Biochemistry, University of California San Diego, La Jolla, California

Paul M. Thompson Laboratory of Neuro Imaging, Department of Neurology, University of California Los Angeles School of Medicine, Los Angeles, California

Arthur W. Toga Laboratory of Neuro Imaging, Department of Neurology, University of California Los Angeles School of Medicine, Los Angeles, California

Maria Toledo-Rodriguez Brain Mind Institute, Lausanne, Switzerland

Dennis A. Turner Departments of Surgery (Neurosurgery) and Neurobiology, Duke University Medical Center, Durham, North Carolina

Roger Unwin San Diego Supercomputer Center and the Departments of Bioengineering and Chemistry and Biochemistry, University of California San Diego, La Jolla, California

David C. Van Essen Department of Anatomy and Neurobiology, Washington University School of Medicine, St. Louis, Missouri

John Darrell Van Horn The fMRI Data Center, Dartmouth Brain Imaging Center, Center for Cognitive Neuroscience, Dartmouth College, Hanover, New Hampshire

Wubin Weng Department of Computer Science, Columbia University, New York, New York

John W. Wilbur National Center for Biotechnology Informatics, National Library of Medicine, National Institutes of Health, Bethesda, Maryland

Raimond L. Winslow Center for Cardiovascular Bioinformatics and Modeling, The Johns Hopkins University School of Medicine and Whiting School of Medicine, Baltimore, Maryland

Warren G. Young Neurome, Inc., La Jolla, California

Hong Yu Department of Computer Science, Columbia University, New York, New York

COMPUTER SCIENCE
MEETS NEUROSCIENCE

PART

I

Databasing the Brain. Edited by Stephen H. Koslow and Shankar Subramaniam
ISBN 0-471-30921-4 © 2005 John Wiley & Sons, Inc.

Database Needs of Neuroscience: Schema and Design

Floyd E. Bloom, M.D. and *Warren G. Young, Ph.D.*

CONTENTS

Databasing the Brain. Edited by Stephen H. Koslow and Shankar Subramaniam
ISBN 0-471-30921-4 © 2005 John Wiley & Sons, Inc.

1.1 INTRODUCTION

Over the last decade, many initiatives have attempted to address the ongoing issues in the overlapping disciplines of neuroscience and computer sciences (NIMH Human Brain Project—Huerta et al., 1993; Koslow and Huerta, 1997) (NIH Brain Molecular Anatomy Project—Gibbons, 1999). New programs have been started, current ones enhanced, and groups called to meet and share their experiences and look for potential areas of collaboration and seeding of new ideas. Many of these are taken from a research perspective, sometimes as a highly focused effort on behalf of a research group's ongoing program. Very few discuss (a) the challenges facing an entire company as a whole and (b) the efforts to integrate both research and business needs with a uniform solution.

Neurome, Inc. (La Jolla, CA) is a neuroinformatics-focused murine-centric biotechnology company founded by current and formerly academic neuroscientists. Here the opportunity existed to create a completely new information system to address the needs of the firm's research groups as well as the important needs of the company's business objectives. At Neurome, we have the responsibility of addressing the needs of our clients, maintaining their data securely and independently, but at the same time increasing the value of the data by providing insight and knowledge by comparison with other known data in the research knowledgebase of both published and internal information. Business efforts are tied into the data, from the raw data that emerges from the laboratory instrumentation, to the derived interpretations that are delivered as papers, reports, and internal presentations. New areas of concern emerge when the entire corporate entity needs to be addressed. The integration of business and research as a unified solution for investors, clients, institutions, and academic collaborators required a different way of thinking and handling of data, as well as consideration for the time-sensitive phasing of the development of the software to implement this solution.

On the research side, we have developed means to integrate images from microscopes into a database backend. Each image is a cross section of a mouse brain and ranges in size from 30 MB to 24 GB. Each of the images is one part of a three-dimensional (3D) model containing up to 200 such images. The brain models are then segmented in two-dimensional (2D) and 3D into smaller structures representing known brain regions. Both the core research and the needs of the clients determine the extent and list of brain regions that we analyze. Additional research adds valuable annotations to these structural frameworks. The experimental data ranges from whole brain models, to brain regions, to nuclei and even to the level of the cell or subcellular component.

The huge volumes of data from research and business need to be integrated back into an easy-to-access system that serves three main purposes: (1) generating reports for the clients; (2) contributing to the core knowledgebase at Neurome for internal research programs; and (3) leveraging the business units to attract clients (big pharma and biotech) and financiers (investors, venture capitalists, and financial institutions).

To address the large volumes of data that we expect to acquire for research and business, we focused on designing database schema using industrial-strength and proven technology. This chapter will highlight some of the key concepts in the database effort, relate it to real world testing and tuning, and discuss some of the decisions we had to make in order to provide the best technological database backend as a model for new biotech companies in the emerging industry of commercialized neuroscience research.

1.2 BRAIN DATABASES

We have been addressing the issue of "brain databases" for decades now, both here at Neurome (previously in our academic laboratories) and elsewhere across the globe (Bowden and Martin, 1988, 1995; Bloom and Young, 1994, 1999, 2000; Bloom et al., 1997). Questions such as "What are the primary data of morphologic research?" and "How are observations replicated?" are continuously asked. To a microscopist, largely trained through the centuries as a subjective observer of events, the compiled observations need to be converted into a form that is presentable to colleagues, but in a format that is readily comparable to the work of others. The scientific method requires that these observations be verified by replication in order to confirm or reinterpret the conclusions of a study. Modern approaches apply rigorous methods for acquiring quantitative morphological information, provide a much more realistic and reasonable basis for the delivery of the conclusions of the observations, and support clean methods of rigorously and quantitatively comparing data among the observers. As natural outgrowths of applying technology to many of the disciplines of the neurosciences, more sophisticated methods of data collection, data reduction, and the development of information as well as knowledge from these data begins to emerge (Bloom and Young, 1994, 1999; Bloom, 1995).

A brain atlas is a research tool that neuroscientists regard as a graphical database in which the database entities are "places" listed in an index that refer to a given "plate" or series of drawings in one, two, or all three of the standard planes of section. If you already know the name of the brain structure whose location you want to see, this database is highly useful. It is not useful if the name does not exist in the database table, or if you don't know what structure you are looking at in the microscope or schematic. One of the biggest problems in neuroanatomy is the lack of standards in nomenclature, or perhaps the bigger problem of scientists not wanting to adhere to any emerging standard. As Bowden and Dubach point out in their chapter elsewhere in this volume and in previous publications (Bowden and Martin, 1988, 1995), most information in neuroscientific databases is referenced to specific brain structures through shared, but not yet interchangeable nor architecturally compatible, taxonomies. Through these implicit terminological architectures, neuroscientists record and communicate information about discrete structures by their consensus names (nomenclature) and by visual images (the boundary coordinates of the structures in species-specific brain atlases). They (in this volume) describe the current status of attempts to (a) adapt these neuroscientific databases for interoperability via the World Wide Web and (b) adapt their requirements to attain the goal of full interoperability among large confederated databases.

Atlas users acknowledge other recurring problems: how to relate the schematics and images in the atlas to the strain, age, and sex of the animal they are using. How to create a new view that more closely matches the plane or angle of section that is being analyzed. How to get deeper information than is available from the standard atlas, such as learning more about a specific brain location, its cells, transmitters, connections, and possible functions. In most cases, a separate external reference system has to be used. There are efforts underway that try to merge electronically delivered atlases with annotations and other referential material (Martin and Bowden, 1996; Williams, 2003).

1.3 TYPES OF DATA

Three basic terms need to be discussed when talking about databases, whether they are databases on nervous systems or nucleic acid sequences: data, information, and knowledge. Data assembled with the proper concept could be interpreted as information, which, when validated and extended, can contribute a level of understanding that is called knowledge. A database needs to manage the data so that the entities can be reassembled along demonstrated conceptual relationships. Knowledge expands from this when the relationships suggest conclusions that can be verified by new experiments. Few scientists can be expert in every area in which they need to extract knowledge, and especially if this involves new relationships that they may not be aware of.

In the world of scholarly analysis, there are four levels of data to be collected and organized in a brain database: (1) raw data, such as from experiments gathered and written in a laboratory notebook (Pittendrigh and Jacobs, 2002); (2) reports, in which the raw data have been reduced to consistent observations and combined with interpretations (Felleman and Van Essen, 1991; Healey et al., 1997; Skoufos et al., 1999, 2000; Crasto et al., 2002); (3) knowledge, in which experts have collected, compared, reduced, and interpreted the reports into a consensus (Shepherd et al., 1997, 1998); and (4) encyclopedic, in which the main principles and characteristics are stated for introducing others to the topic (Bloom et al., 1989; Bloom and Young, 1993).

The issue of data structures begins to take on more relevance when designing database systems that can accommodate all four levels of data under control by user selection. As the database becomes increasingly populated with data and as the query tools become more powerful to go from level 1 to level 4, new neuroscientists can draw reliable conclusions and extract answers from heuristic questions. Conclusions based on the raw data can be compared at the report level within the community. The database can combine reports together to create a sharing of them as knowledge, offering a means to bring in new reports on a comparative level and see if interpretations of the data follow a norm or whether departures from emerging norms deserve follow up and replication. Finally, the database can interact and share data with other data systems, such as those from publishers, or other data systems from other companies, universities, or laboratories, and gate data back and forth based on known tabular elements. The interface to such an amalgamation of data systems would function encyclopedically, providing key facts and concepts, links to other knowledge bases, to the reports contained within, and finally to the raw data itself in hand or to be collected.

1.3.1 Neuroscience Data Entities

Nevertheless, consensus in terminology is required in order to begin to implement a database that can attempt to address the four levels of data outlined above. All major forms of brain structural research needs to be considered—from noninvasive imaging of the whole brain (by metabolic, blood flow, proton density, or hemoglobin oxygenation), to functional assessments by electrophysiology (from EEG recordings to patch-clamp single-ion channel analyses) to conventional microscopy (light, laser, two-photon, and electron). The types of data that are acquired from these forms of research are considered multimodal—graphics, text, video, audio, and so on—and cover the life science domains of the animal's anatomy, biochemistry, physiology, pharmacology, and behavior and must be categorized by species, and in some cases strain contextual constraints. Some of the discrete aspects that need to be considered are as follows:

- *Molecular.* Data on neurotransmitters, their synthetic and catabolic enzymes, receptors, and transductive mechanisms; their regulated ion channels; the sites and molecules to which drugs bind; other neuron-restricted molecular markers for structural, secretory, and metabolic functions (e.g., all those properties recognized by gene markers or monoclonal antibodies).
- *Cellular.* Attributes critical in categorizing circuitry to or from specific cells in specific locations would include their functional classes, sizes, shapes, locations, dendritic and axonal arbors, and nuclear or laminar locations; their developmental stage of origin; their survival during senescence.
- *Circuitry.* Afferent and efferent connections between nuclear or laminar neuronal locations, their trajectories, their density; the forms of synaptic connections made and their relative numbers; the degree to which the projections are reciprocal, topographic, collateralized, or specific; the methods by which the pathways were defined; the "functions" of such circuitry (excitatory, inhibitory, or conditional), and the transmitter(s) and receptor(s) attributable to it.
- *Systems.* The functional systems (e.g., sensory, motor, endocrine, visceral regulatory, emotion) to which a given set of cells and circuits has been attributed by physiological or behavioral experiments; the methods used.
- *Behavioral.* Relationships between nuclei or cortical regions and their circuits with distinctive forms and stages in behaviors, or of lesions that incapacitate certain forms of behaviors.
- *Cognitive/Emotional.* Using noninvasive structural or functional imaging of the human subject performing higher-order mental tasks, the sites to which such operations can be attributed; the alterations in specific pathologic states which may be shown to incapacitate such higher-order functions.

1.4 A PROTOTYPE DATABASE

We developed a prototype brain database a few years ago called the "Brain Browser™" (Bloom et al., 1989; Bloom and Young, 1993). The initial goal was to create an electronic system to manage atlas oriented data for the rodent brain and to include the entities described above. These entities were expressed in a database that could provide links or relationships among themselves. The basic data structure in the Brain Browser™ were a series of atlas plates derived from a commonly available hardcopy form of a rat brain atlas (Paxinos and Watson, 1986), onto which software tools were designed to permit the fast location of up to 400 known "places" in the rat brain or more specifically a 180-g female Sprague–Dawley rat. Because the location of the places were known, relationships could be expressed between two or more places, to define "circuits" of places that expressed the then-known qualitative connectivity patterns in the rat brain. Finally, all of the data entities described above could also be entered for any newly studied place in the database. The idea was to allow the experimentalist to use this system as a template for the inclusion of raw data, from which could be derived reports from experiments, leading to additional knowledge with summaries from a more global perspective and, finally, to community data-sharing (Bloom et al., 1997; Bloom and Young, 2000). With the inclusion of reference information from the literature, the knowledge begins to emerge as encyclopedic.

We learned quite a bit from this exercise, including the resistance of established neuroscientists to taking on new software, climbing the learning curve, especially for systems that don't quite match their current needs. Nevertheless, the multimedia aspects, pseudo-relational and hierarchical architecture, and the ability to input new data and to share and search the data have yet to be approached by any subsequent neuroscience databases.

When we started developing a new system at Neurome, we recognized the value of what Brain Browser™ tried to accomplish. But we also understood that there were additional complex requirements needed for the business logic, and Brain Browser™ lacked state-of-the-art sophisticated viewing tools. In harvesting reviews of the literature to populate the database of circuitry in Brain Browser™, it also became clear that the literature was filled with structural degeneracy; that is, some authors report cell-to-cell circuitry across regions, while others refer to the data on a region-to-region basis, and with all levels of intermediate precision. Therefore we created a hierarchical outline of the structures of the rat brain, starting with the six classically defined, ontogenetically derived terms for the segments of the central nervous system, namely: spinal cord, medulla, ponscerebellum, midbrain, diencephalon, and telencephalon. Each of these major segments are subdivided into progressively smaller structurally defined divisions, until the lowest level is reached, namely the specified collections of neurons contained within a specific nucleus or layer; in some instances, these segments are subdivided into geographically restricted subsets or subsets of the subsets (ad infinitum) of these nuclei or layers. The hierarchical organization is useful to understand the relationships between the levels of structures encompassed by a given named structural entity. It is now possible to speak of structures at five general levels of the hierarchy:

1. The six levels of the neuraxis
2. Regions within levels
3. Cortical (cytoarchitectonic) areas and sets of nuclei in subcortical structures
4. Specific brain "places" defined as the subcortical nuclei or cortical laminae (as well as subsets of these when sufficiently specifiable)
5. Cell elements within these "places"

1.4.1 The Brain Place

One advantage of this hierarchical outline is its tolerance for degeneracy when adding to or searching for encoded structures. For example, scientists looking at the same structure can disagree on the boundaries of known places in the brain. This effect also reveals itself as an interindividual variation within strains and clearly presents itself among different species. One can liken this to looking at the cities from the sky above and trying to discern the city, county, and state borders. Imagine doing this for a 3D object. To prevent the endless dissection of structures into atomic units where you "can't see the forest for the trees," we have decided to settle on the level of the "place" as the most logical subdivision in which to begin to ascribe biochemical, physiological, pharmacological, and behavioral attributes. We have used this metaphor in Brain Browser™, where a place is operationally defined as "a circumscribed set of neurons whose location, name, and properties are defined in the database." The supra-unitary structures in which the basic unit is nested in the hierarchy build up from this level, to define areas, regions, and neuraxis levels. Smaller clusters of neurons within them, and the circuits to which they contribute, are then defined as a subset within this unit. By establishing *that the basic unit of the database is the smallest collection of neurons whose named location* is defined, an operational scheme is created. This scheme appears to serve usefully as a basis for each of the related entities, at each of the hierarchical levels of organization. The scheme also readily accommodates the addition of newly defined structural units, as well as forming a logical basis for the translation of named structural subunits from one species brain to another. Each place has attributes associated with it that can be compiled as to the neurons within the place, their morphological attributes, their known transmitters, receptors, and other molecular markers, their known developmental time of origin, and the functional supra-systems of the brain to which this place has been related.

1.4.2 Circuits among Places

As mentioned above, circuits are connections from one group of neurons in one place to the neurons in another place. The actual "connections" occur on a cell-to-cell level, but many of these connections define the more global aspect of a pathway or circuit. The interneuronal communication and the connections of chains of neuronal locations into functionally defined "systems" are more functionally meaningful aspects of brain structure. The circuit properties are symbolic of the major elements of the data explosion of the past three decades. Simply put, there are so many brain circuits being defined by the powerful and sensitive methods of orthograde and retrograde transport and immunohistochemistry that even highly experienced anatomists cannot keep them all accurately in mind. Hence, the anatomists may not be able to recognize immediately the importance of a newly reported afferent or efferent connections of a structure or the patterns of cells and fibers that express epitopes for a newly defined marker. New circuits, new transmitters, new forms of receptors, and new forms of synaptic signal transduction all require establishment to define the qualitative properties of a pathway. In addition, anticipating that genetic manipulations in transgenic mice may offer means to model some human neuropathologies, we have emphasized quantitative morphometric assessments and sought means to compare brains at all the above-noted levels of resolution within and across strains.

Our original rat-only Brain Browser™ had data on slightly less than a thousand brain circuits, which comprise the more-or-less classical pathways of neuroanatomy, based on relatively recent comprehensive review articles (Björklund et al., 1987), and often with references to the original report data indirectly referenced through these reviews. The circuitry data have the facility for defining the methods used to establish these circuits, and the user can personally rate the reliability or acceptability of these data for subsequent analysis if data conflicts arise. In many circuitry records, Brain Browser™ shows only the places that are linked, because the specific cells within those places that form the sources and target of the circuits were not yet known, and the same caveats held for transmitters, receptors, or their transductive mechanisms.

The current Neurome mouse-centered database system accommodates this notion of brain places and circuits, now tailored to our commercial situation in an industrial-strength approach. While we find that the hardware is faster and smaller, and the technologies slightly better in terms of development tools, delivery platforms, compatibility, and data rendering, the general concepts underlying the brain database are the same. There are still basic tenets to the types of entities that need to be implemented in the database schema. There are the same attributes that need to be stored for each entity. And there is still a lot of data out there that have yet to be stored in a competent, easy-to-use database system.

1.5 THE NEUROME TECHNOLOGIES

Neurome controls a series of proprietary technologies developed by the founders. The Neurome technologies are used to produce, collect, and integrate accurate 3D quantitative, spatial, and volumetric data on gene expression within the brain and to correlate that data with the developing wealth of learning on the architecture and functions of brain structures, circuits, and cells. The technologies include:

1. *MiceSlice™*. Defined workflows used throughout the laboratory for the handling, sectioning, collection, and histological staining of all brain tissue. Standard stereological principles are employed to carefully determine the minimum number of sections to analyze, while maintaining high efficiency and accurate reproducibility. This provides the foundation material for the development of standardized experimental protocols.

2. *NeuroZoom™*. Computerized microscope instrumentation control for precise extraction, normalization, and storage of quantitative data from microscope images of the brain.

Ultra-high-resolution digital brain sections are seamlessly assembled from thousands of microscope image tiles.

3. *BrainArchive™*. A comprehensive database of neuroinformation and an electronic brain "atlas" for archiving, integration, and comparison of brain structure and circuitry data, with the use of tools in morphometrics, stereology, and image processing.

4. *BrainPrints™*. Drawing upon the prodigious amounts of data in the Neurome database, digital profiles can be produced for comparison of quantitative, spatial, and volumetric data among different transgenic models of mice.

These technologies dictate a certain organization to the research effort at Neurome. The research units that contribute to overall data workflow and storage into Neurome's database are depicted in Figure 1.1.

The research internal units correspond to the Neurome technologies: MiceSlice™, NeuroZoom™, BrainArchive™, and BrainPrint™. They are considered internal units because they are inside the Neurome firewall and contribute freely to the overall data workflow. The internal units have complete access to all data in the workflow. Since these operate mostly within a firewall, security to such data is rigidly controlled.

The external units represent entities external to Neurome: collaborators (academic, commercial), contractors, and consumers. Collaborators are more search-oriented than contractors, who usually would be receiving our data and their interpretations under a contract for specific business goals. Both collaborators and contractors have bilateral access to the data flow. They can contribute as well as consume data. Consumers have a unidirectional access to data. While they generally take out data, they do not contribute any data to Neurome's database. External units have access to a subset of the data through the workflow. Since these units are external to the firewall, there are restrictions placed on the method of access to the data and the data itself that are reachable.

1.5.1 Internal Research Units

MiceSlice™

The MiceSlice™ unit primarily prepares physical tissue sections that are used in the NeuroZoom™ and BrainArchive™ units. The workflow is carefully controlled so that three brains may be physically cut and stained simultaneously. Figure 1.2A shows one of the workflows to produce immunohistochemically stained sections or in situ hybridization imaged series from aligned brains before sectioning. All histochemical procedures are done under controlled conditions to reduce variability among brain datasets. Figure 1.2B shows a portion of the basic workflow used to record various metadata for each brain section into the Neurome database, as well as the pertinent data itself that is captured in the next research units (see below: NeuroZoom™ and BrainArchive™). As brains are sectioned and stained according to specific experimental protocols, they are entered automatically by instrumentation into Neurome's database for analysis. Each of the conditions for each animal and tissue slice also has associated collection parameters and individual sets of neuroanatomic metadata describing attributes such as species, animal, age, sex,

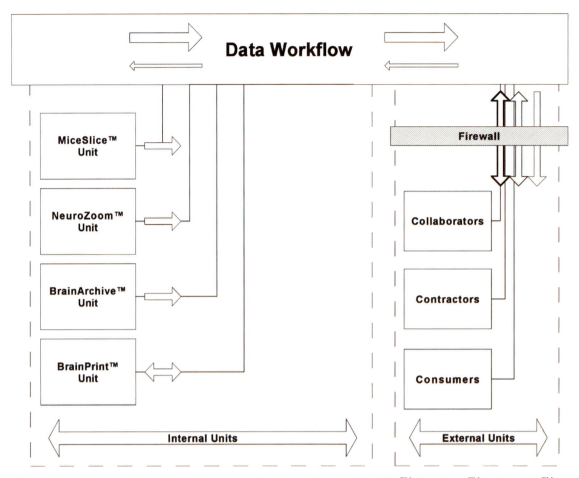

Figure 1.1 The research internal units correspond to the Neurome technologies: MiceSlice™, NeuroZoom™, BrainArchive™, and BrainPrint™. The data that each internal unit produces flow among each other inside the firewall, and strict rules determine access to and from the external units, which are comprised of Neurome's collaborators, contractors, and consumers.

condition, and whether kept in-site or delivered to collaborators or contractors, as well as where (see below: Animal Tracking Database).

NeuroZoom™

Although this unit is the basic means by which the data to populate the database are acquired, we describe it only briefly. Much of this software is based on the original NeuroZoom™ software developed under the Human Brain Project consortium project (Bloom et al., 1997; Nimchinsky et al., 1998a,b; Bloom and Young, 1999). The NeuroZoom™ unit consists of instruments that are responsible for the acquisition of neuroanatomic data, usually in the form of topographical maps or reduced quantitative information datasets. The sources of the data are brain tissue sections. Data are expressed in two and three dimensions within the spatial context of the tissue, relative to normals stored in Neurome's database.

Figure 1.3A shows the basic microscope instrument setup that is controlled with NeuroZoom™ software. Up to eight standard glass slides are mounted on the motorized microscope stage. Each slide holds five to eight cross sections of mouse brain tissue. The slides are automatically analyzed for the location of every section. The sections are then automatically imaged at very high resolution directly into Neurome's database. Figure 1.3B shows

the individual tiles that are imaged as they correspond to each microscope field of view. The tiles are reassembled automatically into a single large-scale mosaic (Figure 1.3C). Finally, the NeuroZoom™ unit aligns the sections by rotating and translating them into 3D registration to produce a single 3D dataset of the mouse brain (Figure 1.3D).

Banks of microscopes are controlled 24 hours a day, 7 days a week, acquiring images from mice of different strains, ages and histological stains. The system is highly scalable. The NeuroZoom™ unit runs software in parallel on the mainframe computer and uses a distributed computing grid during intense computations, such as during 3D alignment operations.

BrainArchive™

The BrainArchive™ unit reorders the data from the NeuroZoom™ unit so that it is available as a 3D model and can be rendered and displayed in a 3D navigational display system. Using a mixture of manual and automated contouring tools, the boundaries of the known places in the standard neuroanatomic nomenclature of our mouse brain are outlined and entered into Neurome's database. The raw data are thus organized into usable spatial models or brain atlases. Three-dimensional models, morphometric data, and other comparative analyses are then extracted from these atlases.

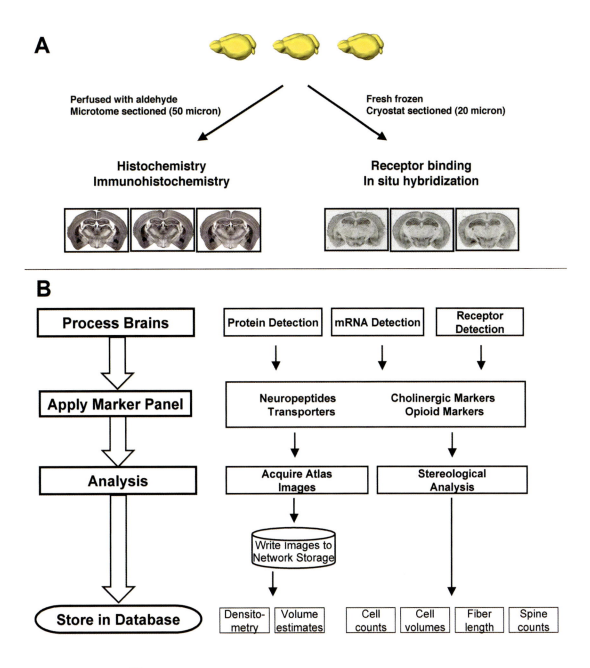

Figure 1.2 MiceSlice^{TM} unit. Figure 1.2A is one of the workflows to produce immunohistochemically stained sections or in situ hybridization imaged series from aligned brains before sectioning. All histochemical procedures are done under controlled conditions to reduce variability among brain datasets. All background metadata for each section are recorded in the Neurome database. Figure 1.2B shows that as brains are sectioned and stained according to specific experimental protocols, they are entered automatically by instrumentation into Neurome's database for analysis.

Figure 1.4 shows a Nissl-stained brain, collected from about 60 sections and reassembled by computer. Figure 1.4A shows one of the original cut sections in the coronal plane. Figure 1.4B is the reassembled view in the sagittal plane. Figure 1.4C shows the same brain in the horizontal plane. Figure 1.4D shows the CA1 and CA2 region of the hippocampus rendered in color against the three orthogonal planes in a 3D view.

The rigorous controls in the MiceSlice^{TM} unit as well as computer-controlled registration of 2D and 3D coordinate systems in the NeuroZoom^{TM} unit provide for accurate correlations

between two different imaging modalities, such as magnetic resonance microscopy (MRM, Figure 1.4E) and bright-field microscopy (Nissl stain, Figure 1.4F). In fact, we have found such good correlation that the standard method of brain analysis at Neurome is now implemented with sectioned brain tissue using light and laser microscopy systems, instead of noninvasive imaging techniques, such as MRM.

Contours that are stored in Neurome's database are applied to sections imaged under different staining conditions. Figure 1.4G shows a darkfield-imaged section captured by the NeuroZoom^{TM}

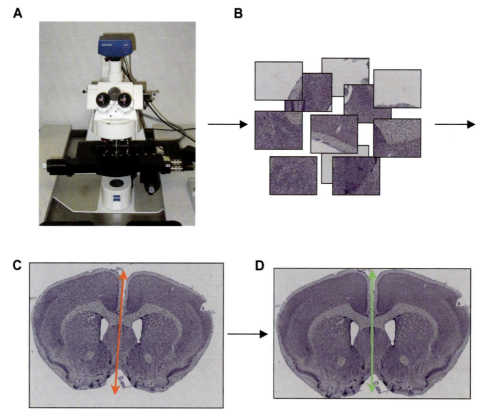

Figure 1.3   NeuroZoom™ unit. (**A**) Basic microscope instrument setup that is controlled with NeuroZoom™ software. Up to eight standard glass slides are mounted on the motorized microscope stage. Each slide holds five to eight cross sections of mouse brain tissue. The slides are automatically analyzed for the location of every section. The sections are then automatically imaged at very high resolution directly into Neurome's database. (**B**) Individual tiles that are imaged as they correspond to each microscope field of view. The tiles are reassembled automatically into a single large-scale mosaic (**C**). Finally, NeuroZoom™ aligns the sections by rotating and translating them into 3D registration to produce a single 3D dataset of the mouse brain (**D**).

unit. A contour from Neurome's database is applied to outline the hippocampus. The same contour is then applied to the thioflavin-stained image of the same section (Figure 1.4H) to extract the region of interest (Figure 1.4I). Finally, the signal arising from plaques is extracted with a variety of image processing and filtering operations (Figure 1.4J).

BrainPrint™

The BrainPrint™ unit analyzes the experimental core datasets originating from the NeuroZoom™ unit and the BrainArchive™ unit and identifies those attributes that are useful for developing profile information that correspond to certain characteristics under various transgenic conditions.

Figure 1.5A shows the 3D dataset of a Nissl-stained series produced by the MiceSlice™ unit, acquired and aligned by the NeuroZoom™ unit, ordered by the BrainArchive™ unit, and analyzed for regional profile data with the BrainPrint™ unit. In this case, the subregions of the hippocampus, as outlined in the upper left of the figure, are analyzed for α-7 nicotinic acetylcholine receptor as measured in nCi/mg. This is displayed in the upper right side of the figure, with the colors correlated to a pseudo-color chart (not shown) showing the relative amounts of signal.

Figure 1.5B is an example of a reduction of another signal acquired into Neurome's database by the BrainPrint™ unit that produces a series of values for normal (C57BL/6) mouse strain and compares this to the same stain conditions for the experimental transgenic (Tx) mouse strain. The difference delta (Δ) is displayed as a color-coded profile and linked to the known places in the standard neuroanatomic nomenclature of Neurome's brain database. In this example, the places that show a measured difference are red nucleus, medial geniculate, and the CA1 subregion of the hippocampus. This combination of data constitutes a "profile," or a BrainPrint™ signature of the region that is specific to this brain dataset. Profiles from the Neurome database are easily compared across different conditions and strains once the data are acquired from the other units.

The BrainPrint™ unit creates much of the data that are used in the content section (see below) and that are repurposed for other units. Many external databases will also be consulted in this unit's actions, bringing together experimental, archival, and legacy data, thus exposing itself—or emerging—as the encyclopedic level of data.

1.5.2 External Units

All external units are in front of the Neurome firewall. The movement of data to and from these units is strictly controlled. The

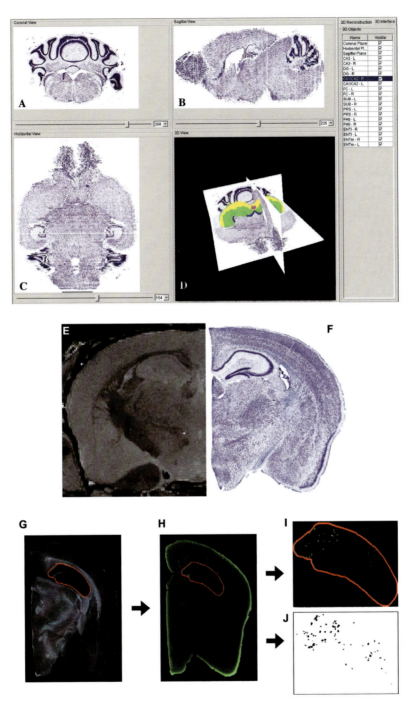

Figure 1.4   BrainArchive™ unit. (**A**) One of about 60 original cut Nissl-stained sections in the coronal plane. These have been reassembled by computer into a 3D model. (**B**) Reassembled view in the sagittal plane. (**C**) The same brain in the horizontal plane. (**D**) shows the CA1 and CA2 region of the hippocampus rendered in color against the three orthogonal planes in a 3D view. There is good correlation between two different imaging modalities: magnetic resonance microscopy (MRM) (**E**) and Bright-field imaging with Nissl stain (**F**). (**G**) Darkfield-imaged section captured by NeuroZoom™ . A contour from Neurome's database is applied to outline the hippocampus. The same contour is then applied to the thioflavin-stained image of the same section (**H**) to extract the region of interest (**I**). Finally, the signal arising from plaques can be extracted with a variety of image processing and filtering operations (**J**).

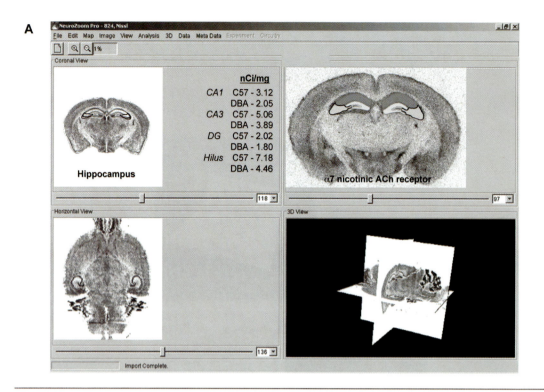

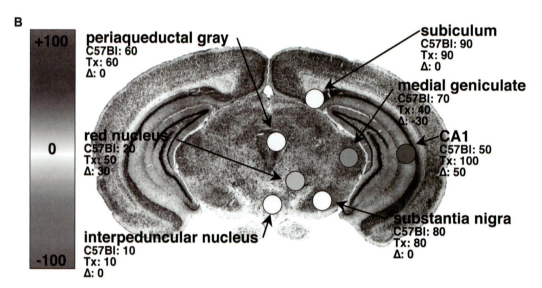

Figure 1.5 BrainPrint™ unit. (**A**) The 3D dataset of a Nissl-stained series produced by the MiceSlice™ unit, acquired and aligned by the NeuroZoom™ unit, ordered by the BrainArchive™ unit, and analyzed for regional profile data with the BrainPrint™ unit. In this case, the subregions of the hippocampus, as outlined in the upper left of the figure, are analyzed for α-7 nicotinic acetylcholine receptor as measured in nCi/mg. This is displayed in the upper right side of the figure, with the colors correlated to a pseudo-color chart (not shown) showing the relative amounts of signal. (**B**) Example of a reduction of another signal acquired into Neurome's database by the BrainPrint™ unit that produces a series of values for normal (C57BL/6) mouse strain and compares this to the same stain conditions for the experimental transgenic (Tx) mouse strain. The difference delta (Δ) is displayed as a color-coded profile and linked to the known places in the standard neuroanatomic nomenclature of Neurome's brain database. This combination of data constitutes a "profile," or a BrainPrint™ signature of the region that is specific to this brain dataset. Profiles from the Neurome database can be easily compared across different conditions and strains once the data are acquired from the other units.

precise nature of what is externally viewable, reviewable, and so on, is a function of each client's needs. Whenever possible, data review will be performed in-house, within the firewall.

1.6 CONTENT MANAGEMENT SYSTEM

The core operation at Neurome is to deal with neuroscience data. Those data are accumulated in a linear manner, by conducting experiments using our tools and technologies. Data are also contributed from partners and other collaborators. In all cases, "data in" produces a "data out" effect. However, it is not the data itself that forms any one product. Instead, the data may be restructured or recombined in new ways that add exponentially to their individual value. This information or content is more valuable than the data itself. In this case, the comment "The whole is greater than the sum of the parts" rings true.

It is important to realize that the database system being implemented is really an enterprise-wide content management system. Content must be managed within all parts of the development, maintenance, and extension of the database architecture. In order to support this approach, the content management system is structured with a component architecture where data entities are stored in minimalist form and operated on at the lowest level by basic tools to support the business as outlined by the research units above. This system has extensible markup language (XML)

validation (Bray et al., 1998), user interfaces, logging, storage, and versioning. The diagram in Figure 1.6 shows the interaction with regard to data movement and access among the various internal research units and several core units that arise from these interactions.

In addition to the internal research units (MiceSlice[TM], NeuroZoom[TM], BrainArchive[TM], and BrainPrint[TM]), three new modules are apparent: content database, content manipulation, and content use/reuse. Note that the MiceSlice[TM] unit spans vertically from the lowest point (experimental layer) to the highest point (product layer) because MiceSlice[TM], one of the Neurome technologies, can be externalized as a product, as well as providing a data source for Neurome's other units (NeuroZoom[TM] and BrainArchive[TM]). The data bus under the MiceSlice[TM], NeuroZoom[TM], and BrainArchive[TM] units is a private bus supporting the communication of data for those units. Data acquired from the MiceSlice[TM], NeuroZoom[TM], and BrainArchive[TM] units move into the experimental data store, which is a temporary holding area until pulled into content database.

The content database is the main database store. Located in this are the physical devices and the table structures to hold the basic object entities that comprise data. Content manipulation has full access to all data objects in the content database. Content manipulation is accessed by the BrainPrint[TM] unit and feeds mainly into content use/reuse. The BrainPrint[TM] unit uses content manipulation to refine content from data until they can be

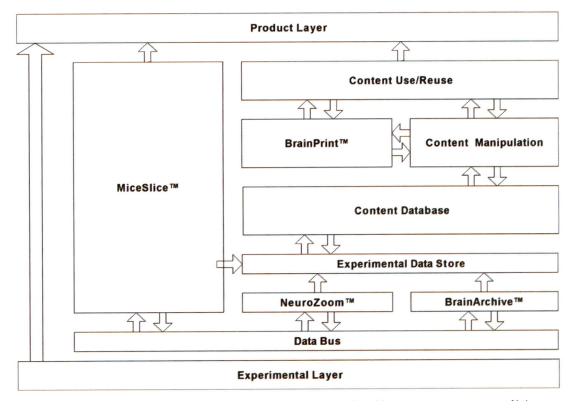

Figure 1.6   The data from the various internal research units create an enterprise-wide content management system. Various approaches are taken to restructure the data into usable modules for refactoring content into the forms and products that Neurome can deliver to the end-user. Research oriented raw data is at the bottom, and filters through the various components of the content management system and eventually are promoted to the product layer. As an example, the histology module can be externalized as a product, packaging tissue sections with image data from the content database (which receives input from the imaging and microscopy modules). The result of this arrangement into a content management system is to create a flexible arrangement of data acquisition, analysis, and presentation.

moved into a content use/reuse area that repurposes the majority of the data for export or sale to customers, clients, contractors, and Neurome's own internal use. Content use/reuse restructures the data pool in the content database by using content manipulation and readies data, content, and information for products and repurposing. The result of this arrangement into a content management system is to create a flexible arrangement of data acquisition, analysis, and presentation. The content management system is comprised of three basic parts as outlined in the previous figure: (1) content database, (2) content manipulation and (3) content use/reuse.

The content database has the very basic entities expressed in the technology that supports databases, namely, binary large objects (BLOBs) for the images and XML data for general data and metadata, including vector data. The content database also contains stored procedures and database triggers used on the data itself for XML validation, user interfaces, data version, logging, reporting, security, and so on. All the images, data points, references, experimental data, metadata, and other types that we collect, maintain, and store in the Neurome database system are located in the content database. The content database is a large, scalable, and reliable database system and fully integrated with the other parts of Neurome's technologies. XML is used throughout all of the modules of the content management system due to its platform-independence, openness, extensibility, human readable code, and self-describing nature (Bray et al., 1998).

Content manipulation is accessible by the BrainPrint™ unit because it supports data editing, approval, insertion, organization, characterization, and association. Typically, there is extensive input from external database systems from our partners for validation of experimental data. Other forms of legacy databases are linked into this unit. This content manipulation system is the central module that touches all parts of the content management system. This is oftentimes a more abstract layer than a physical component of the content management system. However, tools exist in various other units or layers that contribute to this layer indirectly.

Content use/reuse takes the filtered and organized data from content manipulation and "repackages" it for new purposes (repurposing). This is not necessarily a physical product, but might be more a dynamic action imposed on the workflow by software control systems. It can act, for example, as the interface to the external world, providing a multitude of different "looks" at the content database. As with content manipulation, this is oftentimes a more abstract layer than a physical component of the architecture. However, tools also exist in various other units or layers that contribute to this layer indirectly.

Collectively, the content management system marshals data to and from the content database. It resides mostly in the traditional middle tier of a multi-tiered system. At Neurome, the content management system is exposed to most users, internal and external to Neurome, by its portal system: NeuroPortal™. NeuroPortal™ will bring to all users content not just from within Neurome's database, but from content on the Internet that would normally be available by visiting those data sources directly (e.g., via the Web). The internal and external content would be integrated by the content management system and navigated with NeuroPortal™.

1.7 CONTENT DATABASE SCHEMA

The content database is built on the relational schema of the Oracle 9i database. Therefore, the hierarchical approach to building a complex database must be implemented by having reference records in the same table. Furthermore, table design must be minimalist, with the preponderance of functionality provided by scaling out the number of tables and their specific attributes. Once the schema is designed, additional interfaces from either Oracle or other vendors can be used to add value to the application development, such as Oracle's object-relational libraries.

The content database is currently divided up functionally into broad logical groupings as:

1. Neuroanatomy database
2. Atlas database
3. Animal tracking
4. Device database
5. Experimental database
6. Histochemistry database
7. Standard operating procedure (SOP) database
8. Portal database
9. Image database
10. Security database
11. Business database

A discussion of all of these database issues is beyond the scope of this chapter. For the purposes of showing examples of various approaches to modeling neuroscience data into Neurome's database schema, the following will be discussed: neuroanatomy, atlas, animal tracking and images.

1.7.1 Neuroanatomy Database

The Neuroanatomy database contains the basic tables for representing brain places, circuits, physiology, and pharmacology. These data form the vast majority of the core part of neuroscience knowledge and represent much of the intellectual property at Neurome. There are 20 main tables and 30 tables used for associations. The main tables are: animal, cell, cell_model, cell_structure, circuit, circuitpharm, density, envelope, fiber_tree, ionchannel, place, place_link, place_name, receptor, species, strain, structure_physics, system_level, transmitter, and transporter. Many of these tables refer to other tables by key association and build up more sophisticated combinations of attributes that describe neuroanatomic entities. This will be described in more detail in the sections below on place-to-place relationships.

It is important to note that while the tables here represent neuroanatomic concepts and structures, the groupings of these into families represent brain atlases, and these are reflected by the tables in the atlas database. None of the metadata stored in any of the tables in the neuroanatomy database are considered ephemeral or experimental. Those data are stored in the experimental database. Only validated and certified data go into the neuroanatomy database. Much of the data that are pre-populated into the tables of the neuroanatomy database are from knowledge assimilated from the literature, already considered a source

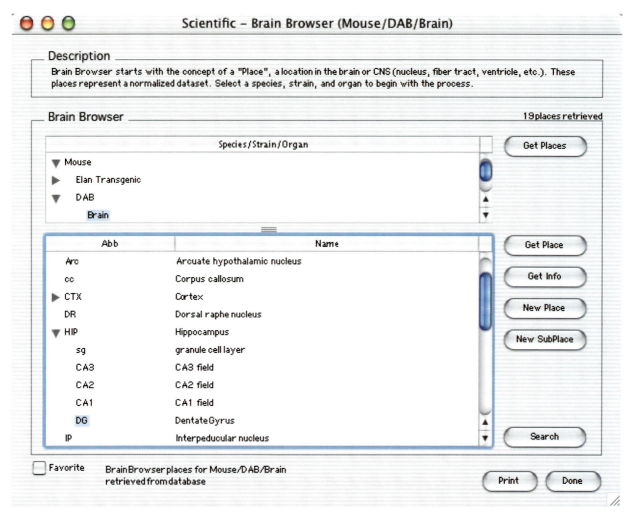

Figure 1.7 A representative screen shot of the display of species/strain/organ, and the hierarchical view of the "places" within that organ. Selecting any place can retrieve additional information from the database, such as circuits, pharmacology, and so on. This example shows the species as "mouse," the strain as "DBA," and the organ as "brain." The selected place is "DG" (dentate gyrus) of the hippocampus.

of validated data in a sense. These data are then certified by Neurome anatomists before being added into the neuroanatomy database. This basic model is based on Brain Browser™ as described earlier in this chapter and the basic entity is the "place." Relationships between places define network-based circuits and hierarchy-based parent–child order.

Places

Every structure in the brain has a place. Figure 1.7 shows the selected species as "mouse," the selected strain as "DBA," the selected organ as "brain." Note also in the lower field that the places known for this combination are displayed, with the place "hippocampus" expanded to reveal its subplaces, such as the "dentate gyrus."

What is not obvious is that running behind the scenes here is a fully schematized system in the Neurome database that transported the brain organization to this application—Neurome's NeuroPortal™—via fully descriptive XML. The operations are more readily visualized in the rendered XML browser shown in Figure 1.8.

The same XML stream is expressed as a hierarchically organized tree (XML tree). Note the clean progression from species (mouse), to strain (DBA), to organ (brain), to places (all places), to one place (hippocampus), and finally to a subplace within the hippocampus (dentate gyrus). The value "dentate gyrus" is actually an attribute and displayed as data associated with this place. We use XML because of its self-descriptive and thus self-supporting open source nature. The XML vocabulary is work-in-progress at Neurome and will describe all the entities in use in the content management system.

The place table contains records of all places among all brains of all species and strains. The structure of the table is quite simple, containing foreign keys to other tables for place names, notes, animals, and structure physics. In general, every place of each combination of animal species, strain, sex, and age will have physical attributes associated with it. Such attributes are volume, 3D location, surface area, shape, texture, granulation, clustering indices, and the like. Every record for each animal combination may be different, and this schema supports that possibility. All of the "physics" that are used to describe every structure are

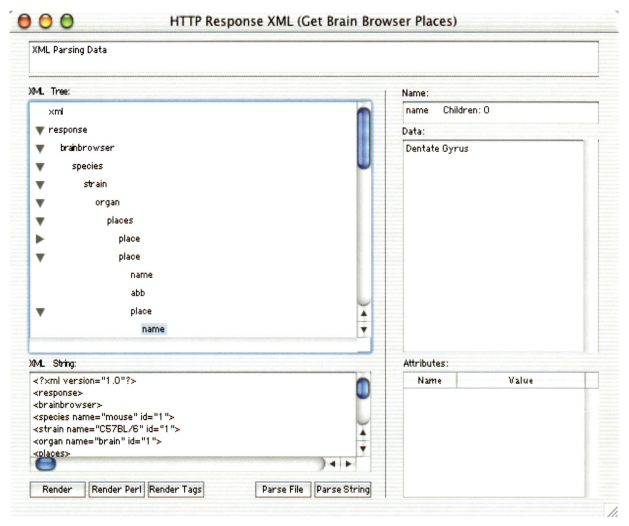

Figure 1.8   All data are transported to and from user applications, such as Neurome's NeuroPortal™ access application as fully descriptive XML. The XML hierarchy is shown here in this browser. The same hierarchy of species, strain, organ, hippocampus, and dentate gyrus is displayed here as in the previous figure. In this browser, only the XML tags are displayed (i.e., species, strain, organ, places, place, name, etc.).

stored in a table named "structure physics." The use of these descriptions gives the brain hierarchical and network models (see below) a 3D spatial orientation. This becomes important when linking in details gathered for brain atlas databases.

Place-to-Place Relationships

Places in Neurome's database schema can be related to other places through one or both of two relationships: hierarchical envelopes and network connections.

Hierarchical Envelopes. The brain is hierarchical in nature (Bowden and Martin, 1995). One can start with the brain itself as the main structure and begin subdividing it into smaller and smaller pieces. The brain, for example, can be roughly divided into these major "places": spinal cord, medulla, pons, midbrain, diencephalon, and telencephalon. Each of those places can be divided further. For example, the diencephalon can be divided into the epithalamus, the thalamus, the metathalamus, the subthalamus, and hypothalamus. The hypothalamus can be divided into the anterior hypothalamus and the posterior hypothalamus. Eventually, a

tree-like hierarchy develops where all parent–child relationships are known.

The names and abbreviations of these standard places in the brain hierarchy are being developed at Neurome by neuroscience consultants for mouse and rat. There will not be a perfect relationship between the two species because the brains are different. In this case, a translation table will be developed so that any place of one species can be related to another place in the other species. The database tables express their relationship to one another in this hierarchy. This hierarchy does not change substantially over time, unless new knowledge requires changes in the nomenclature. The hierarchical database now lends itself to the addition of other data and relationships that express the network connections that are found in the brain.

The hierarchical envelopes are mostly used to express the hierarchies seen in the organization of places. In the XML example above, the brain hierarchically envelops all of the known places, which includes the hippocampus as a place, which then hierarchically envelops all of its subregions, which includes the dentate gyrus as one of its subplaces. There are intrinsic parent–child relationships. For example, the hippocampus is the parent of the dentate gyrus; likewise, the dentate gyrus is the child of the hippocampus. This enveloping method can continue on, down to the subcellular level if so desired, but since the "place" was chosen as the normative neuroanatomic database entity, it stops at that point, and other attributes are associated back to the places through their own individual tables.

Network Connections. The network relationships in the brain are expressed as "circuits." Circuits are connections of one class of neural cells to another class of neural cells. Neurons communicate with one another by sending chemical information from cell to cell. A cell-to-cell connection is one fascicle of a place-to-place circuit. Cells may be clustered together in places, and the circuit domain may consist of multiple cells of the same type, communicating to a cluster of cells of a same or different type in another place.

The Neuroanatomy database can express these circuits with a mechanism similarly used to express hierarchical envelopes. Whereas in the hierarchical organization of the brain there are parent and child relationships between places, a network connection has source and target places. Source places, also known as the origin of the circuit, communicate or project to target locations. Every source location has an association with a cell type. The cell type is a table in the Neuroanatomy database. Every target location also has an association with a cell type. Every cell, depending on whether it is a source or target cell, has transmitters, receptors, transporters, and ion channels as the basic pharmacological makeup of the cell. Source cells traditionally have transmitters, because these chemical structures are usually created in the source area of the circuit, transported by a transporter chemical structure down the axon of the neuron to the target location where the transmitter is released and attaches to the target cell's receptor, a chemical structure that chemically and molecularly binds to the transmitter. Once the target cell receives the transmitter, it can then continue the progression of the signal to its target, thus becoming the source of the subsequent circuit.

Circuits are to be defined as one class of cells communicating with one or more neurotransmitters to another class of cells. If there is a one-to-many, or a many-to-one arrangement, this is more than one circuit. In the neuroanatomy database, circuits are related as another form of link. Many places in the brain can have target locations, each of which receives a circuit. The same place can be source locations for others circuits. Since the circuit is also built around the "place" concept, the hierarchical organization in the database, pre-populated with standard nomenclature, serves as an ideal framework on which to enter in this network or circuit information. Much of the circuit information is derived from the scientific literature, or from ongoing experiments at Neurome. The circuitry database at Neurome now contains over 11,000 circuitry data items from both archival Brain Browser™ data on the rat and other early research efforts in the authors' laboratories at The Scripps Research Institute (La Jolla, CA).

Circuits

As discussed above, circuits are the connections between the places, or more accurately between the cells defined and located in the places. Figure 1.9 shows an example of the circuits

(afferent and efferent) to and from the nucleus of the solitary tract (NTS).

Clicking on any place, either on the afferent side (known as source or origins of the circuit) or on the efferent side (known as targets of the circuit), shifts the focus of the window to that newly selected place. The central area of the screen shows additional metadata on the place: cells, transmitters, receptor subtypes, system levels, and birth dates of the cells. Each of these are selected from pull-down menus, showing dozens of standard data that can be selected for each metadata category. Note next to "Multipolar" there is a notation "1 of 21". This means that there are 21 discrete types of cells listed for NTS, with the first one being shown as "Multipolar." Selecting the second cell type will display its own set of metadata for the transmitters, receptor subtypes, system levels, and birth dates of the cells.

The circuit table contains the circuit records. Each circuit has a name, description, projection density, source cell, target cell, birth date, and structure physics. The circuit's projection path is stored in the structure physics table. The circuit coordinate table contains the actual 3D pathway of the circuits. Similar to the use of the structure physics records for places, the use of this table will give a strong 3D spatial orientation to circuits. If the details are strong enough, the actual navigation down the cellular fibers will be viewable.

Another table of interest for circuits is the circuit pharmacology table that contains foreign keys to: transmitter, transmitter level, receptor, receptor level, receptor density, transporter, transporter density, transporter localization, ion channel, secondary messenger, and topography. One record is descriptive of a circuit's pharmacology. This record is then associated back to a record in the circuit table.

Pharmacology

As discussed above in Brain Browser™, there are various forms of metadata associated with the main neuroanatomic entities of places and circuits. In the Neurome schema, several database tables deal with lower-level entities in the neuroanatomy database, such as transmitters, transporters, receptors, animals, species, sex, age, and so on. Each of these tables have specific attributes that express their functionality and are sometimes related to primary key records through various forms of one-to-many association table, or directly to the primary key record through its foreign key if it is a one-to-one relationship.

For example, there is a single table for all the neurotransmitters in all brains in the neuroanatomy database. However, every cell of each circuit may have one or more neurotransmitters associated with it. Association tables make the one-to-many association that is needed in this kind of data model. The same situation applies to receptors, transporter, and ion channels. Many entities have relationships with one another. For example, transmitters and receptors are usually discussed in pairs, much like a lock and key. There is a receptor entity for each transmitter entity. The Neurome schema will allow for infinite levels of transmitter–receptor pairings because a single cell can contain more than two transmitters–receptors.

Figure 1.10 shows an example of one pharmacology entity, the neurotransmitter "acetylcholine," in its detail window. Various attributes about acetylcholine are available in the pharmacology tab, such as the manufacturer (chemical/reagent supply

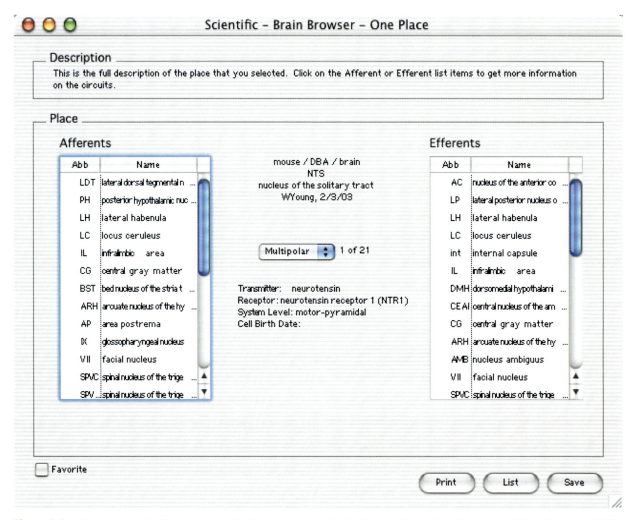

Figure 1.9   This is an example of the browser used to display neuroanatomic circuits to and from a place [i.e., nucleus of the solitary tract (NTS)]. Information on the place itself (NTS) is located in the center of the window. Other places leading into the NTS, known as afferent connections, are in the scrolling field on the left side. Other places leading out of the NTS, known as efferent connections, are in the scrolling field on the right side.

company). This general information is linked to the ordering system of Neurome's NeuroPortal™. The currently selected tab is the relationship section, used to show relationships to other neuroanatomic entities. For example, this neurotransmitter is located in the glossopharyngeal nucleus (as well as elsewhere, but this is an example screen shot), has an afferent connection to the nucleus of the solitary tract (again, an example), and is located in motor neuron cells. The bottom part of the window shows how relationships to other entities are created. In this case, the system level for this transmitter in the motor neuron is being related to the motor-pyramidal system. The other tabs are used for images related to acetylcholine and references from the literature (external to Neurome's database).

1.7.2 Atlas Database

Neurome has implemented an atlas creation and display system within its database. This system comprises a collection of high-resolution images and a searchable database of labeled structures that is used to facilitate morphometry and accurate localization and analysis of gene expression. The atlas system is fully integrated with all other Neurome's databases and serves as a visual gateway to quantitative data on gene expression.

The atlas database contains the basic tables for representing atlases. There are four main tables and one table used for associations. The main tables are: atlas_image, atlas_page, atlas_plane, and atlas_set. Atlases are organized into atlas sets. The atlas set table contains all of the sets of atlases in the Neurome database. These may be divided into finer details such as by species, strain, sex, and age. Each atlas set is then divided in atlas pages, similar to the organization of a hardcopy atlas. Each atlas page also has an atlas plane of orientation assigned to it. An atlas is equated to a "snapshot" of the brain, and thus an atlas set, representative of this snapshot, uses as its discriminators species, strain, sex, and age. The plane of section is not a discriminator, but rather are just view models into the same brain.

Atlas-based images are a special form of regularly formatted images that can be used for the display of brain atlases. Images would form the backdrop or the actual brain section acquired by scanning the sections with a microscope. The commercial atlas on the C57BL/6 and 129/Sv mouse (Hof et al., 2000) is installed

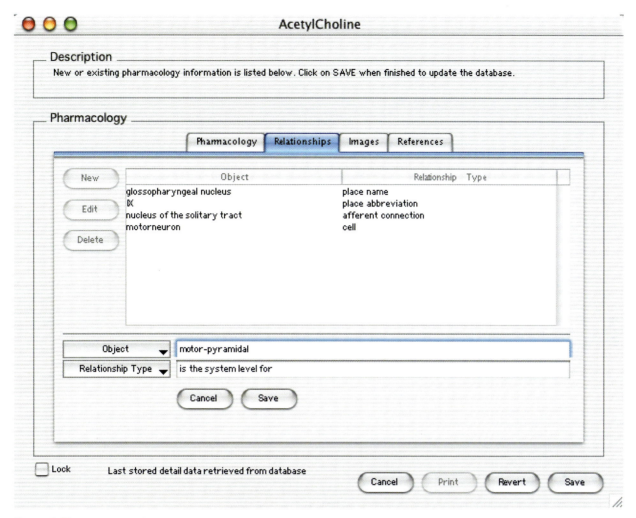

Figure 1.10 An example of one pharmacology entity, the neurotransmitter acetylcholine, in its detail window. Various attributes about acetylcholine are available in the pharmacology tab, such as the manufacturer (chemical/reagent supply company). This general information is linked to the ordering system of Neurome's NeuroPortal™. The currently selected tab is the relationship section, used to show relationships to other neuroanatomic entities. For example, this neurotransmitter is located in the glossopharyngeal nucleus (as well as elsewhere, but this is an example screen shot), has an afferent connection to the nucleus of the solitary tract (again, an example), and is located in motor neuron cells. The bottom part of the window shows how relationships to other entities are created. In this case, the system level for this transmitter in the motor neuron is being related to the motor-pyramidal system. The other tabs are used for images related to acetylcholine and references from the literature (external to Neurome's database).

in Neurome's database system and displayed through Neurome's NeuroPortal™ application.

Figure 1.11 shows the database being searched for all sections matching the place "secondary somatosensory cortex"; 2531 database entries for the mouse strain C57BL/6 were searched when this process was completed. This figure is a snapshot in time, so at that time, 126 matches were already made with 16 sections collected for display. After the searching process completes, the sections that contain the place "secondary somatosensory cortex" may be selected from a pick list and displayed independently. Figure 1.12 shows the mouse brain section at Bregma −0.90 mm.

The Nissl-stained section is on the left, and the corresponding vector map is on the right side. The vector map shows all of the known places and their boundaries as entered by Neurome's anatomists. These places correspond to the standard taxonomy that is used for this species (mouse). Note that although

"secondary somatosensory cortex" was the search string, all the sections containing this place are displayed. The scale menu at the bottom left can be used to scale up the images (both Nissl and vector). The magnification can be increased to very high levels if the images have been acquired by Neurome's NeuroZoom™ unit.

In summary, the atlas database contains images of brain sections cut in standard planes of section. The real sections are imaged at high magnification with various stains. The edge boundaries of standard places in the taxonomy are outlined on top of the real sections. The places are located and normalized to a variety of coordinate systems, and they are placed inside a query system. Any section of any animal of many stains can be recalled, with structural schematics available on top, and annotations and links to many other data sources (i.e., references, raw data, experiments, other knowledge or information data stores).

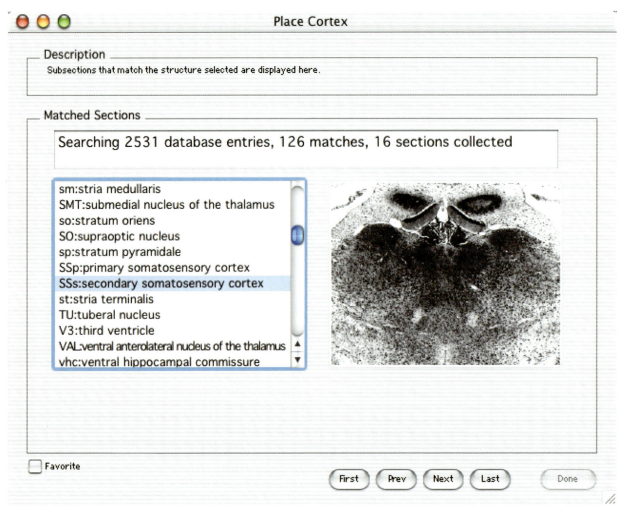

Figure 1.11 This is a snapshot of the searching process for the place "secondary somatosensory cortex" in the Hof mouse brain atlas contained in the Neurome atlas database (Hof et al., 2000). When the searching process completes, the sections that contain the place "secondary somatosensory cortex" may be selected from a pick list and displayed independently.

1.7.3 Animal Tracking Database

Neurome has created a complete animal tracking database. The use of computers provides the proper accuracy and integrity that is required for our business and provide a degree of blindedness that is needed to remove investigator bias from the scientific research. The basic goals of the animal tracking database are to (1) store the valuable and necessary information regarding all the animals used in an experiment, (2) guarantee proper documentation and facilitate the recording of information by scientists, (3) outline workflow and create projections for when an experiment may be completed from a technical/procedural standpoint, and (4) be able to generate reports by experimental study (e.g., immunohistochemistry or MRM).

The animal tracking database contains the basic tables for all animal tracking data. There are eight main tables and two tables used for associations. The main tables are: animal_dissection, animal_housing, animal_order_group, animal_organ, animal_organ_data, animal_perfusion, animal_project, and animal_mrm. Animals are tracked completely, especially since

Neurome performs contract research. Nearly every aspect of an animal is monitored and recorded.

The animal organ data table contains data on certain groups such as pre-sectioning, sectioning, and usage. The animal order group table has attributes indicating order information: requester, request date, order date, delivery date, check in date, the vendor, the species, the strain, the genotype, sex, arrival age, animal housing identity, and the number of animals requested.

Figure 1.13 shows an example of some of the individual animal detail that is recorded, specifically for the perfusion process.

Tabs at the top of the detail section for this animal select different parameters recorded during the tracking process:

- *General Statistics.* Animal ID number, a barcode value to track all of the animal's assets, the vendor, date of birth, birth weight, the date, age, and weight at sacrifice, project(s) that the animal will be used in, animal technicians making the observations.
- *Perfusion.* Animal technician performing the perfusion, date of perfusion, body weight, ratings for liver clearing and stiffness of carcass, the type, dose, and time of anesthetic, the type,

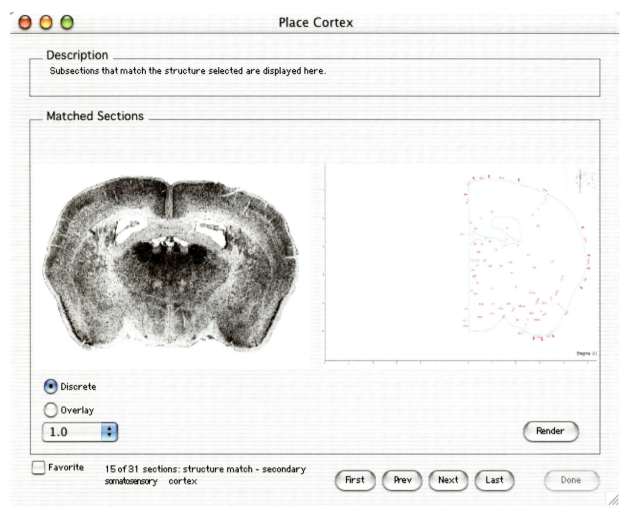

Figure 1.12 This is the mouse brain section at Bregma −0.90 mm. The Nissl-stained section is on the left, and the corresponding vector map is on the right side. The vector map shows all of the known places and their boundaries as entered by Neurome's anatomists. These places correspond to the standard taxonomy that is used for this species (mouse). Note that although "secondary somatosensory cortex" was the search string, all the sections containing this place are displayed. The scale menu at the bottom left can be used to scale up the images (both Nissl and vector).

volume, and time of rinse, and the type, volume, and time of fixative.

- *Dissection.* Animal technician performing the dissection, date of dissection, time to finish, post-fix duration, type of resting solution, whether the brain was frozen and, if so, the date when frozen, and images of the dissection process.
- *Pre-Sectioning Data.* Brain volume as measured by a plethysmometer, weight, and time in 15% or 30% sucrose.
- *Sections.* Date of sectioning and animal technician performing the sections, the sectioning type (microtome, vibratome, cryostat), sectioning thickness, number of sections.
- *Stains.* Any number of stains applied to the sections, when applied, by which histologist, and the range of sections.
- *Images.* Any images pertaining to the animal or the brain or other organs of the animal.
- *References.* This lists any references in the literature that would be useful for this animal and its tracking. The journal name, the authors, the article title, and the page range are all stored and linked to this animal.

1.7.4 Image Database

Images comprise the majority of the data in the Neurome database. Some examples of images have already been discussed in the atlas database. Other types of images may be rendered models. For example, Figure 1.14 is the hippocampus (A), cerebellum (B), and whole brain (C) rendered from a mouse brain using MRI. Figure 1.14D is the 3D surface reconstructions within a 3D volumetric MRM file. Details of these models and the data analysis are in a previous publication (Redwine et al., 2003).

Other types are models are reconstructions from mapped data. Figure 1.15 shows the 3D reconstruction of amyloid beta distribution in the mouse brain. Figure 1.15A shows serial coronal sections immunostained with 3D6. Figure 1.15B shows the amyloid deposits segmented by simple thresholding. Figure 1.15C is the 3D reconstruction of amyloid in red viewed from the posterior aspect of the brain with the hippocampus in yellow. Figure 1.15D shows the lakes and ribbons of amyloid beta in the frontal cortex and olfactory bulb. Figure 1.15E shows the amyloid beta

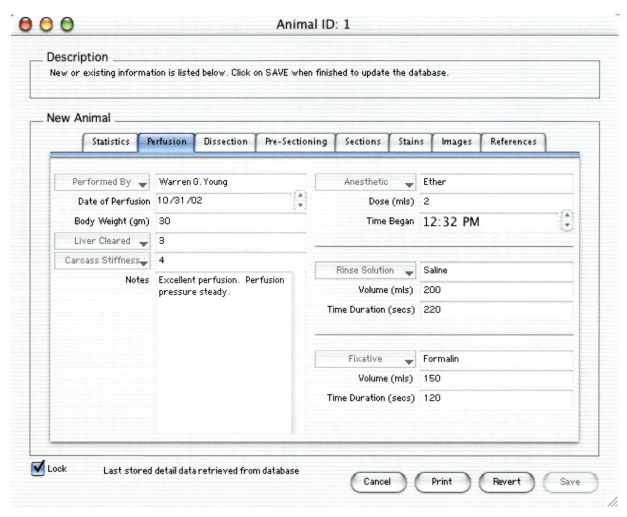

Figure 1.13 Animal detail is available for any animal tracked through Neurome's animal tracking database, a part of the overall database schema. Data available are: general statistical information (animal ID, barcode, vendor, date of birth, date, age, and weight at sacrifice, etc), perfusion data (animal technician, date of perfusion, body weight, liver clearing and carcass stiffness ratings, anesthetic type and dose, rinse type, volume and time, and fixative type, volume and time), dissection data (animal technician, date of dissection, time to finish, post-fix duration, type of resting solution, date when frozen, images of dissection), pre-sectioning data (brain volume, weight, and time in sucrose), sections (date of sectioning, animal technician, sectioning type, thickness, and number of sections, stains (kind, histologist and range), images (any images to view concerning the animal and its procedures), and references (links to literature references).

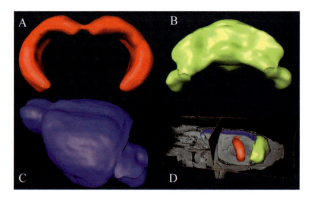

Figure 1.14 Three-dimensional surface reconstructions generated from image segmentation of the hippocampus (**A**), as well as the cerebellum (**B**) and whole brain (**C**). (**D**) The 3D surface reconstructions within a 3D volumetric MRM file. Note that the brain is undissected and still lies within the mouse head. Modified from Redwine et al. (2003).

as sheets in the rostral part of the dentate gyrus. Figure 1.15F displays a magnified view of the amyloid beta lakes and ribbons in the dentate gyrus extending into the retrosplenial cortex. Details of these models and the data analysis are in a previous publication (Reilly et al., 2003).

Experiment-based images are those that come from other technologies being used at Neurome. For example, the NeuroZoom™ unit could be used to acquire images from transmitted light and laser confocal microscopes. The images acquired from Neurome's instrumentation are of a very high optical resolution. With completely automated microscopy systems we routinely captured images of each tissue section at 264×, with a 1-in-4 series comprising approximately 14 GB of data. This is actually at the low end of the available resolution scale that Neurome is capable of acquiring. Each of the individual serial sections is actually automatically acquired mosaics. Small tiles representing a single microscope field of view back are seamlessly "stitched" back into a large mosaic image. The mosaic can be rendered as a whole section down to cellular detail.

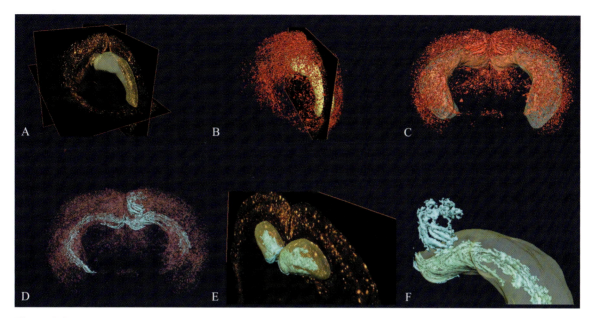

Figure 1.15   A 3D reconstruction of amyloid beta distribution. (**A**) Serial coronal sections immunostained with 3D6 were imaged and compiled into a 3D data file. A surface reconstruction of the hippocampus is shown in yellow. (**B**) Amyloid beta deposits were segmented by thresholding and are displayed as a 3D reconstruction (red). A surface reconstruction of the hippocampus (yellow) and a single coronal section are shown for orientation. (**C**) A 3D reconstruction of amyloid beta (red) viewed from the posterior aspect of the brain with the hippocampus shown as transparent yellow. Note the extensive deposition in the neocortex and hippocampus, along with the central lucency representing the midbrain and caudate-putamen with punctate amyloid beta visible in the frontal cortex and olfactory bulb. (**D**) Large lakes and ribbons of amyloid beta (cyan) were identified by automated detection of contiguous structures within the 3D reconstruction of amyloid beta (shown as transparent red; same angle of view as C). (**E**) Amyloid beta sheets (cyan) are visible in the rostral part of the dentate gyrus, shown against a single coronal section, with the surface reconstruction of the hippocampus in transparent yellow. (**F**) Magnified view of the amyloid beta lakes and ribbons (cyan) in the dentate gyrus (within the transparent yellow hippocampal surface) and extending into the retrosplenial cortex (above). Modified from Reilly et al. (2003).

Figure 1.16 shows an example of the scale of the resolution we maintain in our database schema.

1.7.5 Other Databases

There are many other databases and many more tables in the Neurome database that will not be discussed in detail in this chapter. For example, the business database is involved in the maintenance of the business entities (employees, vendors, other companies, clients, contractors, collaborators, lab and record books, company documents, company literature, projects, literature, profiles, reviews, shareholders, employee vacations, etc.).

The portal database deals with company presentations and technology demonstrations. The device database contains all of the tables that control and maintain all company instrumentation, most notable the research instruments such as microscopes, cameras, imagers, barcode scanners, and so on. The experiment database contains the schema needed to create, organize, execute, maintain, and analyze scientific experiments. These tables describe the very large images that are acquired from the microscope and organize them into datasets that relate back to the animal's tissue under study. The SOP database contains all of the standard operating procedures in use at Neurome, describing all aspects of all units, research, and business. These are linked heavily into the experiment database when laboratory procedures

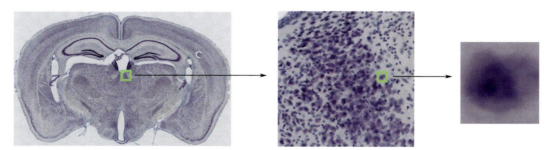

Figure 1.16   Very high resolution images are stored in Neurome's database by accurately controlling microscope instrumentation and using sophisticated techniques to stitch the images back together as a complete mosaic. The images available for viewing in the atlas module of Neurome's NeuroPortal™ viewing and access application ranges from macroscopic views of the entire section, to regions of cells, and finally to cellular and even subcellular levels of resolution.

are involved. Finally, the security database is used to track, audit, monitor, and safeguard against intrusion and malicious activities.

1.8 SUMMARY

All together, the Neurome database has been designed to provide a flexible, but powerful, data backend. A sophisticated schema, relational in design, object-oriented in use, and open and extensible with XML, has created a system that satisfies much of what a new commercialized neuroscience research company needs to have in order to reduce costs, maximize efficiency, and garner the kind of trust and confidence that we anticipate our partners will want to see. Neurome is trying to manage many kinds of data, not just the traditional laboratory data, but data as they are turned into knowledge and information by managing them as content, linking them to Neurome's commercial endeavors (such as real products: MiceSlice™ for tissue sections, NeuroZoom™ for software controlling instrumentation, BrainArchive™ for atlases, BrainPrints™ for comparative profiles, etc.). On the business side, potential inventions or emerging intellectual properties are constantly being identified and evaluated. Numbers are constantly derived from the use of resources to create forecast models, for financials, inventory, and research project planning. All of these data are placed into the enterprise-wide content management system that researchers, computer scientists, business management, and administration can easily access.

However, our initial basic scientific requirements vary only modestly from those days when the same scientists were based in academia: understanding enough from the validated, certified reports of ourselves and others when new observations departed from the literature and providing sufficient clues to devise new experimental validations or extensions. In this murine postgenomic era, our task will be to determine where genes are expressed within the mouse brain and how such expression differs across strains including new transgenic knock-in or knock-out mice. Two main objectives have been singled out for the highest priority: (1) accelerating the pace of making calls on the mapping of expression sites within the context of prior literature on cellular roles in system circuitry functions and (2) the ability to compare quantitative data on gene expression maps across groups of individual mice. These high-priority data collection and analysis streams lie both behind and on top of the utilization of the data organized in the database. It is our perspective that only precise, rigorously collected, and precisely analyzed data merit inclusion in a database from which to draw conclusions on possible pathologies, screening of novel interventions, or demonstration of the earliest detectable departures from the norms. No matter how good and facile a database system may be, it can be no better than the data that are in it.

References

Björklund, A., Hökfelt, T., and Swanson, L. eds. (1987). *The Handbook of Chemical Neuroanatomy*, Vol. 5, Part I. Elsevier, Amsterdam.

Bloom, F.E. (1995). Neuroscience-knowledge management: Slow change so far. *Trends Neurosci.* **18**(2), 48–49.

Bloom, F.E., and Young, W.G. (1993). *The Brain Browser for Windows*. Academic Press, New York.

Bloom, F.E., and Young, W.G. (1994). New solutions for neuroscience communications are still needed. *Prog. Brain Res.* **100**, 275–281.

Bloom, F.E., and Young, W.G. (1999). The multi-dimensional database requirements of brain information. In *Handbook of Molecular-Genetic Techniques for Brain and Behavioral Research* (W.E. Crusio and R.T. Gerlai, eds.). Elsevier, Amsterdam.

Bloom, F.E., and Young, W.G. (2000). Electronic collaboration: Implications for neurosciences. In *Electronic Collaboration in Science* (S.H. Koslow and M.F. Huerta, eds.), Chapter 6, pp. 113–142. Erlbaum, Hillsdale, NJ.

Bloom, F.E., Young, W.G., and Kim, Y. (1989). *The Brain Browser*. Academic Press, New York.

Bloom, F.E., Young, W.G., Nimchinsky, E.A., Hof, P.R., and Morrison, J.H. (1997). Neuronal vulnerability and informatics in human disease. In *Neuroinformatics: An Overview of the Human Brain Project* (S.H. Koslow and M.F. Huerta, eds.), pp. 83–124. Erlbaum, Hillsdale, NJ.

Bowden, D.M., and Martin, R.F. (1988). A hypercard glossary of macaque and human neuroanatomical nomenclature. *Soc. Neurosci. Abstr.* **14**, 1256.

Bowden, D.M., and Martin, R.F. (1995). NeuroNames brain hierarchy. *NeuroImage* **2**, 63–83.

Bray, T., Paoli, J., and Sperberg-McQueen, C.M. (1998). Extensible Markup Language (XML) 1.0, eds. Available at: http://www.w3.org/TR/REC-xml

Crasto, C., Marenco, L., Miller, P., and Shepherd, G. (2002). Olfactory Receptor Database: A metadata-driven automated population from sources of gene and protein sequences. *Nucleic Acids Res.* **30**(1), 354–360.

Felleman, D.J., and Van Essen, D.C. (1991). Distributed hierarchical processing in the primate cerebral cortex. *Cereb. Cortex* **1**, 1–47.

Gibbons, P. (1999). Brain Molecular Anatomy Project Resource Center. Available at: http://grants1.nih.gov/grants/guide/notice-files/not99-120.html

Healey, M.D., Smith, J.E., Singer, M.S., Nadkarni, P.M., Skoufos, E., Miller, P.L., and Shepherd, G.M. (1997). Olfactory Receptor Database (ORDB): A resource for sharing and analyzing published and unpublished data. *Chem. Senses* **22**, 321–326.

Hof, P.R., Young, W.G., Belichenko, P., Bloom, F.E., and Celio, M.R. (2000). *Comparative Cytoarchitectonic Atlas of the SV129 and C57Bl/6 Mouse Brain*. Elsevier, Amsterdam.

Huerta, M.F., Koslow, S.H., and Leshner, A.I. (1993). The Human Brain Project: An international resource. *Trends Neurosci.* **16**, 436–438.

Koslow, S.H., and Huerta, M.F., eds. (1997). *Neuroinformatics: An Overview of the Human Brain Project*. Erlbaum, Mahwah, NJ.

Martin, R.F., and Bowden, D.M. (1996). A stereotaxic template atlas of the macaque brain for digital imaging and quantitative neuroanatomy. *NeuroImage* **4**, 119–150.

Nimchinsky, E.A., Hof, P.R., Young, W.G., Bloom, F.E., and Morrison, J.H. (1998a). NeuroZoom software: An all-purpose quantitative tool for analyzing biological structures. *BECON BioEngineering NIH Symposium, Building the Future of Biology and Medicine*.

Nimchinsky, E.A., Hof, P.R., Young, W.G., Bloom, F.E., and Morrison, J.H. (1998b). NeuroZoom software: Development, validation, and neurobiological application. *FASEB J.* **12**, A628.

Paxinos, G., and Watson, C. (1986). *The Rat Brain in Stereotaxic Coordinates*, 2nd ed. Academic Press, San Diego, CA.

Pittendrigh, C.S., and Jacobs, G.A. (2003). NeuroSys: An electronic laboratory notebook and semi-structured database. *Neuroinformatics* **1**(2), 167–176.

Redwine, J.M., Kosofsky, B., Jacobs, R.E., Games, D., Reilly, J.F., Morrison, J.M., Young, W.G., and Bloom, F.E. (2003). Dentate gyrus volume is reduced before onset of plaque formation in PDAPP mice: A magnetic resonance microscopy and stereologic analysis. *Proc. Natl. Acad. Sci. U.S.A.* **100**(3), 1381–1386.

Reilly, J.F., Games, D., Rydel, R.E., Freedman, S., Schenk, D., Young, W.G., Morrison, J.M., and Bloom, F.E. (2003). Amyloid deposition in the hippocampus and entorhinal cortex: Quantitative analysis of a transgenic mouse model. *Proc. Natl. Acad. Sci. U.S.A.* **100**(8), 4837–4842.

Shepherd, G.M., Healy, M.D., Singer, M.S., Peterson, B.E., Mirsky, J.S., Wright, L., Smith, J.E., Nadkarni, P.M., and Miller, P.L. (1997). SenseLab: A project in multidisciplinary, multilevel sensory integration. In *Neuroinformatics: An Overview of the Human Brain Project* (S.H. Koslow and M.F. Huerta, eds.), pp. 21–56. Erlbaum, Hillsdate, NJ.

Shepherd, G.M., Mirsky, J.S., Healy, M.D., Singer, M.S., Skolufos, E., Hines, M.S., Nadkarni, P.M., and Miller, P.L. (1998). The Human Brain Project: Neuroinformatics tools for integrating, searching and modeling multidisciplinary neuroscience data. *Trends Neurosci.* **21**, 460–468.

Skoufos, E., Healey, M.D., Singer, M.S., Nadkarni, P.M., Miller, P.L., and Shepherd, G.M. (1999). Olfactory Receptor Database: A database of the largest eukaryotic gene family. *Nucleic Acids Res.* **27**, 343–345.

Skoufos, E., Marenco, L., Nadkarni, P.M., Miller, P.L., and Shepherd, G.M. (2000), Olfactory Receptor Database: A sensory chemoreceptor resource. *Nucleic Acids Res.* **28**(1), 341–343.

Williams, R. (2003). The Mouse Brain Library. Available at: http://www.mbl.org/main.html

Web Resources

Bray, T., Paoli, J., and Sperberg-McQueen, C.M. (1998). Extensible Markup Language (XML) 1.0, eds. Available at: http://www.w3.org/TR/REC-xml

Gibbons, P. (1999). Brain Molecular Anatomy Project Resource Center. Available at: http://grants1.nih.gov/grants/guide/notice-files/not99-120.html

Williams, R. (2003). The Mouse Brain Library. Available at: http://www.mbl.org/main.html

Neuroanatomical Nomenclature and Ontology

Douglas M. Bowden, M.D. and *Mark Dubach, Ph.D.*

CONTENTS

Databasing the Brain. Edited by Stephen H. Koslow and Shankar Subramaniam
ISBN 0-471-30921-4 © 2005 John Wiley & Sons, Inc.

2.1 INTRODUCTION

Nomenclatures and ontologies are essential tools of informatics. They are key to the fourth great advance in the efficiency of human communication. The first step was simply the evolution of *language* as a means to convey complex thoughts. While language represented a uniquely powerful advance, communication was limited to individuals within earshot of each other. The second advance was the development of *writing*, to convey symbolic information, and *drawing*, to convey visual impressions, to individuals in different places and future times. Third was the invention of *printing* and *libraries*, which increased by orders of magnitude the number of people who could be reached by information in the form of words and drawings. Now *information technology*, particularly the World Wide Web, brings us to an age in which information in vastly larger amounts will become immediately available to anyone on the planet with a computer. Whereas the office shelves of the typical neuroscientist may offer immediate access to 50,000 pages of information, the computer on the desk offers access to several billion pages. Informatics is the scientific entree to an age in which time and distance can be eliminated as obstacles to human communication.

Informatics is an applied science that draws on principles of several basic disciplines to achieve more efficient communication. For neuroscience, it enhances communication by applying principles of computer science, linguistics, and cognitive psychology to the knowledge structure of the discipline. The roles of neuroanatomical nomenclature and ontology in this process depend on the kinds of obstacle one aspires to eliminate. Two approaches are common. The first uses primarily nomenclature and principles of computer science to create internet links to virtual books and libraries. Libraries, publishers, and book sellers can speed communication considerably by eliminating the immense expenditure of time required to acquire hardcopy publications. Researchers save the time required to travel to libraries. The second approach, in which ontology supplements nomenclature, applies computer technology to organize information in database formats that, in addition, can virtually eliminate the search time necessary to locate information and to assemble it in a useful format. The power of the latter approach, which is now pursued intensively in a few scientific fields, such as genomics and proteomics (Biology Workbench, 2002; Chapter 9, this volume), is only beginning to be recognized in neuroscience (Kötter, 2001). Search Google (2003) for a genomics term such as "DNA sequence for type 1 adenylyl cyclase" and you get 2810 listings, including a direct link to the enzyme's GenBank ID (IDentification code) prominently displayed on page one. This is good. Search for "connections of the olivary pretectal nucleus" and you get 78 listings, but you search in vain for an unambiguous link to a concise, comprehensive list of afferent and efferent pathways. The difference is attributable to lack of adherence to a common ontology in neuroscience.

If neuroscience is a laggard in the application of informatics to communication about the brain, it is not for lack of need (National Institute on Drug Abuse [NIDA], 2002). Neuroscientists, like their colleagues in other disciplines where the numbers of scientists have grown into the tens of thousands and technological advances, have sped the flow of data to a torrent, are threatened with inability to know what is known, much less to integrate that knowledge into a satisfactory conceptual framework. But, if necessity is the mother of invention, feasibility is the father, and neuroscience is not genomics. In genome projects the first informatics needs were to develop fast software that would reconstruct genes from fragmentary overlapping sequences and to develop Web-accessible databases of gene sequences where investigators could make comparisons and bank new sequences for others to view. While these goals may have provided a significant challenge to computer science, the challenge in terms of nomenclature and ontology was minuscule. The basic concepts were four functionally distinct nucleotides of DNA, each of which had a unique name that was universally accepted by the genetics community, and linear sequences of nucleotides that could be specified in a one-dimensional conceptual model readily represented in digital format.

By contrast, the basic concepts of neuroanatomy are some 1000 cell groups with potentially distinct functions and multiple names for each; their spatial relationships are expressed in a three-dimensional conceptual model based on visual features that, until recently, were not readily classified in digital format, and their functional relationships are expressed by an equally large and different vocabulary referenced to a different conceptual model in which structures are grouped on the basis of their connections. One need not look beyond problems of neuroanatomical nomenclature and ontology to see why Google cannot provide a direct link to a concise, comprehensive list of connections of the pretectal olivary nucleus. The necessary nomenclature and ontologies are rapidly emerging, but the codification of knowledge about the brain is not sufficiently implemented to allow retrieval of much information at that level of detail.

2.2 NOMENCLATURE

Language gives humans the capacity to communicate about *entities*, which in neuroscientific discourse usually reduces to objects, such as the brain, and events, such as neuronal firing patterns and behavior. Scientists communicate about entities in terms of *concepts*. As primates endowed with hypertrophic visual, auditory, and vocal systems, scientists communicate concepts largely by auditory and visual channels using *words* and *images*, respectively. The capacities and limitations of the human as a transmitter, repository, and receiver of verbal and visual information determine goals and constraints in the development of informatics tools for more rapid, accurate, and complete communication.

Entities are the subjects of scientific study and reporting. When the cognoscenti palpate an elephant, entities are the elephant and its parts. Concepts, on the other hand, are the cognitive representations established in the brains of the cognoscenti as they palpate, and words are the linguistic codes by which they transmit and receive each others' concepts. The job of the scientist is ultimately to identify and represent entities in terms of useful concepts that they can share with others. The job of the informaticist is to analyze the structure of scientific nomenclature and concepts in order to construct an ontology. An ontology is a codification of the relationships of words to concepts that information storage and transmission systems need to maximize the efficiency of scientific communication.

Neuroanatomical nomenclature, the essential core of a neuroscience ontology, is here regarded as the complete set of words that refer to structures in the nervous system. It is a subset of the neuroscientific nomenclature, which includes names for methods, events, neurochemicals, genes, and other entities involved in neural functions. The neuroanatomical nomenclature includes all words that neuroscientists, clinicians, and teachers use to describe and store information about the physical organization of the nervous system. It is composed largely of nouns—that is, the names of neural structures, which are the entities of neuroanatomy.

The challenge of codifying the nomenclature of neuroanatomy is substantial. The structural concepts of neuroanatomy, which represent neuronal cell groups and combinations of cell groups, number in the thousands, enough to support a rich stream of communication and an entire subdiscipline of anatomy. Unfortunately for communicators the names for those concepts number in the many thousands, an average of six English and Latin names per concept for the average "classical" structure (Table 2.1; Bowden and Dubach, 2002a).

The historical reasons for redundancy are clear. When *classical neuroanatomy*—that is, the classification of brain structures on the basis of topology and stains for gray and white matter—passed through its exponential growth phase in the late nineteenth and early twentieth centuries, neuroanatomists developed a systematic Latin nomenclature to allow efficient communication among individuals who spoke different languages (International Anatomical Nomenclature Committee [IANC], 1983). At the same time, clinicians in various countries developed nomenclatures in their own languages. Many hybrid terms emerged in each language as authors adapted the Latin names to their own language. Sometimes they spelled Latin names according to the rules of their own language (e.g., "Sulcus precentralis" for "Sulcus praecentralis") or changed the order of nouns and adjectives to conform to the syntax of their own language (e.g., "precentral sulcus") or translated the Latin roots into their own language [e.g., "Brückenkerne," German translation of "pontine nuclei" ("bridge kernels" in English)]. Most of the terms in every language resulted from pidginization with Latin, although some of the multiple names for structures were derived from entirely different roots. Field H1, thalamic fasciculus, and Area tegmentalis, pars dorsalis, for example, all represent the same structure (Table 2.1).

While terminological redundancy is understandable, it imposes a much greater waste of time and effort on students

and practitioners of neuroscience than is commonly recognized. Students who learn a few hundred concepts to master the basic structure of the brain must spend many additional hours learning some of the thousands of additional names that they will need if they are to recognize the concepts in their reading. Psychologists, molecular biologists, and other scientists who come to the study of the brain from other disciplines are discouraged from thinking about the implications of their findings in terms of brain function because of the effort necessary to become oriented to the brain literature. Even neuroscientists must keep several textbooks and atlases at hand when reading outside their area of expertise, and authors of neuroscientific papers often find themselves compelled to define concepts by multiple terms and provide a glossary to assist their readers. Finally, the lack of a comprehensive nomenclature and ontology prevents the creators of databases from indexing information at the level of anatomical detail necessary for efficient retrieval. This is the single greatest obstacle to interoperability of neuroscientific databases.

A person seeking information about a brain structure on the Web ordinarily has three basic ways to locate it. The first is through a *menu* or a *site map* within a website. In this kind of approach the computer simulates a library directory or the table of contents of a book. The searcher must find websites where the information is likely to exist, intuit the category or location in the website where the information is stored, and navigate through pages of the website to see if it is there. Most neuroscientific websites [e.g., sites supported by the Human Brain Project (HBP, 2002)] and publishers' websites [e.g., Neuroscion (http://www.neuroscion.com/features.html)] are currently organized on this principle.

The second way is through a *scrollable index* within a website that lists all of the structure names in a website's nomenclature. Again, seekers of information must first find a website where the information exists. There they scroll to the name for the structure of interest adopted by the website and click to view a definition or image. Examples of such sites include the Talairach Daemon (TD, 2002) and The Whole Brain Atlas (WBA, 1999).

The third way to access neuroanatomical information is through a *search box*. The user types the name of a structure into the box. The computer (server) of the website or search engine analyzes the entry, searches its files, and most commonly displays a list of possible websites or publications where the information may exist. Informaticists have developed a number of software tools based simply on nomenclature to reduce the amount of knowledge of websites and website organization that a user must bring to the search. For example, the Brain Architecture Management System (BAMS, 2003) provides a search box through which one can scan the nomenclatures of seven brain atlases representing five species (human, macaque, cat, rat, and mouse) for atlases that use a particular structure name (Figures 2.1 and 2.2).

Search engines, such as Google, represent an all-purpose application of the search box method to the entire Web. These are websites that scan the Web as a whole and index other websites on the basis of the words they contain. Google has indexed several billion web-pages according to the words contained in the first portion of each. When a user submits a character string, Google searches its index for pages that contain the words in the string. It displays a list of sites and pages organized such that

TABLE 2.1 Field H1 Neuroanatomical nomenclature

English	Latin
H1 bundle of Forel	Area subthalamica tegmentalis,
H1 field of Forel	pars dorsomedialis
Tegmental area H1	Area tegmentalis H1
Thalamic fasciculus	Area tegmentalis, pars dorsalis
Forel's field H1	Campus Forelli (pars dorsalis)
Field H1	Fasciculus thalamicus
	Fasciculus thalamicus hypothalami
	Forelli campus I

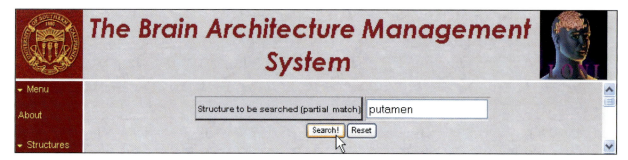

Figure 2.1 A user has entered "putamen" into the search box of the Brain Architecture Management System; when they click "Search" the system will display links to several sources of information about the structure (Figure 2.2) (BAMS, 2003).

websites that have been selected most frequently by other users and extracts in which the words appear closest to each other are displayed first.

As an informatics tool for neuroscience, the search engine provides a good start, but no more. The list of 78 websites and pages returned from our search for "connections of the olivary pretectal nucleus" represents a remarkable sifting of relevant sources from three billion possible pages. Nevertheless, one

can spend several hours searching through 78 websites looking for a summary table of connections. One cannot be sure such a table exists, or, if it exists, that its location is in the list. Perhaps it does not appear in the list because the person who created it used different words. A Google search for "connectivity of olivary pretectal nuclei (*instead of nucleus*)," for instance, returns only 11 citations, and "hodology (*synonym of connectivity*) of olivary pretectal nuclei" returns none. Thus,

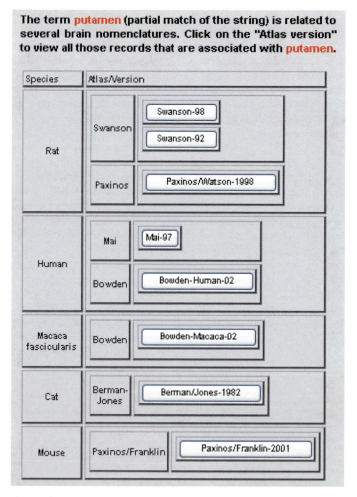

Figure 2.2 A search box query (Figure 2.1) yields a list of seven brain atlases in five species that use the term "putamen." If the atlas is on the Web, one can navigate to view the structure there; if not, one can view a full citation of the source and a tabulation of information extracted from it (BAMS, 2003).

a search based on character strings analyzed exclusively on the basis of space and punctuation characters can only take one so far. The bare words as character strings carry a certain amount of information necessary to locate other information, but often not enough.

Other websites apply more complex nomenclature-based tools to narrow the search; they extract more information from the character string. For example, PubMed, a Web-based service of the National Library of Medicine (PubMed, 2003), identifies the words in a character string and converts the characters to lowercase. It then converts each word to its uninflected "normalized word" form—that is, strips the word of characters such as the "s" or "es" of plural forms, the "apostrophe s" of possessive forms, and so on. If the term consists of several words, the computer strings the normalized words together, deletes conceptually unnecessary words such as the "the" and "of" of "ventromedial nucleus of the hypothalamus," alphabetizes the remaining words and arranges them, separated by spaces, to create a *normalized string* for the term, for example, "hypothalamus nucleus ventromedial." This string is used to search the database, which in PubMed is a collection of citations, key words, and abstracts of several million articles representing the world's medical literature of recent decades. PubMed also uses a number of other strategies to squeeze the maximal amount of information from user-supplied character strings. If one submits the string "connections of olivary pretectal nucleus" to PubMed, one receives a list of 10 citations, many fewer than from Google, but much more focused. The number of false positives is reduced enough that the user can determine rapidly from the abstracts that the collated information does not exist in these references. So search time is reduced. However, the information has not been found.

When information seekers submit words to Google or PubMed, they are providing the computer clues to the location of the information they want. When Google displays links to websites or PubMed displays citations of articles, they usually are not providing the actual information that the user seeks, but clues as to where it might be located. Google provides links to websites with pages containing the words of interest. It may also provide an indented paragraph in which the search words appear highlighted, embedded in phrases with a few preceding and succeeding words to show the context. PubMed returns the title, authors, journal name, and usually an abstract of the article that provides searchers a summary of the kind of information to be found if they retrieve the article. Once one arrives at the destination, one can use the FIND function of the browser to search the document and locate specific terms.

While the Web reduces to zero the effect of distance on information display time and nomenclature-based tools can substantially reduce the time required to locate the information, tools that are strictly nomenclature-based still require the knowledge seeker to spend large amounts of time finding information and assembling it into a useful format. With the current state of neuroscientific database development, the example of a search for connections of the olivary pretectal nucleus illustrates the limits of search capability based on nomenclature. For the computer to assist further, it must know more about the meanings of words. It must know about the concepts to which words refer and the relationships among those concepts. That kind of information is contained in an ontology.

2.3 ONTOLOGY

According to current thinking, human language emerged evolutionarily not as a steady accumulation of disconnected words that gradually became organized into complex grammatical forms, but as a brain function with all of the syntactic and semantic capacities of modern languages (Pinker, 1994). The *syntax* of a language is a codification of rules by which linguists attempt to capture the regularities of word order, parts of speech, prefixes, suffixes, and so on, that convey information in a given language. *Semantics* is a specification of the links of words to the concepts they represent. The computer can speed the search for information substantially if it can relate the words in a search string to concepts in a database. If the computer knows the knowledge structure of a discipline—that is, the relations of words to concepts and the relations of concepts to each other—it can build a bridge between the character string and the knowledge base that will conduct users much closer to the information they seek. Ontology specifies such relationships.

Some definitions of "ontology" are very broad: "the science or study of being; a doctrine of the being and relations of all reality" (OED, 2003). Others restrict its meaning to the expression of relationships by words and symbols: "human-readable text describing what the names mean, and formal axioms that constrain the interpretation" (Gruber, 1993). Since much neuroanatomical knowledge is most informatively communicated by images as representations of concepts, a more inclusive definition is in order. For our purposes, a *neuroanatomical ontology* consists of human-readable text and human-recognizable visual images specifying relationships between the words of a nomenclature and the operationally defined structural concepts they represent (Bowden and Dubach, 2002a).

From a linguistic point of view, the ontology that relates most neuroanatomical nomenclature to the brain is relatively simple. All of the words are nouns or adjective–noun combinations that serve as nouns. So the syntactic complexity of the character strings that represent brain structures is minimal. Second, the semantic information in neuroanatomical terms refers to objects that are conceptualized as visual entities. (Though, in an exception that perhaps proves the rule, the first written reference to the brain was to a tactile entity. It was recorded by an Egyptian battlefield physician writing recommendations for triage of warriors with penetrating head wounds. It suggested that the examiner should palpate the open wound and finding "something therein throbbing (and) fluttering under thy fingers . . . and he [the patient] discharges blood from both his nostrils . . . an ailment not to be treated" (Gross, 1997).)

An ontology is defined for a particular domain of knowledge. Neuroanatomical ontology in the broadest sense applies to structural concepts of the nervous system on a scale extending from the organ level (neuraxis = brain + spinal cord) down to the level of organelles and even molecules specific to neural tissue (Martin et al., 2001). More narrowly, it refers to the topological anatomy of the nervous system which, in the brain, extends from the level of brain, as a suborgan of the neuraxis, down to the level of subnuclei and cortical layers. Substructures of the brain down to the level of gyrus or nucleus tend to have unique names, such as "superior frontal gyrus" or "abducens nucleus," and can be considered *primary structures*. Beyond that level, structures have generic names, and one must specify a context in the form "of the . . ." along with the name to produce an unambiguous term for the structure, such as "molecular layer of the dentate gyrus" or "caudal part of the spinal trigeminal nucleus." In mammalian brains, as opposed to some invertebrates, structures below the layer/subnucleus level (i.e., individual cells) do not have names but are assigned to classes (e.g., "pyramidal cells") on the basis of structural features. A neuroanatomically unambiguous term for the cells of a structure thus takes the form "pyramidal cells of the external pyramidal layer of area 4 of Brodmann." These characteristics of the nomenclature reflect the way neuroscientists think and communicate about the brain and have implications for the organization of a database in which information is referenced to neuroanatomical structure.

2.4 BRAIN ATLASES

Names communicate concepts. Since the concepts of brain structure are visual, the core of a neuroanatomical ontology is a brain atlas that defines by illustration the relations of names in the nomenclature to visual concepts of structures as represented in the atlas. Brain atlases vary greatly in the kinds of information that they present (Bowden, 2000a; Swanson, 2000a; Chapter 23, this volume). Atlases for teaching neuroanatomy commonly present surface views and cross sections of the brain with labeled arrows pointing to the centers of structures. Atlases for locating sites for implantation of cannulas for intracerebral injections or electrodes for stimulation and recording may be less detailed but provide scales for use with stereotaxic instruments. Brain atlases for informatics—for use in mapping anatomical data into image databases and for quantitative structural analyses—require a number of additional features (Table 2.2; Bowden, 2000b).

The most distinctive characteristics of an atlas in which structural concepts are used for spatial indexing of data are the definition of structures by boundaries and the comprehensive segmentation of images into primary structures. Every point in the brain is identified unambiguously with a structure and only one structure at the finest level of segmentation (Hohne et al., 1992). In the past decade, atlases that meet these requirements have become available in printed format for several of the species most studied by neuroscientists, including the mouse (Paxinos and Franklin, 2001), the rat (Paxinos and Watson, 1998), and the macaque (Paxinos et al., 2000). Some are available

in digital format on compact disk, including the mouse (Hof et al., 2000), the rat (Swanson, 1998), the macaque (Martin and Bowden, 2000), and the human (Mai et al., 2003). On the Web, one has access without subscription to a digital stereotaxic atlas of the mouse brain segmented at low granularity (47 volumetric structures; GenePaint, 2003; Visel et al., 2002) and a macaque brain segmented at a higher granularity (476 volumetric structures; BrainInfo, 2002; Bowden and Dubach, 2002b). Each of these atlases provides the basic elements of a neuroanatomical ontology of the species it represents. All provide a name and, in principle, a bounded visual concept for each primary structure of a comprehensively segmented brain, albeit at different levels of granularity. The most thoroughly segmented atlases, which show the boundaries between layers in the cortex and between subnuclei, show on the order of 1000 structures at the most granular level; atlases segmented only to the level of primary structures show about half that number. Some atlases go a step further toward defining an ontology by specifying common synonyms and by defining a hierarchy of structures and superstructures that models their organization within the brain (Swanson, 1998; Martin and Bowden, 2000; BrainInfo, 2002).

2.5 CONCEPTUAL RELATIONSHIPS

Models are overarching concepts that represent scientists' understanding of how the underlying entities relate to one another (Chapters 20–22, this volume). Most neuroanatomical terms belong to three basic kinds of model: partitive hierarchies, categorical hierarchies and systemic models.

2.5.1 Partitive Hierarchies

In primate neuroanatomy the model that accommodates the largest number of terms is a *partitive hierarchy* based on the classical Nomina Anatomica (IANC, 1983). The Encephalon division of the Nomina Anatomica divides the brain into three parts: forebrain, midbrain, and hindbrain, each of which is subdivided into parts that are further subdivided through as many as nine levels to the level of primary structures. The Nomina Anatomica provides the skeleton for a model of the brain as a visual entity, like a three-dimensional puzzle that can be taken apart in successively smaller pieces down to the level of primary structures.

TABLE 2.2 Comparison of conventional atlases and informatics-oriented atlases of the brain

Conventional Brain Atlas	Informatics-Oriented Brain Atlas
Printed format	Digital format
Structures designated by labels at centers	Structures defined by boundaries
Many areas undefined	Comprehensively segmented
Revised editions at 10- to 20-year intervals	Continuously updated
Approximate mapping of data to structures	Unambiguous mapping of data to structures
Few tools for manual mapping of data	Many tools for automated mapping of data
Stereotaxis referenced to a few cranial or commissural landmarks	Spatial indexation (mapping) referenced to hundreds of local anatomical landmarks
No mechanism for display of mapped data	Overlays for comparison of multimodal mapped data
Query by atlas-specific structure names	Query by any name or by spatial coordinates
High resolution in two dimensions	High resolution in three dimensions

Virtually all modern textbooks of neuroanatomy organize information about the brain according to the basic model set forth in the Nomina. The Nomina was developed to establish a common Latin nomenclature for concepts that an international committee of neuroanatomists could agree should be represented. It was not intended to be a comprehensive nomenclature, and so it only covers 416 concepts, about 50% of the concepts described and illustrated in comprehensive textbooks. Some textbook authors have undertaken to extend the partitive hierarchy in the Nomina to include many primary structures not covered there. Authors who have extended the Nomina approach most systematically include Crosby et al. (1962), Carpenter (1991), Gray et al. (1995), Schiebler et al. (1999), and Kahle (2001).

While the Nomina-based hierarchy accounts for the largest single grouping of neuroanatomic concepts, primary structures of the brain are also grouped in many smaller partitive hierarchies to model relationships according to other models of brain organization. For example, the term "basal ganglia" has two very common and many less common definitions (Anthoney, 1994). The Nomina definition, which originated in the earliest days of brain dissection, groups structures into a four-level hierarchy on the basis of proximity within the forebrain. The definition most widely used in clinical circles, on the other hand, groups structures on the basis of connectivity and mutual involvement in clinical syndromes; so in addition to forebrain nuclei, it includes the substantia nigra, a midbrain structure. Similarly, the definition of the hippocampus in the Nomina Anatomica, which is a subdivision of a temporal lobe structure in the human, is different from a multiple-level definition based on comparative anatomy and developmental studies that include the supracallosal gyrus and paraterminal gyrus of the limbic lobe (Stephan, 1975). And the nuclei of the hypothalamus are grouped by two equally time-honored approaches, by coronal planes into anterior, intermediate, and posterior regions and by sagittal planes into periventricular, medial, and lateral zones (Anthoney, 1994). A comprehensive neuroanatomical ontology must be capable of linking the same structure to any number of hierarchies.

Printed textbooks and brain atlases are essentially neuroanatomical ontologies in illustrated narrative format. Many are organized on the partitive principle. The most comprehensive ontology coded in digital format for computerized indexing, search, and retrieval of information about the brain is NeuroNames (Bowden and Martin, 1995; Bowden and Dubach, 2002a). Based on a survey of more than 70 neuroanatomical textbooks, atlases, and articles, NeuroNames contains 879 concepts related to one another by the classical partitive hierarchy. The survey identified 592 concepts of *primary volumetric structures* in the brains of primates (human and macaque combined); about 300 of the primary volumetric structures were identified as having homologs in the rat brain (Bowden and Martin, 1997). The primary structures are grouped into a hierarchy of 112 *superstructures* (e.g., frontal lobe and pontine tegmentum) and associated with 175 *superficial structures* (e.g., lateral sulcus and inferior olive). A database that references information to structures in the classical partitive hierarchy of the brain can provide very specific and comprehensive answers to questions of the form, "Where is structure X?" and, when paired with databases containing other kinds of information, to questions like, "What is known about structure X? What structures and data have been mapped to it?"

Despite limitations (see "Systemic Models" below), the classical nomenclature and partitive ontology play an inescapable role in communication about location in the brain. They form an essential key to the neuroscientific literature of the past and provide the framework to which all more informative nomenclatures and ontologies are referenced. The 500+ classically defined primary structures of the human brain are the landmarks whereby everyone from the neuromolecular biologist to the practicing clinician recognizes, remembers, and communicates about the location of neural circuits, lesions, and specific structural and functional markers.

In addition to structures in the classical hierarchy, NeuroNames currently defines some 1200 further concepts in relation to those structures. These include cortical areas defined by architectonics, alternative segmentations and groupings of subcortical primary structures, embryonic precursors to adult structures, and so on. The classical structures are the lab scientist's link to the existing literature and the clinician's link to the multiple subdisciplines of neuroscience (Bowden and Martin, 1997; Sato et al., 1993). NeuroNames provides the core nomenclature for indexing BrainInfo, a neuroanatomic knowledge base on the Web (BrainInfo, 2002; Bowden and Dubach, 2002b). The terminology of the NeuroNames hierarchy has been incorporated into the Foundational Model of Anatomy, a systematic nomenclature and representation of concepts and relationships that describe the physical organization of the body (Martin et al., 2001). It is made available for use in other neuroanatomically referenced systems as a source vocabulary of the National Library of Medicine's Unified Medical Language System (NLM, 2003; Hole and Srinivasan, 2002) and is also distributed to individual laboratories on compact disk through the University of Washington (WaNPRC, 2002).

2.5.2 Categorical Hierarchies

Primary structures are also related in *categorical hierarchies*, commonly known as "-is a- hierarchies" (Rosse et al., 1998a,b; Martin et al., 2001). In a categorical hierarchy structures are grouped on the basis of common features. A structure inherits features of its category and adds new features of its own. The neuraxis is an "organ with organ cavity." As an organ, it shares characteristics of other organs as "self-contained units of macroscopic anatomy morphologically distinct from other such units." As an "organ with organ cavity," it shares characteristics of other organs with cavities, such as the heart and stomach, but is also distinct in that, for example, it "consists of gray matter and white matter." The brain, which shares those characteristics with the neuraxis as a whole, is considered an "organ part."

In another categorical hierarchy (Figure 2.3), all neuroanatomical entities of the neuraxis are represented in a categorical hierarchy as volumetric or superficial; volumetric structures can be parenchymal or fluid; parenchymal structures can be predominantly gray matter or white matter, and so on. Such a model is useful in generating visual displays of the brain in which different kinds of structure are represented by different colors. Indexation of information about the brain in both categorical and partitive models enables the computer to provide comprehensive answers to queries such as, "What nuclei are found in the midbrain tegmentum?" or "What are superficial features of the frontal lobe?"

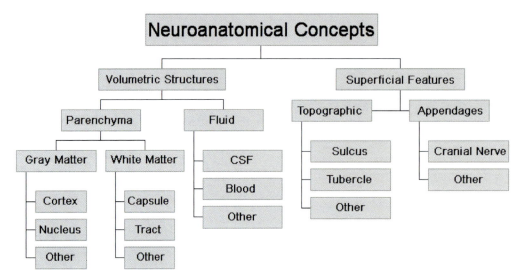

Figure 2.3   A categorical hierarchy classifies concepts on the basis of common characteristics.

2.5.3 Systemic Models

Finally, primary structures of the brain are commonly, and from a functional viewpoint most usefully, related in *systemic models*. In such models, structures are grouped into anatomically overlapping circuits and networks that mediate physiological and behavioral functions. The partitive and categorical hierarchies that group structures on the basis of proximity and topographic features are necessary parts of a comprehensive neuroanatomical ontology, but their utility in communication about brain function is limited (Bowden and Martin, 1997). While Nomina Anatomica is accepted as the most authoritative reference for neuroanatomical nomenclature, most neuroscientists recognize its limitations as a medium for communication about the functional organization of the brain.

Different ontologies are based on different methods of partitioning the brain, and methods of partitioning can help or hinder communication about the subject matter. The *classical nomenclature* of Nomina Anatomica is based on a partitioning method that dismembers pathways into nuclei, tracts, and terminal areas. Because of this methodology, Nomina Anatomica more often than not assigns the cell bodies and the axonal projections of a pathway to different structures. This makes for an extremely awkward linguistic medium for communicating about structure–function relationships. Consider, for example, trying to describe the role of motor cortex in the control of arm movements using the phrase "tract that runs from the precentral gyrus through the cerebral white matter, internal capsule, cerebral crus, corticospinal tract of pons, and pyramidal tract to the ventral horn of the spinal cord" instead of "corticospinal tract."

Whereas the primary concept of partitive brain models is the primary structure, the primary concept of systemic models is the connection between one brain structure or "place" in the brain and another (Bloom et al., 1990). While the nomenclature for systemic models is less standardized than that for the classical partitive model, the concepts fit two general categories, topological models and hodological (connectivity) models. In topological models, neuroanatomists have a nomenclature for tracts, fasciculi, bundles, and so on, that connect the classical cortical and nuclear structures, as in the "corticospinal tract" example above.

This nomenclature, which references concepts defined operationally by dissection and the gray/white dichotomy, is the core content of most neuroanatomy textbooks. The structural names are derived from the classical nomenclature, so the connections they represent are easily recognized by a person who knows the classical nomenclature, and their relation to function is sufficiently precise to meet communication needs of neurologists and neurosurgeons. Most clinical syndromes in which anatomical site is a determining factor, such as traumatic injuries, tumors, and infections, involve lesions that affect all of the axons in a given area without regard to their source and target structures. More specific knowledge as to the internal structure of the area is largely academic. Most important from the point of view of diagnosis, therapy, and prognosis are the signs, symptoms, and location of the lesion in a classical sense.

From a research viewpoint, however, the classical topological model and related nomenclature have long been inadequate vehicles of communication. As understanding of brain function has progressed from a concept of "brain centers" to "neural systems," the functional unit of the brain has become a set of similar neurons located together and having similar connections. Since the classical tracts and fasciculi evident by dissection typically contain connections in both directions between diverse groups of cells that are located in many different structures, a systemic model in which connecting structures are limited to classical tract concepts cannot reference functional systems in sufficient detail.

A *hodological nomenclature* has developed based on *connectivity*. Here key words are "pathway," "projection," "afferent," "efferent," "myelinated," "unmyelinated," and synonyms thereof. Such terms refer to connections between cell groups defined at the level of nuclei, subnuclei, and layers in architectonically defined cortical areas. Connections are defined between "source" and "target" cell groups which may be defined on the basis of neurotransmitters or gene products rather than the classical Nissl and myelin stains. They can be categorized as unidirectional or reciprocal. They can also be strung together to form "*circuits*," where successive connections link back to a given source. Cell groups that project to more than one other group become "nodes" of more extensive "*networks*." Since the same subnucleus can contain cells of different classes that project to, or receive projections from, specific classes of cell in other areas, and since the same axon can send collaterals by different pathways to more than one cell group on the same or opposite side of the brain, systems modelers have developed names and symbols for ever more precise concepts to enable unambiguous communication about connections in neural networks (Arbib, 1995). A database that indexes information according to partitive, categorical, and hodological models can provide comprehensive answers to queries such as "What nuclei in the brainstem receive projections from the olivary pretectal nucleus?" and "Are there superficial features that mark these nuclei?"

As models of the brain evolve, they become more abstract. Nonanatomical attributes come to outweigh anatomical attributes in importance. Whereas in the classical partitive model nonanatomical attributes—like lesion effects, neurophysiological responses, and classes of cell present—are indexed and stored in a database as attributes of primary structures in the classical nomenclature, in connectivity models the nodes are more likely to be stored as attributes of subnuclei, architectonic areas, and specific layers thereof. This gives rise to the need for an ontology that can handle multiple overlapping pathways that project from multiple cell groups through a common tract to other cell groups.

Brain Browser (Bloom et al., 1990), a HyperCard application distributed on floppy disks before the appearance of the Web, is a neuroanatomically referenced database of information on the rat brain. Each connection is stored separately as an attribute of the place where it originates (efferent connection) and as an attribute of the place where it terminates (afferent connection). A "place" can be any locational concept the data enterer wishes to name—a primary structure, superstructure, layer, nucleus, or a cell group characterized by expression of a particular gene. Some attributes of the connection, such as cells of origin, neurotransmitters, or cotransmitters, are linked to the place of origin; others, such as receptors and second messengers, are linked to the target. Connections are displayed visually by acronyms for the source and target places separated by an arrow pointing toward the target—for example, NTS → DMTg for a projection from nucleus of tractus solitarius to the dorsomedial tegmental area. The Brain Browser database was distributed containing some 1000 connections. Designed to be expanded and used by an individual scientist or laboratory, the database contained little ontologic information, such as codification of synonyms or relationships among the places joined by different connections.

BAMS (2003) also provides information on the Web about connectivity in the rat brain. More than 5000 connections are indexed according to a systemic model in which the important features of a connection are the structure where it originates and the structure where it terminates. Responding to a query about afferent connections of the caudatoputamen, the system produces a table where the cell groups of origin of 17 pathways to the structure are listed (Figure 2.4).

CoCoMac (2003a; Stephan et al., 2001) is another website that provides connectivity information to the neuroscientific community. Continually updated, it focuses on connections in the macaque brain. Again, information is linked primarily to the structures of origin and termination of each connection. An ambitious goal of CoCoMac is to allow visitors to the website to conduct automated analyse of the database to identify connectivity patterns based on studies by many investigators using a variety of methods. Considerable attention is given to coding attributes of structures in sufficient detail to allow the users to combine and compare data across studies. The structural attributes available for comparison include the spatial precision with which structures of origin and termination are defined, the degree of overlap between structures, the methods by which structures and connections were demonstrated, and other indices of the strength of evidence for their existence.

The result is an exhaustive ontology as well as nomenclature of terms that have been used in publications about connectivity in the adult macaque brain. A systematic extraction of information from hundreds of reports published from the early 1900s to the present has yielded acronyms and terms for some 1000 structures, largely architectonic areas of the cerebral cortex and layers thereof. CoCoMac's authors have indexed some 5700 connections based on 1700 tracer injections (Kamper et al., 2002). The spatial relations of structures are reported in a tabular format that allows one, for example, to see that area FL in the cortical map of von Bonin and Bailey (1947) is identical to area 25 of

Afferent structure	Number of reports	Collator(s)
SNc	4	Gully Burns
SNr	4	Gully Burns
VTA	3	Gully Burns
RR	1	Gully Burns
MT	1	Gully Burns
GP	1	Gully Burns
GPm	1	Gully Burns
SI	1	Gully Burns
SGN	1	Gully Burns
SPF	1	Gully Burns
MGm	1	Gully Burns
LP	1	Gully Burns
MGv	1	Gully Burns
MGd	1	Gully Burns
PF	1	Gully Burns
TM	2	Larry Swanson
PPN	1	Larry Swanson

Figure 2.4 A table listing structures in the rat brain that project to the caudatoputamen includes a unique acronym for the structure of origin of each afferent pathway, the number of studies in which the connection was reported and the person who entered the information into the database (BAMS, 2003).

Brodmann (1909), a substructure of area 24 of Brodmann (1905), a superstructure of area 14 of Walker (1940), and overlaps area 24 of Walker (1940). Each statement of the spatial relationships among overlapping areas named by different authors is annotated with the citation of the literature source and page number. Connectivity relationships revealed by tracer techniques are stored for retrieval in a source–target format. Each report of a connection is annotated as to experimental conditions (e.g., subject characteristics, postmortem time), injections (e.g., tracer substance, concentration, volume, injection method, localisation), and quantitative data (e.g., exact numbers of labeled cells). Connections are represented in tabular form and in graphical image formats (Figure 2.5). CoCoMac's comprehensive nomenclature and ontology allow virtually all data on cortical connectivity in the macaque, published in 300 sources over the course of a century, to be communicated in a matter of minutes. Depending on the number of connections of interest, the same query would take hours, days, or even weeks to answer by conventional book and library mechanisms.

CoCoMac codes and indexes structures by the acronyms that authors commonly use to label cortical areas and subcortical nuclei—for example, "2" for area 2 of Brodmann (1909), or "FL" for area FL of von Bonin and Bailey (1947). In addition to displaying connectivity data in a variety of formats and images on the Web, CoCoMac distributes the nomenclature and ontology of acronym–concept relationships, including synonyms and homologies, in a variety of formats for use in other systems and for interoperability of such systems with CoCoMac (2003b).

2.6 EQUIVALENT REPRESENTATIONS OF CONCEPTS

2.6.1 Controlled Vocabularies

Different authors use different terms and images to represent the same concept. Thus, a neuroanatomical ontology can enhance communication by codifying synonymous relationships among terms with a common meaning. Synonyms and multiple images used to illustrate a given neuroanatomical structure are *equivalent representations* of the same concept. The way in which a database handles equivalent representations has great practical significance.

A computerized database handles the representations of concepts by assigning every name, every image, and every concept a unique ID (identification code). Relationships are stored as links between IDs. New information about a structure is stored by links to its IDs. Under some conditions, it may be desirable to allow users of a database to use any accepted synonym or image to represent a structure. To this end, one could theoretically create a database in which every new piece of information about the structure is linked to each of the synonyms and illustrated in relation to every image of the structure. The number of links in such a database, however, would grow excessively as new synonyms, images, and concepts are added. A system based on this design would violate a basic principle of efficient database design, namely, that of *scalability*; one should be able to expand the database by orders of magnitude beyond its initial capacity without modification of the original design.

Untangling the brain

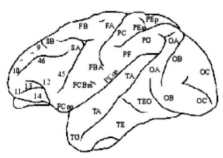

To overcome the problem of divergent brain maps we developed ORT (Objective Relational Transformation), an algorithmic method to convert data in a coordinate-independent way based on logical relations between areas in different brain maps.

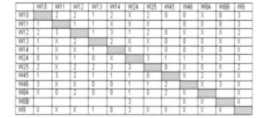

CoCoMac (Collations of Connectivity data on the Macaque brain) is our approach to produce a systematic record of the known wiring of the primate brain. The main database contains details of hundreds of tracing studies in their original descriptions. Further data are continuously added.

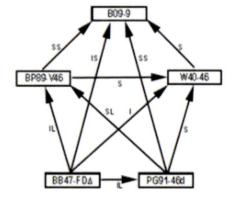

We use CoCoMac data to analyse the organisation of the cerebral cortex, and to establish its structure-function relationships. This includes multi-variate statistics and computer simulation of models that take into account the real anatomy of the primate cerebral cortex.

Figure 2.5 The Start Page of CoCoMac, a web repository of data on connectivity in the macaque brain that is based on a comprehensive nomenclature and ontology of cortical areas published in the neuroscientific literature from 1900 to the present (modified from CoCoMac, 2003b).

For database managers, the simplest solution to the synonyms problem is a *controlled vocabulary*. One of the names for each structure is selected as its standard term, and all data related to the structure are indexed to that term. The first controlled vocabulary for neuroanatomical concepts was the Latin nomenclature of the Nomina Anatomica. An example of a vocabulary developed specifically for digital database management is a controlled vocabulary of neuroanatomical terms posted by the Center for Bioinformatics (CBIL, 2002) and based on the Mouse Gene Expression Database of the Jackson Laboratory supplemented with human terms from Gray's anatomy text (Gray et al., 1995). Another is the set of default terms from NeuroNames (WaNPRC, 2002) used by BrainInfo (2002) to represent the classical hierarchy of brain structure. An informatics tool, Braintree (Toga et al., 1996) enables users of neuroscientific databases to "prune and graft" a comprehensive hierarchy, like that of NeuroNames, to create alternative hierarchies to guide quantitative analyses and for display purposes while preserving the underlying nomenclature.

A controlled vocabulary has the advantage of assuring that information is indexed by a person who knows it best at the time it is entered into the database. Communication is much more efficient if the problem of search time is resolved by precise indexation at the time of data entry than by burdening numerous users with the need to conduct independent searches through multiple imprecisely indexed databases. Controlled vocabularies, however, have the disadvantage that, unless different databases use identical vocabularies, they prevent interoperability. Furthermore, all potential contributors and users must know the standard terms specific to each database they use. Determining a system's name for a concept that the user has in mind can be time-consuming, and most systems currently do not explicitly define or illustrate the standard terms in their controlled vocabularies. A system can improve the efficiency of communication by helping a user identify its standard term for the concept the user has in mind. The number of words a person can recognize is some 10 times larger than the number of words the same person can recall and use in speech and writing (Pinker, 1994). A more usable database design for neuroscience, therefore, links new information about a structure to a single "*default term*," and it links the default term with all of its synonyms. One can search in one's own language for one's own preferred term and be linked through the default term to the available information. Likewise, such a database maps spatial information to coordinates in a *canonical brain* or "*template atlas*." The mapped data can be displayed at will as a set of overlays on a template drawing in the atlas. A unique link between the default term in the nomenclature and the outlines of the corresponding structure in the template atlas allows efficient storage and communication of information about the structure. NeuroNames (Bowden and Dubach, 2002a), for example, allows visitors to the BrainInfo (2002) website to query its database using any accepted neuroanatomical term in English, Latin, and five other languages (Bowden and Dubach, 2002b). This function is intended to aid scientists, clinicians and students who read English, but whose knowledge of neuroanatomy is largely in their native language. They can enter the database using any term with which they are familiar, learn immediately the website's default English term, and navigate to the information they seek.

Another tool made possible by inclusion of synonyms in an ontology is a software routine that recognizes truncated words. Both the accuracy of recording and the efficiency of information retrieval are increased if the computer can guide the user through a nomenclature that includes every term a user might use and provide prompts to the terms by which information is indexed in the database. One need not know the full synonym or how to spell it to find information about it. With an interactive system programmed to respond to queries that contain the least information necessary to identify a concept, one can enter the first few letters of any word from any name for the structure. The computer returns a list of synonyms that contain words starting with those letters together with the system's default terms for the structures. When the searcher recognizes the intended term and selects it, the computer proceeds to display the desired information. For example, the searcher may recall only that the name starts with "dark-" and sounds Slavic; they enter "dark" in a search box, and the computer returns synonyms with "nucleus of Darkschewitsch" in five different forms. Each is paired with the system's default spelling, which the searcher clicks to see a central directory of information about the nucleus (BrainInfo, 2002).

Template Atlases

A *template atlas* plays an analogous role in communication by images to the role of the default vocabulary in communication by words. Structural concepts are represented as boundaries mapped to the standard coordinate space of a canonical brain atlas of the species. The *templates*, pages in the atlas, may be based on the brain of one individual representative of the species or they may show the mean location of boundaries in a "*probabilistic* atlas" based on a number of individuals. For sites near a boundary, a probabilistic atlas provides the probabilities that the site is located in one or the other of the two structures that meet at the boundary. Such an atlas is highly useful for blind stereotaxic procedures, such as introducing an electrode into the brain without an MRI or other image to guide the implantation.

For *spatial indexation* of data, template atlases based either on representative brains or on probabilistic atlases are equally appropriate from an informatics viewpoint provided all of the structures common to the species are present. (A template atlas of the human cortex poses particular problems, because many sulci and gyri are so variably present that no template atlas is likely to be appropriate for mapping and comparing data from all individuals.) Except for use in blind stereotaxis, the exact coordinates of boundaries between structures in a template atlas are not as important as the degree to which the structures shown, and their spatial relationships, are representative of the species. The primary role of the template atlas is not portrayal of data but efficient linkage of the data via the atlas to overlapping data mapped from other sources. Several template atlases are now available for some species (see Section 2.2.3). Each represents the authors' best judgment as to landmark boundaries characteristic of the species. The utility of such atlases for databasing image information depends upon the extent to which they meet the criteria listed in Table 2.2.

Just as any synonym can serve as the standard term in a controlled vocabulary, any template that shows the boundary of a structure in a representative location relative to other structures is a legitimate tool for mapping spatial information. Just as linking new text information to the default name for a structure defines its relation to other concepts in the same nomenclature, mapping new data to a "default" template illustrates its spatial relationship to all other information mapped to the same template. Data from many sources are compared by mapping them to appropriate locations relative to boundaries in the template. Investigators have used this approach to present data from rodent studies in a standard image format for decades. Among the earliest applications was that of Ungerstedt (1971), who plotted the locations of noradrenergic, dopaminergic, and serotonergic pathways on standard templates from the rat brain atlas of König and Klippel (1963). Later German and Bowden (1974) were able to test hypotheses regarding the relationship of intracranial self-stimulation sites to the monoaminergic pathways by mapping some 900 sites reported in more than 20 studies by different authors to templates from the same atlas. In recent years more detailed atlases of the rat brain have been widely used to display spatial data in standard format (Paxinos and Watson, 1998; Swanson, 1998). Currently, templates in digital format are available on compact disk for the human (Mai et al., 1998), the macaque (Martin and Bowden, 2000), the rat (Swanson, 1998), and the mouse (Hof et al., 2000).

The linkage of a neuroanatomical nomenclature with a template atlas enables communication in image formats of information indexed according to various models of brain structure. A picture is worth a thousand words, but until the advent of the Web, a picture cost more to publish than a thousand words. As a result, the information that neuroscientists communicated in image format was largely limited to representative photomicrographs, summary diagrams and the like. With advent of the Web and powerful image processing software, it has become as easy and economical to distribute original neuroanatomic data in images as to distribute text descriptions. For indexing image data, greatest precision is achieved by reference to coordinates in a standard three-dimensional space, a template atlas. For linking the visual concept of data to other data and to a nomenclature, the greatest precision is achieved by mapping it to a template atlas.

The fundamental unit in a *neuroanatomical ontology* is the link of a default term in the nomenclature with a primary structure in a template atlas. Superstructures have unique names and are defined as parents of groups of primary structures; internal structures have generic names and are defined as children of primary structures. The *granularity* of an ontology in terms of nomenclature is measured by the smallest unit to which information can be indexed by name. For the brain, the smallest functional unit in anatomical discourse is a group of cells with common neurochemical characteristics and connectivity, which commonly reduces to a particular layer of an architectonically defined cortical area or to a subnucleus. In terms of spatial indexing, the granularity is determined by the resolution of the template atlas on which the ontology is based. Since all digital atlases currently available are collections of two-dimensional sections taken through the brain, the granularity of an ontology based on them depends on the distance between sections and the level to which the template atlas is segmented. Atlases based on sections taken at close intervals and segmented to the level of cortical layers and subnuclei define the brain with higher granularity than atlases based

on sections taken at greater intervals and less thoroughly segmented.

Spatial indexation of data to different template atlases poses an obstacle to interoperability similar to that posed by indexing textual information with different controlled vocabularies. Ideally, different image databases would map spatial data to a common atlas. In practice, however, different databases use templates from different atlases. Interoperability depends upon the incorporation of tools for "warping" data from one standard atlas to another. German and Bowden (1974) found self-stimulation sites mapped to two stereotaxic rat brain atlases, König and Klippel (1963) (K&K) and Pellegrino and Cushman (1967) (P&G). In order to include sites mapped to P&G templates in analyses of overlap between stimulation sites and the monoamine pathways mapped to the K&K atlas, they developed formulas for translation, rotation, and scaling of coordinates in P&G to the stereotaxic space of K&K. A number of image processing tools now enable the warping of image data in three-dimensional stereotaxic space (Toga, 1999). They will be useful for creating interoperable digital image databases that are not based on common template atlases.

2.7 OBSOLETE MODELS AND DEFAULT VOCABULARIES

Designing and populating a knowledge base to include every neuroanatomical model that has ever been described, while possible in theory, would be overwhelming and unnecessary, in practice. The neuroanatomical literature includes at least five conceptual models of the brain itself (Swanson, 2000b); extrapolate that to the hundreds of lower-level groupings of structures that various authors have regarded as significant, and one recognizes the magnitude of the challenge. The ontologist must restrict the domain of concepts and conceptual models accommodated by a database.

The models included in a database ontology will depend upon the purpose of the database. If the purpose is to provide a framework for organizing information useful to scientists and clinicians in the present and future, many models can be excluded on the basis that they have been superseded by models based on more definitive research. For example, Aristotle's dual brain model as composed of a large anterior brain (the cerebral hemispheres) and a small posterior brain (the cerebellum) has played no role in neuroscientific theory for more than a century. Similarly, models based on the concept that the cerebral cortex is essentially a thick protective covering to insulate the ventricles, the sites where the important mixing of "humors" takes place, are fascinating but only of historical interest.

The selection of models to incorporate into a neuroanatomical ontology can be a great scholarly challenge. On the other hand, the selection of words to be accommodated in the nomenclature is more straightforward. English and Latin neuroanatomical terms appearing in publications since the late nineteenth century probably number less than 10,000. The effort to codify them can be measured in person-years rather than decades and is, in fact, well underway. The NeuroNames nomenclature (Bowden and Dubach, 2002a) currently contains some 8000 English and Latin names of neuroanatomical structures. About 5000 of those are

Figure 2.6   Common English and Latin names for the olivary pretectal nucleus. Note that this term appears twice in the list of common English and Latin Names, because the same term is used in atlases of the rat brain and the macaque brain.

so little used in neuroscientific discourse that a PubMed search for them in articles between 1975 and 2000 yields no citations. Such terms are largely of historical interest and need not appear in the ontology of many brain databases. Unless a searcher is interested in synonyms per se, including them can be a distraction. When queried for synonyms, BrainInfo, which contains the entire NeuroNames nomenclature, first displays English and Latin names that have appeared at least once in PubMed in the last 25 years. From there visitors can elect to view all English and Latin names or all names in several languages (Figure 2.6).

The working nomenclature and ontology of neuroscience is continuously evolving. Just as some names and models become obsolete, others emerge. A database that is intended to accommodate the results of studies into the future must be designed to accept new synonyms for old concepts and names for new concepts. Up to now the identification and codification of terms has been performed by individuals with basic neuroanatomic knowledge who scan the literature for new terms and concepts. This is a very time-consuming process that becomes less and less efficient as the codified nomenclature grows and the number of new terms per hundred pages of text declines. A *compound term extractor* (Srinivas et al., 2003) is a software tool that, paired with a comprehensive nomenclature of existing terminology, can potentially speed the identification of new terms. Such a tool scans atlases and articles for words and compound terms that are nouns or noun phrases. The several thousand terms extracted from a particular publication can be screened against the existing nomenclature and reduced to a small enough number that a database curator can quickly review them and, as appropriate,

add them to the nomenclature as synonyms for existing concepts or labels of new concepts.

Even if the nomenclature adopted for a given database includes only terms used in recent decades, the ontologist is likely to need a default term for each concept for indexing and for communication among users of the database. Selection of default terms is a challenge. Neuroscientists would feel best served if the default terminology of every database were the terminology that they use in their own work. Since that is not possible, the ontologist must select default terms for concepts that will be most acceptable to the largest number of individuals likely to use the database and then provide tools for customizing displays to taste.

The most systematic database designers adopt a vocabulary from an authoritative atlas, textbook, or the UMLS of the National Library of Medicine (NLM, 2003) and modify it as necessary to accommodate the concepts in their database. The UMLS includes concepts from more than 100 source vocabularies representing all fields of medicine. For its own internal communication, the NLM designates a default term, or "preferred term," for each concept. Source vocabularies are ranked in a precedence hierarchy based on the number of concepts represented in each vocabulary, frequency of update by the organizations that maintain them, and adherence to principles of ontology definition that determine breadth of application (NLM, 2003). Having ranked the source vocabularies, UMLS developers determine the source of highest rank that includes the concept and assigns that source's term as the preferred term. The preferred term is used in UMLS communications about the concept—for example, in defining the concept and its relation to other concepts in the UMLS ontology.

The authors of the UMLS recommend that database developers in a specific discipline such as neuroscience select from the source vocabularies the vocabulary most appropriate to their use. The NLM's criteria for ranking source vocabularies may not be appropriate for a specific field. The largest, highest priority vocabularies of the UMLS are the NLM's own Metathesaurus and Medical Subject Headings (MeSH), the Diagnostic and Statistical Manual of Mental Disorders (DSM), and the Systematic Nomenclature of Human and Veterinary Medicine (SNOMED). These nomenclatures are rooted for the most part in clinical models that incompletely represent the organization of current concepts in a particular discipline. Thus, they may not be the best sources of default terms for a database intended to serve the community of scientists, students, and clinicians in a particular area.

Most developers of controlled vocabularies and default nomenclatures for digital databases draw standard terms from the terminology of one or a few authoritative texts or atlases. For example, the Mouse Brain Atlas (MBL, 2002) draws terms from the atlas of Franklin and Paxinos (1997). Digital atlases of the rat brain and human brain distributed on compact disk have been produced by professional neuroanatomists who developed their own comprehensive nomenclatures (Swanson, 1998; Mai et al., 1998). Two Web-based, interactive atlases, the human atlas of the Digital Anatomist (DA) (2002) and the macaque atlas of BrainInfo (2002), use the default terminology of NeuroNames (Bowden and Martin, 1995; Bowden and Dubach, 2002a). The English terms in NeuroNames were adopted largely from Carpenter and Sutin (1983), Crosby et al. (1962), and Paxinos (1990), with lesser numbers from some 70 other textbooks and atlases. The NeuroNames criteria for selecting a default term from among multiple synonyms were: (1) Avoid eponyms; for example, use "hippocampus," not "Ammon's horn"; (2) Avoid awkward Latin spellings; for example, use "locus ceruleus," not "locus coeruleus"; (3) avoid duplication of terms; if the same name is used for two different structures, select a different name for one of them; (4) avoid numeric designations; for example, use "superior frontal gyrus," not "Gyrus F1"; and (5) short is better than long; for example, use, "amygdala," not "amygdaloid nuclear complex" (Bowden and Dubach, 2002a). The NeuroNames nomenclature is made available for incorporation into computerized databases and websites that convey information about the brain. It is distributed in the form of Microsoft Excel spreadsheets that provide for each structure a default English term, a default Latin term, a unique acronym for use in labeling structures, a unique identification number (ID), and the ID of the parent in the classical hierarchy. It also provides the web address of the central directory of each structure in the BrainInfo website for use by database designers who wish to produce a system to be interoperable with BrainInfo (WaNPRC, 2002).

2.8 INTEROPERABILITY

The ultimate aim of neuroinformatics as an applied science is to make all information about the brain immediately accessible to everyone regardless of their location or the location of the repository where the information resides. The ultimate neuroanatomical ontology of the mammalian brain would be a universal nomenclature that includes all terms in all languages. The nomenclature would be linked to structures in three-dimensional template atlases of all species used in neuroscientific research and segmented to the layer/subnuclear level. Extended to the temporal dimension, such a system would include template atlases of each species at each stage of embryonic, fetal, and postnatal development (Toga et al., 2001; Toga, 2002).

No such system exists today. Nevertheless, some neuro-website developers are developing databases that will be compatible with such a comprehensive system. Current efforts at interoperability take three forms. The first and most common is a list of *links to websites*, hyperlinks to the home pages of other websites that contain information likely to be of interest to its visitors.

Greater specificity is achieved by *links to Web pages*, hyperlinks to the very page in another website where desired information is known to exist. BrainInfo (2002), which is primarily a repository of image information about the macaque brain, also serves as a portal to hundreds of illustrations of human and mouse brain structures located at other places on the Web. For example, BrainInfo presents a visitor seeking information about a structure like the hippocampus a Central Directory for information about the hippocampus (Figure 2.7).

If the visitor clicks "Show It!" a display appears that provides links to six different kinds of image showing the hippocampus (Figure 2.8). Then, if the visitor clicks the thumbnail that illustrates a dissection of the dorsal aspect of the hippocampus in the human, BrainInfo displays a page of the Digital Anatomist (DA, 2002) where that illustration is located (Figure 2.9). Or, if the visitor wishes to view a three-dimensional image of the hippocampus and related structures rotating in space, BrainInfo displays a videoclip at the Laboratory of Neuroimaging (LONI, 2002) website (Figure 2.10).

Links to Web pages represent a powerful advance toward the ideal of eliminating search and retrieval time for users of detailed information that is located at multiple, widely dispersed databases. Unfortunately, however, it requires more time and effort to maintain than the average website database manager is able to provide. The links by which the original website sends a visitor to other websites are the web addresses (URLs) of pages in those websites. Those links must be programmed manually by a "domain expert," or curator, at the original website. That curator must explore the Web for sites that have pages of suitable text and images. For each website found, the curator must record its address in the original website's database. Visitors only have access to the small amount of information to which the curator has created specific links. If the linked website is reorganized, which is common for actively growing sites, the Web addresses of pages may change. The curator must regularly search for broken links and correct them. If a linked site changes its modus operandi such that Web pages are generated "on the fly" from data in the database rather than residing in static pages, the pages may no longer have a unique address. In this case the original website can no longer send visitors directly to the information of interest. Thus, while the links to Web pages approach can meet the needs of the person seeking information, the efficiency of establishing and maintaining interoperability is severely limited.

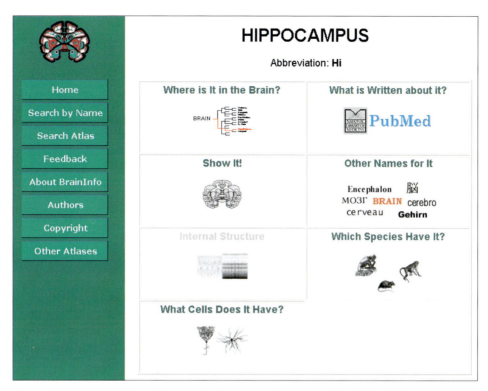

Figure 2.7   BrainInfo's Central Directory for the structure "hippocampus" (BrainInfo, 2002).

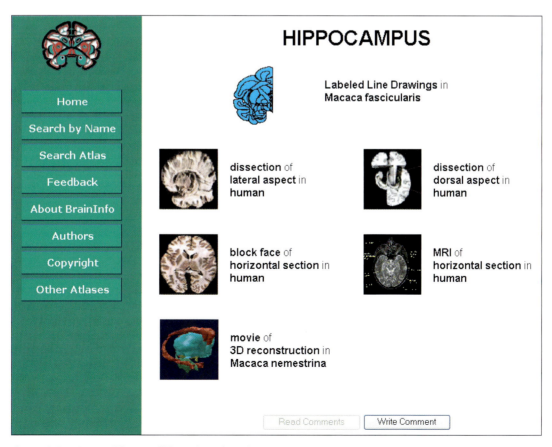

Figure 2.8   Thumbnail images of illustrations of the hippocampus available to a visitor to BrainInfo (2002).

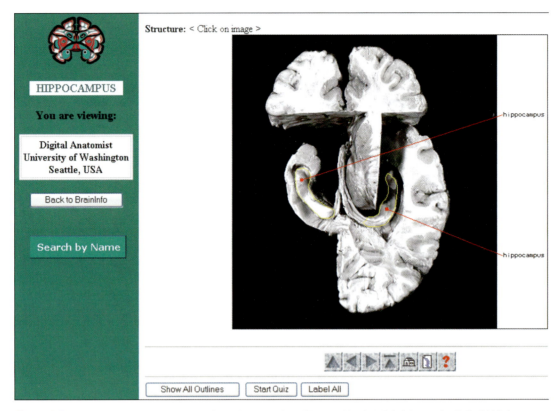

Figure 2.9 A dissection of the hippocampus in the human brain as illustrated by the Digital Anatomist (DA) (2002) in BrainInfo (2002).

In the Web-address-based approaches to interoperability, only the original website needs to possess an ontology. Its curator identifies the structural concepts and relationships expressed in text and images at other sites and creates appropriate links. When BrainInfo sends a visitor to a site that uses a term different from the NeuroNames default term, it sends along instructions to help the visitor locate the precise information of interest. For instance, if the visitor is looking for the hippocampus, and the hippocampus is labeled "Ammon's horn" at the second site, the message

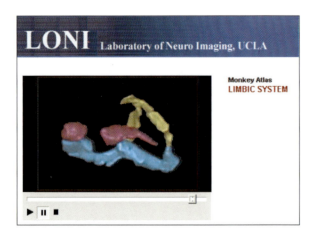

Figure 2.10 Rotational model of the bilateral hippocampus (blue) together with the amygdala (red) and the posterior columns of the fornix (yellow) (LONI, 2002).

"Hippocampus: look for Ammon's horn" appears in the Brain-Info sidebar.

The most powerful form of interoperability is that of *federated databases* in which multiple websites with related databases function symmetrically (Chapter 4, this volume). A visitor to any site can retrieve and analyze data from other sites seamlessly without regard to the physical location of the data. This kind of interoperability requires a common ontology that includes all of the concepts in all of the participating databases.

Users of the different databases do not have to use the same names for structures to search each other's systems, but all systems must link their nomenclatures to an identical set of concepts in each other's databases. The definitions of structural concepts and the models programmed into the systems must have unique database IDs through which other sites can reach information of common interest.

Several neuroscientific websites are developing databases and ontologies necessary to operate as federated databases. They include databases for neurophysiological data (BrainML, 2002; Chapter 3, this volume), connectivity data from the macaque in text format (CoCoMac, 2003a; Stephan et al., 2001), connectivity data and surface-based atlases in the macaque and human (CARET, 1999 VEL (1999)), and illustrations of neuroanatomical structures by atlas overlays in the macaque (BrainInfo, 2002; Bowden and Dubach, 2002b). These and others are developing ontologies appropriate to the subject matter of their databases and are using the Web's extended mark-up language (XML), or a variant thereof, that will enable websites that share ontologies to access raw data at their sites.

2.9 CONCLUSIONS

1. Nomenclatures and ontologies are essential tools of informatics; they integrate principles of linguistics, cognitive psychology, and computer science to facilitate scientific communication through space and time.

2. The nomenclature of neuroanatomy consists of all terms used to communicate concepts based on the physical structure of the nervous system.

3. Since neuroanatomical concepts are for the most part visual, they are communicated by images as well as by words.

4. A neuroanatomical ontology specifies relationships among neuroanatomical concepts, including the names and image representations of structures in template atlases, which are brain atlases specially designed for informatics applications.

5. A neuroanatomical ontology also includes overarching neuroanatomical concepts in terms of models, the most common of which are (a) partitive hierarchies, such as the classical hierarchy of brain structures defined in the Nomina Anatomica; (b) categorical hierarchies, such as the classification of brain structures as predominantly gray matter, white matter, or fluid; and (c) systemic models that, for example, define relations among brain structures in terms of their connections.

6. The communicative power of Web browsers, search engines, portals, and databased websites depends on the kinds of ontologic tools they employ. The most powerful applications of informatics technology will be seen in federated databases that adopt a common ontology encompassing all names and concepts by which data are indexed in the participating databases.

ACKNOWLEDGMENTS

The authors appreciate comments on this discussion provided by Cornelius Rosse and the many contributions of Evan Song to development of TSrainInfo. The manuscript was prepared with the support of grants LM-006243 and RR-00166 from the National Institutes of Health to the University of Washington. The authors make NeuroNames available for incorporation into computerized databases and websites conveying information about the brain. It is distributed at no cost in the form of Microsoft Excel spreadsheets. For more information, contact dmbowden@u.washington.edu.

References

Anthoney, T.R. (1994). *Neuroanatomy and the Neurologic Exam: A Thesaurus of Synonyms, Similar-Sounding Non-Synonyms, and Terms of Variable Meaning*. CRC Press, Boca Raton, FL.

Arbib, M.A. (1995). *The Handbook of Brain Theory and Neural Networks*. MIT Press, Cambridge, MA.

BAMS (2003). *Brain Architecture Management System*. University of Southern California, Los Angles (http://brancusi.usc.edu/bkms/brain/search_bname_con.php).

Biology Workbench. (2002). *Biology Workbench*. San Diego Supercomputing Center, University of California at San Diego (http://workbench.sdsc.edu).

Bloom, F.E., Young, W.G., and Kim, Y.M. (1990). *Brain Browser: A HyperCard Application for the Macintosh*. Academic Press, San Diego, CA.

Bowden, D.M. (2000a). History of brain atlases for nonhuman primates. In *Primate Brain Maps: Structure of the Macaque Brain*, (R.F. Martin and D.M. Bowden, eds.), pp 1–11. Elsevier, Amsterdam.

Bowden, D.M. (2000b). Atlases for informatics and quantitative neuroanatomy. In *Primate Brain Maps: Structure of the Macaque Brain*. (R.F. Martin and D.M. Bowden, eds.), pp. 12–22. Elsevier, Amsterdam.

Bowden, D.M., and Dubach, M.F. (2002a). NeuroNames 2002. *Neuroinformatics* **1**, 43–60.

Bowden, D.M., and Dubach, M.F. (2002b). BrainInfo: An online interactive brain atlas and nomenclature. In *Neuroscience Databases: A Practical Guide*, (R. Kötter, ed.), pp. 259–273. Kluwer Academic Publishers, Boston, MA.

Bowden, D.M., and Martin, R.F. (1995). NeuroNames brain hierarchy. *NeuroImage* **2**, 63–83.

Bowden, D.M., and Martin, R.F. (1997). A digital Rosetta Stone for primate brain terminology. In *Handbook of Chemical Neuroanatomy*, (F.E. Bloom, A. Björklund, and T. Hökfelt, eds.), Vol. 13, Part 1. Elsevier, Amsterdam.

BrainInfo (2002). *BrainInfo*. Neuroscience Division, National Primate Research Center, University of Washington, Seattle (http://braininfo.rprc.washington.edu).

BrainML (2002). *BrainML*. Laboratory of Neuroinformatics, Weill Medical College of Cornell University, New York (http://brainml.org).

Brodmann, K. (1905). Beiträge zur histologischen Lokalisation der Grosshirnrinde: Dritte Mitteilung: Die Rindenfelder der niederen Affen. *Psycho. Neuro.* **4**:(5/6), 177–226.

Brodmann, K. (1909). *Beschreibung der einzelnen Hirnkarten. IV. Kapitel in Vergleichende Lokalisationslehre der Grosshirnrinde*. Barth, Leipzig. [English translation: L.J. Garey, *Brodmann's 'Localisation in the Cerebral Cortex*, Imperial College Press, London, 1999]

CARET (1999). *CARET and Surface-Based Atlases*. Van Essen Laboratory, Washington University, St. Louis, MO (http://stp.wustl.edu/resources/caretnew.html) accessed March, 2003.

Carpenter, M.B. (1991). *Core Text of Neuroanatomy*. Williams & Wilkins, Baltimore, MD.

Carpenter, M.B., and Sutin, J. (1983). *Human Neuroanatomy*. Williams & Wilkins, Baltimore, MD.

CBIL (Computational Biology and Informatics Laboratory). (2002). Center for Bioinformatics, University of Pennsylvania, Philadelphia (http://www.cbil.upenn.edu/anatomy.php3).

CoCoMac (2003a). *Collations of Connectivity Data on the Macaque Brain*. Heinrich Heine University, Duesseldorf (http://www.mon-kunden.de/cocomac).

CoCoMac (2003b). *Untangling the Brain*. Heinrich Heine University, Duesseldorf (http://www.cocomac.org).

Crosby, E.C., Humphrey, T., and Lauer, E.W. (1962). *Correlative Anatomy of the Nervous System*. Macmillan, New York.

DA (2002). *Digital Anatomist*. University of Washington, Seattle (http://www9.biostr.washington.edu/da.html).

Franklin, K.B.J., and Paxinos, G. (1997). *The Mouse Brain in Stereotaxic Coordinates*. Academic Press, San Diego, CA.

GenePaint (2003). *GenePaint*. Department of Molecular Embryology, Max-Planck Institute of Experimental Endocrinology, Hannover, Germany (http://genepaint.org/Frameset.html).

German, D.C., and Bowden, D.M. (1974). Catecholamine systems as the neural substrate for intracranial self-stimulation: An hypothesis. *Brain Res.*, **73**, 381–419.

Google (2003). Website (http://google.com); General review (http://www.searchengineshowdown.com/features/google/review.html); Technical review (http://www.google.com/technology/pigeonrank.html).

Gray, H., Williams, P.L., and Bannister, L.H. (1995). *Gray's Anatomy: the Anatomical Basis of Medicine and Surgery*. Churchill-Livingstone, Edinburgh.

Gross, C.G. (1997). From Imhotep to Hubel and Wiesel: The story of the visual cortex. In *Cerebral Cortex* (K.S. Rockland, J.H. Kaas, A. Peters, and EG Jones, eds.), Vol. 12, Chapter 1, Plenum Press, New York.

Gruber, T. (1993). What is an ontology? (http://www-ksl.stanford.edu/kst/what-is-an-ontology.html).

HBP (2002). *Human Brain Project*. National Institute of Mental Health, Bethesda, MD (http://www.nimh.nih.gov/neuroinformatics/researchgrants.cfm).

Hof, P.R., Young, W.G., Bloom, F.E., Belichenko, P.V., and Celio, M.R. (2000). *Comparative Cytoarchitectonic Atlas of the C57BL/6 and 129/Sv Mouse Brains*. Elsevier, Amsterdam.

Hohne, K.H., Bowmans, M., Riemer, M., Schubert, R., Tiede, U., and Lierse, W. (1992). A volume-based anatomical atlas. *IEEE Comput. Graphics Appl.*, pp. 72–78.

Hole, W.T., and Srinivasan, S. (2002). Adding NeuroNames to the UMLS Metathesaurus,. *Neuroinformatics* **1**, 61–64.

International Anatomical Nomenclature Committee (IANC). (1983). *Nomina Anatomica*, 5th ed. Williams & Wilkins, Baltimore, MD.

Kahle, W. et al. (2001). Nervensystem und Sinnesorgane. *Taschenatlas der Anatomie*, Vol. 3. Thieme, Stuttgart.

Kamper, L., Bozkurt, A., Rybacki, K., Geissler, A., Gerken, I., Stephan, K.E., and Kötter, R. (2002). An introduction to CoCoMac-Online. In *Neuroscience Databases: A Practical Guide* (R. Kötter, ed.), pp. 155–169. Kluwer Academic Publishers, Boston, MA.

König, J.F.R., and Klippel, R.A. (1963). *The Rat Brain: A Stereotaxic Atlas of the Forebrain and Lower Parts of the Brain Stem*. Williams & Wilkins, Baltimore, MD.

Kötter, R. (2001). Neuroscience databases: Tools for exploring brain structure-function relationships. *Philos. Trans. R. Soc. London, Ser. B* **356**, 1111–1120.

LONI (2002). *Laboratory of Neuroimaging*. University of California at Los Angeles (http://www.loni.ucla.edu/).

Mai, J., Assheuer, J., and Paxinos, G. (2003). *Atlas of the Human Brain*, 2nd Edition (book with CD-ROM). Academic Press, San Diego / Elsevier, Amsterdam.

Martin, R.F., and Bowden, D.M. (2000). *Primate Brain Maps: Structure of the Macaque Brain*. Elsevier, Amsterdam.

Martin, R.F., Mejino, J.L.V., Bowden, D.M., Brinkley, J.F., and Rosse, C. (2001). Foundational model of neuroanatomy: Implications for the Human Brain Project, *Proc. Am. Med. Inf. Assoc. Symp, 2001*, pp. 438–442.

MBL (2002). *Mouse Brain Library*. University of Tennessee, Memphis (http://www.nervenet.org/mbl/atlas170/atlas170_frame.html).

National Institute on Drug Abuse (NIDA). (2002). *Setting Priorities for Molecular Neuroanatomy in the Postgenomic Era*. NIDA, Bethesda, MD (http://www.drugabuse.gov/MeetSum/Postgenomic.html).

Neuroscion (2002). *Neuroscion*. Elsevier Science, London (http://www.neuroscion.com/features.html).

NLM (2003). *Unified Medical Language System: Knowledge Sources*, 14th ed., January Release 2003AA. National Library of Medicine, Bethesda, MD.

OED (2003). *Oxford English Dictionary* (http://dictionary.oed.com/cgi/entry/00163666).

Paxinos, G., ed. (1990). *The Human Nervous System*. Academic Press, San Diego, CA.

Paxinos, G., and Franklin, K.B.J. (2001). *The Mouse Brain in Stereotaxic Coordinates*, 2nd ed. Academic Press, San Diego, CA.

Paxinos, G., and Watson, C. (1998). *The Rat Brain in Stereotaxic Coordinates*, 4th ed. Academic Press, San Diego, CA.

Paxinos, G., Huang, X.-F., and Toga, A.W. (2000). *The Rhesus Monkey Brain in Stereotaxic Coordinates*. Academic Press, San Diego, CA.

Pellegrino, L.J., and Cushman, A.J. (1967). *A Stereotaxic Atlas of the Rat Brain*. Appleton-Century-Crofts, New York.

Pinker, S. (1994). *The Language Instinct: How Mind Creates Language*. Harper Collins, New York.

PubMed (2003). *PubMed*. National Library of Medicine, Bethesda, MD (http://www.ncbi.nlm.nih.gov/PubMed).

Rosse, C., Mejino, J.L., Modayur, B.R., Jakobovits, R., Hinshaw, K.P., and Brinkley, J.F. (1998a). Motivation and organizational principles for anatomical knowledge representation: The Digital Anatomist symbolic knowledge base. *J. Am. Med. Inf. Assoc.* **5**(1), 17–40.

Rosse, C., Shapiro, L.G., and Brinkley, J.F. (1998b). The Digital Anatomist Foundational Model: Principles for defining and structuring its concept domain. *J. Am. Med. Inf. Assoc. Symp. Suppl.*, pp. 820–824.

Sato, L., McClure, R.C., Rouse, R.L., Schatz, C.A., and Greenes, R.A. (1993). Enhancing the Metathesaurus with clinically relevant concepts: Anatomic representations. *Proc. 16th Annu. Symp. Comput. Appl. Med. Care*, pp. 388–391.

Schiebler, T.H., Schmidt, W., and Zilles, K. (1999). *Anatomie*. Springer-Verlag, Berlin.

Srinivas, P.R., Gusfield, D., Mason, O., Gertz, M., Hogarth, M., Stone, J., Jones, E.G., and Gorin, F.A. (2003). Neuroanatomical term generation and comparison between two terminologies. *Neuroinformatics* **1**(2), 177–192.

Stephan, H. (1975). Allocortex. In *Handbuch der mikroskopischen Anatomie des Menschen* (W. Bargmann, ed.), Vol. 4, Part 9. Springer-Verlag, Berlin.

Stephan, K.E., Kamper, L., Bozkurt, A., Burns, G.A.P., Young, M.P., and Kötter, R. (2001). Advanced database methodology for the Collation of Connectivity data on the *Maca*que brain (CoCoMac). *Philos. Trans. R. Soc. London, Ser. B.* **356**, 1159–1186 (http://www.cocomac.org).

Swanson, L.W. (1998). *Brain Maps: Structure of the Rat Brain – A Laboratory Guide with Printed and Electronic Templates for Data, Models and Schematics*. Elsevier, Amsterdam (http://www.elsevier.nl/homepage/sah/brainmaps/doc/intro_frm.html).

Swanson, L.W. (2000a). A history of neuroanatomical mapping. In *Brain Mapping: The Systems* (A.W. Toga, ed.), Chapter 3. pp. 77–109. Academic Press, San Diego, CA.

Swanson, L.W. (2000b). What is the brain? *Trends Neurosci.* **23**, 519–527.

TD (2002). *Talairach Daemon*. Research Imaging Center, University of Texas Health Science Center at San Antonio (http://biad73.uthscsa.edu/research/body.html).

Toga, A.W. ed. (1999). *Brain Warping*. Academic Press, San Diego, CA.

Toga, A.W. (2002). Imaging databases and neuroscience. *Neuroscientist* **8**, 423–436.

Toga, A.W., Thompson, P.M., Holmes, C.J., and Payne, B.A. (1996). Informatics and computational neuroanatomy. *J. Am. Med. Inf. Assoc. Symp. Suppl.*, pp. 299–303.

Toga, A.W., Thompson, P.M., Mega, M.S., Narr, K.L., and Blanton, R.E. (2001). Probabilistic approaches for atlasing normal and disease-specific brain variability. *Anat. Embryol.* **204**, 267–282.

Ungerstedt, U. (1971). Stereotaxic mapping of monoamine pathways in the rat brain. *Acta Physiol. Scand.* **82** (Suppl. 367), 1–48.

VEL (1999). *Human Atlas*, HTML slide show. Van Essen Laboratory, Washington University, St. Louis, MO (http://stp.wustl.edu/caret/slides/human/img3.html).

Visel, A., Ahdid, J., and Eichele, G. (2002). A gene expression map of the mouse brain. In R. Kötter, (ed.), *Neuroscience Databases*. Chapter 2, pp. 19–35. Kluwer Academic Publishers, Boston, MA.

von Bohin, G., and Bailey, P. (1947). *The Neocortex of Macaca mulatta*. University of Illinois Press, Urbana.

Walker, A.E. (1940). A cytoarchitectural study of the pre-frontal area of the macaque monkey. *J. Comp. Neurol.* **73**, 59–86.

WaNPRC (2002). *NeuroNames 2002: The NeuroNames Brain Hierarchy Default Names and Synonyms*, compact disk (dmbowden@u.washington.edu). Neuroscience Division, Washington National Primate Research Center, Seattle.

WBA (1999). *Whole Brain Atlas*. Harvard Medical School, Boston, MA (http://www.med.harvard.edu/AANLIB/cases/caseM/case.html).

Web Resources

BAMS (2003). *Brain Architecture Management System.* University of Southern California, Los Angles (http://brancusi.usc.edu/bkms/brain /search_bname_con.php).

Biology Workbench. (2002). *Biology Workbench.* San Diego Supercomputing Center, University of California at San Diego (http:// workbench.sdsc.edu).

BrainInfo (2002). *BrainInfo.* Neuroscience Division, National Primate Research Center, University of Washington, Seattle (http:// braininfo.rprc.washington.edu).

BrainML (2002). *BrainML.* Laboratory of Neuroinformatics, Weill Medical College of Cornell University, New York (http://brainml.org).

CARET (1999). *CARET and Surface-Based Atlases.* Van Essen Laboratory, Washington University, St. Louis, MO (http://stp.wustl .edu/resources/caretnew.html).

CBIL (Computational Biology and Informatics Laboratory). (2002). Center for Bioinformatics, University of Pennsylvania (http://www. cbil.upenn.edu/anatomy.php3).

CoCoMac (2003a). *Collations of Connectivity Data on the Macaque Brain.* Heinrich Heine University, Duesseldorf (http://www.monkunden.de/cocomac).

CoCoMac (2003b). *Untangling the Brain.* Heinrich Heine University, Duesseldorf, (http://www.cocomac.org).

DA (2002). *Digital Anatomist.* University of Washington, Seattle (http:// www9.biostr.washington.edu/da.html).

GenePaint (2003). *GenePaint.* Department of Molecular Embryology, Max-Planck Institute of Experimental Endocrinology, Hannover, Germany (http://genepaint.org/Frameset.html).

Google (2003). Website (http://google.com): General review (http:// www.searchengineshowdown.com/features/google/review.html); Technical review (http://www.google.com/technology/pigeonrank. html).

Gruber, T. (1993). What is an ontology? (http://www-ksl.stanford.edu/ kst/what-is-an-ontology.html).

HBP (2002). *Human Brain Project.* National Institute of Mental Health, Bethesda, MD (http://www.nimh.nih.gov/neuroinformatics /researchgrants.cfm).

LONI (2002). *Laboratory of Neuroimaging.* University of California at Los Angeles (http://www.loni.ucla.edu/).

MBL (2002). *Mouse Brain Library.* University of Tennessee, Memphis (http://www.nervenet.org/mbl/atlas170/atlas170_frame.html).

National Institute on Drug Abuse (NIDA). (2002). *Setting Priorities for Molecular Neuroanatomy in the Postgenomic Era.* NIDA, Bethesda, MD (http://www.drugabuse.gov/MeetSum/Postgenomic.html).

Neuroscion (2002). *Neuroscion,* Elsevier Science, London (http://www .neuroscion.com/features.html).

OED (2003). *Oxford English Dictionary* (http://dictionary.oed.com/cgi /entry/00163666).

PubMed (2003). *PubMed.* National Library of Medicine, Bethesda, MD (http://www.ncbi.nlm.nih.gov/PubMed).

Stephan, K.E., Kamper, L., Bozkurt, A., Burns, G.A.P., Young, M.P., and Kötter, R. (2001). Advanced database methodology for the Collation of Connectivity data on the Macaque brain (CoCoMac). *Philos. Trans. R. Soc. London, ser. B* **356**, 1159–1186, (http://www.cocomac.org).

Swanson, L. W. (1998). *Brain Maps: Structure of the Rat Brain – A Laboratory Guide with Printed and Electronic Templates for Data, Models and Schematics.* Elsevier, Amsterdam (http://www.elsevier .nl/homepage/sah/brainmaps/doc/intro_frm.html).

TD (2002). *Talairach Daemon.* Research Imaging Center, University of Texas Health Science Center at San Antonio (http://biad73.uthscsa .edu/research/body.html).

VEL (1999). *Human Atlas,* HTML slide show. Van Essen Laboratory, Washington University, St. Louis, MO (http://stp.wustl.edu/caret /slides/human/img3.html).

WaNPRC (2002). *NeuroNames 2002: The NeuroNames Brain Hierarchy Default Names and Synonyms,* compact disk (dmbowden@u. washington.edu). Neuroscience Division, Washington National Primate Research Center, Seattle.

WBA (1999). *Whole Brain Atlas.* Harvard Medical School, Boston, MA (http://www.med.harvard.edu/AANLIB/cases/caseM/case.html).

Neuroinformatics for Neurophysiology: The Role, Design, and Use of Databases

Daniel Gardner, Ph.D., Michael Abato, Kevin H. Knuth, Ph.D., and
Adrian Robert, Ph.D.

CONTENTS

Databasing the Brain. Edited by Stephen H. Koslow and Shankar Subramaniam
ISBN 0-471-30921-4 © 2005 John Wiley & Sons, Inc.

3.1 SCOPE AND PURPOSE

This chapter has a threefold purpose. It is intended as a guide to users of our neurodatabases, as a stimulus to other domains or communities within neuroscience who may wish to implement data sharing of neurophysiologic signals or related neuroscience data, and as a framework toward developing a compatible set of interfaces for exchange of neuroscience data, queries, and analytic tools. The guide illustrates the scope and ease of use of a contemporary resource for neurophysiology. For a stimulus, we summarize the potential utility of neurodatabases and emphasize design choices that promote and facilitate their use to maximize the information and insight available from neurophysiology data. The framework, collegially directed toward neuroinformatics developers, presents our in-development work on BrainML, designed as an extensible standard and interface for interoperability among disparate domains of neuroscience. Each of these is a component directed toward synthesizing available information concerning the function of neurons and neural systems.

Earlier reports on these projects have appeared in journal article (Gardner et al., 2001a,c) and abstract (Abato and Gardner, 1999; Abato et al., 2002; DeBellis et al., 2000; Gardner, 2001; Gardner and Erde, 1997a,b; Gardner and Fins, 2003; D. Gardner et al., 1998, 1999, 2000a,b, 2001a,b,c, 2002, 2003a; Knuth et al., 1999; Robert et al., 2003; Xiao et al., 2002) form.

3.2 NEUROPHYSIOLOGY DATA: ARCHIVING THE BRAIN'S SIGNALS

3.2.1 The Scope of Neural Data Processing

Contemporary neurophysiology is broad in scope, investigating a wide range of questions with multiple techniques that produce many types of data and require disparate analyses and interpretations. Sensor-based brain data including time-series neurophysiological records (extracellular spike trains and slow potentials, patch-clamp channel and whole-cell currents, EEG, MEG) convey and report information processing in the brain, complementing imaging and functional imaging methods such as PET, MRI, and fMRI.

A major task of neuroscience is decoding how the brain encodes information. One of the goals is an understanding of the neural code, that pattern of activity in each neuron of the brain that encodes and transmits sensation, motor commands, cognition, and memory. Some of this information, transmitted in a rapidly changing time scale with great parallelism, is in patterns of action potentials. For representative work in this field, see Abeles and Gerstein (1988), Bialek et al. (1991), Gerstein and Aertsen (1985), Johnson et al. (2001), Knight (1972), Mainen and Sejnowski (1995), Perkel et al. (1967), Reich et al. (2001), Rieke et al. (1997), Shadlen and Newsome (1998), and Wilson (1999). Other information, with greater parallel extent but varying less rapidly, is expressed via the strengths of synapses. Summed synaptic potentials contribute as well to several classes of electrical signals. For a review of synaptic function, see Cowan et al. (2001).

Among other questions, neurophysiology is concerned with functional differences between distinct classes of neurons and between similar neurons processing disparate information. It is concerned as well with the relations between neuronal signaling and properties of the state of the nervous system and the organism, including sensation, motor behavior, other functions, and learning. For a broad review, see Chapters 12, 13, 17, 19, 20, and 26 in this volume (see also Gardner, 1993; E.P. Gardner et al., 1999; Gray and Singer, 1989; Harris-Warrick et al., 1992; Markram et al., 1997; Segev et al., 1995; Victor and Purpura, 1996). Neurophysiology also properly informs and is informed by neuronal and network models (Beeman et al., 1997; Dayan and Abbott, 2001; Gerstner and Kistler, 2002; Hines, 1989; Koch and Segev, 1998; Ziv et al., 1994). Neither questions nor methods are static; as new techniques are developed and new questions explored, the scope of neurophysiology expands.

3.2.2 Neuroinformatics: Complementing Publications, Genomics, Proteomics, and Physiomics

This breadth and evolution of neurophysiology present challenges for neuroinformatics, which comprises technical methodologies and data description schemas designed to facilitate sharing of a universe of neuroscience data. Neuroinformatics properly includes methods to archive, exchange, and analyze actual data records, complementing and going beyond the information obtainable from static journal figures. The goal is to provide insights into fundamental properties of neuronal ensembles in the same manner as archives of protein and nucleic acid sequences and structures have revolutionized molecular biology (Baxevanis, 2003; Benson et al., 2003; Bernstein et al., 1977; Westbrook et al., 2003). Parallel informatic efforts addressing cellular, subcellular, and ensemble state and function—the proteome or physiome—can inform and be informed by neurophysiologic neuroinformatics.

3.2.3 Data Sharing and the Power of Reanalysis

In March 2003, the U.S. National Institutes of Health released a final policy requiring all high-direct-cost (> US$500,000/yr) grant proposals submitted in October 2003 or after to specify a plan for sharing data acquired by the project (NIH, 2003). With this new policy and related implementation guidelines for data sharing, we note the following:

- The *goals* and *scope* that have characterized neuroinformatics for the past decade are now applicable to an expanded set of areas, communities, techniques, and investigators.
- The *significance* of neuroinformatics has been enhanced.
- The *perspective* of neuroinformatics has broadened.

Neurophysiology databases can help answer central problems of neuroscience by enabling data sharing (Gardner et al., 2003b; Datasharing.net, 2003). Archiving and delivery of neurophysiologic datasets in neurodatabases or peer-to-peer networks allows users to download the actual datasets so that they can be meta-analyzed or analyzed using additional techniques or algorithms. Such methods expand the reach of each investigator by providing tools to allow locally recorded data from individual labs to be made available for display and comparison and reanalysis remotely, establishing a virtual oscilloscope capable of reporting from a potentially global set of microelectrodes. Neurodatabases and parallel methods will thus permit individual investigators or collaborative groups to:

- Archive, exchange, and reanalyze data from remote labs in parallel with data from local sensors
- Link disparate but related experimental datasets
- Link experimental data to the quasi-data output of models
- Optimize the use of experimental animals

The ability to bring to bear such expanded techniques and viewpoints on datasets that previously were only available to a single lab may advance such goals of neurophysiology as:

- Correlating neuronal signals with the function, location, or type of the neuron or region from which they originate
- Extracting information from neuronal signals, and distinguishing informational from purely intrinsic features of signals
- Shaping and testing models of neural function both locally and remotely
- Designing and testing hypotheses about neural function, including neural coding
- Formulating hypotheses similarly informed by networks of related data and hypotheses
- Relating multiple levels of brain function—and beyond brain function to genetic and genomic, expression, and other physiologic data as well
- Integrating neurophysiology data to the function of the organism, to behaviors including sensation, action, and learning

Our laboratory is developing databases to collect and archive neurophysiologic recordings from brain neurons and regions, as part of a thrust funded by the Human Brain Project (Huerta et al., 1993; Koslow and Huerta, 1997; Pechura and Martin,

1991; Shepherd et al., 1998). These databases, resembling yet transcending sequence archives, are designed to exchange and deliver actual brain recordings worldwide, permitting further analysis and comparison to other local or global data. As such work is just beginning, there are few existing standards for data interchange, and a parallel effort is directed toward designing compatible yet extensible standards for interoperability.

In this work, we have tried to (a) adopt design goals that facilitate use by neurophysiologists and (b) leverage enabling methods of contemporary informatics. In the following sections we briefly describe these design goals and methods and present our ongoing projects.

3.3 NEUROINFORMATICS FOR NEUROPHYSIOLOGISTS

Neuroscience data are richer and more complex than sequence data, which can be searched and expressed as simple text strings. Formats for neurophysiology data include two- or three-dimensional images, arrays of digitized sensor data such as electrode current or voltage, and histograms of signal amplitudes or durations. As many of the chapters in this volume emphasize, methods addressing such data and serving neuroinformatics need to transcend those developed for sequence or structure bioinformatics.

3.3.1 The Publication Model for Accessible Databases

Neurophysiologists organize and present data according to each of several models that differ in volume, scope, and type of organization, depending upon the nature and volume of the data, and the purposes to which it can be put. Examples include such disparate entities as lab notebooks, computer backup media, abstracts, and posters. From these familiar models, we advocate that neuroinformatic resources adopt a model whose scope, organization, utility, and methodology are drawn from publications. Gardner et al. (2001a) have discussed at length the goals and method of publication-model databases that accept, archive, and deliver neurophysiologic datasets with conventions and techniques designed to emulate those of journal articles. Figure 3.1 summarizes similarities between such neurodatabases and publications, as well as ways in which these databases transcend and complement journal articles, books, and book chapters.

Briefly, a publication model echoes the selection and organization of reports in the scientific literature by specifying characteristic or defining data, rather than the all-encompassing sets of data found in modalities such as lab notebooks. Publications present data using a visual organization of figures and panels, and we favor a corresponding visual grammar for presentation of datasets from databases. Publications combine data figures with accompanying text, and database datasets should be linked to textual and numeric metadata to aid both description and selection. As timing of publications is recognized as asserting precedence, so too can time-stamped database submissions establish priority (Gardner et al., 2001a, 2003b).

Publications are established media for scientific communication, with common methods for distribution, archiving, and

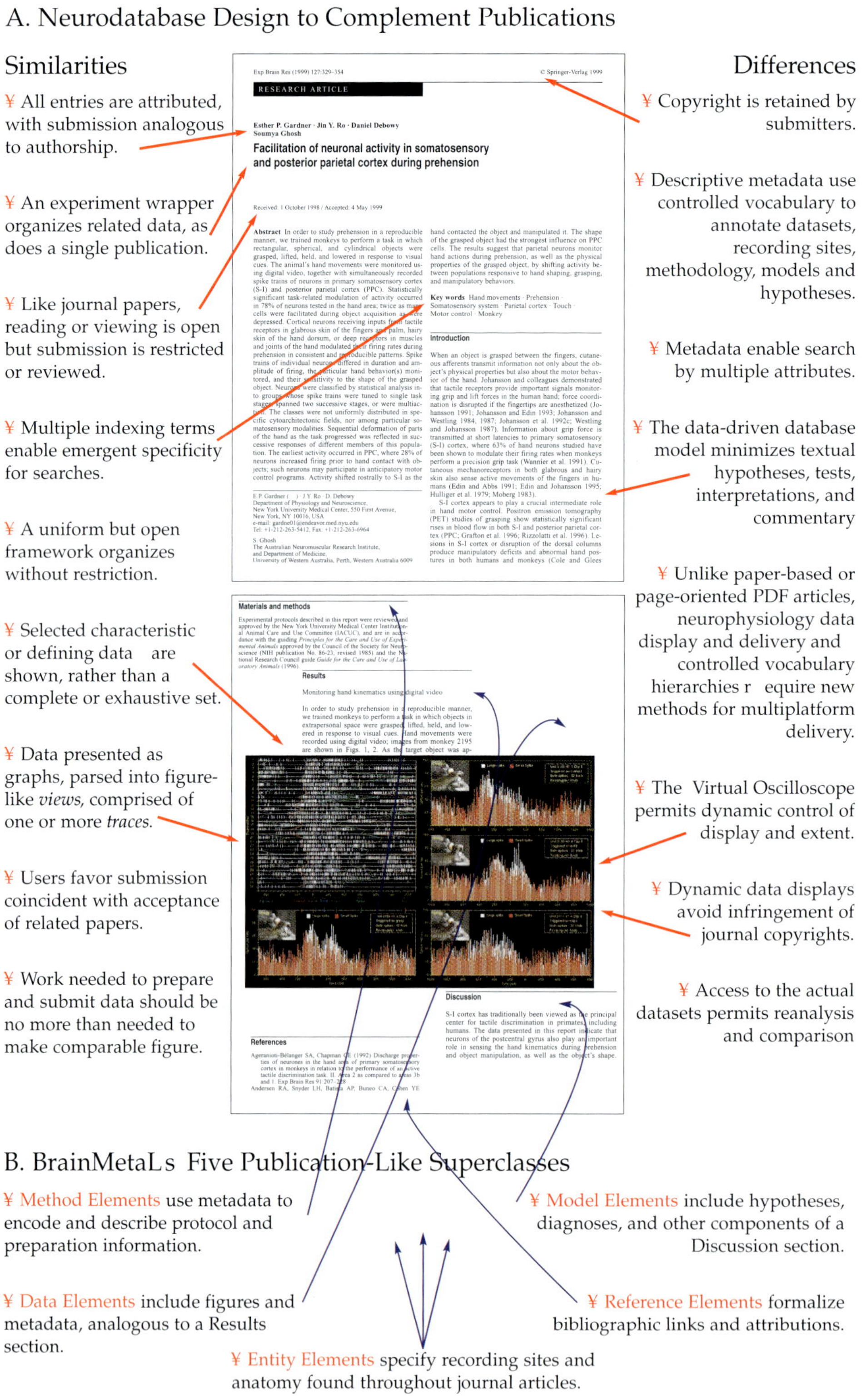
A. Neurodatabase Design to Complement Publications
Similarities
¥ All entries are attributed, with submission analogous to authorship.
¥ An experiment wrapper organizes related data, as does a single publication.
¥ Like journal papers, reading or viewing is open but submission is restricted or reviewed.
¥ Multiple indexing terms enable emergent specificity for searches.
¥ A uniform but open framework organizes without restriction.
¥ Selected characteristic or defining data are shown, rather than a complete or exhaustive set.
¥ Data presented as graphs, parsed into figure-like views, comprised of one or more traces.
¥ Users favor submission coincident with acceptance of related papers.
¥ Work needed to prepare and submit data should be no more than needed to make comparable figure.
Differences
¥ Copyright is retained by submitters.
¥ Descriptive metadata use controlled vocabulary to annotate datasets, recording sites, methodology, models and hypotheses.
¥ Metadata enable search by multiple attributes.
¥ The data-driven database model minimizes textual hypotheses, tests, interpretations, and commentary
¥ Unlike paper-based or page-oriented PDF articles, neurophysiology data display and delivery and controlled vocabulary hierarchies r equire new methods for multiplatform delivery.
¥ The Virtual Oscilloscope permits dynamic control of display and extent.
¥ Dynamic data displays avoid infringement of journal copyrights.
¥ Access to the actual datasets permits reanalysis and comparison
Exp Brain Res (1999) 127:329–354
© Springer-Verlag 1999
RESEARCH ARTICLE
Esther P. Gardner · Jin Y. Ro · Daniel Debowy
Soumya Ghosh
Facilitation of neuronal activity in somatosensory and posterior parietal cortex during prehension
Materials and methods
Results
Discussion
References
B. BrainMetaLs Five Publication-Like Superclasses
¥ Method Elements use metadata to encode and describe protocol and preparation information.
¥ Model Elements include hypotheses, diagnoses, and other components of a Discussion section.
¥ Data Elements include figures and metadata, analogous to a Results section.
¥ Reference Elements formalize bibliographic links and attributions.
¥ Entity Elements specify recording sites and anatomy found throughout journal articles.

Figure 3.1 Neurophysiology database design can reflect and complement publications. (**A**) The publication model for a neurodatabase is illustrated by a journal article that reports findings archived at neurodatabase.org. (*Left*) Conventions familiar from journal articles can inform neurodatabase design, promote utility, and enhance ease of use. (*Right*) Neuroinformatic techniques can transcend paper or PDF formats with dynamic views, access to actual datasets, and advanced searchable metadata. (**B**) Five BrainMetaL superclasses for methods, models, references, entities, and data partially reflect a publication model as well. Excerpts from Gardner et al. (1999b) used by permission of E.P. Gardner and Springer-Verlag.

citation. To prove their worth, neuroinformatic resources must therefore enhance or supplement the information available via the literature. On-screen figures can transcend their published counterparts by enabling transforms beyond the static two-dimensionality of the printed page, including animation, rotation, and selection via focused expansion or contraction of coordinates. More significant is the availability of datasets themselves, supplementing graphical figures based on the data and thereby enabling reanalysis. Articles are locatable using a restricted set of identifiers: (a) bibliographic entries such as authors and title and (b) a usually small set of general keywords. Informatic techniques enhancing publication-model databases can expand this through the use of defined attributes and metadata schemas for descriptors and search terms.

3.3.2 Java2 and XML Technologies Supporting Neuroinformatics

Contemporary neurophysiology is the joint effort of multiple laboratories that acquire and analyze brain data with the assistance of local computers, for which multiple standards are in widespread use. Each hardware architecture and related operating system is largely or partially distinct and is characterized by specific conventions, tools and applications, and adherents. To serve neurophysiology and promote interoperability as broadly as possible, neuroinformatics should recognize and serve this diversity.

Contemporary computer languages such as Java and XML span multiple architectures and operating systems. They thus aid the construction of tools allowing ready access to, and utilization of, data archives from any networked computer, including Macs or machines utilizing Windows, Linux, or any of several varieties of Unix. These languages can be used to design, implement, and distribute state-of-the-art tools to support general methods for data entry, storage, visualization and search, each designed to aid investigator-directed retrieval and analysis of information.

The Java language and associated JVM virtual machines (Gosling and McGilton, 1996; Joy et al., 2000; Java, 2003) provide a near-uniform environment within disparate hardware and software environments. Recent enhancements to Java including Java2 and Java Web Start provide a true multiplatform environment that combines capability with versatility and generality. As a result, neuroinformatic tools can be designed to a single standard and yet target and enable any of several architectures interchangeably. (Similar independence allows resources to design and implement server tools and methods that can be migrated to newer architectures.)

The Extensible Markup Language (XML) is an emerging standard that presents a protean structure and offers a powerful set of related tools for data description (Cover, 2003; XML, 2003). Defined quite generally, in XML:

- <tags> specify meaning </tags>
- <nested><tags> arrange hierarchies </tags></nested>
- <schemas_or_DTDs> verifiably assemble sets of tags </schemas_or_DTDs>

Such generality allows tags to be defined and arranged at will, imposing a requirement that the definitions and arrangements be codified and made available, allowing intelligent reuse and avoiding conflict or contention between identical or overlapping definitions. Neuroinformatic information including datasets, metadata, queries, algorithms, and transforms all can be defined, marked up, and transmitted using XML. As these definitions can vary among resources, additional methods are needed to rationalize XML representations of the structure and neuroscience content of different resources and so make them interoperable. One such method, our BrainML, is discussed below, following presentation of our cortical neurophysiology database.

3.4 TOWARD THE GLOBAL MICROELECTRODE: neurodatabase.org

Adopting these design goals and informatic methods, we have implemented a database of mammalian cortical neurophysiology to accept, archive, and deliver actual reanalyzable author-submitted datasets such as spike trains. The database, available at http://neurodatabase.org, incorporates the publication model for data selection and storage. Underlying the database are an intuitive data model, data structures, and methods that aid data sharing. Each dataset is annotated by controlled-vocabulary metadata that enable search by either broad or focused criteria (Gardner et al., 2001a,c).

Conformant to the principle that neurodatabases should be available to the widest range of users, and interoperable with the widest range of other resources, we have designed this database with XML for neuroscience markup and Java for multiplatform delivery and technology evolution. Query, display, upload, and acquisition of data and metadata is by Java user tools that provide open access for Macintosh OS/X, Windows, Solaris, or Linux, with migration to a Java Web Start model for ease of access. An extensive Java midlayer provides public methods for linking client and database while shielding users from the database engine and other internals. The midlayer also converts internal representations to XML for communication with users, and it serves Java user tools and controlled-vocabulary trees for metadata. Figure 3.2 provides a schematic view of this three-tier arrangement.

3.4.1 Easing the Users' Path

Technological sophistication is secondary to utility for neuroinformatic resources and utility depends strongly upon incentive, protective, and reward structures. Unlike publications, which enjoy widespread use and ready acceptance, databases have yet to develop consistent conventions for submission, use, citation, and reward.

To facilitate and encourage database use and data sharing, we have incorporated protections and incentives into the design of neurodatabase.org. Database access requires agreement with a statement on appropriate use of data, which includes the requirement to acknowledge sources and a reminder that extensive re-use of data may favor a collaboration (Figure 3.3). For submitters, ease of data submission and perceived benefits of submission should match or better those provided by conventional

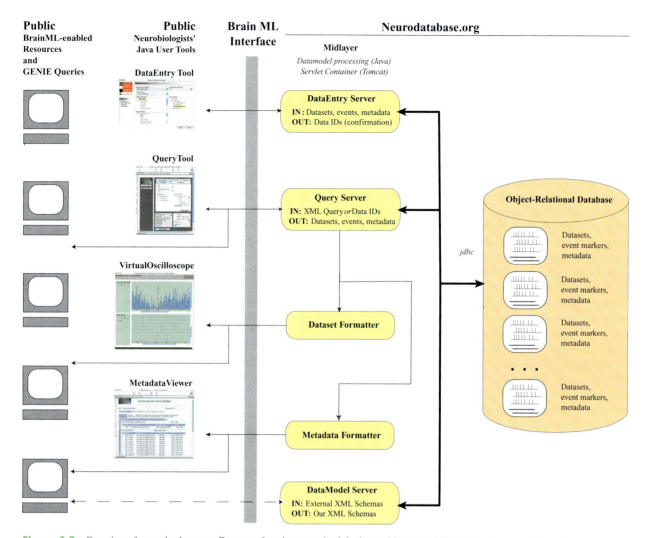

Figure 3.2   Overview of neurodatabase.org. For ease of use by neurophysiologists and interoperability with other neuroinformatic resources, neurodatabase.org adopts a three-tier organization. Communication with the object-relational database itself (*right*) is mediated by a Java midlayer that includes server and formatter processes (*center*). Communication with the Java user tools for data entry, query, and viewing of data and metadata (*near left*), as well as delivery of the tools themselves, is via a BrainML-enabled interface (*gray bar*). Using multiplatform Java, user tools provide identical functionality on Macintosh OS/X, Windows, linux, Solaris, and other unix distributions. Upon completion of the DataModel Server component of the midlayer, and extension of the interface, neurodatabase.org will also mediate queries and schema exchanges with other neuroinformatic resources (*far left*).

journal publication. User-friendly tools such as our guided mode DataEntryTool (discussed below) provide one such incentive. Another we offer is the projected availability of a parallel resource to develop and test a range of analytic algorithms as well as a computational engine to apply such algorithms to spike train and other data, thus adding value for submitters.

At neurodatabase.org, users see an open repository with universal access and intuitive tools. These include a QueryTool for searching (Figure 3.4), a Virtual Oscilloscope (Figure 3.5) that downloads and dynamically displays actual datasets, a Metadata Viewer (Figure 3.6) for descriptive annotation of datasets, and a DataEntryTool (Figure 3.7) for direct data entry by submitters. As noted above, all user tools are Java2-based, free, and multiplatform, and they are distributed by our application servers to most common standard networked platforms. Complementary tools to provide upload and analysis from within popular applications are being developed under a separate project by Bruxton Corporation (Figure 3.8).

3.4.2 Targeting Queries Using Neurophysiologic Criteria

Laboratory notebook descriptors often use local knowledge and conventions. Such specialized terms as date, trial, experimenter, animal tag, or dataset acquisition or sequence number are not useful identifiers for a public database. Neuroscientists searching for information need global search terms or descriptors that specify what outside investigators would want to know about the data, and would be likely to use to describe it. Such global terms can also inform or enable methods for interoperability between resources, as discussed below.

Our Java QueryTool (Figure 3.4) permits browsing or directed queries using descriptive globally useful metadata attributes that are delivered in real time from the database. The metadata include types of data, recording techniques and experimental protocols, neurons and other recording sites, and publication data. Such terms are properly neurophysiologic, not neuroinformatic,

neurodatabase.org
Cortical Neuron Database
© 2001 Weill Medical College of Cornell University
This Human Brain Project / Neuroinformatics Resource
is funded by the National Institute of Mental Health
and the National Institute of Neurological Disorders and Stroke

What do I do? What will I
see?

About the tools

What do I need?

Reload access
conditions

To access the database,
view conditions at right; if
necessary, reload them.

To enter the database, please read, acknowledge, and comply with these conditions:

- Each dataset and metadata description archived in this database remains the intellectual property of the individuals, laboratories, or organizations responsible for the recording, processing, annotation, and submission of the attributed data.
- Use of these data requires recognition of contributions of the above parties. For published datasets, this must include citation of literature references accompanying datasets. For unpublished datasets, this should include a citation of the form: (*investigator(s) name(s)*, databased dataset(s)). Extensive re-use requires explicit permission of the submitter; in some cases, an agreed-upon collaboration may be appropriate.
- We also ask that re-use of any data from this site include as well an acknowledgment such as: " Data used in this study were delivered via **neurodatabase.org** --a neuroinformatics resource funded by the Human Brain Project."

For database access, please signify approval: I acknowledge these conditions and my use of any data from this database will be compliant.

Figure 3.3   Neurodatabase.org explicitly recognizes rights of data submitters. To access databases at neurodatabase.org requires agreement with this statement on appropriate use of data. Such steps toward protection of data can reassure potential submitters and thus increase data sharing.

and largely intuitive, familiar, self-defining, and useful to the researchers for whom the database is intended. Extensive use of controlled vocabulary eases the users' path not only by directing searches to terms used to describe database datasets, thus providing guidance and specificity, but also through a fast, familiar point-and-click interface. With such a broad set of descriptors, and recognizing the diversity possible descriptors, most attributes are optional. Most terms are presented in a tree, with increasingly specific descriptors deriving from more broad ones. The underlying Hierarchic Attribute–Value schema allows construction of multi-term queries that can incorporate both broad and targeted searches of specific attributes describing the database contents (Gardner et al. 2000b, 2001c); for comparison with Entity–Attribute–Value schema, see Chapter 20 in this volume, as well as Nadkarni et al., (1999).

Just as bibliographic searches return references to publications rather than text or figure fragments, so database searches using the QueryTool return unified sets of data that were grouped into a single submission at the time of data entry. In our schema, these are called *experiments*, and they comprise linked datasets and associated descriptive and search metadata. Just as papers and other articles are divided into figures and text sections, so experiments have a logical, intuitive, visual grammar, presenting one or more *views*, each comprising related individual datasets, called *traces*, on common axes.

3.4.3 Dynamic Viewing Using the Virtual Oscilloscope

To extend views of data available as print or static image-based figures, the display methods provided present an interactive and dynamic view of neurophysiology data, providing additional incentives by enabling the user to control interactively selection and scaling of data, with or without metadata. Our Java

Virtual Oscilloscope includes methods to download and cache the datasets themselves and to construct dynamically created displays with user-selectable sweep and extent (Figure 3.5). This dynamic display also presents views of data that do not duplicate figures in the published literature, thus avoiding potential copyright infringement.

The underlying data model, object-relational database (Codd, 1970; Rowe and Stonebraker, 1987), and user tools include support for acquisition, storage, and delivery of both spike train and continuous records, as well as behavioral or stimulus event time stamps. Building on the generality of the data model and Virtual Oscilloscope, we are implementing dynamic display styles designed individually for specific classes of datasets characterizing an expanded range of techniques and preparations, including sampled patch-clamp and voltage-clamp time-series and related histogram data as well as the time-stamped records for spike trains.

3.4.4 Relating Data Using Linked Metadata

Interpretation of neurophysiology records is dependent on understanding from what neurons or structures they were recorded, using what techniques, and under what experimental or behavioral conditions or constraints. For this reason, presentation of data via the Virtual Oscilloscope is accompanied by descriptive textual metadata that annotate each experiment, view, and trace. These are presented for convenience in a separate window by the Metadata Viewer. Figure 3.6 provides an example of such information for a cortical experiment.

Linked to experiment, view, or trace, as appropriate, the Metadata Viewer specifies recording methodologies (e.g., extracellular single electrode) and neurophysiological data descriptions (e.g., spike train trace). Additional metadata specify submitters, related literature references, recording site, and protocols.

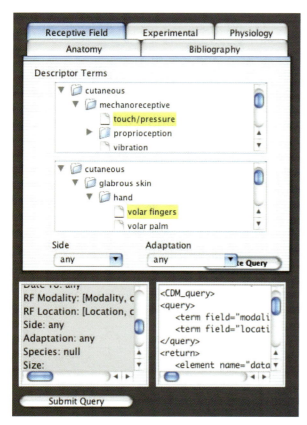

Figure 3.4   The Java QueryTool enables neurodatabase search by multiple controlled vocabulary descriptive attributes. Shown are two panes displaying hierarchic trees allowing selection of controlled vocabulary terms specifying receptive field modality and location. Tabs permit selection of additional bibliographic, anatomic, experimental, and physiologic criteria. Any combinations of several attributes may be specified by users. Queries are assembled in text (*lower left*) and BrainML (*lower right*) forms before transmission to neurodatabase.org. Following posting of queries, a list of experiments matching search terms is returned and displayed in the lower right pane (not shown).

Multiple attributes are available to describe neurons and recording sites. Reflecting neurophysiologic practice and the variable assignability of precise locations to electrode placement, we provide a set of optional attributes for location. For cortical neurons, these include cytoarchitectural and functional areas, depth, and cell class. As the initial target domain for neurodatabase.org was somatosensory neurophysiologists, we include as well experimentally determined functional neuronal response profiles. These include both background firing patterns and receptive fields, which are specified via location and modality. As many cortical receptive fields are broad or multimodal, individual neurons may be described by more than one set of attributes.

Most of the attributes cited above are database-served controlled vocabulary, whereas others such as experiment, view, and trace labels are implemented as free text. Specification by controlled vocabulary enhances ease of use by precisely, uniformly specifying descriptive and query metadata. The Hierarchic Attribute–Value implementation enables selectable specificity queries, supporting both (a) domain experts who wish to target highly specific datasets and (b) the broader group of neuroscientists whose interests and search strategies are likely to

be more general. Such a scheme additionally anticipates future expansion, providing enhanced precision without obsolescence.

3.4.5 Facilitating Entry of Neurophysiologic Data

Preparing and submitting sharable data to a database, like preparing and submitting a manuscript for publication, requires time and effort. In each case, reducing the effort required relative to the perceived reward promotes submission. For this reason, our latest Java DataEntryTool is designed to minimize the work needed to enter data into neurodatabase.org. For example, although the underlying data model is derived from the standards and practices of neurophysiology, it includes much that is technical and implementation-dependent. The DataEntryTool guides users through data selection and metadata specification, consistent with our data model without requiring users to be explicitly familiar with it. The tool presents a logical series of windows, with prompts and explanations, to assist users in organizing datasets and selecting appropriate metadata (Figure 3.7). The DataEntryTool presents the same controlled vocabulary trees as the QueryTool, ensuring that identical attributes, identical terms, and the same syntactic structure is presented to both submitters of data and users requesting sharable data.

3.4.6 Reanalysis of Neural Datasets

Perhaps the most significant feature of neurophysiologic databases such as neurodatabase.org—and the most significant advantage over presentation of data via the literature—is the provision of actual datasets, adding value to the data by enabling reanalysis. Our current Java-based tools allow database data to be downloaded via XML. A set of additional tools will allow database users to transfer data to, and thereby analyze data in, any of several standard software packages. A complementary tool set will allow users to assemble, format, edit, and annotate data collected by existing data acquisition programs and to submit these data to neurodatabase.org. These additional analysis and acquisition tools are being developed by Bruxton Corporation, with our collaboration, and will be made available without charge to users of neurodatabase.org. A prototype of the type of data transfer these tools will enable is shown in Figure 3.8.

3.5 TOWARD INTEROPERABLE NEUROINFORMATICS

3.5.1 The Challenges of Interoperability

Neuroinformatic and bioinformatic resources now online or under development differ in purpose; some are databases, some are tool repositories, and some are modeling sites. Both data types and underlying data models differ. Each resource adopts a data model, syntax, and semantics that cover a specific domain of interest, and these domains span multiple levels, questions, techniques, preparations, and data types. Within the Human Brain Project, and as shown by many examples throughout this volume, resources include libraries of fMRI images and MEG recordings, dendritic tree morphologies, three-dimensional EM

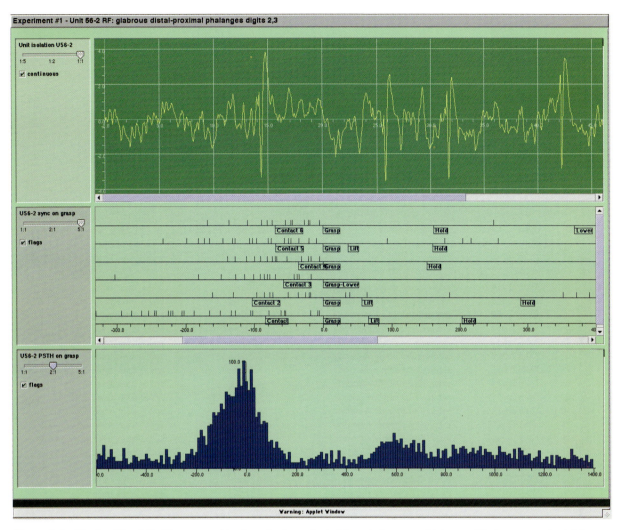

Figure 3.5 The Java Virtual Oscilloscope provides dynamic views of actual datasets from neurodatabse.org. Displays are constructed from XML-wrapped datasets delivered via the database midlayer and interface. Transcending static journal figures or other images, such views can be expanded, contracted, or scrolled to allow an overview of selected datasets or a close examination of a portion of the data. This example presents three distinct database datatypes, illustrating as well the intuitive experiment>view>trace data hierarchy. (**Top view**) A single trace displays the raw extracellular electrode record to show unit isolation. Virtual Oscilloscope style specifies a familiar oscilloscope-like color and format. (**Middle view**) Six traces of spike replicas are accompanied by metadata flags providing behavioral context to each trace. (**Bottom view**) A peristimulus time histogram plots distribution of spike times around a behavioral event. Data from the laboratory of E.P. Gardner.

reconstructions of spine arrangements on dendrites, conductance properties of ion channels underlying synaptic currents, classes of olfactory receptors, spike trains of cortical neurons, atlases of neurotransmitters in mouse brains, and flat maps of the cerebral cortices of humans and monkeys.

Illustrating this scope, Figure 3.9 presents an array of bioinformatic data types and levels of analysis. Within this array, many species and techniques have specific descriptors and data formats, so that different data types and sets of metadata attributes can be required, even for a particular intersection of type and level. Moreover, even when attributes are similar, the meaning of the terms themselves can vary. Some data, such as patient records, require stringent techniques for de-identification.

Gardner et al. (2001c) have noted: "Beyond the community database focused on one technique or preparation, the true challenge of interoperability is coordinating such disparate informa-

tion resources." This requires agreement on technical and lexical standards to (a) describe data, queries, tools, models, and hypotheses and (b) allow them to be referenced, exchanged, transferred, or converted. The need for interoperability transcends neuroscience, as neurophysiology resources and schemas should properly link to an expanding universe of bioinformatics resources and information.

3.5.2 Interoperability Through Interfaces

Toward interoperability, we support a model in which compatible interfaces implement interoperability between, and federation of, resources. Properly designed interfaces shield users from resource internals, while making available in compatible format both structure and descriptors of input queries and output data (Figure 3.10).

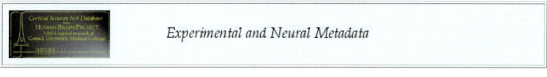

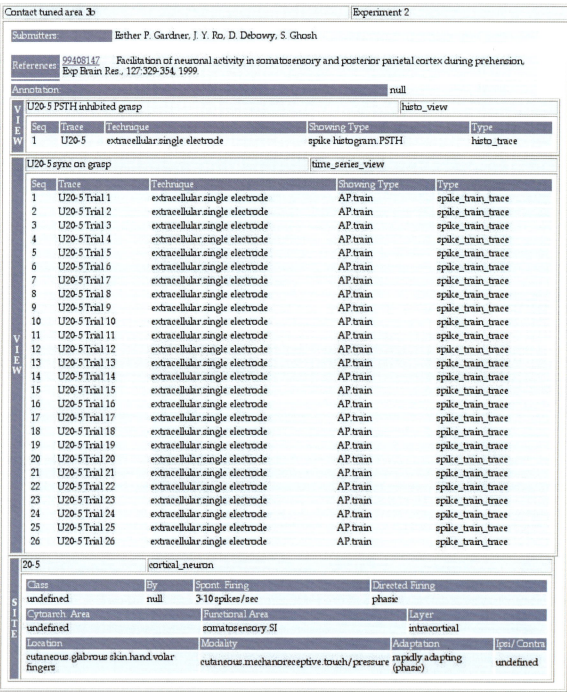

Figure 3.6 Complementing dynamic data displays via the Virtual Oscilloscope, the Metadata Viewer presents textual and numeric metadata useful for interpretation and reanalysis of neurophysiology data. Reference data provide a literature citation, including a live PubMed link, and site data characterize the cortical neuron from which the data were recorded. Each of two views provides label and type data, as well as labels and controlled-vocabulary descriptors for individual traces. The example shown here is a different experiment from that displayed in Figure 3.5.

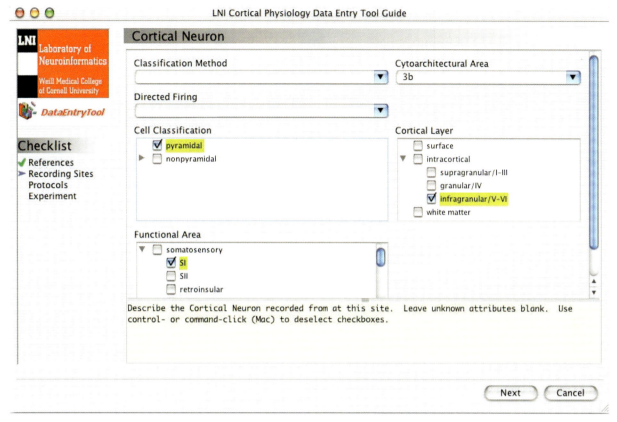

Figure 3.7   Neurodatabase.org minimizes the effort for neurophysiologists to share data and select descriptive metadata. To encourage use of neuroscience databases, ease of data submission and perceived benefits should equal or exceed those provided by conventional journal publication. User-friendly tools such as our guided mode DataEntryTool provide one such incentive. On any Java2-equipped local computer, the DataEntryTool allows submitters to organize their datasets as for figures, annotate by selecting descriptive controlled vocabulary metadata, and upload data and metadata to a compliant neurodatabase via BrainML, Java, and a midlayer DataServer that adds date and author tags. The sample screen shown here guides input of controlled-vocabulary metadata, some hierarchic, characterizing a cortical neuron from which data is to be archived.

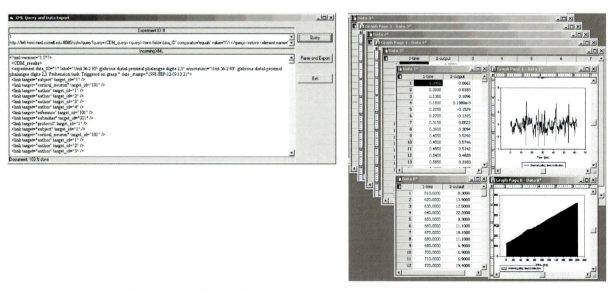

Figure 3.8   Cooperatively developed tools aid analysis of shared data from neurodatabase.org. (**Left**) BrainML-transmitted datasets from neurodatabse.org. (**Right**) Tools under development by Bruxton Corporation for open distribution to database users enable transfer of database data to third-party analytic programs, such as SPSS SigmaPlot, as well as upload of sharable data from third-party acquisition programs (not shown).

	Gene	Protein	Cell	Organ	Subject
Sequence	comparable	comparable			relatable
Expression			relatable	relatable	relatable
Structure		same	relatable	varies	varies
Sensor data			largely same	technique-dependent	technique-dependent
Image		same	technique-dependent	varies	varies
Chart note					unique

Figure 3.9 Interoperability interfaces need to relate data of disparate types across multiple levels. To support interoperability, such interfaces must accommodate and relate neuroscience data of many types (rows) obtained from systems at many levels (columns). The chart summarizes identical, comparable, relatable, variable, and unique combinations of type and level.

Interfaces Need Universal yet Nonrestrictive Standards

Such interfaces should ideally enable information exchange not just with dedicated user tools, as for our neurodatabase.org, but with any compatible present or to-be-developed resource. This, in general, requires standardization upon data formats that are broadly interpretable across platforms, operating systems, and data types. It also requires development and standardization of a rich semantics for descriptors that cover both neuroscience anatomy and data types, as well as an appropriate range of experimental techniques, procedures, and practices. In addition, the complexity and diversity of experimental neuroscience requires a defined syntax as well as semantics. This can aid targeted searches that rely on more than simple term matching, such as for trees or hierarchies, as well as specifying and parsing datasets. Given this complexity, and the speed with which new data and concepts appear, such standards need to be both comprehensive enough to form a basis for interoperability and yet expandable.

As noted above, we have used XML for database-client transmission of queries, data, and metadata for neurodatabase.org, and we favor the use of this widely adopted versatile standard more generally as a syntactic layer for interfaces.

Interfaces Should Specify Both Detectors and Selectors

To make present and planned neuroinformatic resources interoperable with other resources as well as with researchers in other domains of neuroscience, each resource should be designed to enable both users and other resources to know what information is available and how to make use of it.

This goal is aided by differentiating interoperable descriptors into two classes. Detectors, common to most or all entries in a focused resource, specify the scope and domain of what the resource supplies and how it can be obtained. Another set of terms, selectors, are used to select specific datasets, or tools, or models from those available. Both categories can include such relevant descriptors as anatomical locations, hypotheses, references, and methodology. Both types of terms should serve as well to (a) link multiple data types from multiple levels and (b) extend within-neuroscience interoperability to link to genomic and other resources outside of neuroscience.

3.5.3 Metadata Enable Interoperable Interfaces

Design Goals for Metadata

Many neuroscience data types are not self-describing, as sequences are, thus making the role of metadata more crucial. Goals for designing metadata to complement neurophysiologic data include:

- Metadata can properly include both textual and numeric components.
- Terms and values should be consistent with community usage.
- Design must have the breadth to span the database domain.
- Identifiers should ideally be accessible and useful both to specialists within a domain and to those outside the area of specialization.
- Optional and alternate categories of metadata provide flexibility.
- Schemes should interoperate and be generalizable to other domains of neuroscience and to non-neuroscience data.
- It should be possible to extend metadata without obsolescence.
- Design should permit efficient implementation to speed queries.
- Methods should transcend specific architectures or technology.

The Semantics of Neuroscience are Imprecise

Interoperability between such resources thus requires agreement on lexical and technical standards to (a) describe data, queries, tools, models, and hypotheses and (b) allow them to be referenced, exchanged, transferred, or converted. However, each domain of neuroscience has sets of preferred and specialized vocabularies; this specialization often extends to individual laboratories as well. Words and phrases thus have multiple meanings, and similar terms are rarely truly synonymous, revealing ambiguity that affects descriptive and search metadata used to characterize and specify entries in neuroscience databases.

Syntax, Context, or Ontologies Can Focus Descriptors

Thus terms in isolation are often insufficient as neuroscience descriptors, and additional structure or methods are of use for

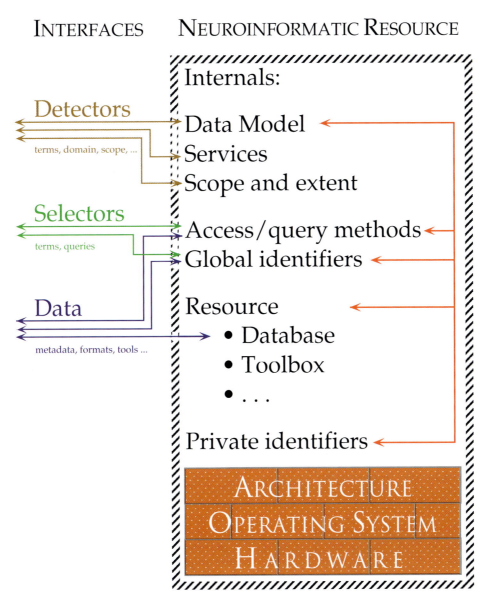

INTERFACES   NEUROINFORMATIC RESOURCE

Figure 3.10 Interfaces mediate open access and shield internals of neuroinformatic resources. Compatible interfaces expose detectors for identifying resources, selectors for specific search, and data served by the resource. Interfaces should be either standardized or traceable to a defined schema. Ideally, interfaces map without exposing the internal structure, architecture, or private identifiers of resources, and without limitations imposed by particular operating systems. Such interfaces, designed for interoperability with compatible resources, can complement, rather than replace, native access methods.

focus or disambiguation. Context and structure can be used to extract significance, as well as to disambiguate similar terms and relate different ones, and these can be derived from syntactic specifications or extracted from free text.

Where descriptors are unstructured free text with ad-hoc terminology, ontologies may be useful to parse or translate text into standard sets of identifiers and relations. Ontologies can specify or disambiguate terms, but the role of ontologies in neuroinformatics depends, in part, on how we select the level of knowledge the ontologies will specify and the scope of neuroscience to be covered. These, in turn, influence the methodology and tools employed for ontology development as well as the choice of implementers to specify, design, maintain, and extend the ontologies.

Drawing in part on work on medical and other informatic and controlled-vocabulary approaches, we favor specification via controlled vocabularies (see, for example, Chapter 5 in this volume, as well as Cimino, 1998, 2000; Lindberg et al., 1993). In our preferred scheme, specific attribute slots are associated with specific datatypes, giving our neuroscience vocabulary terms the required has-a context. Reflecting the need to be sometimes precise, sometimes general, such attribute slots are filled with a specific, focused vocabulary hierarchy that adds is-a syntax and selectable specificity. Thus datatypes are characterized by multiple short sets of selectable values. Ontologies may additionally be used to place and relate controlled vocabulary terms and to provide neuroscience terms with more general context.

Either ontologies or namespaces may be used to specify or relate interconvertible dialects, serving heterogeneity.

One Size Won't Fit All

The above goals call for interoperability throughout and beyond neuroscience. They recognize the utility of XML. They specify design requirements for syntactic and semantic specifications for metadata that promote ease of use and avoid ambiguity while serving both domain experts and more general users. These goals acknowledge the present complexity and inevitable evolution of resources, datatypes, and descriptors. Given this scope, we believe that no XML-derived data description language can be detailed enough to serve focused needs within an area, comprehensive enough to link related but distinct investigations, and broad enough to anticipate future expansion.

3.6 BRAINML: A DEVELOPING INTERFACE AND FUNCTIONAL ONTOLOGY FOR NEUROSCIENCE

To fulfill these goals, to provide a functional standard underlying interoperability between resources, and to serve as an interface enabling archiving, query, and exchange of neuroscience data and the metadata that describe them, we are designing and codifying the neuroscience markup language BrainML. Rather than a formal ontology for a field as complex and dynamic as neuroscience, BrainML combines a scheme for defining simple, usable functional ontologies, a data model, and tools for synthesizing multiple levels of structure and function. BrainML derives jointly from the standards and practices of neuroscience and informatics. From neuroscience, BrainML adopts a research-friendly structure as well as sets of familiar descriptive attributes, arranged in multiple focused hierarchies. It is not restricted to one level, such as the cell, nor to one technique, such as simulation. It incorporates neuroanatomy, but recognizes the utility of other descriptors as well. From informatics, BrainML

takes and extends a series of tools, design choices, and standards, including XML, XML Schema, and XML namespaces. Each of these, widely adopted, is hardware-, software-, and vendor-independent.

BrainML is open, to encourage community development, and nonformal, for efficiency. By nonformal, we mean that it uses implicit rather than formal semantic definitions of terms, and the definitions relate to the attributes they describe and to the position of a term in a hierarchy of related values.

We project that communities associated with specific preparations or techniques will be able to use BrainML to implement ontologies and data models that appropriately describe their domain of neuroscience while also maintaining interoperability with other domains whose data, methods, models, and entities may differ. BrainML schemas should additionally be compatible with other XML-derived languages such as NeuroML (Goddard et al., 2001).

BrainML has a layered architecture, illustrated in Figure 3.11 and described further below. The data, queries, tools, and models that users exchange, and resources supply, are instance documents shown on the top layer of Figure 3.11. BrainML specifies their neuroscience information using semantic tags referencing domain-specific data models, but in a way that is designed to be interoperable with other resources. These specifications rely upon sets of definitions of neuroscience descriptors that form BrainML type libraries. Many of these libraries will include hierarchic lexicons that map trees of descriptors to specific attributes, enabling both broad and focused searches as well as extensibility. Toward generality, BrainML is built on the metalanguage BrainMetaL, a substrate defining basic types and structures (Abato et al., 2002; Xiao et al., 2002). The lexical structure of BrainML, BrainMetaL, and instance documents is defined using XML Schema and XML namespaces, which provide rich content typing and layered extensibility. Each of these derives its syntax from XML. As noted above, XML and its derivations, widely adopted, are hardware-, software-, and vendor-independent, and this independence and layered architecture aid interoperability, extensibility, and maintenance.

Figure 3.11   BrainML is a designed as a layered suite. Neuroinformatic resources archive, supply, exchange, and utilize instance documents such as datasets, queries, models, and tools. The architecture of BrainML is designed to enable resources to describe these instance documents with reference to defined schemas and thus facilitate interoperability. Instance documents are layered upon BrainML-defined data models, type libraries, and lexicon dictionaries. These BrainML components provide neuroscience content to, and build upon, the abstract semantics of our BrainMetaL metalanguage. BrainML and BrainMetaL, in turn, rely on standards of the informatics and computer science communities, including lexical and syntactic structures specified using XML and XML Schema.

We are developing BrainMetaL, the core of BrainML and several specific-type libraries for neurophysiology, and we seek collaborative coalitions to develop BrainML lexicons to serve additional areas of neuroscience. We additionally encourage other groups with complementary expertise to encode compatible anatomy and neurogenetic schemas. For examples, see Chapters 10, 14, 18, 21, and 22 in this volume; also see Brinkley et al., 1997; Felleman and Van Essen, 1991; Jones, 1986).

3.6.1 BrainML Is Derived from XML and XML Schema

For XML to advance interoperability, additional standards are needed to rationalize XML representations for neuroscience data and make them interoperable. We use XML Schema to define BrainML and BrainMetaL syntax publicly so that other resources can use our data files or queries and integrate them with their own. XML Schema (2003) transcends several of the limitations of document-type definitions (DTDs). It has broad support and an expanding toolset including validators and test suite collections. XML Schema defined types can be provided in libraries and used or extended for datasets as well as queries, and namespaces can be used to rationalize similar schemas or extensions of schemas with new capabilities. XML Schema also provides layered extensibility, in conjunction with namespaces, which allow resources to coordinate derived, inherited and complementary ontologies. BrainML schemas can make type-based references and utilize common facilities including descriptions and attributes, validate and reference one another, and cross-compare or analyze common datatypes. Finally, XSL transform allows translation between compatible BrainML-derived schemas.

3.6.2 BrainML Is a Layered Suite

We note above that design goals seem incompatible with a comprehensive single data description language. Such a unitary canonical XML Schema or DTD would likely be neurobiologically inadequate, sociologically implausible, and evolutionarily inflexible. Conversion between disparate XML representations would plausibly require hand-coded translators, one-to-one negotiation between pairs of resource developers, or ontology-based mediators. As no single such model can serve such a wide field, or accommodate evolution of techniques and data types, we have adopted a distinct solution: BrainML is structured as a suite, based upon a metalanguage designed to enable interoperable schemas and interfaces. Layering promotes ease of development, with both (a) a metalanguage that provides the basics and (b) specific compatible schemas to fit the needs of particular communities of neuroscience.

3.6.3 BrainMetaL: A Substrate for Interoperable Data Description

To serve as a functional standard supporting BrainML, underlying interoperability between developing resources in neuroscience, biophysics, and physiology, and to provide a layer of structure and abstract semantics enabling development of compatible data description languages, ontologies, and interfaces, we are developing the BrainML metalanguage BrainMetaL.

We are designing BrainMetaL as the simplest set of abstractions that can be implemented directly and can provide a substrate for the broad set of domain-specific dictionaries, data models, and data-type libraries that comprise BrainML, as well as future specializations. Such abstractions must be basic yet sufficiently comprehensive to accommodate physical, chemical, psychological, and clinical correlates of biophysics or neuroscience, enabling broad applicability for BrainML.

BrainMetaL Abstracts Core Scientific Data Descriptors

BrainMetaL specifies the extensible abstract core for BrainML and includes classes for:

- Data, entities, methods, models, and references
- Standard scientific quantities, units, and prefixes
- Hierarchic lexical structures to organize controlled vocabularies of descriptors

The architecture of BrainMetaL is designed to enable neuroinformatic and other biomedical resources to layer data models, libraries, and methods for interoperability on our language and technique development efforts, which in turn rely on standards of the informatics and computer science communities. As noted above, instance documents—queries, data, and tools—are described using semantic XML tags defined for specific domains by languages such as BrainML (Figure 3.11).

Quintessence. Quintessence encompasses five classes of abstract elements designed to describe any data resource. Brain-MetaL defines symbolic or image-based data elements: basic datasets and wrappers that BrainML can specialize for neurophysiologic datasets; quasi-data such as simulation results; metadata; and our experiment or view wrappers. Entity elements include things with physical rather than informational existence, abstractable as having structure and location, with names and possibly coordinates. These include anatomical structures that may be defined at any granularity, or by any coordinate scheme, such as dendrites, surface-based locations, regions of interest, and even subjects. Molecular species are also entity elements. Techniques and protocols are method elements, which also include experimental conditions and simulation engines. The reference element superclass is offered for bibliographic information: references, publications, and authors. And model elements cover hypotheses, diagnoses, and simulations.

Components of this quintessence schema abstract both attributes and structure of their respective domains. Thus data and method elements are intuitively organized bottom-up, declaring successive wrapper classes for atomic or basic units of data and manipulation. The more lexically dependent reference, model, and entity elements, in contrast, use a similarly intuitive top-down abstract schema that accommodates both broad categories and more specific instances. The five types vary as well in their relative utility to specify descriptors, ontologies, or search terms, or to exchange instantiated data. Data elements are largely instantiable and exchangeable. Method elements are mixed; data analysis algorithms and tools are instantiable and exchangeable, but protocols and others are descriptive. Model element simulations and parameter sets are instantiable, whereas hypotheses or diagnoses can be any of instantiable, exchangeable, or descriptive. Most entities are descriptive, excepting macromolecular sequence or structure datasets. Finally, references are largely descriptive but could be exchangeable (for bibliographic resources, for example).

Quantities and Units. BrainMetaL also defines base SI quantities, units, and exponents, supporting named or unnamed combinations without restricting rule sets.

Lexicon Structure. Also included in BrainMetaL are base abstractions for the syntactic framework, but not the semantic content, of hierarchic controlled vocabularies. Semantic dictionaries and neuroscience vocabularies can be developed and specified by BrainMetaL-derived BrainML schemas. Figure 3.12 shows both the BrainMetaL schema definition of lexicon structure and an example BrainML lexicon for functional areas of primate cortex.

3.6.4 BrainML Enables Interoperable Queries and Exchangeable Datasets

BrainML Extends BrainMetaL with Neuroscience Content

Fundamental elements and types defined by BrainMetaL allow neuroinformatic resources to design specialized yet compatible data models, libraries, and methods for interoperability. BrainML can be used to define structures and specify semantics. Structures and content of data models, datasets and their formats, queries, analytic tools, model specifications, and algorithms can be constructed from quintessence and other BrainMetaL elements. Utilizing the BrainMetaL lexicon definitions, BrainML can also specify the semantic content of attributes describing any of these BrainML-defined structures (Figure 3.12).

Data Wrappers

The derivation and use of data element provides an example of compatible extensibility. The BrainMetaL-defined XML Schema element <data-element> and its underlying complex type <data-element-type> are recursively defined to include

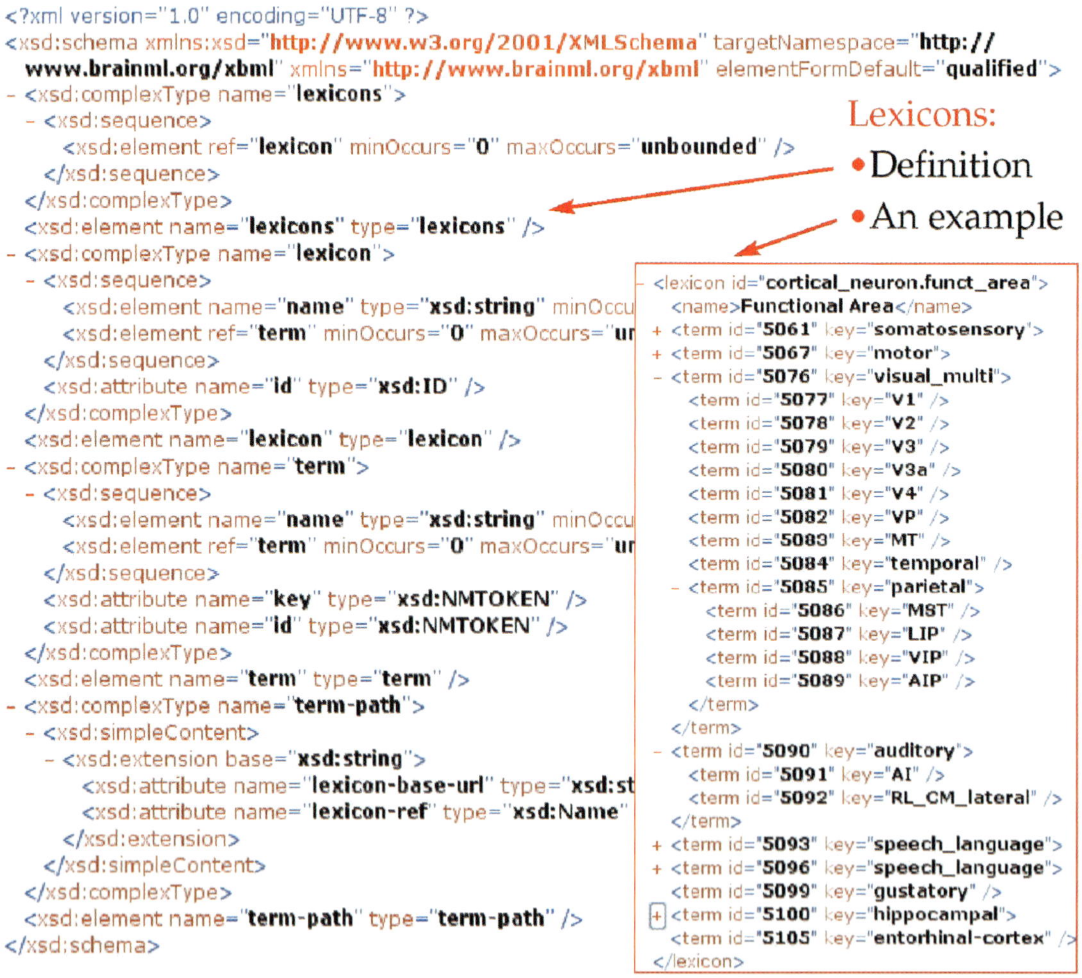

Figure 3.12   BrainML controlled-vocabulary lexicons: definition and example. BrainMetaL uses XML Schema to define a top-level structure for lexicons: hierarchic attribute-value trees of descriptive terms modifying specified attributes. As shown in the inset, BrainML, GENIE, and compatible schemas use this structure to specify focused sets of controlled vocabularies for neuroscience, here for cortical functional area.

<data-container> and <data-element> elements (BrainML.org, 2003). These are logically coherent collections built in turn on arrangements of *n*-dimensional combinations of *datoms*: data points (as XML-described and wrapped elements) or existing defined data formats. Data-element, data-container, and dataset may thus be defined as frameworks enabling BrainMetaL-compatible schemas to define a broad set of hierarchic data collections to be specified and exchanged.

Lexicon Semantics; Lexicon Extensibility

Although BrainML schemas and structures may be composed of any valid XML type—permitting, for example, free-text values for elements—we favor the use of lexicons to specify controlled-vocabulary (CV) hierarchies. Such controlled vocabularies, user-selected via graphical user interface forms presenting trees of terms for selection, reduce or eliminate the need for resource moderators to review submitter-supplied free-text terminology. Such defined terms also aid interoperability among resources or domains.

BrainML semantic attributes and values thus potentially form a functional ontology for neuroscience. Where attribute values of BrainML-compatible ontologies are arranged in defined CV hierarchies, specialized descriptors may be derived from more general ones. Attributes and hierarchies together provide a set of implicit is-a and has-a operators that form a functional ontology. Our current BrainMetaL lexicon definition specifies a simple tree, although this could be generalized to a directed graph. Describing neurobiological elements using multiple attributes appears to be able to serve similar knowledge concepts as a directed acyclic graph implementing multiple inheritance.

Specific Domains Require Targeted Sets of Attributes or Values

Different attribute sets are often required for similar entities characterizing differing domains, preparations, or techniques. As an example, Figure 3.13 presents lists of attributes specifying two classes of neurons: molluscan identified cells and mammalian cortical units. Some attributes are common to both classes, and they can share the same lexicon for possible values. Other attributes appear common to both classes, but the semantic values are preparation-specific and therefore the lexicons are distinct; the location of the neuron's receptive field is an obvious instance. Other attributes are specific to one or the other class of neuron.

Sample BrainML Schemas and Exchangeable Datasets

Concurrent with the codification of BrainMetaL, we are developing specialized functional BrainML representations for somatosensory cortex and for invertebrate electrophysiology, both to serve those communities and as test cases for interoperable schema development. Figure 3.14 presents a noncanonical example of a BrainMetaL-derived BrainML schema for a spike train data trace from neurodatabase.org.

We further recognize that for BrainML representations to continue to serve neuroscience, they cannot remain static. This evolution has two components. First, continued monitoring of the utility of lexicons, with leaf terms in shallow trees converted to branch elements and expanded with more specific descriptors. As Figure 3.15 shows, lexicon design makes this

Molluscan Neuron:		Cortical Neuron:	
identifier		cell_class	classif_method
species	ganglion	subject_taxonomy	
coordinates	appearance	coordinates	coord_scheme
connects-to	relative_to	funct_area	cytoarch_area
innervation		cortical_layer	
RP	spont_AP	directed_firing	
receptive_field_modality		receptive_field_modality	
receptive_field_location		receptive_field_location	
motor_behavior		motor_behavior	
releases	expresses	releases	expresses
homolog-of			

Figure 3.13 Specific preparations or techniques require targeted sets of BrainML attributes and values. As shown for molluscan and mammalian cortical neurons, sets of descriptive attributes may vary. In this example, some attributes are common with a common set of values, others use the same attribute but populate its lexicon with distinct values (not shown), and still other attributes are specific to one or another cell type.

process relatively straightforward while maintaining the selectable specificity inherent in lexicon structure. More difficult are syntactic extensions to the data model, data formats, or other BrainML structures. Here versioning via namespaces may prove useful.

Brain ML will be expanded to encompass other resources being developed in neuroscience, biophysics, and physiology. It is important to note that BrainML does not obsolete existing data models. By providing an additional interface, it works in parallel with existing tools that have been designed for existing resources. BrainML can additionally express data models represented in Java, universal modelling language, DTDs, or resource description format, as well as XML Schema. Multiple definitions—current and evolving, overlapping but incompletely congruent—can be ordered, and versions rationalized, by the use of namespaces.

We also propose how BrainML-compatible ontologies might be coordinated between linked communities and domains of neuroscience. In this scheme, neuroscientists select concepts, neuro-ontologists construct definitions, computer scientists develop tools to parse and interconvert descriptors and query terms, and neuroinformaticists, GUI designers, and users build intuitive, ontology-shielding tools. These communities will need to define methods for updating ontologies as neuroscience evolves.

3.7 GENIE: COMPLEMENTARY INTEROPERABLE PEER-TO-PEER DATA SHARING

The imminent scheduled requirement for data sharing does not provide time for most areas of neuroscience to develop centralized databases such as those we have provided in our initial phase and described above. For some preparations, techniques, and communities, centralized databases such as ours will be supplemented by informal or mediated networks for sharing distributed data. As a consequence, there is a need for rapidly deployable peer-to-peer networks for data sharing.

To aid this effort, and support interoperability between centralized and distributed resources, we are developing a new methodology for peer-to-peer serving and searching

```
<?xml version="1.0" encoding="UTF-8" ?>
- <document xmlns="http://www.brainml.org/genie-x1" xmlns:xbml="http://
    www.brainml.org/xbml" xmlns:xsi="http://www.w3.org/2001/XMLSchema-
    instance" xsi:schemaLocation="http://www.brainml.org/genie genie-x1.xsd">
  - <time-series-trace>
      <label>U56-2</label>
      <horizontal-units>ms</horizontal-units>
      <vertical-units ref="mV" />
      <time-rate>0.045</time-rate>
      <recording-technique>extracellular.single electrode</recording-technique>
      <data-class>AP.multiunit'</data-class>
      <stimulus-response>response</stimulus-response>
    - <data-trace order="1">
        <datum>-0.0662231</datum>
        <datum>-0.0183105</datum>
        <datum>0.1095581</datum>
        <datum>0.005188</datum>
        <datum>-0.1528931</datum>
        . . .
      </data-trace>
    </time-series-trace>
  - <xbml:prefixed-unit id="mV">
      <xbml:name>millivolt</xbml:name>
      <xbml:si-prefix-ref>http://www.brainml.org/brainmetal/
        units.xml#milli</xbml:si-prefix-ref>
      <xbml:si-unit-ref>http://www.brainml.org/brainmetal/
        units.xml#volt</xbml:si-unit-ref>
    </xbml:prefixed-unit>
  </document>
```

Figure 3.14   BrainML describes exchangeable datasets. In this example of an instance document, a BrainML-transmitted cortical neuron trace, spike times are <datum> elements within the defined <time-series-trace> element, which also includes defined tags and values for relevant metadata. Such BrainML-defined tags compactly and verifiably specify structure and metadata values of exchangeable datasets and other instance documents, enabling data exchange among compatible systems.

heterogeneous neuroscience data—GENIE, the Generalized Extensible Neuroscience Internet Examiner. GENIE is a self-organizing peer-to-peer web designed for ease of use and rapid deployment to exchange heterogeneous and distributed datasets,

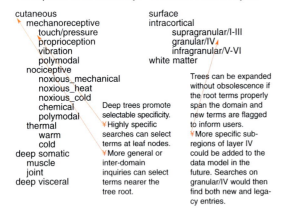

Figure 3.15   BrainML lexicons enable selectable-specificity searches and extension without obsolescence. Lexicon terms are intuitively represented as trees, allowing either highly specific searches or queries that maximize broad returns. Lexicon hierarchies can be expanded without rendering prior entries obsolete, as areas evolve and require enhanced specificity.

whether stored locally as flat files or in databases. Prototypes of the simple header files specifying the scope of a GENIE resource as provided in Figure 3.16. Local data archives can serve searchable metadata descriptors, or hybrid schemes under development by other groups are possible, in which metadata but not data are maintained by centralized servers. Search agents can be made general enough to accept any data model or descriptors, and the web architecture enables recursive searches.

Befitting a cooperative scheme, GENIE uses platform-independent XML, standard protocols, and common open-source software. Although designed to be enhanced by BrainML, GENIE can use other XML-based or free-text schemas for domains of neuroscience that have not yet had BrainML lexicons developed. It may be of use to coordinate future development efforts with other Human Brain Project groups that are devising similar schemes, toward interoperability or possibly a synthesis.

3.8 NOT A CONCLUSION BUT AN EVOLUTION

The success of BrainML will depend on how well it serves as an interface describing the data models, ontologies, or schemas of neuroscience individual resources, as well as their instance documents such as queries and datasets, while also maintaining

A. Sample header for a /genie file:

```
<html>
<head>
    <title>NeuroPeer Database</title>
    <meta name="detectors"
    content="genie,visual_multi,optical.neurointrinsic,
    nonhuman.primate, neuroinformatix" scheme="brainml">
    <meta name="robots"content="genie,neuroinformatix,brainml">
</head>
. . .
```

B. Sample /hosts file:

```
<?xml version="1.0" encoding="UTF-8"?>
<genie-hosts>
    <host>neurodatabase.org</host>
    <host>neocortex.med.cornell.edu</host>
    <url>http://datasharing.net/hosts</url>
    . . .
</genie-hosts>
```

Figure 3.16   GENIE is designed to facilitate interoperable peer-to-peer data sharing. For local archives or digital lab notebooks, GENIE is being developed to enable BrainML-compatible hosting of focused or compact data from a single lab, workgroup, or community. (**A**) This sample HTML header for the /genie file served by a local archive enables search agents to recognize a GENIE-compatible server with specific content and defined data model. The local server can use any of HTML, Java, or Javascript. (**B**) Another local server file specifies a list of compatible genie-hosts that can recursively contain or reference additional lists.

interoperability with other domains whose data, methods, and models may differ.

We present a series of test criteria for BrainML and Brain-MetaL that will evaluate how well they will aid interoperability of developing neuroinformatic resources:

- What data, tools, and models need to be made available, and can these be readily described and exchanged via BrainML?

- What descriptive metadata are needed to characterize such data, tools, or models, and can such metadata be derived compatibly from BrainMetaL types?

- Can these descriptive metadata be readily parsed into selector and detector classes?

- What other domains of neuroscience need to be linked to particular neurophysiological resources?

- Are BrainMetaL's abstractions enabling without being excessively restricting?

- Is the knowledge domain of neuroscience adequately served by BrainMetaL and BrainML abstractions?

- Will hierarchic controlled vocabulary lexicons facilitate mapping of neuroanatomy terms?

- Will BrainML and BrainMetaL aid development of non-native interfaces for databases and other resources?

- How well do BrainMetaL-derived types map to those of other neurobiological or related scientific markup languages?

As BrainML and related schemas progress, they will face challenges and requirements that are already clear. It will be necessary to rationalize or recognize different XML representations for brain data and make them interoperable. This within-neuroscience interoperability will have to be extended to link to

genomic and other resources outside of neuroscience. This will require that logical data models map data from levels including genetic, molecular, tissue, and clinical.

This chapter reports a set of works in progress and urges development of complementary methods for neuroscience data sharing, many of which are reported in other chapters in this volume. It therefore ends not with a conclusion, but a projection. An array of centralized and distributed neuroinformatics resources, tools, and methods is now under development. Coupled to an emerging consensus of the value of data sharing, this array will greatly advance (a) our understanding of neurophysiologic substrates of brain function, modulation, and learning and (b) our ability to relate neurophysiologic findings to those in genomics, proteomics, and physiomics.

3.8.1 Datasharing.net

To continue to advance these perspectives, we have established as a community resource datasharing.net (2003), complementing our neurodatabase access through neurodatabase.org (2003) and BrainML and BrainMetaL schemas at brainml.org (2003).

ACKNOWLEDGMENTS

We thank Robert DeBellis, Steven M. Erde, Thomas White, and Youping Xiao for their contributions to the development work that made these projects possible. Esther P. Gardner supplied sample data and many valuable comments on this report, as well as the work on which it is based. Jonathan Victor's perspectives were invaluable for information-based and other techniques for analyses of sharable data. Consultant-collaborators for the somatosensory cortical database, including Kenneth O. Johnson, Robert H. LaMotte, and Harold Burton, provided insightful comments. Daniel T. Brown, Joseph Grote, and Guohua Liu of Bruxton Corporation are developing additional tools for reanalysis. This Human Brain Project/Neuroinformatics research is supported by NIMH and NINDS via MH57153, with additional support from MH60538, past NINDS support from NS36043, and past NSF support from BIR/DBI-9506171.

References

Abato, M., and Gardner, D. (1999). Common data model—A data type definition for biophysical data exchange. *Biophys. J.* **76**, A197.

Abato, M., Xiao, Y., Knuth, K.H., and Gardner, D. (2002). BrainMetaL: Underlying interoperable neuroscience markup languages. *Soc. Neurosci. Abstr.* **28** (Program No. 610.15).

Abeles, M., and Gerstein, G.L. (1988). Detecting spatiotemporal firing patterns among simultaneously recorded single neurons. *J. Neurophysiol.* **60**, 909–924.

Baxevanis, A.D. (2003). The molecular biology database collection: 2003 update. *Nucleic Acids Res.* **31**(1), 1–12.

Beeman, D.E., Bower, J.M., De Schutter, E., Efthimiadis, E.N., Goddard, N., and Leigh, J. (1997). The GENESIS simulator-based neuronal database. In *Neuroinformatics: An Overview of the Human Brain Project* (S.H. Koslow and M.F. Huerta, eds.), pp. 57–81. Erlbaum, Mahwah, NJ.

Benson, D.A., Karsch-Mizrachi, I. Lipman, D.J., Ostell, J., and Wheeler, D.L. (2003). GenBank. *Nucleic Acids Res.* **31**(1), 23–27.

Bernstein, F.C., Koetzle, T.F., Williams, G.J.B., Meyer, E.F., Brice, M.D., Rodgers, J.R., Kennard, O., Shimanouchi, T., and Taumi, M. (1977). Protein data bank: A computer-based archival file for macromolecular structures. *J. Mol. Biol.* **112**, 535–542.

Bialek, W., Rieke, F., de Ruyter van Steveninck, R., and Warland, D. (1991). Reading a neural code. *Science* **252**, 1854–1857.

BrainML.org (2003). *BrainML and BrainMetaL Schema Documents.* Laboratory of Neuroinformatics, Weill Medical College of Cornell University, New York (http://brainml.org).

Brinkley, J.F., Myers, L.M., Prothero, J.S., Heil, G.H., Tsudura, J.S., Maravilla, K.R., Ojemann, G.A., and Rosse, C. (1997). A structural information framework for brain mapping. In *Neuroinformatics: An Overview of the Human Brain Project.* S.H. Koslow, and M.F. Huerta, (eds.), pp. 309–334 + plate 16. Erlbaum, Mahwah, NJ.

Cimino, J.J. (1998). Desiderata for controlled medical vocabularies in the twenty-first century. *Methods Inf. Med.* **37**, 394–403.

Cimino, J.J. (2000). From data to knowledge through concept-oriented terminologies: Experience with the Medical Entities Dictionary. *J. Am. Med. Inf. Assoc.* **7**, 288–297.

Codd, E.F. (1970). A relational model of data for large shared data banks. *Commun. ACM* **13**, 377–387.

Cover, R., ed. (2003). Core standards: Extensible Markup Languages (XML). *Cover Pages: Online Resource for Markup Language Technologies.* (http://xml.coverpages.org/xml.html).

Cowan, W.M., Südhof, T.C., and Stevens, C.F., eds. (2001). *Synapses.* Johns Hopkins, Baltimore, MD.

Datasharing.net (2003). Laboratory of Neuroinformatics, Weill Medical College of Cornell University, New York (http://datasharing.net).

Dayan, P., and Abbott, L.F. (2001). *Theoretical Neuroscience.* MIT Press, Cambridge, MA.

DeBellis, R., Abato, M., Erde, S.M., Knuth, K.H., and Gardner, D. (2000). Common data model extensions for exchanging simulations and comparing recorded to computed datasets. *Soc. Neurosci. Abstr.* **26**, 2001.

Felleman, D.J., and Van Essen, D.C. (1991). Distributed hierarchical processing in the primate cerebral cortex. *Cereb. Cortex* **1**, 1–47.

Gardner, D., ed. (1993). *The Neurobiology of Neural Networks.* MIT Press, Cambridge, MA.

Gardner, D. (2001). A Common Data Model enabling and federating electrophysiology databases. *Biophys. J.* **80**(1), 33a.

Gardner, D., and Erde, S.M. (1997a). Database schemas for electrophysiological datasets may complement genomic and protein sequences. *Biophys. J.* **72**, A118.

Gardner, D., and Erde, S.M. (1997b). Neurodatabases implement evolving neurophysiology data structures using an object-relational model. *Soc. Neurosci. Abstr.* **23**, 1048.

Gardner, D., and Fins, J.J. (2003). Neuroscience data sharing II. Ethical perspectives. *Soc. Neurosci. Abstr.* **29** (Program No. 428.6).

Gardner, D., Abato, M., and Erde, S.M. (1998). Open neurodatabase access by java-mediated controlled–vocabulary neuron terms and experimental protocols. *Soc. Neurosci. Abstr.* **24**, 134.

Gardner, D., Abato, M., DeBellis, R., Erde, S.M., Knuth, K.H., and White, T. (1999). Common data model 2000: Open methods for neuroscience data description and interchange. *Soc. Neurosci. Abstr.* **25**, 1910.

Gardner, D., Abato, M., Knuth, K.H., DeBellis, R., and Erde, S.M. (2000a). A dynamic publication model for data-driven neurodatabases implements interoperability for neuroscience information resources. *Soc. Neurosci. Abstr.* **26**, 2001.

Gardner, D., Knuth, K.H., DeBellis, R., and Abato, M. (2000b). Hierarchical attribute-value schema enables extensible controlled vocabularies systematizing neuroscience databases. *Biophys. J.* **78**, 218A.

Gardner, D., Abato, M., Knuth, K.H., DeBellis, R., and Erde, S.M. (2001a). Dynamic publication model for neurophysiology databases. *Philos. Trans. R. Soc. London, Ser. B.*, **356**, 1229–1247.

Gardner, D., Abato, M., Knuth, K.H., White, T., DeBellis, R., and Gardner, E.P. (2001b). BrainML: An extensible interoperability schema and functional ontology for neuroscience. *Soc. Neurosci. Abstr.* **27** (Program No. 21.10).

Gardner, D., Knuth, K.H., Abato, M., Erde, S.M., White, T., DeBellis, R., and Gardner, E.P. (2001c). Common data model for neuroscience data and data model interchange. *J. Am. Med. Inf. Assoc.* **8**, 17.

Gardner, D., Xiao, Y., Abato, M., Knuth, K.H., and Gardner, E.P. (2002). BrainML and GENIE: Neuroinformatics schemas for neuroscience data sharing. *Soc. Neurosci. Abstr.* **28** (Program No. 610.14).

Gardner, D., Abato, M., Knuth, K.H., and Xiao, Y. (2003a). BrainML: Neuroinformatics for data sharing. *Biophys. J.* **82**, 305a.

Gardner, D., Toga, A.W., Ascoli, G.A., Beatty, J., Brinkley, J.F., Dale, A.M., Fox, P.T., Gardner, E.P., George, J.S., Goddard, N., Harris, K.M., Herskovits, E.H., Hines, M., Jacobs, G.A., Jacobs, R.E., Jones, E.G., Kennedy, D.N., Kimberg, D.Y., Mazziotta, J.C., Miller, P., Mori, S., Mountain, D.C., Reiss, A.L., Rosen, G.D., Rottenberg, D.A., Shepherd, G.M., Smalheiser, N.R., Smith, K.P., Strachan, T., Van Essen, D.C., Williams, R.W., and Wong., S.T.C. (2003b). Towards effective and rewarding data sharing. *Neuroinformatics* **1**, 289–295.

Gardner, E.P., Ro, J.Y., Debowy, D., and Ghosh, S. (1999). Facilitation of neuronal activity in somatosensory and posterior parietal cortex during prehension. *Exp. Brain Res.* **127**, 329–354.

Gerstein, G.L., and Aertsen, A. (1985). Representation of cooperative firing activity among simultaneously recorded neurons. *J. Neurophysiol.* **54**, 1513–1527.

Gerstner, W., and Kistler, W. (2002). *Spiking Neuron Models: Single Neurons, Populations, Plasticity.* Cambridge University Press, London and New York.

Goddard, N.H., Hucka, M., Howell, F., Cornelis, H., Shankar, K., and Beeman, D. (2001). Towards NeuroML: Model description methods for collaborative modelling in neuroscience. *Philos. Trans. R. Soc. London, Ser. B.* **356**, 1209–1228.

Gosling, J., and McGilton, H. (1996). *The Java Language Environment— A White Paper*, Technical Report. Sun Microsystems, Mountain View, CA (http://java.sun.com/docs/white/langenv/).

Gray, C.M., and Singer, W. (1989). Stimulus-specific neuronal oscillations in orientation columns of cat visual cortex. *Proc. Natl. Acad. Sci. U.S.A.* **86**(5), 1698–1702.

Harris-Warrick, R.M., Marder, E., Selverston, A.I., and Moulins, M., eds. (1992). *Dynamic Biological Networks: The Stomatogastric Nervous System.* MIT Press, Cambridge, MA.

Hines, M. (1989). A program for simulation of nerve equations with branching geometries. *Int. Biomed. Comput.* **24**, 55–68.

Huerta, M.F., Koslow, S.H., and Leshner, A.I. (1993). The Human Brain Project: An international resource. *Trends Neurosci.* **16**, 436–438.

Java (2003). *The Source for Java Technology.* Sun Microsystems, Mountain View, CA (http://java.sun.com).

Johnson, D.H., Gruner, C.M., Baggerly, K., and Seshagiri, C. (2001). Information-theoretic analysis of neural coding. *J. Comput. Neurosci.* **10**(1), 47–69.

Jones, E.G. (1986). Connectivity of the primate sensory-motor cortex. In *Cerebral Cortex*, (E.G. Jones and A. Peters, eds.), Vol. 5, pp. 113–183. Plenum Press, New York.

Joy, B., Steele, G., Gosling, J., and Bracha, G., eds. (2000). *Java(TM) Language Specification*, 2nd ed. Addison-Wesley, Reading, MA.

Knight, B.W. (1972). Dynamics of encoding in a population of neurons. *J. Gen. Physiol.* **59**, 734–766.

Knuth, K.H., Abato, M., White, T., and Gardner, D. (1999), Generalizability of common data model 2000 to representation and exchange of data for advanced neuroimaging techniques. *Soc. Neurosci. Abstr.* **25**, 1910.

Koch, C., and Segev, I., eds. (1998). *Methods in Neuronal Modeling: From Ions to Networks*, 2nd ed. MIT Press, Cambridge, MA.

Koslow, S.H., and Huerta, M.F., eds. (1997). *Neuroinformatics: An Overview of the Human Brain Project.* Erlbaum, Mahwah, NJ.

Lindberg, D.A.B., Humphreys, B.L., and McCray, A.T. (1993). The unified medical language system. *Methods Inf. Med.* **32**, 281–291.

Mainen, Z.F., and Sejnowski, T.J. (1995). Reliability of spike timing in neocortical neurons. *Science* **268**, 1503–1506.

Markram, H., Lubke, J., Frotscher, M., and Sakmann, B. (1997). Regulation of synaptic efficacy by coincidence of postsynaptic APs and EPSPs. *Science* **275**, 213–215.

Nadkarni, P.M., Marenco, L., Chen, R., Skoufos, E., Shepherd, G., and Miller, P. (1999). Organization of heterogeneous scientific data using the EAV/CR representation. *J. Am. Med. Inf. Assoc.* **6**(6), 478–493.

National Institutes of Health (NIH). (2003). *NIH Data Sharing Policy.* NIH, Washington, DC (http://grants.nih.gov/grants/policy/data_sharing/).

Neurodatabase.org (2003). *Cortical Neuron Database.* Laboratory of Neuroinformatics, Weill Medical College of Cornell University, New York (http://neurodatabase.org).

Pechura, C.F., and Martin, J.B., eds. for Institute of Medicine (U.S.) Committee on a National Neural Circuitry Database (1991). *Mapping the Brain and Its Functions: Integrating Enabling Technologies Into Neuroscience Research.* National Academy Press, Washington, DC.

Perkel, D.H., Gerstein, G.L., and Moore, G.P. (1967). Neuronal spike trains and stochastic point processes. II. Simultaneous spike trains. *Biophys. J.* **7**(4), 419–440.

Reich, D.S., Mechler, F., and Victor, J.D. (2001). Independent and redundant information in nearby cortical neurons. *Science* **294**, 2566–2568.

Rieke, F., Warland, D., de Ruyter van Steveninck, R., and Bialek, W. (1997). *Spikes: Exploring the Neural Code.* MIT Press, Cambridge, MA.

Robert, A., Abato, M., Knuth, K.H., and Gardner, D. (2003). Neuroscience data sharing I. Interfaces, incentives, and internals for interoperability. *Soc. Neurosci. Abstr.* **29**, (Program No. 428.7).

Rowe, L.A., and Stonebraker, M.R. (1987). The POSTGRES data model. *Proc. 13th, Intl. Conf Very Large Databases.*

Segev, I., Rinzel, J., and Shepherd, G.M., eds. (1995). *The Theoretical Foundation of Dendritic Function: Selected papers of Wilfrid Rall with Commentaries.* MIT Press, Cambridge, MA.

Shadlen, M.N., and Newsome, W.T. (1998). The variable discharge of cortical neurons: Implications for connectivity, computation, and information coding. *J. Neurosci.* **18**(10), 3870–3896.

Shepherd, G.M., Mirsky, J.S., Healy, M.D., Singer, M.S., Skoufos, E., Hines, M.S., Nadkarni, P.M., and Miller, P.L. (1998). The Human Brain Project: Neuroinformatics tools for integrating, searching and modeling multidisciplinary neuroscience data. *Trends Neurosci.* **21**, 460.

Victor, J.D., and Purpura, K.P. (1996). Nature and precision of temporal coding in visual cortex: A metric-space analysis. *J. Neurophysiol.* **76**(2), 1310–1326.

Westbrook, J., Feng, Z., Chen, L., Yang, H., and Berman, H.M. (2003). The protein data bank and structural genomics. *Nucleic Acids Res.* **31**(1), 489–491.

Wilson, H.R. (1999). *Spikes, Decisions, and Actions: The Dynamical Foundations of Neuroscience.* Oxford University Press, New York.

Xiao, Y., Abato, M., Knuth, K.H., and Gardner, D. (2002). BrainMetaL: An abstract semantic metalanguage for developing and interfacing data description languages. *Biophys. J.* **81**, 168A.

XML (2003). W3C Architecture Domain. Extensible Markup Language (XML). Available at http://www.w3.org/XML

XML Schema (2003). W3C Architecture Domain. XML Schema. Available at http://www.w3.org/XML/Schema

Ziv, I., Baxter, D.A., and Byrne, J.H. (1994). Simulator for neural networks and action potentials: Description and application. *J. Neurophysiol.* **71**, 294–308.

Web Resources

BrainML.org (2003). *BrainML and BrainMetaL Schema Documents.* Laboratory of Neuroinformatics, Weill Medical College of Cornell University, New York (http://brainml.org).

Cover, R., ed. (2003). Core standards: Extensible Markup Languages (XML). *Cover Pages: Online Resource for Markup Language Technologies* (http://xml.coverpages.org/xml.html).

Datasharing.net (2003). Laboratory of Neuroinformatics, Weill Medical College of Cornell University, New York (http://datasharing.net).

Gosling, J., and McGilton, H. (1996). *The Java Language Environment—A White Paper,* Technical Report, Sun Microsystems, Mountain View, CA (http://java.sun.com/docs/white/langenv/).

Java (2003). *The Source for Java Technology,* Sun Microsystems, Mountain View, CA (http://java.sun.com).

National Institutes of Health (NIH). (2003). *NIH Data Sharing Policy.* NIH, Washington, DC (http://grants.nih.gov/grants/policy/data_sharing/).

Neurodatabase.org (2003). *Cortical Neuron Database.* Laboratory of Neuroinformatics, Weill Medical College of Cornell University, New York (http://neurodatabase.org).

XML (2003). W3C Architecture Domain. Extensible Markup Language (XML). Available at: http://www.w3.org/XML

XML Schema (2003). W3C Architecture Domain. XML Schema. Available at http://www.w3.org/XML/Schema

Persistent Collections

Reagan W. Moore, Ph.D.

CONTENTS

Databasing the Brain. Edited by Stephen H. Koslow and Shankar Subramaniam
ISBN 0-471-30921-4 © 2005 John Wiley & Sons, Inc.

4.1 INTRODUCTION

Scientific disciplines generate massive amounts of data from observations, simulations, and analyses (Boisvert and Tang, 2001; Moore et al., 1997). The data are typically annotated with semantic tags, organized in collections, and published in a digital library (Chen, 2001). The data may be distributed across multiple geographically separated sites that are managed on separate administration domains by a data grid (Moore et al., 1999). The relevance of the data may be of a sufficiently long time period that the underlying storage technologies may change multiple times, requiring the use of a persistent archive to manage technology evolution (Moore et al., 2000). These three environments—digital libraries, data grids, and persistent archives—are integrated in the concept of a persistent collection. The ability to publish, share, and preserve data is accomplished by building a collection that can be managed independently of the choice of supporting infrastructure through use of basic data management abstractions. We will look at the concepts that form the abstraction mechanisms, the integration of the concepts into digital libraries, data grids, and persistent archives, as well as the formation of persistent collections that manage data for arbitrarily long periods of time.

Persistent collections require the ability to build infrastructure-independent representations for both (a) the digital entities within the collection and (b) the collections themselves (Ludäscher et al., 2001a). The infrastructure-independent representations correspond to levels of abstraction within the supporting software infrastructure that can be used to interface to access protocols, to application programming interfaces (APIs), and to encoding formats that change over time. The types of abstraction levels will be defined for managing not only data, but also information and knowledge. Examples of systems that implement these abstractions will be presented based upon the NIH Biomedical Informatics Research Network (Web Resource No. 5), the NLM Visible Embryo project (Web Resource No. 36), and the NIH Joint Center for Structural Genomics project (Web Resource No. 17).

4.2 DIGITAL ENTITIES

A persistent collection manages the data, information, and knowledge content of selected digital entities. A digital entity consists of a string of bits on which a data format is imposed. Examples of digital entities include images, binary files, URL strings, SQL command strings, applications, file system directories, database tables, and even collections. Each type of digital entity is interpreted using an appropriate data format. Images are interpreted using image formats such as tiff (Tag Image File Format) (Web Resource No. 33) or gif (CompuServe Graphics Interchange Format) (Web Resource No. 10). Binary files may be interpreted using Hollerith format statements in Fortran codes, and URLs and SQL commands are interpreted as ASCII strings.

Digital entities are manipulated based upon a data model (Moore, 2002). The data model defines (a) the semantics that are used to label structures within the digital entity and (b) the operations that are used to manipulate the structures. The manipulation of complex digital entities such as SQL command strings, applications, directories, and collections requires a characterization of the operations that can be performed on the digital entity. The operations, in turn, depend upon their implementation in a supporting infrastructure. Thus SQL command strings are interpreted by databases, applications depend upon the system calls supported by operating systems, directories depend upon file systems, and collections depend upon databases. Managing a persistent collection depends upon not only the infrastructure needed to create a collection, but also upon the infrastructure needed to interpret the digital entities that are stored in the collection. A persistent collection is created by working with characterizations of digital entities, as well as with characterizations of the systems that are used to manage and interpret the digital entities.

4.2.1 Characterizing Data, Information, and Knowledge

Digital entities contain data, information labels, and knowledge relationships. Persistent collections require the ability to organize, store, and manipulate the data, information, and knowledge content of digital entities. The characterization of a digital entity requires a careful definition of each of these components. We use the following simple definitions to differentiate between data, information, and knowledge (Moore, 2000):

- *Data* correspond to the bits (zeroes and ones) that comprise a digital entity.
- *Information* corresponds to any semantic label associated with the digital entity. The label assigns semantic meaning to structures within the digital entity, and it provides a context for discussing the structures. Instead of referring to, say, the thirtieth through fortieth bits in a digital entity, a label can be applied to the bits to associate a meaning for that structure within the digital entity.
- *Knowledge* corresponds to any relationship that is defined between semantic labels. Examples are semantic/logical relationships between the meaning of information labels, spatial/structural relationships for mapping structures within digital entities to coordinate systems, temporal/procedural relationships for mapping causal ordering to operations that can be applied to the structures, and functional/algorithmic relationships between the procedures that are applied to the structures.

Data are typically managed as files in storage repositories, or binary large objects (blobs) in databases, or objects in real-time object ring buffers. An immediate challenge for the management of data is that the mechanisms used to interact with each of these types of storage repositories are different. The access mechanisms used to retrieve a digital entity vary from file system calls, to SQL commands, to methods on objects. A storage repository abstraction provides a uniform access mechanism for all types of storage systems.

Information is typically managed as attributes in a database or information repository. The semantic label defines the attribute name. The associated bits define the attribute value. The combination of attribute name and attribute value is referred to as metadata. Again, multiple types of information repositories are used

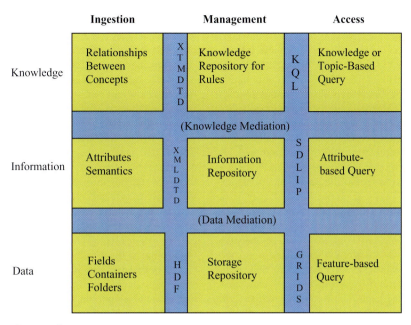

Figure 4.1   Characterization of management systems for data, information, and knowledge.

to manage and organize metadata, from XML files in file systems, to relational tables in relational databases, to XML structures in XML databases, to objects in object-oriented databases. An information repository abstraction is used to create a uniform access mechanism for all types of information repositories.

Knowledge is typically organized in concept spaces or ontologies that are stored in knowledge repositories. An ontology organizes relationships between semantic labels. An example is the use of "is a" and "has a" logical relationships to define whether two terms are semantically equivalent or subordinate (Gupta et al., 2000; Ludäscher et al., 2001b). Graphical Information Systems manage spatial relationships to facilitate the overlay of maps and images. Data processing pipelines manage procedural relationships to specify the order in which processes are applied to digital entities. Feature extraction systems apply algorithmic relationships to identify, for example, a genome, based on an expected sequence of bases. A knowledge repository abstraction is used to create a uniform access mechanism for all types of knowledge repositories and knowledge manipulation systems.

4.2.2 Managing Data, Information, and Knowledge

As seen above, persistent collections manage characterizations of the data, information, and knowledge content of a set of digital entities. The characterizations define a context within which the digital entities can be interpreted and manipulated. The information repository abstraction defines not only operations to manipulate information recorded about individual digital entities, but also operations to manipulate the collection that is housed in the information repository. Collection manipulation includes the import and export of metadata through use of XML files, schema extension for adding or deleting metadata, and automated generation of SQL commands based on specification of operations on

attribute names and values. The collection is viewed as a digital entity that can be managed and manipulated. Persistent collections provide the ability to preserve collections as well as the digital entities organized by the collection.

The persistent collection infrastructure is shown in Figure 4.1 as a three-by-three matrix of the mechanisms used to ingest, manage, and access the data, information, and knowledge content of digital entities (Moore, 2000, 2001). Infrastructure components are represented as yellow rectangles. The abstraction mechanisms for integrating the infrastructure components are represented by the blue grid. The lower row of the matrix defines the data handling systems. The middle row defines the information handling systems, and the top row defines the knowledge handling systems. The columns of the matrix define the ingestion, management, and access mechanisms. Here ingestion refers to the mechanisms and standards used to add digital entities to the persistent collection. Management refers to the storage and preservation of the digital entities, and access refers to the discovery and retrieval of digital entities.

The rectangles in the left column of the persistent collection matrix represent mechanisms for characterizing the data, information, and knowledge content of digital entities. In the ingestion process, operations are performed upon the digital entity to transform its structure into a standard form for long-term preservation. The transformation can be physical, in which the encoding format is changed to a standard encoding format. This process is called migration (Rajasekar et al., 1999). The transformation can also be abstract, in which a characterization of the relationships that are present within the digital entity is created (Moore, 2000). The relationships are organized in a digital ontology that defines the causal order in which bits are structured into words, words are mapped into arrays, and arrays are mapped onto coordinate systems. The digital ontology also records the relationships for assigning time stamps and semantic labels. Operations on the digital ontology correspond to emulation, the process of displaying a digital entity in its original form.

The data bits of a digital entity are structured into fields that are stored in files. The files are physically aggregated into containers, and the containers are logically organized into folders. Multiple types of containers are used for physically organizing files. The hierarchical data format (HDF) stores information about the data format and data models directly within the container, and it provides I/O libraries for extracting and manipulating individual digital entities within the container (Web Resource No. 13). The storage resource broker (SRB) data grid provides support for both the physical aggregation of files and the logical organization of files, by storing the associated attributes in a metadata catalog (Web Resource No. 20; Rajasekar and Moore, 2001).

The use of containers to aggregate digital entities is mandated by interactions with archival storage systems, which write digital entities to removable media (magnetic tape, optical tape, CD-Rom). They work most efficiently when dealing with large files. For example, the archival storage system at the San Diego Supercomputer Center stores 680 terabytes of data (a terabyte is a trillion bytes or a thousand gigabytes) in only 18.8 million files. The average file size is 36 megabytes. Projects that want to store smaller files into the archive use the SRB data grid technology to aggregate the files into containers before they are archived. Storing the 2-micron All Sky Survey of astronomical images required the aggregation of 5 million image files into 147,000 containers before the containers were written to tape. Each image was 2 megabytes in size, resulting in an average container size of 300 megabytes (Web Resource No. 1).

The information content within the digital entities can be annotated using a standard syntax, such as the Extensible Markup Language (XML) (Web Resources No. 38), and the tags can be applied as annotations within the digital entity. Systems such as the Extensible Scientific Interchange Language (XSIL) use the XML syntax to annotate the format of the structures within the digital entity (Web Resource No. 39). Alternatively, the metadata tags can be extracted from the digital entity for import into an information repository. The advantage of the latter is the ability to then support discovery and browsing of the digital entities within the collection. The advantage of the former is that the information content can be associated directly with the digital entity, without requiring the maintenance of an information repository. The reference model for an open archival information system (OAIS) (Web Resource No. 24) specifies the metadata attributes that should be associated with a digital entity to describe its provenance (origin of the digital entity), its properties (format), and its context (descriptive metadata). The metadata attributes and the digital entity-comprise an archival information package (AIP) that can then be preserved. The creation of the AIP for a digital entity is part of the ingestion process.

The knowledge content within a digital entity must also be characterized upon ingestion. A standard syntax such as the resource description framework (RDF) is used to annotate the internal relationships (Web Resource No. 27). RDF represents relationships between the semantic labels as tuples, matching a subject with an object through a predicate. Both the subject and object can be semantic labels that are applied to the digital entity, and the predicate is the defined relationship. RDF uses XML as the interchange syntax for encoding a set of RDF relationships.

The rectangles in the middle column of the persistent collection matrix represent instances of repositories. Digital entities are stored as files in storage repositories such as file systems and archives, and they are also stored in databases as binary large objects or "blobs."

Information is stored as metadata in databases. Knowledge is stored as relationships in a knowledge repository. The repository instances can be distributed across multiple sites and can include multiple types of repository. A persistent collection can be distributed across file systems, archival storage systems, and databases. Software infrastructure, such as data grid technology, is needed to integrate the multiple repositories into a consistent data management system.

The rectangles in the right column represent the standard query mechanisms used for discovery and access to each type of repository. Feature-based query of the storage repository corresponds to application of a template to decide if a feature is present within a digital entity. This requires knowing the name of the physical file in order to apply the template.

Query-based access of the information repository corresponds to specifying logical relationships between attribute names and operations on attribute values to discover a relevant digital entity. A mapping is then done from a logical name that represents the digital entity to the physical file name. The researcher is expected to know the names of the attributes, as well as the correct way to interpret an attribute value. For example, an attribute for a time stamp may be interpreted as a month/day/year date, or as the number of seconds since a specific point in time. The application of operations to the attribute values assumes that the same characterization is applied to all digital entities.

Knowledge-based access of the knowledge repository corresponds to queries on concept spaces. The concept spaces use terms present within a scientific discipline textbook to describe physical quantities relevant to the discipline. A mapping is made from these terms to the attributes used by a particular database. This makes it possible to query a database without having to know the names of the attributes used in the database (Ludäscher et al., 2000). An example is the emerging topic map ISO standard (Web Resource No. 34) which makes it possible to define terms that will be used in a scientific discipline, and then map from these terms to the attribute names implemented within a local database. Topic maps define logical relationships between semantic terms to define whether two terms are semantically equivalent or subordinate. For the most general query mechanism, one would like to be able to map from a concept discovered by browsing through a concept space, to a database attribute that represents that concept, to a logical name for a relevant digital entity, and then to the physical file stored in a data repository.

Persistent collections use multiple levels of naming indirection to support discovery, from a concept space, to a database schema, to a logical name, to a physical file name. When digital entities are ingested into the persistent collection, consistent naming mechanisms need to be used that will be relevant to the discovery mechanisms. A persistent collection uses the capabilities of grid technology (represented as the grid that connects each of the rectangles in Figure 4.1) to build a consistent environment and manage the multiple levels of naming indirection.

Data grid technology provides the concept of a logical name space. Logical names are used to create common user-defined names that are independent of the physical file names. The logical names are equivalent to global persistent identifiers. The logical name is mapped to the physical file names in the storage repositories, based upon administrative metadata that are kept in an information repository. The metadata attributes that are managed by the information repository are also mapped to the logical name, making it possible to discover a digital entity by queries on the metadata. Data grids use the logical name space as the "foreign key" for integrating all types of metadata. The services that are supported by the data grid generate administrative metadata that are also mapped onto the logical name space. The administrative metadata correspond to state information that is generated by the application of a grid service. Examples are the location of a digital entity within a container, the location of a container on a storage repository, the access controls for the digital entity, the date the digital entity was created, and so on.

The grid in the persistent collection matrix that interconnects the rectangles in Figure 4.1 represents the interoperability mechanisms provided by grid technology. The lower row of the grid represents the data mediation level. An example is the San Diego Supercomputer Center Storage Resource Broker (Baru et al., 1998b; Web Resource No. 32). The SRB implements a logical name space to support global persistent identifiers, containers for the physical aggregation of digital entities, a collection hierarchy for the logical organization of digital entities, replicas for multiple copies of digital entities on different storage repositories, and a storage repository abstraction for the access and manipulation of digital entities. The storage repository abstraction defines the permitted operations that can be performed across all of the storage repositories, and it includes standard Unix file system operations. The SRB storage repository abstraction provides a uniform interface to files systems, archives, real-time object ring buffers, FTP sites, and binary large objects (blobs) in databases.

The upper row of the grid in the persistent collection matrix represents the knowledge mediation systems used to map from the concepts described in a knowledge space to the attribute names used in a collection (Web Resource Nos. 3, 15; Baru et al., 1999; Gupta et al., 2000). An example is the SDSC model-based mediation system used to interconnect multiple data collections (Ludäscher et al., 2000). The concept space is based on the terms used by a discipline to describe relationships between physical quantities. Examples are (a) spatial relationships between the named components of an embryo and (b) structural relationships used in anatomy to characterize the evolution of an embryo. The concept space is typically drawn as a directed graph with the links representing the logical relationships and the nodes representing the terms. In topic maps, the concepts are represented as topics, the relationships are represented as associations between topics, and the mappings to instances of the topics in databases are represented as occurrences. The knowledge mediation system makes it possible to map from a concept space to multiple, independent collections that have disjoint sets of metadata attributes.

The left column of the grid in the persistent collection matrix represents the encoding standards used to describe the data, information, and knowledge content of each type of digital entity. The encoding standards that are applied at ingest of material into a persistent collection should correspond to the encoding formats expected by the access mechanisms. Transformative migrations are used to convert to the desired standard encoding format. At the data level, standard encoding formats include SRB containers, the Hierarchical Data Format version 5 (HDF), and tar (tape archive) files. At the information level, the standard encoding format is the Extensible Markup Language syntax (XML). The XML encoding may be organized in Document-Type Definition (DTD) files or in XML Schema. At the knowledge level, multiple choices are available as markup languages, including RDF, Topic Maps, entity relationship diagrams, and so on. Relationships in a Topic Map can be organized in an XML Topic Map Document-Type Definition (XTM DTD).

The right column of the grid in the persistent collection matrix represents the standard access mechanisms that can be used to interact with a repository. For data access, a standard set of operations is provided by the Unix file system (open, close, read, write, seek, stat, sync). For information access, the Simple Digital Library Interoperability Protocol (SDLIP) provides a standard way to retrieve results from search engines (Web Resource No. 29). The Open Archives Initiative (OAI) provides a standard way to extract metadata from an information repository (Web Resource No. 23). For knowledge access, there are multiple existing mechanisms, including the Knowledge Query Manipulation Language (KQML), for interacting with a knowledge repository (Web Resource No. 18).

The interoperability mechanisms between information and data and between knowledge and information represent the levels of abstraction needed to implement a persistent collection. As noted above, persistent collections are built on the capabilities provided by digital libraries, data grids, and persistent archives. Digital libraries are focused on the central row of the persistent collection matrix, as well as on the manipulation and presentation of information related to a collection (Brunner et al., 2000). The digital library community is developing metadata standards for the description of compound digital entities, including documents, multi-media, and images. The METS (Metadata Encoding and Transmission Standard) standard is used to define metadata that is exchanged between digital libraries (Web Resource No. 21). Digital libraries are now being integrated with data grids to support access to digital entities that are located on a remote storage repository. Digital libraries are also being built using logical name spaces to make it possible to create a personal digital library that points to material that is stored somewhere else on the web, as well as to material that is stored on your local disk. The digital library community is starting to work with concept spaces to support knowledge-based access. Web interfaces represent one form of knowledge access systems (the upper right-hand corner of the grid), in that they organize the information extraction methods and organize the presentation of the results for a particular discipline. Examples are portals such as the Biology Workbench that are developed to tie together interactions with multiple web sites and applications, to provide a uniform access point (Web Resource No. 4).

Data grids are focused on the lower two rows of the grid, and they seek to tie together multiple storage systems through use of a logical name space (Baru et al., 1998a; Web Resource No. 12). Computational grids provide mechanisms to support job execution across multiple compute engines (Web Resource No. 11). The integration of the capabilities of data grids and computational grids is done through the definition of web-based services. The Open Grid Services Architecture (OGSA) defines basic access services for files, databases, and computational batch queues (Web Resource No. 25). The OGSA services environment is based upon (a) the Simple Object Access Protocol (SOAP) for transferring service parameters (Web Resource No. 31) and (b) the Web Service Definition Language (WSDL) for describing the services (Web Resource No. 37). Each grid service creates state information that can be mapped onto the logical name space. The logical name space then becomes the naming convention that is used to tie together the results of computations executed within the computational grid, with the persistent collection environments.

Persistent archives manage technology evolution by using storage and information repository abstractions for interacting with different versions of repository infrastructure. There is a very close relationship between data grids and persistent archives. Data grids make it possible to interact with heterogeneous systems interconnected by networks. Persistent archives use the same technology to migrate data from an old technology to a new technology. At the point in time when the migration occurs, the persistent archive has access to both versions of the technology, and hence can use data grid mechanisms to perform the migration. Persistent archives are differentiated from data grids in the types of state information they must maintain. Persistent archives manage state information related to the archival processes of appraisal, accession, arrangement, description, preservation, and access. The additional metadata attributes are related to provenance information and authenticity information. Typical provenance metadata are described by the Dublin Core metadata schema (Web Resource No. 7). Authenticity information includes digital signatures to ensure that digital entities are not corrupted, access controls to restrict migration invocations to authorized archivists, audit trails to track operations done on the digital entities, and location information to ensure all copies can be found.

A persistent collection uses all of the capabilities discussed above. This includes the ability to create persistent digital entities, organize digital entities into collections, manage both the collection and the digital entities across evolving technology infrastructure, and support advanced discovery mechanisms on the resulting collections. Fortunately, generic solutions exist for managing the data and information levels of the persistent collection matrix. Knowledge characterization, ingestion, management, and access, however, remain research issues, with solutions tailored to specific scientific disciplines.

4.3 CREATION OF PERSISTENT COLLECTIONS

The management functions of a persistent collection are very similar to those of a persistent archive. A description of the archival processes for persistent archives is given in Moore (2003). The archival functions provide standard mechanisms for loading new digital entities into the collection, for transforming the encoding formats to accepted standards, for extracting relevant metadata to support discovery, for creating the archival forms that are stored, and for accessing the collection. The archival processes represent explicit tasks that should be applied in the creation of a persistent collection. Given the close coupling between data grids and persistent archives, one can also build persistent collections upon data grid technology. For each

archival process, we list the processing steps that are also needed within persistent collections.

4.3.1 Appraisal

Appraisal is the decision process for whether digital entities should be included in a persistent collection. Typical decision parameters are the uniqueness of the digital entity, its relationship to other digital entities, and its relationship to the criteria under which the persistent collection is being assembled. Data grids allow the curator of the persistent collection to get a quick overview of the digital entities that have already been ingested. The metadata associated with the other digital entities are queried to identify overlaps, as well as areas where additional information is needed. The metadata are also assessed to determine the types of descriptive and provenance information that are needed for the digital entities. If this information is not available for the new material, then they possibly are not suitable for inclusion in the persistent collection.

4.3.2 Accessioning

Accessioning is the controlled import of digital entities. Data grids provide a logical name space that supports the registration of digital entities into collection/subcollection hierarchies. The logical name space is decoupled from the underlying storage systems, making it possible to reference digital entities without moving them. It is possible to represent a digital entity by a handle (such as a URL), register the pointers into the logical name space, and organize the pointers independently of the physical digital entities. Data grids put digital entities under management control, such that automated processing can be done across an entire collection. Data grids provide mechanisms that can be used to validate data models, extract metadata, and authenticate the identification of the submitter. Data grid capabilities that are used to manage the ingestion of digital entities into a persistent collection under process control are typically implemented as remote proxies that can be executed at the storage repository that holds the digital entity.

4.3.3 Arrangement

Arrangement is the process of organizing digital entities. Data grids use collections to define the context to associate with data entities. The context includes provenance information describing the processes under which the data entities were created, attributes used to support information discovery to identify an individual data entity, and relationships that can be used to determine whether associated attribute values are consistent with implied knowledge about the collection, or represent anomalies and artifacts. An example of a knowledge relationship is the range of permissible values for a given attribute, or a list of permissible values for an attribute. The context management also is used to control the level of granularity associated with the organization of the data entities into collections/subcollections. Containers, the digital equivalent of a cardboard box, are used to physically aggregate data entities. Data grids also provide support for logical organization of digital entities into collection hierarchies.

4.3.4 Description

Description is the process of extracting metadata to load into a collection to support future queries. The ability to identify derived data products is based on persistent logical identifiers that are independent of the local storage system file names. For persistent collections, this includes the ability to provide persistent logical identifiers for the data entities stored within the data collections. The global name space may be organized into a collection/subcollection hierarchy with each subcollection supporting unique metadata. Each subcollection is described by an extensible set of attributes that can be defined independently of other subcollections.

4.3.5 Preservation

Preservation is the process of creating archival forms of digital entities and then managing their loading onto a storage repository. Data grids manage replicas of digital entities, replicas of collection attributes, and replicas of collections. The replicas can be located at geographically remote sites, ensuring safety from local disasters. Data grids provide a consistent environment, which guarantees that the administrative attributes used to identify digital entities always remain consistent with migrations performed on the data entities. The consistent state is extended into a persistent state through management of the information encoding standards used to create platform independent representations. The ability to migrate from an old representation of an information encoding standard to a new representation leads to persistent management of digital entities.

Data grids track all operations done on each data entity, to be able to guarantee that the information and knowledge content of each digital entity have only been transformed by approved curators. Audit trails record the dates of all transactions and the names of the persons who performed the operations. Digital signatures and checksums are used to verify that between transformation events the digital entity has remained unchanged. The mechanisms used to accession records can be reapplied to validate the integrity of the digital entities after they have been migrated. Data grids also support versioning of digital entities, making it possible to store explicitly the multiple versions of a record that may be received. The version attribute can be mapped onto the logical name space as both a time-based snapshot of a changing record and as an explicitly named version.

4.3.6 Access

Access is the process of discovery and retrieval of digital entities. Data grids provide transport mechanisms for accessing data in a distributed environment that spans multiple administration domains. This includes support for moving data and metadata in bulk, while authenticating the user across administration domains. Data grids also provide multiple roles for characterizing the allowed operations on the stored data, independently of the underlying storage systems. Users can be assigned the capabilities of a curator, with the ability to create new subcollections, or annotator with the ability to add comments about the digital entities, or submitter, with the ability to write data into a specified subcollection, or public user, with the ability to read selected subcollections. Data grids provide the mechanisms needed to access data distributed across heterogeneous data resources. Data grids implement servers that map from the protocols expected by each proprietary storage repository to the storage repository abstraction. This makes it possible to access digital entities through a standard interface, no matter where it is stored.

4.4 PERSISTENT COLLECTION FUNCTIONALITY REQUIREMENTS

The requirements for a persistent collection can be expressed in general as "transparencies" that mask the heterogeneity of software infrastructure. Examples include digital entity name transparency, data location transparency, platform implementation transparency, encoding standard transparency, and authentication transparency for single sign-on systems. The "transparencies" are also used to manage technology evolution. Implementations exist in data grids for at least four key functionalities or transparencies that simplify the complexity of accessing distributed heterogeneous systems:

1. *Name Transparency.* The ability to identify a desired digital entity without knowing its name can be accomplished by queries on descriptive attributes, organized as a collection. Persistent collections of digital entities map from unique attribute values to a global, persistent identifier.

2. *Location Transparency.* The ability to retrieve a digital entity without knowing where it is stored can be accomplished through use of a logical name space that maps from the global, persistent identifier to a physical storage location and physical file name. If the data grid owns the digital entities (digital entities are stored under the data grid Unix ID), the administrative attributes for storage location and file name can be self-consistently updated every time the digital entity is moved.

3. *Platform Implementation Transparency.* The ability to retrieve a digital entity from arbitrary types of storage systems can be accomplished through use of a data grid that provides a storage repository abstraction. The data grid maps from the protocols needed to talk to the storage systems to the operations defined by the storage repository abstraction. Every time a new type of storage system is added to the persistent collection, a new driver is added to the data grid to map from the new storage access protocol to the data grid data transport protocol. Similar platform transparency is needed for the information repository in which the persistent collection stores the collection context. An information repository abstraction is defined for the set of operations needed to manipulate a catalog in an information repository, or database.

4. *Encoding Standard Transparency.* The ability to display a digital entity requires understanding the associated data model and encoding standard for information. If infrastructure-independent standards are used for the data model and encoding standard (nonproprietary, published formats), a persistent

collection can use transformative migrations to maintain the ability to display the digital entities. The transformative migrations will need to be defined only between the infrastructure-independent data model standards.

The infrastructure that supports the above transparencies exists in multiple data grid implementations. When one examines the data grid implementations, it is possible to identify over 150 different capabilities that have been implemented to facilitate the management of data and information in distributed environments (Moore, 2003). The challenge is defining the minimal set of capabilities that should be provided by a data grid for implementing a viable persistent collection. The fundamental capabilities can be categorized as logical name space, storage repository abstraction, information repository abstraction, and distributed resilient scalable architecture.

A set of core capabilities has been defined in Table 4.1. The list includes the essential capabilities that simplify the management of collections of digital entities while the underlying technology evolves. The use of each capability by one of the six archival processes is indicated. The columns are labeled as App. (Appraisal), Acc. (Accessioning), Arr. (Arrangement), Desc. (Description), Pres. (Preservation), and Ac. (Access).

Possibly the unique capability that must be present in a persistent collection is the ability to maintain a consistent environment. This implies an environment in which only authorized actions can take place. Every operation within the persistent archive must be tracked, and the corresponding metadata must be updated to guarantee consistency of the metadata. This can be most easily implemented by having the digital entities stored under the control of the data grid. This forces access to be done through the data grid, making it possible to track all operations that are done

TABLE 4.1 Mapping of the core capabilities of data grids to archival processes

Core Capabilities	App.	Acc.	Arr.	Desc.	Pres.	Ac.
Storage repository abstraction		X	X		X	X
Storage interface to at least one repository		X	X		X	
Standard data access mechanism		X	X		X	X
Standard data movement protocol support		X	X		X	X
Containers for data		X	X		X	
Logical name space	X	X	X	X	X	X
Registration of files in logical name space	X	X	X	X	X	
Retrieval by logical name		X			X	X
Logical name space structural independence from physical name space	X	X	X	X	X	X
Persistent handle		X	X	X	X	X
Information repository abstraction	X	X	X	X	X	X
Collection owned data	X	X	X	X	X	X
Collection hierarchy for organizing logical name space		X	X	X		
Standard metadata attributes (controlled vocabulary)		X	X	X	X	X
Attribute creation and deletion		X	X	X	X	
Scalable metadata insertion		X	X	X	X	
Access control lists for logical name space to control who can see, add, and change metadata		X	X	X	X	X
Attributes for mapping from logical file name to physical file names			X		X	X
Encoding format specification attributes		X		X		X
Data referenced by catalog query						X
Containers for metadata		X	X	X	X	
Distributed resilient scalable architecture	X	X	X	X	X	X
Specification of system availability		X			X	X
Standard error messages		X	X	X	X	X
Status checking		X	X	X	X	X
Authentication mechanism	X	X	X	X	X	X
Specification of reliability against permanent data loss	X				X	
Specification of mechanism to validate integrity of data and metadata		X	X		X	X
Specification of mechanism to ensure integrity of data and metadata		X	X		X	X
Virtual data grid		X	X	X	X	X
Knowledge repositories for managing collection properties			X	X	X	X
Application of transformative migration for encoding format		X	X	X	X	X
Application of archival processes		X	X	X	X	X

on the digital entities, from transformative migrations, to media migrations, to replication, to accesses.

The choice of core capabilities is an opportunistic definition of the mechanisms that are now available through data grids. Many of the core capabilities can be implemented as procedural policies on current-file-system-based storage repositories, without using data grid technology. For example, authenticity can be managed by defining a set of user IDs that are allowed to write to the archive. One can then require that the defined set of persons manually enter characterizations of all operations that they perform. In practice, this approach would be labor-intensive. The list of core capabilities is intended to minimize the labor associated with organizing, managing, and evolving a persistent collection.

A question is whether the levels of abstraction associated with data grids are consistent with operations in persistent collections. One can think of a data grid as the set of abstractions that manage differences across storage repositories, information repositories, knowledge repositories, and execution systems. Data grids also provide abstraction mechanisms for interacting with the objects that are manipulated within the grid, including digital entities (logical namespace), processes (service characterizations or application specifications), and interaction environments (portals). The data grid approach can be defined as a set of services, and the associated APIs and protocols can be used to implement the services. The data grid is augmented with portals that are used to assemble integrated work environments to support specific applications or disciplines. An example is the Biology Workbench, which provides mechanisms to interact with traditional biological applications (Web Resource No. 4). A major question is whether a persistent archive is better implemented as a data grid, incorporating the required functionality directly into the grid, or as a portal, with the required authenticity and management control implemented as an application interface.

Data grids have been implemented for multiple scientific disciplines, including the high-energy physics community, the earth systems science community, the bioinformatics community, the biology community, the astronomy community, the education community, and the ecology community.

4.5 CONSISTENCY CONSTRAINTS

One of the major challenges in development of infrastructure to support persistent collections is the management of consistency constraints on the execution of grid services and archival processes. In the case of the SRB the consistency management between the services is done at the server level, using the MCAT metadata catalog to provide transaction-level consistency management (Web Resource No. 20). But the problem with the SRB is that these interactions and consistency management constraints are hard-coded and hence are not able to adapt dynamically to new user requirements. The addition of a new grid service, or new set of distributed state information, requires that the software be rewritten to specify additional consistency constraints. A current research topic is the development of an architecture that relieves the user of maintaining consistency and transactional capabilities but that also allows the flexibility of using independent services while tailoring composite services to the needs of a project or user group without breaking the coherency of the system.

We provide a few examples where dynamic constraints will be helpful. For example, in the Biomedical Informatics Research Network (Web Resource No. 5), a requirement is the use of a sophisticated access control system that is consistent with HIPAA (*Health Insurance Portability and Accountability Act of 1996*) requirements for how medical data can be shared and moved (Web Resource No. 14). In effect, HIPAA access constraints are applied dynamically to each file. One can add a label to each digital entity to specify whether or not HIPAA access requirements must be applied. The services that can be performed on the digital entities can check whether or not HIPAA access requirements are to be enforced. However, when new services are added to the persistent collection, a mechanism is needed to specify how the new service will interact with prior services to preserve HIPAA access constraints. In the SRB, these constraints are hard-coded in software, requiring that the code be rewritten each time a new service is added. The ability to characterize dynamically the consistency constraints is needed to facilitate the addition of new services.

Another data grid example is the desire for tailoring audit trails for different projects: Some would like to track access by groups, whereas others want to track access by individuals. Dynamic constraints are needed to allow the choice of audit trails to be changed over time. Similar problems occur in the persistent collection domain in which access controls may be imposed for a time period greater than a specific storage repository lifetime. The access controls have to follow the data as they migrate onto new storage systems. This same requirement is a key feature in federation of collections when specifying which collection is in charge of metadata consistency. In the digital library domain, metadata consistency is needed while accommodating multiple types of curatorship permissions, and when re-purposing collections of digital objects without breaking current usage patterns (Boisvert and Tang, 2001). Digital library systems that are currently available [e.g., ADL (Web Resource No. 2), Berkeley ELIB (Web Resource No. 9), U Michigan DL (Web Resource No. 35), etc.] are hard-coded to support explicit schema and services. Changing them to accommodate new features becomes rather difficult.

Hence, to solve the diverse problems faced by data grids (SRB, Globus, and others), digital libraries, and persistent archival systems, a coherent framework is needed for consistency management in a dynamically changing environment.

Data grids can be viewed as systems that manage and manipulate consistency constraints on mappings of distributed state information onto a logical name space. Digital libraries add mappings to manage user-defined metadata to support discovery and browsing. Persistent archives add mappings to manage the authenticity of the deposited digital entities (Moore, 2000). The three types of data management systems can be viewed as defining multiple levels of aggregation semantics upon collections of digital entities. At the same time, each level of aggregation is managed by a set of constraint relationships. We recognize multiple levels of aggregation and associated ontologies, shown in Table 4.2, that organize the constraint relationships (Moore, 2002).

We note that a *digital entity* can be represented as a set of relationships imposed on bits or bytes: for example, file, table, structured object, and so on. The *digital ontology* specifies the order in which the relationships should be imposed to correctly interpret

TABLE 4.2 Ontology hierarchy for constraint-based collections

Aggregation Level	Aggregation Sets	Ontologies
Federation	Operations on virtual organizations	Federation ontology
Virtual organization	Sets of services	Grid ontology
Service	Operations on collections	Service ontology
Collections	Sets of digital entities	Collection ontology
Digital entity	Relationships on bits	Digital ontology

the digital entity. A *collection* represents a (possibly ordered) set of digital entities. A collection can be viewed as imposing a semantic mapping on digital entities by associating a logical name to each digital entity along with descriptive attributes (metadata). The *collection ontology* organizes semantic relationships between the metadata attributes, structural relationships between attributes and database tables, and procedural relationships for metadata updates. Examples of relationship constraints include ingestion criteria (e.g., check for anonymization in clinical data), automated metadata extraction templates, and so on. A (grid) *service* can be viewed as operations (read methods and functions) on collections. Each service maps distributed state information about the operations onto the logical name space of the entities in the collection. The *service ontology* specifies temporal and structural constraints on updates to the distributed state information. Examples of such constraints include automatic data placement strategies to meet user demands, multiple access control mechanisms as required by user communities, and so on. A *virtual organization (vo)* is a set of services on associated collections along with registries for discovery of services. A *grid ontology* specifies the order in which grid services must be applied (and the order of the distributed state information update) that is needed to maintain consistency. Examples of such constraints include consistency maintenance between replication services and metadata registry services, recovery mechanisms for synchronization of metadata, and so on. Finally, the *federation* of virtual organizations requires mappings between the grid ontologies to ensure consistency between grids. The *federation ontology* specifies the relationships that are enforced between the grid ontologies. Examples of such constraints include assignment of ownership of metadata update and access controls, hand-over policies for data movement, and so on. A collection management system that is able to impose dynamic constraints can be used to implement, manage, and apply any one or all of these ontologies.

The current research activities related to persistent collections are focused on the creation of a collection management system that implements dynamic constraints as a generic mechanism for managing each of these levels of aggregation of digital entities. The manipulation of the ontologies requires (a) specification of a standard syntax for describing relationships and (b) the specification of a language for manipulating the syntax. The Common Logic standard (Web Resource No. 6) and the Ontology Web Language standard (Web Resource No. 26) are emerging as appropriate systems for the management of dynamic consistency constraints. The expectation is that subsets of each of these emerging standards can be used to implement a constraint-based persistent collection management system.

4.6 PERSISTENT COLLECTION EXAMPLES

The data grid community has been developing software infrastructure to support distributed data collections for many scientific disciplines, including high-energy physics, chemistry, biology, earth systems science, and astronomy. These systems are in production use, managing data collections that contain millions of digital entities and aggregate terabytes in size. It is noteworthy that across the many implementations, a common approach is emerging. There are a number of important projects in the areas of data grids and grid toolkits. They include the Storage Resource Broker data grid described here (Web Resource No. 32), the European DataGrid replication environment (based upon GDMP, a project in common between the European Data-Grid and the Particle Physics Data Grid, and augmented with an additional product of the European DataGrid for storing and retrieving metadata in relational databases called Spitfire and other components) (Web Resource No. 8; Stockinger et al., 2001), the Scientific Data Management data grid from Pacific Northwest Laboratory (Web Resource No., 30), the Globus toolkit (Web Resource No. 11), the SAM Sequential Access using Metadata data grid from Fermi National Accelerator Laboratory (Web Resource No. 28; Terekhov and White, 2000), the Magda data management system from Brookhaven National Laboratory [Web Resourece No. 19], and the JASMine data grid from Jefferson National Laboratory (Web Resource No. 16). These systems have evolved as the result of input from their user communities for the management of data across heterogeneous, distributed storage resources.

EGP, SAM, Magda, and JASMine data grids support high-energy physics data. In addition, the European Data Grid provides access to distributed data for earth observation and computational biology communities. The SAM data grid is a file-based data management and access layer between storage and data processing layers that is used to support the D-Zero high-energy physics experiment at Fermi National Accelerator Laboratory. Magda is a distributed data management system used to support the ATLAS high-energy physics experiment. Magda serves as a file catalog and an automated file replication tool between CERN (European Organization for Nuclear Research) and Brookhaven National Laboratory mass stores and the United States ATLAS grid testbed sites. JASMine is a Mass Storage System Manager that is designed to meet the needs of current and future high-data-rate experiments at Jefferson Laboratory.

The SDM system provides a digital library interface to archived data for PNL and manages data from multiple scientific disciplines. The Globus toolkit provides services that can be composed to create a data grid. The SRB data handling system is used in projects for multiple US federal agencies, including the NASA Information Power Grid (digital library front end to archival storage), the DOE Particle Physics Data Grid (collection-based data management for the BaBar experiment), the National Library of Medicine Visible Embryo project (distributed data collection), the National Archives Records Administration (persistent archive), the NSF National Partnership for Advanced Computational Infrastructure (Web Resource No. 22) (distributed data collections for astronomy, earth systems science, and neuroscience), the Joint Center for Structural Genomics (data grid), and the National Institute of Health Biomedical Informatics Research Network (data grid).

The seven data management systems listed above included not only data grids, but also distributed data collections, digital libraries, and persistent archives. However, at the core of each system was a data grid that supported access to distributed data. By examining the capabilities provided by each system for data management, we can define the core capabilities that all the systems have in common. The systems that have the most diverse set of user requirements tend to have the largest number of features. Across the seven data management systems, a total of 152 capabilities were identified. The SRB supports 90% of the capabilities. About one-third of the capabilities (50) have been implemented in at least five of the data grids. This set was further reduced to the 34 capabilities that are listed in Table 4.1 (Moore, 2003).

The common data grid capabilities that are emerging across all of the data grids include implementation of a logical name space that supports the construction of a uniform naming convention across multiple storage systems. The logical name space is managed independently of the physical file names used at a particular site, and a mapping is maintained between the logical file name and the physical file name. Each data grid has added attributes to the name space to support location transparency (access without knowing the physical location of the file), file manipulation, and file organization. Most of the grids provide support for organizing the data files in a hierarchical directory structure within the logical namespace, as well as support for ownership of the files by a community or collection ID.

The logical name space attributes typically include the replica storage location, the local file name, and user-defined attributes. Mechanisms are provided to automate the generation of attributes such as file size and creation time. The attributes are created synchronously when the file is registered into the logical name space, but many of the grids also support asynchronous registration of attributes. Most of the grids support synchronous replica creation, and they provide data access through parallel I/O. The grids check transmission status and support data transport restart at the application level. Writes to the system

are done synchronously, with standard error messages returned to the user. The error messages provided by the grids to report problems are quite varied. Although all of the grids provided some form of an error message, the number of error messages varied from less than ten to over 1000 for the SRB. The grids statically tuned the network parameters (window size and buffer size) for transmission over wide area networks. Most of the grids provide interfaces to the GridFTP transport protocol.

The most common access APIs to the data grids are a C++ I/O library, a command line interface, and a Java interface. The grids are implemented as distributed client-server architectures. Most of the grids support federation of the servers, enabling third-party transfer. All of the grids provide access to storage systems located at remote sites including at least one archival storage system. The grids also currently use a single catalog server to manage the logical name space attributes. All of the data grids provide some form of latency management, including caching of files on disk, streaming of data, and replication of files.

4.6.1 Storage Resource Broker

The Storage Resource Broker implements the persistent collection capabilities that are listed in Section 4.4. The SRB is structured as shown in Figure 4.2, with explicit layers for each of the transparencies, or abstraction layers.

An access abstraction is used to support a wide variety of application programming interfaces. The consistency management layer is hard-coded in software and ensures that all metadata stored in the MCAT catalog are updated (Web Resource No. 20). The consistency management includes locks when writing to files stored in containers, synchronization flags for managing updates to replicas, access control lists for data, and access control lists for metadata. All operations are done relative to the logical name space, ensuring that when a file is moved between storage repositories, the same access controls will continue to apply. An information repository abstraction is used to support manipulation of collections stored in databases, including DB2, Oracle, Sybase, Postgresql, Informix, and SQLServer. A storage repository abstraction is used to support manipulation of digital

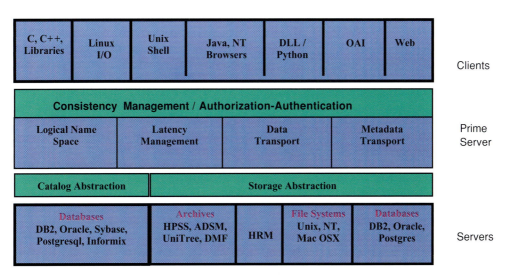

Figure 4.2   The SDSC Storage Resource Broker data grid.

entities that reside in archives, in file systems, in Hierarchical Resource Managers, in object ring buffers, and as blobs within databases.

The SRB is used as a data grid for the sharing of data, a digital library for the publication of data, and a persistent archive for the preservation of data. Fourteen projects are currently using the SRB to store over 50 terabytes of data at SDSC, comprising over 7.8 million files.

4.6.2 Biomedical Informatics Research Network

The BIRN project uses the Storage Resource Broker to implement a data collaboration environment to federate access to multiple independent data collections within the neuroscience community. Three major collaborations are supported within BIRN: the development of virtual data grid technology by a team located at the University of California, San Diego, with activities at the School of Medicine (SOM), the Center for Research on Biological Structures (CRBS), the San Diego Supercomputer Center (SDSC), and the California Institute for Telecommunications and Information Technology [Cal-(IT)2]; the study of animal (mouse) models of human diseases with activities at Duke University, UCLA, Caltech, and UCSD; and the study of brain morphometry with activities at two research groups at Harvard Medical School, one at Duke University, one at John Hopkins University, and a group at UCSD.

At each participating site, a "Grid Brick" is installed that provides the CPU processing power, the disk cache, and the network connectivity that is needed to federate access to the existing collections. The Grid Bricks run grid technology for managing job execution, and the Storage Resource Broker to implement a data sharing environment. Replicas of datasets are made into a storage repository at SDSC to support large scale data manipulations. A MCAT metadata catalog is used to implement a digital library and metadata that describes the digital holdings. Researchers apply access control lists to their data and to their metadata to meet privacy requirements. Parallel I/O is used to move data between the sites. The goal of the effort is to support directed cross-correlation of studies between the sites. By the end of FY 2003, over 4 terabytes of data is expected to be registered into the data grid.

4.6.3 Visible Embryo

The project entitled "Human Embryology Digital Library and Collaboratory Support Tools" uses the Storage Resource Broker to implement a digital library of embryo images for use as an education resource in embryology courses and for clinical science applications. The project is part of the Next Generation Internet Initiative and is funded by the National Library of Medicine. The project's purpose is to demonstrate how leading-edge information technologies in computation, visualization, collaboration, and networking can expand capabilities in science and medicine for developmental studies, clinical work, and teaching. Images are generated at the Armed Forces Institute of Pathology in Washington, DC, sent over wide area networks to a collection housed at the San Diego Supercomputer Center, and accessed by researchers at Johns Hopkins University, Oregon Health Science

University, the University of Illinois at Chicago, and George Mason University.

The images are typically 15 megabytes in size, and are aggregated into containers for replication into archives at SDSC and Caltech. A web-accessible digital library containing information about the images is maintained at SDSC. The total amount of data stored is 43,000 files comprising over 650 gigabytes of data. The SRB functions both as a data grid for distributed image sharing and as a digital library for the publication of the images onto the web.

4.6.4 Joint Center for Structural Genomics

The Joint Center for Structural Genomics (JCSG) is a consortium of California scientific research organizations with the goal of determining the 3D structure of up to 2000 proteins. In the project, state-of-the-art high-throughput technology is being developed for the generation of x-ray diffraction data, the organization of the data into a collection, and the determination of protein structures at high resolution. The main organizations in the JCSG consortium are: The Scripps Research Institute (TSRI)/Genomics Institute for the Novartis Research Foundation (GNF); the San Diego Supercomputer Center (SDSC) at University of California, San Diego (UCSD); and the Stanford Synchrotron Radiation Laboratory (SSRL, a Division of the Stanford Linear Accelerator Center, SLAC) at Stanford University.

The Storage Resource Broker is used to manage the movement of data from SLAC into an archive at SDSC and the registration of the data into a collection managed in the MCAT metadata catalog. The project currently stores 1.66 terabytes of data comprising over 236,000 files in the SRB data grid. An SRB server is installed at SLAC to manage the movement of data from SLAC to SDSC. The MCAT catalog runs on an Oracle database at SDSC. The SRB functions both as a data grid for managing ingestion of data from a remote resource and as a digital library for the registration of the digital entities into a collection.

4.7 SUMMARY

Persistent collections are the combination of digital libraries for the publication of digital entities, data grids for the sharing of digital entities, and persistent archives for the preservation of digital entities. Persistent collections are based on access abstractions for storage and information repositories. A logical name space is used to provide global persistent identifiers that are location independent. State information that results from the application of data grid services or archival processes is mapped onto the logical name space and managed in a metadata catalog. Persistent collections are built on top of data grids. While data grids are being used quite successfully in support of data sharing, publication, and preservation, additional capabilities will be needed in the future to simplify the integration of new services and support federation of independent data grid implementations. The new capabilities can be expressed as the characterization, management, and manipulation of knowledge relationships. The immediate application is the specification of consistency constraints for the federation of independent grid environments.

ACKNOWLEDGMENTS

The ideas presented here were developed by members of the Data and Knowledge Systems group at the San Diego Supercomputer Center. Michael Wan developed the data management systems, Arcot Rajasekar developed the information management systems, Bertram Ludasecher and Amarnath Gupta developed the logic-based integration and knowledge management systems, and Ilya Zaslavsky developed the spatial information integration and knowledge management systems. Richard Marciano created digital library and persistent archive prototypes. The data management system characterizations were only possible through the support of Igor Terekhov (Fermi National Accelerator Laboratory), Torre Wenaus (Brookhaven National Laboratory), Scott Studham (Pacific Northwest Laboratory), Chip Watson (Jefferson Laboratory), Heinz Stockinger and Peter Kunszt (CERN), Ann Chervenak (Information Sciences Institute, University of Southern California), and Arcot Rajasekar (San Diego Supercomputer Center). This research was supported by NSF NPACI ACI-9619020 (NARA supplement), NSF NSDL/UCAR Sub-award S02-36645, NSF I2T EIA9983510, DOE SciDAC/SDM DE-FC02-01ER25486, and NIH BIRN-CC3~P41~RR08605-08S1. The views and conclusions contained in this document are those of the authors and should not be interpreted as representing the official policies, either expressed or implied, of the National Science Foundation, the National Archives and Records Administration, or the U.S. government.

References

Baru, C., Moore, R., Rajasekar, A., Schroeder, W., Wan, M., Klobuchar, R., Wade, D., Sharpe, R., and Terstriep, J. (1998a). A data handling architecture for a prototype federal application. *6th Goddard Conf. Mass Storage Syst. Technol.*, *1998*.

Baru, C., Moore, R., Rajasekar, A., Wan, M. (1998b). The SDSC storage resource broker. *Proc. CASCON'98 Conf.*, Toronto, Canada, *1998*.

Baru, C., Chu, V., Gupta, A., Ludascher, B., Marciano, R., Papakonstantinou, Y., and Velikhov, P. (1999). XML-based information mediation for digital libraries. *ACM Conf. Digital Libr., 1999*, exhibition program.

Boisvert, R., and Tang, P. (2001). The architecture of scientific software. In *Data Management Systems for Scientific Applications*, pp. 273–284. Kluwer Academic Publishers, Boston, MA.

Brunner, R., Djorgovski, S., and Szalay, A. (2000). Virtual observatories of the future. *Astrono. Soc. Pac. Conf. Ser.*, Vol. 225, pp. 257–264.

Chen, C. (2001). Global digital library development. *Knowledge-based Data Management for Digital Libraries,* pp. 197–204. Tsinghua University Press, Beijing, China.

Gupta, A., Ludäscher, B., and Martone, M.E. (2000). Knowledge-based integration of neuroscience data sources. *12th Int. Conf. Sci. Stat. Database Manage. (SSDBM)*, Berlin, Germany, 2000.

Ludäscher, B., Gupta, A., and Martone, M.E. (2000). Model-based information integration in a neuroscience mediator system, *26th Int. Conf. Very Large Databases (VLDB)*, Cairo, Egypt, 2000, demonstration track.

Ludäscher, B., Marciano, R., and Moore, R. (2001a). Towards self-validating knowledge-based archives. 11th International Workshop on Research Issues on Data Engineering (Ride-Dm 2001), Heidelberg, Germany 1–2 April 2001, IEEE Computer Society, *RIDE-DM* pp. 9–16.

Ludäscher, B., Gupta, A., and Martone, M.E. (2001b). Model-based mediation with domain maps. *17th Int. Conf. Data Eng. (ICDE)*, Heidelberg, Germany, 2001.

Moore, R. (2000). Knowledge-based persistent archives. *Proc. Conserv. Doc. Inf. Aspetti Orga. Tec.*, Rome, Italy, *2000*.

Moore, R. (2001). Knowledge-based Grids. *Proce. 18th IEEE Symp. Mass Storage Syst. 9th Goddard Conf. Mass Storage Syst. Technol.*, San Diego, CA, 2001.

Moore, R. (2002). The San Diego Project: Persistent objects. *Proc. Workshop XML Preserv. Lang.*, Urbino, Italy, *2002*.

Moore, R. (2003). Persistent archive concept paper. *Global Grid Forum 8*, Seattle, WA, 2003.

Moore, R., Baru, C., Bourne, P., Ellisman, M., Karin, S., Rajasekar, A., and Young, S., Information based computing. *Proce. Workshop Res. Dir. Next Gener. Internet, 1997.*

Moore, R., Baru, C., Rajasekar, A., Marciano, R., and Wan, M. (1999). Data intensive computing. In *The Grid: Blueprint for a New Computing Infrastructure*, (I. Foster and C. Kesselman, eds.) Morgan Kaufmann, San Francisco, CA. pp. 105–129.

Moore, R., Baru, C., Rajasekar, A., Ludäscher, B., Marciano, R., Wan, M., Schroeder, W., and Gupta, A. (2000). Collection-based persistent digital archives. Parts 1 & 2. *D-Lib Maga.*, April/March 2000 (http://www.dlib.org/)

Rajasekar, A., and Moore, R. (2001). Data and metadata collections for scientific applications. *High Perf. Comput. Network. (HPCN 2001)*, Amsterdam, 2001, pp. 72–80.

Rajasekar, A., R., Marciano, and Moore, R. (1999). Collection based persistent archives. *Proc. 16th IEEE Symp. Mass Storage Syst.*, *1999*, pp. 176–184.

Stockinger, H., Rana, O., Moore, R., and Merzky, A. (2001). Data management for grid environments. *High Perform. Comput. Network. (HPCN 2001)*, Amsterdam, *2001*, pp. 151–160.

Terekhov, I., and White, V. (2000). Distributed data access in the sequential access model in the D0 Run II data handling at Fermilab. *Proce. 9th IEEE Int. Symp. High Perform. Distrib. Comput.*, Pittsburgh, PA, *2000*.

Web Resources

2MASS – Two Micron All Sky Survey (http://www.ipac.caltech.edu/2mass/).

ADL, Alexandria Digital Library, University of California, Santa Barbara (http://www.alexandria.ucsb.edu/adl.html); AFCS—Alliance for Cell Signaling (http://www.afcs.org).

AMOS—Active Mediators for Information Integration (http://www.dis.uu.se/~udbl/amos/amoswhite.html).

Biology—The Biology Workbench (http://www.ncsa.uiuc.edu/News/Access/Releases/96Releases/960516.BioWork.html).

BIRN—The Biomedical Informatics Research Network (http://www.nbirn.net).

Common Logic Standard, proposed ISO standard (http://cl.tamu.edu/).

Dublin Core Metadata Initiative (http://dublincore.org/).

EDG—European Data Grid (http://eu-datagrid.web.cern.ch/eu-datagrid/).

ELIB, UC Berkeley Digital Library Project (http://elib.cs.berkeley.edu/).

Gif—CompuServe Graphics Interchange Format (http://ptolemy.eecs.berkeley.edu/eecs20/sidebars/images/gif.html).

Globus—The Globus Toolkit (http://www.globus.org/toolkit/).

Grid Forum Remote Data Access Working Group (http://www.sdsc.edu/GridForum/RemoteData/).

HDF—Hierarchical Data Format (http://hdf.ncsa.uiuc.edu/).

HIPPA, Health Insurance Portability and Accountability Act of 1996 (http://www.hep-c-alert.org/links/hippa.html).

I2T—The Information Integration Testbed project (http://www.sdsc.edu/DAKS/I2T).

Jasmine—Jefferson Laboratory Asynchronous Storage Manager (http://cc.jlab.org/scicomp/JASMine/).

JCSG—Joint Center for Structural Genomics (http://www.jcsg. org/).

KQML—Knowledge Query Manipulation Language (http://www.cs.umbc.edu/kqml/papers/).

Magda—Manager for Distributed Grid-based Data (http:// atlassw1.phy.bnl.gov/magda/info).

MCAT—The Metadata Catalog (http://www.npaci.edu/DICE/ SRB/mcat.html).

METS—Metadata Encoding and Transmission Standard (http:// www.loc.gov/standards/mets/).

NPACI Data Intensive Computing Environment Thrust Area (http://www.npaci.edu/DICE/).

OAI—Open Archives Initiative (http://www.openarchives.org/).

OAIS—Reference Model for an Open Archival Information System (OAIS). ISO draft, 1999 (http://www.ccsds.org/ documents/pdf/CCSDS-650.0-R-1.pdf).

OGSA—Open Grid Services Architecture (http://www.globus. org/ogsa/).

OWL—Ontology Web Language, from the Web Ontology working group of the W3C (http://www.w3.org/2001/sw/WebOnt/).

RDF—Resource Description Framework (RDF). W3C Recommendation (www.w3.org/TR/).

SAM—Sequential data Access using Metadata (http://d0db.fnal. gov/sam/).

SDLIP—Simple Digital Library Interoperability Protocol (http:// www-diglib.stanford.edu/~testbed/doc2/SDLIP/).

SDM—Scientific Data Management in the Environmental Molecular Sciences Laboratory (http://www.computer.org/conferences/ mss95/berard/berard.htm).

SOAP—Simple Object Access Protocol (http://www.w3.org/TR/ SOAP/).

SRB—The Storage Resource Broker Web Page (http://www.npaci. edu/DICE/SRB/).

Tiff, Tag Image File Format (http://www.libtiff.org/).

Topic Maps—ISO/IEC FCD 13250, 1999 (http://www.ornl.gov/ sgml/sc34/doocument/0058.htm).

U Mich DL, University of Michigan Digital Library Project (http://www.si.umich.edu/UMDL/).

Visible Embryo Project, Human Embryology Digital Library and Collaboratory Support Tools, part of the Next Generation Internet Initiative and funded by the National Library of Medicine (http://netlab.gmu.edu/visembryo.htm).

WSDL—Web Service Definition Language (http://www.w3. org/TR/wsdl).

XML—Extensible Markup Language (http://www.w3.org/XML/).

XSIL—Extensible Scientific Interchange Language (http://www. cacr.caltech.edu/SDA/xsil/).

Entity-Attribute-Value Database Approaches for Heterogeneous, Evolving Neuroscience Data

Perry L. Miller, M.D., Ph.D., Luis Marenco, M.D.,
Gordon M. Shepherd, M.D., Ph.D., and Prakash M. Nadkarni, M.D.

CONTENTS

Databasing the Brain. Edited by Stephen H. Koslow and Shankar Subramaniam
ISBN 0-471-30921-4 © 2005 John Wiley & Sons, Inc.

5.1 INTRODUCTION

Looking to the future, there will be increasing needs in the neurosciences to store, analyze, and integrate many types of diverse heterogeneous data. These data include (1) two-, three-, and four-dimensional images of structures ranging from the brain as a whole to individual neurons and to subcellular components, (2) electrophysiological recordings, (3) data from behavioral instruments such as psychological questionnaires, (4) clinical patient data, (5) genomic data including sequence, gene expression, and protein expression data, (6) computer models of diverse neuroscience phenomena, and (7) the results of many different neuroscience experiments. In addition, the data collected in the neurosciences often evolve over time as the underlying experimental techniques themselves evolve.

The need to be able to store and analyze data that are heterogeneous and that evolve over time poses an interesting set of challenges for the computer systems that are used to store the data. There are a number of approaches that could be used to attempt to accommodate such data.

One approach is to allow the data to be stored in a very loosely structured format—for example, in a set of linked Web pages—that allow each research group generating data flexibility to include whatever data they want and to impose whatever Web-based structure they choose. A disadvantage of this approach is that it may be difficult, if not impossible, to allow the data to be searched, analyzed, and integrated automatically in a robust fashion.

A second approach is to store the data in a conventional relational database. This approach has a number of advantages, as discussed later in this chapter, but does introduce a number of potential inflexibilities. For example, it can be difficult to add new types of data without significant reprogramming. It can also be difficult to allow the types of data stored to evolve over time in an organized, graceful fashion.

As part of the national Human Brain Project, we have developed a set of neuroscience databases for the Yale SenseLab project (Shepherd et al., 1998; Miller et al., 2001). SenseLab is exploring a number of neuroinformatics research questions using the olfactory system as a model domain. We originally built SenseLab's databases using a conventional relational database approach. In recent years, however, we have used the Sense-Lab project as a vehicle to explore a more flexible approach to database design which we call EAV/CR (Entity–Attribute–Value with Classes and Relationships). We believe that the EAV/CR database approach has potential utility in many areas of neuroscience, as well as in many other areas of the biosciences in which heterogeneous, evolving data must be stored and analyzed.

The EAV/CR approach is based on the EAV (Entity–Attribute–Value) approach, which has previously been used in

a number of successful clinical information systems. The EAV approach is also similar to "row modeling," which has been used in computer science to handle sparse matrices (tables). EAV/CR augments the basic EAV approach by providing the ability to handle complex data items (*classes*) and with the ability to explicitly store and manipulate *relationships* between data items (Nadkarni et al., 1999; Marenco et al., 1999). We believe that these augmentations are particularly useful in handling certain of the complexities of neuroscience data.

As described in detail later in this chapter, a key feature of the EAV (and EAV/CR) approach centers around the use of *metadata* (data describing data). The processing of an EAV database is guided by metadata stored in a data library that contains a variety of different data describing each data item. For each data item, the metadata includes information such as (1) the name of the data item, (2) the data type, (3) allowable upper and lower bounds for numeric data items (if appropriate), (4) allowable values for enumerated data items, and (5) the various data attributes associated with each data class (for example, that the class "neuron" has the attributes "receptor," "transmitter," "channel"). Whenever a data item is accessed (created, modified, or retrieved), the metadata are consulted to allow the data item to be accessed appropriately. Metadata can also be used to guide the display of data items—for example, by specifying how they are to be formatted on the screen.

Because the metadata are used every time a data item is accessed, it is easy to add a new data item. A description of the new data item (i.e., metadata describing the item) is merely added to the data library. No programming is required. In contrast, in a conventional relational database, adding a data item requires adding a new column to a table and then modifying the programs that access that table. In the EAV approach, the name of the new data item is stored (as *data*) in the data library. In a relational database, the name of a new data item is hardwired in as the name of a new column of a table. The EAV approach achieves its flexibility because the data library is not just a file to be read by human database programmers, but contains machine-processible metadata that is read by the computer itself each time a data item is accessed.

This chapter first discusses various background issues to help place the EAV and EAV/CR approaches into perspective. It then describes the EAV/CR approach, and how it is implemented in SenseLab, with a particular focus on the use of metadata. Finally, the chapter discusses the current status, limitations, and possible future directions for the EAV/CR approach.

5.2 RELEVANT BACKGROUND INFORMATION

This section discusses a number of issues to help place the EAV/CR approach into perspective. There are several approaches that might be taken to handle the challenges posed by complex heterogeneous neuroscience data. These include:

1. Non-database approaches
2. Conventional relational databases
3. EAV-based databases
4. Hybrid databases

5.2.1 Nondatabase Approaches

Using a database requires that one (a) analyze the data that one is planning to include in considerable detail and (b) impose a significant degree of structure on that data. If one's goal is to allow many different research groups to make whatever data they have available to one another—for example, via the Web—by far the easiest and most flexible way to do this is to use a non-database approach.

One widely used approach in this regard, of course, is to use text. Text allows essentially unlimited flexibility to describe a set of data, how it was developed, its nuances, idiosyncrasies, and limitations, and so on. A modern variation of this approach is to use hypermedia (e.g., hypertext that includes images and other nontextual information). Using hypermedia, one can (a) create an interconnected set of Web pages that contain a great deal of diverse neuroscience experimental data and (b) make that data widely available.

The advantage of such a loosely structured approach is that it provides great flexibility to store many different types of data and to easily add new, unanticipated types of data. Because the data are so loosely structured, however, they can be very difficult to search and analyze in an organized way. This is particularly true if one wishes to analyze the data automatically on a large scale in a systematic fashion.

5.2.2 Conventional Relational Databases

Conventional relational databases offer a number of advantages over a loosely structured approach. Here data are stored in defined tables with "linking tables" that define relationships between data elements in different tables. The process of defining an appropriate set of relational tables for a set of complex heterogeneous data can be a time-consuming task that requires an intimate understanding of (1) the data items and their interrelationships, (2) how the data are likely to be queried and analyzed, and (3) how relational database technology can best be used to model the data for these anticipated uses.

Advantages of using database technology include:

1. *Helping to Maintain Data Accuracy.* Each data item typically has an associated data type and may also have defined, permissible values. The database system will enforce these constraints, not allowing the user to enter data that violate these conditions.

2. *Helping Maintain "Referential Integrity."* Databases typically contain complex interrelated information. For example, if data about a purchase refer to a customer, then information about that customer must exist in the database. A database system maintains such "referential integrity" by ensuring that all items referred to are contained in the database. Maintaining referential integrity is important both when adding new information and when editing or removing existing information.

3. *A Structured Query Language (SQL).* Another key feature of a relational database system is that it provides a structured query language (SQL) that allows the user to search and analyze the data in a powerful, well-defined fashion.

4. *Query Optimization.* Database systems typically also provide a number of tools and capabilities to facilitate query optimization so that queries can be performed as efficiently as possible. These capabilities include (a) the ability to *index* selected database fields to facilitate efficient searching and (b) the ability to *order* the performance of the various operations needed to carry out a query to enhance rapid processing.

5.2.3 EAV-Based Approaches

The EAV approach has previously been used very successfully in a number of clinical information systems. These include the Help system developed in Utah (Huff et al., 1991, 1994), the Columbia Presbyterian clinical data repository (Friedman et al., 1990; Johnson et al., 1990), and Trial/DB, a database for clinical research developed at Yale (Nadkarni et al., 1998). Figure 5.1 illustrates, in a simplified example, how the use of a conventional relational database differs from the use of the EAV approach.

Figure 5.1a shows a conventional table containing several data items for different patients. In this relational table, each row contains a patient name followed by values for various clinical data items: hemoglobin (Hb), hematocrit (Hct), white blood cells (Wbc), and platelets (Plts) expressed in thousands. Note that in this conventional approach, the data element names are an integral part of the database itself, "hardwired" in as the names of the table columns. As more data items are added (e.g., to store additional types of clinical data), additional columns are added to the table.

Figure 5.1b shows the same data represented in simplified EAV format. In contrast to the relational table, the EAV table has only three columns: Entity, Attribute, and Value. Each row contains a single "fact." (1) The Entity column contains the patient name. (2) The Attribute column contains the data item name (e.g., "Hb," "Hct") *stored as data* (not hardwired into the database). (3) The Value column contains the value of that data item for that patient.

To help make the use of metadata more concrete, we show below the metadata that might be entered into an EAV data library defining the data item "Hemoglobin":

Data item name: Hb

Display name: Hemoglobin

Data type: real number

Significant digits: ##.#

Allowed range: 0–35

Normal range: 10–16

UMLS CUI: C0518015

SNOMED ID: P3-34100

Display format: simple text box

As can been seen, these metadata contain a variety of information about the data item that could help to check the accuracy of the data (the data type and allowed range), to display the data item (the display name, significant digits, normal range, and display format), and to link the data item to other information resources (the Unified Medical Language System Concept identifier and the SNOMED identifier).

As discussed previously, *advantages* of the EAV approach compared to the relational approach include greater flexibility to accommodate heterogeneous data that evolves over time. Potential *disadvantages* are discussed later in the chapter, but include

(A)

Name	Hb	Hct	Wbc	Plts
Robert	12.2	38.5	12.2	70
Mary	14.1	41.2	6.2	120

∞
∞
∞

(B)

Entity	Attribute	Value
Robert	Hb	12.2
Robert	Hct	38.5
Robert	Wbc	12.2
Robert	Plts	70
Mary	Hb	14.1
Mary	Hct	41.3
Mary	Wbc	6.2
Mary	Plts	120

∞
∞
∞

Figure 5.1   A simplified example contrasting a conventional relational table to the EAV approach.

potential performance inefficiencies when accessing EAV data for broad queries across multiple entities.

Before concluding this discussion of the EAV approach, it is worth making two comments.

1. The EAV approach is typically implemented within a relational database framework. For example, in our work we usually use Oracle. As a result, the real distinction made in this chapter is between (a) the "conventional" relational approach and (b) the EAV approach. An EAV database itself is typically built on top of an underlying relational database engine.

2. The description of the EAV approach in this section has been deliberately *simplified* to allow us to better convey the basic concepts involved. In practice, the implementation of an EAV system can be considerably more complex. For example, the description of an Entity may be involve considerably more than just a name. It may also include qualifying information such as date and time. Also, the information actually stored in the Entity and Attribute fields of an EAV database are typically pointers into other tables (*not* the names of the data items as shown in Figure 5.1). For the Attribute column, these are pointers into the data library (to the metadata entry that defines that data item).

5.2.4 Hybrid Approaches

In building practical operational database systems, it is often logical to combine the EAV approach (for certain of the data) with conventional relational tables for other data. For example, in Trial/DB (a clinical research database we have developed using EAV techniques), we store demographic information about the patient (e.g., name, address, telephone number, social security number, etc.) in relational format. Using relational tables makes this information more efficient to retrieve and process, and it does not cause problems because the set of data items used is not expected to change. (The values of a given item may change, of course—for example, when a patient changes address. But the set of data elements stored is not likely to change and is constant across all the patients in the database despite the fact that they are in different clinical studies that collect very different clinical data.)

Another reason to store certain data in conventional relational tables arises in the biosciences because high-throughput techniques are increasingly generating massive amounts of very *homogeneous* data. For example, the use of microarray technology to study gene expression can result in a huge amount of data from many different experiments in which the set of high-throughput data items collected is identical (describing the microarray image intensity corresponding to the expression of tens of thousands of genes in a single experiment). It is much more efficient to store this massive amount of homogeneous data in a relational table than to use EAV techniques. On the other hand, storing the data required to describe the very diverse biological *samples* used in different microarray experiments is well-suited for EAV if the database will contain data from many different types of microarray experiments.

5.3 TECHNICAL DETAILS AND METHODOLOGY

5.3.1 The EAV/CR Approach

As described above, the EAV approach has been previously used in several clinical databases. In these systems, there is typically one principal type of entity, the patient (typically modified by factors such as date and time). There are thousands of data items that are potentially relevant to a patient's care. These include

Class	Entity	Attribute	Value
NCR	ncr-12	Neuron	smn (spinal motor neuron)
NCR	ncr-12	Receptor	gaba-receptor
NCR	ncr-12	Compartment	ded
Neuron	smn	Organism	vertebrate
Neuron	smn	Category	principal neuron
Receptor	gaba-receptor	Parent	amino acid receptors
Receptor	gaba-receptor	Agonist	gaba
Compartment	ded	Parent	de

Figure 5.2   A simplified example illustrating the EAV/CR approach, as described in the text.

many different laboratory test results, many findings on different x-ray procedures and other diagnostic procedures, many findings when the doctor conducts a comprehensive physical examination of the patients, and a very wide range of possible symptoms that a patient might describe. Any given patient will only have a very small subset of this vast amount of data that is relevant to their problems and that need to be recorded in a clinical data system.

If one attempts to store all these data using a straightforward conventional approach (with tables that have columns for each possible data item), a number of problems arise. First, one needs a huge number of tables, since a relational database system typically limits the number of columns in a table (often to a maximum of 256). In addition, the vast majority (probably well over 99%) of the table entries would be empty (irrelevant to a given patient). Thus the data are very *sparse*. As a result, a straightforward conventional database approach would be very inefficient since it would result in a massive waste of computer storage. An advantage of using the EAV approach to store these data is that only those "facts" that are relevant to the patient need to be recorded.

When one attempts to adapt the EAV approach to bioscience data, there is no longer a single main "entity" that one wants to describe. Whereas in a clinical database important data may center primarily around patients, in the biosciences, these are many different entities about which one needs to collect data. In SenseLab, these entities include neurons, receptors, transmitters, channels, and so on. No one entity is predominant. In addition, the relationships between the various entities are also important and are often themselves experimentally derived data.

Because of issues such as these, when we first started to adapt the EAV approach (which we had previously used in the clinical arena) to deal with the complexity of neuroscience data, it became clear that the basic EAV approach could be productively enhanced in a number of ways. The EAV/CR approach was developed to better handle some of the new challenges posed by the neuroscience domain. The EAV/CR approach enhances the EAV approach in two fundamental ways.

1. *Classes*. In the EAV approach, each EAV triple includes a single value which may be a number, a text field, an enumerated choice, and so on. In the biosciences, however, it became clear that values are often complex entities themselves. As a result, in the EAV/CR approach we allow values to be either simple or complex. A complex value is defined by a *class* and consists of several attributes, each of which has a value, any of which may in turn be defined by a class. (We show examples of such classes later in this section.)

2. *Relationships*. In a conventional relational database, relationships are built into the database, either as the names of column headers, or in the relational bridge tables that link data items in one table to related data items in another table. In the biosciences, however, relationships between biological entities are potentially important experimental data (e.g., that a particular protein *binds* to a particular receptor). As a result, we designed the EAV/CR approach so that such relationships could be explicitly stored. New relationships can be added to the system merely be entering a description of them in the data library.

Figure 5.2 provides a simplified example of how data is stored using the EAV/CR approach. In addition to the three EAV columns, we now have a fourth column, "Class." In this example, we see four classes. One class, Neuronal Compartment Receptor (NCR), has the other three classes for values for its three attributes. An instance of the NCR class specifies a relationship between a particular neuron, a particular compartment, and a particular receptor (indicating that that receptor as been experimentally identified in that compartment of that neuron).

1. In this example, a specific instance of the NCR class (*ncr-12*) specifies that a *gaba-receptor* has been identified in the *Ded* compartment of the *spinal motor neuron* (SMN). This information is stored in rows 1–3 of Figure 5.2.

2. Rows 4–5 of Figure 5.2 store information about the specific instance (SMN) of the class "Neuron." This class has two attributes ("Organism" and "Category"), each of which has a single value describing the SMN.

3. Rows 6–7 of Figure 5.2 store similar information about gaba-receptor, which is an instance of the Receptor class and has two attributes ("Parent" and "Agonist") describing it.

4. The final row contains similar information describing the Ded compartment.

The key difference between the EAV/CR approach and the EAV approach is that each EAV triple is stored as an independent fact, whereas using EAV/CR, one defines a set of explicitly related facts. In Figure 5.2, all of the data are explicitly related (since they are grouped into a hierarchical structure defined by the four classes). One advantage of this approach is that the related facts can be more readily accessed and analyzed as a group. A second advantage is that each class is defined by metadata which indicate, for example, which attributes are allowed to describe each class. In this way, the metadata can help maintain the logical consistency of the database and can also help in the retrieval and display of the data. It can also help the database designer in maintaining the database as it evolves over time.

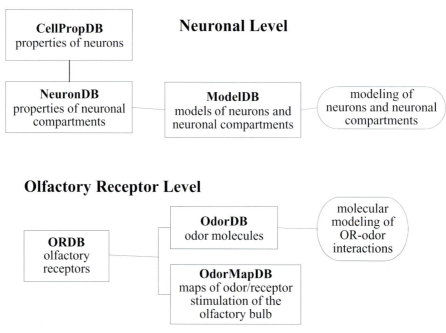

Figure 5.3 A schematic overview showing the six SenseLab databases, along with related computer modeling activities, at two different levels: the neuron level and the olfactory receptor level.

In Section 5.4 of this chapter, we show examples of how SenseLab's metadata are structured and how they can be used to facilitate display, editing, and querying of the SenseLab databases.

5.3.2 Performance Efficiency

The EAV approach achieves its flexibility at the cost of certain potential inefficiencies. Because EAV data are stored in one long table, the task of accessing and analyzing that data becomes more complex. One facet of this complexity concerns the code that a programmer must write to store, access, display, and analyze the data. This coding task can be facilitated if preprogrammed routines are provided to help accomplish these tasks.

A fundamental issue concerns runtime efficiency—for example, the time required to process of query. Chen et al. (2000) explored certain EAV-related efficiency issues in the context of Trial/DB. This study suggests that "entity-centered" queries (queries that retrieve data about a single entity, such as a patient) run reasonably efficiently. On the other hand, "attribute-centered" queries (queries that analyze data across many entities) can be significantly slower in an EAV system. There are a number of ways to help deal with these potential inefficiencies.

1. One approach is to place data of different datatypes into separate EAV tables. This allows the data to be much more efficiently indexed, and therefore more efficiently retrieved.

2. Appropriate indexing of the EAV tables by the underlying database system can also help improve efficiency.

3. An approach that has been adopted in several clinical EAV systems has been to maintain an automatically generated relational copy of the database that can be used for broad, attribute-centered queries that would run too slowly on the EAV system itself.

5.4 CURRENT APPLICATIONS

The EAV/CR approach was developed in the context of the SenseLab project at Yale. SenseLab is a neuroinformatics research project supported by the national Human Brain Project that focuses on the olfactory system as a pilot domain. (See Figure 5.3.) SenseLab research is divided into three broad categories: (1) building databases and related tools to support neuroscience research, (2) computer-based modeling of neurons, of neuronal compartments and circuits, and of selected molecules, and (3) basic neuroinformatics research performed in the context of SenseLab whose results are used to enhance the capabilities of SenseLab as a whole.

5.4.1 The Six SenseLab Databases

SenseLab currently includes six databases that are in different stages of maturity. Three databases focus on the olfactory receptor:

1. *ORDB* (Olfactory Receptor Database) contains information about olfactory receptors (ORs) and other closely related receptors. ORDB serves as a national repository for this information and has been operational for several years. As of September 2002, ORDB contains data describing over 4000 receptors.

2. *OdorDB* is a pilot database that contains information about ligands (odor molecules) that have been used experimentally to interact with ORs.

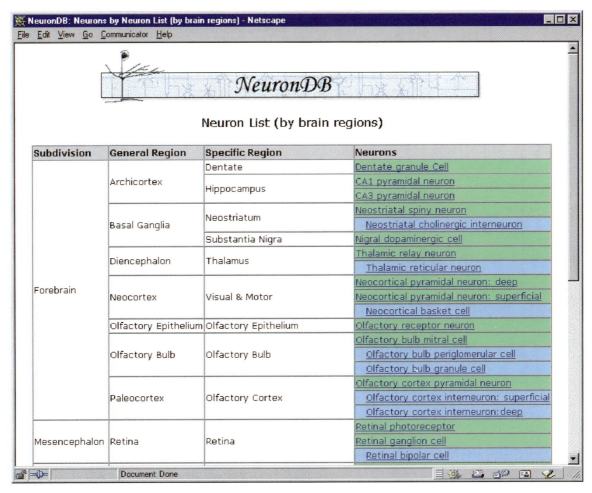

Figure 5.4   A NeuronDB screen that allows the user to select a neuron of interest.

3. *OdorMapDB* is a prototype database currently being developed that contains maps of the patterns of olfactory bulb stimulation for various odor molecules and ORs.

The other three databases, which are currently in pilot operational use, focus on the neuron:

1. *CellPropDB* contains information about membrane properties (receptors, transmitters, and channels) that have been experimentally identified in different neurons.

2. *NeuronDB* contains the same information as CellPropDB, with the information localized to specific neuronal compartments.

3. *ModelDB* contains a growing number of computer models of neurons, of neuronal compartments and components, and of neuronal circuits. These models can be downloaded over the Internet and run on a user's machine. As of September 2002, ModelDB contains over 75 models.

Detailed descriptions of SenseLab can be found in Miller et al. (2001) and Shepherd et al. (1998). To help illustrate the nature of SenseLab's databases, this section gives a brief overview of NeuronDB. NeuronDB contains information about membrane properties (receptors, transmitters, and channels), localized to different neuronal compartments of specific neurons.

The data stored in NeuronDB is taken from the neuroscience literature.

Figure 5.4 shows a NeuronDB screen that allows the user to select a particular neuron of interest. If the olfactory mitral cell is chosen, Figure 5.5 shows the NeuronDB screen that can be used to request information about different compartments of that cell. This screen shows (1) an actual picture of a mitral cell, (2) a schematic "canonical form" outlining the various neuronal compartments used by NeuronDB, and (3) a menu that allows the user to request various data at different levels of detail. For example, if the user requests data "plus references/notes," about the "Dad" (distal apical dendrite) compartment, for "All Properties," Figure 5.6 shows a portion of the information retrieved. This information contains annotated information about various receptors, currents (channels), and transmitters that have been experimentally identified in the Dad compartment of the mitral cell. By scrolling down the Web page in Figure 5.6, additional information is available, including all the annotation references (which also provide automated links to PubMed).

NeuronDB is designed to serve as a pilot national repository for neuronal properties similar to the national role played (on a much larger scale) by GenBank for sequence data. NeuronDB provides rapid access to information that otherwise would be

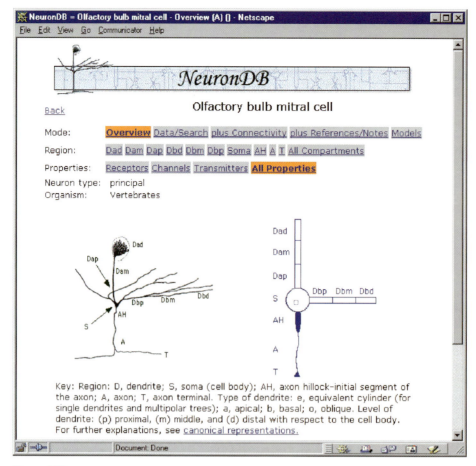

Figure 5.5   A NeuronDB screen that allows the user to request data about a mitral cell neuron.

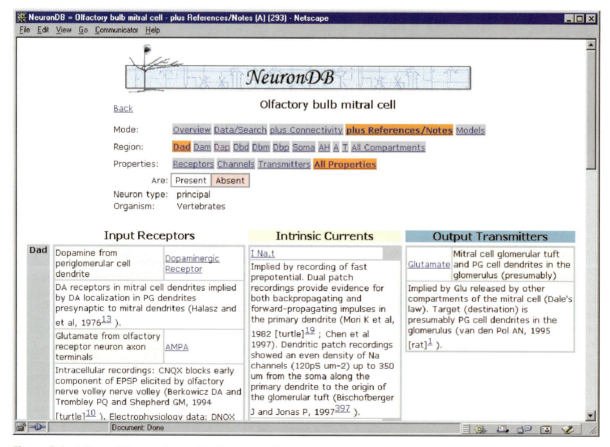

Figure 5.6   A NeuronDB screen that provides data requested by the user about the distal apical dendrite (Dad) compartment of the mitral cell.

extremely time-consuming and unreliable to extract manually from the literature. The database can be searched in a variety of ways. For example, in Figure 5.6, clicking on the name of a receptor will provide a list of all other cells that have that receptor in the Dad compartment.

5.4.2 The EAV/CR Framework and Metadata Underlying SenseLab

All six of SenseLab databases are stored within a single EAV/CR system. The EAV/CR metadata defines each database and each of the components of that database. This section shows successive levels of this metadata starting at the top metadata level (that defines each database as a whole) and moving down through the various levels of metadata defining classes and attributes.

1. Figure 5.7 shows a SenseLab screen designed to help the database designer start to browse or manage the EAV/CR metadata. This screen lists the six available SenseLab databases.
2. Figure 5.8 shows metadata that define the NeuronDB database itself.
3. Figure 5.9 shows metadata listing all the classes that have been defined for NeuronDB.
4. Figure 5.10 shows metadata that define the "Neuronal Compartment Receptor" class within the NeuronDB database.

5. Figure 5.11 shows metadata listing the attributes that have been defined for the Neuronal Compartment Receptor class.
6. Figure 5.12 shows the metadata that define the "Receptor" attribute of NeuronDB.

In this way, SenseLab uses metadata to define the various data elements and objects at each level of an EAV/CR database. These metadata can be inspected and edited by a database designer when creating or refining a SenseLab database. The metadata are also consulted constantly by the routines that access and display the data to the user. In the next two sections, we show examples of how this is done.

5.4.3 SenseLab's Metadata-Driven Interfaces for Browsing and Editing

One way that SenseLab uses the EAV/CR metadata is to construct screens automatically for use in viewing and editing the data. For example, Figure 5.13 shows two interfaces automatically created using the metadata for ORDB. Figure 5.13a shows an interface for *viewing* data describing an olfactory receptor (OR). Figure 5.13b shows a different interface constructed automatically from the same metadata for *editing* that data. Because these interfaces are constructed automatically from the metadata, if metadata describing a new data element are added, these two interfaces would automatically adjust to the new definition of an OR, without any programming required at all.

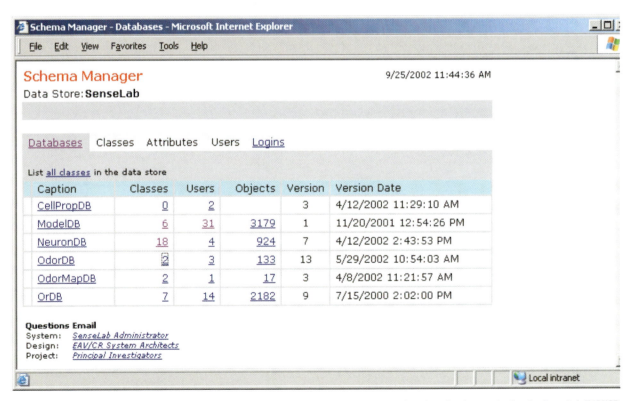

Figure 5.7   The topmost level of SenseLab's metadata, which list the logical databases that have been implemented using the SenseLab EAV/CR database system.

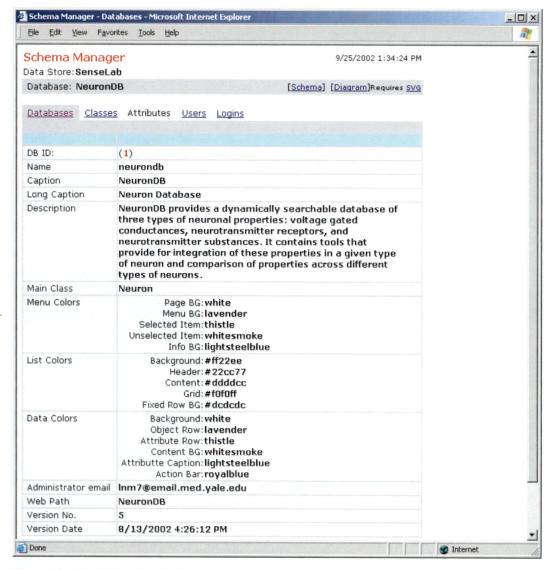

Figure 5.8 Metadata describing the NeuronDB database.

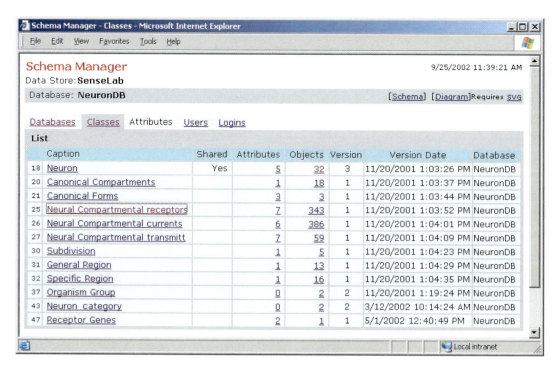

Figure 5.9 This screen is generated from metadata and lists the classes that have been defined for NeuronDB.

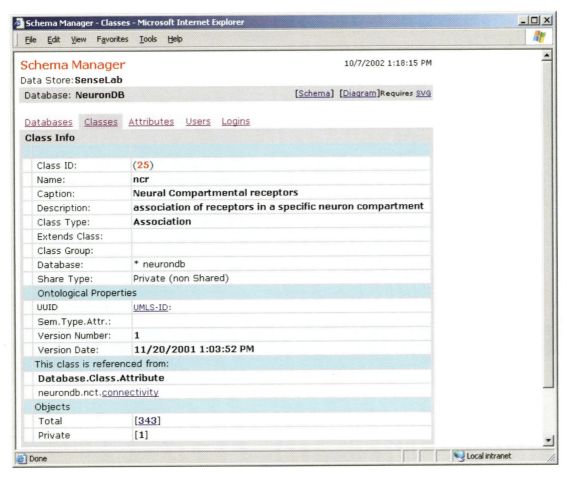

Figure 5.10   Metadata describing the Neural Compartment Receptor (NCR) class.

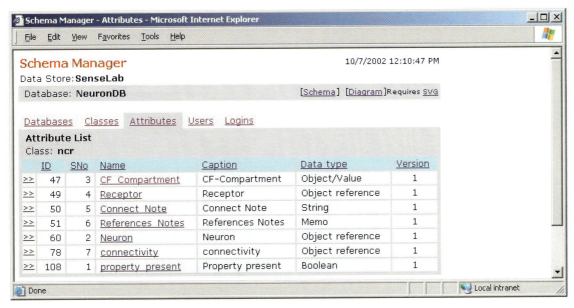

Figure 5.11   A listing of the attributes that have been defined for NeuronDB (metadata).

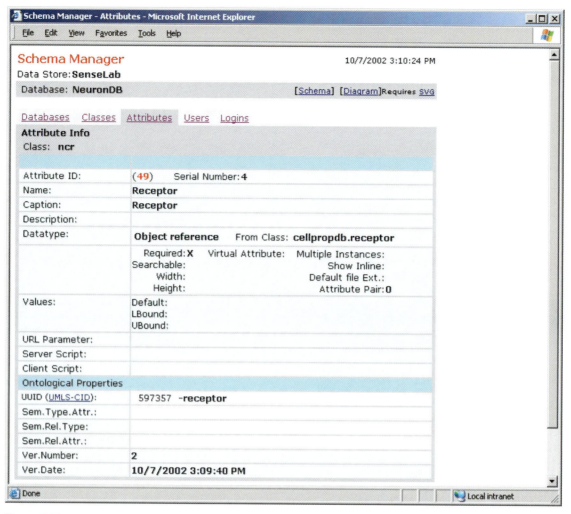

Figure 5.12 Metadata describing Receptor attribute.

(A)   (B)

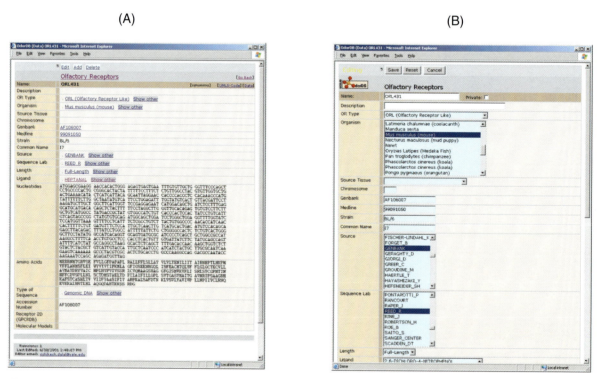

Figure 5.13 Two screens that have been automatically generated from ORDB metadata, one screen for viewing the data about an OR and one for editing that data.

5.4.4 SenseLab's Metadata-Driven Query Interface

Figure 5.14 shows a metadata-driven interface for querying an EAV/CR database, using ORDB as an example. The top of the screen lists the various attributes associated with an ORDB Olfactory Receptor. The user can specify (in the "Value" column) specific values to search for in each field. If the user is unsure what value to use, he merely clicks on the appropriate "LookUp" button, and the system will search the actual data (not the metadata) to see what values have actually been used for that attribute and will display a choice list from which the user can make a

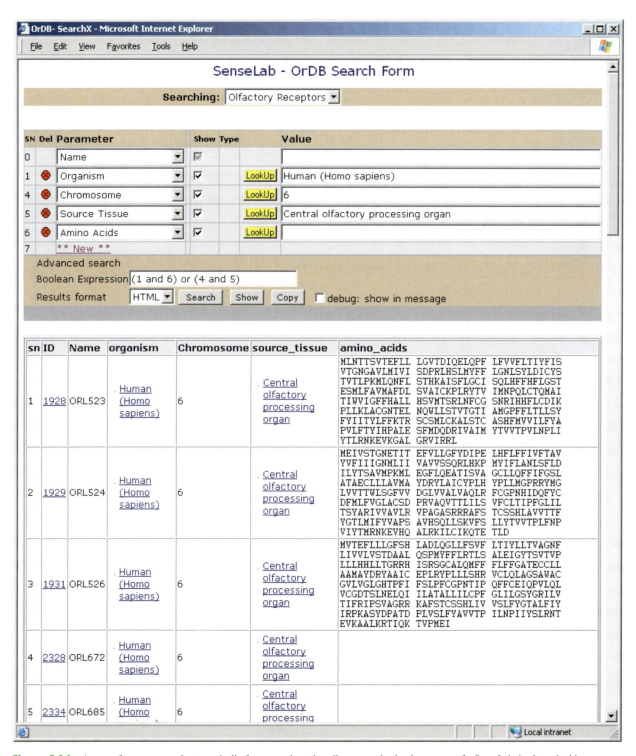

Figure 5.14 A query form, generated automatically from metadata, that allows complex boolean query of a SenseLab database, in this case ORDB, as described in the text.

selection. In the "Show" column, the user checks each field that should be displayed in the results returned. In the "Boolean Expression" field below this list, the user may specify a complex combination of the conditions for the search. If this field is left blank, the values specified are "anded" together in the search. At the bottom of the screen, we see the results returned from this search.

Thus the EAV/CR routines that construct this interface consult not only the metadata (to construct the screen as a whole), but also the primary data stored in the database itself (to list values that are actually present in the database for possible use in searching) to provide both flexibility and focus. Consulting the metadata provides flexibility. Consulting the actual data allows the interface to list those choices that will actually retrieve data. In addition, if desired, it can list those options that are most likely to be of interest based on recent use of the database.

5.5 LIMITATIONS

This chapter describes our work at Yale to develop and explore the EAV/CR approach on a research basis. To help place this work into a broader perspective, it is useful to describe certain limitations:

1. Limitations of the EAV/CR approach itself
2. Limitations of our work, which has explored EAV/CR in a limited domain as a research project

5.5.1 Limitations of the EAV Approach

This chapter has already discussed advantages and limitations of the EAV approach. On the one hand, the EAV approach provides a great deal of flexibility in representing heterogeneous, evolving data. It is also well-suited for storing sparse data. A major advantage is that the data stored can evolve over time gracefully, without the need for constant reprogramming.

On the other hand, the EAV approach imposes several challenges. One challenge is that it can be much more difficult for a programmer to write the code to retrieve and manipulate data than with a relational database. This problem can be addressed by creating parameterized routines to facilitate this manipulation. Another challenge concerns performance efficiency. As discussed previously, retrieving data about a single entity can be performed quite efficiently in an EAV database. Performing a query that requires an integrated analysis of data about many entities, however, can be considerably slower with EAV than with a relational database. One previous approach to this problem has been to create an automatically generated relational copy of the database which can handle such queries more efficiently.

In summary, there are tradeoffs in flexibility and performance when deciding whether to use a conventional or EAV approach. Deciding which approach to use can depend on the nature of the data and the type of queries anticipated. In practice, a hybrid approach may often be a good solution.

5.5.2 Limitations of Our Exploration of the EAV/CR Approach

Members of our research group have previously used the EAV approach to build Trial/DB, a clinical research system that has been in operational use at Yale and elsewhere for a broad range of different clinical studies. The EAV/CR approach was developed as part of the SenseLab project and has been used only for that project. At present the six SenseLab databases (described above) all run in a single integrated EAV/CR framework. Several of the SenseLab databases (e.g., ORDB, NeuronDB, CellPropDB, and ModelDB) already serve a role as pilot national resources and are used regularly by researchers outside of Yale.

At the same time, our implementation of EAV/CR undergoes periodic refinement as we explore how it could be enhanced to provide additional capabilities. As a result, the EAV/CR approach is not currently implemented as a database management system that can be used broadly by many other research groups. To allow such broad use, EAV/CR would need to be taken over by an operational support group who could refine it as a product, document it for broad use, and maintain its distributed use over time.

5.6 FUTURE IN NEUROSCIENCE

As one looks to the future in the field of neuroinformatics, one can readily envision a large, growing, evolving federation of neuroscience databases. Each database may contain a range of different types of experimental data. Much of these data will be sparse and heterogeneous. At the same time, some of the data will involve massive amounts of very homogeneous data produced by high-throughput techniques, such as microarray gene expression experiments.

There will be many informatics-related challenges in pursuing this vision of a federation of databases. One major challenge will involve exploring how best to combine conventional database techniques with more flexible approaches such as EAV and EAV/CR. The best solutions may often involve a hybrid of these techniques, as well as other techniques including loosely structured approaches. Our current pilot work with EAV/CR is one step in exploring how the components of an ultimate federation of neuroscience databases might best be designed.

ACKNOWLEDGMENTS

This work was supported in part by NIH grant P01 DC04732 and by NIH grants P20 LM07253 and T15 LM07056 from the National Library of Medicine.

References

Chen, R.S., Nadkarni, P.M., Marenco, L., and Miller, P.L. (2000). Exploring performance issues for a clinical database organized using an entity–attribute–value representation. *J. Am. Med. Inf. Assoc.* **7**, 475–487.

Friedman, C., Hripcsak, G., Johnson, S., Cimino, J., and Clayton, P. (1990). A generalized relational schema for an integrated clinical patient database. *Proc. 14th Symp. Comput. Appl. Med. Care*, Washington, DC, *1990*, pp. 335–339.

Huff, S.M., Berthelsen, C.L., Pryor, T.A., and Dudley, A.S. (1991). Evaluation of a SQL model of the HELP patient database. *Proc. 15th Symp. Comput. Appl. Med. Care*, Washington, DC, *1991*, pp. 386–390.

Huff, S.M., Haug, D.J., Stevens, L.E., Dupont, C.C., and Pryor, T.A. (1994). HELP the next generation: A new client-server architecture. *Proc. 18th Symp. Comput. Appl. Med. Care*, Washington, DC, *1994*, pp. 271–275.

Johnson, S., Cimino, J., Friedman, C., Hripcsak, G., and Clayton, P. (1990). Using metadata to integrate medical knowledge in a clinical information system. *Proc. 14th Symp. Comput. Appl. Med. Care*, Washington, DC, *1990*, pp. 340–344.

Marenco, L., Nadkarni, P.M., Skoufos, E., Shepherd, G.M., and Miller, P.L. (1999). Neuronal database integration: The Senselab EAV data model. *Proc. 1999 Am. Med. Inf. Assoc. Annu. Fall Symp.* Washington, DC, *1999*, pp. 102–106.

Miller, P.L., Nadkarni, P.M., Singer, M., Marenco, L., Hines, M., and Shepherd, G. (2001). Neuroinformatics research at Yale in support of the Human Brain Project: SenseLab. *J. Am. Med. Inf. Assoc.* **8**, 34–48.

Nadkarni, P.M., Brandt, C., Frawley, S., Sayward, F., Einbinder, R., Zelterman, D., Schacter, L., and Miller, P.L. (1998). ACT/DB: A client-server database for managing entity-attribute-value clinical trials data. *J. Am. Med. Inf. Assoc.* **5**, 139–151.

Nadkarni, P.M., Marenco, L., Chen, R., Skoufos, E., Shepherd, G.M., and Miller, P.L. (1999). Organization of heterogeneous scientific data using the EAV/CR representation. *J. Am. Med. Inf. Assoc.* **6**, 478–493.

Shepherd, G.M., Mirsky, J.S., Healy, M.D., Singer, M.S., Skoufos, E., Hines, M.S., Nadkarni, P.M., and Miller, P.L. (1998). The Human Brain Project: Neuroinformatics tools for integrating, searching, and modeling multidisciplinary neuroscience data. *Trends Neurosci.* **21**, 460–468.

Data Grids

Ian Foster, Ph.D.

CONTENTS

Databasing the Brain. Edited by Stephen H. Koslow and Shankar Subramaniam
ISBN 0-471-30921-4 © 2005 John Wiley & Sons, Inc.

6.1 INTRODUCTION

Neuroscience has become data-intensive, computational, and collaborative. Experimental studies continue to be vitally important, but the "laboratory" within which the researcher works to obtain new understanding and knowledge must increasingly extend beyond the bench to encompass geographically distributed databases, computational services, and experimental facilities—and, in many cases, to connect the researcher with distributed, often multidisciplinary, teams (Committee on a National Collaboratory—National Research Council, 1993).

The growing importance of informatics is, in a broad sense, the theme of this book. My purpose in this chapter is to address those aspects of this theme that relate specifically to geographical distribution. To this end, I introduce the reader to the motivation, design, technical details, applications, and likely evolution of the so-called Grid technologies (Foster and Kesselman, 2004) that are emerging as a basis for large-scale resource sharing within the sciences. I describe both (a) the basic Grid middleware infrastructure that addresses foundational issues such as authentication, authorization, discovery, data transport, and secure access and (b) various services and tools that build on this foundation to address specific concerns such as data sharing and federation, access to computational services, and collaboration. I illustrate my presentation with examples from a variety of disciplines.

The title of this chapter, "Data Grids," is chosen to parallel this book's focus on data-intensive applications. However, this term should not be read as implying any distinction between the Grid technologies used for data access and analysis and those used for remote access to computation or instrumentation. In practice, any substantial data analysis necessarily requires access to the full range of Grid services, including computing, network, and security.

6.2 BACKGROUND

I introduce the technical challenges that have motivated the development of Grid technologies and then sketch the evolution of those technologies.

6.2.1 Resource Sharing in Virtual Organizations

The broad deployment and adoption of the Internet and high-speed networks make it possible for researchers to access distributed resources and collaborate with remote colleagues in ways not previously possible. In addition, rapid improvements in sensor, storage, simulation, imaging, and other related technologies make remote resources and multidisciplinary teams increasingly critical to scientific progress. The technical and methodological consequences of these interrelated causes and effects have been explored first within dedicated Gigabit testbeds (Catlett, 1992; Catlett and Smarr, 1992) and more recently on larger scales (Stevens et al., 1997; Brunett et al., 1998; Eickermann and Hommes, 1999; Johnston et al., 1999; Beiriger et al., 2000). Applications have included distributed computing for large-scale data analysis (pooling of compute power and storage), the federation of distributed datasets, collaborative visualization of large scientific datasets, interpersonal collaboration (Stevens, 2004), and coupling of scientific instruments with remote users, computers, and archives to increase functionality as well as availability (Bonkalski et al., 1998; Johnston, 1999; Hadida et al., 2000).

Experience with these applications suggests a cross-cutting requirement for middleware infrastructure that can enable coordinated resource sharing and problem solving in dynamic, multi-institutional settings. This sharing concerns not primarily simple file exchange but rather direct access to computers, software, data, services, and other resources. This sharing is, necessarily, highly controlled, with resource providers and consumers defining clearly and carefully just what is shared, who is allowed to share, and the conditions under which sharing occurs. A set of individuals and/or institutions defined by such sharing rules form a *virtual organization* (VO) (Foster et al., 2001).

VOs can vary greatly in terms of scope, size, duration, structure, distribution and capabilities being shared, community, and sociology, as we illustrate with three examples.

- A collection of data centers that maintain, in a consistent format and subject to standard policies, a collection of online repositories of neural imaging data, plus data mining services that allow authorized participants to search this federated collection (Ellisman and Peltier, 2004).
- The entire neuroscience community, which is far less organized and mutually trusting, but can benefit from registry, authentication, resource access, and other mechanisms for locating and accessing data, computational, and other services.
- Two co-authors sharing data for a research article.

Despite this diversity of application scenarios, we can identify a broad set of common concerns and requirements. We need to establish and maintain flexible sharing relationships capable of expressing collaborative structures such as client–server and peer-to-peer along with more complex relationships such as brokered sharing via intermediaries. We also need high levels of control over how shared resources are used, including fine-grained access control, delegation, and application of local and global policies. We need basic mechanisms for discovering, provisioning, and managing varied resources, ranging from programs, files, and data to computers, sensors, and networks, so as to enable time-critical or performance-critical collaborations. We also need support for diverse usage modes, ranging from single user to multi-user and from performance-sensitive to cost-sensitive and hence embracing issues of quality of service, scheduling, co-allocation, and accounting.

6.2.2 The Data Challenge

Resource sharing is motivated in particular by the demands of increasingly large data collections. Digital data are now fundamental to all branches of science and engineering and play a major role in medical research and diagnosis. Increasingly, these data are organized as shared and structured collections held in databases, XML documents, and structured assemblies of binary files. Driven by advances in simulation and sensor technology, data collections have grown to the extent that multiple terabyte-sized, and soon petabyte-sized, collections of data are becoming prevalent (Hey and Trefethen, 2003).

The sheer size of datasets being generated makes the interpretation of the data in any one collection challenging. Analysis may demand teraflops of compute power and require access to terabytes of data distributed across millions of binary files and multiple databases. Furthermore, analysis may require complex series of processing steps, each generating intermediate data products of size comparable to the input datasets. These intermediate data products need to be stored, either temporarily or permanently, and made available for discovery and use by other analysis processes.

The advent of ubiquitous network connectivity and the scale of modern challenges, such as deciphering the function of all genes in a large number of species, have led to widespread collaboration in the creation, curation, publication, management, and exploitation of these structured collections. Substantial advances can often be achieved by combining information from multiple data sources. For example, astronomers are building virtual observatories that allow data collected at different frequencies (x-ray, radio, optical, infrared, etc.) and at different times to be combined to discover new properties of the universe (Szalay and Gray, 2001). Similarly, functional genomics requires comparison between species and integration with protein biochemistry and crystallography databases, laboratory phenotypical data, and population studies. Consequently, data analysis must both deal with issues of large-scale computation and data movement and provide mechanisms to integrate information from diverse, geographically distributed structured collections.

As collections increase in scale and number, it becomes impractical to arrange for integration of data access and analysis into application workflows through ad hoc schemes. More structured mechanisms for discovering, accessing, analyzing, and integrating data become critical if these important assets are to be shared and exploited to their potential.

6.2.3 Evolution of Grid Technologies

The development of Grids has been spurred and enabled by the staged development of increasingly sophisticated and broadly used technologies. As illustrated in Figure 6.1, early experiments with "metacomputing" (Catlett, 1992; Catlett and Smarr, 1992; Grimshaw et al., 1994; Eickermann and Hommes, 1999; Messina, 1999) worked primarily with custom tools or specialized middleware that emphasized message-oriented communication between computing nodes.

The transition from metacomputing to Grid computing occurred in the mid-1990s with the introduction of middleware designed to function as wide-area infrastructure to support diverse online processing and data-intensive applications. Systems such as the Storage Resource Broker (Baru et al., 1998), Globus Toolkit® (Foster and Kesselman, 1998), Condor (Litzkow et al., 1988; Frey et al., 2002c), and Legion (Grimshaw and Wulf, 1997) were developed primarily for scientific applications and demonstrated at various levels of scale on a range of applications. Other developers attempted to leverage the middleware structure being developed for the World Wide Web by using HTTP servers or Web browsers as Grid computing platforms (Baratloo et al., 1996; Brecht et al., 1996; Vahdat et al., 1998). However, these systems did not gain significant use, partly because the middleware requirements for distributed information systems such as the Web are different from those for Grid applications.

Globus Toolkit

By 1998, the open source Globus Toolkit (GT2) (Foster and Kesselman, 1998) had emerged as a de facto standard software infrastructure for Grid computing. GT2 defined and implemented protocols, APIs, and services used in hundreds of Grid deployments worldwide. By providing solutions to common problems such as authentication, resource discovery, resource access, and data movement, GT2 accelerated the construction of real Grid applications. And by defining and implementing "standard" protocols and services, GT pioneered the creation of interoperable Grid systems and enabled significant progress on Grid programming tools. This standardization played a significant role in spurring the subsequent explosion of interest, tools, applications, and deployments.

Open Grid Services Architecture

As interest in Grids continued to grow, and in particular as industrial interest emerged, the importance of true standards increased. In particular, 2002 saw the emergence of the Open Grid Services Architecture (Foster et al., 2002a) (OGSA), a community standard with multiple implementations—including the OGSA-based GT version 3 (Welch et al., 2003) released in 2003. Building on and significantly extending GT2 concepts and technologies, OGSA aligns Grid computing with broad industry initiatives in service-oriented architecture and Web services.

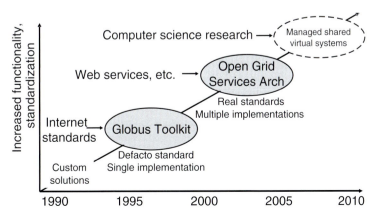

Figure 6.1   Evolution of Grid technology.

In addition to defining a core set of standard interfaces and behaviors that address many of the technical challenges introduced above, OGSA provides a framework within which can be defined a wide range of interoperable, portable services. OGSA thus provides a foundation on which can be constructed a rich Grid technology ecosystem comprising multiple technology providers. Thus, we see, for example, major efforts underway to develop data access and integration services (Paton et al., 2002; Atkinson et al., 2004).

Concurrent with these developments, we see a growing recognition that large-scale development and deployment of Grid technologies is critical to future success in a wide range of disciplines in science, engineering, and the humanities, along with increasingly large investments within industry in related areas. While research and commercial uses can have different concerns, they also have much in common, and there are promising signs that the required technologies can be developed in a strong academic–industrial partnership. The current open source code base and emerging open standards provide a solid foundation for the new open infrastructure that will emerge from this work.

6.2.4 Service Orientation, Integration, and Virtualization

Three related concepts are key to an understanding of the Grid and its contemporary technologies and applications: service orientation, integration, and virtualization.

A service is an entity that provides some capability to its clients by exchanging messages. A service is defined by identifying sequences of specific message exchanges that cause the service to perform some *operation*. By thus defining these operations only in terms of message exchange, we achieve great flexibility in how services are implemented and where they are located. A service-oriented architecture is one in which all entities are services and thus any operation that is visible to the architecture is the result of message exchange.

By encapsulating service operations behind a common message-oriented service interface, service orientation isolates users from details of service implantation and location. For example, a storage service might present the user with an interface that defines, among other things, a "store file" operation. A user should be able to invoke that operation on a particular instance of that storage service without regard to how that instance implements the storage service interface. Behind the scenes, different implementations may store the file on the user's local computer, in a distributed file system, on a remote archival storage system, or in free space within a department desktop pool—or even select from among such options depending on context, load, amount paid, or other factors. Regardless of implementation approach, the user is aware only that the requested operation is executed—albeit with varying cost and other qualities of service, factors that may be subject to negotiation between the client and service. In other contexts, a *distribution framework* can be used to disseminate work across service instances, with the number of instances of different services deployed varying according to demand (Appleby et al., 2001; Graupner et al., 2002).

While a service implementation may directly perform a requested operation, services may be *virtual*, providing an interface to underlying, distributed services, which in turn may be virtualized as well. Service virtualization also introduces the need for *service integration*. Once applications are encapsulated as services, application developers can treat different services as building blocks that can be assembled and reassembled to adapt to changing business needs. Different services can have different performance characteristics, and in a virtualized environment, even different instances of the *same* service can have different characteristics. Thus new distributed system integration techniques are needed to achieve end-to-end guarantees for various qualities of service.

6.3 TECHNICAL DETAILS AND METHODOLOGY

I now turn to a more detailed description of specific Grid technologies and factors that may guide their selection and use within scientific communities such as neuroscience. I structure this presentation in terms of (a) specific technical challenges that the working scientist may encounter and (b) the tools that may be used to address these challenges.

6.3.1 Data-Oriented Services

Perhaps the most compelling motivation for Grid technologies is a need to access remote data and/or to provide remote colleagues with remote access to local data. Depending on context, issues can include discovery, access control, secure and reliable data movement, performance, and server-side processing. In addition, data may need to be integrated from multiple sources. I describe some of the basic building blocks that can be deployed to address these challenges and also present in the next section a case study of a data access and integration system, the Earth System Grid.

Discovery
Users often require the ability to discover data based on metadata attributes rather than knowledge of name or location. Key issues for the discovery of data (and indeed of any resource in a distributed system) include the description, publishing, indexing, and updating of relevant attributes (i.e., metadata). The most challenging issues often relate to the definition of metadata conventions that allow a community to share information using common vocabulary and semantics, issues that are discussed in other chapters. However, the secure and reliable organization of metadata is also an important concern.

The metadata catalog (MCAT) provided by the Storage Resource Broker (SRB) (Baru et al., 1998) represents one approach to maintaining metadata. MCAT is used for storage and discovery of logical and physical metadata attributes in a relational database, as well as for enforcing access control and providing consistency among replicated data stored in SRB servers. Another approach is to decompose metadata services further—for example, storing only relationships between attributes and logical (location-independent) collection names, as in the metadata catalog service (MCS) (Chervenak et al., 2003). MCS decouples metadata management from other distributed data management functions. Thus, data may be maintained anywhere and in any format, and access control and replication are handled by distinct services.

MCAT and MCS both provide a general metadata schema, and both are extensible to allow the specification of attributes specific to a particular application community, in addition to standard schemas for data characterization and discovery, such as the Dublin Core produced by the digital library community. In other contexts, ontologies and other technologies for capturing application-specific knowledge can be vitally important (Goble et al., 2004).

Moving Data Securely and Efficiently

Data movement between storage systems or between programs and data storage is a fundamental operation, providing the foundation for replication, caching, and bulk data access. Large datasets and wide-area data transfer delays can make data transfer efficiency important. Data may be filtered before transfer, effectively running a query or analysis to extract only a desired subset of the data. Many applications are concerned with the reliability of data movement, and thus we may wish to maintain state on outstanding data transfer operations, retry failed transfers, and/or restart interrupted transfers.

GridFTP (Allcock et al., 2002) defines a standards-based, extensible data transfer protocol that provides a uniform interface to various storage systems, including hierarchical storage systems, disk systems, and storage brokers. It provides a fundamental data access and data transport service for distributed systems.

GridFTP functionality includes both features supported by the FTP standard and a number of extensions. FTP was chosen as a base because of its widespread use on the Internet and because the FTP protocol provides a well-defined architecture for protocol extensions and supports dynamic discovery of the extensions supported by a particular implementation. In addition, various groups have already added extensions through the IETF, some of which address Grid requirements. To facilitate interoperation with other Grid services, GridFTP services use the widely deployed Grid Security Infrastructure for robust and flexible authentication, integrity, and confidentiality.

GridFTP can be used both to access specific data values and to move data blobs from point to point. To facilitate its use as a data access protocol, server-side processing allows for the inclusion of user-written code that can process the data prior to transmission or storage. Partial file retrieval is included by default. Third-party control of data transfer allows a user or application at one site to initiate, monitor, and control a data transfer operation between two other sites.

GridFTP supports (a) parallel data transfers that use multiple TCP streams between a source and destination (Figure 6.2)—useful over dedicated but lossy networks—and (b) striped data transfers for use when data objects are interleaved across multiple hosts. Large TCP windows are used to improve the efficiency of file transfer for large files, reducing the overhead of control message and TPC synchronization. Also, because large transfers can take a long time and hence are more susceptible to failure, GridFTP contains (a) mechanisms for reliable and restartable data transfer and (b) user-provided fault recovery algorithms to handle failures such as transient network and server outages.

Enhancing the Reliability of Data Movement

The user who has a large number of data transfers to perform must necessarily be concerned with submitting, monitoring, and managing these many transfers. In this context, a *reliable data transfer service* can be a useful tool. Such a system augments the basic data transfer services described previously (e.g., GridFTP) with enhanced failure semantics and reliability. It maintains the state of outstanding data transfer operations, monitors the progress of transfers, and attempts to restart failed transfers. Thus, a user can submit a large number of simultaneous data transfer operations and monitor and manage the status of each.

Figure 6.3 depicts an OGSA-compliant reliable file transfer (RFT) service, developed as part of the Globus Toolkit, which allows monitoring and control of third-party data transfer operations between two GridFTP servers. A client issues a request to an RFT factory, which instantiates an RFT service instance. This RFT instance controls the transfer and stores the state of the transfer in persistent storage (in this case a database). The

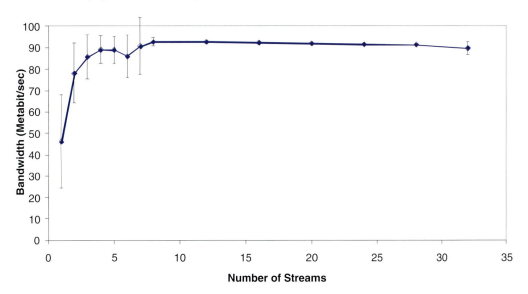

Figure 6.2   Effect of multiple streams on network transfer performance between Chicago and Los Angeles. Data courtesy of B. Allcock.

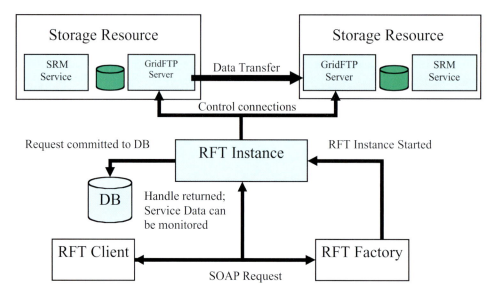

Figure 6.3 The Globus Toolkit's reliable file transfer service.

RFT instance communicates with two storage resources running GridFTP servers as well as storage resource managers. The RFT instance initiates a third-party transfer from the source to the destination GridFTP server and monitors the status of the transfer, updating the state describing the transfer in the database. If the transfer fails because the client or one of the storage resources fails, then the transfer state in the RFT database is sufficient to resume or restart the interrupted transfer when all resources become available.

Access to Structured Data

Data access functionality in GridFTP is primarily directed toward file-oriented structured data with access primitives to return subsections of files. Although FTP's extended data commands support "get" operations with complex specifications, a higher-level query interface can provide more uniform access to data sources.

Standard interfaces for accessing structured data such as ODBC and JDBC typically establish a session and then submit a series of query statements within some transaction regime. Each submission obtains a response, either a result set or a status response indicating whether the execution succeeded or failed. These interfaces make an application program independent of some aspects of the database to which it connects, and they support remote connection.

Using these standard interfaces as a baseline, the Data Access and Integration (DAI) Services Working Group of the Global Grid Forum has developed specifications for data access and integration (Antonioletti et al., 2003) appropriate for the Grid environment. The UK e-Science Programme project OGSA-DAI has constructed "plug-and-play" data access and integration components (Antonioletti and Jackson, 2003), the first version of which was released in conjunction with the Globus Toolkit 3 (GT3) in June 2003. OGSA-DAI assumes an OGSA-compliant architecture and provides a simple set of composable components.

As illustrated in Figure 6.4, the client uses a data registry first to locate a Grid data service factory (GDSF) service capable of generating the required access and integration facilities.

The information returned allows the client to choose an appropriate GDSF and activate it using its Grid service handle (GSH). The client then asks that GDSF to produce a set (here one) of Grid data services (GDSs) that provide the required access to data resources. A GDS may be the data resource itself or a proxy for that data resource, as illustrated here. The client then requests the GDS to perform a sequence of operations, such as update, query, bulk load, and schema edit. To enable an open-ended range of data models and operations, the required operations are requested by using a request document, which specifies a sequence of activities such as database operations defined using standard query languages and data delivery. Additional components required for data translation are included in the DAI architecture.

Data Replication

One reason for moving data is to create *replicas* to reduce access latency, maintain local control over necessary data, and/or improve reliability and load balancing. Thus, we need services for replicating data, locating existing replicas, selecting among available replicas, and proactively replicating data items to satisfy demand.

A replication management service is responsible for creating replicas and potentially supports selection among replicas. For example, Reptor (Guy et al., 2002) provides for the management of data replication operations, controlling the copy of an existing data item and its registration with a replica location service and metadata catalog. An optimization component selects from among existing replicas and picks the best location for creation of new replicas.

The problem of locating replicas when the number of files, replicas, and replica locations are large is challenging. The GT3 replica location service (Chervenak et al., 2002), developed by the Globus and EU DataGrid projects, provides interested parties with replica location information. This service uses soft-state update protocols to simplify the protocol and increase reliability, and it transmits summaries rather than complete information to reduce communication requirements.

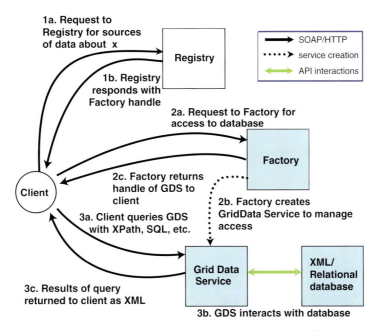

Figure 6.4 The OGSA-DAI architecture, showing the steps involved in retrieving data from a remote database.

6.3.2 Accessing Remote Computers and Software

A second common reason for considering Grid technologies is a need for either more specialized or greater quantities of computing resources than are available on a local computer. A variety of systems exist for providing secure access to remote computers, for harnessing distributed computing resources, and for organizing distributed computations. We review some of these systems here.

Secure and reliable access to remote computers for purposes of job submission can be achieved via the Grid Resource Access and Management (GRAM) protocol and service provided by the Globus Toolkit (Czajkowski et al., 2004). This component defines a client-side API and network protocol that a client can use first to submit, and subsequently to monitor and manage, jobs on any remote computer that runs a GRAM server. Grid Security Infrastructure authentication and authorization mechanisms are used, and GRAM status information can be accessed via standard information service mechanisms.

GRAM mechanisms can be accessed directly from user programs or from the command line via programs such as "globus-job-submit." However, they are most commonly used via other more application-oriented tools. We describe several such tools in the following.

Workflow systems such as DAGman (Thain and Livny, 2004) focus on the coordination of components written in traditional programming languages. DAGman allows the user to specify directed acyclic graphs in terms of tasks that must be performed and dependencies among those tasks; it then schedules the execution of those tasks on available computers, using GRAM for remote submission. Nimrod-G (Abramson et al., 2000) and the AppLeS Parameter Sweep Template (Casanova et al., 2000) are specialized to parameter sweep applications in which a single application must be run many times while varying specified parameters.

Portal toolkits (Thomas et al., 2001; Aloisio and Cafaro, 2002; Novotny, 2002; Russell et al., 2002) support remote access to either computers or applications from "thin clients" such as Web browsers or Java applications. These systems have been used to construct many successful Grid portals, some application-specific and some designed to provide access to a particular class of resources.

MPICH-G2 (Karonis et al., 2003) is a Grid-aware version of the public-domain Message Passing Interface (MPI) (Gropp et al., 1999). It provides a standards-based message-passing environment for Grid environments, allowing a client to interact with a remote server using MPI mechanisms, and/or for the distribution of a single computation over several distributed computers.

NetSolve (Casanova and Dongarra, 1997) and Ninf-G (Tanaka et al., 2002) are examples of systems designed to provide remote access to software.

6.3.3 Virtual Data: Integrating Data and Computation

Many applications involve a series of data-processing and data analysis steps. The construction and subsequent execution of such operations can pose difficult planning, scheduling, and monitoring challenges. One approach to managing this complexity is to adopt a *virtual data* abstraction (Foster et al., 2002b), in which desired data collections are specified by name or attributes, and underlying services construct the processing steps, taking into consideration preexisting intermediate results, data location, computational resource availability, and so forth. For example, the Chimera virtual data system provides a language for specifying virtual data products, an interpreter that maps from a virtual data description to a plan for creation of the data, catalogs for recording provenance information for newly created data items, and Grid infrastructure that can discover whether the data item exists or can initiate production of the data item (Foster et al., 2002b).

Once an execution graph is produced, either as the result of an explicit query or by a virtual data system, one must map the execution of the operations onto available services and resources. Issues here include planning the computation (Deelman et al., 2003), perhaps breaking it into a series of data movement and computational tasks; selecting the best storage and computational resources to perform each task and scheduling them; and monitoring progress, including recovery from task failures. The Virtual Data Toolkit developed within the GriPhyN project addresses these issues.

We note that for any derived data product, it is important to record information not only about the data but about how the data were produced; this is called *provenance* information (Buneman et al., 2001). Such information can include exact information about the analysis or simulation programs that produced a data item and the inputs to those programs. Although provenance data are another form of metadata, they present unique challenges with respect to how they are structured, obtained, and used to reconstruct data analysis chains (Foster et al., 2002b).

6.4 APPLICATION EXAMPLES

We review two applications to provide a flavor for how Grid technologies can be combined and used to construct useful systems. See also Chapter 7 for a description of the Biomedical Informatics Research Network (Ellisman and Peltier, 2004) and other neuroscience applications of Grid computing.

6.4.1 Earth System Grid

To illustrate how data services can be combined to deliver integrated data access capabilities, we present a case study: the Earth System Grid (ESG) (Allcock et al., 2001). ESG seeks to enable geographically distributed teams of climate researchers to develop new knowledge from massive, distributed data holdings and to share results with a wider community. The amount of data produced by climate simulation models is growing rapidly

with increasing data resolution and computational capabilities. For example, a single 100-year simulation of the community climate simulation model (CCSM) produces approximately 0.75 terabytes of data. Improved data resolution is expected to increase output size to approximately 11 terabytes.

ESG is developing a production system for delivering data to climate scientists, sophisticated metadata services for climate data discovery, and filtering data servers to reduce data transfer requirements. The output datasets of climate model simulations are stored in terabyte- and eventually petabyte-scale data archives, currently using the high-performance storage system (HPSS). Hierarchical storage resource manager (HRM) middleware is used to manage HPSS storage, schedule data requests, stage data from tape storage to disk cache, manage caches, and copy multiple data files among storage locations (Shoshani et al., 2003).

ESG datasets are accessed and filtered by using the OPeNDAP data access protocol system, widely used in the climate modeling community. The OPeNDAP server implementation has been modified to provide authentication of users via the Grid Security Infrastructure, enhanced performance through the GridFTP data transfer protocol, and authorization using the Globus Toolkit's Community Authorization Service (Pearlman et al., 2002). This use of OPeNDAP as a gateway to Grid data services allows ESG to provide high-performance access to replicated and distributed datasets to end-user analysis tools without modification.

Metadata services address the automatic extraction and generation of metadata as well as data discovery and query based on metadata attributes. A common ESG metadata schema has been defined, as has the relationship of this schema to other schemas, including federal standards, the NASA Global Change Master Directory, digital library standards such as the Dublin Core, and metadata schemas used by climate researchers in other nations. ESG metadata are stored in the MCS metadata catalog service mentioned previously, and the replica location service is used to track data replicas. An ESG Web portal provides access to published climate datasets and metadata and manages workflow for accessing the various components of the ESG infrastructure.

Figure 6.5 shows some of the services deployed at various sites in support of ESG functions. Each data provider site runs an

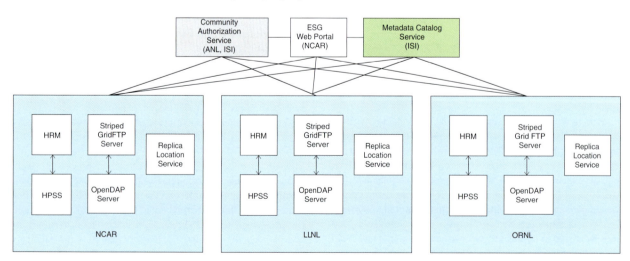

Figure 6.5   Some of the components deployed in the ESG infrastructure.

HRM, GridFTP, and replica location service to enable publication of, and access to, its data resources. Other services support discovery, authorization, and access via Web interfaces.

6.4.2 Virtual Data and the Sloan Digital Sky Survey

The Sloan Digital Sky Survey (SDSS) is a digital imaging survey that will, by the end of 2005, have mapped a quarter of the sky in five colors with a sensitivity two orders of magnitudes greater than previous large sky surveys. SDSS data are being made available online as both a large collection (∼10 TB) of images and a smaller set of catalogs (∼2 TB), containing measurements on each of 250,000,000 detected objects. SDSS is just one example of a growing set of digital sky survey projects that are creating an unprecedented international, distributed multipetabyte collection of digital astronomical data (Szalay and Gray, 2001).

A recent experiment (Annis et al., 2002) (Figure 6.6) showed how this online data can be integrated with distributed computing and storage resources to perform computationally intensive analyses of unprecedented scale. The goal was to process Sloan data to identify galaxy clusters, the largest gravitationally dominated structures in the universe. The approach taken was to use the Chimera and Pegasus systems mentioned earlier to manage the data derivation process and the GriPhyN Virtual Data Toolkit to orchestrate distributed resources. First, the galaxy cluster identification workflow shown on the left of Figure 6.6 was expressed in declarative terms as Chimera virtual data descriptions. A

request for a particular set of derived products then generated a directed acyclic graph of many thousands of computational tasks (Figure 6.6 shows a small example), which were scheduled over computational clusters at four sites across the United States.

6.5 FUTURE

Dramatic increases in data volumes, network capacity, and problem complexity are demanding new approaches to scientific problem solving based on multidisciplinary collaboration and the integration of distributed resources. Grid technologies address key technical challenges—not, for the most part, by providing vertically integrated systems that can be applied "off the shelf," but rather by providing tools for authentication, authorization, metadata-based discovery, replica location, high-performance transfer, workflow management, computation scheduling, and system monitoring and management that can be combined in various ways to develop domain-specific distributed data integration and analysis solutions. The Earth System Grid, Biomedical Informatics Research Network, and Sloan sky survey analysis application are three examples of such integrated systems.

In the immediate future, Grid technology development will complete the integration with Web services started with the initial OGSA specifications. Building on this base, we can then expect to see emerge a complete set of distributed system integration and management services. In addition, domain-specific tools will be developed that support various data integration and analysis scenarios. Longer term, we will see an expanding set

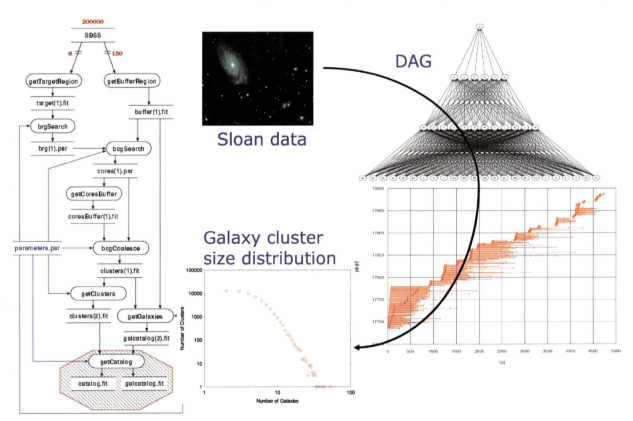

Figure 6.6   *Clockwise from left*: The workflow associated with a galaxy cluster detection algorithm, input data, the directed acyclic graph (DAG) generated by Chimera; execution on a small cluster; and final output.

of interoperable services and systems that address scaling to both larger numbers of entities and smaller device footprints, increasing degrees of virtualization, richer forms of sharing, and increased qualities of service via a variety of forms of active management. This work will draw increasingly heavily on the results of computer science research in such areas as peer-to-peer (Crowcroft et al., 2004), knowledge-based (Berners-Lee et al., 2001; Goble et al., 2004), and autonomic (Horn, 2001) systems.

Opportunities for neuroscience applications of Grid technologies are many. Federation of diverse brain image databases can enable new insights via large-scale data mining. Federation of specialized computational services can enable pooling of computational expertise (and resources) provided by different groups. Coupling of scientific instrumentation and computers can enable more accurate imaging. Collaboration technologies such as Access Grid can enable more effective interpersonal communication. And network access to all of these resources can enable international cooperation, as well as the participation of groups from smaller universities and laboratories in the research enterprise.

Ultimately, Grid services are an enabling technology, and thus the speed and extent of their impact on neuroscience depend on how quickly and effectively the methodologies and practices of the neuroscience community evolve to embrace a highly networked world.

ACKNOWLEDGMENTS

This work was supported in part by the Mathematical, Information, and Computational Sciences Division subprogram of the Office of Advanced Scientific Computing Research, Office of Science, U.S. Department of Energy, under Contract W-31-109-Eng-38.

References

Abramson, D., Giddy, J., and Kotler, L. (2000). High performance parametric modeling with Nimrod/G: Killer application for the global grid? *Proc. Int. Parallel Distrib. Process. Symp. (IPDPS)*, Cancun, Mexico.

Allcock, W., Foster, I., Nefdova, V., Chervenak, A., Deelman, E., Kesselman, C., Lee, J., Sim, A., Shoshani, A., Drach, B., and Williams, D. (2001). High-Performance Remote Access to Climate Simulation Data: A Challenge Problem for Data Grid Technologies, SC2001. ACM Press, Baltimore, MD.

Allcock, W., Bester, J., Bresnahan, J., Chervenak, A., Foster, I., Kesselman, C., Heder, S., Nefedova, V., Quesnel, D., and Tuecke, S. (2002). Data management and transfer in high-performance computational grid environments. *Parallel Comput.* **28**(5), 749–771.

Aloisio, G., and Cafaro, M. (2002). Web-based access to the grid using the grid resource broker portal. *Concurr. Comput.: Pract. Experi. Spec. Issue Grid Comput. Environ.*

Annis, J., Zhao, Y., Voeckler, J., Wilde, M., Kent, S., and Foster, I. (2002). *Applying Chimera Virtual Data Concepts to Cluster Finding in the Sloan Sky Survey*, SC2002. ACM Press, Baltimore, MD.

Antonioletti, M., and Jackson, M. (2003). OGSA-DAI Product Overview. The abbreviation is Open Grid Architecture Data Acess and Integration (OGSA-DAI) www.ogsa.dai.org.uk/downloads/docs/OGSA-DAI-USER-M3-Product-Overview.pdf.

Antonioletti, M., Atkinson, M. et al. (2003). Grid data service specification. *Global Grid Forum.*

Appleby, K., Fakhouri, S., Fong, L., Goldsmith, G., Kalantar, M., Krishnakumar, S., Pazel, D., Pershing, J., and Rochwerger, B. (2001). Oceano—SLA based management of a computing utility. *7th IFIP/IEEE Int. Symp. Integr. Network Manage.*

Atkinson, M., Chervenak, A., Kunszt, P., Narang, I., Paton, N. W., Pearson, D., Shoshani, A., and Watson, P. (2004). Data access, integration, and management. *The Grid: Blueprint for a New Computing Infrastructure.* Morgan Kaufmann, San Mateo, CA.

Baratloo, A., Karaul, M., Kedem, Z., and Wyckoff, P. (1996). *Charlotte: Metacomputing on the Web. 9th Int. Conf. Parallel Distrib. Comput. Syst.*

Baru, C., Moore, R., Rajasekar, A., and Wan, M. (1998). The SDSC storage resource broker. 8th Annu. IBM Cent. Adv. Stud. Conf. Toronto, Canada.

Beiriger, J., Johnson, W., Bivens, H., Humphreys, S., and Rhed, S. (2000). Constructing the ASCI grid. *9th IEEE Int. Symp. High Perform. Distrib. Comput.*

Berners-Lee, T., Hendler, J., and Lassila, O. (2001). The Semantic Web. *Sci. Am.* **284**(5), 34–43.

Bonkalski, J., Anderson, R., Jones, S., and Zaluzec, N.J. (1998). Bringing telepresence microscopy and science collaboratories into the class room. *TeleConf. Mag.* **17** (9).

Brecht, T., Sandhu, H., Shan, M., and Talbot, J. (1996). ParaWeb: Towards world-wide supercomputing. *Proc. 7th ACM SIGOPS Euro. Workshop Syst. Support Worldwide Appl.*

Brunett, S., Czajkowski, K., Fitzgerald, S., Foster, I., Johnson, A., Kesselman, C., Leigh, J., and Tuecke, S. (1998). Application experiences with the Globus Toolkit. *7th IEEE Int. Symp. on High Perform. Distrib. Comput.*

Buneman, P., Khanna, S., and Tan, W.-C. (2001). Why and where: A characterization of data provenance. *Int. Conf. Database Theory.*

Casanova, H., and Dongarra, J. (1997). NetSolve: A network server for solving computational science problems. *Int. J. Supercomput. Appl. High Perform. Comput.* **11**(3), 212–223.

Casanova, H., Obertelli, G., Berman, F., and Wolski, R. (2000). *The AppLeS Parameter Sweep Template: User-Level Middleware for the Grid.* SC'2000. www.sc2000.org.

Catlett, C. (1992). In search of gigabit applications. *IEEE Commun. Mag.* April, pp. 42–51.

Catlett, C., and Smarr, L. (1992). Metacomputing. *Commun. ACM* **35**(6), 44–52.

Chervenak, A., Deelman, E., Foster, I., Guy, L., Hoschek, W., Iamnitchi, A., Kesselman, C., Kunst, P., Ripenu, M., Schwartzkopf, B., Stockinger, H., Stockinger, K., and Tierney, B. (2002). Giggle: A framework for constructing sclable replica location services. *SC'02: High Perform. Network. Comput.*

Chervenak, A., Deelman, E. et al. (2003). *A Metadata Catalog Service for Data Intensive Applications.* Information Sciences Institute, University of Southern California.

Committee on a National Collaboratory—National Research Council. (1993). *National Collaboratories—Applying Information Technology for Scientific Research.* National Academy Press, Washington, DC.

Crowcroft, J., Moreton, T., Pratt, I., and Twigg, A. (2004). Peer-to-peer technologies. *The Grid: Blueprint for a New Computing Infrastructure,* 2nd ed. Morgan Kaufmann, San Mateo, CA.

Czajkowski, K., Foster, I., and Kesselman, C. (2004). Resource and service management. *The Grid: Blueprint for a New Computing Infrastructure,* 2nd ed. Morgan Kaufmann, San Mateo, CA.

Deelman, E., Blythe, J., Gil, Y., Kesselman, C., Mehta, G., Vahi, K., Blackburn, K., Lazzurini, A., Arbtree, A., Caranaugh, R., and Koranda, S. (2003). Mapping abstract workflows onto grid environments. *J. Grid Comput.* **1**(1).

Eickermann, T., and Hommes, F. (1999). Metacomputing in a gigabit testbed west. *Workshop on Wide Area Networks and High Performance Computing,* pp. 119–129. Springer-Verlag, Berlin.

Ellisman, M., and Peltier, S. (2004). Medical data federation: The Biomedical Informatics Research Network. *The Grid: Blueprint for a New Computing Infrastructure,* 2nd ed. Morgan Kaufmann, San Mateo, CA.

Foster, I., and Kesselman, C. (1998). Globus: A metacomputing infrastructure toolkit. *Int. J. Supercomput. Appl.* **11**(2), 115–129.

Foster, I., and Kesselman, C., eds. (2004). The Grid: Blueprint for a New Computing Infrastructure, 2nd ed. Morgan Kaufmann, San Mateo, CA.

Foster, I., Kesselman, C., and Tuecke, S. (2001). The anatomy of the grid: Enabling scalable virtual organizations. *Int. J. High Perform. Comput. App.* **15**(3), 200–222.

Foster, I., Kesselman, C., Nick, J. M., and Tuecke, S. (2002a). Grid services for distributed systems integration. *IEEE Comput.* **35**(6), 37–46.

Foster, I., Voeckler, J., Wilde, M., and Zhao, Y. (2002b). Chimera: A virtual data system for representing, querying, and automating data derivation. *14th Intl. Conf. Sci. Stat. Database Manag.* Edinburgh, Scotland.

Frey, J., Tannenbaum, T., Foster, I., Livny, M., and Tuecke, S. (2002c). Condor-G: A computation management agent for multi-institutional grids. *Cluster Comput.* **5**(3), 237–246.

Goble, C.A., De Roure, D., Shadbolt, N. R., and Fernandes A. A. A. (2004). Enhancing services and applications with knowledge and semantics. *The Grid: Blueprint for a New Computing Infrastructure,* 2nd ed. Morgan Kaufmann, San Mateo, CA.

Graupner, S., Kotov, V., Trinks, H., and Andrzejak, A. (2002). *Control Architecture for Service Grids in a Federation of Utility Data Centers,* HP Labs, Palo Alto, CA.

Grimshaw, A., Weissman, J. et al. (1994). Metasystems: An approach combining parallel processing and heterogeneous distributed computing systems. *J. Parallel Distrib. Comput.* **21**(3), 257–270.

Grimshaw, A. S. and Wulf, W. A. (1997). The Legion vision of a worldwide virtual computer. *Commun. ACM* **40** (1), 39–45.

Gropp, W., Lusk, E., and Skjellum, A. (1999). *Using MPI: Portable Parallel Programming with the Message Passing Interface.* MIT Press, Cambridge, MA.

Guy, L., Kunszt, P., Laure, E., Stockinger, H., and Stockinger, K. (2002). Replica management in data grids. *Global Grid Forum 5.*

Hadida, M., Kadobayashi, Y., Lamont, S., Braun, H.-W., Fink, B., Hutton, T., Kamrath, A., Mori, H., and Ellisman, M. (2000). Advanced networking for telemicroscopy. *10th Annu. Internet Soci. Conf.* Yokohama, Japan.

Hey, A. J. G., and Trefethen, A. (2003). The data deluge: An e-science perspective. Grid Computing: Making the Global Infrastructure a Reality. (F. Berman, G. C. Fox, and A. J. G. Hey, eds.). Wiley, New York.

Horn, P. (2001). *The IBM Vision for Autonomic Computing.* IBM. Available at www.research.ibm.com/autonomic/manifesto.

Johnston, W. E. (1999). Realtime widely distributed instrumentation systems. The Grid: Blueprint for a New Computing Infrastructure. (I. Foster and C. Kesselman, eds.), pp. 75–103. Morgan Kaufmann, San Mateo, CA.

Johnston, W. E., Gannon, D. and Nitzberg, B. (1999). Grids as production computing environments: The engineering aspects of NASA's Information Power Grid. *8th IEEE Int. Symp. High Perform. Distrib. Comput.*

Karonis, N., Toonen, B., and Foster, I. (2003). MPICH-G2: A grid-enabled implementation of the Message Passing Interface. *J. Parallel Distrib. Comput.* **63**(5), 551–563.

Litzkow, M.J., Livny, M., and Mutka, M. W. (1988). Condor—A hunter of idle workstations. *8th Int. Conf. Distrib. Comput. Syst.*, pp. 104–111.

Messina, P. (1999). Distributed supercomputing applications. *The Grid: Blueprint for a New Computing Infrastructure,* pp. 55–73. Morgan Kaufmann, San Mateo, CA.

Novotny, J. (2002). The grid portal development kit. *Concurr. Comput. Pract. Exper.* **14** (13–15), 1145–1160.

Paton, N. W., Atkinson, M. P., Dialani, V., Pearson, D., Storey, T., and Watson, P. (2002). *Database Access and Integration Services on the Grid.* U.K. National e-Science Center.

Pearlman, L., Welch, V., Foster, I., Kesselman, C., and Tuecke, S. (2002). A community authorization service for group collaboration. *IEEE 3rd Int. Workshop Policies Distrib. Syst. and Networks,* IEEE Computer Society Press, Los Alamitos, CA.

Russell, M., Allen, G., Daves, G., Foster, I., Seidel, E., Novotny, J., Shalf, J., and Laszewski, G. von (2002). The astrophysics simulation collaboratory: A science portal enabling community software development. *Cluster Comput.* **5**(3), 297–304.

Shoshani, A., Sim, A., and Gu, J. (2003). Storage resource managers: Essential components for the grid. In *Resource Management for Grid Computing.* (J. Nabrzyski, J. Schopf, and J. Weglarz, eds.), Kluwer Academic, Norwell, MA.

Stevens, R. (2004). Group-oriented collaboration: The Access Grid collaboration system. *The Grid: Blueprint for a New Computing Infrastructure*, 2nd ed. Morgan Kaufmann, San Mateo, CA.

Stevens, R., Woodward, P., De Fanti, T., and Catlett, C. (1997). From the I-WAY to the national technology grid. *Commun. ACM* **40**(11), 50–61.

Szalay, A., and Gray, J. (2001). The world-wide telescope. *Science* **293**, 2037–2040.

Tanaka, Y., Nakada, H., Sekiguchi, S., Suzumura, T., and Matsuoka, S. (2002). Ninf-G: A reference implementation of RPC based programming middleware for grid computing. *J. Grid Comput.* **1**(1), 41–51.

Thain, D., and Livny, M. (2004). Building reliable clients and services. *The Grid: Blueprint for a New Computing Infrastructure*, 2nd ed. Morgan Kaufmann, San Mateo, CA.

Thomas, M., Mock, S., Boisseau, J., Dahan, M., Hueller, K., and Sutton, S. (2001). The GridPort Toolkit Architecture for building grid portals. *10th IEEE Int. Sympo. High Perform. Distribu. Comput.* IEEE Computer Society Press, Los Alamitos, CA.

Vahdat, A., Anderson, T., Dahlin, M., Culler, D., Belani, E., Eastham, P., and Yoshikawa, C. (1998). WebOS: Operating system services for wide area applications. *7th IEEE Int. Symp. High Perform. Distrib. Comput.*

Welch, V., Siebenlist, F., Foster, I., Bresnahan, J., Czajkowski, K., Gawor, J., Kesselman, C., Heder, S., Pearlman, L., and Tuecke, S. (2003). Security for grid services. *12th IEEE Int. Symp. High Perform. Distrib. Comput.*

Building Grid-Based Resources for Neurosciences

Maryann E. Martone, Ph.D., Steven T. Peltier M.S., and
Mark H. Ellisman, Ph.D.

CONTENTS

Databasing the Brain. Edited by Stephen H. Koslow and Shankar Subramaniam
ISBN 0-471-30921-4 © 2005 John Wiley & Sons, Inc.

7.1 INTRODUCTION

A continuing challenge to structural biologists is the understanding of structures on the scale of 1 nm^3 to 10 μm^3, a dimensional range that encompasses macromolecular complexes, organelles, and multicomponent structures like synapses. Such structures have traditionally been difficult to study because they fall in the resolution gap between technologies, spanning X-ray crystallography, electron microscopy, and light microscopy. Structures at this scale represent the heart of information processing in the nervous system and provide a bridge between the molecular information being assembled at one end of the biological continuum and the large-scale brain mapping being performed at the other. These structures will have to be solved if the results of the molecular revolution, the protein products of sequenced genomes, are to be situated in their proper subcellular, cellular, and tissue contexts.

Technological improvements in both light and electron microscopic imaging have led to major advances in our ability to fill in missing information between the molecular and cellular realms. At the light microscopic level, computational image restoration techniques and the optical sectioning capabilities of the confocal and multiphoton microscopes are increasing the resolution achievable with optical-based imaging techniques. At the electron microscopic level, electron tomography is revealing new information about the three-dimensional (3D) ultrastructure of tissues, cells, and macromolecular complexes (as reviewed in McEwen and Frank, 2001; Medalia et al., 2002). Using tomography, a 3D reconstruction is obtained from a series of two-dimensional (2D) projections with the potential of mapping the location and segregation of functionally important macromolecules at much higher resolution than can be achieved by serial sectioning (Martone et al., 2000; Bohm et al., 2000; Medalia et al., 2002; Harlow et al., 2001). Recognizing its unique potential to bridge the molecular with cellular realms, electron tomography was recently named "runner up breakthrough of the year" by *Science* magazine.*

Efforts are underway to streamline the process for electron tomography and other high resolution 3D techniques to increase the rate at which we can acquire and analyze these data. Despite these advances, this type of work remains a time- and resource-intensive process requiring access to specialized instruments, computational resources and analysis tools. In the following sections, we describe some our informatics tools and approaches for increasing the speed and efficiency with which tomographic data can be produced and for providing broader access to tomographic resources and data. These tools take advantage of advances in so-called "grid" technologies, where distributed and heterogeneous instruments, computational resources, and data can be accessed and utilized through high-speed networks and specialized services as if they were a single system. We describe how these approaches are being integrated into the Biomedical Informatics Research Network (BIRN), a large-scale, multi-institution project to establish general data grid infrastructure in support of biomedical science, including a multiscale database federation.

Science 2002, 298(5602):2297–2303.

7.2 THE TELESCIENCE PORTAL: A GRID-BASED RESOURCE FOR ELECTRON TOMOGRAPHY

Electron tomography is an excellent example of a driving application that can benefit from the tight integration of computation, data, and visualization resources. It is computationally, data-, time-, and labor-intensive and requires high-performance, interactive visualization to guide data product refinement and analysis. Users must acquire an initial series of anywhere from 60 to 200 2D projections by imaging the specimen at regular tilt increments on the microscope. These images are then aligned, normalized, and back-projected into a 3D volume, which must be further segmented to extract relevant information. In most instances, the completion of a single tomographic reconstruction occupies several days from the initial acquisition of the raw data through the analysis. Throughout this process an exorbitant amount of data is generated which must be managed and ultimately archived.

The Grid is defined as an infrastructure for the integrated, collaborative use of computational resources, networks, databases, and scientific instruments owned and managed by multiple organizations (Foster et al., 2001; Foster, 2002). The Globus Project (http://www.globus.org) and the Storage Resource Broker (SRB) are two examples of projects actively developing what has been termed "grid middleware"—that is, a collection of software modules and interoperable services that can be used to build flexible and persistent environments of tightly integrated physical resources and applications. As part of the Globus project, researchers at the Information Science Institute (ISI) at the University of Southern California and Argonne National Laboratory (ANL) have developed services spanning four key areas: security, information services, resource management, and data management. The Grid Security Infrastructure (GSI), in particular, is present in all aspects of the Grid, and it provides authentication and authorization services using public key certificates. The SRB, a project led by the San Diego Supercomputer Center, provides a uniform interface for connecting to heterogeneous data resources over a network and accessing replicated datasets (Rajasekar et al., 1999; Rajasekar and Moore, 2001). When used in conjunction with the MetaData Catalog (MCAT), the SRB provides a means for accessing datasets and resources through querying their attributes instead of knowing their physical names and/or locations.

With Grid services, organizations lacking robust infrastructure can utilize the distributed resources necessary to enhance and accelerate specialized and complex processes like electron tomography. Like the power grid from which it gets its name, a biologist should in theory be able to " . . . plug into the grid like an appliance is plugged into an outlet and use resources available on the grid transparently" (Lacroix, 2003). In actuality, the burden of administering and accessing heterogeneous resources through the grid is still quite significant for the noncomputer scientist. For example, beyond the installation of the software on the resources, Globus requires users to obtain and maintain accounts on individual machines and manage security certificates on a regular basis. As each new resource is added to the Grid, increased administrative burden can be passed to the user. Rather than the dazzling vision of vast networks of resources at one's fingertips promoted by grid proponents, the actual infrastructure may appear to a biologist as an administrative quagmire that would do little to increase productivity.

The Telescience Portal (http://telescience.ucsd.edu) was created specifically to hide the complexity of the grid from the end-user while providing access to grid-based services for electron tomography. The Portal is a centralized interface to a fully integrated conglomeration of tools, infrastructure, and services necessary for a biologist to perform end-to-end electron tomography from any Internet-capable location (Peltier et al., 2003). The Portal provides a workflow architecture to manage the many steps involved in the tomographic process along with access to online resource scheduling, remote instrumentation, parallel tomographic grid-based reconstruction, a suite of visualization, segmentation, and image processing tools, access to heterogeneous distributed file systems for data archiving, transparent deposition of data products into databases of cellular structure, and collaborative telecommunication utilities for shared "whiteboard" image annotations and "chatting" between multiple remote researchers. All of these services are managed by a simple Web interface and require only a single username and password (Figure 7.1)

The Telescience Portal is built using the Gridport toolkit, developed at the San Diego Supercomputer Center (http://gridport.npaci.edu). Gridport streamlines the process of incorporating key Globus services and SRB functions into Web applications through the development of Web-friendly wrappers. In doing so, GridPort provides a simple mechanism for Web portal developers to use grid tools like Globus to build an abstraction layer over specific architectures of the grid, providing uniform access to heterogeneous resources such as instrument, computational, and data resources from a platform-independent web interface.

The Telescience Portal impacts the process of electron tomography in four key areas:

1. *Remote Instrumentation*. Through the Portal, users may utilize our specialized intermediate voltage electron microscopes specially configured for biological electron tomography (Hadida-Hassan et al., 1999). Our "Telemicroscopy" system allows multiple remote researchers, students, and observers at distributed sites to conduct collaborative microscopy experiments with little or no previous training.

2. *Workflow Management*. The Telescience Portal operates around the concept of the "tomography workflow." The tomography workflow is composed of the sequence of steps required to acquire, process, visualize, and extract useful information from a 3D volume. Each reconstruction associated with an experiment is assigned a unique reconstruction identifier. This identifier is used by the Portal to track the progress of that reconstruction through the workflow and to supplement the mechanism for organizing data.

3. *Computational Resources*. The Telescience Portal provides a dramatically simplified interface for coordinating and utilizing a globally distributed collection of heterogeneous computational resources. By using parallel implementations of our back-projection algorithms, 3D volumes taking hours to

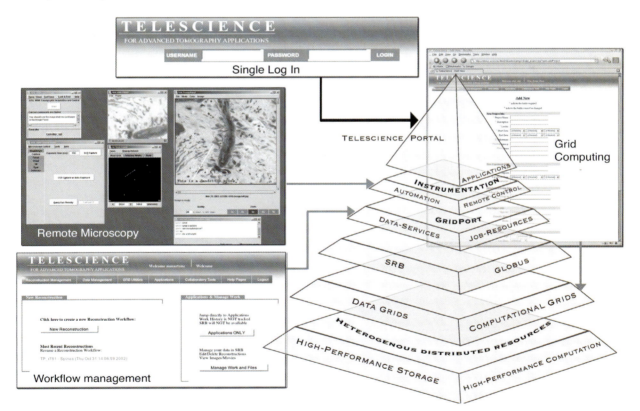

Figure 7.1   Illustration of the Telescience architecture. The user is presented upon log-in with a series of web pages that represent elements of the tomography workflow that transparently access underlying grid services. The architecture is designed in a modular, layered fashion (pyramid) so that additional features, resources, and technologies can be incorporated without bringing the other layers offline.

compute on stand-alone workstations can be computed on the Telescience grid in minutes using multiple processors from workstations, clusters, and supercomputers operating across institutional boundaries and domains. Through the Portal, researchers have launched remote jobs computed simultaneously over resources located in Taiwan, in Osaka, and across the United States (Peltier et al., 2003).

4. *Data Management.* The Telescience Portal further incorporates a number of tools and technologies for securely managing and accelerating the flow of data from acquisition to refinement to data product deposition. Globus, GridFTP, the SRB, and the Cell-Centered Database (described in Section 7.3) are all tightly integrated to streamline the flow of data from instruments, between applications and across distributed resources.

Preliminary research conducted with the Portal has demonstrated the ability of this integrated system to minimize the time, labor, and frustration commonly associated with the electron tomography process. Although the portal was developed around electron tomography, it is generally applicable to any resource intensive process. The Telescience Portal demonstrates how grid-based applications can increase the rate at which we acquire data in the resolution gaps most critical to computational neuroscientists, where molecules fit into their subcellular contexts and where cells fit into larger networks. Through the end-to-end workflow management, access to distributed computation, and the extended availability of data and visualization resources, the tomography process is streamlined and accelerated, increasing the productivity of individual researchers. At the same time, the Web-based tools and interface broaden access to this specialized technology, engaging larger numbers of researchers to utilize the technique.

7.3 THE CELL-CENTERED DATABASE

Techniques like electron tomography are generating large amounts of exquisitely detailed data on cells and their macromolecular organization that need to be exposed to the greater scientific community. The creation of structured shared data repositories in the form of Web-accessible databases has been a driving force behind the genomic revolution. These resources serve not only to organize and manage molecular data being created by researchers around the globe, but also to provide the starting point for data mining operations to uncover interesting information present in the large amount of sequence and structural data. While more than 300 databases are available for molecular information from genes and proteins (Baxevanis, 2003), far fewer resources exist for the type of cellular and subcellular information produced using light and electron microscopy. As part of the Cell-Centered Database (CCDB) project, we are addressing this need by developing a database for 3D light and electron microscopic information (Martone et al., 2002a, 2003).

The CCDB (http://ncmir.ucsd.edu/CCDB) contains structural and protein distribution information derived from confocal, multiphoton, and electron microscopy, including correlated microscopy. Its main mission is to provide a means to make high-resolution data derived from electron tomography and high-resolution light microscopy available to the scientific

community, situating itself between whole-brain imaging databases such as the MAP project (MacKenzie-Graham et al., 2003) and protein structures determined from electron microscopy, NMR spectroscopy and x-ray crystallography (e.g., the Protein Data Bank and EMBL). The CCDB also is serving as a research prototype for investigating new methods for representing imaging data in a relational database system so that powerful data mining approaches can be employed for the content of imaging data. The CCDB data model also addresses the practical problem of image management for the large amounts of imaging data and associated metadata generated in a modern microscopy laboratory. Finally, as described below, the CCDB will serve as a source of microscopic information in the BIRN database federation. As such, the data model of the CCDB needs to ensure that data within the CCDB can be related to data taken at different scales and modalities.

The CCDB is built on an object-relational framework using Oracle 9i and is available online at http://ncmir.ucsd.edu/CCDB. The data model of the CCDB was designed around the process of 3D reconstruction from 2D micrographs, capturing key steps in the process from experiment to analysis. The types of imaging data stored in the CCDB are quite heterogeneous, ranging from large-scale maps of protein distributions taken by confocal microscopy to 3D reconstruction of individual cells, subcellular structures, and organelles (Figure 7.2). The CCDB can accommodate data from tissues and cultured cells regardless of tissue of origin, but because of our emphasis on the nervous system, the data model contains several features specialized for neural data. For each dataset, the CCDB stores not only the original images and 3D reconstruction, but also any analysis products derived from these data, including segmented objects and measurements of quantities such as surface area, volume, length, and diameter.

The CCDB was designed as a grid-based resource and utilizes the SRB as a distributed file management system (Figure 7.3). The descriptive and derived data are managed centrally by Oracle, while the image files are managed by the Storage Resource Broker (SRB). The CCDB acts as a single client of the SRB/MCAT system, so that separate SRB authentications and accounts do not have to be obtained for each user. The CCDB is also interfaced with the Telescience Portal so that data obtained and analyzed through the Portal may be directly deposited into the CCDB, utilizing its function as an imaging data management system. The CCDB can also be queried directly from the Portal, allowing users to compare their data to previous tomography studies or to obtain imaging parameters used in other studies.

The full power of image databases like the CCDB will not be realized until we have the means to query images based not only on descriptive attributes but also on their contents. The development of computer algorithms to identify and extract image features in image data is advancing (Sinha et al., 2002), but it is unlikely that any algorithm will be able to match the skill of an experienced microscopist for many years. The CCDB project is taking two approaches to allow the querying of the content of multi-resolution 3D data. The first approach, currently supported by the CCDB data model, is to store the results of segmentations and analyses performed by individual researchers on the datasets stored in the CCDB. The CCDB allows each object segmented from a reconstruction to be stored as a separate object in the database along with any quantitative information derived from

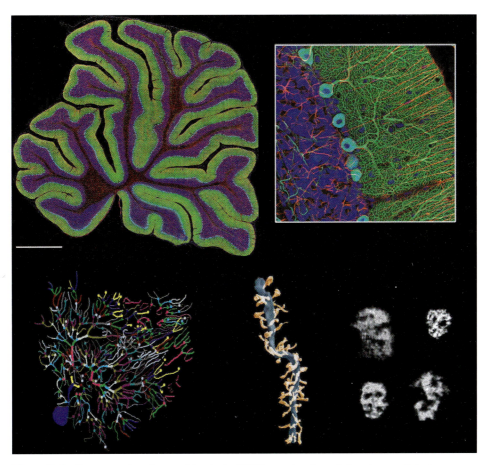

Figure 7.2   Some examples of types of data in the CCDB, spanning the scale from large-scale brain maps from light microscopy (*upper-left*) taken at near to the resolution limit of light microscopy (*upper-right*), to reconstructions of filled neurons (lower left), to subcellular structures like neuronal spiny dendrites (lower middle) and postsynaptic densities (lower right) determined by electron tomography. Scale bar in upper left panel = 1 mm.

it. The list of segmented objects and their morphometric quantities provides a means to query a dataset based on (a) features contained in the data such as object name (e.g., dendritic spine) or (b) quantities such as surface area, volume, and length. For example, users may issue queries to search for neurons based on properties exported from NeuroExplorer, the analysis program supplied with Neurolucida such as number of primary dendrites.

While storing the results of quantitative analyses is useful, this method limits possible queries to those quantities that have been stored with a particular dataset. For true data mining of imaging data to occur, we must be able to exploit information in the database that is not explicitly represented in the schema (Lacroix, 2003). To address this issue, we are developing specific data types around certain classes of segmented objects contained

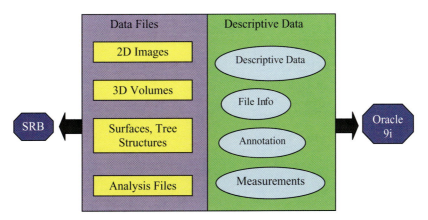

Figure 7.3   The data management structure of the CCDB. See text for description.

in the CCDB. For example, we are creating a "surface data type" to allow users to query directly the original surface data. The properties of the surfaces can be determined through very general operations at query time, which allows the user to query on characteristics not explicitly modeled in the schema—for example, dendrites from striatal medium spiny cells where the diameter of the dendritic shaft shows constrictions of at least 20 % along its length. In this example, the schema doesn't contain any explicit indication of the shape of the dendritic shaft, but these characteristics can be computed as part of the query processing. Additional data types are being developed for volume data and protein distribution data. A data type for tree structures generated by Neurolucida has recently been implemented.

7.3.1 Querying the CCDB

The CCDB provides a query form upon log-in to issue simple queries on attributes such as cell type, protein, and reconstruction technique. An advanced query capability is also provided that allows the user to design a custom query form based on more detailed project, subject, and anatomical information. The results of a query are presented to the user in the form illustrated in Figure 7.4. In the CCDB, a 3D reconstruction is viewed as one interpretation of a set of raw data that is highly dependent on the specimen preparation and imaging methods used to acquire it. Thus, a single record in the CCDB consists a set of raw microscope images and any volumes, images, or data derived from it, along with a rich set of methodological details. These derived products include reconstructions, animations, correlated volumes, and the results of any segmentation or analysis performed on the data. By presenting all of the raw data, reconstructed data, and processed data with a thorough description of how the specimen was prepared and imaged, researchers are free to extract additional content from micrographs that may not have been analyzed by the original author or employ additional alignment, reconstruction, or segmentation algorithms to the data.

The CCDB makes low-resolution 2D views and animations of each dataset available through the Web interface. A general-purpose 3D web viewing tool for CCDB data, JViewer

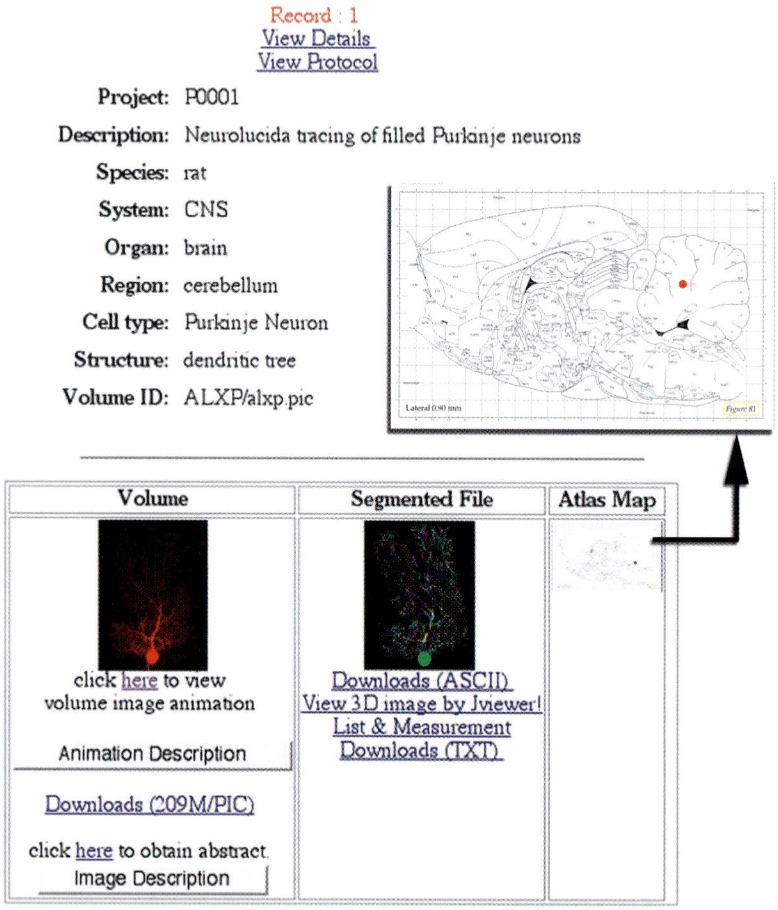

Figure 7.4 Query result from the CCDB. All available image types are displayed as thumbnails. Higher-resolution views may be viewed by clicking on the image, and animations may be viewed by clicking on the indicated link. Some types of segmented objects may be viewed and manipulated using JViewer, a Java-based 3D image viewer. Other types of information available for some datasets include measurement files and atlas maps (enlarged in inset) showing the location of the data in terms of a standard brain atlas. Full-resolution 3D datasets may be obtained by clicking on the "Download" link, if the user has permission for that dataset.

(http://ncmir.ucsd.edu/doc/JViewer/), is available for platforms with Java, Java 3D, and Java Web Start technologies installed. JViewer displays neuronal branching structures created using Neurolucida, surface data created using Synu (Hessler et al., 1992), VRML files, and segmented data created using Xvoxtrace, a manual segmentation tool for tomographic data developed in our laboratory. Users may also download the full-resolution imaging data for any type of data (e.g., raw data, 3D reconstruction, segmented volumes), available for a particular dataset. Each of these types usually consists of a set of related files (e.g., a set of tilt images), and thus all files comprising a given category of information are bundled together as an archive (tar) file so that they may be downloaded as a single file. A set of tools available for viewing and manipulating these datasets is listed on the CCDB website under "Visualization and Analysis Tools."

The CCDB was designed to make information available to the public, and it was also designed as a data management system for unpublished data. In order to protect data before they are ready to be made public, we have implemented multiple levels of access to data using the security features of Oracle. Access privileges may be set separately for each table, allowing flexible design of the amount of data to be exposed to a particular user. The CCDB currently supports three levels of access. For public data, users may view all descriptive data, view Web-based movies and 2D images, and download original data files within the usage agreement of the CCDB. For semi-private data, users may still view a limited set of descriptive information, 2D images, and animations, but may not download the original data without receiving permission from the owner of the data. For private data, nonauthorized users receive no indication that the data are in the database. The owner of the data may reserve sole access or may create user groups who have access to private data.

7.4 FEDERATION OF BRAIN DATA: SEMANTIC AND SPATIAL MARK-UP OF THE CCDB

The CCDB occupies a unique niche in biological databases, with its emphasis on multi-resolution 3D microscopy. On its own, however, it covers only a small portion of the biological spectrum. A major goal of informatics research is to create computer-based approaches to allow scientists to integrate and relate data in resources like the CCDB to that obtained at different scales, experimental systems, and subdisciplines (Kotter, 2001; Fox and Lancaster, 2002). As will be described in a later section, the CCDB is participating in a large-scale database federation through the BIRN project.

One of the well-recognized roadblocks to integration of disparate data resources in the biological sciences concerns reconciling semantic differences among sources (Lacroix, 2003). Scientific terminology, particularly neuroanatomical nomenclature, is vast, nonstandard, and confusing. Anatomical entities may have multiple names (e.g., caudate nucleus and *nucleus caudatus*), the same term may have multiple meanings [e.g., spine (spinal cord) versus spine (dendritic spine)], and, worst of all, the same term may be defined differently by different scientists (e.g., basal ganglia). To minimize semantic confusion and to situate cellular and subcellular data from the CCDB in a larger context, the CCDB is mapped to several shared knowledge sources in the

form of ontologies and spatial atlases. An ontology is essentially a set of terms and the relationships between them and provides one means for communities to formalize a shared understanding of a field (Lacroix, 2003).

Semantic Mark-up of the CCDB. Concepts in the CCDB are being mapped to the Unified Medical Language System (UMLS), a large metathesaurus and knowledge source for the biomedical sciences (Humphreys et al., 1998). Ontologies like the UMLS and Gene Ontologies (http://www.geneontology.org/) can mitigate some of the semantic difficulties described above by assigning each concept in the ontology a unique identifier. All synonymous terms can then be assigned the same ID. For example, the UMLS ID number for the synonymous terms Purkinje cell, cerebellar Purkinje cell, and Purkinje's corpuscles is C0034143. Thus, regardless of which term is preferred by a given individual, if they share the same ID, they are asserted to be the same. Conversely, even if two terms share the same name, they are distinguishable by their unique IDs. In the example given above, spine (spinal cord) = C0037949 while spine (dendritic spine) = C0872341. Ontologies can, in theory, accommodate multiple definitions of the same term. For example, a definition of basal ganglia according to researcher A may have a different ID than one according to researcher B. Thus, ontologies do not require the imposition of a standard set of terms, they only require that the definition of a concept be explicit.

Concepts within ontologies are linked together by a set of relationships. These relationships may be simple "is a" and "has a" relationships; for example, Purkinje cell is a neuron, neuron has a nucleus, or may be more complex (Gupta et al., 2003). In some systems—for example, the BIRN mediator (see below) and the TAMBIS system (Stevens et al., 2003)—the knowledge contained in the ontology is utilized in the formulation of queries. For example, from the above statements, a search algorithm could infer that "Purkinje cell has a nucleus" if the ontology is encoded in a form that would allow such reasoning to be performed. Because the knowledge required to link concepts is contained outside of the source database, the CCDB is relieved of the burden of storing exhaustive taxonomies for individual datasets. For example, for a Purkinje cell dataset, the CCDB does not index the data under cerebellar cortex because the knowledge that the Purkinje cell is part of the cerebellar cortex is contained in the ontology. We deliberately chose to store minimal amounts of information in the database itself because the state of knowledge, even in neuroanatomy, changes over time. By including this type of information in the database itself, the information can quickly become obsolete. In contrast, ontologies and semantic networks can be modified easily to reflect new views of neuroanatomy without requiring revision of the database. Figure 7.5 shows a hypothetical example. In this case, let us say that the CCDB contains data on the amygdala. In the early twentieth century, the amygdala was often considered to be one of the basal ganglia (View 1). By the late twentieth century, it was generally considered to be functionally part of the limbic system (View 2). Through the ontology, the image data in the CCDB "follows" the amygdala, regardless of the semantic network in which it resides. In this way, databases like the CCDB can elaborate their data model around a particular technique, in this case microscopy, but can still be queried as if it were a database of brain anatomy.

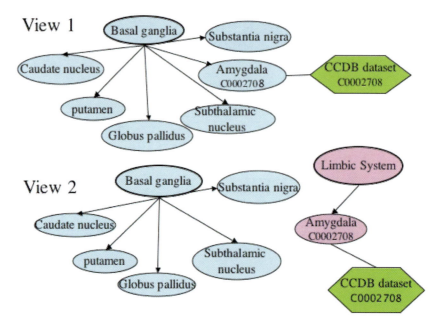

Figure 7.5   Example of how concepts can be recombined in new semantic networks. All arrows represent the "has a" relationship. Any data tagged to that concept are now part of a new semantic network.

The UMLS has recently incorporated the NeuroNames ontology (Bowden and Dubach, 2002) as a source vocabulary. Neuronames is a comprehensive resource for gross brain anatomy in the primate. However, for the type of cellular and subcellular data contained in the CCDB, the UMLS does not contain sufficient detail. As part of the CCDB and BIRN projects, we are developing ontologies for areas like neurocytology and neurological disease in more depth (Gupta et al., 2003). These ontologies are being built on top of the UMLS, utilizing existing concepts wherever possible and constructing new semantic networks and concepts as needed.

Spatial Mark-up: The Smart Atlas Tool. The majority of imaging data in the CCDB is referenced to a higher level of brain organization by registering their location in the coordinate system of a standard brain atlas. Placing data into an atlas-based coordinate system provides one method by which data taken across scales and distributed across multiple resources can be reliably compared to one another (Van Essen et al., 2001; Bjaalie, 2002; Brevik et al., 2001; Fox and Lancaster, 2002; Van Essen, 2002). Through the use of computer-based atlases and associated tools for warping and registration, scientists can move beyond the use solely of terminological description providing the means to express the location of anatomical features or signals in terms of a standardized coordinate system. While there may be disagreement among neuroscientists about the identity of a brain area giving rise to a signal, its location in terms of spatial coordinates is at least quantifiable. The expression of brain data in terms of atlas coordinates also allows it to be transformed spatially to provide alternative views that may provide additional information—for example, flat maps or additional parcellation schemes (Van Essen, 2002). Finally, because individual experiments can study only a few aspects of a brain region at one time, a standard coordinate system allows the same brain region to be sampled repeatedly to allow data to be accumulated over time.

To aid in the spatial registration of data in the CCDB, we have created the Spatial Mark-up and Rendering Atlas Tool ("Smart Atlas"), described in further detail in Martone et al. (2002b). Because much of the data in the CCDB consists of high-resolution light and electron microscopic data, most of the datasets can be represented as a point or polygon on a single 2D slice through a standard brain atlas. Using the Smart Atlas tool, the user draws a circle or polygon representing the approximate location in the atlas from which the data were taken. For larger images that cannot be represented as a point (e.g., the large-scale brain maps shown in Figure 7.2), images are scaled and warped to the atlas coordinate system. A graphical tool, Ratpax, developed by S.P. Lamont in our laboratory, was created to aid in scaling and warping 2D brain slices to a common coordinate system (Martone et al., 2003). The Smart Atlas is currently under additional development for use as a query interface for data in the CCDB. Users may issue queries such as "Display the locations of all filled neurons in cerebellum" (Figure 7.6). Clicking on the location will return data registered to that location.

7.5 CREATING A BIOMEDICAL DATA GRID: THE BIOMEDICAL INFORMATICS RESEARCH NETWORK

The CCDB is one example of the type of neuroimaging resource driving the creation of the BIRN project. The BIRN (http://www.nbirn.net) is a recently launched initiative from the National Institutes of Health to address specifically some of the issues involved exploiting the capabilities of grid-based infrastructure in support of biomedical research. The BIRN is developing and evolving the hardware, software, and protocols necessary to *share* and *mine* data for both basic and clinical research. Central to the project is the establishment of a

Figure 7.6   The Smart Atlas Interface. Using this tool, a standard brain atlas, in this case the Paxinos and Franklin (2000) mouse brain atlas, is converted into a dynamic data registration and query tool. Users may register the spatial location of datasets by drawing the location on a standard atlas plane. The location of registered data can be displayed (green circles). Clicking on a location retrieves the record from the CCDB containing detailed information on that particular cell.

scalable infrastructure consisting of advanced networks, federated distributed data collections, computational resources, and software technologies, to meet the evolving needs of investigators who have formed test bed scientific collaborations. The immediate objectives of the BIRN Project are to:

1. Establish a stable, high-performance network, linking key NIH Biotechnology Centers and General Clinical Research Centers.

2. Develop technologies to federate multiple data collections or databases from distributed partnering centers.

3. Allow collaborative data mining of these federated collections or databases.

4. Employ distributed computational resources to facilitate collaborative visualization, data refinement, and analysis.

5. Address project-wide issues relating to reliability, quality-of-service, performance, scalability, security, and ownership.

6. Build a stable software and hardware communications infrastructure to allow the coordination of large studies across sites.

The development of the BIRN is driven by three neuroimaging test bed activities focused on (1) studying disease states and relationships to human brain morphology (Brain Morphometry BIRN), (2) functional imaging research of schizophrenia (FIRST BIRN), and (3) multiscale analysis of mouse models of disease (Mouse BIRN). By pooling domain expertise, specialized research facilities, instrumentation resources,

advanced applications, and regional information, these investigators are tackling disease studies of greater scope and complexity than are independently possible. Each test bed serves as a guide for the development of a persistent infrastructure to facilitate collaborative biomedical research across multiple disciplines. Neuroimaging was chosen as the driver for the BIRN because it is one of the most rapidly advancing fields in biological science, generating very large amounts of complex data. In addition, through pioneering initiatives like the Human Brain Project, the community has been developing an impressive array of neuroinformatics tools and data caches (Toga, 2002).

7.6 FEDERATION OF BRAIN DATA: THE BIRN MEDIATOR

At the core of the BIRN project is the use of a series of distributed databases as the means by which data are shared and queried by participating scientists and which will serve as a persistent archive of BIRN data for subsequent data mining. Each of the BIRN participants is expected to create and maintain a database tailored to the type of specialized data produced at that site. These databases will span multiple technologies, scales, and species. The primary motivation for integrating data from multiple BIRN sites is to gain a deeper understanding of a scientific problem than would have been possible with any individual site's data.

Although each of the databases is being designed separately, they will be linked together or "federated" so that they may be cross queried.

The federation approach is attractive for many reasons, not the least of which is that it maintains the independence of individual database efforts. Scientists can design a specialized database encapsulating their particular area of expertise, and maintain control of the primary data, while still making it available to other researchers. Each source is wrapped and registered to a federation engine called the mediator. A federation model also allows new types of data to be brought into the BIRN without having to modify significantly either the original or existing resources. For example, the CCDB was created independently of the BIRN project, yet it will participate in the BIRN data federation as part of the Mouse BIRN project. However, in order to be successful, data federation requires that there are elements in common across sources. While this may be the case with databases containing largely similar information (e.g., gene sequence databases), for most scientific databases this assumption is not met. For example, the database being established at Duke University for whole-brain rodent MRI may not share any tables in common with the CCDB. Despite the lack of common semantic links, most neuroscientists can easily relate these two data sources at the conceptual level. In fact, neuroscientists usually can navigate with relative ease from the level of individual molecules to cells to brains to behavior and across experimental disciplines, because they possess the requisite knowledge to conceptually relate data at each level.

To address the challenging problem of integration of multiscale and multimodal data sources, the BIRN is building upon a novel mediator integration paradigm developed in a collaboration between computer scientists at the San Diego Supercomputer Center and neuroscientists in our laboratory through the National Partnership for Advanced Computational Infrastructure (NPACI) program sponsored by the National Science Foundation (Gupta et al., 2000; Ludäscher et al., 2001, 2003). This paradigm exploits expert knowledge contained in ontologies or other knowledge sources as the necessary "glue" to link together heterogeneous neuroscience data, a system that we call "knowledge-guided or model-based mediation." In this paradigm, connections between database elements do not have to be direct, but may be inferred through reasoning operations performed on knowledge sources registered to the mediator at time of query. As a simple example, a scientist posing a query to the mediator for information on mouse cerebellum would retrieve gross anatomical information on the cerebellum from a database established at Duke for MRI brain volumes and information on Purkinje cell structure from the CCDB. The mediator would perform this join because it accessed an ontology of brain anatomy with the relationships "Cerebellum has a cerebellar cortex; cerebellar cortex has a Purkinje cell layer; Purkinje cell layer has a Purkinje cell." From these relationships, the mediator infers that "Cerebellum has a Purkinje cell."

The mediation architecture operates on top of relational and XML databases, as well as Web-based information and is designed to be flexible, scalable, and powerful. Queries are issued against a "mediator," a virtual database that combines the individual data sources in meaningful ways. This combination is achieved using "Integrated View Definitions" that describe how the mediator represents the source databases. A user submits a query to the mediator (or more precisely, to one of the integrated views the mediator can expose). A mediation engine then evaluates the queries with the domain knowledge metadata, a potentially compute-intensive process, and retrieves query results and/or "handles" for relevant datasets. The dataset handles are then used to search and retrieve the actual data. Additional details about the mediation system can be found in Gupta et al. (2000), Ludäscher et al. (2001, 2003), and Martone et al. (2002b).

7.7 CONCLUSIONS

Projects like Telescience, the CCDB, and the BIRN are stretching the boundaries of information technology infrastructure, enriching the Global Grid movement by providing necessary "application pull" from several biomedical domains. Although application-driven, each project represents a true crossdisciplinary partnership involving biomedical scientists and computer scientists. The field of grid computing is still in its infancy. During this period of rapid change, any infrastructure developed runs the risk of soon becoming obsolete. However, even in these early stages, projects such as BIRN are necessary to ensure that the technologies being developed are informed by the requirements of biomedical science—for example, in the area of medical privacy and security. Conversely, biologists will have to become sufficiently proficient in grid technologies to take advantage of the opportunities they afford. The emerging cyberinfrastructure promises both to provide the individual researcher with increased access to data and resources and to foster increased levels of collaboration between communities of neuroscientists, increasing the scope and power of research studies. Through this new collaborative framework, researchers will move ever closer to the grand goal in neuroscience research: to understand how the interplay of structural, chemical, and electrical signals in nervous tissue give rise to behavior.

ACKNOWLEDGMENTS

This work was supported by NIH grants from NCRR RR04050, RR08605, and the Human Brain Project DA016602, as well as NSF grants supporting the National Partnership for Advanced Computational Infrastructure NSF-ASC 97-5249 and MCB-9728338. The Biomedical Informatics Research Network is supported by NIH grants RR08605-08S1 (BIRN Coordinating Center) and RR043050-S2 (Mouse BIRN).

References

Baxevanis, A.D. (2003). The Molecular Biology Database Collection: 2003 update. *Nucleic Acids Res.* **31**, 1–12.

Bjaalie, J.G. (2002). Opinion: Localization in the brain: New solutions emerging. *Nat. Rev. Neurosci.* **3**, 322–325.

Bohm, J., Frangakis, A.S., Hegerl, R., Nickell, S., Typke, D., and Baumeister, W. (2000). From the cover: Toward detecting and identifying macromolecules in a cellular context: Template matching applied to electron tomograms. *Proc. Natl. Acad. Sci. U.S.A.* **97**, 14245–14250.

Bowden, D.M., and Dubach, M.F. (2002). NeuroNames 2002. *Neuroinformatics* **1**, 43–59.

Brevik, A., Leergaard, T.B., Svanevik, M., and Bjaalie, J.G. (2001). Three-dimensional computerized atlas of the rat brain stem precerebellar system: Approaches for mapping, visualization, and comparison of spatial distribution data. *Anat. Embryol.* **204**, 319–332.

Foster, I. (2002). The grid: A new infrastructure for 21st century science. *Phys. Today* **55**(2), 42–47.

Foster, I., Kesselman, C., and Tuecke, S. (2001). The anatomy of the grid: Enabling scalable virtual organizations. *Int. J. Supercomput. App.* **15**, 200–222.

Fox, P.T., and Lancaster, J.L. (2002). Opinion: Mapping context and content: The BrainMap model. *Nat. Rev. Neurosci.* **3**, 319–321.

Gupta, A., Ludaescher, B., and Martone, M.E. (2000). Knowledge-based integration of neuroscience data sources. *Proc. 12th Int. Conf. Sci. Stat. Database Manage. (SSDBM'00).*

Gupta, A., Ludascher, B., Grethe, J.S., and Martone, M.E. (2003). Towards a formalization of a disease specific ontology for neuroinformatics. *Neural Networks* (in press).

Hadida-Hassan, M., Young, S.J., Peltier, S.T., Wong, M., Lamont, S., and Ellisman, M.H. (1999). Web-based telemicroscopy. *J. Struct. Biol.* **125**, 235–245.

Harlow, M.L., Ress, D., Stoschek, A., Marshall, R.M., and McMahan, U.J. (2001). The architecture of active zone material at the frog's neuromuscular junction. *Nature (London)* **409**, 479–484.

Hessler, D., Young, S.J., Carragher, B.O., Martone, M.E., Lamont, S., Whittaker, M., Milligan, R.A., Masliah, E., Hinshaw, J.E., and Ellisman, M.H. (1992). Programs for visualization in three-dimensional microscopy. *NeuroImage* **1**, 55–67.

Humphreys, B.L., Lindberg, D.A., Schoolman, H.M., and Barnett, G.O. (1998). The Unified Medical Language System: An informatics research collaboration. *J. Am. Med. Inf. Assoc.* **5**, 1–11.

Kotter, R. (2001). Neuroscience databases: Tools for exploring brain structure-function relationships. *Philos. Trans. R. Soc. London, Ser. B* 356:1111–1120.

Lacroix, Z. (2003). Issues to address while designing a biological information system. In *Bioinformatics: Managing Scientific Data* (Z. Lacroix and T. Crifchlow, eds.), pp. 75–108. Morgan Kaufmann, San Francisco, CA.

Ludäscher, B., Gupta, A., and Martone, M.E. (2001). Model-based mediation with domain maps. *Proc. 17th Int. Conf. Data Eng.*, Heidelberg, Germany. *2001.*

Ludäscher, B., Gupta, A., and Martone, M.E. (2003). A model-based mediator system for scientific data management. In *Bioinformatics: Managing Scientific* Data (Z. Lacroix and T. Critchlow, eds.), pp. 335–370. Morgan Kaufmann, San Francisco, CA.

MacKenzie-Graham A., Jones, E.S., Shattuck, D.W., Dinov, I., Bota, M., and Toga, A.W. (2003). The informatics of a C57BL/6 mouse brain atlas. *Neuroinformatics* **1**(4), 397–410.

Martone, M.E., Deerinck, T.J., Yamada, N., Bushong, E., and Ellisman, M.H. (2000). Correlated 3D light and electron microscopy: Use of high voltage electron microscopy and electron tomography for imaging large biological structures. *J. Histotechnol.* **23**, 261–270.

Martone, M.E., Gupta, A., Wong, M., Qian, X., Sosinsky, G., Ludascher, B., and Ellisman, M.H. (2002a). A cell-centered database for electron tomographic data. *J. Struct. Biol.* **138**, 145–155.

Martone, M.E., Gupta, A., Ludascher, B., Zaslavsky, I., and Ellisman, M.H. (2002b). Federation of brain data using knowledge guided mediation. In *Neuroscience Databases: A Practical Guide* (R. Kotter, ed.), pp. 275–292. Knopf, New York.

Martone, M.E., Zhang, S., Gupta, S., Qian, X., He, H., Price, D.L., Wong, M., Santini, S., and Ellisman, M.H. (2003). The Cell Centered Database: A database for multiscale structural and protein localization data from light and electron microscopy. *Neuroinformatics* **1**(4), 379–395.

McEwen, B.F., and Frank, J. (2001). Electron tomographic and other approaches for imaging molecular machines. *Curr. Opin. Neurobiol.* **11**, 594–600.

Medalia, O., Weber, I., Frangakis, A.S., Nicastro, D., Gerisch, G., and Baumeister, W. (2002). Macromolecular architecture in eukaryotic cells visualized by cryoelectron tomography. *Science* **298**, 1209–1213.

Paxinos, G., and Franklin, K. B. J. (2001). *The Mouse Brain in Stereotaxic Coordinates.* Second Edition. Academic Press, San Diego CA.

Peltier, S.T., Lin, A.W., Lee, D., Smock, A., Lamont, S., Molina, T., Wong, M., Dai, L., Martone, M.E., and Ellisman, M.H. (2003). The telescience portal for advanced tomography applications. *J. Parallel Distrib. Comput.: Comput. Grids* (in press).

Rajasekar, A., Marciano, R., and Moore, R. (1999). Collection-based persistent archives. *Proc. 16th IEEE Symp. Mass Storage Syst. 1999.*

Rajasekar, A., and Moore, R. (2001). Data and metadata collections for scientific applications. *High Perform. Comput. Network. (HPNC 2001)*, Amsterdam, *2001.*

Sinha, U., Bui, A., Taira, R., Dionisio, J., Morioka, C., Johnson, D., and Kangarloo, H. (2002). A review of medical imaging informatics. *Ann. N. Y. Acad. Sci.* **980**, 168–197.

Stevens, R., Goble, C., Paton, N.W., Bechhofer, S., Ng, G., Baker, P., and Brass, A. (2003). Complex query formulation over diverse information sources in TAMBIS. In *Bioinformatics: Managing Scientific Data* (Z. Lacroix and T. Critchlow, eds.), pp. 189–224. Morgan Kaufmann, San Francisco, CA.

Toga, A. (2002). Neuroimage databases: The good, the bad and the ugly. *Nat. Rev. Neurosci.* **3**, 302–308.

Van Essen, D.C. (2002). Windows on the brain: The emerging role of atlases and databases in neuroscience. *Curr. Opin. Neurobiol.* **12**, 574–579.

Van Essen, D.C., Drury, H.A., Dickson, J., Harwell, J., Hanlon, D., and Anderson, C.H. (2001). An integrated software suite for surface-based analyses of cerebral cortex. *J. Am. Med. Inf. Assoc.* **8**, 443–459.

Visualization in Life Sciences

Donna J. Cox, M.F.A.

CONTENTS

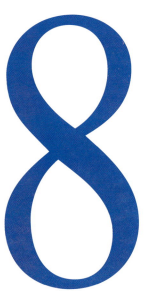

Databasing the Brain. Edited by Stephen H. Koslow and Shankar Subramaniam
ISBN 0-471-30921-4 © 2005 John Wiley & Sons, Inc.

8.1 INTRODUCTION

Computer-mediated visualization is a relatively young technology. During the last half of the twentieth century, the computer graphics imaging (CGI) raster hardware industry matured, three-dimensional (3D) CGI synthesis software advanced, computational science expanded, and a community of developers coalesced around the burgeoning concepts of data-driven scientific visualization (McCormick et al., 1987; Foley and Ribarsky, 1994; Friedhoff, 1989). Here, visualization is defined as the mapping of data to a representation resulting in a digital sensory model, primarily a visual model. This mapping is the transformation of numerical or symbolic information into visual information that people can use to understand, document, synthesize, analyze, hypothesize, and communicate. The primary goal of visualization is insight and decision-making.

In the vast world of information technology, visualization branches into diverse subsystems such as data visualization, scientific visualization, information visualization, and informatics (Ebert et al., 2001). The primary distinguishing factor among these subsystems is the characterization and source of the data; however, all involve the transformation of one domain of information into another domain of visual information. These data include mathematical and scientific simulations, observed data, acquired statistics, instrument recordings, medical imaging, and contextual and symbolic descriptions.

Visualization in the life sciences spans a tremendous range of research and applications that include bioatomics, protein modeling, cellular models, genomic sequencing, medical imaging, and human organ reconstructions. It is beyond the scope of this chapter to cover all visualization techniques in the life sciences. Basic concepts and methods are presented in the process of visualizing primarily numerical data (not contextual data). These methods cross boundaries of data, scientific, and medical visualizations. They include issues of mapping information, reducing visual complexity, and 3D CGI synthetic modeling and rendering. This chapter provides a broad overview and explores areas of particular value to neuroscience.

Historically, visualization was defined as the process of creating and using images and visual models. This process was employed by great scientists. A short history from our rich scientific past to the current information revolution reinforces the idea that the process of visualization contributes to critical thinking and creativity. Visualization is important to the advancement of almost any scientific discipline and is particularly valuable to neuroscience where it will play a key role in providing researchers and practitioners with faster decision-making capabilities and the "bigger" picture of data correlation. Interactive and display technologies are discussed in the context of advanced applications relating to medical imaging and bioinformatics. These applications provide techniques to extract, view, integrate, and explore very large datasets that include instrumentation, statistical, and computational data.

Many challenges still plague the visualization community in the face of extraordinary advances in computing technology because data systems are being produced at a higher rate and with more diversity. Visualization is a tool to help integrate diverse data and bring more insight into complex informational systems; it will play an increasingly important role in neuroscience. As advancing technologies, computing, and digital displays become less expensive and the collective grid infrastructures mature, new horizons expand for visualization in the life sciences.

8.2 BACKGROUND

8.2.1 Historical Threads and Philosophical Ties

Before the computer, hand drawings and diagrams served as conceptualizing, documenting, and didactic tools for early science. Historical scientific illustrations bifurcate into two primary types: those that represented natural phenomena such as drawings of plants and those that represented more abstract concepts such as tree diagrams. Text accompanied the most important and unforgettable illustrative visualizations. As printing technology evolved, the sophistication and proliferation of illustrations evolved.

The Renaissance was the watershed regarding the use and intent of scientific and technology illustrations. The role and influence of visual models upon the process of conception, developing, and communicating scientific knowledge is well documented (Hall, 1960; Peterfreund, 1994; Baigrie, 1996; Kemp, 1996; Allert, 1996; Popper, 1980; Ferguson, 1977). This science history provides a rich interplay among image, science, and the public (Caudill, 1994). During the sixteenth century, botanical, anatomical, and biological books with illustrated text proliferated. Some historians have argued that artists' concepts and imagery helped to form scientific knowledge and methodology during the Renaissance (Shirley and Hoeniger, 1978; Ackerman, 1985; Desantillana, 1969). Copernicus, Kepler, and Descartes created drawings and visual models that not only documented observations but augmented understanding and conception (Holton, 1973). Many historians of science have successfully argued that visuals provided major contributions to the formations of scientific disciplines such as astronomy (Winckler, and Van Heldon, 1992; Cornelis, 2000), geology (Rudwick, 1976), and genetics (Ruse, 1996).

Darwin and the representation of natural systems as tree-form diagrams such as phylogenetic trees are well documented from the nineteenth century (O'Hara, 1996). Many of these early representational devices are incorporated in modern data-driven digital visualizations today (Novacek, 1994; Woolman, 2002). The search for visual structure in science and nature continued to develop during the nineteenth century (Crary, 1990; Alverson, 1991). As technology progressed, scientists employed photographic imagery from instruments such as telescopes and microscopes. By the twentieth century, radiography flourished and coupled with computer image processing techniques. Many scientific fields advanced from the use of radiographs, from medicine to crystallography. Hierarchy Theory grew as an interdisciplinary way to organize and describe forms and processes in nature, physics, and complex systems (Smith, 1981; Whyte, 1949, 1951, 1954; Whyte et al., 1969; Wilson, 1967, 1969; Simon, 1962, 1977; Feekes, 1986; Ahl and Allen, 1996). For example, branching and bifurcation are structural patterns that are ubiquitous in nature, radiographs, scientific and mathematical models, and technological structures. This attempt to organize form and processes directly grew out of the onslaught of technology-mediated scientific images.

Revisionist thinkers in history and philosophy of science such as Thomas Kuhn (1970) and Mary Hesse (1966, 1980) have examined the subculture and discourse of science to reveal metaphorical and paradigmatic practices relating to creativity (Ortony, 1993; Sacks, 1979; Indurkhya, 1992; Headley, 1997; Van Bendegem, 2000; Vandervert, 2001; Forceville, 2002). The use of verbal and visual metaphor and models in science (Bradie, 1998, 1999; Nunez, 2000; Grunfeld, 2000) has been analyzed both as a positive functional element (Hallyn, 2000) and a negative process whereby scientific objectivity is lost to the destructive use of embedded metaphors in gender-biased medical literature (Stepan, 1986; Martin, 1991; Moser, 1993; Keller and Longino, 1997). Issues of funding, media, politics, and the sociology of images have been topics of exploration (Heims, 1987; Lynch and Woolgar, 1990; Buchwald, 1996; Dubow, 2000). Some have recognized and awed at the power of the scientific image (Latour, 1986, 1987; Latour et al., 1992) and have confirmed that our scientific culture is a "visual" culture both historically as well as today. This visual scientific culture came of age in the digital revolution where images and numbers merged.

8.2.2 Digital Visual Revolution: Insight Beyond Numbers

During the twentieth century, data-driven scientific visualization developed in tandem with CGI synthesis after the first vacuum tube computers during the middle of twentieth century (Greenia, 2000). While a small proportion of scientific researchers used these nascent computer technologies to model natural phenomenon, CGI synthetic methods were primarily exploited by the entertainment industry in the 1970s. Hollywood Director George Lucas invested in CGI research, resulting in computer image synthesis algorithms that were enslaved to photo-digital realism and special effects cinema production. Early CGI researchers published many of these foundational techniques at SIGgraph (Special Interest Group in Graphics, Association of Computing Machines) conference proceedings. However, the most advanced CGI algorithms remained the proprietary property of production companies. Most basic research scientists did not have access to advanced CGI algorithms and computers to visualize complex numerical data.

It was not until the dawning of supercomputing that CGI synthetic methods expanded to include techniques particular to scientific problems. In 1985, the National Science Foundation (NSF) established five academic supercomputer centers, and they embarked upon the task of making supercomputing and new technologies available for basic research. Computational science has been called the "third" science and is defined as the scientific mode of numerically modeling theories within digital computers to describe and predict natural phenomenon; and this mode constitutes a new methodology in science (Kaufmann and Smarr, 1993; Bartz, 1998). The solutions to these supercomputer simulations yield billions of numbers and require new methods for humans to view and understand the data.

By 1987, NSF convened a panel to report on Visualization in Scientific Computing (McCormick, et al., 1987). The reaction to the report became a major initiative that unified an industry in the research and advancement of scientific visualization. Eventually, Hollywood-style visualizations gave way

to mathematically rigorous, interactive visualization software (Brown, 1990). Within the last 15 years, data-driven scientific visualization has developed into a language with its conventions and visual methods such as 3D glyphs (Ellson and Cox, 1988) and interactive progressive disclosure techniques. By 1990, new conferences were solely devoted to research for visualization in scientific computing (Patrikalakis, 1991; Rosenblum et al., 1994; Chen et al., 1996).

Most publications are devoted to advancing technical research and focused on mathematical algorithms that make computing more efficient or provide alternate solutions to existing technical problems (ACM SIGgraph Computer Graphics Proceedings, IEEE Visualization Proceedings, IEEE Image Processing Conference Proceedings). Some publications have addressed issues of representation (Tufte, 1983, 1990; Cox, 1988, 1990, 1991; Lynch and Woolgar, 1990), while others have focused on perceptual design approach for information visualization (Gershon, 1990, 1994; Ware, 2000). Many of these basic techniques will be presented in the following sections.

8.3 BASIC METHODOLOGIES AND TECHNIQUES

8.3.1 Varieties of Visualization Experiences

Data visualization is an inclusive term that includes data analysis and a variety of disciplines including science, engineering, geography, business, and finance. Scientific visualization and information visualization have evolved as the two primary subsystems that employ advanced 3D CGI synthetic techniques to display data-driven quantitative and qualitative information. Mostly, scientific visualization provides visual models of natural phenomenon that humans recognize such as physical galaxies, fluid flows, and thunderstorms (Figures 8.1 and 8.2). In general, medical imaging such as computerized axial tomography (CAT) is categorized as scientific visualization and results in digitized slices of data that can be reconstructed into 3D digital visual models (Figure 8.5). In contrast, information visualization provides data-driven graphical models to represent abstract information such as demographics, statistics, genomics (Figure 8.17), and descriptive databases. Both scientific and information visualization are informed by dynamic or static data; both involve mapping one domain of abstract symbolic information into a visual digital model; and both can employ 3D CGI synthetic techniques (Woolman, 2002; Ebert et al., 2001; Banissi et al., 2001). Integration of scientific and contextual information is a valuable strategy for researchers to create "smart" applications that combines descriptive and spatial knowledge discovery (Stoev, and Strasser, 2001; Peltier et al., 2003).

The visualization pipeline is an iterative process and human-computer feedback loop. The visualization expert or scientific user of visualization tools generally preprocesses data, attaches data to modeling schema, and renders data for insight, decision making, and communication. This process is inherently iterative, to explore different aspects of data and refine the visualization. Interactive real-time software applications provide iterative interrogation of datasets (Douglas et al., 2003). Time evolution is a dimension of many datasets; thus visualizations can also take the form of animated sequences of images. However, computational

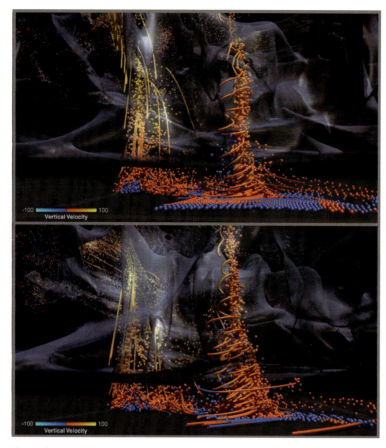

Figure 8.1   High-resolution storm and tornado supercomputer simulations. The animation reveals a swirling tornado to the right that will strengthen as it merges with the larger rotating updraft to the left. Upper figure is animation Frame 7100. A sheet of particles released and traced within the flow field is represented by spherical and stream tube glyphs. The stream tubes represent a short history of selected particles to show flow geometry. When glyphs flow in a positive direction, they are colored red to yellow, with yellow being the highest velocity. Negative velocity is blue to cyan. Lower figure shows frame 7200 in the time evolution of the simulation. Both upper and lower figures show a transparent gray iso-surface, a component scalar field of water droplet and ice. Scientific lead: Robert Wilhelmson (NCSA/UIUC); simulation by Lou Wicker (National Severe Storms Laboratory, NOAA), Matt Gilmore, and Lee Cronce (Department of Atmospheric Sciences, UIUC); visualization by Robert Patterson, Stuart Levy, Matt Hall, Alex Betts, Experimental Technologies (ET), and Donna Cox, Director of ET; copyright NCSA 2003.

constraints, high-resolution, advanced 3D CGI rendering techniques, and multiple-terabyte datasets often require batch mode rendering for animations. Various remote visualization applications and environments are being developed to address these issues.

8.3.2 Mapping Numbers into Pictures

Consider the "higher-level" cognitive process of data mapping to address the most important question in visualization: How can the numbers and loosely correlated facts be designed and transformed into a visual model that makes sense to the viewers? Data mapping is the essential binding of symbolic input to a visual representation according to some convention. Each scientific discipline has its sets of conventions, or in the case of a young science such as neuroscience, these conventions are

evolving. For example, the older discipline of astronomy develops and uses the convention of star catalogues, while young neuroscience is developing the convention of brain atlases. Each scientific community develops consistent descriptive visualization languages to support these conventions, and coherency is one of the primary goals.

The "science" of visualization is based upon the analysis of the human visual and perceptual systems. This approach promises to yield better visualizations that will enable people to better understand more visual information (Ware and Beatty, 1988; Ware, 2000). It is a developing system of guidelines and principles, based upon analytical and calibration methods that include what colors to use (Ware and Beatty, 1988; Gershon, 1990, 1994), how shape forms present information (Beck, 1966), and what affects human perception of visual information. However, visualization is more than perception; it is also a process of interpretation (Gregory, 1990). Experience, habits, cultural

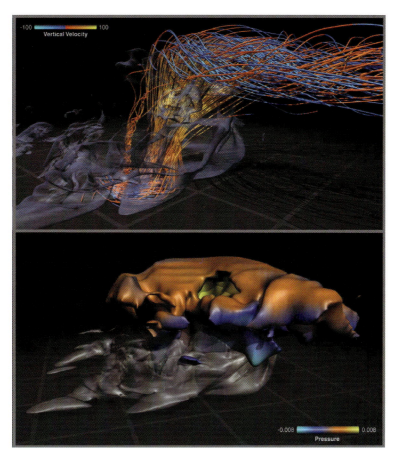

Figure 8.2   High-resolution storm and tornado supercomputer simulations: Upper figure is frame 7200 from an animated visualization. The stream tube glyphs represent the flow geometry from frame 5600–7200, an extended period of the simulation. The transparent gray iso-surface in upper figure shows the component microphysical scalar fields of water droplet and ice. Lower figure shows two component iso-surfaces. The pressure-colored component iso-surface was constructed from two dependent microphysical fields: cloud droplets and ice. This iso-surface has been color mapped with the pressure field sampled on the structured 3D grid. The gray transparent iso-surface in the lower figure is a component of three dependent variables in the 3D grid: rain, snow, and graupel (soft hail). Scientific lead: Robert Wilhelmson (NCSA/UIUC); simulation by Lou Wicker (National Severe Storms Laboratory, NOAA), Matt Gilmore, and Lee Cronce (Department of Atmospheric Sciences, UIUC); visualization by Robert Patterson, Stuart Levy, Matt Hall, Alex Betts, Experimental Technologies (ET), and Donna Cox, Director of ET, NCSA; copyright NCSA 2003.

contingencies, and discipline-specific preferences affect how people see, use, and interpret pictures (Berger, 1977; Berlin and Kay, 1969; Varela et al., 1997; Rosch and Lloyd, 1978; Mulaik, 1995; Lakoff and Johnson, 1980). Choice in the representation and mapping of data affects the interpretation as well as final quality of the visualization.

Glyphs: Icons of Information

The "art" of visualization is the creative translation of data into visual representations called "signs" (Anderson, 1989). This process relates to use of signs in semiology (Hawkes, 1977; Bertin, 1983) and involves the invention of symbolic icons or visual metaphors that are directly bound to data and designed within the constraints of the computer graphics technology (Haber and McNabb, 1990; Haber et al., 1991; Dent-Read et al., 1994). This is a difficult task and often requires the expertise of many people

(Cox, 1990, 1991, 1996; Brown, 1990; Foley and Ribarsky, 1994). Research to incorporate artificial intelligence and automate this representational design process has been demonstrated but can have limitations (Robertson, 1991; Mackinlay, 1986; Ribarsky et al., 1993a,b; Shaw et al., 1999; Ebert, et al., 2000).

Glyphs are data-driven symbolic, iconic, graphical objects that possess attributes such as shape, color, size, position, and orientation in scientific visualization. They are useful to show dependent variables, features, vectors, data correlations, and other abstract information in data (Robertson, 1991; Woolman, 2002). Glyphs are effective visual methods and have become an essential part of many visualization environments (Treinish, 1993; Shaw et al., 1999; Foley and Ribarsky, 1994; Schroeder et al., 1991; Keller and Keller, 1993; SCI, 2003). Signs and icons have been classified in medical imaging (Kergosein, 1991). Figure 8.1 and upper Figure 8.2 exemplify the use of glyphs to understand

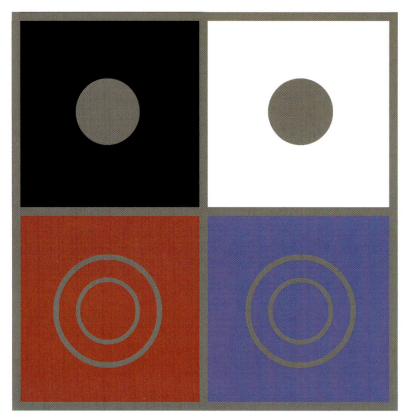

Figure 8.3 Luminance and chroma contrast. Upper figure demonstrates that the perception of identical 50% gray circles are affected by the background color, the gray circle on black appears lighter than the gray circle on white; lower figure demonstrates background complementary chroma contrast where the interior rings are identical in lower left and lower right, but the rings appear bluer in the red square while they appear pinkish in blue square.

turbulent and complicated air flow in an atmospheric simulation (MEAD, 2003). The colored spherical balls and stream tubes in upper and lower Figure 8.1 and upper Figure 8.2 are glyphs that represent vertical velocity particle data in a time-evolving severe thunderstorm simulation. The spherical glyphs indicate positions of tracer particles in the flow of the air currents. Likewise, the stream tube glyphs provide flow geometry. In Figure 8.1, the stream tubes trace a short "history" of motion in the life of individual particles during the time evolution of the simulation. In contrast, upper Figure 8.2 stream tubes trace a long "history" in the life of tracer particles, providing a more complete view of flow geometry.

Color and Mapping

One of the most important and useful, yet difficult, visualization techniques is managing color mapping. Though color can be measured in wavelengths, it is a human perceptual sensation that is affected by subjectivity and cultural interpretations (Berlin and Kay, 1969). A complete theory that satisfies all applications does not exist, although the Opponent Theory of Color has been one of the most useful to visualization experts (Ware, 2000). One major difficulty is the lack of standardization and color constancy in computer displays, projections displays, multimedia, and color reproduction systems (Gershon, 1994). Also, individuals have various levels of perceptual color response to various media. For example, background color affects perception of the

foreground due to perceptual interaction. In upper Figure 8.3, the circles are identical 50% gray, but perceptually the background contrast makes the foreground circles appear different in luminance. In lower Figure 8.3, the interior rings are identical in chroma, but the background induces a complementary contrast and the rings appear bluer in the red square while they appear pinkish in blue square. While luminance predominates in visual perception, it has very different properties than chroma. Indeed, paying attention to color issues is helpful when developing or interpreting visualizations (Rogowitz and Kalvin, 2001; Healey, 1996; Gershon, 1990, 1994).

Using color to provide visual discrimination of data values is a common technique in scientific visualization. A continuous or discrete scalar field with values ranging from x to y is normalized and mapped to a range of indexed values called a color map. This technique is used to show variation on a surface or in a volume. Upper and lower Figure 8.1 shows a velocity field ranging from -100 to $+100$ mapped to an indexed color map. The middle value of zero is the division between positive and negative velocity. The color scale is mapped to the glyphs to indicate whether the air is going up or down. When air flow velocity is up (positive), glyphs and stream tubes are red to yellow. Downward (or negative) air flow is blue to cyan. New techniques in perceptual color maps build upon vision research and are becoming an integral part of visualization (Rogowitz and Kalvin, 2001).

8.3.3 Dealing with Data

Data Sources, Types, and Structures

In general, two primary sources of scientific numerical data are observational and computational. Observational sources include instruments (e.g., telescopes, CAT) and collected or acquired data such as census statistics or textual descriptions. Computational data result from scientific and mathematical models that use digital computers to solve physical equations using approximation methods (Wilhelmson, 1988; Kaufmann and Smarr, 1993). It can be argued that no data is "raw" because all data are filtered and interpreted whether through the orientation of the computational mathematical models or the particular design of the data-gathering instrument. The formats and structures across observed and computed numerical data are diverse and present one of the greatest challenges to visualization (Gallop, 1994; Lévy, 2001). Digital visualization data types include scalar arrays, vectors, meshes, and volumes. Many visualization systems accommodate a variety of data readers and conversion algorithms (Upson and Keeler, 1988; Upson et al., 1989; Schroeder et al., 1992; Fruhaff et al., 1994; SCI, 2003; NCMIR, 2003). Efforts to design common data formats, standards, and metadata tagging have been successful (Rew and Davis, 1990; NCSA, 1999; Kapadia and Yeager, 1999; Folk et al., 2001; Yeh et al., 2001; SRB, 2003) but require adoption within and across disciplines. Visualization methods can employ higher-level contextual data attributes to organize and specify data objects and enable scientific interchange (Hibbard and Dyer, 1992; Rogowitz and Treinish, 1993; Ullmer et al., 1998; SRB, 2003; MCAT, 2003). This attention to data classification fosters a wider variety of data flow visualization applications, discovery environments, and grid applications (Mackinlay, 1986; Baker, 1992; Moore et al., 1998; Reed, 2003). Regardless of these efforts, the variety of cross-discipline scientific data structures often requires customized preprocessing in the visualization pipeline.

The dimension data described here are equal to the number of independent variables, and with structured data the number is three for x, y, z position (Brodlie, 1992; Gallop, 1994). Structured data are defined here as data that are ordered by position such as 3D volumes from MRI (magnetic resonance imaging) or computational 3D grid data. Position is the independent variable, and dependent variables are sampled at position points. Dependent variables can be scalar fields such as density and temperature sampled at points in the 3D structured data. Often, independent variables map to world coordinate systems such as in geosciences data; however, it is also possible to associate data with glyph attributes such as points, lines, and spheres that cannot be projected within a world coordinate system, such as a visualization of a molecular cloud. Finite element simulations are often structured according to the geometry of the material being simulated (Haber and McNabb, 1990).

In many computational simulations, grids are irregular but structured. For example, Figures 8.1 and 8.2 are visualizations of a 3D structured volume of atmospheric computational data. The primary volume dataset is an irregular 3D grid with seven microphysical dependent variables associated with each 3D grid cell. Adaptive mesh refinement (AMR) numerical simulation techniques for computational fluid dynamics (CFD) employ a structured rectilinear grid, but locally adapt producing sets of nested grid cells that have finer spatial and temporal resolutions. Figure 8.4 is an example of time evolution of an AMR cosmological simulation showing a time and spatial domain that varies many orders of magnitude from the large-scale cosmic web of dark matter to small-scale features of star formation and protogalaxies (Norman et al., 1999; Cox, 2000, 2003; Kaufman, 1991; Geller and Huchra, 1989; Ferris, 1982). Upper left and

Figure 8.4 Large-scale to small-scale structure of the universe. Four frames (upper left, upper right, lower left, lower right) from visualization of the evolution of the universe; adaptive mesh refinement (AMR) simulation used 200,000 CPU hours on Origin 2000 and resolves the formation of individual galaxies as well as their web like large-scale distribution in space. Color is used to differentiate gas density, dark matter, birthing stars, and other small-scale features. One continuous camera moves from 300 million light years scale of cosmic web formation (*upper left*) down to a scale of 30 thousand light years of interacting galaxies and star formation (*lower right*). Simulation by Michael Norman, NCSA, Greg Bryon, Princeton, and Brian O'Shea, UIUC. Visualization by Donna Cox, Stuart Levy, and Robert Patterson, NCSA.

upper right of Figure 8.4 shows large-scale structure refining to form filamentary web of gas and stars. Lower left and right of Figure 8.4 are later frames in the simulation and show the birthing stars and protogalaxy interaction as the camera zooms in to view these fine-scale features of the simulation. The AMR numerical technique yields a 3D data structure that is locally unstructured but employs the simplicity of a rectilinear grid (Weber et al., 2001; Berger and Oliger, 1984). Local adaptive refinement on a rectilinear 3D grid produces a greater level of complexity and computational efficiency.

In contrast to structured data, scattered data such as text information (Schumaker, 1976; Nielson, 1993b, 1994; Grzeszczuk et al., 1998) and bioinformatics are not structured and cannot be mapped to a Cartesian or 3D rectilinear grid. Figure 8.17 is a visualization of unstructured datasets yielding probability matrices and relational phylogenetic trees; these data are not ordered according to 3D spatial position. While most sensored instrumentation such as MRI is structured as regular, dense 3D datasets with attributes associated at each point in the volume, some instrumentation data can also be unstructured, scattered data in the form of sparse data samples from sensored roving oceanic ships (Rosenblum and Kamgar-Parsi, 1994; Nielson, 1994).

Reducing Visual Clutter: Progressive Disclosure and Feature Extraction

Most scientific datasets are complex and large. Data generated by supercomputers and large-scale sensors result in multiple-terabyte multidimensional datasets that preclude a visual one-to-one mapping to digital screen space. Visualization techniques map discrete values from an n-dimensional data domain onto pixel colors, but this is largely a dimension-reducing process. Even after the large base dataset has been significantly compressed with standard techniques, the problem remains that on-screen graphics appear too cluttered, noisy, and incomprehensible. Progressive disclosure is an important method to reduce visual clutter and enable continuous in-depth exploration of the dataset. Progressive disclosure is an overall approach that enables the interactive remapping of data, viewing data from arbitrary angles, and progressively revealing datasets at various resolutions and levels of detail. Progressive disclosure employs filtering techniques such as subsampling, polygon reduction, extracting subregions of interest, multiresolution models, and feature visualization.

Subsampling continuous or very large discrete datasets is necessary for most visualization in order to provide an overall, global view with the goal of providing an accurate approximation. Regular or irregular interval subsampling can reduce the underlying 3D data field, visually capture the continuous function of the data (Treinish, 1992; Nielson, 1993b, 1994), and demonstrate efficiency in providing a fast overview. However, subsampling errors and interpolation techniques can lead to misleading artifacts (Carr et al., 2001). Multiresolution modeling, polygon reduction, and feature simplification enable access to levels of detail and enhance interactivity in the visualization pipeline (Walter and Healey, 2001). Advanced multiresolution modeling for high-resolution irregular grids provide methods to zoom into spatially subdivided smaller regions of interest and represent these at higher resolutions and finer scales (Gross, 1994; Zorin et al., 1997; Guskov et al., 1999; Kobbelt et al., 1998; Levin, 1999; Sander et al., 2001; Gavriliu et al., 2001;

Weiler and Ertl, 2001). Wavelet transform methods are useful to reduce data complexity for gridded meshes and volumes (Bonneau, 1998; Bonneau et al., 1996; Muraki, 1995; Bertram et al., 2001; Hubeli and Gross, 2001). Progressively disclosing level of detail enables a smooth transition from courser to finer representations and manages data complexity.

A "feature" is described here as anything that is interesting in the data. "Feature extraction" from data is based upon the assumption that not all data needs to be represented directly; rather, data relationships, attributes, or derived variables will provide important information (Van Walsum et al., 1996; Hubeli and Gross, 2001). The following is an example of feature extraction by computing and visualizing derivative particle trajectories from the primary 3D structured dataset of an atmospheric simulation. A typical method to understanding processes in complicated flows, where salient features may be hidden by turbulent clutter, is to release particles within the flow field and trace the arrangements of the particles. Visualizing these tracer particles as glyphs provides understanding and correlation of flow features within data field quantities. In Figure 8.1 and upper Figure 8.2, the derivative particle trajectories are computed by integrating the velocity field using a fourth-order Runge–Kutta algorithm (Davis and Rabinowitz, 1984). Velocities between grid cells are trilinearly interpolated and registered with other dependent variables. As discussed earlier, Figure 8.1 shows vertical velocity values mapped into two different types of glyphs that trace particles: stream tubes and spherical balls. These glyphs represent extracted features from the primary structured volume dataset. Multidimensional data and unstructured data often require feature extraction to understand complicated embedded processes. The creative use of glyphs and iconic representations with intelligent choices for color mapping, shape, and filtering are necessary to reduce scene clutter (Treinish, 1992, 1993; Durand et al., 2000).

Filtering techniques such as feature and data extraction, normalizing data for interactively remapping color maps, and scaling are typical interactive techniques that help manage visual complexity and enable interactive on-screen data interrogation (Lacroute and Levoy, 1994; Jiang et al., 2001). Often, decoupling compute-intensive techniques is efficient. For example, extracting features such as the above particle trajectories through large time-varying flow fields can be time-consuming due to calculation and disk input–output requirements. With sufficient processing and memory, users can interactively place probes in data, release particles, and visualize the results. Decoupling the particle advection calculation to run on a fast parallel machine with quick access to large datasets enables efficient interactive exploration of particle trajectory visualizations.

Many progressive disclosure techniques depend upon interactive disk data retrieval. Novel data representation schemes and adaptive techniques have been developed to increase efficiency and allow access of structures in an increasing order of smoothness (Machiraju et al., 1998; Pfister et al., 1999). Representing volumetric data as multiresolution tree structures enables faster geometric computation in discretely defined domains (Subramanian and Naylor, 1997; Taubin, et al., 1998). Data compression using discrete cosine transformations for volumetric scalar data can decrease rendering time by a significant factor and enable more efficient geometric computations (Boon-Lock and Liu, 1995; Brunet et al., 1994; Gross, 1994). Likewise, hardware rendering techniques provide increased efficiency for interactive and batch mode volume rendering (Westermann and Ertl, 1998; Weiler and Ertl, 2001; Lum et al., 2002; SCIRun, 2001).

8.3.4 Modeling and Rendering

Raster graphics became widely accepted in the 1970s because hardware became cheaper and easier to use. Raster graphics provides an excellent viewing platform for both 2D data such as medical image scans, 3D CGI image synthesis, and future applications of the life sciences. Creating a 3D CGI synthetic image from data involves modeling and rendering techniques that convert 3D geometric representations into 2D raster image on a digital screen. The following techniques for surfaces and volumes are basic to 3D CGI image synthesis and important to future applications in the life sciences.

3D Surfaces and Volumes

Surface Modeling. Traditionally, surface modeling was called geometric or shape modeling with homogenous interior representations. Surface modeling is the mainstay of 3D CGI synthesis techniques. An object is modeled by a collection of surfaces such as in computer-aided design (CAD). Surfaces can be described as a collection of polygonal meshes (Blinn, 1982; Catmull and Clark, 1978; Schroeder et al., 1992b) or patches. Solid modeling, in contrast to geometric modeling, uses 3D solid primitives such as cones and cubes to describe objects. Generally, surfaces are projected into a 3D Cartesian world coordinate system using isometric or perspective projections.

To view the 3D surface, algorithms scan approximate polygonal meshes to provide screen pixel data values and thus "render" the appearance of the surface. Surfaces are rendered with depth cuing using hidden surface removal techniques to remove invisible polygons that would be obscured by objects in the foreground. Surface graphics is an object-based approach that requires manipulating a display list of geometric objects in order to change the scene view (camera view) and to interact with the geometries. Other surface rendering techniques include shading, coloring, transparency, and texturing of objects (Zhang et al., 1997; Stam, 1999; Lévy, 2001; Neyret and Marie-Paule, 1999; Malzbender et al., 2001). Light sources and shadows provide depth cuing to help the human to perceive 3D placement and appearance of surfaces on screen space. For example, in upper Figure 8.2, the shadows from the stream tubes on the grid plane provide perceptual depth cuing to understand placement of the data. Light model algorithms employ physical laws to calculate shadows, reflections, refraction, and other surface optical appearances.

A myriad of graphics techniques have been developed in pursuit of "realistic" appearances of surface data. Sophisticated algorithms to model the realistic appearance using advanced physics have been developed and could be employed for physically based models, but most of these have not been applied to large scientific datasets until recently (Dorsey et al., 1999; Fedkiw et al., 2001; Jensen and Christensen, 1998; Jensen et al., 2001; Zyda et al., 1997; Riley and Ebert, 2003).

Creating surfaces from structured 3D grid datasets with multiple dependent variables is a useful technique to extract surface features and dynamically explore the data domain as a series of surfaces thresholds, iso-surfaces. A 3D contoured iso-surface is described by specifying an iso-value, a scalar value sampled at 3D grid cells. The surface is extracted by a trilinear interpolation between grid cells, tiling of the triangulated surface, and rendering (Lorensen and Cline, 1987; Schroeder et al., 1994; Nielson and Hamann, 1991; Welch and Witkin, 1994; Stander and Hart, 1997; Kirby and Karniadakis, 2003). The Marching Cubes is a robust contouring algorithm to create 3D surfaces with a constant scalar value from volume datasets (Lorensen and Cline, 1987). The surfaces can then be colored according to a second scalar field sampled at the same location in the 3D structured grid. For example, in lower Figure 8.2, the Marching Cubes algorithm was used to generate two component iso-surfaces from a structured 3D grid. In lower Figure 8.2, the colored iso-surface was constructed from two dependent microphysical variable fields: cloud droplets and ice. This iso-surface has been color mapped with the pressure field sampled at the same location on the structured 3D grid. The lower gray transparent iso-surface is a component of three dependent microphysical variables: rain, snow, and graupel (soft hail). Iso-values and color maps can be interactively manipulated to explore the volumetric data domain as a series of iso-surfaces.

Volume Rendering, Graphics, and Modeling. While surface modeling and graphics are the mainstay of CGI synthesis, it can be argued that scientific visualization futures will focus on volume visualization techniques due to the plethora of collected and computed 3D volumetric data. Volume rendering refers to both the viewing and the rendering of data (Drebin et al., 1988). Volume data are typically graphically represented as a 3D array of voxels: volume elements stored within a 3D raster cubic frame buffer. Voxels represent a scalar data field and can be manipulated, projected, and rendered as intensities (Kaufman, 1990, 1991, 1994; Raufman et al., 1993). Acquired 3D instrumentation data are typically a discrete 3D regular grid of voxels. Each voxel has dependent variable attributes or measured properties associated with it (e.g., density, color). In computed data, many dependent variables may be associated with each voxel.

Interactive volume data techniques include planar slicing and viewing and surface reconstruction to extract features. Volumetric rendering is computationally demanding and requires hardware and software techniques to speed up the process (Westermann and Ertl, 1998; Guthe et al., 2002; Kähler et al., 2002; Prohaska and Hege, 2002; Kanitsar et al., 2002). Unfortunately, instrument data such as CAT is noisy, and direct volume rendering and filtering can distort the volume rendering.

Volume data such as AMR computational output present a challenge to volume rendering since the nested, multi-resolution grids create artifacts in the rendering (Kähler et al., 1999; Norman et al., 1999; Lévy et al., 2001; Kähler and Hege, 2001). Three-dimensional volumetric cells can be converted into a volume of particles and rendered as colored and scalable Gaussian splats (Cox, 1996; Norman et al., 1999). For example, the volumetric gas in lower Figure 8.4 has been converted to a grid of particle and rendered as red Gaussian splats surrounding the protogalaxy stars in the foreground. This approach supports mixing particle data with volumetric 3D data.

Volume graphics is an emerging subfield of volume visualization and is primarily concerned with the synthesis, rendering, and interactivity of volumetric objects that are stored as voxel arrays. Geometric surfaces, objects, and scenes can be scanned and converted into 3D arrays of voxels to render surface properties. Volume graphics for surface rendering eliminates the need to track and manipulate object display lists as in surface modeling.

Volume modeling is the rendering of volumetric data without going through the preprocess of converting volumetric data into voxels (voxelization) and then rendering. For example, a volume dataset can generally be characterized as an integral equation of attribute functions and can directly feed the rendering pipeline, eliminating the voxelization preprocess (Nielson, 1994). A typical direct volume rendering is generated by a one-dimensional transfer function based upon an associated single scalar density at each cell of the dataset. The direct approach to volume models supports the representation of volumes as collections of object attributes and provides an alternative method to identifying and synthesizing 3D objects and their interiors within volumes; this process directly relates to the following applications in segmentation and reconstruction in medical imaging (Nielson, 1993a,b, 1994; Bailey, 2001; Huang et al., 2003).

8.3.5 Segmentation and Reconstruction

In general, 3D CGI segmentation is a technique used to extract meaningful features from data. Model-based segmentation is part of a process to extract meaningful 3D objects from 3D data (Hanrahan, 1993). In medical imaging, structured 2D pixel and 3D voxel datasets from modalities as PET (positron emission tomography), CAT, MRI, and functional MRI are ranges of density values that are undifferentiated in terms of organs, tissues, and substructures. When data are acquired through such medical scanning instruments, the data domain is a collection of homogenous regions and often overlapping density values. Segmentation is the process of sorting these density values and identifying features or objects within the data. Historically, segmentation employed computer vision and data analysis techniques to analyze 2D slices in medical imaging (Tobler, 1978). Model-based segmentation is a technique used to extract 3D features in volume data including computed data from nonmedical disciplines such as geoscience. Isolating meaningful 3D features requires statistical and mathematical techniques to detect edges and boundaries and also involves data subsampling, clustering, registering, and labeling (Cline et al., 1990; Nielson, 1993b, 1999; Bajcsy and Kovacic, 1989; Hu et al., 2003; Huang, 2003; Huang et al., 2003; Lefohn et al., 2003a,b). Segmentation strategies include surface-fitting techniques such as level sets to differentiate and connect homogenous regions (Whitaker et al., 2001; Fedkiw and Osher, 2002; Lefohn et al., 2003a). Level sets employ partial differential equations to deform iso-surfaces, but this is a computationally intensive and often precludes interactivity (Sethian, 1999). While automatic segmentation is desirable, most segmentation is semi-automatic or manual. Interactive segmentation by a skilled user empowered with feature-extraction tools enables data-intensive interrogation and employs her expertise to differentiate subtleties in anatomical features (Höhne and Hanson, 1992; Whitaker et al., 2001).

The 3D CGI reconstruction of segmented regions into surface objects or volumetric renderings provides a powerful visualization tool that supports localization and enhances knowledge about the 3D structures (Sunguroff and Greenberg, 1978; Cline et al., 1987, 1988, 1991; Kikinis et al., 1990, 1991a,b; Höhne et al., 1994; Moharir et al., 1998; J.C. Carr et al., 2001; Bjaale, 2002a; Prohaska and Hege, 2002). The modeling of complex object shapes from structured 3D datasets is a difficult task because standard algorithms that work for simple polygonal surfaces can miss edges or faces in complex natural objects. If the complex object must be described by a set of free-form patches, then global structural information is lost and verification of the integrity of the surface is difficult. Noisy medical imaging modalities such as 3D Ultrasound and high-frequency noise from CT and MR scanners introduce artifacts in conventional algorithms that reconstruct surfaces (Lorensen, 1995; Lengyel et al., 1995; Hahn et al., 2001; Zhang et al., 1997; Zhang et al., 2002; Bartroli et al., 2001; Tory et al., 2001; Preim et al., 2002). Recent research focuses on topological, manifold, and other mathematical and geometric techniques to improve surface approximations in the 3D reconstruction of fine anatomical details such as nerves, vascular systems, and muscle tissue (Kunii and Shinagawa, 1994; Yu and Fessler, 1998; Dong et al., 2001; Fattal and Lischinski, 2001; Brevik et al., 2001; Kanitsar et al., 2002).

8.3.6 Localization and Brain Warping

Localization is identifying and assigning structural and functional locations in anatomical data. Precise location is important to neuroscience researchers trying to understand structural and functional relationships in the brain. Localization accuracy depends on more than computerized 3D CGI segmentation and reconstruction techniques. Accurate registration of data from various sources is paramount; yet, this registration is challenging and complex due to the variety of imaging modalities such as PET and fMRI, differences in individual sizes of brains, lack of precise location information during experiments, and inconsistencies in nomenclature (Bjaalie, 2002a,b; Mazziotta et al., 1995). Registration across modalities and subjects, common reference and coordinate systems, and dynamic atlasing techniques are required for useful neuroscience localization.

Dr. Arthur Toga, Laboratory of Neuro Imaging, University of California at Los Angeles, has comprehensively reviewed and developed warping methods to deform, register, compare, and analyze brain image data (Toga, 1998; Toga et al., 2001). Warping is the geometric transformation of the form and shape of brain image data beyond simple repositioning; it has become an important hybrid tool to study brain structure and function. It includes image processing techniques such as scaling, non linear deformations, and affine transformations. The goal is to provide consistent spatial registration across modalities and subjects. The process results in image datasets for analysis and brain-to-brain comparisons. Since warping transformations provide a higher-order mathematical description of regional differences among brains in individuals and populations of individuals, this information can be used to develop probability and deformation maps for visualization and analysis (Mazziotta et al., 1995). Warping will be important to advance the 3D CGI segmentation and reconstruction visualization techniques of brain data and to increase accuracy in template-based and auto-segmentation applications (Lorensen et al., 1995; Hong et al., 1997; Jolesz et al., 1997; Fried et al.,1999; Nakagohri et al., 1998; Anagnostou et al., 2001; Hurdal et al., 2001; Bartroli et al., 2001; Samsonov et al., 2003).

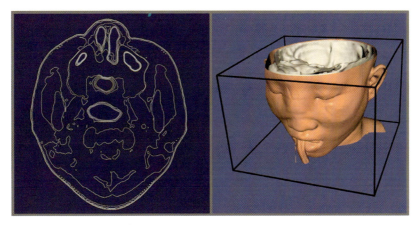

Figure 8.5 The Marching Cubes algorithm (Lorensen and Cline, 1987) generated the two iso-surfaces of skin and bone extracted from a series of 94 CT slice planes at 128^2 resolution. The CT scan slice on the left corresponds to the planar delimitation of the iso-surface rendering on the right. Courtesy of Bill Lorensen, GE's Global Research Center, copyright 2003.

However, warping should not be considered a "black box" technique; it is a powerful tool but depends upon high-quality source data and researchers' understanding of the mathematical transforms (Toga, 1998).

8.4 CURRENT VISUALIZATION APPLICATIONS IN LIFE SCIENCES

Visualization in the life sciences involves a wide variety of applications and display technologies. The following examples demonstrate many of the above methods and introduce how display technologies have and will influence the future of visualization in the life sciences.

8.4.1 VTK and ITK Toolkits

A software toolkit is a system of interoperable algorithmic modules that enables development of customized applications. Many visualization toolkits provide open source and extensible software libraries and interface layers to build and customize

3D visualization applications. For example, The Visualization Toolkit (VTK) is an open source 3D CGI synthesis and image processing software system that is used by thousands of universities and institutions to develop visualization applications (VTK, 2003; Schroeder et al., 1997, 1998; Kitware, 2003). VTK was developed by researchers at GE Global Research Center, and it supports a wide variety of CGI algorithms and techniques such as 2D and 3D scalar, vector, volume and surface modeling, mesh generation and smoothing, data representations, and glyphs. For example, VTK Marching Cubes surface contour algorithm (Lorensen and Cline, 1987) was used to generate the 3D head reconstruction in Figure 8.5 from CT cross-sectional data. Virtual scope techniques such as virtual endoscopy (Lorensen et al., 1995; Jolesz et al., 1997), virtual laryngoscopy (Fried et al., 1999), and virtual colonoscopy are the 3D CGI reconstruction and interactive exploration of medical imaging of anatomical cavities. For example, Figure 8.6 shows a 3D reconstruction of a virtual colon from CT cross-sectional data using VTK (Lorensen et al., 1995; Miller et al., 1991). Figure 8.7 demonstrates VTK's "tissue lens" rendering technique: a mixture of surface and volume rendering. In Figure 8.8 shows multimodal rendering. The anatomy is rendered using a 2D texture mapping volume renderer. The

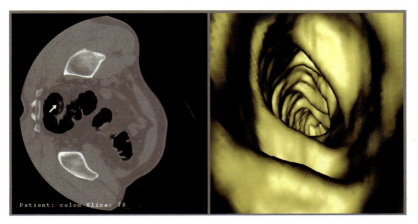

Figure 8.6 Virtual colonoscopy. The 3D reconstruction and interactive exploration of cross-sectional CT data of a colon. Courtesy of Bill Lorensen, GE's Global Research Center, copyright 2003.

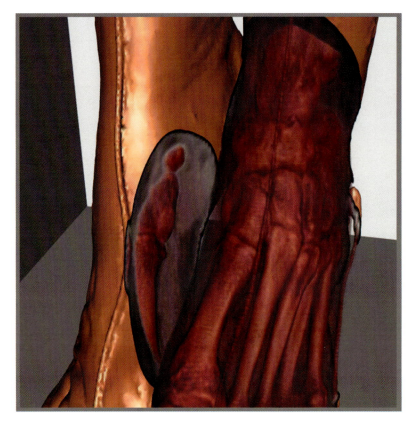

Figure 8.7   VTK "tissue lens." A combination of surface and volume rendering. Courtesy of Bill Lorensen, GE's Global Research Center, copyright 2003.

iso-surfaces are generated with VTK Marching Cubes algorithm (Finnigan et al., 1990).

In general, VTK modules and libraries provide a flexible, robust system and a relatively easy way to build visualization applications due the toolkit's high-level abstraction graphics model. Paraview (Kitware, 2003) is an open source and extensible application that extends VTK to run parallel on distributed and shared memory computational systems, thus providing advanced rendering capacity for large datasets. Paraview provides glyphs, contours, iso-surfaces, vectors, stream tubes, and cutting/slicing/rendering of volumetric data.

The National Library of Medicine Insight segmentation and registration toolkit (ITK) was developed to support the Visible Human Project (ITK, 2003; NLM, 2003). ITK is an open source software system that includes segmentation and registration algorithms. These algorithms include extensive level-set segmentation framework and facilities to do rigid-body and deformable registration techniques. Figure 8.9 shows volume data registration of gray-scale CT data with physical cross sections (color). The data are the Visible Woman dataset from the National Library of Medicine (Lorensen, 1995). The registration was rigid body using ITK's Mutual Information Registration algorithm (Viola and Wells, 1997). The surfaces of the skin and bone are iso-surfaces generated with VTK's Marching Cubes algorithm. The colors are interpolated from the registered physical cross sections.

VTK and ITK are examples of powerful toolkit systems that can accommodate a variety of rendering, segmentation, reconstruction, and registration techniques for the neuroscience community. The primary advantage of open source toolkits is that they encourage flexible community code-building that becomes self-sustaining as developers increase in number. The primary disadvantage of toolkits is that they can take a significant initial investment of computer-literate professional time to customize modules for domain-specific applications.

8.4.2 SCIRun and BioPSE

Dr. Christopher Johnson and research colleagues at the Scientific and Computing and Imaging Institute (SCI), University of Utah, have developed several important visualization and computational problem-solving environments (PSE) for a variety of biomedical and computational applications (SCI, 2003; Johnson, 1998; Johnson et al., 1998, 1999, 2000, 2001, 2002). SCIRun is a data-flow visualization and computational PSE for large-scale datasets and computational modeling. SCIRun is a complete interactive workbench for efficient construction and real time steering of scientific computations and interaction with large-scale datasets (Livnat et al., 2001; Weinstein et al., 1998, 2000). The ability for scientists to control simulation parameters in real time and steer a computation is efficient and insightful. SCIRun is based upon a data-flow programming component model that enables a variety of computational and visualization components to be connected and tightly integrated. SCIRun is the underlying, general-purpose PSE that includes system level codes, libraries, and user interface. BioPSE is a domain-specific package that sits on top and

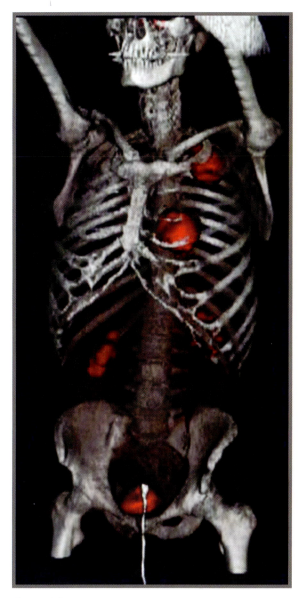

Figure 8.8   Multimodal rendering. The original CT and PET data were generated on a commercial GE PET/CT system called Discovery LS. The anatomy is volume rendered using VTK's 2D texture mapping volume renderer. The iso-surfaces are generated with the VTK Marching Cubes algorithm. Courtesy Bill Lorensen, GE's Global Research Center, copyright 2003.

extends SCIRun by providing bioelectric field modeling, simulation, and visualization. BioPSE includes forward and inverse bioelectric volume conductor problems such as localizing electrically active regions of the brain (Johnson et al., 1993; BioPSE, 2002; Van Uitert and Johnson, 2003; Weinstein et al., 2000). Figure 8.10 shows example screen shots of the modular, data-flow network interface, and visualization windows. A network is assembled by connecting together separate modules. Each module performs a computational or visualization algorithm.

SCIRun supports a variety of geometric modeling, surface mesh generation, and 3D segmentation techniques. Figure 8.11 shows SCIRun visual examples of segmented surface mesh geometries. SCIRun enables scientists to modify geometric

models, interactively change parameters, and mesh adaptation. SCIRun can call ITK's imaging tools, including levelsets, segmentation, and filtering, and then visualize the results with SCIRun's environment. Figure 8.12 is an example rendering of level-set segmentation (Lefohn et al., 2003a,b; Sethian, 1999).

SCIRun also supports direct volume rendering using multidimensional transfer functions (Kniss et al., 2001a,b, 2002a,b; He et al., 1996; Kindlmann and Durkin, 1998). Direct volume rendering employs transfer functions to assign optical properties and the appearance of the voxel. Typical direct volume rendering with 1D transfer functions require trial and error to select the appropriate values. Multidimensional transfer functions are much more complex; thus, they are useful but not widespread due to the complexity of specifying and controlling the transfer functions. SCI researchers have developed an intuitive interface to select multidimensional transfer functions for volume renderings. This interface employs transfer function widgets. A widget is graphical device to control visualization interaction. The specification of the 3D transfer functions correlates with characterizations of data gradient, gradient magnitude, and edge detection. This elegant specification assumes that the most interesting features in a volume are regions of change. The transfer function widget helps guide selections of the multidimensional functions that control the color and opacity of volume objects within the dataset (Kniss et al., 2001a,b, 2002a,b). Other widgets include clipping plane, data probe, and color picker. Figure 8.13 and Figure 8.14 show examples volume renderings and the subtle range of object extraction using 3D transfer functions. At the bottom of each render is a graphical widget for interactive control. The primary disadvantage to multidimensional transfer functions is the increased hardware memory needed to store variables at voxel sample points.

SCIRun and BioPSE support real-time ray tracing, source localization, and other visualization techniques. These frameworks are flexible, extensible, open source and have built-in stereo and link to other display applications. Dr. Johnson and colleague Dr. Steve Parker have developed the SCIRun "big picture" as an integrated PSE and computational workbench. Figure 8.15 shows the complete plan for this integration from data acquisition to real-time intervention.

8.4.3 Visualization Display Applications in Life Sciences

Alternative displays play an increasingly important role in visualization. The examples below demonstrate both high-resolution and mixed-reality display devices that are important contributions to visualization applications.

Bioinformatics and High-Resolution Tiled Display Wall

Figure 8.16 shows an 8192 × 3840 pixel resolution tiled display wall. This scalable tiled display uses 40 PC nodes to drive 40 XGA projectors (NCSA, 2003; Humphreys et al., 2001). It provides a large-scale interaction environment and high-resolution display capabilities where research groups can collaborate and visualize data. Figure 8.17 shows screen shots of the bioinformatics visualization application on the tiled display in Figure 8.16.

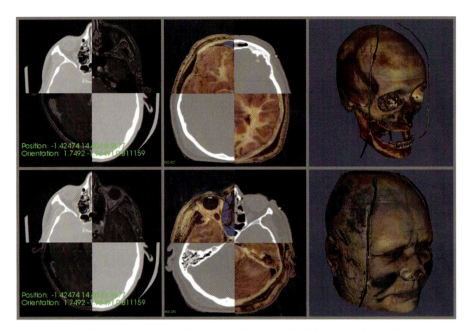

Figure 8.9   Volume registration. The data are of the head of the Visible Woman dataset from the National Library of Medicine (NLM, 2003). The images show registration of physical cross sections (color) with CT data (gray scale). The registration was rigid body using ITK's Mutual Information Registration algorithm. The skin and bone are rendered as iso-surfaces generated with VTK's Marching Cubes algorithm. The colors are interpolated from the registered physical cross-sections. Courtesy of Bill Lorensen, GE's Global Research Center, copyright 2003.

This bioinformatics visualization has greatly benefited from the high resolution of the tile display.

Dr. Eric Jakobsson, Director of Center for Bioinformatics and Computational Biology at the National Institute of General Medical Sciences, and NCSA research colleagues compared all genes in all the completely sequenced prokaryotic genomes to the Transport Commission Database (Saier, 2000; SLBG, 2003). They combined NCBI BLAST (NCBI, 2003) with a parser and matrix generating and visualization program. The full package, NCSA BLAST (E. Jakobsson, S. Natarajan, and Q. Li, unpublished, 2003), is specifically designed to do comprehensive phylogenetic profiling. The above computational comparison of transport proteins to gene sequences in all completely sequenced prokaryotic genomes resulted in an "occurrence matrix" giving the probability of each class of transport protein being represented in each genome. Upper left Figure 8.17 is the wide-view visualization of the derivative "co-occurrence matrix" that shows probabilities for the occurrence of each gene type conditional on the occurrence of each other gene type. Each bright spot represents a high probability that two types of transport genes (represented by the vertical and horizontal axes) interact with each other, based upon the assumption that a high probability of genetic co-occurrence implies a high probability that the gene products will functionally interact (Marcotte et al., 1999). Upper right Figure 8.17 is a close-up view of the "co-occurrence matrix."

Lower left Figure 8.17 shows a wide view of the phylogenetic tree visualization that has been adapted to visualize a mathematical similarity analysis of potassium channels (Palaniappan and Jakobsson, 2004). Potassium channels are a class of proteins that permit movement of potassium ions across cell membranes and are prime determinants of patterned electrical

activity in nerve cells. The tree shows the phylogenetic relationship between prokaryotic (bacterial and archaeal) channels and the channels from two animals, humans and *C. elegans* (a worm). Lower right Figure 8.17 immediately reveals several facts that would have required extensive analysis otherwise. The worm has most of the same classes of channels as the human. For both the human and the worm, there are three major groupings of potassium channels; see lower-right extreme close-up. The prokaryotic channels fit into two of the three groupings while the third grouping developed later in eukaryotes. Far out on the right-hand side of the tree, on a long branch by themselves, is a grouping of five prokaryotic channels that, following the visualization, have been determined to be evolutionary links between potassium channels and two other classes of excitable channels. Lower right Figure 8.17 shows a close-up of the wall-tile visualizations of K-channel phylogenies from prokaryotes, worm, and human. The display high resolution enables one to see (at a glance) evolutionary pathways by which human ion channels originated in bacteria and diversified to provide the complex and subtle variations in electrical signal generation necessary for human thought and other complex functions. The wall-tile display is ideally suited to an analysis of this scale, where researchers want to observe fine details and emergent patterns simultaneously. The disadvantage of a tiled display is the cost and maintenance.

Mixed Realities: Nearly Being There

Virtual reality (VR) has been used for life science visualizations for over 10 years (Brooks et al., 1990; Cruz-Neira et al., 1993; Brady et al., 1995; Bohanan, 2002). VR is an evolving term that generally means the use of a variety of media and CGI synthetic worlds to create an essential and convincing experience. Most

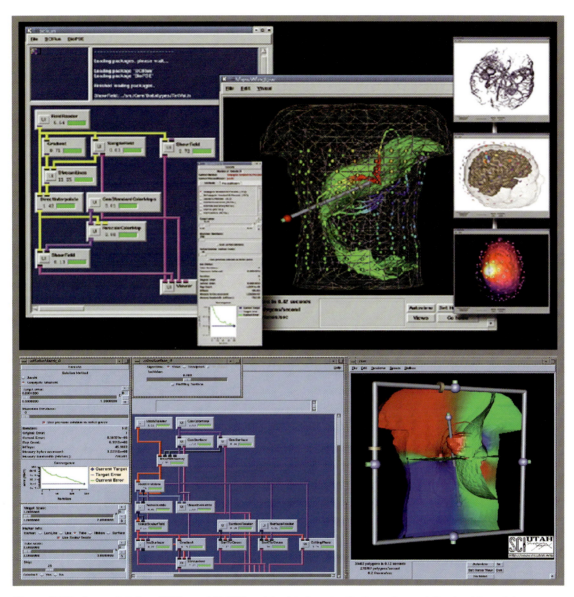

Figure 8.10 Screen shots from SCIRun and BioPSE modular data flow visualization and computational problem solving environments (PSE) showing the module networks and the visualization windows. A network is assembled by connecting together separate modules. Each module performs a computational or visualization algorithm. Courtesy of Scientific Computing and Imaging Institute, University of Utah (SCI, 2003); copyright SCI 2003.

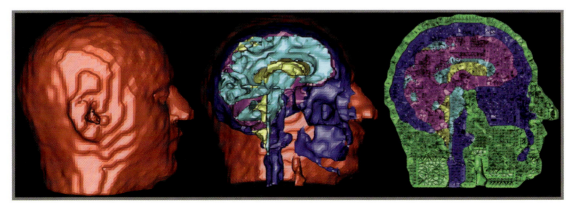

Figure 8.11 SCIRun segmented and rendered surface mesh geometries. Courtesy of Scientific Computing and Imaging Institute, University of Utah (SCI, 2003); copyright SCI 2003.

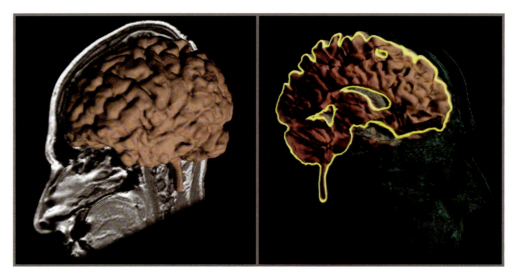

Figure 8.12 An interactive level-set segmentation and rendering. The cerebral cortex of 256 × 256 × 198 MRI volume data. The yellow band indicates the outline of the level-set model on the clipping plane. Courtesy of Scientific Computing and Imaging Institute, University of Utah (SCI, 2003); copyright SCI 2003.

VR display environments track or "sense" the location and interaction of the human user and provide an immersive experience (Sherman and Craig, 2003). VR media can include stereo vision, head-tracking, and hand-held tracking devices for interface control (Czernuszenko et al., 1997; Cruz-Neira et al., 1993; Bryson and Levit, 1991; Bryson, 1994; Cox, 1996; Leigh et al., 1999; Sherman and Craig, 2003). VR enhances the visualization process by increasing visual depth cues, enhances human-computer interaction (Mine et al., 1997; Wartell et al., 1999; Sherman and Craig, 2003), and provides novel strategies for data navigation (Usoh et al., 1999; Foxlin et al., 1998; Cox et al., 2000). VR environments include the CAVE[TM], a room-size, multiperson immersive 3D stereo, rear-screen projection system Figure 8.18. Immersive workbench is a large table-top stereo rear-projection system for VR applications. The effectiveness of VR has been evaluated for its effectiveness (Pausch et al., 1997) and is being used in psychological response experiments (Francis et al., 2003; ISL, 2003). The limitations of VR resolution is improving

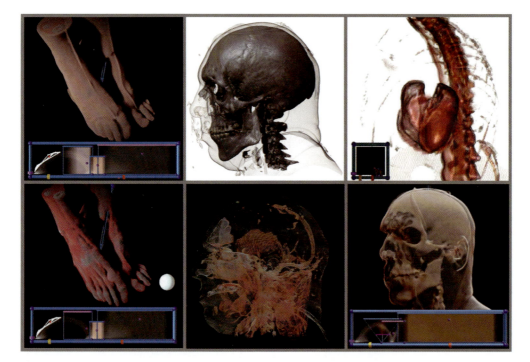

Figure 8.13 SCIRun 3D transfer function volume rendering. These examples show the variety of volumetric object extraction using multidimensional transfer functions. "Widgets," graphical interface specification of multidimensional transfer functions, are shown at the bottom of each volume rendering. Courtesy of Scientific Computing and Imaging Institute, University of Utah (SCI, 2003); copyright SCI 2003.

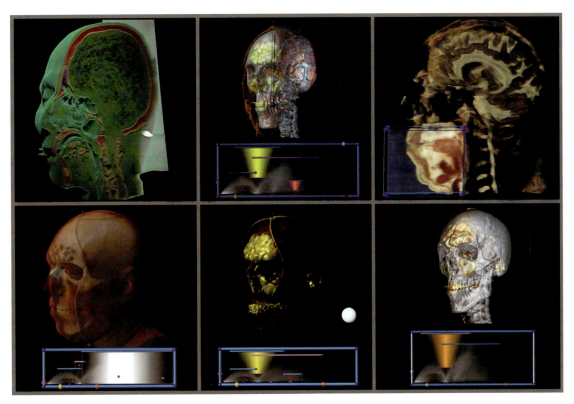

Figure 8.14   SCIRun 3D transfer function volume rendering. These examples show the variety of volumetric object extraction using multidimensional transfer functions. "Widgets," graphical interface specification of multidimensional transfer functions, are shown at the bottom of each volume rendering. Courtesy of Scientific Computing and Imaging Institute, University of Utah (SCI, 2003); copyright SCI 2003.

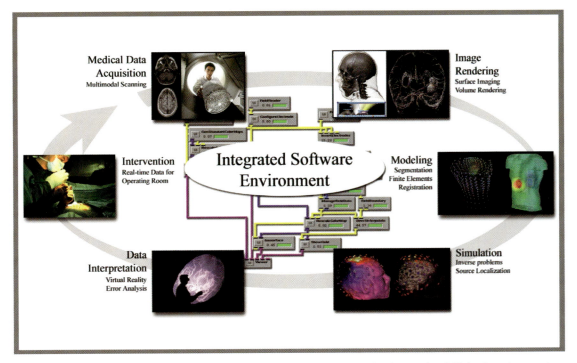

Figure 8.15   Integrated software environment. From data acquisition through real-time intervention, SCI plan for an integrated software environment and scientific workbench. Courtesy of Scientific Computing and Imaging Institute, University of Utah (SCI, 2003); copyright SCI 2003.

Figure 8.16   NCSA's 8192 × 3840 pixel resolution tiled display wall. Each projector tile is driven by a PC. On the left of the tiled display wall is a phylogenetic genetic profile "co-occurrence" matrix; on the right of the tiled display wall is a phylogenetic tree showing major groupings of potassium channels (see Figure 8.17). Photo by Paul Rajlich. Copyright NCSA.

(Deering, 1992), and using commodity equipment is bringing down the cost (Kaczmarski and Zuffo, 2002). However, the biggest problem with VR and other mixed-reality environments is the effort that it takes to make applications work. For example, SCIRun has worked in CAVE and Wall versions in the past. However, the effort to port complex software such as SCIRun to other CAVE implementations often precludes such applications proliferating across a variety of CAVE environments.

VR devices have been expanded beyond the "visual" to include localized audio and haptic force feedback systems. Stereo VR haptic systems are powerful interface strategies for the life sciences (Taylor et al., 1993; Brooks et al., 1990; McNeely et al., 1999; Ikits et al., 2003). For example, the nanomanipulator (CISMM, 2003), co-developed by the departments of computer science and physics at the University of North Carolina, Chapel Hill, has been used by scientists since the early 1990s. The nanomanipulator system provides a VR interface to scanning probe microscopes (SPM) such as scanning tunneling microscopes (STMs) and atomic force microscopes (AFMs). The system couples the microscope to a virtual-reality interface that gives the scientist virtual telepresence on the surface of nanostructures such adenovirus particles and DNA–protein complexes.

Augmented reality (AR) is where the VR application combines virtual synthetic representations with the physical world, such as projecting a CGI synthetic image onto a real space (such

as desk or a room) to perceive and interact with a real and virtual space simultaneously. Alexander Gillet of the Molecular Graphic Lab (MGL), Scripps Institute, developed PyARTK (Gillet, 2003), an interactive computer graphic interface that displays molecular computer graphics concurrently with the physical models. The user manipulates a physical object with his/her hand and views the CGI-augmented surface through a video projection. This interface is built upon the extensible component-based Python environment Pmv developed at MGL and AR-ToolKit, an augmented reality toolkit developed at the University of Washington (ARToolKit, 2003).

Enhanced reality in the surgery room is where the combination of VR, AR, and haptic systems provide a surgical space for the future Figure 8.19. Enhanced reality brings 3D CGI models into the operating room (Lorensen et al., 1993) and provides an alternate method for virtual surgery techniques. For example, a surgeon might perform an operation using minimally invasive techniques where a 3D CGI model provides a virtual view and feedback (Koch et al., 1996).

Telepresence is the application that uses VR to provide an alternate sense of place and enables first-person point-of-view interaction with a physically real environment from a remote location (Sherman and Craig, 2003). Telepresence requires a networking bandwidth that can accommodate real-time interaction without latency. Telepresence, VR, and AR are powerful

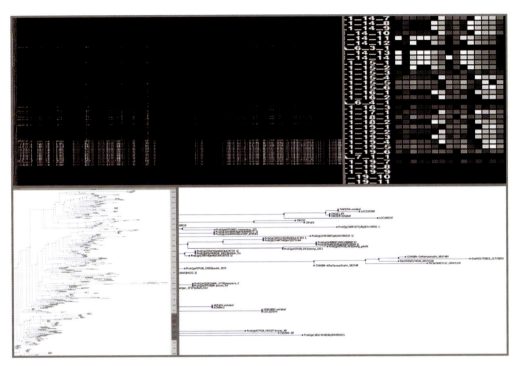

Figure 8.17 Bioinformatics screen shots from Figure 8.16. Upper left is a wide view of the phylogenetic genetic profile "co-occurrence" matrix on left tiled wall (4096 × 3840 pixel resolution); Upper right shows an extreme close-up of the "co-occurrence" matrix; Lower left is a wide view of the phylogenetic tree at (4096 × 3840 pixel resolution); lower right is close-up of phylogenetic tree showing major groupings of potassium channels; Computational biology by Dr. Eric Jakobsson, Ashok Palaniappan, Shreedhar Natarajan; Visualizations by Paul Rajlich, Visualization Group, NCSA.

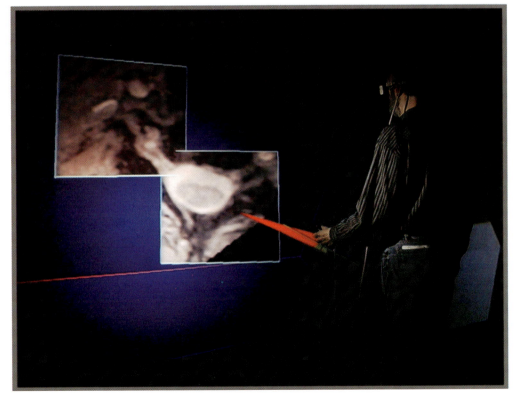

Figure 8.18 Crumbs CAVE application for viewing biomedical data. Developed by Rachael Brady, John Pixton, George Baxter, Pat Moran, Clint Potter, Bridget Carragher, and Andrew Belmont. Alan Craig is viewing an MRI of his skull using Crumbs CAVE application at NCSA (Brady et al., 1995; Sherman and Craig, 2003). Photo Courtesy of William Sherman and Alan Craig.

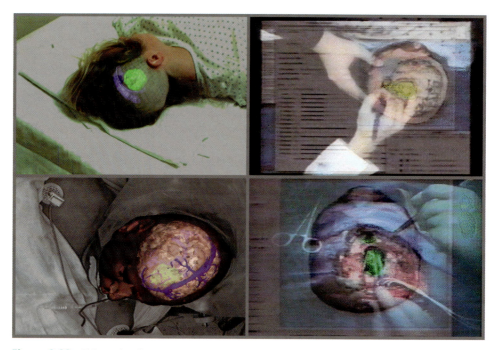

Figure 8.19   Enhanced reality brings 3D CGI models in the surgery room where the combination of virtual reality, augmented reality, and haptic systems enhance a surgical space for the future. Courtesy of Bill Lorensen, GE's Global Research Center, copyright 2003.

strategies over high-speed networks to provide visualization and collaboration on the grid (Leigh et al., 1996; DeFanti et al., 1996).

8.4.4 Remote Interaction and Grid Visualization

The "grid" concept is a system of networked, interconnected technologies such as distributed computational resources, data storage and retrieval, remotely controlled instruments, telepresence, and remote visualization. The goal of the grid is that these specialized services will be provided as a secure, simple, and unified system (Foster and Kesselman, 1999). Grids are part of the larger concept of cyberinfrastructure (Hiltz and Turoff, 1993), the technological glue that enables virtual community interaction and collaboration. Cyberinfrastructure includes systems of portals, teleconferencing, data management and visualization applications, distributed computation, and multiple-discipline grids. Most grids provide domain-specific applications, and many use existing Internet2 connectivity.

Televisualization

Often the computing resources to generate a simulation are not located physically near the person who wants to visualize and view the results. Remote visualization often involves the remote control of visualization processes, visualization displays, and the transportation of the image data to local workstation.

Telepresent visualization (televisualization) is the generation and viewing of digital images and animations remotely by remote control of applications over the grid (Simpson et al., 2002). For example, televisualization can involve navigating and controlling the viewing camera of large-scale displays from remote locations. NCSA researchers controlled and choreographed the 440-person Hayden Planetarium dome, American Museum of

Natural History, New York City, from their CAVE in Champaign, Illinois. They collaborated, virtually choreographed in real time, and visualized significant portions of the public show (Cox, 2000; Green, 2000). Remote data can include video, audio, and other multimedia; and the virtual community can involve not only researchers but the general public as part of educational outreach.

NCMIR, CCDB, and BIRN

Telescience is the process of using telepresence, televisualization, and other grid tools to do science. Dr. Mark Ellisman, Dr. Maryann Martone, and other researchers at the National Center for Microscopy and Imaging Research (NCMIR), Department of Neurosciences, University of California, San Diego, have embraced telescience to provide an international resource for structural biology and neuroscience communities (NCMIR, 2003). NCMIR provides telemicroscopy that includes electron tomography, intermediate high-voltage electron, multiphoton, and confocal microscopes. The high-quality, multiscale resolution of this data acquisition provides excellent precision for 3D CGI segmentation and reconstruction. NCMIR 3D light and electron microscopic resconstructions are reposited at the Cell Centered Database (CCDB) (Martone et al., 2002; Perkins et al., 1997; Shenglan et al., 2003; CCDB, 2003). The Biomedical Informatics Research Network (BIRN) initiative is a test-bed consortium of biomedical researchers who are harnessing grid infrastructure for performing large-scale multresolution imaging studies and providing convenient access and integration of a variety data sources (Moore et al., 1998; SRB, 2003; NPACI, 2003; Martone, 2003; Ellisman, 2003; Peltier et al., 2003).

NCMIR, CCDB, and BIRN provide a suite of image processing/analysis and 3D CGI segmentation and reconstruction visualization tools that include methods for manual contouring, creating iso-surfaces, selecting subsectioning of volumes, and tracing 3D neuronal branching (CCDB, 2003). The tools include a variety of software platforms such as Java, OpenGl, and Windows. The CCDB deposits 3D reconstruction data volumes with links to original raw images and processing details for reconstruction. Segmented visualization objects from 3D volume data are classified into four categories: surface, contour, volume, and tree objects. Each visualization object is labeled with measurement information and indexed to the parent reconstruction (CCDB, 2003).

A primary visualization goal is to span many orders of magnitude in spatial scale from subcellular structures to organs. This "big picture" is similar to astrophysics efforts in Figure 8.4, where a complete universe mapping involves visualizing the range from small-scale features to large-scale structure. This scale of visual integration across multiresolution data is the holy grail of advanced visualization techniques.

In the history of science, the most important visualizations were those with associated descriptive annotations that reinforced the visual models. Intelligent 3D data volumes integrate symbolic and spatial knowledge (Höhne et al., 1994) and are important to brain atlasing. The NCMIR, CCDB, and BIRN initiatives are focused on knowledge-based mediation to navigate the federation of databases and provide intelligent volumes of visual data. These are powerful strategies for the advancement of neuroscience knowledge and the future of visualization (Shepherd et al., 1998).

8.5 LESSONS AND FUTURES

In a broad sense, visualization is the mapping of one domain of information into a model. However, in this process, information is lost. We must remember that the map is not the territory. All data are mediated whether acquired through machines or computed through mathematical models. This fact requires the examination of error ranges and confidence intervals coupled with visualizations (Johnson and Sanderson, 2003). In the future, visualizations must show these important uncertainties and the potential for lost information in the translation process. Historically, "annotated" visual models provided the most important contributions to the formulation of scientific knowledge and communication. Likewise, the future of visualization requires linked knowledge databases that augment and contextualize evolving visualization storehouses (Martone 2003; Razdan et al., 2002).

Historical studies have demonstrated that creativity is inspired and limited by our languages and visual metaphors (Lakoff and Johnson, 1980; Barlow, 1991). For example, computer science has adopted interface metaphors such as "windows," "capture," "navigate," "raw data," and "portal" (Lafrance, 1992). The "brain as a networked computer" is a predominant metaphor (Draaisma, 2000). We need new visualization metaphors and schemas for effective interfaces to visually manage digital information and to integrate the vast range of temporal and spatial scales in neuroscience (Welsh et al., 2001; Hertxmann et al., 2001).

Finally, imagine a future where each individual has a 3D CGI "digital self," including a "digital brain" with a personalized brain atlas that can be registered and monitored as life evolves.

Researchers like Bill Lorensen foresee a "digital body double" that serves as a repository of the individual as he/she ages. This "digital self" could provide a personal test bed for drugs, disease, and surgery and could serve as a bank for genomic histories. CGI techniques will be paramount to the digital self. This type of future will involve federations of vast interdisciplinary communities of researchers across the globe who use advanced 3D data visualization as an integral part of daily inquiry. In the future, visualization will be an important tool to help integrate diverse data across neuroscience and the biological sciences.

8.6 GLOSSARY

Computer Image Synthesis. Using raster computer graphics to create 3D objects from geometric surfaces and to render these surface models with properties such as color, light, and shadows; These techniques have expanded to include volumetric and other modeling procedures.

Chroma. Color described by red–green and yellow–blue channels of perceptual response information; the hue of the color.

Computer Graphics. Communication between people and computers using imagery as opposed to text and numbers. Channel communication into pictures, human ability to recognize patterns

Computerized Tomography. X-ray beam passes through body and is collected with an array of detectors; beam is rotated for a slice or can collate whole-body scans. X-ray info collected is then used by a computer to reconstruct the internal structures resulting in image.

Context Data. Often text that reveals relationships within a visualization data set; it is very useful for expository and explanatory purposes

Data-Flow Visualization. A modular, component, and sequential approach to building a visualization pipeline; data-flow application builders enable the user to build a pipeline by building and connecting modules; data flow is in contrast to monolithic environments.

Depth Cue. Using visual techniques to provide the illusion of depth in spatial CGI representations; these techniques include removing surfaces that would be occluded, shading surfaces using light sources, using size, shape and distance to create the illusion of perspective fall off.

Direct Manipulation Widget. Graphical device to control visualization interaction.

Extensible Software. Software that is open and flexible and enables developers to add functionality to the system.

Electron Tomography. Similar CAT and MRI medical imaging; produces a 3D structural volume from a series of 2D projections.

Feature Visualization. A feature in the data is defined as anything interesting; feature extraction is the application of visualization techniques to make features visible.

Grid. An infrastructure analogue to the "power grid"; the grid concept includes a system of networked, interconnected technologies such as distributed computational resources and grids, data storage and retrieval facilities, telepresence, and remote visualization.

Immersive.   Virtual reality media technology that provides an alternate reality or convincing point of view using CGI synthetic visuals, audio, haptic devices, or other VR media.

Iso-surface.   A CGI surface that represents a constant-valued scalar function.

Iso-value.   The scalar value used to set a threshold and create an iso-surface.

Localization.   The process of identifying and assigning functional and structural locations to anatomical data.

Luminance.   The black-to-white channel of color information; the closeness of a color to black or white; the value of the color.

Magnetic Resonance Imaging (MRI).   An imaging technique based upon the variation in magnetic field in response to radio-wave pulses.

Medical Imaging.   Imaging techniques that employ x rays, gamma rays, high-frequency sound waves, and magnetic fields to produce pictures of the body. The most common are film radiographs.

Metadata.   Data that describes or tags other data; a set of abstractions that describe a variety of data contexts such as data sources, types, formats, applications, linked information such as documents associated with data, and other associated semantic information and knowledge.

Monolithic Visualization Environment.   A traditional visualization application designed for special purpose and has a fixed pipeline. This is in contrast to a data-flow environment that is modular and flexible.

Pixel.   Single picture element of a 2D computer screen display. Resolution of the computer-driven display is specified in 2D pixels (e.g., 1280 × 1024 pixel resolution).

Phylogenetic Profiling.   Comparing patterns of existence among particular genes across different organisms.

Phylogenetic Trees.   Historically, phylogenetic trees were used to show classification and evolutionary processes in species. A branching point on a phylogenetic tree indicates divergence in a single class resulting in two persistent progeny.

Polygonal Mesh.   A connected set of polygonal flat surfaces.

Polygon Reduction.   One of many filtering and data reduction techniques that has the goal of maintaining a sufficient approximation of the original geometric model.

Portal.   An interface to Web-based applications, infrastructure, services, and software tools.

Render.   The process of creating the "appearance" of the data. The term *data visualization* is used here to specify this process of transforming scientific numerical or information data (often called informatics) and to distinguish data visualization from other visualizations such as non-data-driven artwork.

Telepresence.   Closely related to virtual reality to provide an alternate sense of place and enables first-person point of view interaction with a physically real environment from a remote location.

Tele visualization.   The interactive remote control of visualization services that include parameter selection, computation, graphics selection, interactive viewing, and display control and routing of multimedia.

Toolkit.   A software toolkit is a system that includes libraries and interoperable algorithmic modules that enables development of customized applications.

Track.   Using technology to locate the body or parts of the body in real space and using that information to control or provide feedback to virtual reality or computer applications.

Transfer Function.   A function that controls the red, green, blue, and alpha channels of voxel information.

Voxel.   A cubic picture element from a volume that is commonly stored as a 3D array of voxels in a cubic or volume frame buffer for raster displays.

Virtual Reality.   An evolving term that generally means the use of a variety of media and CGI synthetic worlds to create an essential and convincing experience.

Virtual Scopes.   Using medical imaging techniques (e.g., CT or MRI) to record anatomical cavities and computer visualization techniques to virtually dissect and unfold these data.

Wavelet Transform.   Expanding a signal by means of basis functions, wavelets; data can be represented as a wavelet basis; decomposing data into a wavelet basis localizes in both frequency and spatial domains.

ACKNOWLEDGMENTS

Thanks to my Experimental Technology team, who helped during the writing of this chapter. Special thanks go to Lorne Leonard, who suffered through hours of entering references and was essential to help prepare images for print. Special thanks also go to Chris Johnson and Bill Lorensen, who provided inspiration, images, and references via email. Finally, thanks go to Shankar and Stephen for endless patience during the illness of my mother.

References

Ackerman, J. (1985). The involvement of artists in renaissance science. In *Science and the Arts in the Renaissance* (J.W. Shirley and F.D. Hoeniger, eds.). Folger Shakespeare Library, Washington, DC, pp. 102–116.

Advisory Panel on Graphics, Image Processing and Workstations. (1987). *Visualization in Scientific Computing.* National Science Foundation, Washington, DC.

Ahl, V., and Allen, T.F.H. (1996). *Hierarchy Theory: A Vision, Vocabulary, and Epistemology.* Columbia University Press, New York.

Allert, B. (1996). *Languages of Visuality: Crossings between Science, Art, Politics, and Literature.* Wayne State University Press, Detroit, MI.

Alverson, H. (1991). Metaphor and experience: Looking over the notion of image schema. In *Beyond Metaphor: The Theory of Tropes in Anthropology* (J.W. Fernandez, ed.), pp. 94–117. Stanford University Press, Stanford, CA.

Anagnostou, K., Atherton, T.J., and Waterfall, A.E. (2001). 4D volume rendering with the shear warp factorisation: Extensions and quantitative results. In *Fifth International Conference on Information Visualization* (E. Banissi, F. Khosrowshahi, M. Sarfraz, and A. Ursyn, eds.): IEEE Computer Society, Los Alamitos, CA.

Anderson, C. (1989). Images worth thousands of bits of data: Computer artist transform equations into dramatic simulations at Illinois. *Scientist* **3**(3), 1, 16–17.

ARToolKit. (2003). HIT Lab, University of Washington. Web page: http://www.hitl.washington.edu/artoolkit/

Auger, P. (1989). *Dynamics and Thermodynamics in Hierarchically Organized Systems, Applications in Physics, Biology and Economics.* Pergamon Press, Oxford.

Bailey, M. (2001). Visualization viewpoints: Interacting with direct volume rendering. *Comput. Graphics Appl.* **21**(1), 10–12.

Bajcsy, R., and Kovacic, S. (1989). Multiresolution elastic matching. *Comput. Vision, Graphics Image Process.* **46**, 1–21.

Baker, P. (1992). Knowledge-based visualization. *IEEE Visual. '92 Workshop on Autom. Des. Visual.*

Banissi, E., Khosrowshashi, F., Sarfraz, M., and Ursyn, A., eds. (2001). *Fifth International Conference on Information Visualization.* IEEE Computer Society, Los Alamitos, CA.

Barlow, C. (1991). *From Gaia to Selfish Genes: Selected Writings in the Life Sciences.* MIT Press, Cambridge, MA.

Bartroli, A.V., Wegenkittl, R., Konig, A., and Gröller, E. (2001). Nonlinear virtual colon unfolding. *Proc. Visual. 2001*, pp. 411–418.

Bartz, D., ed. (1998). *Visualization in Scientific Computing '98.* Springer, Vienna.

Beck, J. (1966). Effect of orientation and of shape similarity on perceptual grouping. *Percept. Psychophys.* **1**, 300–302.

Berger, J. (1977). *Ways of Seeing.* Penguin Books, New York.

Berger, M., and Oliger, J. (1984). Adaptive mesh-refinement for hyperbolic partial differential equations. *Computat. Phys.* **53**, 484–512.

Berlin, B., and Kay, P. (1969). *Basic Color Terms: Their Universality and Evolution.* University of California Press, Berkeley.

Bertin, J. (1983). *Semilogy of Graphics.* University of Wisconsin Press, Madison.

Bertram, M., Laney, D.E., Duchaineau, M.A., Hansen, C.D., Hamann, B., and Joy, K.I. (2001). Wavelet representation on contour sets. *Proc. Visual., 2001*, pp. 303–310.

BioPSE. (2002). BioPSE: Problem Solving Environment for modeling, simulation, and visualization of bioelectric fields. Web page: http://software.sci.utah.edu/biopse.html.

Bjaalie, J.G. (2002a). Opinion: Localization in the brain: New solutions emerging. *Nat. Rev. Neurosci.* **3**, 322–325.

Bjaalie, J.G. (2002b). Localization in the brain: New solutions emerging. *Neuroscience.* **3**, 323.

Blinn, J. (1982). A generalization of algebraic surface drawing. *ACM Trans. Graphics* **1**, 235.

Bohanan, J. (2002). The human genome in 3D, at your fingertips. *Science* **298**, 737.

Bonneau, G.P. (1998). Multiresolution analysis on irregular surface meshes. *Visual. Proc.*, pp. 365–378.

Bonneau, G.P., Hahmann, S., and Nielson, G. (1996). Blac-wavelets: A multiresolution analysis with non-nested spaces. *Visual. Proc. IEEE*, 43–48.

Boon-Lock, Y., and Liu, B. (1995). Volume rendering of DCT-based compressed 3D scalar data. *Visual. Proc.*, pp. 29–43.

Bradie, M. (1998). Models and metaphors in science: The metaphorical turn. *Protosociology* **12**, 305–318.

Bradie, M. (1999). Science and metaphor. *Bio. Philos.* **14**(2), 159–166.

Brady, R., Pixton, J., Baxter, G., Potter, P., Moran, C.S., Carragher, B., and Belmont, A. (1995). Crumbs: A virtual environment tracking tool for biological imaging. *IEEE Symp. Front. Biomed. Visual.*, pp. 18–25.

Brevik, A., Leergaard, T.B., Svanevik, M., and Bjaalie, J. (2001). Three-dimensional computerized atlas of the rat brain stem precerebellar system: Approaches for mapping, visualization, and comparisons of spatial distribution data. *Anat. Embryol.* **204**, 319–332.

Brodlie, K.W. (1992). *Models for Scientific Visualization. Animation and Scientific Visualization.* British Computer Society Displays Group, London.

Brooks, F., Ouh-Young, J., Blatter, J., and Kilpatrick, P. (1990). Project GROPE—Haptic displays for scientific visualization. *Comput. Graphics Proc. SIGGRAPH Conf., Ser.* 24, No. 4.

Brown, M. (1990). *History of Visualization*, personal communication.

Brunet, P., et al. (1994). Modeling and visualization through data compression. In *Scientific Visualization: Advances and Challenges* (L. Rosenblum, R.A. Earnshaw, J. Encarnacao, H. Hagen, A. Kaufman, S. Klimenko, G. Nielson, F. Post, and D. Thalmann, eds.), pp. 157–170, Academic Press, San Diego, CA.

Bryson, S. (1994). Real-time exploratory scientific visualization and virtual reality. In *Scientific Visualization: Advances and Challenges.* (L. Rosenblum, R.A. Earnshaw, J. Encarnacao, H. Hagen, A. Kaufman, S. Klimenko, G. Nielson, F. Post, and D. Thalmann, eds.), pp. 65–85. Academic Press, London.

Bryson, S., and Levit, C. (1991). The Virtual Wind Tunnel: An environment for the exploration of three dimensional unsteady flows. *Visual., Proc., IEEE Conf. Visual.* (sponsoring organization: IEEE Computer Society Technical Committee on Computer Graphics in cooperation with Association for Computing Machinery/SIGGRAPH).

Buchwald, J. (1996). *Theories of Practice. Stories of Practice.* University of Chicago Press, Chicago.

Carr, H., Möller, T., and Snoeyink, J. (2001). Simplicial subdivisions and sampling artifacts . *Proc. Visual., 2001*, pp. 99–106.

Carr, J.C., Beatson, R.K., Cherrie, J.B., Mitchell, T.J., Fright, W.R., McCallum, B.C., and Evans, T.R., (2001). Reconstruction and representation of 3D objects with radial basis functions. *Comput. Graphics Proc., Annu. Conf. Ser.*, Los Angeles, CA, *2001*, pp. 67–76.

Catmull, E., and Clark, J. (1978). Recursively generated B-Spline surfaces on arbitrary topological meshes. *Comput.-Aided Des.* **10**(6), 239–248.

Caudill, E. (1994). The bishop-eaters: The publicity campaign for Darwin and On the Origin of Species. *J. Hist. Ideas*, pp. 441–460.

CCDB. (2003). Cell Centered Database. Web page: http://donor.ucsd .edu/CCDB/visualize.shtml.

Chen, M., Townsend, P., and Vince, J.A. (1996). High performance computing for computer graphics and visualization. *Proc. Inter. Workshop High Perform. Comput. Comput. Graphics Visual.*

CISMM (Center for Computer Integrated Systems for Microscopy and Manipulation). (2003). Nanomanipulator. Web page: http://www.cs .unc.edu/Research/nano/cismm/nm/index.html

Cline, H.E, Dumoulin, C.L., Lorensen, W.E., Hart, H.R., and Ludke, S. (1987). 3D Reconstruction of the brain from magnetic resonance images using a connectivity algorithm. *Magn. Reson. Imag.* **5**(5), 345–352.

Cline, H.E, Lorensen, W.E., Ludke, S., Crawford, C.R., and Teeter, B.C. (1988). Two algorithms for the three-dimensional construction of tomograms. *Med. Phys.* **15**(3), 320–327.

Cline, H.E, Lorensen, W.E., Kikinis, R., and Jolesz, F. (1990). Three-dimensional segmentation of MR images of the head using probability and connectivity. *J. Comput. Assisted Tomogr.* **14**(6), 1037–1045.

Cline, H.E., Dumoulin, C.L., Lorensen, W.E., Souza, S.P., and Adams, W.J. (1991a). Volume rendering and connectivity algorithms for MR angiography. *Magn. Reson. Med.* **18**, 384–394.

Cline, H.E, Lorensen, W.E., Souza, S., Jolesz, F., Kikinis, R., Gerig, G., and Kennedy, T. (1991b). 3D surface rendered MR images of the brain and its vasculature. *J. Comput.-Assisted Tomogr.* **15**(2) 344–351.

Cornelis, G.C. (2000). Analogical reasoning in modern cosmological thinking. In *Metaphor and Analogy in the Sciences* (F. Hallyn, ed.), pp. 165–180. Kluwer Academic Publishers, Dordrecht, The Netherlands.

Cox, D.J. (1988). Using the supercomputer to visualize higher dimensions: An artist's contribution to scientific visualization. *Leonardo* **21**, 233–242.

Cox, D.J. (1989). The Tao of postmodernism: Computer art, scientific visualization, and other paradoxes. ACM SIGGRAPH '89 Art Show Catalogue, Comput. Art in Context. *Leonardo, Suppl. Issue*, pp. 7–12.

Cox, D.J. (1990). Scientific visualization: Mapping information. *AUSGRAPH '90 Proc.*, pp. 101–106.

Cox, D.J. (1991). Collaborations in art/science: Renaissance teams. *J. Biocommun.* **18**(2), 10–15.

Cox, D.J. (1996). Cosmic voyage: Scientific visualization for Imax film. SIGGRAPH '96 Visual. Proc., p. 129.

Cox, D.J. (2000). Creating the Cosmos: Visualization & Remote Virtual Collaboration. Consciousness Reframed.

Cox, D.J. (2001). The Art of Scientific Visualization, Data Representation, and Renaissance Teams, Informatics SeminarArrowsmith Project. University of Illinois at Chicago.

Cox, D.J. (2003). Algorithmic Art, Scientific Visualization, and Tele-Immersion: An Evolving Dialogue with the Universe. Art, Women and Technology. MIT Press, Cambridge, MA.

Cox, D.J., Patterson, R., and Thiebaux, M., inventors. (2000). Virtual reality 3D interface system for data creation, viewing and editing.

Crary, J. (1990). Techniques of the Observer: On Vision and Modernity in the Nineteenth Century. MIT Press, Cambridge, MA.

Cruz-Neira, C., Sandin, D.J., and DeFanti, T.A. (1993). Surround-screen projection-based virtual reality: The design and implementation of the CAVE. Comput. Graphics: Proc. SIGGRAPH '93.

Czernuszenko, M., Pape, D., Sandin, D., DeFanti, T.A., Dawe, G., and Brown, M. (1997). The ImmersaDesk and Infinity Wall Projection-Based Virtual Reality Displays. Comput. Graphics 32(2), 46–49.

Davis, P.J., and Rabinowitz, P. (1984). Methods of Numerical Integration. Academic Press, Orlando, FL.

Deering, M. (1992). High resolution virtual reality. Comput. Graphics: Proc. SIGGRAPH '92.

DeFanti, T., Brown, M., and Stevens, R. (1996). Virtual reality over high speed networks. IEEE Comput. Graphics Appl. 16(4), 42–43.

Dent-Read, C.H., Klein, G., and Eggleston, R. (1994). Metaphor in visual displays designed to guide action. Metaphor Symb. Act. 9(3), 211–232.

Desantillana, G. (1969). The role of art in scientific renaissance. In Critical Problems in the History of Science (M. Clagett, ed.), pp. 33–65. University of Wisconsin Press, Madison.

Dong, F., Clapworthy, G.J., and Krokos, M. (2001). Volume rendering of fine details within medical data. Proc. Visual. 2001, pp. 387–394.

Dorsey, J., Edelman, A., Legakis, J., Jensen, H.W., and Kohling Pedersen, H. (1999). Modeling and rendering of weathered stone. Comput. Graphics Proc., Annu. Conf. Ser., Los Angeles, CA, 1999, pp. 225–234.

Douglas, C.C., Cole, M., Efendiev, Y., Ewing, R., Ginting, V., Johnson, C.R., Jones, G., Lazarov, R., Shannon, C., and Simpson, J. (2003). Workshop Dyn. Data-Driven Appl. Syst., 2003.

Draaisma, D. (2000). Metaphors of Memory : A History of Ideas about the Mind. Cambridge University Press, Cambridge, MA.

Drebin, R.A., Carpenter, L., and Hanrahan, P. (1988). Volume rendering. Comput. Graphics Proc., Annu. Conf. Ser., Atlanta, GA, 1988, Vol. 22, No. 4, pp. 65–74.

Dubow, J. (2000). "From a view on the world to a point of view in it". Rethinking sight, space and the colonial subject. Interventions 2(1), 87–102.

Durand, F., Drettakis, G., Thollot, J., and Puech, C. (2000). Conservative visibility preprocessing using extended projections. Comput. Graphics Proc., Annu. Conf. Ser., New Orleans, LA, 2000, pp. 239–248.

Ebert, D., Rohrer, R., Shaw, C., Kukla, J., and Roberts, D.A. (2000). Procedural shape generation for multi-dimensional data visualization. Comput. Graphics 24(3), 375–384.

Eck, M., Derose, T., Duchamp, T., Hoppe, H., Lounsbery, M., and Stuetzle, W. (1995). Multiresolution analysis of arbitrary meshes. Comput. Graphics SIGGRAPH, p.173–182.

Ellisman, M. (2003). Multiscale imaging of the nervous system and BIRNing in the neurogrid: Integrating IT for biomedical research. Web page: http://www.nimh.nih.gov/neuroinformatics/martonandel lisman.cfm

Ellson, R., and Cox, D. (1988). Visualization of injection molding. simulation: J. Soc. Comput. Simul. 51(5), 184–188.

Fattal, R., and Lischinski, D. (2001). Variational classification for visualization of 3D ultrasound data. Proc. Visual., 2001, pp. 403–410.

Fedkiw, R., and Osher, S. (2002). Level Set Methods and Dynamic Implicit Surfaces. Springer, Berlin.

Fedkiw, R., Stam, J., and Jensen, H.W. (2001). Visual simulation of smoke. Comput. Graphics Proc., Annu. Conf. Ser., Los Angeles, CA, 2001, pp. 15–22.

Feekes, G.B. (1986). The Hierarchy of Energy Systems, From Atom to Society. Pergamon, Oxford.

Ferguson, E.S. (1977). The mind's eye: Nonverbal thought in technology. Science 197, 827–836.

Ferris, T. (1982). Galaxies. Stewart, Tabori and Chang, Publishers, New York.

Finnigan, P., Hathaway, A., and Lorensen, W. (1990). Merging CAT and FEM. Mech. Eng. 112(7), 32–38.

Foley, J., and Ribarsky, M.W. (1994). Next-generation data visualization tools. In Scientific Visualization Advances and Challenges (L. Rosenblum, R.A. Earnshaw, J. Encarnacao, H. Hagen, A. Kaufman, S. Klimenko, G. Nielson, F. Post, and D. Thalmann, eds.), pp. 103–127. Academic Press, London.

Folk, M., Pourmal, E., Schoof, L., and Miller, M. (2001). Sharable and scalable I/O solutions for high performance computing applications. Web page: http://hdf.ncsa.uiuc.edu/HDF5/papers/SC2001/SC01_tutorial/

Forceville, C. (2002). The identification of target and source in pictorial metaphors. J. Pragm. 34, 1–14.

Foster, I., and Kesselman, C., ed. (1999). The Grid: Blueprint for a New Computing Infrastructure. Morgan Kaufmann, San Francisco, CA.

Foxlin, E., Harrington, M., and Pfeifer, G. (1998). Constellation™: A wide-range wireless motion-tracking system for augmented reality and virtual set applications. Comput. Graph. Proc., Annu. Conf. Ser., Orlando, FL, 1998, pp. 371–378.

Francis, G., Goudeseune, C., Kaczmarski, H., Schaeffer, B., and Sullivan, J.M. (2003). ALICE on the eightfold way: Exploring curved spaces in an enclosed virtual reality theatre. Visual. Math. III.

Fried, M.P., Moharir, V.M., Shinmoto, H., Alyassin, A.M., Lorensen, W.E., Hsu, L., and Kikinis, R. (1999). Virtual laryngoscopy. Ann. Ontol. Rhinol., Laryngol. 108(3), 221–226 .

Friedhoff, R.M. (1989). The Second Computer Revolution Visualization. Harry N. Abrams, NewYork.

Fruhauf, T., Gobel, M., Haase, H., and Karlsson, K. (1994). Design of a flexible monolithic visualization system. In Scientific Visualization: Advances and Challenges (L. Rosenblum, R.A. Earnshaw, J. Encarnacao, H. Hagen, A. Kaufman, S. Klimenko, G. Nielson, F. Post, and D. Thalmann, eds.), pp. 265–285. Academic Press, London.

Gallop, J. (1994). Underlying data models and structures for visualization. In Scientific Visualization: Advances and Challenges (L. Rosenblum, R.A. Earnshaw, J. Encarnacao, H. Hagen, A. Kaufman, S. Klimenko, G. Nielson, F. Post, and D. Thalmann, eds.), pp. 239–250. Academic Press, San Diego, CA. Limited.

Gavriliu, M., Carranza, J., Breen, D.E., and Barr, A.H. (2001). Fast extraction of adaptive multiresolution meshes with guaranteed properties from volumetric data. Proc. Visual., 2001, pp. 295–302.

Geller, M., and Huchra, J. (1989). Mapping the universe. Science 246, 885, 897–904.

Gershon, N.D. (1990). Visualization of three-dimensional image processing of position emission tomography (PET) images. Proc. IEEE Visual. '90 Conf. Comput. Soci. Press, pp. 120–130.

Gershon, N.D. (1994). From perception to visualization. In Scientific Visualization: Advances and Challenges (L. Rosenblum, R.A. Earnshaw, J. Encarnacao, H. Hagen, A. Kaufman, S. Klimenko, G. Nielson, F. Post, and D. Thalmann, eds.), pp. 129–139. Academic Press, London.

Gillet, A., PyARTK. Web page: http://www.scripps.edu/pub/olson-web/pyartk/pyartk.html

Grande, L., and Rieppel, O. (1994). *Interpreting the Hierarchy of Nature: From Systematic Patterns to Evolutionary Process Theories.* Academic Press, San Diego, CA.

Green, K. (2000). *Passport to the Universe: Intergalactic Travel Is Still Science Fiction.* (But a new animated presentation at the American Museum of Natural History's Rose Center for Earth and Space gives visitors the most accurate depiction available of what it might be like out there.) National Center for Supercomputing Applications, Champaign, IL.

Greenia, M. (2000). *History of Computing,* 2nd ed., CD-ROM. Lexikon Services, San Francisco, CA.

Gregory, R.L. (1990). *Eye and Brain. The Psychology of Seeing.* Princeton University Press, Princeton, NJ.

Gross, M.H. (1994). Subspace methods for visualization of multiresolution data sets. In *Scientific Visualization: Advances and Challenges,* (L. Rosenblum, R.A. Earnshaw, J. Encarnacao, H. Hagen, A. Kaufman, S. Klimenko, G. Nielson, F. Post, and D. Thalmann, eds.), pp. 171–186. Academic Press, San Diego, CA.

Grunfeld, J. (2000). Mathematics and metaphor. *Prima-Philosophia* **13**(4), 323–332.

Grzeszczuk, R., Terzopoulos, D., and Hinton, G. (1998). NeuroAnimator: Fast neural network emulation and control of physics-based models. *Comput. Graph. Proc., Annu. Conf. Ser.,* Orlando, FL, *1998,* Vol. 24, pp. 9–20.

Guskov, I., Sweldens, W., and Schroer, P. (1999). Multiresolution signal processing for meshes. *Comput. Graphics Proc., Annu. Conf. Ser.,* Los Angeles, CA, *1999,* pp. 325–334.

Guthe, S., Wand, M., Gonser, J., and Straber, W. (2002). Interactive rendering of large volume data sets. *Visual. Proc.,* pp. 53–60.

Haber, R.B., and McNabb, D.A. (1990). Visualization idioms: A conceptual model for scientific visualization systems. In *Visualization in Scientific Computing* (G. Nielson et al., eds.), IEEE Press, New York, pp. 263–270.

Haber, R.B., Lucas, B., and Collins, N. (1991). A data model for scientific visualization with provisions for regular and irregular grids. *Proc. IEEE Visual. '91,* pp. 298–305.

Hahn, H., Preim, B., Selle, D., and Peitgen, H.-O. (2001). Visualization and interaction techniques for the exploration of vascular structures. *Proc. Visual., 2001,* pp. 395–402.

Hall, A.R. (1960). *The Scientific Revolution: 1500–1800: The Formation of the Modern Scientific Attitude.* Beacon Press, Boston.

Hallyn, F., ed. (2000). *Metaphor and Analogy in the Sciences.* Kluwer Academic Publishers, Dordrecht, The Netherlands.

Hanrahan, P. (1993). *Future Directions in 3D Medical Imaging. SIGGRAPH 93.* ACM Press, New York.

Hawkes, T. (1977). *Structuralism and Semiotics.* University of California Press, Berkeley.

He, T.S., Hong, L.C., Kaufman, A., and Pfister, H.P. (1996). Generation of transfer functions with stochastic search techniques. *Proc. IEEE Visual., 1996,* pp. 227–234.

Headley, C.R. (1997). Platonism and metaphor in the texts of mathematics: Godel and Frege on mathematical knowledge. *Man-and-World* **30**(4), 453–481.

Healey, C.G. (1996). Choosing effective colors for data visualization. *IEEE Visual. '96,* pp. 263–270.

Heims, S.J. (1987). *John Von Neumann and Norbert Wiener: Form Mathematics to the Technologies of Life and Death.* MIT Press, Cambridge, MA.

Hertxmann, A., Jacobs, C.E., Oliver, N., Curless, B., and Salesin, D.H. (2001). Image analogies. *Comput. Graphics Proc., Annu. Conf. Ser.,* Los Angeles, CA, *2001,* pp. 327–340.

Hesse, M. (1966). *Models and Analogies in Science.* University of Notre Dame Press, Notre Dame, IN.

Hesse, M. (1980). *Revolutions and Reconstructions in the History of Science.* Indiana University Press, Bloomington.

Hibbard, W., Paul, C., and Dyer, B. (1992). Display of scientific data structures for algorithm visualization. *Proc. IEEE Conf. Visual. '92,* pp. 139–144.

Hiltz, S.R., and Turoff, M. (1993). *The Network Nation.* MIT Press, Cambridge, MA.

Höhne, K.H., and Hanson, W.A. (1992). Interactive 3D segmentation of MRI amd CT volumes using morphological operations. *Comput.-Assisted Tomogr.* **16**(2), 285–294.

Höhne, K.H., Pommert, A., Riemer, M., Schiemann, T., Schubert, R., and Tiede, U. (1994). Medical volume visualization based on "Intelligent volumes." In *Scientific Visualization: Advances and Challenges* (L. Rosenblum, R.A. Earnshaw, J. Encarnacao, H. Hagen, A. Kaufman, S. Klimenko, G. Nielson, F. Post, and D. Thalmann, eds.), pp. 21–35. Academic Press, San Diego, CA.

Holton, G. (1973). *Thematic Origins of Scientific Thought.* Harvard University Press, Cambridge, MA.

Hong, L.C., Muraki, S., Kaufman, A., Bartz, D., and He, T. (1997). Virtual voyage: Interactive navigation in the human colon. *Comput. Graphics Proc., Annu. Conf. Ser.,* Los Angeles, CA, *1997,* pp. 27–34.

Hu, J., Razdan, A., Farin, G., and Nielson, G.M. (2003). Segmenting linear parts using layered region growing. *3D MODELLING Symp.* 8th ed., pp. 1–8.

Huang, A. (2003). *3D Biomedical Image Segmentation & Visualization: A Shape-Based Approach.* May 7, 2003 Ph.D. Thesis.

Huang, A., Nielson, G., Razdan, A., Farin, G., Capco, D., and Baluch, P. (2003). Line and net pattern segmentation using shape modeling. *Visual. Data Anal., 2003 (VDA 2003).*

Hubeli, A., and Gross, M. (2001). Multiresolution feature extraction for unstructured meshes. *Proc. Visual., 2001,* pp. 287–294.

Humphreys, G., Eldridge, M., Buck, I., Stoll, G., Everett, M., and Hanrahan, P. (2001). WireGL: A scalable graphics system for clusters. *Comput. Graphics Proc. Annu. Conf. Ser.,* Los Angeles, CA, *2001,* pp. 129–140.

Hurdal, M.K., Kurtz, K.W., and Banks, D.C. (2001). Case study: interacting with cortical flat maps of the human brain. *Proc. Visual., 2001,* pp. 469–472.

Ikits, M., Brederson, J.D., Hansen, C.D., and Johnson, C.R. (2003). A constraint-based technique for haptic volume exploration. *Proc. IEEE Visual., 2003,* pp. 263–269.

Indurkhya, B. (1992). *Metaphor and Cognition: An Interactionist Approach.* Kluwer Academic Publishers, Dordrecht, The Netherlands.

ISL. (2003). Web page: http://www.isl.uiuc.edu/

ITK. (2003). National Library of Medicine Insight Segmentation and Registration Toolkit. Web page: http://www.itk.org

Jakobsson, E., Natarajan, S., and Li, Q. (2003). NCSA Blast.

Jensen, H.W., and Christensen, P.H. (1998). Efficient simulation of light transport in scenes with participating media using photon maps. *Comput. Graphics Proc., Annu. Conf. Ser.,* Orlando, FL, *1998,* pp. 311–320.

Jensen, H.W., Durand, F., Stark, M.M., and Simon, P. (2001). A physically-based night sky model. *Comput. Graphics Proc., Annu. Conf. Ser.,* Los Angeles, CA, *2001,* pp. 399–408.

Jiang, T.-Y., Ribarsky, M.W., Wasilewski, T., Faust, N., Hannigan, B., and Parry, M. (2001). Acquistion and display of real-time atmospheric data on terrain. In *Data Visuaulization 2001* (D. Ebert, J.M. Favre, and R. Peikert, eds.), pp. 15–24. Springer-Verlag, Vienna and New York.

Johnson, C.R. (1998). Computer visualization in medicine. *Nat. Forum Fall,* pp. 17–21.

Johnson, C.R., and Sanderson, A.R. (2003). A next step: Visualizing errors and uncertainty. *IEEE Comput. Graphics Appl.* **23**(5), 6–10.

Johnson, C.R., MacLeod, R.S., and Matheson, M.A. (1993). Computational medicine: Bioelectric field problems. *IEEE Comput.,* pp. 59–67.

Johnson, C.R., Berzins, M., Zhukov, L., and Coffey, R. (1998). SCIRun: Applications to atmospheric diffusion using unstructured meshes. In

Numerical Methods for Fluid Dynamics VI (M. J. Baines, ed.). Oxford University Press, London and New York.

Johnson, C.R., Parker, S.G., Hansen, C., Kindlmann, G.L., and Livnat, Y. (1999). Interactive simulation and visualization. *IEEE Comput.* **32**(12), 59–65.

Johnson, C.R., Parker, S.G., and Weinstein, D. (2000). Large-scale computational science applications using the SCIRun problem solving environment. *Supercomputer 2000.*

Johnson, C.R., Livnat, Y., Zhukov, L., Hart, D., and Kindlmann, G. (2001). *Computational Field Visualization. Mathematics Unlimited— 2001 and Beyond*, Vol. 2, pp. 605–630.

Johnson, C.R., Parker, S., Weinstein, D., and Heffernan, S. (2002). Component-based problem solving environments for large-scale scientific computing. *Concurr. Comput.: Pract. Experi.* **14**.

Jolesz, F.A., Lorensen, W.E., Shinmoto, H., Atsumi, H., Nakajima, S., Kavanaugh, P., Saiviroonporn, P., Seltzer, S.E., Silverman, S.G., Phillips, M., and Kikinis, R. (1997). Interactive virtual endoscopy. *Am. J. Res.* **169**, 1229–1235.

Kaczmarski, H., and Zuffo, M.K. (2002). Commodity clusters for immersive projection environments. *SIGGRAPH 2002, SIGGRAPH Course No. 47: SIGGRAPH 2002 Course Notes.*

Kähler, R., and Hege, H.C. (2001). *Interactive Volume Rendering of Adapative Mesh Refinement Data.*, Internal Report, ZIB-Report 01–30.

Kähler, R., Cox, D., Patterson, R., Levy, S., Hege, H.C., and Abel, T. (2002). Rendering the first star in the universe - A case study. *Proc. Visual., 2002*, pp. 537–539.

Kanitsar, A., Fleischmann, D., Wegenkittl, R., Felkel, P., and Groller, M. E. (2002). CPR - Curved planar reformation. *Visual. Proc.*, pp. 37–44.

Kapadia, A., and Yeager, N. (1999). Performance evaluation of HDF version 4 Cunking facility when used for subsetting pathfinder data. Web page: http://hdf.ncsa.uiuc.edu/apps/dods/perfeval/report.html.

Kaufman, A. (1990). Volume visualization tutorial. *Visualization Proc.*

Kaufman, A., ed. (1991). *Volume Visualization.* IEEE Computer Society Press, Los, Alamitos, CA.

Kaufman, A. (1994). Trends in volume visualization and volume graphics. In *Scientific Visualization: Advances and Challenges* (L. Rosenblum, R.A. Earnshaw, J. Encarnacao, H. Hagen, A. Kaufman, S. Klimenko, G. Nielson, F. Post, and D. Thalmann, eds.), pp. 3–19. Academic Press, San Diego, CA.

Kaufman, A., Cohen, D., and Yagel, R. (1993). Volume graphics. *Computer* **26**(7), 51–64.

Kaufmann, W.J., III, and Smarr, L. (1993). *Supercomputing and the Transformation of Science.* Scientific American Library, United States.

Keller, E.F., and Longino, H.E. (1997). *Feminism & Science.* Oxford University Press, New York.

Keller, P.R., and Keller, M.M. (1993). *Visual Cues: Practical Data Visualization.* IEEE Society Press, Manning Publication, Los Alamitos, CA.

Kemp, M. (1996). Temples of the body and temples of the cosmos: Vision and visualization in the Vesalian and Copernican revolution. In *Picturing Knowledge: Historical and Philosophical Problems Concerning the use of Art in Science* (B.S. Baigrie, ed.), pp. 40–85. University of Toronto Press, Toronto.

Kergosien, Y.L. (1991). Generic sign system in medical imaging. *IEEE Comput. Graphics Appli.* **11**(5), 46–65.

Kikinis, R., Jolesz, F.A., Gerig, G., Sandor, T., Cline, H.E., Lorensen, W.E., Halle, M., and Benton, S.A. (1990). 3D morphometric and morphologic information derived from clinical brain MR images. *3D Imaging in Medicine*, pp. 441–454. Springer-Verlag, Berlin.

Kikinis, R., Jolesz, F.A., Lorensen, W.E., Cline, H.E., Stieg, P., and Black, P. (1991). Three-dimensional reconstruction of skull base tumors from MRI data for neurosurgical planning. *Soc. Magn. Reson. Med.*, p. 752.

Kindlmann, G., and Durkin, J.W. (1998). Semi-automatic generation of transfer functions for direct volume rendering. *IEEE Sympo. Vol. Visual.*, pp. 79–86.

Kirby, R.M., and Karniadakis, G.E.M. (2003). De-aliasing on non-uniform grids: Algorithms and applications. *J. Comput. Phys.* **191**, 267–282.

Kitware. (2003). Para view: Parallel visualization application. Web page: http://www.paraview.org/HTML/Index.html.

Kniss, J.M., Kindlmann, G., and Hansen, C. (2001a). Interactive volume rendering using multi-dimensional transfer functions and direct manipulation widgets. *Proc. Visual., 2001*, pp. 255–262.

Kniss, J.M., McCormick, P.S., McPherson, A., Ahrens., J., Painter, J., Keahey, A., and Hansen, C. (2001b). Interactive texture-based volume rendering for large data sets. *IEEE Comput. Graphics Appl.* **21**(4), 52–61.

Kniss, J.M., Kindlmann, G., and Hansen, C. (2002a) Multidimensional transfer functions for interactive volume rendering. *IEEE Trans. Visual. Comput. Graphics* **8**(3), 270–285.

Kniss, J.M., Premoze, S., Hansen, C., and Ebert, D. (2002b). Interactive translucent volume rendering and procedural modeling. *IEEE Visual., 2002.*

Kobbelt, L., Campagna, S., Vorsatz, J., and Seidel, H.P. (1998). Interactive multi-resolution modeling on arbitary Meshes. *Comput. graphics Proc., Annu. Conf. Ser.*, Orlando, FL, *1998*, pp. 95–104.

Koch, R.M., Gross, M.H., Carls, F.R., von Buren, D.F., Fankhauser, G., and Parish, Y.H. (1996). Simulating facial surgery using finite element methods. In *Comput. Graphics Proc., Annu. Conf. Ser.*, New Orleans, LA, *1996*, pp. 421–428.

Kuhn, T. (1970). *The Structure of Scientific Revolutions.* University of Chicago Press, Chicago, IL.

Kunii, T., and Shinagawa, Y. (1994). Research issues in modeling complex object shapes. In *Scientific Visualization: Advances and Challenges* (L. Rosenblum, R.A. Earnshaw, J. Encarnacao, H. Hagen, A. Kaufman, S. Klimenko, G. Nielson, F. Post, and D. Thalmann, eds.), pp. 495–502. Academic Press, San Diego CA.

Lacroute, P., and Levoy, M. (1994). Fast volume rendering using a shear-warp factorization of the viewing transformation. *Comput. Graphics Proc., Annu. Conf. Ser.*, Orlando, FL, *1994*, pp. 451–458.

Lafrance, M. (1992). Excavation, capture, collection, and creation: Computer scientists' metaphors of eliciting human expertise. *Metaphor Symb. Act.* **7**(384), 135–156.

Lakoff, G., and Johnson, M. (1980). *Metaphors We Live By.* University of Chicago Press, Chicago, IL.

Latour, B. (1986). Visualization and cognition: Thinking with eyes and hands. Knowledge and Society: *Stud. Sociol. Cult., Past Present* **6**, 1–40.

Latour, B. (1987). *Science in Action.* Harvard University Press, Cambridge, MA.

Latour, B., Mauguin, P., and Teil, G. (1992). A note on socio-technical graphs. *Soc. Stud. Sci.* **22**, 33–57.

Lefohn, A.E., Kniss, J.M., Hansen, C.D., and Whitaker, R.T. (2003a). *Interactive Deformation and Visualization of Level Set Surfaces Using Graphics Hardware*, UUCS-03-005. University of Utah School of Computing, Salt Lake City.

Lefohn, A.E., Cates, J.E., and Whitaker, R.T. (2003b). *Interactive, GPU-Based Level Sets for 3D Brain Tumor Segmentation*, UUCS-03-004. University of Utah School of Computing, Salt Lake City.

Leigh, J., Johnson, A.E., Vasilakis, C.A., and DeFanti, T.A. (1996). Multi-perspective collaborative design in persistent networked virtual environments. *IEEE Virtual Reality Annu. Int. Symp. (VRAIS), 1996*, pp. 253–260.

Leigh, J., Johnson, A., Brown, M., Sandin, D., and DeFanti, T. (1999). Visualization in teleimmersive environments. *IEEE Comput.*, pp. 67–73.

Lengyel, J., Greenburg, D.P., and Popp, R. (1995). Time-dependent three-dimensional intravascular ultrasound. *Comput. Graphics Proc., Annu. Conf. Ser.*, Los Angeles, CA, *1995*, 465–474.

Levin, A. (1999). Interpolating nets of curves by smooth subdivision surfaces. *Comput. Graphics Proc., Annu. Conf.* Ser., Los Angeles, CA, *1999*, pp. 57–64.

Lévy, B. (2001). Constrained texture mapping for polygonal meshes. *Comput. Graphics Proc., Annu. Conf. Ser.,* Los Angeles, CA, *2001*, pp. 417–424.

Lévy, B., Cauman, G., Conreaux, S., and Cavin, X. (2001). Circular incident edge lists: A date structure for rendering complex ultrastructured grids. *Proc. Visual. 2001*, pp. 191–198.

Livnat, Y., Hansen, C.D., Parker, S.G., and Johnson, C.R. (2001). Isosurface extraction for large-scale datasets. *Proc. Sci. Visual. Dagstuhl'2000*, pp. 59–60.

Lorensen, W. (1995). *Marching Through the Visible Man.* IEEE Press, New York.

Lorensen, W.E., and Cline, H.E. (1987). Marching cubes: A high resolution 3D surface construction algorithm. *Comput. Graphics* **21**(3), 163–169.

Lorensen, W., Cline, H., Nafis, C., Kikinis, R., Altobelli, D., and Gleason, L. (1993). Enhancing reality in the operating room. *Visualization '93*, pp. 410–415.

Lorensen, W., Jolesz, F.A., and Kikinis, R. (1995). The exploration of cross-sectional data with a virtual endoscope. *Interact. Technol. New Health Paradigm*, pp. 221–230.

Lum, E.B., Ma, K.-L., and Clyne, J. (2002). A hardware-assisted scalable solution for interactive volume rendering of time-varying data. *Visual. Proc.*, pp. 286–301.

Lynch, M., and Woolgar, S. (1990). *Representation in Scientific Practice.* MIT Press, Cambridge, MA.

Machiraju, R., Zhifan, Z., Fry, B., and Moorhead, R. (1998). Structure-significant Representation of structured datasets. *Visual. Proc.*, pp. 117–132.

Mackinlay, J. (1986). Automating the design of graphical presentations of relational information. *ACM Trans. Graphics* **5**(2), 110–141.

Malzbender, T., Gelb, D., and Wolters, H. (2001). Polynomial texture maps. *Comput. Graphics Proc., Annu. Conf. Ser.*, Los Angeles, CA, *2001*, 519–528.

Marcotte, E.M., Xenarios, I., Van der Bliek, A.M., and Eisenberg, D. (1999). Localizing proteins in the cell from their phylogenetic profiles. *Proc. Natl. Acad. Sci. U.S.A.* **97**(22), 12115–12120.

Martin, E. (1991). The egg and the sperm: How science has constructed a romance based on stereotypical male-female roles. *Signs: J. Women Cult. Soc.* **16**, No. 3.

Martone, M. (2003). Navigation through multiresolution brain imaging data using knowledge-based mediation. Web page: http://www.nimh .nih.gov/neuroinformatics/martoneandellisman.cfm

Martone, M.E., Gupta, A., Wong, M., Qian, X., Sosinsky, G., Ludaescher, B., and Ellisman, M.H. (2002). A cell centered database for electron tomographic data. *J. Struct. Biol.* **138**, 145–155.

Mazziotta, J.C., Toga, A.W., Evans, A.C., Fox, P., and Lancaster, J.A. (1995). A probablistic atlas of the human brain: Theory and rationale for its development. *NeuroImage* **2**, 89–101.

MCAT (2003). The Metadata Catalog. Web page: http://www.npaci .edu/DICE/SRB/mcat.html

McCormick, B.H., DeFanti, T.A., and Brown, M.D. (1987). Visualization in scientific computing. *Comput. Graphics* **21**, No. 6.

McNeely, W.A., Puterbaugh, K.D., and Troy, J. (1999). Six degrees-of-freedom haptic rendering using voxel sampling. *Comput. Graphics Proc., Annu. Conf. Ser.*, Los Angeles, CA, *1999*, pp. 401–408.

Mead (2003). Modeling Environment for Atmospheric Discovery (MEAD). Web page: http://www.ncsa.uiuc.edu/expeditions/MEAD/

Miller, J.V, Breen, D.E., Lorensen, W.E, O'Bara, R., and Wozny, M.J. (1991). Geometrically deformed models: A method for extracting closed geometric models from volume data. *Comput. Graphics* **25**(3), 217–226.

Mine, M.R., Brooks, F.P., and Sequin, C.H. (1997). Moving objects in space: Exploiting proprioception in virtual-environment interaction. *Comput. Graphics Proc., Annu. Conf. Ser.*, Los Angeles, CA, *1997*, pp. 19–26.

Moharir, V.M., Fried, M.P., Vernick, D.M., Janecka, I.P., Zahajsky, J., Hsu, L., Lorensen, W.E., Anderson, M., Wells, W.M., Morrison, P., and Kikinis, R. (1998). Computer-assisted three-dimensional reconstruction of head and neck tumors. *Laryngoscope* **108**, 1592–1598.

Moore, R.W., Prince, T.A., and Ellisman, M. (1998). Data-intensive computing and digital libraries. *Communi. ACM* **41**(11), 56–62.

Moser, S. (1993). Gender stereotyping in pictorial reconstructions of human origins. In *Women in Archaeology: A Feminist Critique* (H. Smith and L. DuCros, eds.). Australian National University Press, Canberra, pp. 80–95.

Mulaik, S.A. (1995). The metaphoric origins of objectivity, subjectivity, and consciousness in the direct perception of reality. *Philos. Sci.* **62**(2), 283–303.

Muraki, S. (1995). Multiscale volume representation by a DoG wavelet. *Visual. Proc.*, pp. 109–116.

Nakagohri, T., Jolesz, F.A., Okuda, S., Asano, T., Kenmochi, T., Kainuma, O., Tokoro, Y., Aoyama, H., Lorensen, W.E., and Kikinis, R. (1998). Virtual pancreatoscopy of mucin-producing pancreatic tumors. *Comput.-Aided Surg.* **3**, 264–268.

NCBI. (2003). National Center for Biotechnology Information. Web page: http://www.ncbi.nlm.nih.gov/BLAST/

NCMIR. (2003). National Center for Microscopy and Imaging Research. Web page: http://ncmir.ucsd.edu/

NCSA. (1999). HDF, Hierarchical Data Format Web Site. Web page: http://hdf.ncsa.uiuc.edu

NCSA. (2003). Tiled display wall. Web page: http://www.ncsa.uiuc.edu/ Projects/AllProjects/Projects83.html

Neyret, F., and Marie-Paule, C. (1999). Pattern-based texturing revisited. *Comput. Graphics Proc., Annu. Conf. Ser.,* Los Angeles, CA, *1999*, pp. 235–242.

Nielson, G.M. (1993a). Modeling and visualizing volumetric and surface-on-surface data. In *Focus on Scientific Visualization* (H. Hagen, H. Mueller, and G. Nielson, eds.), pp. 219–274. Springer, Berlin.

Nielson, G.M. (1993b). Scattered data modeling. *IEEE Comput. Graphics Appli.* **13**, 60–70.

Nielson, G.M. (1994). Research issues in modeling for the analysis and visualization of large data sets. In *Scientific Visualization: Advances and Challenges* (L. Rosenblum, R.A. Earnshaw, J. Encarnacao, H. Hagen, A. Kaufman, S. Klimenko, G.M. Nielson, F. Post, and D. Thalmann, eds.), pp. 143–155. Academic Press, San Diego, CA.

Nielson, G.M., and Hamann, B. (1991). The asymptotic decider: Removing the ambiguity in marching cubes. In *IEEE Visualization '91* (G.M. Nielson and L.J. Rosenblum, eds.), pp. 83–91. IEEE Computer Society Press, Los Alamitos, CA.

NLM. (2003). *The Visible Human Project*®. National Library of Medicine (NLM), Washington, DC.

Norman, M., Shalf, J., Levy, S., and Daues, G. (1999). Diving deep: Data-management and visualization strategies for adaptive mesh refinement. *Comput. Sci. Eng.* **1**(4), 22–32.

Novacek, M.J. (1994). Morphological and molecular inroads to phylogeny. In *Interpreting the Hierarchy of Nature: From Systematic Patterns to Evolutionary Process Theories* (G. Lance and R. Olivier Grande, eds.), pp. 85–131. Academic Press, San Diego, CA.

NPACI. (2003). Web page: http://www.npaci.edu/DICE/Pubs/

Nunez, R.E. (2000). Conceptual metaphor and the embodied mind: What makes mathematics possible? In *Metaphor and Analogy in the Sciences* (F. Hallyn, ed.), pp. 125–145. Kluwer Academic Publishers, Dordrecht, The Netherlands.

O'Hara, R. (1996). Representations of the natural system in the nineteenth century. In *Picturing Knowledge: Historical and Philosophical Problems Concerning the use of Art in Science* (B.S. Baigrie, ed.), pp. 164–183. University of Toronto Press, Toronto.

Ortony, A. (1993). *Metaphor and Thought*. Cambridge University Press, Cambridge, London and New York.

Palaniappan, A., and Jakobsson, E. (2004). Understanding classification, evolution, and structural conservation of K+ channels via human genome-wide analysis of permeation pathway. *Biophysical Journal* (submitted for publication).

Patrikalakis, N.M., ed. (1991). *Scientific Visualization of Physical Phenomena*. Springer-Verlag, Tokyo.

Pausch, R., Proffitt, D., and Williams, G. (1997). Quantifying immersion in virtual reality. *Comput. Graphics Proc., Annu. Conf. Ser.*, Los Angeles, CA, *1997*, pp. 13–18.

Peltier, S.T., Lin, A.W., Lee, D., Mock, S., Lamont, S., Molina, T., Wong, M., Dai, L., Martone, M.E., and Ellisman, M.H. (2003). The telescience portal for advanced tomography applications. *J. Parallel Distrib. Comput.: Comput. Grids.*

Perkins, G., Renken, C., Martone, M.E., Young, S.J., Ellisman, M., and Frey, T. (1997). Electron tomography of neuronal mitochondria: Three-dimensional structure and organization of cristae and membrane contacts. *J. Struct. Bio.* **119**, 260–272 .

Peterfreund, S. (1994). Scientific models in optics: From metaphor to metonymy and back. *J. Hist. Ideas* **55**(1), 59–73.

Pfister, H., Hardenbergh, J., Knittel, J., Lauer, H., and Seiler, L. (1999). The volume pro real-time ray-casting system. *Comput. Graphics Proc., Annu. Conf. Ser.*, Los Angeles, CA, *1999*, pp. 251–260.

Popper, D. (1980). To reason by means of images: J.J. Thomason and the mechanical picture of nature. *Ann. Sci.* **37**, 31–57.

Preim, B., Tietjen, C., Spindler, W., and Peitgen, H.-O. (2002). Integration of measurement tools in medical 3d visualizations. *Visual. Proc.*, pp. 21–28.

Prohaska, S., and Hege, H.-C. (2002). Fast visualization of plane-like structures in voxel data. *Visual. Proc.*, pp. 29–36.

Razdan, A., Rowe, J., Tocheri, M., and Sweitzer, W. (2002). Adding semantics to 3D digital libraries. *5th Int. Conf. Digital Libr. (JCDL 2002)*.

Reed, D.A. (2003). Grids, the TeraGrid, and beyond. *IEEE Comput.* **36**(1), 62–68.

Rew, R.K., and Davis, G. (1990). NetCDF: An interface for scientific data access. *IEEE Comput. Graphics Appl.* **10**(3), 76–82.

Ribarsky, M.W., Ayers, E.J., and Mukherjea, S. (1993a). Using glyphmaker to create customized visualizations of complex data, GIT GVU-93–25.

Ribarsky, M.W., Hodges, L.F., Minsk, R., and Bruchez, M. (1993b). Visual representations of complex, discrete multivariate data. *Visual Comput.*

Riley, K., and David, E. (2003). Visually accurate weather visualization. Web page: https://engineering.purdue.edu/people/kirk.j.riley.1/weatherproject

Robertson, P.K. (1991). A methodology for choosing data representations. *IEEE Comput. Graphics Appl.* **11**(3), 56–67.

Rogowitz, B.E., and Kalvin, A.D. (2001). The "Which Blair Project": A quick visual method for evaluating perceptual color maps. *Proc. Visual., 2001*, pp. 183–190.

Rogowitz, B.E., and Treinish, L. (1993). Data structures and perceptual structures. *SPIE Proc.: Hum. Vision, Visual Process. Digital Display*, ACM vol 1913 p. 102–105.

Rosch, E., and Lloyd, B. (1978). *Cognition and Categorization*. Wiley, New York.

Rosenblum, L., and Kamgar-Parsi, B. (1994). Progress and problems in ocean visualization. In *Scientific Visualization: Advances and Challenges* (L. Rosenblum, R.A. Earnshaw, J. Encarnacao, H. Hagen, A.

Kaufman, S. Klimenko, G. Nielson, F. Post, and D. Thalmann, eds.), pp. 435–454. Academic Press, San Diego, CA.

Rosenblum, L., Earnshaw, R.A., Encarnacao, J., Hagen, H., Kaufman, A., Klimenko, S., Nielson, G., Post, F., and Thalmann, D., eds. (1994). *Scientific Visualization: Advances and Challenges*. Academic Press, San Diego, CA.

Rudwick, M.J.S. (1976). The emergence of a visual language for geological science 1760–1840. *Hist. Sci.* **14**, 149–195.

Ruse, M. (1996). Are pictures really necessary? The case of Sewall wright's 'Adaptive Landscapes.' In *Picturing Knowledge: Historical and Philosophical Problems Concerning the use of Art in Science* (B.S. Baigrie, ed.), pp. 303–337. University of Toronto Press, Toronto.

Sacks, S. (1979). *On Metaphor*. University of Chicago Press, Chicago, IL.

Saier, M. (2000). A functional-phylogenetic classification system for transmembrane solute transporters. *Microbiol. Mol. Biol. Rev.* **64**, 354–411.

Samsonov, A.A., Kholmovski, E.G., and Johnson, C.R. (2003). Image reconstruction from sensitivity encoded MRI data using extrapolated iterations of parallel projections onto convex sets. *SPIE Med. Imag.* (accepted for publication).

Sander, P.V., Snyder, J., Gortler, S.J., and Hoppe, H. (2001). Texture mapping progressive meshes. *Comput. Graphics Proc., Annu. Conf. Ser.*, Los Angeles, CA, *2001*, pp. 409–416.

Schroeder, W., Volpe, C., and Lorensen, W. (1991). *The Stream Polygon: A Technique for 3D Vector Field Visualization*, pp. 126–132. IEEE Press, New York.

Schroeder, W., Lorensen, W., Montanaro, G., and Volpe, C. (1992a). *Visage: An Object-Oriented Scientific Visualization System*, pp. 219–261. IEEE Press, New York.

Schroeder, W., Zarge, J., and Lorensen, W. (1992b). Decimation of triangle meshes. *Comput. Graphics* **26**(2), 65–70.

Schroeder, W., Lorensen, W., and Linthicum, S. (1994). Implicit modeling of swept surfaces and volumes. *Visualization*, pp. 40–45.

Schroeder, W., Martin, K., and Lorensen, W. (1997). *The Visualization Toolkit: An Object-Oriented Approach to 3D Graphics*. Prentice Hall, Upper Saddle River, NJ.

Schroeder, W., Martin, K., and Lorensen, W. (1998). *The Visualization Toolkit, An Object-Oriented Approach to 3D Graphics*. Prentice Hall, Upper Saddle River, NJ.

Schumaker, L. (1976). Fitting surfaces to scattered data. In *Approximate Theory II* (C.K. Chui, L.L. Schumaker, and G.G. Lorentz, eds.), pp. 203–268. Academic Press, New York.

SCI (2003). Scientific Computing and Imaging Institute, University of Utah. Web page: http://www.sci.utah.edu/.

SCIRun (2001). SCIRun: A Scientific Computing Problem Solving Environment. Web page: http://software.sci.utah.edu/scirun.html

Sethian, J.A. (1999). *Level Set Methods and Fast Marching Methods Evolving Interfaces. Computational Geometry, Fluid Mechanics, Computer Vision, and Materials Science*. Cambridge University Press, London and New York.

Shaw, C., Hall, J., Ebert, D., and Roberts, A. (1999). Interactive lens visualization techniques. *IEEE Visual. '99*, pp. 155–160.

Shenglan, Z., Price, D., Qian, X., Gupta, A., Ellisman, M.H., and Martone, M.E. (2003). A cell centered database (CCDB) for multiscale microscopy data management. *Microsc. Microanal., 2003*.

Shepherd, G., Mirsky, J., Healy, M., Singer, M., Skoufos, E., Hines, M., Nadkarni, P., and Miller, P. (1998). The Human Brain Project: Neuroinformatics tools for integrating, searching and modeling multidisciplinary neuroscience data. *Trends Neurosci.* **21**, No. 11.

Sherman, W., and Alan, C. (2003). *Understanding Virtual Reality: Interface, Application, and Design*. University of California, Berkeley.

Shirley, J.W., and Hoeniger, F.D., ed. (1978). *Science and the Arts in the Renaissance*. Folger Shakespeare Library, Washington, DC.

Simon, H. (1962). The architecture of complexity. *Proc. Am. Philos. Soc.* **106**, 467–482.

Simon, H. (1977). What computers mean for man and society. *Science* **195**, 1186–1191.

Simpson, J., Sanderson, A., Luke, E., Balling, K., and Coffey, R. (2002). Collaborative remote visualization. *IEEE/ACM SC2002 Conf.*

SLBG. (2003). SLBG. Web page: http://tcdb.ucsd.edu/tcdb/background.php

Smith, C.S. (1981). *A Search for Structure: Selected Essays on Science, Art, and History.* MIT Press, Cambridge, MA.

SRB. (2003). The Storage Resource Broker Web Page. Web page: http://www.npaci.edu/DICE/SRB/

Stam, J. (1999). Diffraction shaders. *Comput. Graphics Proc., Annu. Conf. Ser.*, Los Angeles, CA, *1999*, pp. 101–110.

Stander, B.T., and Hart, J.C. (1997). Guaranteeing the topology of an implicit surface polygonization for interactive modeling. *Comput. Graphics Proc. Annu. Conf. Ser.*, Los Angeles, CAL, *1997*, pp. 279–286.

Stepan, N.L. (1986). Race and gender: The role of analogy in science. *ISIS* **77**, 261–277.

Stoev, S., and Strasser, W. (2001). A case study on interactive exploration and guidance aids for visualizing historical data. *Proc. Visual. 2001*, pp. 485–488.

Subramanian, K.R., and Naylor, B.F. (1997). Converting discrete images to partitioning trees. *Visual. Proc.*, pp. 273–288.

Sunguroff, A., and Greenberg, D. (1978). Computer generated images for medical applications. *Comput. Graphics Proc., Q. Rep. SIGGRAPH-ACM SIGGRAPH '78 Proc.*, *1978*, pp. 196–202.

Taubin, G., Gueziec, A., Horn, W., and Lazarus, F. (1998). Progressive forest split compression. *Comput. Graphics Proc., Annu. Conf. Ser.*, Orlando, FL, *1998*, pp. 123–132.

Taylor, R., Robinett, W., Chi, V., Brooks, F., and Wright, W. (1993). The nanomanipulator: A virtual reality interface for a scanning tunnelling microscope. *Comput. Graphics Proc. SIGGRAPH Conf. Ser.*, pp. 45–62.

Thiebaux, M. (1997). Virtual director: Steering scientific visualization with virtual camera choreography. Masters Thesis, Electronic Visualization Lab (EVL), University of Illinois at Chicago.

Tobler, W.R. (1978). Comparing figures by regression. *Comput. Graphics Proc., Q. Rep. SIGGRAPH-ACM SIGGRAPH '78 Proc.*, *1978*, pp. 193–195.

Toga, A.W., ed. (1998). *Brain Warping.* Academic Press, New York.

Toga, A.W., Thompson, P.M., Mega, P.M., Narr, K.L., and Blanton, R.E. (2001). Probablistic approaches for atlasing of normal and disease-specific brain variability. *Anat. Embryol.* **204**, 267–282.

Tory, M., Röber, N., Möller, T., Celler, A., and Atkins, M.S. (2001). 4D space-time techniques: A medical imaging case study. *Proc. Visual., 2001*, pp. 473–476.

Treinish, L.A. (1992). Unifying principles of data managment for scientific visualization. *Animation Sci. Visual.* British Computer Society Displays Group, London.

Treinish, L.A. (1993). Inside multidimensional data. *Byte* **18**(4), 132–133.

Tufte, E.R. (1983). *The Visual Display of Quantitative Information.* Graphics Press, Cheshire, CT.

Tufte, E.R. (1990). *Envisioning Information.* Graphics Press, Cheshire, CT.

Ullmer, B., Hiroshi, I., and Glas, D. (1998). mediaBlocks: Physical containers, transports, and controls for online media. *Comput. Graphics Proc., Annu. Conf. Ser.*, Orlando, FL, *1998*, pp. 379–386.

Upson, C., and Keeler, M. (1988). VBUFFER: Visible volume rendering. *Comput. Graphics Proc., Annu. Conf. Ser.*, Atlanta, GA, *1988*, Vol. 22, No. 4, pp. 59–64.

Upson, C., Faulhaber, T., Kamnis, D., Laidlaw, D., Schlegel, D., Vroom, J., Gurwitz, R., and Van Da, A. (1989). The application visualization system: A computational environment for scientific visualization. *IEEE Comput. Graphics Appl.*, pp. 30–42.

Usoh, M., Arthur, K., Whitton, M.C., Steed, R., Bastos, A., Slater, M., and Brooks, F.P., Jr. (1999). Walking > walking-in-place > flying, in virtual environments. *Comput. Graphics Proc., Annu. Conf. Ser.*, Los Angeles, CA, *1999*, pp. 359–364.

Van Bendegem, J.P. (2000). Analogy and metaphor as essential tools for the working mathematician. In *Metaphor and Analogy in the Sciences.* (F. Hallyn, ed.), pp. 105–123. Kluwer Academic Publishers, Dordrecht, The Netherlands.

Vandervert, L.R. (2001). How working memory and cognitive modeling functions of the cerebellum contribute to discoveries in mathematics. *New Ideas Psycho.*

Van Uitert, R., and Johnson, C.R. (2003). Influence of brain conductivity on magnetoencephalographic simulations in realistic head models. *25th Annu. Int. Conf. IEEE Eng. Med. Biol. Soc.*

Van Walsum, T., Post, F.H., Silver, D., and Post, F.J. (1996). Feature extraction and iconic visualization. *IEEE Trans. Visual. Comput. Graphics* **2**(2), 111–119.

Varela, F.J., Thompson, E., and Rosch, E. (1997). *The Embodied Mind: Cognitive Science and Human Experience.* MIT Press, Cambridge, MA.

Viola, P., and Wells, W.M., III. (1997). Alignment by maximization of mutual information. *Int. J. Comput. Vision* **24**(2), 137–154.

VTK. (2003). Visualization Toolkit (VTK). Web page: http://public.kitware.com/VTK/

Walter, J., and Healey, C. (2001). Attribute preserving dataset simplification. *Proc. Visual. 2001*, pp. 113–120.

Ware, C. (2000). *Information Visualization: Perception for Design.* Academic Press, San Diego, CA.

Ware, C., and Beatty, J. (1988). Using color dimensions to display data dimensions. *Hum. Factors* **20**(2), 127–142.

Wartell, Z., Hodges, L.F., and Ribarsky, W. (1999). Balancing fusion, image depth and distortion in stereoscopic head-tracked displays. *Comput. Graphics Proc., Annu. Conf. Ser.*, Los Angeles, CA, *1999*, pp. 351–358.

Weber, G.H., Kreylos, O., Ligocki, T.J., Shalf, J.M., Hagen, H., Hamann, B., and Joy, K.I. (2001). Extraction of crack-free isosurfaces from adaptive mesh refinement data. In *Data Visuaulization 2001.* (D. Ebert, J.M., Favre, and R. Peikert, eds.), pp. 25–34. Springer-Verlag, Vienna and New York.

Weiler, M., and Ertl, T. (2001). Hardware-software-balanced resampling for the interactive visualization of unstructured grids. *Proc. Visual., 2001*, pp. 199–206.

Weinstein, D.M., Potts, G., Tucker, D., and Johnson, C.R. (1998). The SCIRun inverse EEG pipeline—a modeling and simulation system for cortical mapping and source localization. 1st *Int. Conf. Med. Image Comput. Comput.-Assisted Intervention.*

Weinstein, D.M., Zhukov, L., and Johnson, C.R. (2000). An inverse EEG problem solving environment and Its applications to EEG source localization. *NeuroImage, Suppl.*, p. 921.

Welch, W., and Witkin, A. (1994). Free-form shape design using triangulated surfaces. *Comput. Graphics Proc., Annu. Conf. Ser.*, Orlando, FL, *1994*, pp. 247–256.

Welsh, T., Mueller, K., Zhu, W., Volkow, N., and Meade. J. (2001). Graphical strategies to convey functional relationships in the human brain: A case study. *Proc. Visual. 2001*, pp. 481–484.

Westermann, R., and Ertl, T. (1998). Efficiently using graphics hardware in volume rendering applications. *Comput. Graphics Proc., Annu. Conf. Ser.*, Orlando, FL, *1998*, pp. 169–177.

Whitaker, R., Breen, D., Museth, K., and Soni, N. (2001). Segmentation of biological volume datasets using a level set framework. *Vol. Graphics*, pp. 249–263.

Whyte, L.L. (1949). *The Unitary Principle in Physics and Biology.* Holt, New York.

Whyte, L.L. (1951). *Aspects of Form: A Symposium on Form in Nature and Art.* Percy Lund Humphries.

Whyte, L.L. (1954). *Accent on Form: An Anticipation of the Science of Tomorrow.* Harper, New York.

Whyte, L.L., Wilson, A., and Wilson, D. (1969). *Hierarchical Structures.* American Elsevier, New York.

Wilhelmson, R. (1988). *High-Speed Computing: Scientific Applications and Algorithm Design.* University of Illinois Press, Urbana.

Wilson, A.G. (1967). A hierarchical cosmological model. *Astron. J.* **72**, 326.

Wilson, A. (1969). Hierarchical structure in the cosmos. In *Hierarchical Structures* (L.L. Whyte, A. Wilson, and D. Wilson, eds.), pp. 113–134. American Elsevier, New York.

Winckler, M.G., and Van Helden, A. (1992). Representing the Heavens: Galileo and visual astronomy. *Isis* **83**, 195–217.

Woolman, M. (2002). *Digital Information Graphics.* Watson-Guptill Publ., New York.

Yeh, P.-S., Serafino, W.-X., Lowell, M., Kobler, B., and Menasce, D. (2001). Implementation of CCSDS lossless data compression in HDF. *Earth Sci. Technol. Conf.*

Yu, D.F., and Fessler, J.A. (1998). Edge-preserving tomographic reconstruction with nonlocal regularization. *IEEE Int. Conf. Image Process.*, pp. 29–33.

Zhang, H.-S., Manocha, D., Hudson, T., and Hoff, K.E., III. (1997). Visibility culling using hierarchical occlusion maps. *Comput. Graphics Proc., Annu. Conf. Ser.*, Los Angeles, CA, *1997*, pp. 77–88.

Zhang, Y.-W., Rohling, R., and Pai, D.K. (2002). Direct surface extraction from 3D freehand ultrasound images. *Visual. Proc.*, pp. 45–52.

Zorin, D., Schroder, P., and Sweldens, W. (1997). Interactive multiresolution mesh editing. *Comput. Graphics Proc., Annu. Conf. Ser.*, Los Angeles, CA, *1997*, pp. 259–268.

Zyda, M., Cox, D., Katz, W., Larson-Mogal, U., Lypaczewski, P., Pausch, R., Singer, A., and Weisman, J. (1997). *Modeling and Simulation: Linking Entertainment and Defense.* National Academy Press, Washington, DC.

Biology Workbenches

Andreia Maer, Ph.D., Brian Saunders, Ph.D., Roger Unwin, and Shankar Subramaniam, Ph.D.

CONTENTS

Databasing the Brain. Edited by Stephen H. Koslow and Shankar Subramaniam
ISBN 0-471-30921-4 © 2005 John Wiley & Sons, Inc.

9.1 INTRODUCTION

9.1.1 What Is a Workbench?

Workbench systems provide a framework for integrating different applications/tools and databases into a single, cohesive, intuitive interface. This framework facilitates the traffic of data to and from the subapplication. The Workbench is responsible for providing a mechanism to allow the users to select/input what objects (e.g. sequences, images) they wish to pass to the application and what parameters they wish to use. The Workbench also handles the presentation/display of the application output in a way that makes the visualization easier, as well as extracting any objects from the output that could be useful and lead to further analysis. The primary goal of a workbench system is to provide computational analysis support to the users.

9.1.2 Problem-Solving Environment

Workbenches consist of a number of problem-solving environments. Primary interface into a workbench will be the real-time interaction interface. This is the system that allows a user to put their data into the system as well as manipulate using included tools. After this problem-solving environment comes the pipeline analysis environment. This environment allows the user to define a set of tools through which the stream of objects to be analyzed will be passed. This pipeline can consist of any number of tools and may span multiple machines and architectures. As the user sets up the pipeline, he will not only decide on the tools, and the order to execute them, but also the parameters, the output filters and destinations, and the process flows. The interactive problem-solving environment provides an interface for the construction and configuration of these batched pipeline systems. Lastly in the problem solving environments is the custom presentation system. This is a system that allows a user to present the results of his analysis in a polished human readable format complete with a customized search interface.

9.1.3 How Will They Be Useful in Biology?

The largest barrier in the utilization of extant data and bioinformatics tools by experimental biologists today is the scarcity of easy-to-use interoperable mechanisms. There is an increased need of strong infrastructure foundations to deal with large volume of heterogeneous data, and of complex computational tools along with easy to use Web interfaces. In addition to that, the databases are very diverse in terms of organization, content, and format, and there are few repositories of these databases that allow querying across multiple databases over the Web. Often the biologists have to manipulate and re-format the output of one program/tool in order to make the visualization of the results more useful and easier, or in order to submit it for analysis by another program.

An ideal workbench analysis environment would provide *interoperability* between programs and databases to present the user with a *single unified interface* and make available computational resources which were previously unavailable. The current level of communication infrastructure that would be utilized is the Web, which provides access to almost every important database. The access to the tools would be *seamless*, with automatic reformatting of the data to suit the analysis program input. It should also be able to process the output of one program to create input for the next program in pipeline. The interface should be designed in such a way that is easy to use and biologist-friendly.

9.2 ARCHITECTURE AND DESIGN OF WORKBENCHES

The biggest challenge toward creation of such workbenches (database and analysis systems) lies in integrating various databases and analysis tools. In addition to *interoperability*, one may also want the system to be *dynamic*, in that it updates regularly and does routine bookkeeping and automatic analysis. The need for a robust infrastructure results in some general design principles for the implementation. First, the reliability and efficiency of data processing are enhanced by minimizing the human intervention in routine steps, such as querying databases. Therefore, as much *automation* as is reasonably possible should be implemented. Second, the need for open access to the database via the Internet suggests that the interface should be built with *Web-browser-based* design. Third, one should expect that the pertinent data would increase in a nonlinear scale and, therefore, make the system *scalable* to effectively manage changes in data volume without extensive software modifications. Fourth, one should also expect that with the development of novel experimental protocols, new data types might be acquired. Therefore, one should also try to make the system *extensible*, to accommodate the creation and processing of new data types.

We will describe briefly different types of architectures first, then the components of the architectures (databases, tools, interfaces).

9.2.1 Types of Architecture

Nonclient-server Architectures

Mainframe Architecture. With mainframe software architectures all intelligence is within the central host computer. Users interact with the host through a terminal that captures keystrokes and sends that information to the host. One of the advantages of Mainframe is that Mainframe software architectures are not tied to a hardware platform. User can interact using PCs and UNIX workstations.

Limitation of mainframe software architectures is that they do not easily support graphical user interfaces (GUI) or access to multiple databases; produce substantial network traffic; require a complex operating system; are expensive to maintain. However, in the last few years, mainframes have found a new use as a *server* in client-server architectures.

File-Sharing Architecture. The original PC networks were based on file-sharing architectures, where the server downloads files from the shared location to the desktop environment. The requested user job is then run (including logic and data) in the desktop environment. A disadvantage of file-sharing architectures is that they could only work efficiently if shared usage is low, update contention is low, and the volume of data to be transferred is low. One of the advantages is that they used PCs, which could be hooked up to LANs (local area networks) and supported GUI's. The client-server architecture emerged as a result of the limitations of file sharing architectures.

Single-Tier Architecture. Single-tier architecture has the interface, business logic, and database highly coupled. All of the data and all of the program logic typically ran off of the same machine. Any changes made to one program require changes being made to all others. Pros: ease of administration; no network traffic between Application and Database servers; ease of maintenance; lower hardware costs. Cons: it lacks flexibility; scalability is achieved by investing in single, costly scalable server.

Client-server Software Architecture. Client-server software architecture is a versatile, message-based and modular infrastructure that is intended to improve *usability, flexibility, interoperability,* and *scalability* as compared to centralized, mainframe, time-sharing computing. A *client* is defined as a requester of services, and a *server* is defined as the provider of services. A single machine can be both a client and a server depending on the software configuration. As a result of the limitations of file-sharing architectures, the client-server architecture emerged. This approach introduced a database server to replace the file server. Using a relational database management system (DBMS), user queries could be answered directly. The client-server architecture reduces network traffic by providing a query response rather than total file transfer. It improves multi-user updating through a GUI front end to a shared database. In client-server architectures, Remote Procedure Calls (RPCs) or standard query language (SQL) statements are typically used to communicate between the client and server.

Two-Tier Architectures. Two-tier architectures consist of three components distributed in two layers: client and server. The three components are as follows:

- User System Interface (such as session, text input, dialog, and display management services)
- Processing Management (such as process development, process enactment, process monitoring, and process resource services)
- Database Management (such as data and file services)

The two-tier design allocates the user system interface exclusively to the client. It places database management on the server and splits the processing management between client and server, creating two layers. With two-tier client-server architectures, the user system interface is usually located in the user's desktop environment and the database management services are usually in a server that is a more powerful machine that services many clients. The database management server provides stored procedures and triggers. The two-tier client-server architecture is a good solution for distributed computing when work groups are defined as a dozen to 100 people interacting on a LAN simultaneously.

The two-tier architecture has a number of limitations. When the number of users exceeds 100, performance begins to deteriorate. This limitation is a result of the server maintaining a connection via "keep-alive" messages with each client, even when no work is being done. A second limitation of the two-tier architecture is that implementation of processing management services using vendor proprietary database procedures restricts flexibility and choice of DBMS for applications. Finally, current implementations of the two-tier architecture provide limited flexibility in moving (repartitioning) program functionality from one server to another without manually regenerating procedural code.

Two-tier software architectures are used extensively in non-time critical information processing where management and operations of the system are not complex, where the transaction load is light and there is minimal operator intervention required. The two-tier architecture works well in relatively homogeneous environments with processing rules (business rules) that do not change very often and when workgroup size is expected to be fewer than 100 users, such as in small businesses.

Three-Tier Architectures. The three-tier architecture (also referred to as the multi-tier architecture) emerged to overcome the limitations of the two-tier architecture. In the three-tier architecture, a middle tier was added between the user system interface client environment and the database management server environment. This middle tier provides process management where business logic and rules are executed and can accommodate hundreds of users (as compared to only 100 users with the two-tier architecture) by providing functions such as queuing, application execution, and database staging. The three-tier architecture is used when an effective distributed client-server design is needed that provides (when compared to the two-tier) increased performance, flexibility, maintainability, reusability, and scalability, while hiding the complexity of distributed processing from the user. There are a variety of ways of implementing this middle tier, such as transaction processing monitors, message servers, or application servers. The middle tier can perform queuing, application execution, and database staging. For example, if the middle tier provides queuing, the client can deliver its request to the middle layer and disengage because the middle tier will access the data and return the answer to the client. In addition, the middle-layer adds scheduling and prioritization for work in progress. The three-tier client-server architecture has been shown to improve performance for groups with a large number of users (in the thousands) and improves flexibility when compared to the two-tier approach. The added modularity makes it easier to modify or replace one tier without affecting the other tiers. Separating the application functions from the database functions makes it easier to implement load balancing. Flexibility in partitioning can be a simple as "dragging and dropping" application code modules onto different computers in some three-tier architectures. These characteristics have made three-layer architectures a popular choice for Internet applications and netcentric information systems.

Sometimes, the middle tier is divided in two or more units with different functions; in these cases the architecture is often referred as multilayer. This is the case, for example, of some Internet applications. These applications typically have light clients written in HTML and application servers written in C++ or Java. Because the gap between these two layers is too big to link them together, there is an intermediate layer (Web server) implemented in a scripting language instead. This layer receives requests from the Internet clients and generates HTML using the services provided by the business layer. This additional layer provides further isolation between the application layout and the application logic. It should be noted that recently, mainframes have been combined as servers in distributed architectures to provide massive storage and improve security.

Building three-tier architectures is complex work. Programming tools that support the design and deployment of three-tier architectures do not yet provide all of the desired services needed to support a distributed computing environment. A potential problem in designing three-tier architectures is that separation of user interface logic, process management logic, and data logic is not always obvious. Some process management logic may appear on all three tiers. The placement of a particular function on a tier should be based on criteria such as the following: ease of development and testing; ease of administration; scalability of servers; performance (including both processing and network load). A limitation with three-tier architectures is that the development environment is reportedly more difficult to use than the visually oriented development of two-tier applications.

Two-tier client-server architectures are appropriate alternatives to the three-tier architectures under the following circumstances: (a) when the number of users is expected to be less than 100 or (b) for non-real-time information processing in non-complex systems that requires minimal operator intervention.

Distributed/collaborative enterprise computing is seen as a viable alternative, particularly if object-oriented technology on an enterprise-wide (numerous smaller systems or subsystems) scale is desired.

Three-Tier Architecture with Transaction Processing Monitor Technology. The most basic type of three-tier architecture has a middle layer consisting of Transaction Processing (TP) monitor technology. The TP monitor technology is a type of message queuing, transaction scheduling, and prioritization service where the client connects to the TP monitor (middle tier) instead of the database server. The transaction is accepted by the monitor, which queues it and then takes responsibility for managing it to completion, thus freeing up the client. TP monitor technology also provides the ability to update multiple different DBMSs in a single transaction; connectivity to a variety of data sources including flat files, nonrelational DBMS, and the mainframe; the ability to attach priorities to transactions; robust security.

Using a three-tier client-server architecture with TP monitor technology results in an environment that is considerably more scalable than a two-tier architecture with direct client to server connection. For systems with thousands of users, TP monitor technology (not embedded in the DBMS) has been reported as one of the most effective solutions. A limitation to TP monitor technology is that the implementation code is usually written in a lower-level language (such as COBOL) and is not yet widely available in the popular visual toolsets.

Three-Tier Architecture with Message Server. Messaging is another way to implement three-tier architectures. Messages are prioritized and processed asynchronously. Messages consist of headers that contain priority information, along with the address and identification number. The message server connects to the relational DBMS and other data sources. The difference between TP monitor technology and message server is that the message server architecture focuses on intelligent messages, whereas the TP Monitor environment has the intelligence in the monitor and treats transactions as dumb data packets. Messaging systems are good solutions for wireless infrastructures.

Three-Tier Architecture with an Application Server. The three-tier application server architecture allocates the main body of an application to run on a shared host rather than in the user system interface client environment. The application server does not drive the GUIs; rather it shares business logic, computations, and a data retrieval engine. Advantages are that with less software on the client there is less security to worry about, applications are more scalable, and support and installation costs are less on a single server than maintaining each on a desktop client. The application server design should be used when security, scalability, and cost are major considerations.

Three-Tier Architecture with an ORB Architecture. Currently, industry is working on developing standards to improve interoperability and determine what the common Object Request Broker (ORB) will be. Developing client-server systems using technologies that support distributed objects holds great promise, because these technologies support interoperability across languages and platforms, as well as enhancing maintainability and adaptability of the system. There are currently two prominent distributed object technologies: Common Object Request Broker Architecture (CORBA) and Component Object Model (COM). Industry is working on standards to improve interoperability between CORBA and COM/DCOM.

Distributed/Collaborative Enterprise Architecture. The distributed/collaborative enterprise architecture emerged in 1993. This software architecture is based on Object Request Broker (ORB) technology, but goes further than the Common Object Request Broker Architecture (CORBA) by using shared, reusable business models (not just objects) on an enterprise-wide scale. The benefit of this architectural approach is that standardized business object models and distributed object computing are combined to give an organization flexibility to improve effectiveness organizationally, operationally, and technologically. An enterprise is defined here as a system comprised of multiple business systems or subsystems. Distributed/collaborative enterprise architectures are limited by a lack of commercially available object orientation analysis and design method tools that focus on applications.

In the following, we will describe components of a three-tier workbench: the databases and tools, the interfaces, and the middle tier (Figure 9.1).

9.2.2 Types of Database Systems

All of external data repositories could be automatically fetched into a central repository using a "mirroring" program that uses standard ftp protocols and runs as a daemon on the server (e.g., Biology Workbench). The local databases could be flat files, relational, XML or object, object-relational, or a combination of them.

Flat Files

The flat-file style of database is ideal for small amounts of data that need to be human readable or edited by hand. Essentially all they are made up of is a set of strings in one or more files that can be parsed to get the information they store; this is great for storing simple lists and data values, but it can get complicated when one tries to replicate more complex data structures. That's not to say that it is impossible to store complex data in a flat-file database; just that doing so can be more costly in time and processing power compared to a relational database.

Relational

The relational databases such as MySQL, Microsoft SQL Server, and Oracle have a much more logical structure in the way that it stores data. Tables can be used to represent real-world objects, with each field acting like an attribute. The "relation" comes from the fact that the tables can be linked to each other—for example, a cross-reference—to provide more information about the author. These relations can be quite complex in nature, and they would be hard to replicate in the standard flat-file format.

One major advantage of the relational model is that, if a database is designed efficiently, there should be no duplication of any data, helping to maintain database integrity. This can also represent a huge saving in file size, which is important when dealing with large volumes of data. Having said that, joining large tables to each other to get the data required for a query can be quite heavy on the processor; so in some cases, particularly when data are read only, it can be beneficial to have some duplicate data in a relational database. Relational databases also have functions "built in" that help them to retrieve, sort, and edit the data in many different ways. These functions save script designers from having to worry about filtering out the results that they get, and so they can go quite some way to speeding up the development and production of Web applications.

In most cases, one would want his/her database to support various types of relations; such databases, particularly if designed correctly, can dramatically improve the speed of data retrieval as well as being easier to maintain. Ideally, one will want to avoid the replication of data within a database to keep a high level of integrity; otherwise, changes to one field will have to be made manually to those that are related. A single connection to a relational database can access all the tables within that database. Relational databases have functions that can sort and filter the data so the results that are sent to the script are pretty much what one needs to work with. It is often quicker to sort the results before they are returned to the script than to have them sorted via a script; few scripting languages are designed to filter data effectively, and so the more functions a database supports, the less work a script has to do. If one is only working with a small amount of data that are rarely updated, then a full-blown relational database solution can be considered overkill. Flat-file databases are not as scalable as the relational model, so if one is looking for a suitable database for more frequent and heavy use, then a relational database is probably more suitable.

The relational databases may be easier to easier/faster to search/query/index than the flat files ones. One could use Web services to get information from remote databases (using HTML or XML), but the process can be slow (if there are too many users) and more complex queries could be trickier to write; one also has to parse the data that he gets back in order to display it.

Object

Object DBMSs add database functionality to object programming languages. They bring much more than persistent storage of programming language objects. Object DBMSs extend the semantics of the C++, Java, and so on, object programming languages to provide full-featured database programming capability, while retaining native language compatibility.

A major benefit of this approach is the unification of the application and database development into a seamless data model and language environment. As a result, applications require less code and use more natural data modeling, and code bases are easier to maintain. Object developers can write complete database applications with a modest amount of additional effort. In contrast to a relational DBMS where a complex data structure must be flattened out to fit into tables or joined together from those tables to form the in-memory structure, object DBMSs have no performance overhead to store or retrieve a Web or hierarchy of interrelated objects. This one-to-one mapping of object programming language objects to database objects has two benefits over other storage approaches: It provides higher-performance management of objects, and it enables better management of the complex interrelationships between objects. This makes object DBMSs better suited to support applications that have complex relationships between data.

Object database systems combine the classical capabilities of relational database management systems (RDBMS), with new functionalities assumed by the object-orientedness. The traditional capabilities include: secondary storage management; schema management; concurrency control; transaction management and recovery; query processing, access authorization, and control; safety, security. New capabilities of object databases include: complex objects; object identities; user-defined types; encapsulation; type/class hierarchy with inheritance; overloading, overriding, late binding, polymorphism; computational and pragmatic completeness of programmers' interfaces.

The object model used by the object database is a perfect match for XML. Object database management systems extend the object programming language with transparently persistent data, concurrency control, data recovery, associative queries, and other database capabilities. Transparent persistence in object database products refers to the ability to directly manipulate data stored in a database using an object programming language. This is achieved through the use of intelligent caching.

Consider an object database when one has a business need for high performance on complex data. A common characteristic of complex data is that they are represented by a graph, in which the nodes can lack unique identification. Many-to-many relationships are a second sign of complex data.

Because an ODBMS stores exactly the same object model that is used at the application level, both development and maintenance costs can be reduced. With an ODBMS, there is no need to develop and maintain two data models: an object model in the application and a relational model stored in the database. This is not needed because an ODBMS uses the same object model as the application; develop and maintain mapping between the relational and the object models.

Object-Relational

Object-relational database management systems (ORDBMSs) add new object storage capabilities to the relational systems at the core of modern information systems. These new facilities integrate management of traditional fielded data, complex objects such as time-series data, and diverse binary media such as audio, video, images, and applets. By encapsulating methods with data structures, an ORDBMS server can execute complex analytical and data manipulation operations to search and transform multimedia and other complex objects.

As an evolutionary technology, the object-relational (OR) approach has inherited the robust transaction- and performance-management features of its relational ancestor and the flexibility of its object-oriented cousin. Database designers can work with familiar tabular structures and data definition languages (DDLs) while assimilating new object-management possibilities. The most important new object/relational features are user-defined types (UDTs), user-defined functions (UDFs), and the infrastructures (indexing/access methods and optimizer enhancements) that support them.

9.3 WORKBENCH COMPONENTS

The system's infrastructure usually has a three-tier architecture: a client/presentation server, an application server, and a database server. The middle server (application server) is not required, and some workbenches use a two-tier architecture (client–server model). In these cases, the server does not present a user interface; instead its sole purpose is to serve a computing server and data repository. This allows for customization in the interface and for a selection between a thin client and a thick client. A thin client is small enough to be downloaded from the workbench server each time it is needed, while a thick client would have so much functionality that its downloading it each time would not be efficient. These client–server workbenches are easy to integrate in a pipeline. Finally, there are the primitive workbenches with a single-tier (or mixed model) architecture (like GENESIS— http://www.genesis-sim.org/GENESIS) in which basically the user downloads and installs the program and runs it from the command line and directly talks to the server.

9.3.1 Interfaces

The interfaces (front-end) to a workbench system handle the following: presenting the user with a way to select and input the data they wish to analyze; controlling mechanisms (being able to change the parameters) as well as presenting the output in a consistent, easy-to-understand, beautified format; tying into specific client side systems for the presentation of data that does not readily fit into the constraints of the systems user interface. The interfaces should be able to handle different types of data, like text, graphics, 2D and 3D images, time-series data, and so on.

9.3.2 Databases

The database server (back-end tier) (e.g., Oracle) would be connected to the client (front-end tier) Web browser or a stand-alone application (e.g., Java Swing) through an application server (middle tier) (e.g., Oracle Application Server OAS) (where the business logic is taken care of and applications are run) (Figure 9.1).

SQL*Loader could be used to load data into Oracle tables. For most databases, parsers will have to be written to get the

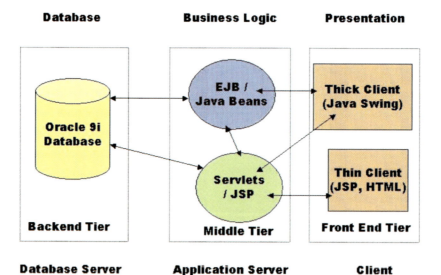

Figure 9.1 Diagram of an example of a three-tier architecture. An Oracle 9i database server is connected through a middle tier, Oracle Application Server (OAS) 9i, to a client Web browser or a stand-alone application using Java Swing. In the middle tier, Java Servlets, Java Server Page (JSP), Java Beans, and/or Enterprise Java Beans (EJB) are used. Servlets/JSP allow easy navigation through firewalls while EJB/Java Beans allows the client to call the server using intuitive method names, obviates the need for XML parsing, and automatically gives remote access and load-balancing. They also allow query of the relational database, creation of an XML model, and export to the client for display purposes.

data content in a form that can be loaded and indexed. In the presentation (front-end) one can use various Java tools/GUI applications for better data display and sorting.

9.3.3 Middle Tier

The reason for a middle tier is to isolate the client tier from changes in the database by forcing communication through a consistent interface featuring the object for which we know are present in our system, but for which the schema still occasionally changes. The use of a middle tier can also allow both Java Swing and Web clients to efficiently obtain information from the database. The middle tier takes care of the business logic and database access for the clients. In the Biology Workbench, the middle tier is composed of the core, the tool wrappers and master wrapper, and API program which talk to the tools and the databases. In the middle tier the necessary format conversions for I/O needed for different tools takes place, as well the running of the queries and of the tools themselves. The middle tier could be based on Enterprise Java technology, which provides common services to the applications, ensuring that the applications are reasonably portable. The specifications cover many areas including:

- **HTTP Communication.** A simple interface is presented for the interrogation of requests from Web browsers and for the creation of the response.
- **HTML Formatting.** Using Java Server Pages (JSP) to create dynamic Web pages.
- **Database Communication.** Java Database Connectivity (JDBC), a standard interface for talking to databases from application code.
- **Database Encapsulation.** the Enterprise Java Beans (EJB) define a way to declare a mapping between application code and database tables using an XML file.
- **Naming Services.** The Java Naming and Directory interface (JNDI) defines a way for application code to consistently obtain references to remote object (i.e., in another tier) based on names defined in the XML file.

9.3.4 Types and Nature of Tools

Workbenches define a set of object types that they will handle—for example, sequence, alignment, 2D or 3D image, structure, graphics, text, and so on. Tools that work on similar data types, or objects, are grouped together to provide interoperability. Certain applications can convert an object of one type to another type, allowing the object to evolve, to be analyzed at a different level. Tools within a workbench system can also consist of a simulation environment. The tools can also provide dynamic views of data (time-series and histogram data, e.g., Neurodatabase—http://www.neurodatabase.org/). Sequence analysis tools would include the most common ones like BLAST and FASTA packages, multiple sequence alignment tools such as CLUSTALW and MSA, domain searching tools such as HMMPFAM, RPSBLAST, and FINGERPRINTS (e.g., Biology Workbench—http://workbench.sdsc.edu; Interpro—http://www.ebi.ac.uk/interpro), finding ORFs, designing primers

(PRIMER3), finding orthologs, getting the translation or the coding region, determining the enzyme cutting sites (TACG), predicting transmembrane domains or secondary structure (TMHMM, CHOFAS, GREASE, GOR4, PredictProtein), gene annotation, protein–protein interactions, and so on. Other tools deal with structural 3D protein display, alignments and similarity searches (e.g., Cn-3D, VAST Search, NCBI—http://www.ncbi.nih.gov/; SWISS-MODEL, ExPASy— http://www.expasy.ch/). Finally other tools deal with text/literature data mining. Some other tools can be very specialized, such as microarray analysis tools or the ones that deal with neuronal recordings.

9.3.5 Miscellaneous

A common representation language is needed, especially in exchanging, representing, and developing data models used by different simulation/analysis tools (e.g., Neurodatabase, Systems Biology Workbench—http://www.cds.caltech.edu/erato/). This would help addressing the interoperability between different tools and databases. It is usually based on XML because of its portability.

In addition, one can also have a user account system, which saves the user's data for future reference or further analysis. The users' data that are taken from local databases (such as sequences) can be stored locally or can be stored in a form of pointers.

With the advantages of Web services, a new design model for workbenches emerges, that of distributed workbench system, where services are offered by individual machines and are coordinated and presented by one or more servers. With this paradigm, the possibility of having multiple workbench systems utilizing the same underlying compute engines is possible. Pipelining systems would be able to utilize these Web services wrapped tools. A Web-services-based workbench would be better able to scale as demand is increased. New tools would be able to be added easily without needing to alter the underlying workbench core.

9.4 EXAMPLES OF EXISTING WORKBENCHES

Various biological workbenches have been created in order to fulfill some of those needs. Some are predominantly sequence analysis workbenches, but could also do some structure analysis—like Biology Workbench (Subramaniam, 1998), NCBI (Ostell and Kans, 1998), KEGG (Ogata et al., 1999), Genotator (Harris, 2000), Interpro (Apweiler et al., 2000) (http://www.ebi.ac.uk/interpro), and ExPASy (http://www.expasy.ch) (Gasteiger et al., 2003). Others do mostly gene prediction—such as GeneMachine (Makalowska et al., 2001) (http://genemachine.nhgri.nih.gov/) and Alfresco (Jareborg and Durbin, 2000) (http://www.sanger.ac.uk/Software/Alfresco), which use, among other applications, Genscan (Burge and Karlin, 1997). Some are dealing with experimental neuroscience data—like Neurodatabase (Gardner et al., 2000) and Neuronal Time-Series Analysis (NTSA) Workbench. Some are more specialized, like the Dicty Workbench and Stanford Microarray database (Gollub et al., 2003). Some are mostly simulation/

modeling workbenches—like GENESIS (Bower et al., 2003) (http://www.genesis-sim.org/GENESIS), the Electrotonic Workbench (http://www.neuron.yale.edu/neuron), which was created using the simulation program NEURON (Hines, 1993), Systems Biology Workbench (Hucka et al., 2001), and Neurodatabase.

Below, we describe a few examples of biology workbenches, some which deal with sequence analysis, such as Biology Workbench, and some which deal with neuronal recordings, such as Neurodatabase.

9.4.1 Biology Workbench

Biology Workbench (http://workbench.sdsc.edu) was a first big step in the direction of eliminating to a significant extent most of the problems mentioned above. Biology Workbench integrates many various sequence databases and tools under a uniform user Web-based interface (Subramaniam, 1998). The interface greatly reduces the amount of computer knowledge needed to use the tools and search databases. The point-and-click interface eliminates the necessity of converting file formats. The sequence data are moved seamlessly between different tools. Biology Workbench has a set of specialized scripts that connect the user's browser to a collection of databases (information sources) and application programs (tools). The scripts are specialized for each database and application. Functionally, they transform the interface for each object, whether database or application program, into a common Web-based form that permits them to be seamlessly interconnected. By reducing all the formats to a common point-and-click Web interface, the users are freed from the necessity of knowing details of the object formats. Also the scripts work through the interfaces very rapidly, so various operations can be done quickly. Reasonable default parameters for the various operations are built into the Web interface. However, the

knowledgeable user can easily adjust parameters for search and analysis via Web menus.

The Biology Workbench was designed to be analysis centric where a multiplicity of databases are integrated and analysis programs are wrapped to operate on the databases. The Biology Workbench uses publicly available databases in flat file formats. While this approach is adequate for dealing with sequence analysis, flat file structure for databases fails to capture complex and rich relationships between different types of data.

The Biology Workbench has been freely available to the national community since 1996 (Subramaniam et al., 1996; Subramaniam, 1998). The Biology Workbench is presently available freely to the research community from the San Diego Supercomputer Center (http://workbench.sdsc.edu). The Biology Workbench uses CGI scripts and programs to integrate databases and tools into the Web environment. Querying, analyzing, and computing can, therefore, be done on a server from any client that is connected to the Internet and has a Web access (Figure 9.2).

All necessary object manipulation tools are embedded in Perl and C programs, which pass the correctly formatted objects, such as protein and DNA sequences and setup parameters to the analysis programs. The Web interface of the Biology Workbench provides a uniform user interface that most researchers are already familiar with, thus reducing the amount of computer knowledge needed to utilize the tools and perform database searches.

The Biology Workbench is a set of Common Gateway Interface (CGI) applications written in Perl (Wall and Schwartz, 1991) and C languages The Workbench is accessible from any client which is Web-browser-compatible and is networked. Any Web browser, such as Netscape Navigator or Microsoft Internet Explorer, can be used.

The core of the Biology Workbench is a C program responsible for handling, among other tasks, the communication between

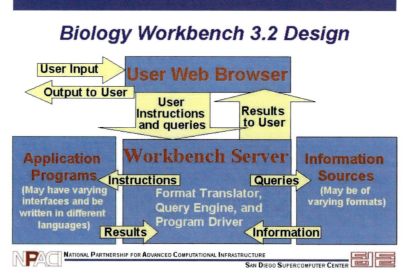

Figure 9.2   Biology Workbench 3.2 Design. The user inputs his data, instructions, and queries via a Web browser interface, which then are relayed to the Workbench server. The Workbench Server performs necessary format translations, drives the application programs, queries the information sources (flat file databases), and then returns the results of the applications and queries to the user via web browser.

the tools and the Web server. Tool wrappers require standard subroutines to communicate with the Workbench core. The tool or module wrappers are written in Perl and provide the "interface" between the tools and the user. In the input, the wrappers do the sequence format conversion, set up the parameters, and execute the program. Then, the wrappers parse the output of the tool into a "prettier" form, extract the information, and identify sequences that can be imported.

Most of tools currently integrated into the Biology Workbench were originally stand-alone, command-line, textual input/output programs. PERL, with its incredibly strong text handling features, is perfect for handling and parsing the textual I/O from the tool. The HTML code necessary can be easily generated to reformat the data for submission to the next tool. Some tools require more involved HTML programming, allowing the user to view and select sequences for importing them back to the Workbench. The HTML environment also allows for enhanced output presentation that can aid the user's interpretation of their results.

Most of the relevant public biological sequence databases are accessible from the Biology Workbench. The databases (flat files) are federated by identification of common objects and indexing each database based on these objects. A database-mirroring program has also been developed for the Biology Workbench, which fetches databases as they are updated in their primary repositories. The indexing and creation of searchable and blastable sequence database files is done locally each time a change in the retrieved databases is perceived by the system.

A novel database management system (Ndjinn), which allows rapid integration of any heterogeneous database for querying and searching, has been developed in the Workbench group. Ndjinn is basically a free-text based flat-file indexing and retrieval system, similar to SRS. The merit of this scheme lies in the versatility in assimilating any type of record-based (text) data into an indexed scheme whereby querying is possible. Further, the querying is made meaningful by ranking the results in order of repeat occurrence of search words. The Workbench permits nested Boolean queries. The results are presented in a visually enhanced and logical field-based format. In addition, hyperlinks are provided to cross-reference databases.

Users' data (sequence information) can be stored using a user accounts system, in session files. Sessions can be used for recalling previously performed analysis as well as comparison with newer data analysis. The sessions can be updated, erased, or stored.

The Biology Workbench contains most of the commonly used, freely available analysis modules for the protein and nucleic acid sequences. "Bread-and-butter" modules for biologists such as BLAST (including PSI-BLAST and RPS-BLAST) (Altschul et al., 1990, 1997; Altschul and Koonin, 1998) and FASTA, other programs for global and local multiple sequence alignment such as CLUSTALW and MSA, some phylogenetic analysis tools (like the PHYLIP package), secondary structure and transmembrane domain prediction tools, and multiple sequence profile and statistical analysis tools are all a point-and-click away on the Workbench (Figure 9.3). A complete list of available tools is too exhaustive to be listed and can be accessed at http://workbench.sdsc.edu.

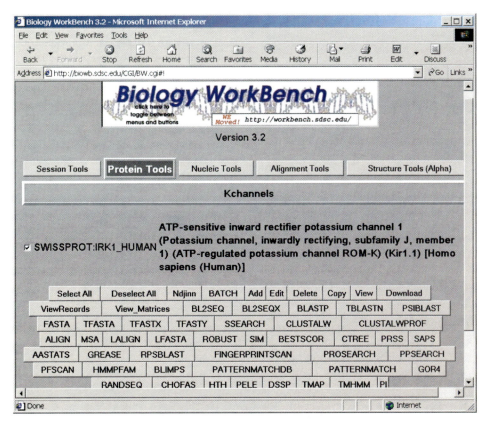

Figure 9.3   A screen snapshot of the Biology Workbench 3.2 Protein Tools.

9.4.2 KEGG—Kyoto Encylopedia of Genes and Genomes

KEGG (http://www.genome.ad.jp/kegg/) is a knowledge base for systematic analysis of gene functions, linking genomic information with higher-order functional information (Ogata et al., 1999). KEGG consists of the PATHWAY database, which contains computerized knowledge on molecular interaction networks such as pathways and complexes and graphical representations of cellular processes; the GENES database, which is a collection of the genes and proteins generated by genome sequencing projects; and the LIGAND database, which contains information about chemical compounds, enzyme molecules and enzymatic reactions. In addition, a limited amount of experimental data for microarray gene expression profiles and yeast two-hybrid systems are stored in the EXPRESSION and BRITE databases, respectively. The data objects in the KEGG databases are represented as graphs and various computational methods are developed to detect graph features that can be related to biological functions. KEGG provides Java graphics tools for browsing genome maps, comparing two genome maps and manipulating expression maps, as well as computational tools for sequence comparison, graph comparison, and path comparison.

9.4.3 National Center for Biotechnology Information (NCBI)

Established in 1988 as a national resource for molecular biology information, the NCBI creates public databases, conducts research in computational biology, develops software tools for analyzing genome data, and disseminates biomedical information— all for the better understanding of molecular processes affecting human health and disease (Jennuth, 2000, Wheeler et al., 2003). (http://www.ncbi.nih.gov/)

NCBI assumed responsibility for the GenBank DNA sequence database in 1992. NCBI staff builds the database from sequences submitted by individual laboratories and by data exchange with the international nucleotide sequence databases, European Molecular Biology Laboratory (EMBL), and the DNA Database of Japan (DDBJ). In addition to GenBank, NCBI supports and distributes a variety of databases for the medical and scientific communities (Ostell and Kans, 1998; Wheeler et al., 2003). These include Genbank, the Online Mendelian Inheritance in Man (OMIM), the Molecular Modeling Database (MMDB) of 3D protein structures, Conserved Domain Database (CDD), Clusters of Orthologous Groups (COGs), the Unique Human Gene Sequence Collection (UniGene), RefSeq/LocusLink (Pruitt et al., 2000), a Gene Map of the Human Genome, the Taxonomy Browser, HomoloGene, EST database, and SNPs database. The protein entries in the Entrez search and retrieval system have been compiled from a variety of sources, including SwissProt, PIR, PRF, PDB, and translations from annotated coding regions in GenBank and RefSeq. Genomes of over 800 organisms can be found in the genome database, representing both completely sequenced organisms and those for which sequencing is in progress.

Entrez is NCBI's search and retrieval system that provides users with integrated access to sequence, mapping, taxonomy, and structural data (Schuler et al., 1996, Geer and Sayers, 2003). Entrez also provides graphical views of sequences and chromosome maps. A powerful and unique feature of Entrez is the ability to retrieve related sequences, structures, and references. The journal literature is available through PubMed, a Web search interface that provides access to over 11 million journal citations in MEDLINE and contains links to full-text articles at participating publishers' Web sites.

BLAST, a program for sequence similarity searching, was developed at NCBI, and it is instrumental in identifying genes and genetic features (Altschul et al., 1997). BLAST can execute sequence searches against the entire DNA database in less than 15 seconds. The BLAST software includes PHI-BLAST, organism-specific blast, and RPS-BLAST. Additional software tools provided by NCBI include: COGnitor and Homologene (for finding potential orthologs), Conserved Domain (CD) Search, Cn3D (A three-dimensional structure and sequence alignment viewer for NCBI structure databases), VAST Search (a structure–structure similarity search), Open Reading Frame Finder (ORF Finder), Electronic PCR, and the sequence submission tools, Sequin and BankIt. All of NCBI's databases and software tools are available from the WWW or by FTP.

9.4.4 Stanford Microarray Database (SMD)

The Stanford Microarray Database (http://genome-www.stanford.edu/microarray) serves as a microarray research database for Stanford investigators and their collaborators (Gollub et al., 2003). In addition, SMD functions as a resource for the entire scientific community, by making freely available all of its source code and providing full public access to data published by SMD users, along with many tools to explore and analyze this data. SMD currently provides public access to data from more than 3500 two-color, spotted DNA microarrays, including data from more than 85 publications. SMD provides online tools for browsing and selecting experimental data, assessing data quality, filtering by individual spot characteristics and by expression pattern, and analyzing data via hierarchical clustering or self-organizing maps—along with extensive help and tutorials on how to use these tools. The most recent tools provide a data selection by gene (Expression History tool) and data visualization and quality assessment (Array Color tool)—which provide also statistical analysis of the variances.

9.4.5 Dicty Workbench

This website provides a comprehensive *Dictyostelium discoideum* genome annotation and analysis portal for *Dicty* researchers (http://dictyworkbench.sdsc.edu/). The *Dicty* Workbench is a specialized *Dictyostelium* sequence and function annotation database that is coupled to the Biology Workbench. The *Dicty* database is an Oracle object-relational database with the query interface, Seeker, developed locally in San Diego Supercomputer Center. The framework for the database was abstracted from the Biology Workbench development and from the Alliance for Cellular Signaling (AfCS) Projects (Gilman et al., 2002). The HMM-predicted ORFs are annotated and provide with links to the HMPFAM, RPS-BLAST, and other BLAST results (which are updated regularly).

9.4.6 Systems Biology Workbench

The ERATO Systems Biology Workbench (SBW) (http://www .cds.caltech.edu/erato) is a modular, broker-based message-passing framework for simplified application intercommunications (Hucka et al., 2001). SBW enables interactions and exchange between software tools for computational biology. The systems biology community needs information standards if the models are to be shared, evaluated, and developed cooperatively. They proposed the Systems Biology Mark-up Language (SBML) as a common representation language for storing biochemical models and representing biochemical reaction networks. SBML is based on XML, and it contains structures for representing compartments, species, and reactions, as well as optional unit definitions, parameters, and rules (constraints). SBML is open and free and is a software-independent language for describing models common to research in many areas of computational biology, including cell signaling pathways, metabolic pathways, gene regulations, and others.

This software framework allows different heterogeneous application components, written in diverse programming languages and running on different platforms, to communicate and use each other's data and algorithmic capabilities. SBW enables applications (potentially running on separated, distributed computers) to communicate via simple network protocol. The interfaces to the system are encapsulated in client-side libraries that are provided for different programming languages (Java, C, Python, etc.)

9.4.7 Neurodatabase

This Internet database, a component of the Human Brain Project, contains author-submitted physiological datasets, like spike trains and raw spike records, metadata, and neuron physiological descriptions (http://neurodatabase.org) (http://brainml.org). Its goal is to enable the neurophysiology resources with the utility and the ease of use of sequence databases. Universal multiplatform Web/Java tools provide dynamic views of the data. The database itself is implemented using an object-relational database engine. They also have a Java Virtual Oscilloscope and Metadata Viewer.

Data requests from the client (who needs only conventional Java-enabled Web browsers such as IE or Netscape) are constructed by their Java query tool, which builds metadata tags for submission and constructs searches from controlled-vocabulary sets.

The authors have developed a Common Data Model (CDM) that aids interoperability and database federation, data exchange, and portability (Gardner et al., 2000). CDM is an intuitive data model for neurophysiologists to search for, acquire and analyze datasets. In CDM, experiments are segmented into views (analogous to figure panels) consisting of one or more traces; metadata accompany datasets at all levels. The CDM model abstracts and relates generalized elements for data (including wrappers and metadata), site, reference, model, and method. CDM data descriptions use a custom-designed Biophysical Description Mark-up Language (BDML) extending open standard Resource Description Framework and XML with neurophysiological vocabulary and formats optimized for data exchange. BDML mediates the interoperability of data models by specifying the intersection of the data models. Data models, metadata, and dataset formats of a wide range of neuroscience data can be derived from these generalized elements and described using CDM.

9.4.8 The University of Southern California Brain Project (USCBP) Repository of Experimental Data (RED)

The USCBP is storing, retrieving, and analyzing experimental neuroscience data in a unified manner (http://www-hbp.usc.edu/). The database component (NeuroCore) is a novel extendible object-relational database implemented in Informix (as ORDBMS). Informix universal server supports standard SQL, and it extends the standard relational database model with object-oriented features including the ability to add new user-defined data types and functions, built-in support for large objects as data types (e.g., images, audio, video), and support of object-oriented concepts such as inheritance and function overloading. For storing neurophysiological time-series data, a time-series data type has been defined and implemented. An online notebook provides a laboratory independent standard for viewing, storing, and retrieving data across the Internet. The notebook interface was developed using Java and HTML specifications.

They emphasize that there will not be a single monolithic database that will store all neuroscience data; rather, there will be a federation of databases. In order to connect and access a federation of databases, certain "hooks" have been included in the core database schema to foster such a communication.

9.4.9 Neuronal Time-Series Analysis (NTSA) Workbench

The focus of the NTSA is to handle information concerning neuronal recordings, and stimulus and behavioral representations of those (http://soma.npa.uiuc.edu/isnpa). These data can be viewed as instances of a general data type, namely, time-series data. This information system has three interrelated components. The first is a database component for organized storage of large datasets and efficient data search and retrieval. A second component is the conventions for data representation and transfer so that the experimental records can be processed independently of the format in which they are stored. The third component is a suite of tools for data filtering, display, analysis, and visualization.

9.4.10 Genotator

Genotator is a workbench for automated sequence annotation and annotation browsing (Harris, 1997, 2000) (http://www .fruitfly.org/~nomi/genotator/). The back end runs a series of sequence analysis tools, handling the various input and output formats conversions required by the tools. The results of the analysis can be viewed with the interactive graphical browser. Genotator's display helps users identify regions of interest. Genotator consists of three components—a set of sequence analysis programs, a database, and a graphical browser—as well as the "glue" that links these components. Genotator's database exists in a Unix directory hierarchy of tabular flat files. Genotator can be run via command-line or with an easy-to-use GUI. They are working on a client-server version that will enable remote users to annotate sequences via a flexible, transparent distributed architecture.

9.5 CHALLENGES IN THE DESIGN AND IMPLEMENTATION OF THE WORKBENCHES

One of the most difficult challenges that workbench designers face is to provide the interfaces and tie in tools and databases in order to achieve interoperability for high-throughput analysis. Another challenge is to create a system that is powerful, yet easy to use by the biologists.

9.6 SUMMARY OF USE OF WORKBENCHES

Workbenches provide a consistent and intuitive interface, which encapsulates the common tasks of the researchers into a common work environment. The environment starts out with reasonable program default settings, but should allow the user to change them, to customize their environment.

Workbenches greatly increase the speed of the daily tasks of researchers, and they enable researchers to look at their data in new ways and to analyze them with different tools without having to worry about format changes and ease the introduction of new tools.

Information Sources

Client/Server Frequently Asked Questions. http://www.faqs.org/faqs/client-server-faq/

Dickman, A. (1995). Two-tier versus three-tier applications. *Inf. Week* **553**, 74–80.

Edelstein, H. (1994). Unraveling client/server architecture. *DBMS* **7**(5), 34(7).

Edwards, J. (1999). *3-Tier Client/Server at Work*. Wiley, New York.

Fournier, R. (1998). *A Methodology for Client/Server and Web Application Development*. Prentice Hall, Yourdon Press, Upper Saddle River, NJ.

Gallaugher, J., and Ramanathan, S. (1996). Choosing a client/server architecture. A comparison of two-tier and three-tier systems. *Inf. Syst. Manag. Mag.* **13**(2), 7–13.

N-Tier Articles: Two-tiered Client/Server Limitations. http://n-tier.com/articles/cslimits.html

Orfali, R., Harkey, D., and Edwards, J. (1999). *Client/Server Survival Guide*, 3rd ed. Wiley, New York.

Software Architectures (Client/server, two tier, three tier)—Carnegie Mellon Software Engineering Institute. http://www.sei.cmu.edu/str/descriptions

References

Altschul, S.F., and Koonin, E.V. (1998). Iterated profile searches with PSI-BLAST—a tool for discovery in protein databases. *Trends Biochem Sci.* **23**(11), 444–447.

Altschul, S.F., Gish, W., Miller, W., Myers, E.W., and Lipman, D.J. (1990). Basic local alignment search tool. *J. Mol. Biol.* **215**(3), 403–410.

Altschul, S.F., Madden, T.L., Schaffer, A.A., Zhang, J., Zhang, Z., Miller, W., and Lipman, D.J. (1997). Gapped Blast and PSI-BLAST: A new generation of protein database search programs. *Nucleic Acids Res.* **25**, 3389–3402.

Apweiler, R., Attwood, T.K., Bairoch, A., Bateman, A., Birney, E., Biswas, M., Bucher, P., Cerutti, L., Corpet, F., Croning, M.D., Durbin, R., Falquet, L., Fleischmann, W., Gouzy, J., Hermjakob, H., Hulo, N., Jonassen, I., Kahn, D., Kanapin, A., Karavidopoulou, Y., Lopez, R., Marx, B., Mulder, N.J., Oinn, T.M., Pagni, M., Servant, F., Sigrist, C.J., Zdobnov, E.M.; InterPro Consortium. (2000). InterPro—an integrated documentation resource for protein families, domains and functional sites. *Bioinformatics* **16**(12), 1145–1150.

Bower, J.M., Beeman, D., and Hucka, M. (2003). The GENESIS Simulation System. In Arbib, M.A. (Ed.), *The Handbook of Brain Theory and Neural Networks*, 2nd ed. MIT Press, Cambridge, MA.

Burge, C., and Karlin, S. (1997). Prediction of complete gene structures in human genomic DNA. *J. Mol. Biol.* **268**, 78–94.

Gardner, D., Abato, M., Knuth, K.H., DeBellis, R., and Erde, S.M. (2000). A dynamic publication model for data-driven neurodatabases implements interoperability for neruroscience information resources. *Soc. Neurosci. Abstr.* **26**,: 748.15.

Gasteiger, E., Gattiker, A., Hoogland, C., Ivanyi, I., Appel, R.D., and Bairoch, A. (2003). ExPASy: The proteomics server for in-depth protein knowledge and analysis. *Nucleic Acids Res.* **31**, 3784–3788.

Geer, R.C., and Sayers, E.W. (2003). Entrez: Making use of its power. *Brief Bioinf.* **4**(2), 179–184.

Gilman, A.G., Simon, M.I., Bourne, H.R., Harris, B.A., Long, R., Ross, E.M., Stull, J.T., Taussig, R., Bourne, H.R., Arkin, A.P., Cobb, M.H., Cyster, J.G., Devreotes, P.N., Ferrell, J.E., Fruman, D., Gold, M., Weiss, A., Stull, J.T., Berridge, M.J., Cantley, L.C., Catterall, W.A., Coughlin, S.R., Olson, E.N., Smith, T.F., Brugge, J.S., Botstein, D., Dixon, J.E., Hunter, T., Lefkowitz, R.J., Pawson, A.J., Sternberg, P.W., Varmus, H., Subramaniam, S., Sinkovits, R.S., Li, J., Mock, D., Ning, Y., Saunders, B., Sternweis, P.C., Hilgemann, D., Scheuermann, R.H., DeCamp, D., Hsueh, R., Lin, K.M., Ni, Y., Seaman, W.E., Simpson, P.C., O'Connell, T.D., Roach, T., Simon, M.I., Choi, S., Eversole-Cire, P., Fraser, I., Mumby, M.C., Zhao, Y., Brekken, D., Shu, H., Meyer, T., Chandy, G., Heo, W.D., Liou, J., O'Rourke, N., Verghese, M., Mumby, S.M., Han, H., Brown, H.A., Forrester, J.S., Ivanova, P., Milne, S.B., Casey, P.J., Harden, T.K., Arkin, A.P., Doyle, J., Gray, M.L., Meyer, T., Michnick, S., Schmidt, M.A., Toner, M., Tsien, R.Y., Natarajan, M., Ranganathan, R., Sambrano G.R.; Participating investigators and scientists of the Alliance for Cellular Signaling. (2002). Overview of the Alliance for Cellular Signaling. *Nature (London)* **420**, 703–706.

Gollub, J., Ball, C.A., Binkley, G., Demeter, J., Finkelstein, D.B., Herbert, J. M., Hernandez-Boussard, T., Jin, H., Kaloper, M., Matese, J. C., Schroeder, M., Brown, P.O., Botstein, D. and Sherlock, G. (2003). The Stanford Microarray Database: Data access and quality assessment tools. *Nucleic Acids Res.* **31**(1), 94–96.

Harris, N.L. (1997). Genotator: A workbench for sequence annotation. *Genome Res.* **7**(7), 754–762.

Harris, N.L. (2000). Annotating sequence data using Genotator. *Mol. Biotechnol. Nov.* **16**(3), 221–232.

Hines, M. (1993). NEURON—A program for simulation of nerve equations. In *Neural Systems: Analysis and Modeling.* (F. Eeckman, ed.), pp. 127–136, Kluwer Acad. Publ. Norwell, MA.

Hucka, M., Finney, A., Sauro, H., Bolouri, H., Doyle, H. and Kitano, H. (2001). The ERATO Systems Biology Workbench: Architectural evolution. *Proc. 2nd Int. Conf. Syst. Biol. (ICSB2001).*

Jareborg, N., and Durbin, R. (2000). Alfresco—A workbench for comparative genomic sequence analysis. *Genome Res.* **10**(8) 1148–1157.

Jennuth, J.P. (2000). The NCBI. Publicly available tools and resources on the Web. *Methods Mol. Biol.* **132**, 301–312.

Makalowska, I., Ryan, J.F., and Baxevanis, A.D. (2001). GeneMachine: Gene prediction and sequence annotation. *Bioinformatics* **17**(9), 843–844.

Ogata, H., Goto, S., Sato, K., Fujibuchi, W., Bono, H., and Kanehisa, M. (1999). KEGG: Kyoto Encyclopedia of Genes and Genomes. *Nucleic Acids Res.* **27**(1), 29–34.

Ostell, J.M., and Kans, J.A. (1998). The NCBI data model. *Methods Biochem. Anal.* **39**, 121–144.

Pruitt, K.D., Katz, K.S., Sicotte, H., and Maglott, D.R. (2000). Introducing RefSeq and LocusLink: Curated human genome resources at the NCBI. *Trends Genet.* **16**(1), 44–47.

Schuler, G.D., Epstein, J.A., Ohkawa, H., and Kans, J.A. (1996). Entrez: Molecular biology database and retrieval system. *Methods Enzymol.* **266**, 141–162.

Subramaniam, S., Fenton, J., Jakobsson, E., Jamison, C., Stupar, M. and Unwin, R. (1996). The Biology Workbench on the World Wide Web. 4th International Conference on Intelligent Systems in Biology. St. Louis, Missouri.

Subramaniam, S. (1998). The Biology Workbench—a seamless database and analysis environment for the biologist. *Proteins* **32**(1), 1–2.

Wall, L. and Schwartz, R.F. (1991). Programming Perl. Publ. O'Reilly and Associates, Inc., Sebastopol, CA.

Wheeler, D.L., Church, D.M., Federhen, S., Lash, A.E., Madden, T.L., Pontius, J.U., Schuler, G.D., Schriml, L.M., Sequeira, E., Tatusova, T.A., and Wagner, L. (2003). Database resources of the National Center for Biotechnology. *Nucleic Acids Res.* **31**(1), 28–33.

SYSTEM APPROACHES

PART

II

Databasing the Brain. Edited by Stephen H. Koslow and Shankar Subramaniam
ISBN 0-471-30921-4 © 2005 John Wiley & Sons, Inc.

Genome to Disease

Iiris Hovatta, Ph.D. and Carrolee Barlow, M.D., Ph.D.

CONTENTS

Databasing the Brain. Edited by Stephen H. Koslow and Shankar Subramaniam
ISBN 0-471-30921-4 © 2005 John Wiley & Sons, Inc.

10.1 INTRODUCTION

Evolving genomics technologies offer neuroscientists tools to tackle more complex questions than ever before. A great deal of current neurobiological research assumes that the rich complexity of human thought and emotion can be reduced to a simple, linear relation between individual genes and behaviors (reviewed in Hamer, 2002). This oversimplified model ignores the complex gene expression networks and environmental influence that are predicted to regulate most behavioral traits. However, neurobiologists are equipped to think in terms of complex systems and using genomic methods that can probe these complexities should be relatively easy. Most of the new genome-wide analysis methods concentrate on evaluating the global gene expression or protein abundance differences. However, RNA-based methods are currently more popular due to their relative ease and the methodological difficulties associated with proteomic approaches. DNA microarrays enable the detection of gene expression profiles from small quantities of brain tissue, or even single cells. This ability is essential for most neurobiological applications because the brain is an extremely complex tissue where neighboring cells might have very different gene expression patterns. However, simply measuring gene expression may not be enough to unravel the molecular underpinnings of a complex behavior or a disease state.

Taking advantage of an increasing amount of DNA sequence information in the context of datasets on gene expression enables us to combine expression studies with sequence analysis and to move from phenomenology to causation. Since the sequence of the human and mouse genomes are known, it is relatively simple to look at the single-nucleotide polymorphisms (SNPs) in the coding regions of the differentially expressed genes and identify SNPs that may either directly or indirectly lead to the expression differences. However, little is known about cis-acting elements outside the coding region of the gene or trans-acting elements that regulate gene expression levels, and identification of such SNPs is currently time-consuming. New approaches to solve these methodological shortcomings have been recently developed, and it is anticipated that they will help in identification of susceptibility genes for complex diseases.

Multifactorial or complex diseases are likely to be caused by some combination of multiple genetic and environmental factors. While a great number of genes that cause monogenic disorders are known, success in mapping susceptibility genes for multifactorial diseases has been limited due to several confounding factors, such as genetic heterogeneity, unknown mode of inheritance, classification of the phenotype, phenocopies, reduced penetrance, and high frequency of disease-predisposing alleles. By narrowing the definition of a disease or restricting the patient population, it is often possible to work with a trait whose inheritance pattern is closer to Mendelian and is more likely to be homogeneous (Lander and Schork, 1994). One can focus on the clinical phenotype and restrict the study to cases with extreme symptoms, or one can choose patients with early onset of the disease. It might also require that an index case with at least two first-degree relatives affected with a given disease be used. Another way of decreasing the genetic heterogeneity is to collect the patients and families from isolated populations such as the Finnish population (Peltonen et al., 2000), the Old Order Amish (McKusick, 1973), or the Ashkenazi Jews (Risch et al., 1995), among many others. In isolated populations the number of

predisposing genes (and mutations in the predisposing genes) is likely to be smaller than in more mixed populations, which makes isolated populations useful in mapping both simple monogenic and complex diseases. Modern gene mapping methods including linkage maps and marker sets and the availability of the human genome sequence have made the mapping of monogenic diseases a fast and straightforward process; however, the same methods have not worked very well for mapping complex diseases.

The success of mapping genes for monogenic diseases led geneticists to think that mapping complex diseases would be possible using the same methods. After a decade of complex disease gene mapping, the number of genes found by linkage or association analyses is very small. It is postulated that there are multiple reasons to account for the failure. For example, the genetic heterogeneity and the number of genes predisposing to these diseases have been underestimated, and the sizes of study materials have generally been too small. In addition, better statistical methods to analyze complex interactions are needed. In the case of many complex diseases involving the brain, there are no measurable biological markers for the disease. In the case of mental disorders, diagnoses are based on clinical symptoms and interviews of the patients rather than biochemical or other, more quantitative measures. In addition, several different disease categories exist. For example, a different set of genes likely predispose to paranoid and catatonic schizophrenia, two separate diseases similar yet different. Finally, the role of environmental factors cannot be controlled in the analysis. These initial disappointments have led geneticists to use new methods, such as microarrays, in order to find genes predisposing to complex diseases. Recently, several studies using human brain samples and either oligonucleotide or cDNA microarrays have been published (Mirnics et al., 2000; Chabas et al., 2001; Hakak et al., 2001; Ho et al., 2001; Loring et al., 2001; Vawter et al., 2001; Whitney et al., 2001). There is also an ongoing search for relevant animal models for complex diseases. The major advantage of working with animals is that the environment can be entirely controlled, and one can concentrate on phenotypic differences resulting from genetic differences.

In this chapter, we will emphasize the use of animal models in conjunction with new genomic methodology in trying to understand complex neurobiological diseases and behavior, and we will review some recent research using these techniques. The nature of the brain tissue calls for special care in the study design, quality control, and data analysis and interpretation. We will discuss the precautions that need to be taken into account when performing gene expression profiling with brain tissue, and we will provide practical advice on how to set up, carry out, and analyze such experiments. We will give several examples of the studies underway in our laboratory and in others and hope they will benefit the reader.

10.2 RELEVANT BACKGROUND INFORMATION

10.2.1 Overview of Microarrays

DNA arrays work by hybridization of labeled RNA or DNA in solution to DNA molecules attached at specific locations on a surface. The concept of DNA microarrays is more than a decade old (Freimert et al., 1989; Gress et al., 1992) and several different techniques emerged in mid 1990s (Schena et al., 1995; DeRisi et al., 1996; Lockhart et al., 1996). Currently available arrays offer increased sensitivity and a larger number of target genes. The two most widely used microarray types for gene expression analysis are Affymetrix oligonucleotide and cDNA microarrays. In addition, other types of microarrays exist, such as oligonucleotide arrays from Agilent Technologies.

Production of the different types of arrays varies greatly. Commercially available Affymetrix oligonucleotide arrays were developed by Lockhart and colleagues, and they are based on single-color hybridization. They are produced by in situ synthesis of oligonucleotides using photolithography, and $\sim 10^7$ copies of each selected oligonucleotide (usually 20–25 nucleotides in length) are synthesized base by base on a glass surface. Typically, multiple probes per gene are placed on the array (11–20 pairs) (Lockhart et al., 1996). cDNA arrays are based on two-color hybridization and are produced by robotic deposition of nucleic acids (PCR products or plasmids) onto a glass slide. Approximately 1 ng of material is deposited, and usually a single longer (up to 1000 bp) double-stranded DNA probe is used for each target gene. These arrays are available commercially, but a large number of laboratories or microarray core facilities print their own cDNA arrays (Schena et al., 1995; DeRisi et al., 1996). Agilent Technologies produce oligonucleotide arrays that are based on in situ ink-jet synthesis of single 60-nucleotide-long probes (Hughes et al., 2001). While these arrays are less variable than cDNA arrays, they are based on two-color hybridization like cDNA arrays. In all cases, probes are usually designed from sequence located nearer to the 3' end of the gene, and different probes can be used for different exons to enable the detection of variant splice forms (Lockhart and Winzeler, 2000).

Both cDNA arrays and oligonucleotide arrays have advantages and disadvantages. The strong advantage of Affymetrix oligonucleotide arrays is that they are designed for single-sample, single-color hybridizations. Because these arrays are prepared in a highly reproducible industrial process, appropriately triaged data from any array of a given type used in a consistent way can be compared in detail to that from any other array of the same type, without the need for two-color labeling and co-hybridization of samples (Figure 10.1). This enables building of large databases where information of hundreds of experiments can be stored, retrieved, and compared to each other. In addition, Affymetrix arrays have multiple oligonucleotide features called probe sets per gene, which enable more reliable detection of the transcript in question. The main advantage of the cDNA arrays are their versatility. Gene sets can be changed, expanded, or modified to include, for example, variant splice forms or newly discovered genes. They are also of modest cost, but because of variability in their synthesis, individual arrays cannot be compared with each other as reliably as in the case

| | Single-color hybridization | | | | Two-color cohybridization | | | |
| | | | n=20 (n=200) | | | | n=20 (n=200) | |
	No. samples	No. arrays	No. samples	No. arrays	No. sample	No. arrays	No. samples	No. arrays
Simple pairwise	2n	2n	40 (400)	40 (400)	2n	n	40 (400)	20 (200)
All pairwise	2n	2n	40 (400)	40 (400)	2n	n(2n-1)	40 (400)	780 (79,800) [#]
Using common reference					2n + ref	2n	40 + ref (400 + ref)	40 (400) [*]

[*] Unlike with the single-color system this is not an absolute signal, but relative only to common reference.

[#] Common reference for CNS may not be possible and in those cases the direct comparison must be performed.

Figure 10.1 Calculation of the number of samples, arrays, and experimental measurements. The figure shows a comparison between the number of samples and arrays required to produce a comparable number of explicit pairwise comparisons for single-color and two-color hybridization methods. As shown, with a highly reproducible single-color method, $2n$ measurements (or n pairs) lead to a total of $n(2n - 1)$ pairwise comparisons ($2n$ measurements taken two at a time). To reliably detect subtle expression changes with arrays designed for two-color hybridizations, it is generally necessary to co-hybridize each pair of samples to the same physical array. In this case, $n(2n - 1)$ pairwise co-hybridization measurements on this many different arrays must be performed. The ratio of the number of two-color to single-color experimental measurements is then $(2n - 1)/2$, or approximately n. Expression measurements need to be performed at least in duplicate, and sometimes independent triplicates or quadruplicates are recommended, depending on the consistency of the sample source. So, for 20 different measurements (e.g., 20 brain regions) done in duplicate, 40 samples must be prepared and 40 single-color experiments are needed to generate the full set of 780 pairwise comparisons in silico. By contrast, 780 samples and 780 actual two-color cohybridization experiments would be required to generate a comparable set of explicit pairwise comparison. With 200 different brain regions, 400 single-color measurements suffice, whereas 79,800 two-color measurements would be required to generate the complete set of all possible pairwise comparison. Of course, it may not be necessary to perform all pairwise measurements for all purposes. For example, in time-course studies, comparisons to the initial time point may be sufficient, and simple pairwise comparisons may suffice when samples come in natural pairs (e.g., sets of disease tissue and normal tissue from the same individuals). To reduce the number of measurements required with large sets of two-color hybridizations, alternative experimental designs are often employed. One technique is to use a common reference sample for all co-hybridization measurements. In this approach, direct physical comparisons between samples are not made, but each sample is compared to the reference, and then various computational methods are used to make comparison between samples, based on the explicit reference-based measurements. Commonly used referencing strategies employ a "universal" reference, or a reference sample made from pools of the experimental samples. This is a reasonable approach, but it does require the production of a very large amount of the labeled reference material. In addition, any spot that gives a very low signal (or too large a signal) with the reference sample can be rather uninformative. Most importantly, sensitivity to subtle differences between real samples is sacrificed when they have to be determined indirectly through a common reference, rather than directly in an explicit co-hybridization measurement. This issue does not arise with single-color measurements on highly reproducible oligonucleotide arrays because each experiment is quantitative in an absolute sense and can be compared in silico to any other experiment performed in the same way on the same chip type. Therefore, single-color measurements on this type of chip have distinct advantages when a significant number of different sample types are involved, when new samples will continue to be added to the data set, and when direct comparisons of any sample to any other are required in order to detect subtle yet important changes in gene expression.

of oligonucleotide arrays. Furthermore, samples cannot be retrieved from cDNA array hybridization, there is no universal reference for brain samples, and the quality control of the data and the determination of the false-positive rate are difficult to perform. In order to maximize sensitivity to subtle expression changes and low-abundance messages, while keeping false positives to a minimum, direct and detailed pairwise comparisons of the hybridization patterns between independent samples need to be performed. The specifics of how to accomplish this are different depending on whether Affymetrix oligonucleotide arrays or cDNA arrays are used (Figure 10.1). More detailed information about microarrays can be found in recent methods books that have been written focusing on technical issues in neurogenomics (Chin and Moldin, 2001; Geschwind and Gregg, 2002).

10.2.2 Animal Models in Gene Mapping

A growing body of evidence has been obtained for the existence of genetic loci involved in the etiology of many complex diseases utilizing human linkage analysis and animal models. Experimental allergic encephalomyelitis (EAE) in rodents may be a relevant animal model for multiple sclerosis (MS). Human syntenic regions to murine loci predisposing to EAE (Sundvall et al., 1995) have been tested as candidate regions for genetic susceptibility of MS in Finnish families (Kuokkanen et al., 1996), and several markers on one of the three tested regions showed evidence for linkage in human families. Recently, two studies have used cDNA arrays to study genes involved in EAE in rodents (Chabas et al., 2001; Whitney et al., 2001) and in human MS patients (Whitney et al., 1999). Experimental crosses and inbred strains of mice offer an ideal setting for the genetic dissection of some diseases and phenotypes. With the opportunity to study hundreds of meioses from a single set of parents, the problem of genetic heterogeneity disappears, and far more complex genetic interactions can be probed than is possible in human families.

10.2.3 Quantitative Trait Loci (QTL)

Studies of inbred strains of mice have allowed the mapping of many QTLs associated with a range of abnormal behaviors. The task of identifying the gene(s) underlying the QTLs is typically accomplished using standard genetic techniques to narrow the chromosomal region, followed by an attempt to identify the specific gene or genes responsible for the phenotype (Figure 10.2). However, this latter step is often difficult and time-consuming. Recently, with the increase in the number of genes discovered, the main focus is on testing candidate genes rather than discovery of new genes in mapped regions. Given the difficulties in identifying the genes responsible for the phenotype, gene expression profiling may be extremely useful in identifying or establishing the role of a particular set of genes mapping to the region. We and others have proposed that gene expression analysis can be used to complement QTL analysis (Karp et al., 2000; Sandberg et al., 2000; Luo and Geschwind, 2001), and the recently published sequence information of the mouse genome will benefit this approach as well.

10.2.4 Advances in Mouse Genetics

Progress in understanding the neurobiology of behavior has increased rapidly in recent years due to several major advances, most notably advances in mouse genetics. The rapid pace at which new genes are being identified, the availability of the DNA sequence for the mouse, the extensive analysis of inbred strains of mice, and the use of induced mutations has created new tools for investigating the molecular and genetic basis of behavioral phenotypes. The use of inbred strains of mice has been critical in providing new information about neurobehavioral and neurological phenotypes. Since animals of one strain are genetically identical, inter-strain phenotypic differences are caused by genetic factors (Figure 10.3). Many studies have focused on trying to understand the constellation of genetic variations that account for significant differences between inbred strains of mice. For example, neurogenesis after exposure to an enriched environment differs substantially between the C57BL/6 and 129SvJ mouse strains (Kempermann et al., 1997, 1998). Other studies showed that despite similar seizure susceptibility, various inbred strains showed large differences in neuronal cell death after seizures (Schauwecker and Steward, 1997). At the level of behavior, inbred strains of mice vary greatly in their behavioral response to drugs of addiction including ethanol (Metten et al., 1998). These strains also show marked differences in some types of behavioral testing, such as pre-pulse inhibition, anxiety, and running wheel activity (Crawley et al., 1997; Carter et al., 2001).

We and others have proposed combining behavioral testing in inbred mouse strains with gene expression profiling (Sandberg et al., 2000; Carter et al., 2001; also see Figure 10.10). We are currently using this approach to find genes regulating anxiety. We have chosen six different inbred mouse strains that are being tested in several anxiety-related behavioral paradigms. Concurrently, we have microdissected brain regions that have been shown to be important in regulating anxiety (e.g., amygdala, hippocampus, periaqueductal gray, and cingulate cortex) from the same inbred strains. We are using Affymetrix GeneChips to find differentially expressed genes in these brain regions and mouse strains. By comparing the gene expression patterns of highly anxious and less anxious mouse strains, we will be able to identify genes or pathways regulating anxiety in mice. A major advantage to this approach is the availability of sequence data for five different inbred strains. To find causative changes, gene expression data can be combined with corresponding sequence information. The same genes that regulate anxiety in mice likely regulate anxiety in humans as well, and this information can be applied in studies of human anxiety disorders.

10.2.5 Transgenic Mice

Another notable advance is the use of transgenic technologies to generate targeted mutations in genes (null mutations) or to over- or mis-express genes (transgenics) (for a review, see Steele et al., 1998). Many novel behavioral phenotypes have been observed. These have ranged from mice with decreased or enhanced learning and memory (Abeliovich et al., 1993; Mayford et al., 1995; Tang et al., 1999), to mice with difficulty in performing motor

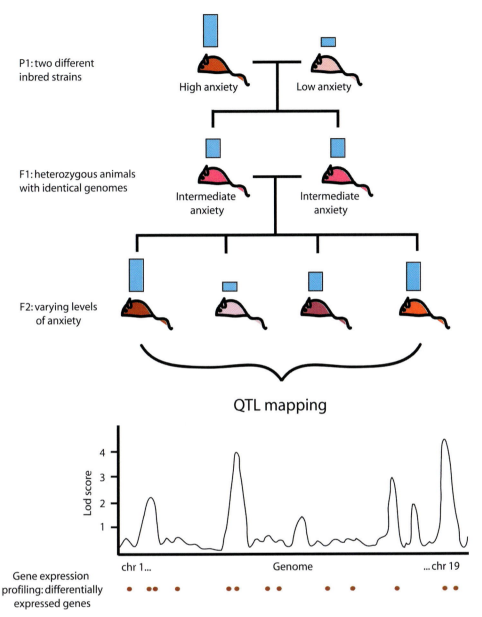

P1: two different inbred strains
High anxiety   Low anxiety

F1: heterozygous animals with identical genomes
Intermediate anxiety   Intermediate anxiety

F2: varying levels of anxiety

QTL mapping

Gene expression profiling: differentially expressed genes

Figure 10.2 A traditional quantitative trait locus (QTL) analysis approach to identify genes involved in complex diseases. First, two phenotypically extreme parental inbred mouse strains are selected. These mice are bred to obtain genetically identical, heterozygote F1 animals that are then crossed to obtain F2 animals that show variable levels of the phenotype. F2 animals are genotyped and phenotyped, and a standard linkage-based analysis is performed to identify chromosomal regions that are likely to harbor genes predisposing to the phenotype. Typically, these genomic regions are large (1–30 cM), and positional cloning of the causative gene can be tedious. By determining gene expression profiles of the F2 animals, one might be able to find candidate genes by simply choosing genes that are differentially expressed or are directly regulated by a differentially expressed gene in animals with diverse phenotypes and that localize on the chromosomal region of a QTL.

tasks (Barlow et al., 1996), mice with sensitivity to drugs of addiction (Phillips, 1997), and many others too numerous to site here.

10.2.6 Modifier Genes

Another interesting recent development has been the observation that modifier genes can greatly impact a particular phenotype.

For example, it has been found that the neurobehavioral phenotype of a particular mouse results not only from a specific alteration induced by a targeted mutation, the mis-expression of a particular gene, or the administration of a particular drug, but also from the effects of modifiers, which may differ significantly based on the genetic background (Crawley, 1996; Crusio, 1996; Dawson et al., 1996; Gerlai, 1996; Lathe, 1996; Morrison et al., 1996; Watanabe et al., 1996). By crossing mutant mice to other strains with differing susceptibility to the particular phenotype,

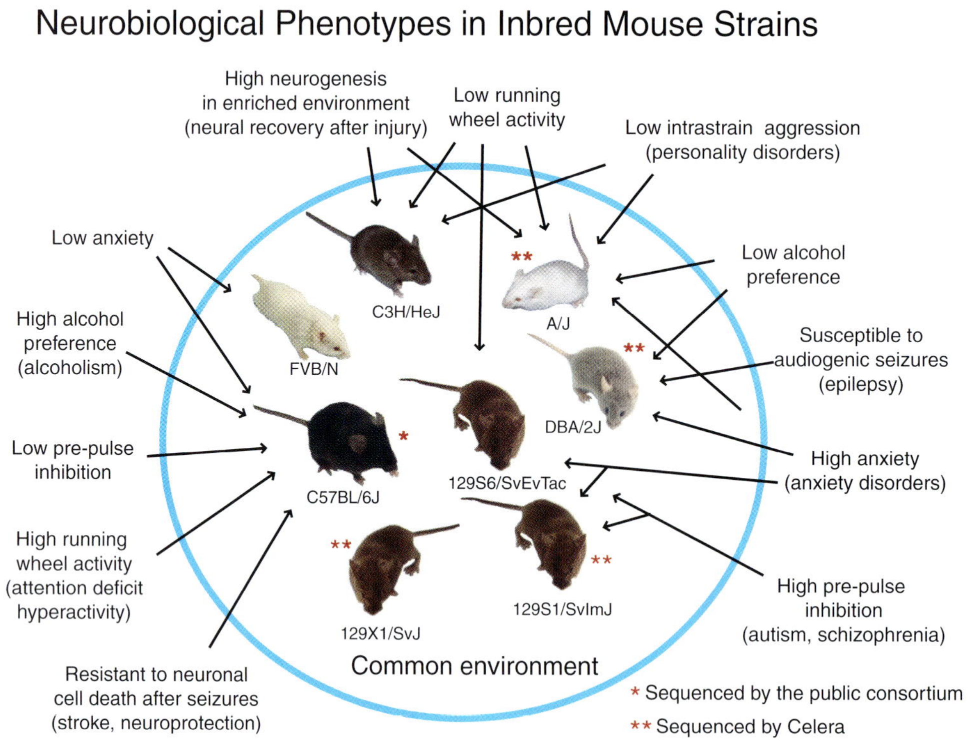

Neurobiological Phenotypes in Inbred Mouse Strains

High neurogenesis
in enriched environment
(neural recovery after injury)

Low running
wheel activity

Low intrastrain aggression
(personality disorders)

Low anxiety

Low alcohol
preference

High alcohol
preference
(alcoholism)

Susceptible to
audiogenic seizures
(epilepsy)

Low pre-pulse
inhibition

High anxiety
(anxiety disorders)

High running
wheel activity
(attention deficit
hyperactivity)

High pre-pulse
inhibition
(autism, schizophrenia)

Resistant to neuronal
cell death after seizures
(stroke, neuroprotection)

Common environment

C3H/HeJ A/J FVB/N DBA/2J C57BL/6J 129S6/SvEvTac 129X1/SvJ 129S1/SvImJ

* Sequenced by the public consortium
** Sequenced by Celera

Figure 10.3 Neurobiological phenotypes of inbred mouse strains. Since animals of one inbred strain are genetically identical and all mice share the same environment, inter-strain phenotypic differences are caused by genetic factors. Inbred strains differ in many neurobiological phenotypes, some of which are listed in the figure. By comparing the genomes of phenotypically extreme strains, we may identify genes regulating that particular phenotype. The genome sequence of the C57BL6/J strain is available in the public domain, and the sequence of 129X1/SvJ, 129S1/SvImJ, A/J, and DBA/2J are available from Celera. Note that the 129X1/SvJ is a contaminated strain and is therefore not commonly used. 129S1/SvImJ and 129S6/SvEvTac are derived from the same congenic strain and are closely related.

it may be possible to elucidate the influence of various gene interactions.

These and other studies demonstrate that the mouse is an excellent model for studying many aspects of behavior with relevance to a wide range of human mental health, neurological disorders, and/or psychiatric disorders. However, it remains to be established how the constellation of modifying genes influences these phenotypes and what areas of the brain are involved. Because behaviors are influenced by many factors ranging from the environment to specific gene interactions, it is increasingly important to consider candidate genes or mutations in light of the multitude of potential modifiers.

10.2.7 The Mouse, Complex Biology, and Gene Expression Profiling

Highly parallel gene expression approaches allow one to look at the global interactions of genes and modifiers and their effects, and it will greatly enhance our ability to define the role of developmental alterations, mutations, and compensatory mechanisms in causing or modifying particular behaviors. Behavioral testing combined with microarray profiling has proven to be a powerful approach to identify genes regulating mouse behaviors, and the

same genes and metabolic pathways most likely, at least partially, regulate the same behaviors in humans as well (Figure 10.4).

When strict procedures are applied during animal and tissue handling, sample preparation, and array hybridization, overall assay variability can be minimized, and the false-positive rate can be kept extremely low. For example, a brain mapping study from our laboratory shows that the false-positive rate, as assessed by independently repeating expression measurements on the hippocampus of different mice, was between zero and three genes per ~13,000 measured, while maintaining sensitivity to rare transcripts and small changes (Sandberg et al., 2000; Lockhart and Barlow, 2001b; Barlow and Lockhart, 2002).

The drawback of using animals as models for human diseases is that it requires a well-characterized animal model for a given disease, and such models do not exist for many psychiatric diseases, such as schizophrenia. However, individual components of complex diseases can be studied separately, and in case of schizophrenia this has been utilized by studying pre-pulse inhibition. Pre-pulse inhibition is the suppression of the normal response to a startling stimulus when that stimulus is immediately preceded by a weak pre-stimulus or pre-pulse (Graham, 1975). Pre-pulse inhibition differs in schizophrenia patients and controls as well as in different inbred mouse strains as measured using acoustic startle in humans and mice.

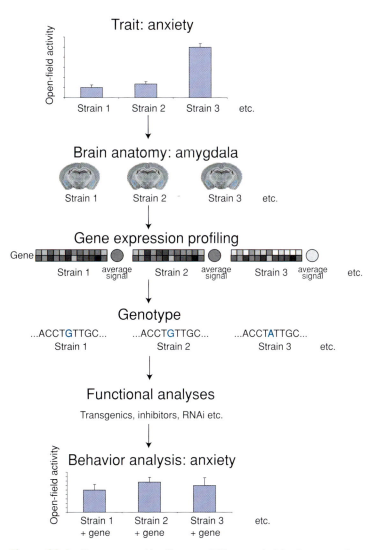

Figure 10.4 Our strategy to identify susceptibility genes in inbred mouse strains. Inbred mouse strains are tested for a given phenotype (e.g., anxiety) to identify the strains with lowest and highest values. Subsequently, the brain regions involved in regulating this phenotype are dissected from the same inbred strains. Total RNA is extracted and the samples are hybridized on Affymetrix GeneChips. Gene expression profiles of the high-anxiety strains and low-anxiety strains are compared to identify genes that are differentially expressed. These genes are then sequenced to find causative SNPs that may lead to observed gene expression differences. Depending on the nature of the gene, different strategies may be used to prove that this gene actually regulates the phenotype in question. Transgenic technologies, protein inhibitors, or RNAi may be utilized. For example, if an enzyme is up-regulated in high-anxiety strains, it might be possible to reverse the phenotype by administering an inhibitor of the enzyme.

10.3 TECHNICAL DETAILS AND METHODOLOGY

10.3.1 Tissue Collection and RNA Extraction

Human Samples

Gene expression profiling has contributed to our understanding of the pathophysiology of many diseases, including cancer. The progress in neurobiological diseases has not been as rapid. In part this is due to the poor availability of good-quality human brain samples (Figure 10.5). Although fresh tissue is always preferred, this is infrequently available. However, the mRNA in most regions of the brain appears to be surprisingly well-preserved, at least up to 48 h post mortem (Harrison et al., 1995; Bahn et al., 2001). It is now clear that agonal status and the freezing of the tissue impact mRNA integrity more than post-mortem interval (reviewed in Bahn et al., 2001). In our hands, samples collected less than 12 h post mortem usually produce consistent results. An essential requirement in studies using human brain samples

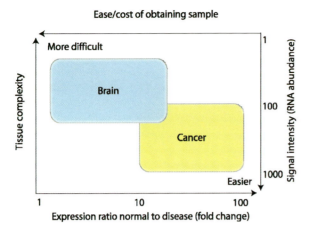

Figure 10.5 Differences in the tissue complexity and gene expression fold change levels in the brain versus cancer tissues. Typically, brain tissue is more complex and has a larger number of different cell types than cancer tissue. In addition, the gene expression fold changes when comparing a normal and a pathological state are fairly small in the brain, while fold changes between a cancer tissue and a corresponding normal tissue are typically large. Therefore the analysis of gene expression profiles from brain tissue requires sophisticated methods to identify very small gene expression differences that might be significant for a phenotype.

is that a sufficient number of experiments be performed across multiple individuals and multiple tissue samples to account for individual variation and possible tissue inhomogeneity.

To minimize the number of hybridizations, it is possible to pool samples of several individuals. It is believed that this helps smooth out individual differences, and gives the average expression behavior of a gene, but requires the use of more individuals. However, this approach masks the underlying distribution of expression levels, and it is not generally advisable when studying human samples. Ideally, it is best to analyze individual samples because it allows identification of noise due to genetic variation or sample artifacts so that true disease associations can be detected.

Samples from Experimental Animals

The advantage of working with mice is that the handling of the animals and the tissue dissections can be performed systematically and consistently. In our experience, this is extremely important in order to minimize sample to sample variations (Lockhart and Barlow, 2001a). In our laboratory, all animals are singly housed for 7 days prior to sacrifice, and the dissections are carried out at specified hours of the day. Dissections are performed on the surface of petri dishes filled with wet ice. Samples are dissected as fast as possible and frozen on dry ice. We try to avoid pooling whenever possible because dissection artifacts from one sample in a pool will invalidate the results of the entire pool. If dissecting small brain regions, such as amydala or cortical subregions, either we pool samples from two to five animals and collect them in RNAlater buffer (Ambion) or we use methods for small sample amplification that do not require pooling. Samples in RNAlater can be stored at room temperature until all of the samples are collected. The extra buffer is then removed, and the samples are frozen on dry ice and stored at $-80°$C.

We perform many brain dissections by hand using a sophisticated dissection microscope and very fine dissection tools. It is important to dissect the exact anatomical structure of inter-

est. Inclusion of too much additional material might "dilute" a weak expression difference seen between two regions, and failure to include all small subnuclei might lead to missing important differences. Laser capture microdissection (LCM) is a method for producing reasonably pure populations of targeted cells from small regions of tissue sections that cannot be obtained otherwise (Bonner et al., 1997). LCM uses a specially designed microscope and an integrated laser to select and collect cells onto a special transfer film. It has been shown that the detailed morphology of the captured cells is maintained and that DNA and RNA of high quality can be extracted from laser-captured cells and used for a range of different analyses, including gene expression monitoring on DNA microarrays (Luo et al., 1999; Sgroi et al., 1999). It is important to use stringent quality control of extracted total RNA. This can be achieved by checking RNA quality on an agarose gel and by an absorption measurement or by using a bioanalyzer (Lockhart and Barlow, 2001a). We have found that RNA of virtually identical quality can be obtained using LCM as compared to that obtained using standard methods (e.g., OD 260/280 of 2.0–2.1). It is clear that gene expression profiling of a brain tissue from humans or model organisms is feasible if the special nature of the brain tissue is taken into account and rigorous quality control is performed (Figure 10.6).

10.3.2 Labeling and Hybridization

In most current implementations of array-based approaches, the RNA or DNA to be hybridized must be labeled prior to the hybridization reaction. The surface bound molecules can then be detected fluorescently and quantitated. cDNA arrays are commonly labeled with fluorescein, Cy3, or Cy5. Labeling with biotin is most common for use with oligonucleotide arrays (Lockhart et al., 1996). The biotin-containing molecules are made fluorescent following hybridization by staining with a streptavidin–phycoerythrin conjugate. With cDNA arrays or oligonucleotide arrays that are still based on two-color fluorescence measures, samples are labeled with two different fluorescent dyes, and the two samples are hybridized on one array. With Affymetrix oligonucleotide arrays, each sample is labeled identically and hybridized independently to different arrays. Signal intensities can then be compared directly between any two or more arrays (Lockhart and Barlow, 2001a).

The standard procedure for mRNA amplification and labeling for oligonucleotide arrays has been described elsewhere (Lockhart et al., 1996; Wodicka et al., 1997). This amplification and labeling procedure is highly reproducible and introduces very little bias in the mRNA population (Lockhart and Barlow, 2001a). This method works routinely with starting material of 5–10 μg of total RNA. It is important that the same amount of total RNA is used for the labeling reactions to avoid false-positive results. However, when small brain regions are dissected, it is often impossible to obtain sufficient RNA without pooling samples from many animals. Nevertheless, pooling introduces additional potential error and should be avoided especially when analyzing human samples or samples from outbred animals. Efficient and reproducible mRNA amplification methods are required, and two methods show significant promise. The first is a PCR-based approach that has been used to make single-cell cDNA libraries (Wang et al., 1989; Jena et al., 1996). We and others currently use another method that does not rely on PCR

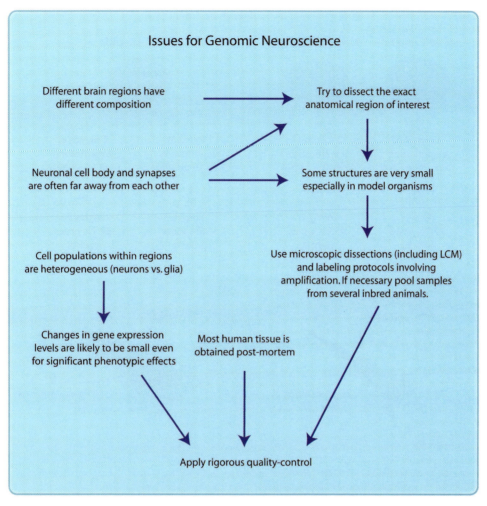

Figure 10.6 Special features of brain tissue. Gene expression profiling from brain tissue is feasible if the nature of the tissue is taken into account and if rigorous quality control is performed. Different brain regions have very different composition, and often the body of the neuron resides far away from the synaptic terminals. Therefore it is critical to dissect the exact region of interest to avoid "diluting" the possibly weak but meaningful gene expression differences between different cell populations. When rigorous quality control is applied, gene expression differences in brain tissue can be trusted.

but on multiple rounds of linear amplification based on cDNA synthesis and a template-directed IVT reaction (Van Gelder et al., 1990; Eberwine et al., 1992; Kacharmina et al., 1999; Luo et al., 1999; Wang et al., 2000). Each round of amplification yields a typical amplification between 100- and 1000-fold (Lockhart and Barlow, 2001a). Each round consists of a cDNA synthesis step, which yields no amplification, plus an IVT step to produce single-stranded RNA. This basic method was developed specifically to characterize mRNA from single live neurons.

We find that the multiple-round cDNA/IVT amplification method produces sufficient quantities of labeled material and is reproducible between independent preparations. In our hands, 50 ng of starting total RNA yields enough labeled cRNA using two-round amplification and 5 ng is enough when using three-round amplification, to produce consistent expression patterns (Figure 10.7). While multiple rounds of amplification produces consistent results in our hands, it is important to note that it introduces some skewing in the expression patterns. Therefore, it is not advisable to make direct comparisons of results obtained from single-round amplifications to results obtained from

two or three rounds of amplifications (e.g., in Microarray Suite software) (Figure 10.7). However, the gene lists obtained from comparisons between one-round and one-round amplification or two-round and two-round amplification can be compared.

An advantage of the Affymetrix oligonucleotide-array-based method is that samples can be hybridized repeatedly on multiple arrays, up to 8–10 times (Wodicka et al., 1997; Lipshutz et al., 1999). This is possible because only a small fraction of the molecules in a sample remain bound to the oligonucleotide probes on the array surface. The vast majority of the molecules are recovered simply by removing the hybridization reaction mixture from the array cartridge after each hybridization reaction. Samples can be stored at -80°C, thawed, and rehybridized to new arrays. It is thus possible to collect expression data using the arrays we have now, but then at a later date when we have new arrays to use, the same samples can be interrogated without having to perform additional tissue collections, RNA extractions, and labeling reactions. As a result, data can be collected at different points of time and still be reliably compared to other datasets. This is extremely important for genomic

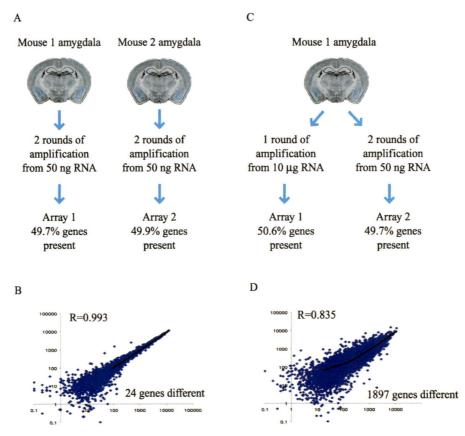

Figure 10.7   Comparison of results from labeling protocols involving one or two rounds of in vitro transcription (IVT). Two 8-week-old C57BL/6J male mice were used for microarray analysis. Amygdalae of two mice were hand-dissected under a dissection microscope and collected in the RNAlater buffer (Ambion). Extra buffer was removed and the samples were snap-frozen on dry ice and stored at −80°C until the total RNA was extracted using the Trizol reagent. (A) fifty nanograms of total RNA from each animal was processed through two rounds of IVT (step 1: reverse transcription; step 2: second strand synthesis; step 3: IVT; step 4: reverse transcription; step 5: second strand synthesis; step 6: IVT). Thirty micrograms of labeled cRNA was hybridized on Affymetrix MG_U74Av2 arrays. Gene expression profiles of two different animals were compared using the Bullfrog program and our standard criteria (a fold change of 1.8 or greater, a qualitative call of "increased," "marginally increased," "decreased" or "marginally decreased," an average difference change greater than 50 fluorescence units, and the probe set of the gene must have been scored as "present" in at least one array). Twenty-four probe sets were observed as differentially expressed (leading to a false-positive rate of 0.20%), and the correlation coefficient of the signal intensity of the probe sets between the two arrays was 0.993 (B). We generally see similar false-positive rates when starting from 10 μg of total RNA and doing only one round of IVT. (C) The same total RNA sample was processed through one round of amplification starting from 10 μg of total RNA or two rounds of amplification starting from 50 ng of total RNA and hybridized on separate arrays. One thousand eight hundred ninety-seven probe sets were scored differentially expressed using our standard criteria leading to a false-positive rate of 15.5%. The correlation coefficient of the signal intensity of the probe sets in the two arrays was 0.835 (D). Two rounds of linear amplification produce skewing of the labeled cRNA product leading to a large number of false positives when two-round amplification results are compared to one-round amplification results. This example shows that it is extremely important to directly compare (e.g., using the Microarray Suite software to generate comparison files) only arrays with the same starting amount of RNA and same number of amplification steps with each other to avoid a large number of false-positive results.

neuroscience because it is often difficult to obtain high-quality brain samples, especially for human studies.

10.3.3 Data Analysis

Assessing Chip and Sample Quality

The scanned arrays should first be inspected visually to assess the quality of the array and to detect any manufacturing errors, such

as scratches or glue spots, and to see that the hybridization pattern is even. Following the basic image analysis, data "triage" should be performed to make sure that the array data are of sufficient quality for further analysis and comparison with other datasets (Figure 10.8). Affymetrix arrays have several internal controls to assess data quality, and there is software that extracts this information easily as discussed later in this section. The factors to monitor include background, noise, overall signal strength, the

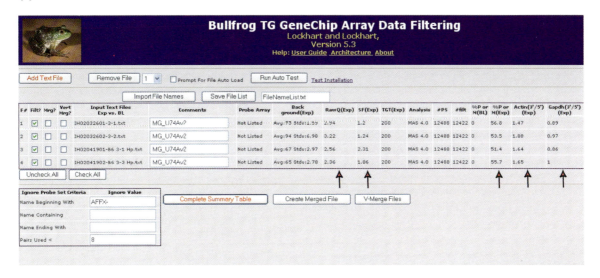

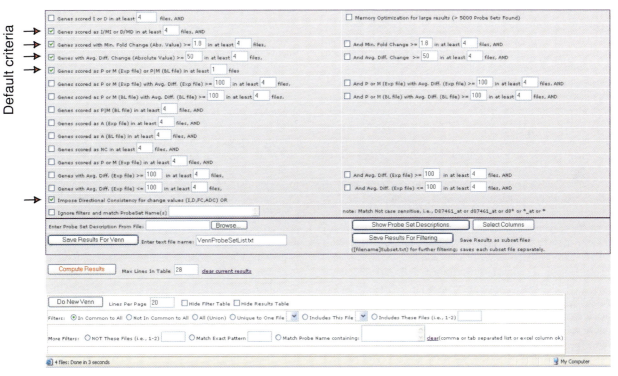

Figure 10.8 Expression experiment sample and data triage using Bullfrog freeware. (A) This screen capture includes a minimal set of sample and data quality measures that should be recorded and checked for in every array-based gene expression experiment. "Background" is the overall, average background signal across the entire array. "Stdv of the background" is the standard deviation of the background across different physical regions and measures the background consistency. "RawQ" is a measure of the noise in the background signal and is used to set minimum thresholds in the data analysis algorithms. "SF" is the linear scaling factor that is applied to equalize the average signal for every array dataset. "TGT" is the target intensity for scaling. "Analysis" tells which algorithm was used for data analysis. "#PS" gives the number of probe sets on the array, and "#filt" gives the number of filtered probe sets. "%P or M" is the percentage of probe sets that scored either present or marginally present (typically 30–60% for mammalian cells). Actin $(3'/5')$ and Gapdh $(3'/5')$ are the ratios of the signals observed for probe sets derived from the $3'$ and $5'$ ends of these abundant transcripts. A value near 1.0 indicates that the two ends are approximately equally represented and that the original mRNA was not significantly degraded. When using no amplification during the labeling, these values should be very close to 1.0 and less than 2.0. Amplification introduces skewing of the mRNA population, and ratios greater than 50 are often seen. However, if the values of all arrays are similar, the number of false-positive findings is not increased significantly (see Figure 10.7). (B) The upper part of this screen capture shows adjustable filters used to identify differentially expressed genes, and the lower part of the screen capture shows the Venn diagram function.

ability to detect spiked bacterial control RNA, the ratio of the 3′ and 5′ signals for actin and Gapdh (glyceraldehyde-3-phosphate dehydrogenase) mRNA (this is a measure of RNA length and quality, and degraded RNA will result in high 3′/5′ ratios because only the region of the mRNA near the 3′ poly-A tail will be amplified and labeled), and the percentage of genes scored as "present." The number of genes reliably detectable (based on statistical "present calls") in the brain tissue is usually 45–55% when a single round of IVT is performed. The number of genes present in other tissues is generally lower, usually 35–55%. The exact quality control criteria that we use in our laboratory can be found elsewhere (Lockhart and Barlow, 2001a). It is critical that the biologist have tools to directly assess the quality of the data and use a general analysis to see if biological or experimental noise exists. Arrays that do not meet established standards need to be discarded from further analysis. Later in the chapter we will describe freeware tools designed to make this task easier for a bench scientist.

Estimating False-Positive Rate

For high-throughput, parallel measurements, data quality is of critical importance if one is attempting to identify with high confidence specific genes that are differentially expressed. The reason is that when monitoring, for example, 10,000 genes, even a low false-positive rate of 1% results in 100 incorrect difference calls, comparable to the number of true changes observed in many types of experiments (Lockhart and Barlow, 2001a). In order to increase confidence in the results, as well as the quantitative accuracy, we perform at least two replicate experiments using independent cell sources, independent sample preparations, and independent arrays. If there are expected sources of additional variation, a larger number of replicates may need to be performed to ensure the same level of confidence in the results. This might be necessary when inbred strains of mice are not used or for any studies that use human tissue or cell lines.

A major advantage of the single-color platform, such as Affymetrix arrays, is that the false-positive and the true-positive rate can be evaluated directly without the need to perform additional experiments. The false-positive rate can be estimated using array data from independent replicate samples where tissue is derived from different individuals or animals. This will control for the biological variance between individuals and for experimental variance that is due to tissue dissection, RNA extraction and labeling, and array hybridization differences. Replicate arrays are compared using exactly the same filtering criteria that will be used in the actual experiment comparing different conditions. The genes that come up as differentially expressed between the replicate samples are false positives, and the false-positive rate can be estimated (Figure 10.7). If inbred mouse strains are used, the false-positive rate is generally less than 0.2% when analyzing two replicate samples. When studying human tissue, cell lines, or outbred animals, three or four replicate samples are needed to achieve a low false-positive rate. The acceptable false-positive rate will vary based on the question being asked and the amount of independent verification the experimenter is willing to perform. Similarly, the number of replicate samples required increases when three-round IVT protocol is used because additional variance is introduced. In the next section we will describe in more detail how data are routinely analyzed in our laboratory (see also Zapala et al., 2002).

Determining If Genes Are Differentially Expressed Between Sets of Arrays

There are many types of analyses that can be performed to extract information from gene expression data. One can ask, for example, how many genes are differentially expressed in two different samples using various filtering criteria. It is important to take into account the nature of the brain tissue and the false-positive rate when designing the analysis approach, and it is to keep in mind that the gene expression differences are going to be subtle in the brain tissue (Figure 10.5). In cancer research or in studies looking at very different cell cycle states, it is not uncommon to find hundreds of genes with fold change values of greater than five. In contrast, in most studies involving the brain, few genes change and most of those have fold change values of less than five, and many genes have a fold change value between 1.2 and 1.8 (Figure 10.5).

Filtering criteria for finding differentially expressed genes should include both quantitative and qualitative criteria. For example, a gene may be considered differentially expressed if it meets the following criteria: relative change (fold change) of 1.8 or greater, a qualitative call of "increased," "marginally increased," "decreased," or "marginally decreased" and a difference in the signal (average difference change or ADC) greater than 50 fluorescence units. All of these criteria have to be consistent in the majority of the comparison files. Also, the probe set for the gene could score as "present" in at least one of the experiments to avoid finding genes that are not expressed in the tissue under study. In the Affymetrix software the qualitative calls are based on statistical and empirical algorithms using several criteria. When used in combination with carefully controlled, independent replicates, this approach produces an acceptably low false-positive rate of less than a few incorrectly assigned calls per 10,000 genes monitored (Lockhart and Barlow, 2001a). If one wishes to find a large number of candidate genes that will be tested with other molecular biology methods, it is possible to loosen the above-mentioned criteria. For example, filtering for genes with a fold change of 1.5 or greater will result in a larger number or candidate genes, but the number of false-positive genes will also increase. Similarly, one might want to require that the fold change and call criteria are met in only half of the experiments. When working with the brain, subtle changes in genes expression might be important and one does not want to miss those genes even if it requires a larger number of follow-up experiments.

A significant obstacle in microarray research has been the inability to easily process experimental data, assess the data quality, manage multiple data sets, and mine the data with user-friendly tools that can be quickly learned and applied for routine analysis by laboratory scientists. Our laboratory, in collaboration with David Lockhart and Daniel Lockhart, developed two HTML-based programs to analyze and filter gene expression data, "Bullfrog" for Affymetrix GeneChips and "Spot" for custom cDNA arrays (Figure 10.8) (Zapala et al., 2002). The programs provide intuitive data filtering tools through an easy-to-use interface. They were designed to help the researcher narrow their search from tens of thousands of gene candidates, to several hundred or fewer that meet specific but adjustable criteria. We periodically update the programs and have them available for Affymetrix outputs. These programs are freeware with an open source code. They have multiple additional functions, but at the moment the user guide is not comprehensive and the user is encouraged to experiment. The basic functions include data quality assessment, the ability to merge data from several arrays to a single experiment, filtering based on multiple adjustable criteria, and a Venn diagram function to easily compare gene lists. The general approach to data analysis is as follows. We might have two replicate cortex samples from wild-type mice (WT) and two replicate cortex samples from knock-out mice (KO). First, absolute files and then all possible pairwise comparison files (WT cortex 1 vs. WT cortex 2, KO cortex 1 vs. KO cortex 2, WT cortex 1 vs. KO cortex 1, WT cortex 2 vs. KO cortex 1, WT cortex 2 vs. KO cortex 2) are generated using Affymetrix software. First the four absolute text files are imported to Bullfrog to assess the overall data quality. If the % present difference between arrays is greater than 5%, data are often skewed and the outlier array should be omitted from the analysis. If all quality measures are met, comparison files are imported to Bullfrog for further quality control and data filtering. The false-positive rate is first determined by comparing the replicate experiments with each other (i.e., WT cortex 1 vs. WT cortex 2 and KO cortex 1 vs. KO cortex 2) using the same filtering criteria that would be used in WT vs. KO comparisons. All genes that come up as differentially expressed between the two replicate samples are false positives. If the false positive rate between WT cortex 1 and WT cortex 2 is 5%, a third replicate is likely needed. If false-positive rates for both WT replicate and KO replicate samples are within acceptable range, one can begin to compare WT samples to KO samples to find genes that are differentially expressed using the filtering function of Bullfrog (Figure 10.8).

After obtaining a list of differentially expressed genes, the researcher wants to know what proteins the genes encode and see their function in a biological context. The internet site of Affymetrix, NetAffx, lists sequence information of all probe sets on their arrays and provides links to appropriate NCBI (National Center for Biological Information) pages. Detailed information about genes and their function and their homologs in other species is easily available. It is also interesting to know if differentially expressed genes fall in the same biological pathways. Some clustering programs, such as GeneSpring, that will be introduced in the next section have a gene ontology function that assigns genes into pathways. Affymetrix, via the NetAffx site, will also add the functionality in the near future. Many useful tools are provided by the Gene Ontology Consortium whose goal is to produce a dynamic controlled vocabulary that can be applied to all organisms even as knowledge of gene and protein roles in cells is accumulating and changing.

Cluster Analysis

Another data analysis approach involves using the observed gene expression behavior to cluster genes together into groups (Eisen et al., 1998; Wen et al., 1998; Alon et al., 1999). However, it is critical to pre-select genes by filtering to avoid clustering noise. We generally use the Bullfrog filtering tool, but other tools such as D-Chip can be used. Thereafter, one can use higher-order analysis tools that exist commercially or as freeware (e.g., GeneSpring, Resolver, Cluster/Treeview, D-Chip). The basic assumption underlying clustering is that genes with similar expression behavior (e.g., increasing and decreasing together under similar circumstances) are likely to be functionally related (see Lockhart and Winzeler, 2000). In this way, genes without

previous functional assignments can be given tentative assignments or assigned a role in a biological process based on the known functions of genes in the same expression cluster. The approach is made more systematic and statistically sound by calculating the probability that the observed functional distribution of differentially expressed genes could have happened by chance. However, whereas this approach has been useful in situations where large changes occur, such as cell cycle in yeast and cancer in humans, this approach remains to be validated in neuroscience. Different methods for clustering are reviewed in Quackenbush (2001).

Clustering can also be used to assess the quality of a dataset by looking how individual arrays are related to each other (Figure 10.9). Sometimes arrays that were processed the same day might cluster together more tightly than arrays that are supposed to be more closely related simply because there was variation in the sample handling. Once this possibility is ruled out, it is possible to fully trust the data obtained from microarray experiments and proceed with asking biological questions.

10.3.4 Data Storage and Sharing

An important consideration is how to store the data obtained from gene expression profiling. One Affymetrix GeneChip expression measurement produces approximately 100 megabytes of data including raw and processed data files. Traditionally, people have made their gene expression data available to the scientific community in the form of Excel spreadsheets and large text files. However, this is not practical, and it is critical to store the data in carefully designed databases that include the gene expression patterns as well as information about tissue collection, handling, and experimental procedures in a format that makes it easy to access and query by a larger number of scientists. Gene expression databases already exist, but they are generally not suitable for neuroscience research. Currently, researchers do not have standardized nomenclature for brain structures they are profiling. Adapting anatomical nomenclature, such as the NeuroNames Hierarchy (Bowden and Martin, 1995), is essential to documentation. In addition, comprehensive information about the details of the study are needed, such as the age and strain of the animals used, experimental treatments they received, RNA extraction and hybridization protocols, and the arrays used. The goal is to develop a Web-accessible database that would not only store data but allow analysis of large amount of GeneChip data from several laboratories and accompanying anatomical and other information in appropriate database architecture. In addition, tools for data storage, querying, linking, visualization, and data analysis are needed to enable collaboration of participating neuroscientists all around the world. It is important that such a tool is in the hands of the bench scientist who is repeatedly asking questions such as, "Where is my gene of interest expressed in the brain and to what extent?" It is not practical to consult a statistician every time an analysis is run when simple statistical tests can be incorporated in the database analysis tools. One difficulty is, How does the academic community find a resource to fund the significant cost of such a database and its maintenance? Currently, there is no public support for a gene expression database for neuroscience and consequently no effective mechanism to ensure data access to entire neuroscience community.

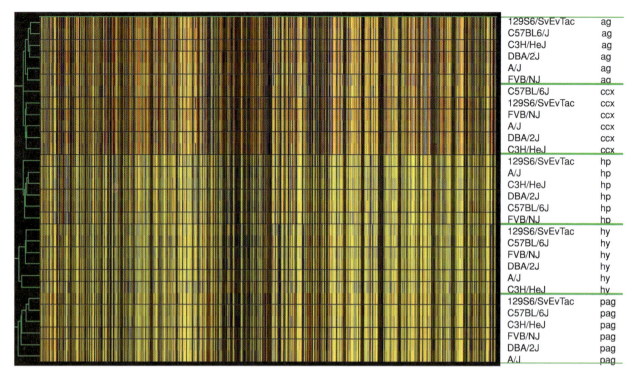

Figure 10.9 A cluster of 60 GeneChips using GeneSpring software's (Silicon Genetics) experiment cluster tool. Five brain tissues—hypothalamus (hy), hippocampus (hp), amygdala (ag), cingulate cortex (ccx), and periaqueductal gray (pag)—were dissected from six mouse strains (129S6/SvEvTac, A/J, C57BL/6J, C3H/HeJ, DBA/2J, and FVB/NJ). Two samples of each region from each strain were obtained. Samples were processed as described in the text and hybridized on Affymetrix MG_U74Av2 arrays. Microarray Suite software was used to generate absolute analysis for each sample. The absolute signal intensities (based on the perfect match–mismatch value) were imported in the GeneSpring. Signal intensities from the replicate samples were averaged, and a clustogram of genes present in at least two experiments was formed showing the relationships of different arrays. As expected, all different brain regions cluster separately. Note that the amygdala and cingulate cortex samples clustered closely together, separate from other tissues, reflecting that these samples were processed through two rounds of in vitro transcription (IVT), unlike the other samples that were processed through a single round of IVT.

For example, the database used by our laboratory has only limited support to host outside users.

We and others are taking advantage of the highly scalable analytical data warehousing systems that have been developed over the past decade for the business and defense sectors, and we are tailoring them for the needs of genomics and biological research (Dennis, 2002). The first version of such a resource, the Web-based TeraGenomics system, is in use in our laboratory. The .cel files from a GeneChip experiment are first loaded into the database together with detailed information about the experiment, such as strain and age of the mouse, dissected tissue, total RNA concentration and purity, and hybridization protocol. It is also possible to include histological images to show exactly which brain area was dissected for an array in question, and the anatomical nomenclature is adapted from the NeuroNames Brain Hierarchy (Bowden and Martin, 1995). The loaded gene expression data are normalized, and the absolute analysis is performed to assess the quality of the data. It is then possible to query the database and select arrays meeting specific criteria. Once arrays for a particular analysis have been selected, one can make any comparison files and filter those according to user specified criteria that are easily modified. It is possible to filter based on "call values" (increased, marginally increased, decreased, or marginally decreased) together with fold change criteria, or use p-value criteria from several different statistical

tests. The comparison files remain in the system for reuse. Once a desired gene list is shown on the screen, one can click on interesting genes and obtain more information through links to the NetAffx site and data imported from other sites. The system supports collaborative research of multiple scientists across multiple laboratories and institutions; array data can be designated as shared or not shared, identified as being associated with one or more research projects, and located by searching on one or more metadata fields of information about the experiment. Other similar types of resources are being developed for other areas of biology (see Stanford Genome Technology and Yale Microarray database websites).

A large database also enables efficient sharing of data between laboratories. Sharing data should make research more efficient and greatly facilitate our understanding of brain function (Koslow, 2002). Completely free and open data sharing is clearly a desirable goal; it is fundamental tenet of good science that data are made available for scrutiny and further analysis by others. This is undeniably critically important for rigorous and efficient scientific progress. Realistically, however, there are a few issues in addition to the purely technical problems. A real but less tangible problem is that many researchers, often with good reason, are reluctant to make all their data available, especially prior to publication. This is because of the potential for misuse by bioinformatics "cherry-picking" in which the data of others are mined and often published without the knowledge or any input from the original authors. This is perhaps an inevitable part of the scientific process, but it is important that researchers are given credit for being the designers of experiments and the generators of primary data. Authors of the original papers or publishers of primary data should not be "scooped," undermined, or otherwise punished by making raw data widely available, and the publication of secondary analyses of primary data should be required to meet the same demanding standards placed on those who publish original work.

10.3.5 Follow-Up Experiments of Expression Profiling Data

Because the false-positive rate of the array experiments can be made sufficiently low, it is not necessary to independently confirm every change for the results to be valid and trustworthy, especially if conclusions are based on the behavior of sets of genes rather than individual genes (Lockhart and Barlow, 2001a). The more conventional, lower-throughput methods are used in a more targeted fashion to complement the broader measurements and to follow up on the genes, pathways, and mechanisms implicated by the array results. Such methods include Northern blots, Western blots, real-time PCR, immunohistochemistry, and in situ hybridization, among others. More detailed follow-up using these methods is recommended if a gene is being chosen, for example, as a drug target, as a candidate for population genetics studies, or as the target for the construction of a transgenic or knock-out mouse.

These methods can provide a different type of information than can array-based methods. In situ hybridization might be a good confirmation method if small hand-dissected brain regions are studied, because information on the anatomical distribution of the gene expression is obtained, and it serves as a control that initial dissections were made correctly. Sometimes it is interesting to obtain information concerning the protein levels rather than mRNA abundance. Immunohistochemistry is an important method, because it reveals information about protein localization which cannot be measured directly using RNA based methods.

10.4 CURRENT APPLICATIONS

10.4.1 Running Wheel Activity in Inbred Mouse Strains as an Example of Combining Behavioral Analysis and Genomics

Our group is interested in combining behavioral analyses of inbred mouse strains with gene expression monitoring using oligonucleotide arrays to identify genes regulating the behavior of interest. When initial candidate genes are found, additional genetic analyses can be performed to further assess the function of these genes. An example is an ongoing study where genes involved in running wheel activity are being sought (Carter et al., 2001; also see Figure 10.10). Inbred mouse strains vary in their amount of voluntary running wheel activity (Ingram et al., 1981; van Praag et al., 1999; Allen et al., 2001). For this study, activity was measured in several inbred strains of mice (Carter et al., 2001). These strains included 129S6/SvEvTac, DBA/2J,

Castaneus EI, C3H/HeSnJ, and C57BL/6J. Additionally, a heterogeneous F2 cross, generated from the mating of the F1 offspring of 129S6/SvEvTac mice with C57BL/6J, was examined. Individual mice were housed in a rat cage equipped with a running wheel, and their activity was monitored. To quantitate the amount of activity, an optical counter was attached to the running wheel and running was measured each day for 7–9 days (Figure 10.10). The low-level runners, 129S6/SvEvTac and the 129/BL6 F2 mice, averaged 2245 and 2145 revolutions per day, respectively. The mid-range runners, DBA/2J and Castaneus EI, averaged 4729 and 10,805 revolutions per day, respectively. The high-level runners, C3H/HeSnJ and C57BL/6J, averaged 11,491 and 14,011 revolutions per day. An interesting observation was that the 129/BL6 F2 mice showed activity levels similar to only one of their parent strains, namely, 129S6/SvEvTac. This suggests that there might only be a small number of genes regulating this behavior in mice.

To determine which genes were differentially expressed between the various strains, gene expression profiling was performed using the striata from the four strains that exhibited the most extreme differences in the level of running wheel activity including 129S6/SvEvTac, 129/BL6 F2, C3H/HeSnJ, and C57/BL6 strains. The striatum was studied because of its known association with movement, and the tissue was dissected from 8-week-old male animals. Different mice were used in the behavioral analyses and gene expression profiling to ensure that the behavioral testing did not have an influence on gene expression patterns and that any gene expression difference identified is due to baseline differences between the strains. The extracted RNA was labeled and hybridized to Mu11KSubA GeneChips from Affymetrix.

This study was designed to find candidate genes involved in running wheel activity, and the analysis resulted in 16 comparisons (all low-level runners compared to all high-level runners, in 2 replicates). The criteria for differential expression included (a) a relative expression change of 1.3 or greater and (b) a difference in the signal (ADC) of greater than 50 fluorescence units. The probe set was scored as "present" in at least one of the 16 measurements, and the pattern was consistent in at least 13 of 16 comparisons. Using these criteria, 51 genes were identified as consistently differentially expressed between the different strains exhibiting low levels and high levels of running wheel activity (Figure 10.10). These genes function in a range of activities including cytoskeletal remodeling, signal transduction, ion transport, and transcription. The Hierarchical Clustering Tool in the GeneSpring software package was used to further evaluate these genes. Thirty genes showed higher expression levels in the low-level runners than in the high-level runners. Nineteen genes were more highly expressed in high-level runners than in low-level runners. Only two genes were difficult to classify as consistently different between low- and high-level runners (Carter et al., 2001). Of course, not all of these genes are expected to have a role in regulating running wheel activity in mice, but they can be considered as good candidates.

The use of less stringent criteria in this study makes it necessary to perform additional follow-up studies using Northern blots, real-time PCR, and in situ hybridizations. One might benefit from QTL analyses that have already shown certain chromosomal regions to contain genes involved in the particular behavior. The strongest candidate genes are those located on these

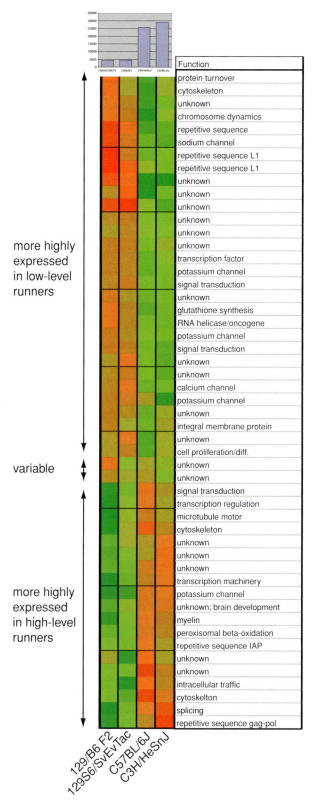

Figure 10.10 Cluster of genes that correlate with running wheel activity. Fifty-one genes were identified that met the criteria of at least a 1.3-fold change and an ADC ≥ 50 in comparisons of the high-level running strains (C3H/HeSnJ and C57BL/6J) to the low-level running strains (129/SvEv and 129/B6 F2). A call of "present" in at least one sample was also required. The expression results for these 51 genes were then subjected to hierarchical clustering using the GeneSpring software program (Silicon Genetics). The results are shown below the summary running wheel activity graph. An increase in gene expression level in a strain, relative to the average expression level across all four strains, is represented in red, and a decrease in expression level is represented in green. The magnitude of expression level change is represented by color intensity, with brighter intensities representing higher expression levels. Thirty genes were found to be more highly expressed in the strains exhibiting low-level running, 19 genes were more highly expressed in high-level runners, and two genes were variable. Known gene functions are listed to the right. Reprinted from Carter et al. (2001) with permission of Elsevier Science, Inc.

chromosomal regions. For the most promising candidates, a screen for polymorphisms and mutations can be performed, and ultimately functional analyses in cell culture systems or construction of transgenic or knock-out animals can be done to further understand the role of the gene in regulating the behavior in interest. This result was meant as an example, and significant recent progress should make this type of analysis easier and faster. For example, DNA sequence information exists for five inbred mouse strains; and by comparing the sequences of differentially expressed genes and their regulatory regions across strains, one might be able to identify cis-acting causative elements.

Information derived from gene expression monitoring in mice might be useful when studying complex human phenotypes. It is usually easier to obtain high-quality brain tissue from mice, and animal handling and tissue collection can be performed in a standardized way. The data obtained from the running wheel activity experiments might help us to better understand some aspects of complex human diseases, such as attention deficit-hyperactivity disorder. We are currently using the same methodology to identify genes regulating anxiety in mouse, and a similar strategy can be applied to virtually any well-characterized phenotype that varies in different inbred mouse strains. By combining the gene expression profiling approach with SNP mapping and QTL analysis, as shown in the recent article (Schadt et al., 2003), it may be possible to identify regulatory elements responsible for the gene expression changes.

10.4.2 Gene Expression Profiling of the Human Brain Tissue

Gene expression profiling coupled with cluster analysis has proven to be useful in classifying several types of cancer. New diagnostic categories have been identified for lymphomas, leukemias, and breast cancer, among others (reviewed in Chung et al., 2002; Liu, 2003). Central nervous system tumors have also been classified according to their cell type of origin (Pomeroy et al., 2002). It is possible to obtain good-quality brain tissue from patients who undergo surgical treatment for brain tumors. However, for many other brain-related diseases it is very difficult, and this has slowed down the use of microarray-based technologies when studying neurological and mental illnesses. However, there is a growing number of studies utilizing DNA array technology for monitoring gene expression levels in the brain tissue of these diseases. As studies of cancer have shown, it might be possible to use microarray profiling and cluster analysis to better define the subtypes of a given disease, identify target genes and pathways, predict optimal therapeutic interventions, and assess pharmacodynamic response. The next sections provide an overview of some studies aiming to resolve the biology behind three complex brain diseases.

Gene Expression Profiling in Schizophrenia Patients
Schizophrenia is a severe and debilitating mental disorder affecting about 1% of the population worldwide. Despite the effort of numerous groups, only a few susceptibility genes have been found using linkage analysis in families with several affected individuals, and it remains unclear how these genes contribute to the pathogenesis of the illness. However, multiple promising chromosomal loci have been found that are likely to harbor genes predisposing to schizophrenia. Recent microarray studies using

schizophrenia patients and matched controls have also provided new candidate genes and pathways (Mirnics et al., 2000, 2001; Hakak et al., 2001; Vawter et al., 2001, 2002; Middleton et al., 2002; Mimmack et al., 2002).

The earliest microarray study by Mimmack et al. (2002) used a custom-made candidate gene cDNA array comprising 300 genes. Using prefrontal cortex tissue from 10 schizophrenia patients and 10 controls, the group found up-regulation of the apolipoprotein L1 gene. Using real-time PCR, expression profiles of other members of the apo L family (apo L2–L6) were investigated, showing that apo L2 and L4 were significantly up-regulated in the brains of schizophrenia patients. Mirnics et al. (2000, 2001) studied six schizophrenia patients and controls using a UniGEM-V cDNA array that contains over 7000 genes. The RNA was obtained from the dorsal prefrontal cortex. By dividing the genes into functional groups, they were able to show that transcript levels were significantly decreased in the schizophrenia patients for a group of genes that were related to the presynaptic secretory machinery (PSYN). In addition, gene expression differences were observed in several other groups, namely, GABA transmission, glutamate transmission, energy metabolism, growth factors, and receptors. The interesting observation was that individual patients had different patterns of the specific members of the PSYN gene group showing decreased expression levels. The most consistently changed individual transcript was regulator of G-protein signaling 4 (RGS4), and this result was verified with in situ hybridization in a larger set of patients. A more recent follow-up study (Middleton et al., 2002) showed that common metabolic pathways were disturbed in schizophrenia patients and that these changes may not be a result of treatment with antipsychotic medication, as assessed in haloperidol-treated monkeys. Vawter et al. (2001, 2002) also used microarrays to study (a) cerebellum and prefrontal cortex samples from eight pooled patients and eight pooled controls and (b) middle temporal gyrus samples from 10 pooled patients and 10 pooled controls. In the follow-up study, three pools, each having five schizophrenia patients and five controls, were used. They applied the NIA-Neuroarray that consists of 1128 clones sorted for brain relevant genes. Synaptic, proteolytic, and signal genes accounted for one-half of the differentially expressed genes, and histidine triad nucleotide-binding protein, ubiquitin conjugating enzyme E2N, and glutamate receptor, ionotropic, AMPA2 showed the greatest down-regulation. Interestingly, some of these genes were the same as those identified by Mirnics et al. In a separate study, Hakak et al. (2001) hybridized samples from 12 schizophrenia patients and 12 controls and an additional four samples from drug-free patients on Affymetrix HuGeneFL arrays that contain probes for over 6000 human genes. The tissue was obtained from the dorsolateral prefrontal cortex. An interesting finding was the identification of five genes whose expression is enriched in myelin-forming oligodendrocytes, all of which were transcriptionally down-regulated in schizophrenia. In contrast, other genes were predominantly up-regulated in expression.

There are several factors that make comparisons between these studies difficult. First, every group used different microarrays that contain different sets of genes. Since each one of them contain only a fraction of all human genes, it is possible that genes that came up in one study were not included in an array of another group. Second, considering the expected heterogeneity of schizophrenia, only a small number of patients were used in each study. This is understandable since obtaining good-quality post-mortem tissue from well-characterized schizophrenia patients is very difficult. Also, the age of the patients differed between the studies. Hakak et al. used samples from old schizophrenia patients that had been hospitalized for long periods. Mirnics et al. used patients that died in middle age. It will be important in the future to use sufficiently large numbers of well-characterized patient and control samples that will be hybridized to similar microarrays containing a larger number of genes to obtain a larger-scale picture of the genes involved in the pathogenesis of schizophrenia. However, these first studies have already given us some valuable candidate genes to test and have pointed to some specific pathways that have not been studied intensively in schizophrenia research. A critical question to ask is, What are the similarities and differences of these datasets and what can be learned when all data are combined? An emerging view seems to be that the gene expression differences between schizophrenia patients and controls are subtle, and a number of genes are down-regulated in schizophrenia patients compared to controls.

Gene Expression Profiling in Alzheimer's Disease

Alzheimer's disease (AD) is a complex disease that accounts for the majority of dementia in elderly population. Three genes predisposing to early-onset familial AD have been described: APP (amyloid beta A4 precursor protein), PSEN1 (presenilin 1), and PSEN2 (presenilin 2) (reviewed in Tanzi and Bertram, 2001). Also, one risk factor, the ApoE4-ε4 allele of the apolipoprotein E gene, has been described in late-onset AD. However, abnormalities in these four genes account for only a fraction of the total genetic predisposition to the disease. Several studies using microarrays have been performed in order to understand the pathophysiology of AD.

Elegant experiments performed by Ginsberg et al. (2000) compared the neurofibrillary tangle-bearing versus normal CA1 neurons aspirated from sections of AD and control brains. Follow-up experiments included reverse Northern blots and immunohistochemistry. Significant reductions in several classes of mRNAs that are known to encode proteins implicated in AD neuropathology, including phophatases/kinases, cytoskeletal proteins, synaptic proteins, glutamate receptors, and dopamine receptors, were detected. In addition, changes in levels of mRNAs encoding proteins not previously implicated in AD were found. Hata et al. (2001) studied neurofibrillary tangle-associated hippocampi with the lesion-free parietal cortex from the same patient, and with the same regions from control individuals. The most up-regulated gene in the four patients studied was calcineurin Aβ mRNA. This result was confirmed by in situ hybridization and RT-PCR. Ho et al. (2001) identified candidate genes whose expression was altered in cerebral cortex of patients with early AD. They used samples extracted from superior temporal gyrus pooled from six controls and five cases of moderate dementia. Among the genes identified were the synaptic vesicle protein synapsin II that plays an important role in neurotransmitter release. Loring et al. (2001) compared affected and unaffected brain regions in nine controls and six AD cases. The affected brain regions include amygdala and cingulate cortex, and the unaffected regions studied were striatum and cerebellum. The most prominent up-regulated physiological correlates of pathology involved chronic inflammation, cell adhesion, cell

proliferation, and protein synthesis. The down-regulated correlates involved signal transduction, energy metabolism, stress response, synaptic vesicle synthesis and function, calcium binding, and cytoskeleton genes. Colangelo et al. (2002) studied pooled hippocampal CA1 samples from six AD patients and six controls using Affymetrix GeneChips. Samples from patients showed a generalized depression in brain gene transcription, including decreases in RNA encoding transcription factors, neutotrophic factors, and signaling elements involved in synaptic plasticity such as synaptophysin, metallothionein III, and metal regulatory factor 1.

Clearly, the problems related to the gene expression profiling in AD are similar to those in schizophrenia. The sample sizes are small, and the methods used vary greatly among different laboratories. However, these initial experiments show that microarray analyses can identify interesting new candidate genes that might help investigators understand the pathophysiology of AD.

Gene Expression Profiling in Multiple Sclerosis

Multiple sclerosis (MS) is a complex autoimmune disease that is characterized by central nervous system lesions that lead to blood–brain barrier breakdown, inflammation, and myelin damage. Experimental allergic encephalomyelitis (EAE) in rodents is an animal model for MS. Whitney et al. have taken advantage of this animal model and compared the gene expression patterns in EAE mice and human MS patients (Whitney et al., 1999, 2001). One of the genes found to be up-regulated in both MS lesions and EAE brains was 5-lipoxygenase, a key enzyme in the biosynthesis of the proinflammatory leukotrienes. Another study using spinal cords of EAE mice at the onset and at the peak of the disease found increased gene expression of immune-related molecules, extracellular matrix and cell adhesion molecules, and molecules involved in cell division and transcription (Ibrahim et al., 2001). In addition, differential regulation of molecules involved in signal transduction, protein synthesis, and metabolism were found. Of the 104 genes with defined chromosomal locations, 51 mapped to known EAE-linked QTLs representing putative candidate genes for susceptibility to EAE.

Another study combined large-scale sequencing of cDNA libraries derived from brains of MS patients, microarray analysis of spinal cords from EAE rats, and studies in osteopontin knockout mice (Chabas et al., 2001). This group showed that osteopontin, a proinflammatory cytokine, is involved in the pathophysiology of MS. Perhaps the most comprehensive and elegant study to date is recent analysis by Lock et al. (2002) using Affymetrix arrays to study MS lesions of various stages from four patients and two controls. Two of the more than 70 genes identified using arrays were pursued further in EAE mice, and manipulation of the two gene products delayed progression of the disease. These two examples demonstrate that by combining large-scale genetic and genomic techniques in mice and humans, one can reveal valuable information about genes involved in complex disease pathophysiology. However, in all cases a shortcoming is that data were not deposited in a public database that would allow comparisons between different studies that had quite different focus. Clearly a database is needed that would provide raw microarray data and all associated information required to use the data from animal and clinical studies to enable efficient comparisons (Tompkins and Miller, 2002).

10.5 LIMITATIONS

The sequencing and annotation of human and mouse genomes is proceeding at a rapid pace. Currently available commercial arrays now cover virtually all identified genes. Affymetrix has a set of two arrays to cover all human genes and has another set of two arrays to cover all mouse genes. As the annotation of human and mouse genomes advance, we will have more comprehensive information about different splice variants, and their expression patterns and new arrays incorporating this information will be developed. Statistical geneticists are working on better tools to efficiently combine RNA-level and DNA-level variation to obtain information on how specific DNA variations lead to gene expression changes, and important steps in this direction have been taken recently (Schadt et al., 2003).

Microarrays measure the differences in mRNA levels. mRNA is only an intermediate on the way to the functional protein product. Changes in protein stability or post-translational protein modifications might only be indirectly reflected in the levels of mRNA. Therefore, protein-based methods are important in complementing the RNA-based methods. One might ask why study mRNA levels at all if it is only an intermediate on the way to produce the functional protein products? One reason is that no scalable protein-based methods are available. They are more difficult, less sensitive, and low throughput. But more importantly, mRNA levels are immensely informative about cell state and the activity of genes; and for most genes, changes in mRNA abundance are related to changes in protein abundance (Lockhart and Winzeler, 2000).

A large number of technical limitations related to microarrays have been solved in the past few years. It seems that the greatest challenge derives from the tissue under study. Levels of gene expression changes are generally smaller in the brain tissue as compared tissues such as cancer (Figure 10.5). In addition, some phenotypes may be the result of multiple, subtle changes in mRNA abundance that will be difficult to detect. It is possible to detect these smaller differences, but to do so reliably requires performing a larger number of independent experiments so that smaller differences can be trusted with reasonable confidence (Lockhart and Barlow, 2001a). It might be necessary to use other RNA- and protein-based methods to confirm these small changes.

10.6 FUTURE IN NEUROSCIENCE

It is clear that a large amount of important information regarding brain-related diseases has already been obtained using genomic approaches, mainly microarrays. Gene expression profiling is helping us find pathways that are important in the pathophysiological mechanisms of diseases, but it will be important to combine RNA-based technology with DNA variation detection in order to find the causative DNA polymorphisms for both cis-acting and trans-acting regulatory elements. Only a minority of gene expression differences are due to polymorphisms in the coding region of the corresponding gene. By comparing genome sequences of several vertebrate organisms, we can identify conserved regulatory regions that lie outside the coding sequences helping the identification of cis-acting elements in particular. However, a large number of regulatory elements are expected to be trans-acting (i.e., not in the physical proximity of the gene whose expression levels it regulates). More sophisticated tools are needed in order to identify such elements, and a method that combines gene expression profiling and QTL mapping has been shown to be successful in several species (Schadt et al., 2003).

In silico genomic methods also show promise and are likely to become important tools. In silico gene mapping can speed up QTL identification tremendously because it only involves the initial genotyping of several inbred mouse strains, but afterwards virtually any phenotype can be analyzed utilizing the same setting (Grupe et al., 2001). This method can be combined with microarray analysis to identify differentially expressed genes that map in the QTL regions, and it has been applied in a study for genetic susceptibility to asthma (Karp et al., 2000).

Microarrays and other genomic techniques provide us with a large amount of interesting data, but to prove that a gene is involved in mediating a certain disease or behavior, functional analyses are needed. Complex diseases are caused by several susceptibility genes whose effects are further regulated by numerous modifier genes. Transgenic or knock-out animals might be useful in showing how one gene affects a phenotype. At the moment, RNA interference or RNAi shows great potential as well. Introduction of double-stranded RNAs into cells can suppress gene expression by mRNA degradation and inhibition of translation (reviewed in Agami, 2002; Shi, 2003). While there are technical challenges to be solved, this method has been shown to work in mouse in vivo (McCaffrey et al., 2002; Carmell et al., 2003; Rubinson et al., 2003; Tiscornia et al., 2003). By suppressing the gene expression of several genes at the same time, it may be possible to study the complex networks of genes in an animal model and get valuable information about the mechanisms of complex diseases in humans.

We believe that an important goal in neurobiology is the development of a combined high-resolution atlas that contains a complete set of quantitative gene expression data for all important regions and cell types in the brain, combined with anatomical information and three-dimensional views in an accessible, navigable environment. The goal is a searchable and queryable database of region-specific and cell-type-specific gene expression profiles linked to the detailed, digitized, three-dimensional brain anatomy to guide us toward systems-level descriptions of complex networks and a global yet detailed view of how genes act in concert to produce particular phenotypes (Barlow and Lockhart, 2002). It would be important to include disease-related data from human and experimental animal studies and clinical trials. This database should not only be a repository for the data, but should also contain tools to query and analyze large sets of data coming from experiments focusing on various fields of neuroscience.

Maybe in the near future we may be able to define subclasses of brain-related diseases or predict optimal therapeutic interventions, much like is possible in cancer research now. For this purpose, we will need (a) coordinated efforts of clinicians and brain banks in order to obtain high-quality post-mortem brain samples from patients suffering from neurological and mental illnesses and (b) databases that allow the storage and analysis of large amounts of genomic data.

ACKNOWLEDGMENTS

Todd Carter is greatly acknowledged for providing the data of running wheel activity and gene expression profiling in inbred mouse strains. David J. Lockhart is acknowledged for his continuous contributions and insights. The Barlow laboratory is acknowledged for fruitful discussions and support. IMC and Teradata are acknowledged for support of the TeraGenomics database.

References

Abeliovich, A., Paylor, R., Chen, C., Kim, J.J., Wehner, J.M., and Tonegawa, S. (1993). PKC gamma mutant mice exhibit mild deficits in spatial and contextual learning. *Cell* **75**, 1263–1271.

Agami, R. (2002). RNAi and related mechanisms and their potential use for therapy. *Curr. Opin. Chem. Biol.* **6**, 829–834.

Allen, D.M., van Praag, H., Ray, J., Weaver, Z., Winrow, C.J., Carter, T.A., Braquet, R., Harrington, E., Ried, T., Brown, K.D., Gage, F.H., and Barlow, C. (2001). Ataxia telangiectasia mutated is essential during adult neurogenesis. *Genes Dev.* **15**, 554–566.

Alon, U., Barkai, N., Notterman, D.A., Gish, K., Ybarra, S., Mack, D., and Levine, A.J. (1999). Broad patterns of gene expression revealed by clustering analysis of tumor and normal colon tissues probed by oligonucleotide arrays. *Proc. Natl. Acad. Sci. U.S.A.* **96**, 6745–6750.

Bahn, S., Augood, S.J., Ryan, M., Standaert, D.G., Starkey, M., and Emson, P.C. (2001). Gene expression profiling in the post-mortem human brain—no cause for dismay. *J. Chem. Neuroanat.* **22**, 79–94.

Barlow, C., and Lockhart, D.J. (2002). DNA arrays and neurobiology—what's new and what's next? *Curr. Opin. Neurobiol.* **12**, 554–561.

Barlow, C., Hirotsune, S., Paylor, R., Liyanage, M., Eckhaus, M., Collins, F., Shiloh, Y., Crawley, J.N., Ried, T., Tagle, D., and Wynshaw-Boris, A. (1996). Atm-deficient mice: A paradigm of ataxia telangiectasia. *Cell* **86**, 159–171.

Bonner, R.F., Emmert-Buck, M., Cole, K., Pohida, T., Chuaqui, R., Goldstein, S., and Liotta, L.A. (1997). Laser capture microdissection: Molecular analysis of tissue. *Science* **278**, 1481–1483.

Bowden, D.M., and Martin, R.F. (1995). NeuroNames brain hierarchy. *NeuroImage* **2**, 63–83.

Carmell, M.A., Zhang, L., Conklin, D.S., Hannon, G.J., and Rosenquist, T.A. (2003). Germline transmission of RNAi in mice. *Nat. Struct. Biol.* **10**, 91–92.

Carter, T.A., Del Rio, J.A., Greenhall, J.A., Latronica, M.L., Lockhart, D.J., and Barlow, C. (2001). Chipping away at complex behavior: Transcriptome/phenotype correlations in the mouse brain. *Physiol. Behav.* **73**, 849–857.

Chabas, D., Baranzini, S.E., Mitchell, D., Bernard, C.C., Rittling, S.R., Denhardt, D.T., Sobel, R.A., Lock, C., Karpuj, M., Pedotti, R., Heller, R., Oksenberg, J.R., and Steinman, L. (2001). The influence of the proinflammatory cytokine, osteopontin, on autoimmune demyelinating disease. *Science* **294**, 1731–1735.

Chin, H.R., and Moldin, S.O. (2001). *Methods is Genomic Neuroscience.* CRC Press, Boca Raton, FL.

Chung, C.H., Bernard, P.S., and Perou, C.M. (2002). Molecular portraits and the family tree of cancer. *Nat. Genet.* **32** (Suppl.), 533–540.

Colangelo, V., Schurr, J., Ball, M.J., Pelaez, R.P., Bazan, N.G., and Lukiw, W.J. (2002). Gene expression profiling of 12633 genes in Alzheimer hippocampal CA1: Transcription and neurotrophic factor down-regulation and up-regulation of apoptotic and pro-inflammatory signaling. *J. Neurosci. Res.* **70**, 462–473.

Crawley, J.N. (1996). Unusual behavioral phenotypes of inbred mouse strains. *Trends Neurosci.* **19**, 181–182; discussion, pp. 188–189.

Crawley, J.N., Belknap, J.K., Collins, A., Crabbe, J.C., Frankel, W., Henderson, N., Hitzemann, R.J., Maxson, S.C., Miner, L.L., Silva, A.J., Wehner, J.M., Wynshaw-Boris, A., and Paylor, R. (1997). Behavioral phenotypes of inbred mouse strains: implications and recommendations for molecular studies. *Psychopharmacology (Berlin)* **132**, 107–124.

Crusio, W.E. (1996). Gene-targeting studies: New methods, old problems. *Trends Neurosci.* **19**, 186–187; discussion, pp. 188–189.

Dawson, V.L., Kizushi, V.M., Huang, P.L., Snyder, S.H., and Dawson, T.M. (1996). Resistance to neurotoxicity in cortical cultures from neuronal nitric oxide synthase-deficient mice. *J. Neurosci.* **16**, 2479–2487.

Dennis, C. (2002). News feature: Information overload. *Nature (London)* **417**, 14.

DeRisi, J., Penland, L., Brown, P.O., Bittner, M.L., Meltzer, P.S., Ray, M., Chen, Y., Su, Y.A., and Trent, J.M. (1996). Use of a cDNA microarray to analyse gene expression patterns in human cancer. *Nat. Genet.* **14**, 457–460.

Eberwine, J., Yeh, H., Miyashiro, K., Cao, Y., Nair, S., Finnell, R., Zettel, M., and Coleman, P. (1992). Analysis of gene expression in single live neurons. *Proc. Natl. Acad. Sci. U.S.A.* **89**, 3010–3014.

Eisen, M.B., Spellman, P.T., Brown, P.O., and Botstein, D. (1998). Cluster analysis and display of genome-wide expression patterns. *Proc. Natl. Acad. Sci. U.S.A.* **95**, 14863–14868.

Freimert, C., Erfle, V., and Strauss, G. (1989). Preparation of radiolabeled cDNA probes with high specific activity for rapid screening of gene expression. *Methods Mol. Cell. Biol.* **1**, 143–153.

Gerlai, R. (1996). Gene-targeting studies of mammalian behavior: Is it the mutation or the background genotype? *Trends Neurosci.* **19**, 177–181.

Geschwind, D.H., and Gregg, J.P. (2002). *Microarrays for the Neurosciences: An Essential Guide.* MIT Press, Boston, MA.

Ginsberg, S.D., Hemby, S.E., Lee, V.M., Eberwine, J.H., and Trojanowski, J.Q. (2000). Expression profile of transcripts in Alzheimer's disease tangle-bearing CA1 neurons. *Ann. Neurol.* **48**, 77–87.

Graham, F.K. (1975). Presidential Address, 1974. The more or less startling effects of weak prestimulation. *Psychophysiology* **12**, 238–248.

Gress, T.M., Hoheisel, J.D., Lennon, G.G., Zehetner, G., and Lehrach, H. (1992). Hybridization fingerprinting of high-density cDNA-library arrays with cDNA pools derived from whole tissues. *Mamm. Genome* **3**, 609–619.

Grupe, A., Germer, S., Usuka, J., Aud, D., Belknap, J.K., Klein, R.F., Ahluwalia, M.K., Higuchi, R., and Peltz, G. (2001). In silico mapping of complex disease-related traits in mice. *Science* **292**, 1915–1918.

Hakak, Y., Walker, J.R., Li, C., Wong, W.H., Davis, K.L., Buxbaum, J.D., Haroutunian, V., and Fienberg, A.A. (2001). Genome-wide expression analysis reveals dysregulation of myelination-related genes in chronic schizophrenia. *Proc. Natl. Acad. Sci. U.S.A.* **98**, 4746–4751.

Hamer, D. (2002). Genetics. Rethinking behavior genetics. *Science* **298**, 71–72.

Harrison, P.J., Heath, P.R., Eastwood, S.L., Burnet, P.W.J., McDonald, B., and Pearson, R.C.A. (1995). The relative importance of premortem acidosis and postmortem interval for human brain gene expression studies: Selective mRNA vulnerability and comparison with their encoded proteins. *Neurosci. Lett.* **200**, 151–154.

Hata, R., Masumura, M., Akatsu, H., Li, F., Fujita, H., Nagai, Y., Yamamoto, T., Okada, H., Kosaka, K., Sakanaka, M., and Sawada, T. (2001). Up-regulation of calcineurin Abeta mRNA in the Alzheimer's disease brain: Assessment by cDNA microarray. *Biochem. Biophys. Res. Commun.* **284**, 310–316.

Ho, L., Guo, Y., Spielman, L., Petrescu, O., Haroutunian, V., Purohit, D., Czernik, A., Yemul, S., Aisen, P.S., Mohs, R., and Pasinetti, G.M. (2001). Altered expression of a-type but not b-type synapsin isoform in the brain of patients at high risk for Alzheimer's disease assessed by DNA microarray technique. *Neurosci. Lett.* **298**, 191–194.

Hughes, T.R., Mao, M., Jones, A.R., Burchard, J., Marton, M.J., Shannon, K.W., Lefkowitz, S.M., Ziman, M., Schelter, J.M., Meyer, M.R., Kobayashi, S., Davis, C., Dai, H., He, Y.D., Stephaniants, S.B., Cavet, G., Walker, W.L., West, A., Coffey, E., Shoemaker, D.D., Stoughton, R., Blanchard, A.P., Friend, S.H., and Linsley, P.S. (2001). Expression profiling using microarrays fabricated by an ink-jet oligonucleotide synthesizer. *Nat. Biotechnol.* **19**, 342–347.

Ibrahim, S.M., Mix, E., Bottcher, T., Koczan, D., Gold, R., Rolfs, A., and Thiesen, H.J. (2001). Gene expression profiling of the nervous system in murine experimental autoimmune encephalomyelitis. *Brain* **124**, 1927–1938.

Ingram, D.K., London, E.D., Reynolds, M.A., Waller, S.B., and Goodrick, C.L. (1981). Differential effects of age on motor performance in two mouse strains. *Neurobiol. Aging* **2**, 221–227.

Jena, P.K., Liu, A.H., Smith, D.S., and Wysocki, L.J. (1996). Amplification of genes, single transcripts and cDNA libraries from one cell and direct sequence analysis of amplified products derived from one molecule. *J. Immunol. Methods* **190**, 199–213.

Kacharmina, J.E., Crino, P.B., and Eberwine, J. (1999). Preparation of cDNA from single cells and subcellular regions. *Methods Enzymol.* **303**, 3–18.

Karp, C.L., Grupe, A., Schadt, E., Ewart, S.L., Keane-Moore, M., Cuomo, P.J., Kohl, J., Wahl, L., Kuperman, D., Germer, S., Aud, D., Peltz, G., and Wills-Karp, M. (2000). Identification of complement factor 5 as a susceptibility locus for experimental allergic asthma. *Nat. Immunol.* **1**, 221–226.

Kempermann, G., Kuhn, H.G., and Gage, F.H. (1997). More hippocampal neurons in adult mice living in an enriched environment. *Nature (London)* **386**, 493–495.

Kempermann, G., Brandon, E.P., and Gage, F.H. (1998). Environmental stimulation of 129/SvJ mice causes increased cell proliferation and neurogenesis in the adult dentate gyrus. *Curr. Biol.* **8**, 939–942.

Koslow, S.H. (2002). Opinion: Sharing primary data: A threat or asset to discovery? *Nat. Rev. Neurosci.* **3**, 311–313.

Kuokkanen, S., Sundvall, M., Terwilliger, J.D., Tienari, P.J., Wikstrom, J., Holmdahl, R., Pettersson, U., and Peltonen, L. (1996). A putative vulnerability locus to multiple sclerosis maps to 5p14-p12 in a region syntenic to the murine locus Eae2. *Nat. Genet.* **13**, 477–480.

Lander, E.S., and Schork, N.J. (1994). Genetic dissection of complex traits. *Science* **265**, 2037–2048.

Lathe, R. (1996). Mice, gene targeting and behavior: More than just genetic background. *Trends Neurosci.* **19**, 183–186; discussion, pp. 188–189.

Lipshutz, R.J., Fodor, S.P., Gingeras, T.R., and Lockhart, D.J. (1999). High density synthetic oligonucleotide arrays. *Nat. Genet.* **21**, 20–24.

Liu, E.T. (2003). Classification of cancers by expression profiling. *Curr. Opin. Genet. Deve.* **13**, 97–103.

Lock, C., Hermans, G., Pedotti, R., Brendolan, A., Schadt, E., Garren, H., Langer-Gould, A., Strober, S., Cannella, B., Allard, J., Klonowski, P., Austin, A., Lad, N., Kaminski, N., Galli, S.J., Oksenberg, J.R., Raine, C.S., Heller, R., and Steinman, L. (2002). Gene-microarray analysis of multiple sclerosis lesions yields new targets validated in autoimmune encephalomyelitis. *Nat. Med.* **8**, 500–508.

Lockhart, D.J., and Barlow, C. (2001a). DNA arrays and gene expression analysis in the brain. In *Methods in Genomic Neuroscience* (H.R., Chin and S.O. Moldin, eds.), pp. 143–169, CRC Press, Boca Raton, TL.

Lockhart, D.J., and Barlow, C. (2001b). Expressing what's on your mind: DNA arrays and the brain. *Nat. Rev. Neurosci.* **2**, 63–68.

Lockhart, D.J., and Winzeler, E.A. (2000). Genomics, gene expression and DNA arrays. *Nature (London)* **405**, 827–836.

Lockhart, D.J., Dong, H., Byrne, M.C., Follettie, M.T., Gallo, M.V., Chee, M.S., Mittmann, M., Wang, C., Kobayashi, M., Horton, H.,

and Brown, E.L. (1996). Expression monitoring by hybridization to high-density oligonucleotide arrays. *Nat. Biotechnol.* **14**, 1675–1680.

Loring, J.F., Wen, X., Lee, J.M., Seilhamer, J., and Somogyi, R. (2001). A gene expression profile of Alzheimer's disease. *DNA Cell Biol.* **20**, 683–695.

Luo, L., Salunga, R.C., Guo, H., Bittner, A., Joy, K.C., Galindo, J.E., Xiao, H., Rogers, K.E., Wan, J.S., Jackson, M.R., and Erlander, M.G. (1999). Gene expression profiles of laser-captured adjacent neuronal subtypes. *Nat. Med.* **5**, 117–122.

Luo, Z., and Geschwind, D.H. (2001). Microarray applications in neuroscience. *Neurobiol. Dis.* **8**, 183–193.

Mayford, M., Wang, J., Kandel, E.R., and O'Dell, T.J. (1995). CaMKII regulates the frequency-response function of hippocampal synapses for the production of both LTD and LTP. *Cell* **81**, 891–904.

McCaffrey, A.P., Meuse, L., Pham, T.T., Conklin, D.S., Hannon, G.J., and Kay, M.A. (2002). RNA interference in adult mice. *Nature (London)* **418**, 38–39.

McKusick, V.A. (1973). Genetic studies in American inbred populations with particular reference to the Old Order Amish. *Isr. J. Med. Sci.* **9**, 1276–1284.

Metten, P., Phillips, T.J., Crabbe, J.C., Tarantino, L.M., McClearn, G.E., Plomin, R., Erwin, V.G., and Belknap, J.K. (1998). High genetic susceptibility to ethanol withdrawal predicts low ethanol consumption. *Mamm. Genome* **9**, 983–990.

Middleton, F.A., Mirnics, K., Pierri, J.N., Lewis, D.A., and Levitt, P. (2002). Gene expression profiling reveals alterations of specific metabolic pathways in schizophrenia. *J. Neurosci.* **22**, 2718–2729.

Mimmack, M.L., Ryan, M., Baba, H., Navarro-Ruiz, J., Iritani, S., Faull, R.L., McKenna, P.J., Jones, P.B., Arai, H., Starkey, M., Emson, P.C., and Bahn, S. (2002). Gene expression analysis in schizophrenia: Reproducible up-regulation of several members of the apolipoprotein L family located in a high-susceptibility locus for schizophrenia on chromosome 22. *Proc. Natl. Acad. Sci. U.S.A.* **99**, 4680–4685.

Mirnics, K., Middleton, F.A., Marquez, A., Lewis, D.A., and Levitt, P. (2000). Molecular characterization of schizophrenia viewed by microarray analysis of gene expression in prefrontal cortex. *Neuron* **28**, 53–67.

Mirnics, K., Middleton, F.A., Stanwood, G.D., Lewis, D.A., and Levitt, P. (2001). Disease-specific changes in regulator of G-protein signaling 4 (RGS4) expression in schizophrenia. *Mol. Psychiatry* **6**, 293–301.

Morrison, R.S., Wenzel, H.J., Kinoshita, Y., Robbins, C.A., Donehower, L.A., and Schwartzkroin, P.A. (1996). Loss of the p53 tumor suppressor gene protects neurons from kainate-induced cell death. *J. Neurosci.* **16**, 1337–1345.

Peltonen, L., Palotie, A., and Lange, K. (2000). Use of population isolates for mapping complex traits. *Nat. Rev. Genet.* **1**, 182–190.

Phillips, T.J. (1997). Behavior genetics of drug sensitization. *Crit. Rev. Neurobiol.* **11**, 21–33.

Pomeroy, S.L., Tamayo, P., Gaasenbeek, M., Sturla, L.M., Angelo, M., McLaughlin, M.E., Kim, J.Y., Goumnerova, L.C., Black, P.M., Lau, C., Allen, J.C., Zagzag, D., Olson, J.M., Curran, T., Wetmore, C., Biegel, J.A., Poggio, T., Mukherjee, S., Rifkin, R., Califano, A., Stolovitzky, G., Louis, D.N., Mesirov, J.P., Lander, E.S., and Golub, T.R. (2002). Prediction of central nervous system embryonal tumour outcome based on gene expression. *Nature (London)* **415**, 436–442.

Quackenbush, J. (2001). Computational analysis of microarray data. *Nat. Rev. Genet.* **2**, 418–427.

Risch, N., de Leon, D., Ozelius, L., Kramer, P., Almasy, L., Singer, B., Fahn, S., Breakefield, X., and Bressman, S. (1995). Genetic analysis of idiopathic torsion dystonia in Ashkenazi Jews and their recent descent from a small founder population. *Nat. Genet.* **9**, 152–159.

Rubinson, D.A., Dillon, C.P., Kwiatkowski, A.V., Sievers, C., Yang, L., Kopinja, J., Zhang, M., McManus, M.T., Gertler, F.B., Scott, M.L., and Van Parijs, L. (2003). A lentivirus-based system to functionally silence genes in primary mammalian cells, stem cells and transgenic mice by RNA interference. *Nat. Genet.* **33**, 401–406.

Sandberg, R., Yasuda, R., Pankratz, D.G., Carter, T.A., Del Rio, J.A., Wodicka, L., Mayford, M., Lockhart, D.J., and Barlow, C. (2000). Regional and strain-specific gene expression mapping in the adult mouse brain. *Proc. Nat. Acad. Sci. U.S.A.* **97**, 11038–11043.

Schadt, E.E., Monks, S.A., Drake, T.A., Lusis, A.J., Che, N., Colinayo, V., Ruff, T.G., Milligan, S.B., Lamb, J.R., Cavet, G., Linsley, P.S., Mao, M., Stoughton, R.B., and Friend, S.H. (2003). Genetics of gene expression surveyed in maize, mouse and man. *Nature (London)* **422**, 297–302.

Schauwecker, P.E., and Steward, O. (1997). Genetic determinants of susceptibility to excitotoxic cell death: Implications for gene targeting approaches. *Proc. Natl. Acad. Sci. U.S.A.* **94**, 4103–4108.

Schena, M., Shalon, D., Davis, R.W., and Brown, P.O. (1995). Quantitative monitoring of gene expression patterns with a complementary DNA microarray. *Science* **270**, 467–470.

Sgroi, D.C., Teng, S., Robinson, G., LeVangie, R., Hudson, J.R., Jr., and Elkahloun, A.G. (1999). In vivo gene expression profile analysis of human breast cancer progression. *Cancer Res.* **59**, 5656–5661.

Shi, Y. (2003). Mammalian RNAi for the masses. *Trends Genet.* **19**, 9–12.

Steele, P.M., Medina, J.F., Nores, W.L., and Mauk, M.D. (1998). Using genetic mutations to study the neural basis of behavior. *Cell* **95**, 879–882.

Sundvall, M., Jirholt, J., Yang, H.T., Jansson, L., Engstrom, A., Pettersson, U., and Holmdahl, R. (1995). Identification of murine loci associated with susceptibility to chronic experimental autoimmune encephalomyelitis. *Nat. Genet.* **10**, 313–317.

Tang, K., Fu, D.J., Julien, D., Braun, A., Cantor, C.R., and Koster, H. (1999). Chip-based genotyping by mass spectrometry. *Proc. Natl. Acad. Sci. U.S.A.* **96**, 10016–10020.

Tanzi, R.E., and Bertram, L. (2001). New frontiers in Alzheimer's disease genetics. *Neuron* **32**, 181–184.

Tiscornia, G., Singer, O., Ikawa, M., and Verma, I.M. (2003). A general method for gene knockdown in mice by using lentiviral vectors expressing small interfering RNA. *Proc. Natl. Acad. Sci. U.S.A.* **100**, 1844–1848.

Tompkins, S.M., and Miller, S.D. (2002). An array of possibilities for multiple sclerosis. *Nat. Med.* **8**, 451–453.

Van Gelder, R.N., von Zastrow, M.E., Yool, A., Dement, W.C., Barchas, J.D., and Eberwine, J.H. (1990). Amplified RNA synthesized from limited quantities of heterogeneous cDNA. *Proc. Natl. Acad. Sci. U.S.A.* **87**, 1663–1667.

van Praag, H., Kempermann, G., and Gage, F.H. (1999). Running increases cell proliferation and neurogenesis in the adult mouse dentate gyrus. *Nat. Neurosci.* **2**, 266–270.

Vawter, M.P., Barrett, T., Cheadle, C., Sokolov, B.P., Wood, W.H., 3rd, Donovan, D.M., Webster, M., Freed, W.J., and Becker, K.G. (2001). Application of cDNA microarrays to examine gene expression differences in schizophrenia. *Brain Res. Bull.* **55**, 641–650.

Vawter, M.P., Crook, J.M., Hyde, T.M., Kleinman, J.E., Weinberger, D.R., Becker, K.G., and Freed, W.J. (2002). Microarray analysis of gene expression in the prefrontal cortex in schizophrenia: A preliminary study. *Schizophr Res.* **58**, 11–20.

Wang, A.M., Doyle, M.V., and Mark, D.F. (1989). Quantitation of mRNA by the polymerase chain reaction. *Proc. Natl. Acad. Sci. U.S.A.* **86**, 9717–9721.

Wang, E., Miller, L.D., Ohnmacht, G.A., Liu, E.T., and Marincola, F.M. (2000). High-fidelity mRNA amplification for gene profiling. *Nat. Biotechnol.* **18**, 457–459.

Watanabe, Y., Johnson, R.S., Butler, L.S., Binder, D.K., Spiegelman, B.M., Papaioannou, V.E., and McNamara, J.O. (1996). Null mutation of c-fos impairs structural and functional plasticities in the kindling model of epilepsy. *J. Neurosci.* **16**, 3827–3836.

Wen, X., Fuhrman, S., Michaels, G.S., Carr, D.B., Smith, S., Barker, J.L., and Somogyi, R. (1998). Large-scale temporal gene expression mapping of central nervous system development. *Proc. Natl. Acad. Sci. U.S.A.* **95**, 334–339.

Whitney, L.W., Becker, K.G., Tresser, N.J., Caballero-Ramos, C.I., Munson, P.J., Prabhu, V.V., Trent, J.M., McFarland, H.F., and Biddison, W.E. (1999). Analysis of gene expression in mutiple sclerosis lesions using cDNA microarrays. *Ann. Neurol.* **46**, 425–428.

Whitney, L.W., Ludwin, S.K., McFarland, H.F., and Biddison, W.E. (2001). Microarray analysis of gene expression in multiple sclerosis and EAE identifies 5-lipoxygenase as a component of inflammatory lesions. *J. Neuroimmunol.* **121**, 40–48.

Wodicka, L., Dong, H., Mittmann, M., Ho, M.H., and Lockhart, D.J. (1997). Genome-wide expression monitoring in Saccharomyces cerevisiae. *Nat. Biotechnol.* **15**, 1359–1367.

Zapala, M.A., Lockhart, D.J., Pankratz, D.G., Garcia, A.J., and Barlow, C. (2002). Software and methods for oligonucleotide and cDNA array data analysis. *Genome Biol.* **3**, SOFTWARE0001.

Web Resources

https://www.affymetrix.com/site/login/login.affx *Affymetrix/NetAffx*

http://www.chem.agilent.com/scripts/PHome.asp *Agilent Technologies Microarrays*

http://www.mrlpublications.com *Bullfrog and Spot*

http://rana.lbl.gov/EisenSoftware.htm *Cluster and TreeView*

http://www.biostat.harvard.edu/complab/dchip/ *dChip*

http://www.geneontology.org/ *Gene Ontology Consortium*

http://www.silicongenetics.com/cgi/SiG.cgi/Products/GeneSpring/index.smf *GeneSpring*

http://www-sequence.stanford.edu/ *Stanford Genome Technology Center*

http://www.teragenomics.com *Teragenomics database*

http://info.med.yale.edu/microarray/ *Yale Microarray Database*

Stimulus-Dependent Regulation of Gene Expression in the Nervous System

Colleen A. McClung, Ph.D. *and* *Eric J. Nestler, M.D., Ph.D.*

CONTENTS

Databasing the Brain. Edited by Stephen H. Koslow and Shankar Subramaniam
ISBN 0-471-30921-4 © 2005 John Wiley & Sons, Inc.

Synaptic-activity-dependent changes in gene expression are responsible for many long-lasting neuronal adaptations leading to cell growth and differentiation, learning and memory, and even complex behaviors. Gene expression is regulated through the activation of signal transduction pathways in response to neural activity or receptor binding by neurotransmitters, hormones, or growth factors. Signaling pathways eventually lead to changes in the amounts or activities of transcription factors or other nuclear proteins that regulate gene expression. The specificity and activity of transcription factors are regulated in a variety of ways. Furthermore, the control and diversity of expressed transcripts are regulated post-transcriptionally through many different mechanisms. To identify target genes of specific transcription factors, or genes that are regulated after a particular treatment paradigm, several methods have been employed including promoter analysis, large-scale gene expression studies, and chromatin immunoprecipitation. This chapter will focus on the transcriptional and post-transcriptional regulation of stimulus-induced gene expression, as well as the methods used to identify target genes of transcription factors.

11.1 OVERVIEW OF THE BRAIN'S SIGNAL TRANSDUCTION PATHWAYS

When certain neurotransmitters or other molecules in the brain are released, they bind to receptors on the surface of other cells leading to a chain of events that regulates virtually all aspects of that cell's functioning. Perhaps the ultimate endpoint of such signaling is the regulation of gene expression. A great deal is now known about the highly complex, interacting pathways that mediate signal transduction in the brain. A comprehensive review of this vast information is beyond the scope of this chapter. Instead, we provide a brief overview only and refer the reader to recent reviews (Hamm, 1998; Hunter, 2000; Berridge, 1997; West et al., 2002; Cantley, 2002).

In the majority of cases, receptors that induce these signal transduction cascades are termed G-protein-coupled receptors due to their interaction with a guanine nucleotide binding protein, or G protein. G proteins are heterotrimeric proteins consisting of an α, β, and γ subunit. The β and γ subunits have a structural role in anchoring the G protein to the plasma membrane, but they also have direct effects on ion channels. The α subunit confers most of the activity on other signaling cascades. In most of these cases, G proteins interact with an enzyme that produces a diffusible second messenger.

The best-characterized second messenger is cAMP. In this pathway, the α subunit of the G protein is bound in its inactive state to a molecule of GDP. Upon activation, it will exchange a molecule of GTP for GDP and activate the G protein. To first approximation, α_s subunits activate the enzyme, adenylyl cyclase, which catalyzes the conversion of ATP to cAMP, while α_i subunits inhibit adenylyl cyclase. In reality, there are nine known forms of adenylyl cyclase, some of which are not regulated in this manner by α_s and α_i, but instead regulated by $\beta\gamma$ subunits as well as by other factors. The most important effector for cAMP is cAMP-dependent protein kinase, or PKA. PKA can phosphorylate numerous cellular proteins and thereby mediate most of the effects of cAMP on neuronal function. Among these substrates for PKA is the transcription factor CREB (cAMP response

element binding protein), which is activated upon phosphorylation and mediates many of the effects of the cAMP pathway on gene expression. CREB is similarly activated by other intracellular pathways, including the calcium-associated and growth-factor-associated pathways (Figure 11.1). Gene expression changes due to CREB are an important contributor to the neural plasticity in the brain responsible for learning and memory, circadian entrainment, and neuron growth and survival (Mayr and Montminy, 2001; Lonze and Ginty, 2002). CREB-dependent changes in gene expression are also implicated in disease states such as depression and drug addiction.

Other neurotransmitter receptors such as N-methyl-D-aspartate (NMDA) glutamate receptors, as well as some G-protein-coupled receptors, regulate intracellular levels of calcium, another important second messenger in the brain. Calcium acts as a cofactor for calcium/calmodulin-dependent protein kinases (CaM kinases) and for protein kinase C (PKC), which, like PKA, phosphorylate numerous transcription factors to regulate gene expression. The phosphoinositol system is also a prominent second messenger pathway. G-protein-coupled receptors that bind to Gq activate the enzyme phospholipase C, which catalyzes the breakdown of phosphatidylinositol into inositol triphosphate (IP_3) and diacylglycerol (DAG). IP_3 mediates the release of calcium from intracellular stores, which can then activate CaM kinases and PKC. DAG is a required cofactor for most forms of PKC. Furthermore, there are many additional signal transduction pathways, which utilize arachidonic acid metabolites as just one example, as well pathways with gaseous second messengers such as nitric oxide and carbon monoxide.

11.2 TRANSCRIPTION FACTORS: REGULATORS OF GENE EXPRESSION

DNA is transcribed into messenger RNA through the actions of an RNA polymerase complex, combined with the binding of transcription factors (Carey and Smale, 2000). Transcription factors can either activate or inhibit gene transcription, in most cases by binding to consensus sequences in the DNA that confer specificity of their actions. These consensus sequences are typically found in the promoters of genes, usually within 500 base pairs of the transcriptional start site, but are sometimes found further upstream or even within the transcribed region of the gene. One example of such a cis-acting sequence is the cAMP-responsive element or CRE site (5′ TGACGTCA 3′), which is bound by CREB family members (Lonze and Ginty, 2002). Another example is Activator Protein 1 (AP-1) site (5′ TGAG/CTCA 3′), which is bound by AP-1 transcription factors (Chimenov and Kerppola, 2001). In some cases, transcription factors can also bind to nonconsensus sites that are closely related. For example, CREB can bind to AP-1 sites, and AP-1 complexes can bind to CRE sites, under certain circumstances (Andersson et al., 2001).

Most transcription factors function as homo- or heterodimers and are influenced by the binding of additional cofactors. Protein interaction occurs through a dimerization domain that forms a specific structure that can be bound by other proteins. One example is the leucine zipper domain, which is shared by many transcription factors, including CREB and AP-1 proteins (Carey and Smale, 2000). This domain forms an α helix in which every seventh residue is a leucine. The leucines line up on the

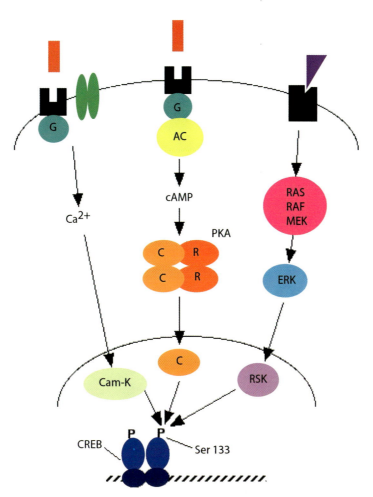

Figure 11.1 Examples of signaling pathways leading to phosphorylation of CREB. Shown are representations of three signal transduction pathways that lead to CREB phosphorylation upon cell stimulation. Release of calcium after receptor binding leads to the activation of calcium/calmodulin-dependent kinases (CaM-K), including CaMK IV, which phosphorylates CREB at ser133. Certain G-protein (G)-coupled receptors can also activate adenylyl cyclase, leading to the dissociation of the catalytic subunits (C) of protein kinase A (PKA) from the regulatory subunits (R). The catalytic subunits translocate to the nucleus to phosphorylate CREB at ser 133. Growth factors can also induce the phosphorylation of CREB through the Ras–Raf–MEK pathway. These proteins activate extracellular signal-related kinases (ERKs), which in turn activate ribosomal S6 kinase (RSK), leading to phosphorylation of ser 133.

face of the helix, allowing for hydrophobic interactions that are stable between the two proteins. Another example is the PAS (PER-ARNT-SIM) domain, which is shared among several transcription factors, including most of the transcription factors that directly regulate circadian rhythms, and allows for dimerization of proteins to form an active transcription factor complex. Examples include the transcription factors PER (period), ARNT (AhR nuclear translocator), Sim1 (single-minded), CLOCK (circadian locomotor outputs kaput), and NPAS2 (neuronal PAS domain protein 2) (Stanewsky, 2003).

Transcription factors are regulated through interactions with binding partners and cofactors in multiple ways. One prominent mechanism, already mentioned, is phosphorylation. Phosphorylation can change the confirmation of the protein to either activate or repress transcription. This can be mediated by changes in its

ability to bind essential cofactors. For example, phosphorylation of CREB by PKA (or other kinases) at serine-133 allows for the binding of CREB with its cofactor, CREB binding protein (CBP) (Chrivia et al., 1993). In contrast, phosphorylation of CREB by CaM kinase II (CaMkII) at a different site inhibits the interaction between CREB and CBP and thereby inhibits transcription (Wu and McMurray, 2001).

Another way that transcription can be regulated is through protein stability and degradation. For example, in the *Drosophila* circadian system, the PER protein binds to the timeless protein, TIM. Both are transcription factors involved in a feedback loop that leads to inhibition of their own transcription. PER is phosphorylated by double time kinase (DBT), which targets PER for degradation. The binding of TIM protects PER from degradation and allows the PER/TIM complex to enter the nucleus and inhibit their own transcription. This type of negative feedback cycle provides the basis for circadian regulation not only in *Drosophila* but also in mammalian cells (Stanewsky, 2003).

There are other examples of transcription factors with unusually high stability, which allows the protein to accumulate over time. The actions of these proteins can change with increasing levels. An example of this is a splice variant of the FosB gene, ΔFosB, which lacks much of the transactivation domain seen in related proteins, while retaining the DNA binding and dimerization domains (Yen et al., 1991). As levels of ΔFosB accumulate in the cell, its action appears to reverse from that of a transcriptional repressor of AP-1, to a transcriptional activator (McClung and Nestler, 2003). This accumulation of ΔFosB occurs in specific brain regions after chronic, but not acute, treatment with drugs of abuse or several other types of chronic perturbations. Because the actions of the protein are very long-lasting, it is thought that ΔFosB is responsible for some of the long-term plasticity leading to drug addiction and other stable forms of neural plasticity (Nestler et al., 2001).

The differential expression of transcription factors among cell types and areas of the brain leads to regulation of gene expression in regionally specific patterns. In addition, transcription factors are controlled by their subcellular localization. In order to function, they need to be in the nucleus, which houses the DNA. Transcription factors can be bound by other proteins and kept in the cytoplasm until the binding protein is removed through a variety of mechanisms. Members of the nuclear receptor family (e.g., steroid hormone receptors, thyroid hormone receptors, vitamin D) are sequestered in the cytoplasm by chaperone proteins. The binding is disrupted by the specific ligand for the receptor (steroid hormone, thyroid hormone, vitamin D, etc.), which frees the receptor to enter the nucleus (Lin et al., 1998). In other cases, the binding protein can be removed via a conformational change (e.g., through phosphorylation), through degradation of the binding protein by local proteases, or targeted degradation through ubiquitination. An example of this mode of regulation is provided by nuclear factor (NF)-κB protein. NF-κB is a transcription factor that is held in the cytoplasm by the inhibitor protein, IκB. Upon cell stimulation, IκB is phosphorylated, then ubiquitinated and destroyed in proteasomes, allowing NF-κB to enter the nucleus (Karin and Ben-Neriah, 2000).

Recently, it has been found that the redox state of the cofactors of transcription factors can either inhibit or enhance DNA binding (Rutter et al., 2001). Brain and muscle ARNT-like protein1 (BMAL1) forms heterodimers with NPAS2 and CLOCK. The activity of these complexes requires nicotinamide adenine dinucleotide (NAD) cofactors. In their reduced state, the redox cofactors NAD(H) and NADP(H) increase the binding of NPAS2/BMAL1 and CLOCK/BMAL1 to DNA, while the oxidized state inhibits binding. In this case, environmental influences like sunlight and food are able to alter the redox state of the cell and influence the regulation of gene expression.

Changes in transcriptional activity also involve changes in chromatin structure (Beaujean, 2002; Felsenfeld and Groudine, 2003). Transcriptional regulatory sites on the DNA can be packaged around histone proteins in nucleosomes, making them inaccessible to transcription factor binding. Histone acetyl-transferase enzymes, or HATs, acetylate the histone proteins and thereby allow greater access of the nearby DNA to regulatory proteins such as transcription factors and the RNA polymerase complex. One such HAT is CBP, which, as mentioned previously, binds to CREB, as well as to several other transcription factors. The CREB/CBP complex is able to relax the chromatin structure through histone acetylation and then activate transcription (Lehrmann et al., 2002). Furthermore, histone deacetylases (HDAC) can reverse the actions of the HATs and inactivate transcription.

There are many transcription factors that are important in neurobiology. CREB, AP-1, and NF-κB are three of the best characterized (Figure 11.2). The CREB family of transcription factors consists of the CREB protein itself, along with the closely related proteins, cAMP response element modulator (CREM), and activating transcription factor 1 (ATF-1) (Lonze and Ginty, 2002; Mayr and Montminy, 2001). All belong to the larger bZIP superfamily of transcription factors, which contain a DNA binding domain and a leucine zipper motif that promotes dimerization with other proteins. The CREB protein also contains other domains, including the kinase-inducible domain or KID, which contains the Ser-133 that, upon phosphorylation, allows for the binding of the transcriptional coactivator, CBP. The mammalian CREB is alternatively spliced into three major products, CREBα, CREBΔ, and CREBβ. CREBα and CREBΔ are the two predominant forms, and both contain the glutamine-rich Q1 and Q2/CAD domains, which can function as constitutive activators in vitro. As well, there are truncated forms of CREBΔ that can function as transcriptional repressors. CREB is expressed ubiquitously and is involved in the regulation of many cellular processes, such as neuronal growth, protection, and plasticity, to name a few.

The AP-1 transcription complex in mammals consists of homodimers and heterodimers of bZIP proteins in the Jun family (c-Jun, JunB, and JunD), Fos family (c-Fos, Fra-1, Fra-2, FosB and its splice variant ΔFosB), Jun dimerization partners (JDP1 and JDP2), and the closely related activating transcription factors (ATF2, LRF1, ATF3, and B-ATF) (Wisdom, 1999; Chimenov and Kerppola, 2001). In addition, a subset of the Maf proteins (v-Maf, c-Maf, and Nrl) can dimerize with c-Jun or c-Fos, while others (MafB, MafF, MafG, and, MafK) can only bind c-Fos. The AP-1 complex binds primarily to the AP-1 site, but it can also bind CRE sites under some conditions. The transcriptional activity and DNA binding ability varies widely among members of the AP-1 complex, leading to great variations in AP-1 activity including some instances of transcriptional repression. Some members of the AP-1 complex, such as c-Fos and c-Jun, can be induced within minutes of cell stimulation. These are termed immediate early genes due to their rapid induction. AP-1 activity is induced in many ways in brain, such as upon cell stimulation by growth factors, cytokines, neurotransmitters, and cellular stressors (e.g., oxidation, irradiation), and is implicated in diverse processes ranging from neuronal growth and plasticity to drug addiction.

NF-κB has important roles in many aspects of neurobiology, including synaptic plasticity, cell survival and apoptosis, and neurodegenerative disorders (Mattson et al., 2001). NF-κB functions as a homo- or heterodimer of proteins from the NF-κB/Rel family. There are five family members that have been identified in mammalian cells: RelA (p65), RelB, c-Rel, p50, and p52. The dimer that is formed between p50 and p65 is the most common and well characterized of the NF-κB factors.

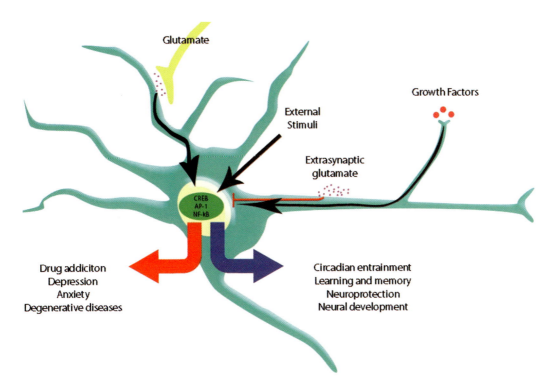

Figure 11.2 Regulation of gene expression by CREB and other transcription factors is important in several neuronal processes. This figure depicts some of the many inputs that lead to transcriptional induction or repression, along with several of the functional outputs in which the regulation of newgene expression plays an important role in the nervous system.

While p50, p52, and p65 are ubiquitous, RelB and c-Rel are expressed predominately in lymphoid tissues. As mentioned previously, prior to cell stimulation, NF-κB proteins are held in the cytoplasm by IκB, which blocks the nuclear localization signal on NF-κB. Three such inhibitory molecules—IκBα, IκBβ, and IκBε—have been reported. After cell stimulation, IκB is phosphorylated by an IκB kinase (IKK) complex, and the IκB is targeted for degradation by proteasomes through its ubiquitination. This allows NF-κB to enter the nucleus and induce gene expression (Mercurio and Manning, 1999). One gene activated by NF-κB is IκBα. IκBα can bind NF-κB and shuttle it back into the cytoplasm, creating a negative feedback loop. Furthermore, the p65 subunit of NF-κB is phosphorylated at serines 529 and 536 in its transactivation domain, leading to increased activity. NF-κB is present in synaptic terminals, where it can be locally activated; presumably, it is then translocated back to the cell nucleus, although local (non-nuclear) functions have not been ruled out. NF-κB is reported to be induced in the brain under several conditions. It is induced in response to glutamate receptor activation in the hippocampus, a region of the brain important in learning and memory (Guerrini et al., 1995). NF-κB is also induced in the nucleus accumbens, an important brain reward region, after chronic cocaine administration (Ang et al., 2001). For these reasons, it is thought that NF-κB is an important mediator of synaptic plasticity. There are many studies that implicate NF-κB as a crucial molecule controlling apoptosis (involving neuronal survival and death) in response to stress or insult to the nervous system. Furthermore, neurodegeneration that results from diseases like Parkinson's disease and Alzheimer's disease may involve the actions of NF-κB (Mattson and Camandola, 2001).

Another unique and important class of transcription factors is the steroid hormone receptors, mentioned earlier (Lin et al., 1998). Unlike most neurotransmitter–hormone receptors, which are G-protein-coupled receptors or ligand-gated channels, steroid hormone receptors are not localized to the plasma membrane but are found in the cell cytoplasm. Steroid hormones, such as glucocorticoids and gonadal steroids, as well as other agents that act via these receptors (e.g., retinoids, vitamin D, thyroid hormone), are lipid-soluble molecules that can easily diffuse across cell membranes. Once a steroid hormone receptor binds its ligand, it translocates into the nucleus and regulates gene expression. Most of these receptors are regulated by the binding of heat shock proteins, which prevent the receptors from entering the nucleus. Ligand binding dissociates the receptor from this complex of chaperone proteins. Once in the nucleus, steroid hormone receptors can regulate transcription directly by binding a hormone response element (HRE) in the regulatory region of certain genes. Alternatively, the receptors can interfere with the actions of other transcription factors such as AP-1 and NF-κB and block transcription. Regulation of gene expression by steroid hormone receptors is very important during development and for proper neuronal functioning throughout adulthood.

11.3 REGULATION OF GENE EXPRESSION BY POST-TRANSCRIPTIONAL MODIFICATION

There are an estimated 30,000–35,000 genes in the human genome and well over 100,000 proteins. The larger number of proteins is due to post-transcriptional modifications. The most common mechanism underlying protein diversity is alternative splicing. In the human genome, over 60% of genes undergo alternative splicing, which can lead to the expression of several distinct proteins from a single gene (Black, 2003). After transcription, a primary RNA transcript contains coding sequences called exons and noncoding sequences called introns (Figure 10.3). Before exiting the nucleus, proteins known as splicing factors bind to the RNA at specific sites, creating RNA protein complexes called spliceosomes. Splicing factors remove intronic sequences while reannealing exon sequences. Sometimes introns or exons are skipped during the splicing process, leading to alternative splice variants that give rise to multiple proteins. Some of the genes encoding transcription factors are subject to alternative splicing. An example is the cAMP response element modulator (CREM), which is a member of the CREB family of transcription factors. The CREM gene is spliced into four variants, two of which act as transcriptional activators (CREMτ and CREMα), while the other two act as transcriptional repressors (S-CREM and ICER) (Lonze and Ginty, 2002). In this case, one gene is spliced into four variants, all four regulate different genes in different ways, leading to more variation in gene regulation.

A less common posttranscriptional modification is RNA editing. Here, single base pair substitutions are made in the RNA after transcription. Therefore, as is the case with alternative splicing, RNA editing creates multiple transcripts from a single gene. There are two known forms of RNA editing. One is A-to-I editing and the other is C-to-U editing. A-to-I editing is mediated by adenosine deaminases acting on RNA (ADARs) and requires a short sequence in an intron that is complementary to an exon sequence containing the adenosine that will be edited (Seeburg, 2002). These complementary sequences fold onto each other to form double-stranded RNA. Interestingly, the only known transcripts that undergo A-to-I editing, besides the ADARs themselves, are membrane proteins in the nervous system that function as ion channels or G-protein-coupled receptors such as the *GluR2* subunit of the AMPA receptor in mammals and the major voltage-activated Na(+) channel polypeptide in *Drosophila, para* (Higuchi et al., 2000; Hanrahan et al., 2000) . Loss of A-to-I RNA editing has very severe consequences. The *Drosophila* ADAR gene is expressed only in the nervous system, and flies with a null mutation in this gene show a range of motor

deficits, abnormal courtship behavior, obsessive cleaning, slow recovery from induced hypoxia, and neurodegeneration with age (Reenan, 2001). In mammals, ADAR is necessary for development, and mutation in either of the two ADAR genes (ADAR 1 and 2) leads to either prenatal or early postnatal lethality (Wang et al., 2000). This is most likely due to the loss of editing in the GluR2 (GluR-B) subunit of AMPA receptors. AMPA receptors are heteromeric proteins that mediate most of the fast excitatory neurotransmission in the brain. The GluR2 subunit confers many unique properties to the AMPA channel, all of which rely on a functionally critical arginine residue that is encoded by the CIG codon, containing an A-to-I substitution (Seeburg, 2002). Interestingly, the A-to-I substitution not only regulates the functional properties of the receptor, but also is involved in the facilitation of the RNA transcript itself. The unedited RNA slows overall RNA splicing leading to lower GluR2 protein levels and problems with cellular trafficking of the GluR2 subunit.

The less common, C-to-U editing also affects transcripts in the nervous system (Blanc and Davidson, 2003). One example is the neurofibromatosis type 1 gene (NF-1). Here a site-specific deamination changes a CGA codon to UGA, generating a stop codon at position 3916. This leads to a truncated protein lacking a region in the 5′ end involved in GTPase activation. C-to-U editing of NF-1 has been documented in peripheral nerve sheath tumors in neurofibromatosis patients. In the liver, C-to-U editing of apolipoprotein B (ApoB) results in a stop codon that creates a truncated protein, ApoB48, that is important in intestinal and hepatic lipid transport.

Another way that gene expression is regulated posttranscriptionally is through RNA stability and transport. At the end of mRNAs, after the translational stop site, is the 3′ untranslated region and the poly A tail. This region contains important regulatory sequences, such as the AUUUA consensus sequence, which is involved in the stability and transport of the mRNA (Grzybowska et al., 2001). An example of a protein family that binds to the 3′ UTR is the STAR family of RNA binding proteins (Lasko, 2003). Some of these proteins are involved in regulating translation, while others such as QKI-5, QKI-6, and QKI-7 bind to a TGE (tra-2 and GLI element) in the 3′ UTR and increase RNA stability. There is also evidence that some of the RNA binding proteins involved in RNA stability and transport are downstream

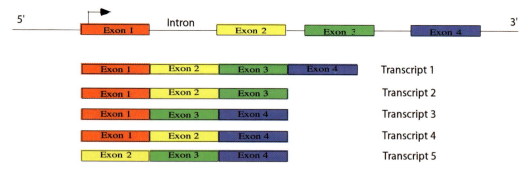

Figure 11.3 Alternative splicing of mRNA leads to multiple products from a single gene. Boxes represent exons, while lines represent introns. The region 5′ of the first exon represents the 5′ regulatory region of the gene which generally contains the promoter sequences, as well as the TATA box (though some cis-regulatory regions reside in introns). The region 3′ of the last exon contains the 3′ untranslated region (UTR) and the region of polyadenylation. Shown are five examples of possible splice variants that may arise from this gene. Each of these splice variants could go on to make a separate protein with a unique function.

of G-protein-mediated signaling cascades, as well as receptor tyrosine kinase cascades involving RasGAP, a negative regulator of the small GTPase, Ras (Tourriere et al., 2001). In addition to binding at the 3′ UTR, there are reports that proteins involved in splicing remain bound to the exon–exon junctions to influence mRNA stability and export into the cytoplasm (Lasko, 2003). The length of the poly-A tail is also very important in regulating RNA stability, though the mechanism of this is unclear.

11.4 IDENTIFYING TARGET GENES OF TRANSCRIPTION FACTORS

Some of the most difficult and problematic studies in molecular biology are those that seek to determine the physiological target genes of specific transcription factors in vivo. One approach is to search for genes in the genome that contain a specific transcriptional regulatory sequence in the genes' promoters. For example, it has been known for some time that CREB binds to CRE sites in gene promoters. Until the recent genomic era with the completion of the human, mouse, and *Drosophila* genomes, the search for potential CRE sites was done on a gene by gene basis. Now, large-scale searches for CRE and other cis-acting elements can be done virtually in silico through a position weight matrix (PWM) or other algorithm, which searches for these cis-acting elements in raw sequence upstream of genes (Heinemeyer et al., 1998). However, because most sites are only 5–25 base pairs, there are large numbers of these sequences, and it is difficult to determine which of these sites are functionally relevant. Moreover, not all CREs are perfect consensus sequences. To try to reduce the number of nonrelevant sites, a comparative genomic approach, or phylogenetic footprinting, has been developed to identify conserved noncoding regions among different species (typically two species), to zero in on important regulatory elements in the 5′ flanking region of genes (Qiu et al., 2003).

This method gives greater confidence in an identified regulatory site, but any gene that has a regulatory region that is not shared among species will be lost. Other methods use a more limited approach and cluster genes first based on functional similarity, expression patterns, or co-regulation of gene expression with expression of a particular transcription factor, and then they search for common cis-acting elements among those genes. All of these in silico methods are important in identifying which genes are potentially regulated by a given transcription factor, but do not ascertain whether they are truly regulated under physiological conditions.

Another large-scale approach to finding target genes of specific transcription factors is to (a) overexpress or repress the transcription factors either in cell culture or in vivo and (b) use an open-ended gene expression system to look at the changes in gene expression that result. Then additional techniques are used to determine if those genes are direct or indirect targets of a specific transcription factor. Expressing genes in cell culture is relatively straightforward and enables the isolation of large quantities of RNA, which is required for some gene expression systems. However, as mentioned before, this approach can ascertain which genes might be regulated in a tissue of interest, but not whether they are in fact regulated under physiological conditions. Some in vivo methods of expression have been developed to allow expression or removal of transcription factors

in the region of the brain or cell type that is important for the neurobiological question under investigation. One such method is to create transgenic mice that carry an engineered transgene to overexpress the transcription factor, or to overexpress a dominant negative antagonist of the transcription factor, to activate or repress transcription, respectively. In recent years, transgenic technology has evolved to enable temporal and spatial expression of these factors in the brain not only to look for target genes of particular transcription factors, but to determine the function of these proteins in particular brain regions during adulthood (Kelz et al., 1999; Sakai et al., 2002).

One such method utilizes the TetOP system, which employs the antibiotic, tetracycline, to control gene expression (Chen et al., 1998). In this system, a line of mice is created that carries the gene for the tetracycline transactivator (tTA), driven by a regionally specific promoter that only expresses in certain tissues or cell types (Figure 11.4). Another line of mice is created that harbors a gene with the tetracycline-responsive promoter (TetOp) flanking the transcription factor of interest. In bigenic mice carrying both transgenes, in the absence of tetracycline, the tTA protein will be made in the cells in which the promoter is functional, and it will bind to the TetOP promoter, driving expression of the transcription factor that is encoded by the transgene. When animals are raised with tetracycline or its derivative, doxycycline (dox), in their drinking water, the tTA protein is inactivated and thus there is no overexpression of the transcription factor. This system allows the experimenter to turn gene expression on or off by adding dox to the drinking water, thus giving regional and temporal control over the transcription factor of interest. The temporal control is especially important since transcription factors have very different roles during development than they do in adulthood, and they regulate different genes as the animal matures. The regional expression is also important since many of these proteins regulate hundreds of genes under different conditions in different cell types. For example, the transcription factor, CREB, is ubiquitous and is involved in many facets of neurobiology. The genes that are regulated by CREB and that lead to learning and memory in the hippocampus may be very different from the CREB-regulated genes involved in drug addiction in the nucleus accumbens.

One of the major problems of this approach has been lack of complete control over the regional expression of the transgenes. Based on the insertion site of a transgene, regional enhancer elements from other genes can influence the amount and sites of its expression. Therefore, it is necessary to generate multiple independent lines and determine the specificity and level of expression in each line. Furthermore, it takes a considerable amount of time (roughly 6 weeks) for the dox to wash out of the system and induce full expression of the transgene. This gradual buildup of transcription factor levels can influence gene expression, which can be an advantage if the goal of a study is to investigate the actions of a protein that accumulates over time. This system is also expressing the gene ectopically and at high levels, so like the cell culture studies, this is a better, but still somewhat artificial, situation.

Viral-mediated gene transfer is another approach that allows temporal and spatial control over transcription factor expression. The virus is physically injected into the brain region of interest, and, depending on the type of virus used, transgene expression can peak anywhere from several days (herpes simplex virus) to

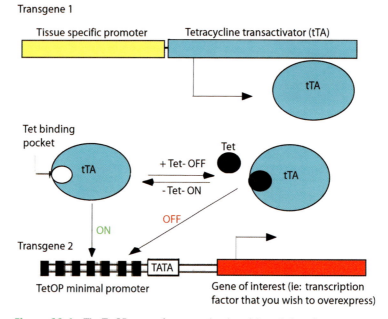

Figure 11.4   The TetOP system for temporal and spatial regulation of transgene overexpression. In transgene 1, a tissue-specific promoter drives the expression of the tetracycline transactivator (tTA), leading to the production of the tTA protein in a regionally specific pattern (based on the promoter). When tetracycline (Tet) is present, the tTA protein is inactive. The protein becomes activated when Tet is removed. This allows the binding of tTA to the TetOP promoter on transgene 2, activating transcription of the transgene. The ability to add or remove Tet at will allows the temporal control of gene expression.

2 weeks (adeno-associated virus) after surgery (Carlezon et al., 2000). The advantages of this system are that this injection can be done in adult animals, and in both mice and rats, while transgenic approaches are only readily available for mice. Rats are easier to use for certain behavioral and electrophysiological experiments, and more RNA can be harvested from the larger tissue samples. In addition, this approach does not require the breeding and housing of large numbers of animals. The regional specificity of injection can be better controlled, since it is based on the site of injection and not the expression of a promoter. One of the disadvantages of the viral vector approach is that some types of viruses (e.g., herpes) drive expression for short periods of time (days) only. Furthermore, viruses are limited by the size of the transgenes they can carry, which influences the choice and number of genes that can be expressed by this technique. The technique also requires intra-cranial surgery and viral infection, which can be toxic to cells and cause considerable stress and even cell death, and not all cells in a given region will become infected. As with the transgenic studies, the levels of expression can be far different than physiological levels, leading to increases or repression of genes that may not normally occur in this system.

Other methods that have been used to knockdown or knockout transcription factors include traditional gene knockout through homologous recombination, antisense oligonucleotides, and RNA interference (RNAi). Gene knockouts in mice are a very successful way to rid the animal of a particular gene. However, in traditional knockouts, the gene is missing throughout development, which often leads to compensation of protein function by other closely related genes or embryonic lethality in the case of genes that are necessary for development. These knockouts

are also genome-wide so they affect all cells in which this gene is expressed. Consequently, an animal may suffer severe problems due to loss of function in organ systems outside the brain. More sophisticated knockout systems have been developed that allow for spatial and even temporal removal of genes. These techniques use the P1 bacteriophage Cre recombinase to induce recombination at *loxP* sites (Yu and Bradley, 2001). A mouse is created that harbors the gene of interest flanked by two *loxP* sites. This mouse can then be crossed with a mouse that expresses Cre recombinase under a regionally specific promoter. Recombination will occur in only those cells that express Cre recombinase, leading to excision of the gene of interest. Furthermore, the TetOp system can be combined with the Cre/lox system to allow for temporal control of the knockout through the addition or removal of dox (Monteggia et al., 2004). However, this system requires extensive breeding efforts over a long period of time. As well, a virus encoding the Cre recombinase can be injected into a specific region of the brain to induce recombination in that region (Scammel et al., 2003).

Studies using intracranially injected antisense oligonucleotides to bind endogenous mRNAs from a particular gene have had mixed results. Some studies show considerable knockdown of gene expression with little toxicity to the cell, while others are less successful. RNA interference technology is a newer technique, which holds great promise for temporally and regionally knocking out gene expression (Scherr et al., 2003). It has been used successfully for some time in worms and flies. In this technique, a small double-stranded RNA molecule sharing a sequence specific to a gene of interest is injected into cells. The cells activate an endogenous mechanism that destroys RNA containing that sequence. Thus far it has been difficult to get this system to function in mammalian neurons, though some recent studies are hopeful.

11.5 HIGH-THROUGHPUT MEASURES OF CHANGES IN GENE EXPRESSION

Comparing the genes that are regulated by overexpression of a transcription factor with those that are repressed when the protein is disrupted can help identify target genes of that transcription factor. There are several high-throughput measures of gene expression that allow the screening of hundreds or even thousands of genes at the same time. Some of the older methods include subtractive library screening and differential hybridization (Sagerstrom et al., 1997). In subtractive library screening, cDNA is synthesized from RNA from the experimental animals (i.e., overexpressing the transcription factor) and is hybridized with RNA from the control group. cDNA that does not form cDNA–RNA hybrids is purified by hydroxyapatite chromatography and used to create a cDNA library. This library should consist of cDNA that is differentially expressed in the experimental tissue. In differential hybridization, a cDNA library is constructed from the tissue of interest, expressed in bacteria, and plated. The resulting colonies are lifted onto filters and screened with probes from two different RNA populations (experimental and control). Then the two filters are examined for differences in levels of expression between the two populations.

Though these techniques represented a step forward in large-scale gene expression studies, there are several disadvantages. First, there is the need to construct cDNA libraries, which are labor-intensive. Also, when a change in expression is identified, the researcher has to go back to the plate and sequence the DNA from the bacterial colony to determine which gene had changed. Both techniques require a large amount of RNA and are not generally sensitive in terms of picking up small changes in expression.

In 1992, a new technique, called differential display RT-PCR (DDRT-PCR), was reported that overcomes some of these limitations (Stein and Liang, 2002). DDRT-PCR consists of two steps. First is the synthesis of cDNA by reverse transcription from RNA isolated from the experimental and control tissues using degenerate, anchored oligo(dT) primers to generate three pools of cDNA (each containing an A, C, or G anchor just 5′ of the poly A tail). The second step is to nonstringently amplify random fragments by polymerase chain reaction (PCR) using the anchored oligo(dT) primer and an arbitrary primer. During the amplification, a radioactive isotope (α-^{35}S) label is incorporated into the fragments so that they can later be visualized. These random fragments are then run on a large polyacrylamide gel to separate the fragments based on size. Then the resulting fragments from the control and experimental groups are compared to determine if there are differences in their levels of expression. Once differences are identified, the band is excised from the gel, extracted by boiling, and reamplified with the original primer sets. The resulting PCR product can be cloned into a vector, expressed in bacteria, and sequenced to determine the identity of the fragment. The product can also be used as a probe for a Northern blot to confirm changes in gene expression by an alternative method, which is important in all gene expression studies.

This method has several advantages over the earlier methods of subtractive and differential hybridization. First, it is easier and less time-consuming. There is also a need for far less RNA then with previous methods. The sensitivity and reproducibility is also increased. The method enables more flexibility in terms of comparing multiple tissue samples to each other. For example, one could examine differences in gene expression over a time course of treatment. Though this method continues to be widely used and has yielded important information, there are some disadvantages. The rate of false positives can be high. The method is still relatively cumbersome, involving multiple steps before a gene is identified. It is also unclear how well low-abundance RNAs are represented.

With the sequencing of the human and mouse genomes, another method has emerged as a way to identify changes in gene expression, namely, serial analysis of gene expression, or SAGE (Scott and Chrast, 2001). SAGE is based on the principle that a short stretch of DNA sequence from the 3′ end is of sufficient complexity to use as a tag to identify a unique gene transcript. In SAGE, mRNA is isolated from experimental and control tissues and bound to poly(dT) magnetic beads. There it is reverse-transcribed to cDNA and digested with an "anchoring" restriction enzyme that cleaves the 5′ end, leaving the 3′ end attached to the bead. A linker sequence is then ligated to the open end of the cDNA. A "tagging enzyme" cuts the cDNA 10 base pairs 3′ to its site in the linker DNA, releasing the linker sequence and 10-base-pair tag from the beads. The tags are then ligated to form ditags and amplified by PCR and serially concatenated to increase the sequencing efficiency. The DNA is sequenced and analyzed using software that is able to identify unique genes by their tag and quantifies the amount of each gene in a given tissue sample. This method has advantages in that it is less time-consuming and easier to use than some of the earlier methods. It is also very sensitive and is better for analyzing low-abundance genes. Some of the disadvantages are its high cost and somewhat low throughput due to the limitation on the number of tags that can be sequenced. Sometimes multiple genes match to the same tag so the identity of the gene cannot be determined. Furthermore, certain genes lack the recognition site for the anchoring enzyme such that these genes do not attach to the linker sequence. Finally, since the method relies on a completed genomic sequence, it can only be used for organisms for which the genome has been sequenced.

11.5.1 DNA Array Technology

The most widely used gene expression system today is the DNA array (reviewed in Heller, 2002; Barlow and Lockhart 2002; Nguyen et al., 2002). The most basic type of array is a macroarray, where cDNAs of selected genes of interest are spotted onto nitrocellulose filters. Because the DNA is physically spotted onto the filter, the identity of every gene on the filter is known. This is a major advantage over previous methods that require multiple steps before the identities of regulated genes could be determined. RNA from the experimental and control groups is isolated, cDNA is synthesized while incorporating a radioactive nucleotide, and the cDNA is hybridized to the filters. The intensity of the individual spots is then used to determine the signal of each gene on the array. Computer programs can compare the two arrays to determine differences in gene expression. DNA macroarrays can be stripped and reprobed numerous times. One disadvantage is that a limited number of cDNAs can be spotted onto the filters, and typically genes chosen for such arrays are known genes, so no novel genes can be identified. Another disadvantage is that this system still relies on the use of radioactivity. Furthermore, variations in hybridization and washing conditions can lead to nonspecific binding of related genes to a cDNA on the array, or insufficient binding of the gene if the conditions are too stringent.

Based on the need for broader-based screens that include more genes, cDNA microarrays were developed. In this technique, cDNAs are spotted onto a glass slide that has been specially treated to bind the cDNA. Using this technology, thousands of genes can be spotted onto a single slide. This technique allows more flexibility in making custom arrays of known genes and expressed sequence tags (ESTs) in which the function of the gene is unknown. For example, an array can be made that contains only genes that have CRE sites in the promoter. In addition, cDNA synthesized from the RNA is typically labeled with different fluorescent markers, thus avoiding the use of radioactivity and enabling cDNA from experimental and control samples to be run simultaneously on the same array. This helps avoid some the potential differences in expression that can come from unequal hybridization and washing conditions. Fluorescent labeling does require the use of expensive scanners and software to analyze levels of fluorescence and determine differences in expression patterns. The arrays are also created through the use of robotic spotters, which can yield arrays that vary widely in the quality of the spots. Many cDNA arrays are now commercially available, which reduces some of the variation in quality when compared with "homemade" arrays. However, cDNA arrays still have problems with specificity, since many genes within a family have very similar sequences and will crosshybridize. Since the cDNA fragments come from PCR products, it is imperative that quality control measures are taken to ensure that only one PCR product is spotted at each site.

To overcome the nonspecificity that is seen when whole cDNAs are spotted onto an array, oligonucleotide arrays have been developed. These display 25- to 70-mer synthetic oligos that are made from a unique region of each gene. These oligos can either be spotted onto a glass slide in a similar manner to the cDNAs, or they can be built onto a silicon surface, as is the case with the Gene Chip™ from Affymetrix. The use of oligonucleotide arrays also allows for independent examination of alternatively spliced products of genes and single-nucleotide polymorphisms (SNPs) within genes. Affymetrix arrays in particular have many advantages over other arrays in that more genes can fit on a single Affymetrix chip than on a spotted array. Multiple sequences from each gene are represented on every Affymetrix chip, as well as single-base-pair mismatched controls for nonspecific hybridization. These measures build confidence in the data and greatly reduce the number of false positives. Unlike spotted arrays, only one RNA sample is run on a single chip and then two chips are compared, which could lead to artifactual results if the conditions are not equal. The Affymetix system has automated the hybridization, washing, and staining of the chips to reduce this problem. The biggest downside of the Affymetrix chip is the cost. Running Affymetrix chips requires a large monetary investment that is not needed for other types of arrays. On the other hand, fewer replications may be needed due to the greater

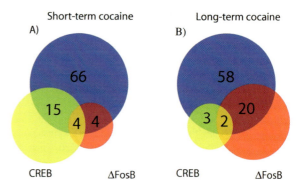

Figure 11.5   Gene expression after cocaine treatment is partially mediated by CREB and ΔFosB. Wild-type mice were given daily i.p. injections of cocaine (10 mg/kg) or saline for 5 days (**A**), or cocaine (15 mg/kg) or saline for 5 days/week × 4 consecutive weeks (**B**), and were used the day after the last injection. RNA was isolated from the nucleus accumbens and subjected to microarray analysis. (**A**) Venn diagram showing the number of genes with shared regulation between short-term cocaine treatment and short-term ΔFosB (1–2 weeks) and CREB (8 weeks) expression. (**B**) Venn diagram showing the number of genes with shared regulation between long-term cocaine treatment and long-term ΔFosB (4–8 weeks) and CREB expression.

confidence in chip data. Another disadvantage is that the binding to oligonucleotides is not as strong as it is to cDNA, so more RNA is needed (though RNA amplification techniques have reduced this problem). There is also the remaining difficulty in sensitivity. Affymetrix chips, like cDNA macro- and microarrays, do not typically detect low-abundance genes with sufficient sensitivity.

The use of these various gene expression systems has greatly accelerated the search for target genes of transcription factors. Furthermore, by comparing the genes that are induced by transcription factor expression, to the genes that become induced after a particular treatment (such as drug treatment), one can begin to identify genes that are regulated by these transcription factors in response to that particular treatment. For example, it has been known for some time that CREB and ΔFosB are induced in the nucleus accumbens after cocaine exposure (Nestler, 2001). However, the genes that are regulated in response to this induction were not known. We have recently identified potential target genes using transgenic mice that express CREB, mCREB (a dominant negative mutant), ΔFosB, or an AP-1 antagonist, ΔcJun, in the nucleus accumbens of adult mice (McClung and Nestler, 2003). Affymetrix microarrays were used to identify genes that were reciprocally regulated by CREB versus mCREB, as well as by ΔFosB versus ΔcJun. The results were then compared with the genes found to be regulated by different cocaine treatment paradigms (Figure 11.5). This approach has succeeded in identifying subsets of genes whose regulation by cocaine is mediated via CREB or ΔFosB.

11.5.2 Other Methods

Each of these gene expression systems discussed above requires verification of the changes in expression by an independent method on independent tissue samples. Examples of methods typically used for this purpose include Northern blotting, RNAse protection assays, quantitative in situ hybridization, and real-time RT-PCR.

After creating a list of potential target genes, the next step is to determine if a transcription factor is indeed bound to the DNA after a particular treatment. This helps determine whether the regulation is direct or indirect. There are a few techniques that can be used for this type of analysis. One is a gel shift assay, which can determine if the transcription factor can bind to promoter sites of specific genes (Molloy, 2000). Another, more powerful assay is the chromatin immunoprecipitation (ChIP) assay, which can be used to determine if the protein is actually bound to the promoter in a particular tissue in vivo (Johnson and Bresnick, 2002). The ChIP assay can also be combined with microarrays as a very powerful "ChIP on Chip" approach as a novel way to search for target genes. This will be described at the end of the chapter. Even if the protein is bound to the promoter, it may not be acting as a transcriptional activator. To address this question, the gene promoter can be cloned into an expression vector containing the gene for firefly luciferase (LUC), or another reporter such as chloramphenicol acetyltransferase (CAT) or β-galactosidase (Naylor, 1999). These constructs can be expressed in cultured cells along with the transcription factor of interest. An increase in reporter activity indicates that the transcription factor is able to induce that gene's promoter, at least in vitro. A major need for the future is to develop equivalent tools assessing gene promoter activity within brain regions of interest in vivo.

11.5.3 Future Directions

One future direction for gene expression studies is at the single-cell level. Though all of the above approaches used to overexpress or repress genes are very powerful, the tissue that is taken for subsequent RNA analysis is still made up of a mixed population of cells that express the gene of interest and those that do not, especially when the tissue is from the brain. The brain is particularly complex, with multiple cell types in a given region. For these reasons, single-cell RT-PCR and single-cell microarrays are being used to determine the actions of transcription factors in specific cell types. Obstacles to this type of analysis have included (a) isolation of the cell of interest and (b) obtaining enough RNA from that cell, or group of similar cells, to run a microarray. Recently, laser capture microdissection has allowed the isolation of single cells from a tissue slice using a laser to either cut around the cell or lift the cell onto a piece of film (Todd et al., 2002). The cell can be identified through the use of a fluorescent tag like green fluorescent protein (GFP). Furthermore, the tissue can even be stained using an antibody for the protein of interest. This use of immunocytochemistry to identify a cell of interest now offers the potential to isolate a cell that is naturally up-regulating a transcription factor in response to a treatment paradigm. For example, by comparing the gene expression pattern in the cells inducing the transcription factor after antidepressant treatment to its neighbors that do not, one can now identify the target genes that are most relevant to this treatment. In addition to laser capture, electrophysiological recording pipettes can be utilized to extract the contents of a cell after recording its activity. The RNA can be then be isolated for gene expression studies. This allows the isolation of RNA specifically from cells that have a particular electrophysiological signature (Liss, 2002).

To generate the amount of RNA needed for microarray analysis, RNA amplification techniques have been developed (Eberwine et al., 1992). This technique, as first described by Jim Eberwine and colleagues, incorporates a T7 RNA polymerase recognition site at the 3′ end during cDNA synthesis. Using this site, new RNA is synthesized from the cDNA by in vitro transcription. Unlike PCR, in vitro transcription is a linear amplification method. This amplification can go through a second round if need be, and the resulting amplified RNA can be used for array analysis. RNA amplification technology represents a major step forward in gene expression technology. However, one problem is that the resulting amplified RNA is not always a good representation of the original RNA population from the cell. Some low-abundance messages drop out, and certain genes amplify to a greater extent than others. However, these genes seem to amplify to the same degree in every population, so as long as the RNA samples from the experimental and control groups are amplified under equal conditions, and the amplification is of equal efficiency, these altered rates of amplification are less of a concern.

Another recently described technique, mentioned above, that holds great promise in identifying target genes of specific transcription factors is the so-called ChIP on Chip or "Chip^2" experiment, which combines the chromatin immunoprecipitation assay with microarray assays (Weinmann and Farnham, 2002). In this case, ChIP is used to isolate DNA fragments that are bound to a specific transcription factor. This technique has a significant advantage over gene expression studies where you are unable to determine whether identified genes are direct or indirect targets of the transcription factor without further analysis. In ChIP, the protein and DNA in chromatin are cross-linked, the DNA is sheared into smaller fragments, and specific DNA–protein complexes are immunoprecipitated with an antibody directed against the transcription factor of interest. The DNA is then isolated from the protein, labeled, and hybridized to a microarray, which comprises promoter sequences of multiple genes. This is a very exciting new approach that has the potential to identify the genes that are bound by the transcription factor in vivo in a high-throughput manner.

11.6 CONCLUSIONS

In conclusion, gene expression is important in regulating cell survival, maintenance, and long-term plasticity in the brain. Genes are regulated downstream of signal transduction pathways in response to neural activity or receptor binding by neurotransmitters, hormones, or growth factors. These signaling pathways eventually signal to the cell nucleus, where changes in the amounts or activities of transcription factors or other nuclear proteins leads to the ultimate changes in gene expression. Searches for consensus transcription factor binding sites in the promoters of genes has allowed the identification of potential target genes for individual transcription factors. Genetic and viral overexpression and repression studies, coupled with high-throughput gene expression systems, have helped determine the target genes that are regulated in specific regions of the brain in vivo. By comparing the genes regulated in these studies with those regulated by various treatment paradigms, one can start to determine the genes involved in regulating complex behaviors or in animal models of psychiatric and neurological disease. Future studies will further characterize the biologically relevant target genes at the cellular and systems level, to help us understand the molecular changes that shape the brain and ultimately help direct us toward more effective treatments, cures, and preventive measures.

References

Andersson, M., Konradi, C., and Cenci, M.A. (2001). cAMP response element-binding protein is required for dopamine-dependent gene expression in the intact but not the dopamine-denervated striatum. *J. Neurosci.* **21**, 9930–9943.

Ang, E., Chen, J., Zagouras, P., Magna, H., Holland, J., Schaeffer, E., and Nestler, E.J. (2001). Induction of nuclear factor-kappaB in nucleus accumbens by chronic cocaine administration. *J. Neurochem.* **79**, 221–224.

Barlow, C., and Lockhart, D.J. (2002). DNA arrays and neurobiology–what's new and what's next? *Curr. Opin. Neurobiol.* **12**, 554–561.

Beaujean, N. (2002). Fundamental features of chromatin structure. *Cloning Stem Cells* **4**, 355–361.

Berridge, M.J. (1997). Elementary and global aspects of calcium signaling. *J. Physiol. (London)* **499**, 291–306.

Black, D.L. (2003). Mechanisms of alternative pre-messenger RNA splicing. *Annu. Rev. Biochem.* **72**, 291–336.

Blanc, V., and Davidson, N.O. (2003). C-to-U RNA editing: Mechanisms leading to genetic diversity. *J. Biol. Chem.* **278**, 1395–1398.

Cantley, L.C. (2002). The phosphoinositide 3-kinase pathway. *Science* **296**, 1655–1657.

Carey, M., and Smale, S.T. (2000). *Transcriptional Regulation in Eukaryotes.* Cold Spring Harbor Laboratory Press, Cold Spring Harbor, NY.

Carlezon, W.A., Jr., Nestler, E.J., and Neve, R.L. (2000). Herpes simplex virus-mediated gene transfer as a tool for neuropsychiatric research. *Crit. Rev. Neurobiol.* **14**, 47–67.

Chen, J., Kelz, M.B., Zeng, G., Sakai, N., Steffen, C., Shockett, P.E., Picciotto, M.R., Duman, R.S., and Nestler, E.J. (1998). Transgenic animals with inducible, targeted gene expression in the brain. *Mol. Pharmacol.* **54**, 495–503.

Chimenov, Y., and Kerppola, T.K. (2001). Close encounters of many kinds: Fos-Jun interactions that mediate transcription regulatory specificity. *Oncogene* **6**, 533–542.

Chrivia, J.C., Kwok, R.P., Lamb, N., Hagiwara, M., Montminy, M.R., and Goodman, R.H. (1993). Phosphorylated CREB binds specifically to the nuclear protein CBP. *Nature (London)* **365**, 855–859.

Eberwine, J., Yeh, H., Miyashiro, K., Cao, Y., Nair, S., Finnell, R., Zettel, M., and Coleman, P. (1992). Analysis of gene expression in single live neurons. *Proc. Natl. Acad. Sci. U.S.A.* **89**, 3010–3014.

Felsenfeld, G., and Groudine, M. (2003). Controlling the double helix. *Nature (London)* **421**, 448–453.

Grzybowska, E.A., Wilczynska, A., and Siedlecki, J.A. (2001). Regulatory functions of 3'UTRs. *Biochem. Biophys. Res. Commun.* **288**, 291–295.

Guerrini, L., Blasi, F., and Denis-Donini, S. (1995). Synaptic activation of NF-κB by glutamate in cerebellar granule neurons *in vitro*. *Proc. Natl. Acad. Sci. U.S.A.* **92**, 9077–9081.

Hamm, H.E. (1998). The many faces of G protein signaling. *J. Biol. Chem.* **273**, 669–672.

Hanrahan, C.J., Palladino, M.J., Ganetzky, B., and Reenan, R.A. (2000). RNA editing of the *Drosophila para* Na(+) channel transcript. Evolutionary conservation and developmental regulation. *Genetics* **155**, 1149–1160.

Heinemeyer, T., Wingender, E., Reuter, I., Hermjakob, H., Kel, A.E., Kel, O.V., Ignatieve, E.V., Ananko, E.A., Podkolodnaya, O.A., Kolpakov, F.A., Podkolodny, N.L., and Kolchanov, N.A. (1998). Databases on transcriptional regulation: TRANSFAC, TRRD and COMPEL. *Nucleic Acids Res.* **26**, 362–367.

Heller, M.J. (2002). DNA microarray technology: Devices, systems and applications. *Annu. Rev. Biomed. Eng.* **4**, 129–153.

Higuchi, M., Mass, S., Single, F.N., Hartner, J., Rozov, A., Burnashev, N., Feldmeyer, D., Sprengel, R., and Seeburg, P.H. (2000). Point mutation in an AMPA receptor gene rescues lethality in mice deficient in the RNA-editing enzyme ADAR2. *Nature (London)* **406**, 78–81.

Hunter, T. (2000). Signaling-2000 and beyond. *Cell* **100**, 113–127.

Johnson, K.D., and Bresnick, E.H. (2002). Dissecting long-range transcriptional mechanisms by chromatin immunoprecipitation. *Methods* **26**, 27–36.

Karin, M., and Ben-Neriah, Y. (2000). Phosphorylation meets ubiquitination: The control of NF-κB activity. *Annu. Rev. Immunol.* **18**, 621–663.

Kelz M.B., Chen J., Carlezon W.A., Jr., Whisler K., Gilden L., Beckmann A.M., Steffen C., Zhang Y.J., Marotti L., Self D.W., Tkatch T., Baranauskas G., Surmeier D.J., Neve R.L., Duman R.S., Picciotto M.R., Nestler E.J. (1999). Expression of the transcription factor deltaFosB in the brain controls sensitivity to cocaine. *Nature* **401**, 272–276.

Lasko, P. (2003). Gene regulation at the RNA layer: RNA binding proteins in intercellular signaling networks. *Sci. STKE*, **179**, 1–8.

Lehrmann, H., Pritchard, L.L., and Harel-Bellan, A. (2002). Histone acetyltransferases and deacetylases in the control of cell proliferation and differentiation. *Adv. Cancer Res.* **86**, 41–65.

Lin, R.J., Kao, H.Y., Ordentlich, P., and Evans, R.M. (1998). The transcriptional basis of steroid physiology. *Cold Spring Harb or Symp. Quant. Biol.* **63**, 577–585.

Liss, B. (2002). Improved quantitative real-time RT-PCR for expression profiling of individual cells. *Nucleic Acids Res.* **30**, e89.

Lonze, B.E., and Ginty, D.D. (2002). Function and regulation of CREB family transcription factors in the nervous system. *Neuron* **35**, 605–623.

Mattson, M.P., and Camandola, S. (2001). NF-kappaB in neuronal plasticity and neurodegenerative disorders. *J. Clin. Invest.* **107**, 247–254.

Mayr, B., and Montminy, M. (2001). Transcriptional regulation by the phosphorylation-dependent factor CREB. *Nat. Rev. Mol. Cell. Biol.* **2**, 599–609.

McClung, C.A., and Nestler, E.J. (2003). Regulation of gene expression and cocaine reward by CREB and ΔFosB. Submitted for publication.

Mercurio, F., and Manning, A.M. (1999). Multiple signals converging on NF-kappaB. *Curr. Opin. Cell. Biol.* **2**, 226–232.

Molloy, P.L. (2000). Electrophoretic mobility shift assays. *Methods Mol. Biol.* **130**, 235–246.

Monteggia L.M., Barrot M., Powell C.M., Berton O., Galanis V., Gemelli T., Meuth S., Nagy A., Greene R.W., Nestler E.J. (2004). Essential role of brain-derived neurotrophic factor in adult hippocampal function. *Proc. Natl. Acad. Sci. USA* **101**, 10827–10832.

Naylor, L.H. (1999). Reporter gene technology: The future looks bright. *Biochem. Pharmacol.* **58**, 749–757.

Nestler, E.J. (2001). Molecular neurobiology of addiction. *Am. J. Addict.* **10**, 201–217.

Nestler, E.J., Barrot, M., and Self, D.W. (2001). DeltaFosB: A sustained molecular switch for addiction. *Proc. Natl. Acad. Sci. USA* **98**, 11042–11046.

Nguyen, D.V., Arpat, A.B., Wang, N., and Carroll, R.J. (2002). DNA microarray experiments: Biological and technological aspects. *Biometrics* **58**, 701–717.

Qiu, P., Qin, L., Sorrentino, R.P., Greene, J.R., and Wang, L. (2003). Comparative promoter analysis and its application in analysis of PTH-regulated gene expression. *J. Mol. Biol.* **326**, 1327–1336.

Reenan, R.A. (2001). The RNA world meets behavior: A-I pre mRNA editing in animals. *Trends Genet.* **17**, 53–56.

Rutter, J., Reick, M., Wu, L.C., and McKnight, S.L. (2001). Regulation of Clock and NPAS2 DNA binding by the redox state of NAD cofactors. *Science* **293**, 510–514.

Sagerstrom, C.G., Sun, B.I., and Sive, H.L. (1997). Subtractive cloning: Past, present, and future. *Annu. Rev. Biochem.* **66**, 751–783.

Sakai, N., Thome, J., Newton, S.S., Chen, J., Kelz, M.B., Steffen, C., Nestler, E.J., and Duman, R.S. (2002). Inducible and brain region-specific CREB transgenic mice. *Mol. Pharmacol.* **61**, 1453–1464.

Scammel, T.E., Arrigoni, E., Thompson, M.A., Ronan, P.J., Saper, C.B., and Greene, R.W. (2003). Focal deletion of the adenosine A1 receptor in adult mice using an adeno-associated viral vector. *J. Neurosci.* **23**, 5762–5770.

Scherr, M., Morgan, M.A., and Eder, M. (2003). Gene silencing mediated by small interfering RNAs in mammalian cells. *Curr. Med. Chem.* **10**, 245–256.

Scott, H.S., and Chrast, R. (2001). Global transcript expression profiling by serial analysis of gene expression (SAGE). *Genet. Eng.* **23**, 201–219.

Seeburg, P.H. (2002). A-to-I editing: New and old sites, functions and speculations. *Neuron* **35**, 17–20.

Stanewsky, R. (2003). Genetic analysis of the circadian system in *Drosophila melanogaster* and mammals. *J. Neurobiol.* **54**, 111–147.

Stein, J., and Liang, P. (2002). Differential display technology: A general guide. *Cell. Mol. Life Sci.* **59**, 1235–1240.

Todd, R., Lingen, M.W., and Kuo, W.P. (2002). Gene expression profiling using laser capture microdissection. *Expert Rev. Mol. Diagn.* **5**, 497–507.

Tourriere, H., Gallouzi, I.E., Chebli, K., Capony, J.P., Mouaikel, J., van der, Geer, P., and Tazi, J. (2001). RasGAP-associated endori-bonuclease G3BP: Selective RNA degradation and phosphorylation-dependent localization. *Mol. Cell. Biol.* **21**, 7747–7760.

Wang, Q., Khillian, J., Gadue, P., and Nishikura, K. (2000). Requirement of the RNA editing deaminase ADAR1 gene for embryonic erythropoiesis. *Science* **290**, 1765–1768.

Weinmann, A.S., and Farnham, P.J. (2002). Identification of unknown target genes of human transcription factors using chromatin immunoprecipitation. *Methods* **26**, 37–47.

West, A.E., Griffith, E.C., and Greenberg, M.E. (2002). Regulation of transcription factors by neuronal activity. *Nat. Rev. Neurosci.* **3**, 921–931.

Wisdom, R. (1999). AP-1: One switch for many signals. *Exp. Cell Res.* **253**, 180–185.

Wu, X., and McMurray, C.T. (2001). Calmodulin kinase II attenuation of gene transcription by preventing cAMP response element-binding protein (CREB) dimerization and binding of the CREB-binding protein. *J. Biol. Chem.* **276**, 1735–1741.

Yen, J., Wisdom, R.M., Tratner, I., and Verna, I.M. (1991). An alternative slice form of FosB is a negative regulator of transcriptional activation and transformation by Fos proteins. *Proc. Natl. Acad. Sci. U.S.A.* **88**, 5077-5081.

Yu, Y., and Bradley, A. (2001). Engineering chromosomal rearrangements in mice. *Nat. Rev. Genet.* **2**, 780–790.

Web Resources

http://www.Affymetrix.com

Cell Signaling Networks in Long-Term Potentiation

Elizabeth A. Grace, Ph.D., Emmanuel M. Landau, M.D., Ph.D.,
Ravi Iyengar, Ph.D., and Robert D. Blitzer, Ph.D.

CONTENTS

Databasing the Brain. Edited by Stephen H. Koslow and Shankar Subramaniam
ISBN 0-471-30921-4 © 2005 John Wiley & Sons, Inc.

12.1 OVERVIEW

Neurons are notable for their capacity to undergo persistent changes in response to brief signaling events, exemplified by the induction of long-term potentiation (LTP) by particular patterns of synaptic stimulation. While a great deal remains to be learned about the components and functioning of the signaling network that underlies LTP, impressive progress has been made in recent years. Recent work has illuminated the physical and functional organization of the network and has provided insights into a prominent property of LTP: the requirement for a remarkably large number of molecules, including scaffolding proteins, cytoskeletal components, and several protein kinases.

Important general features of a signaling pathway include signal amplification, integration, and fidelity of transmission (Sweatt, 1999; Dineley et al., 2001). A common motif in the behavior of signaling pathways is the competition between regulated protein kinase and phosphatase activities directed at common substrates, which determines (1) whether signals are propagated, (2) for how long, and (3) in some cases, the ultimate subcellular distribution of the signal. Scaffolding proteins are essential components of signaling pathways due to their ability to localize the protein kinases and phosphatases to the appropriate compartment, and to present them to their co-localized substrates (Whitmarsh et al., 1998; Weng et al., 1999). The specificity in response to each stimulus naturally follows from the pathway components, but the identity of this set of components is not static; rather, it is shaped by spatial and temporal regulation (Whitmarsh et al., 1998; Weng et al., 1999).

An additional layer of complexity is provided by interactions between different pathways. A signaling pathway that features minimal integration can consist of a simple linear series of reactions. More often, however, the successful transmission of a signal depends on the cell's ability to integrate activity in different signaling pathways when they temporally and spatially overlap, coordinating a unified response. For example, multiple protein kinases are frequently activated by the same external stimuli, recruiting separate signaling pathways that form a signaling network. These networks are integrated when they converge on the same downstream component, funneling their responses into a single, unified set of instructions to the cellular machinery. The integrating component must be able to coordinate a response based on specific combinations of inputs from the separate signaling pathways.

Amplification. Signal amplification occurs when the activity of components increases nonlinearly as the signal moves downstream. This effect may occur at any point within the pathway, and can involve a series of coupled steps resulting in a pronounced signal boost, as exemplified by the MAP kinase cascade (Whitmarsh et al., 1998; Sweatt, 1999). Amplification is especially important if the signal begins at a single cellular location with a high concentration of signaling molecule, and must translocate to a different compartment with a lower concentration, such as from a dendritic spine to the cell body. A mechanism by which amplification occurs is through positive feedback loops, which enhance the activity of upstream effectors.

Compartmentalization. Compartmentalization is an important aspect of signal processing. Microdomains can be established by scaffolding proteins that link effectors with their substrates (Weng et al., 1999; Dorn and Mochly-Rosen, 2002). They can also arise from physical barriers to diffusion, such as the narrow neck associated with many dendritic spines. The microdomains contain unique combinations of signaling enzymes, substrates, and regulatory proteins at different concentrations than those expressed in the cell overall. High, locally restricted concentrations of signaling components, along with the exclusion of negative regulators, can facilitate signaling within microdomains by lowering the threshold for activation. By increasing the ability of effectors to respond to input, compartmentalization could create an environment where signaling pathways are too easily triggered by stimuli that are not biologically relevant. For many signaling pathways, the likelihood of such false positives is reduced by a requirement for the spatially and temporally simultaneous activation of two components before the signal can proceed downstream, a feature known as coincidence detection. Different pathways have different characteristic requirements for the strength of stimulus intensity. Thus, even within the same compartment, the set of active signaling pathways is dictated by the particular pattern of stimulation, and long-term changes cannot occur without the activation of multiple convergent pathways. Another cellular strategy to discourage spurious signaling is the use of negative feedback to curtail upstream pathway activity. Only when the negative feedback is abrogated, often by activity in a converging pathway, can the original signal be sustained.

12.1.1 Long-Term Potentiation as a Model Signaling Network

Long-term potentiation (LTP) is a long-lasting increase in the efficacy of synaptic communication (Bliss and Lomo, 1973). It is thought to be associated with learning and memory, and it may be the cellular analog for them under certain circumstances. Both memory and LTP are associative in nature, long-lasting, and protein synthesis-dependent (Bear, 1997). When induced by synaptic stimulation, LTP conforms to Hebb's postulate that coincident activity between two communicating neurons will increase the strength of their connection (Hebb, 1949). LTP can be induced at excitatory synapses in many telencephalic areas, including the hippocampus, neocortex, amygdala, striatum, and nucleus accumbens. The most intensively investigated of these regions is the hippocampus, a nucleus involved in memory formation, and particularly the CA3–CA1 synapse of hippocampus. In this review of the signaling network that underlies LTP, we primarily will consider data derived from experiments on the CA3–CA1 synapse in both rat and mouse hippocampus. At this synapse, LTP has been induced using a variety of conditioning stimuli. Some of these protocols are purely synaptic, in the sense that presynaptic fibers are electrically stimulated, and the resulting release of glutamate induces LTP. Our description of the LTP signaling network is derived from studies using such protocols, as opposed to those involving the application of exogenous transmitters or other receptor agonists. We will also restrict ourselves with regard to the synaptic locus of the network, focusing on the postsynaptic side. The role of presynaptic plasticity in LTP has generated a great deal of debate among LTP researchers, and it is the subject of ongoing research. Furthermore, the relative inaccessibility of the nerve terminal to experimental manipulations has predictably resulted in an experimental literature that favors an analysis of the postsynaptic network. However, important exceptions to an entirely postsynaptic locus of expression have been observed—for example, when presynaptic vesicular cycling is monitored or when LTP is induced by the application of BDNF (brain-derived neurotrophic factor) (Gottschalk et al., 1998; Zakharenko et al., 2001).

LTP is expressed within a restricted dendritic region, while synaptic connections outside that region remain unaffected. The spatial specificity of LTP reflects the compartmentalized activation of signaling networks that initially affect only that region. However, the maintenance of LTP requires the signaling components translocate beyond the original compartment (i.e., the dendritic spine). Often, the same signaling molecule, typically a kinase, can serve both local and global functions. In reviewing how the LTP signaling network accomplishes these temporally and spatially diverse functions, we will encounter numerous strategies for signal amplification and integration, processes that are expected to figure prominently in any complex signaling network.

Induction of LTP. The first step in the synaptic induction of LTP at the CA3–CA1 synapse is the stimulation-evoked release of glutamate, resulting in postsynaptic membrane depolarization due to the activation of AMPA (α-amino-3-hydroxy-5-methyl-4-isoxazoleprionic acid) receptors (Kirkwood et al., 1993). The AMPA receptor is a ligand-gated channel, primarily permeable to Na^+, which mediates fast excitatory transmission throughout the CNS (McLennan, 1983). The AMPA receptor is also the endpoint for LTP, in which AMPA-mediated currents are potentiated, resulting in more efficient synaptic transmission. The activation of AMPA receptors is not sufficient for LTP induction, however; another type of ionotropic glutamate receptor, the Ca^{2+}-permeable NMDA (*N*-methyl-D-aspartate) receptor, is also required (Malenka and Nicoll, 1999; Luscher et al., 2000; Dineley et al., 2001; Lisman et al., 2002). In the absence of substantial membrane depolarization, glutamate binding to the NMDA receptor results in little Ca^{2+} entry, because Mg^{2+} ions block the channel pore. However, the depolarization that accompanies sufficiently intense AMPA receptor stimulation relieves the Mg^{2+} block, permitting Ca^{2+} entry into dendrites of the postsynaptic (CA1) neuron (Malenka and Nicoll, 1999; Grant and O'Dell, 2001). This requirement for concurrent presynaptic and postsynaptic conditions (i.e., glutamate release and dendritic depolarization) represents the first level of coincidence detection for LTP (Grant and O'Dell, 2001; Rongo, 2002).

Ca^{2+}-Mediated Activation of Multiple Protein Kinases and Phosphatases in the Postsynaptic Region. The rise in postsynaptic Ca^{2+} initiates multiple signal transduction pathways, and the LTP signaling network largely comprises connections between Ca^{2+}-dependent protein kinases and phosphatases. Ca^{2+} signaling is mediated primarily through binding to the protein calmodulin (CaM) (Chin and Means, 2000). CaM is constitutively repressed, but Ca^{2+} binding leads to a conformation shift that then allows CaM to bind to its target helix, which is found on multiple protein kinases and phosphatases (Chin and Means, 2000; Hudmon and Schulman, 2002). The effectors that bind CaM frequently contain auto-inhibitory domains that prevent the catalytic domain from interacting with substrates (Hudmon and Schulman, 2002). CaM binds to the auto-inhibitory domain, relieving the inhibition and releasing the catalytic subunit.

Several protein kinases that are activated by synaptically induced Ca^{2+} influx have been implicated in LTP. Prominent among these is CaMKII (Ca^{2+}/CaM-dependent protein kinase II), the most abundant protein in the postsynaptic density and a central component of the LTP signaling network. Ca^{2+} entry can also activate some isoforms of adenylyl cyclase, generating cAMP (3', 5'-cyclic monophosphate) and stimulating PKA (cAMP-dependent protein kinase) (Xia et al., 1993). Protein kinase C (PKC) is activated by Ca^{2+}, in concert with second messengers released by metabotropic Glu receptor (mGluR) activity (Angenstein and Staak, 1997; Ramakers et al., 1997). A potentially important positive feedback pathway exists between PKC and CaM: PKC can liberate CaM that is otherwise sequestered by the nonphosphorylated form of neurogranin (Ramakers et al., 1997; Pak et al., 2000; Dineley et al., 2001). CaMKII, PKA, and PKC can all increase the activity of an additional protein kinase, MAPK (mitogen-activated protein kinase). Finally, both Ca^{2+}/CaM and PICA indirectly regulate a protein phosphatase, PP1, that has an important modulatory effect on LTP. The roles of these molecules in the LTP signaling network are detailed in later sections of this review. The combined actions of CaMKII, PKA, and PKC (and integration at the MAPK level) orchestrate the LTP-associated post-translational modifications, as well as changes in gene expression (Barria et al., 1997; Malenka and Nicoll, 1999; Lee et al., 2000; Abel and Lattal, 2001; Dineley et al., 2001).

The AMPA Receptor Is a Point of Network Convergence. The postsynaptic LTP signaling network is circular: The AMPA receptor provides the initial stimulus, and LTP is ultimately expressed as an increase in AMPA receptor function. In LTP, the AMPA receptors are thought to be regulated in two independent ways: a change in the receptor's phosphorylation state, and the insertion of new receptors into the synapse. The AMPA receptor is a substrate for several of the protein kinases that are activated by LTP-inducing stimulation: CaMKII and PKC can phosphorylate a residue associated with increased channel conductance, and a second site that modulates open probability is targeted by PKA (Derkach et al., 1999; Banke et al., 2000; Lee et al., 2000). The process of inserting new AMPA receptors into the synapse also depends on CaMKII activity, but through a mechanism unrelated to AMPA receptor phosphorylation (Derkach et al., 1999; Hayashi et al., 2000). The initial synaptic potentiation following LTP-inducing stimulation is likely to reflect AMPA receptor phosphorylation (Soderling and Derkach, 2000; Dineley et al., 2001), but continued potentiation requires the MAPK-dependent insertion of new AMPA receptors in the postsynaptic membrane (Zhu et al., 2002). The insertion is also dependent upon stargazin, a protein that binds to AMPA-R and, through its PDZ C-terminal tail, binds to PSD-95 for membrane insertion (Chetkovich et al., 2002; Schnell et al., 2002). The role of MAPK in AMPA receptor trafficking may be related to the phosphorylation state of stargazin, since its PDZ domain does contain a consensus site for proline-directed protein kinases such as MAPK1/2 (Chetkovich et al., 2002). Interestingly, a threonine site within the stargazin PDZ domain is phosphorylated by PKA, but this prevents insertion, possibly acting as negative feedback (Chetkovich et al., 2002). Within 15 minutes after LTP-inducing stimulation, new AMPA receptors appear within dendritic spines, including those that did not contain AMPA receptors previously (Luscher et al., 2000; Dineley et al., 2001). Synapses that do not contain AMPA receptors are referred to as "silent" synapses, because they contain only NMDA receptors that are not activated by stimulation at resting membrane potentials (Malenka and Nicoll 1999). Insertion of AMPA receptors into silent synapses in the region where the stimulus occurred creates newly functional synapses and a regional potentiation. LTP is impaired when AMPA receptor insertion is disrupted by mutations in the receptor's PDZ domain, which binds to the postsynaptic density (PSD), and by inhibition of NSF (*N*-ethylmaleimide-sensitive fusion protein; an ATPase required for insertion of AMPA receptors into the PSD) (Hayashi et al., 2000; Luscher et al., 2000; Shi et al., 2001). Deletion of the GluR1 subunit of the AMPA receptor, which contains the synaptic localization domain, prevents LTP (Malenka and Nicoll, 1999).

Transcription and Translation. The synapse and its nearby dendrites contain the bulk of the LTP signaling network, but potentiation involves signals that are exported beyond the dendritic compartment. LTP maintenance requires gene transcription, which is regulated following the dendritic export of modulators of the transcription factor CREB (cAMP response element binding protein) and its associate CBP (CREB binding protein) (Frey et al., 1988). CREB converts a decremental form of potentiation into a persistent one, indicating that CREB-mediated transcription is key to the durability of LTP (Barco et al., 2002). The particular genes whose expression are increased in LTP are largely unknown, although one well-documented example is the immediate early gene *Arc* (Steward and Worley, 2002). CREB's role in persistent synaptic potentiation is phylogenetically conserved and has been demonstrated behaviorally in *Aplysia* and *Drosophila*, as well as in rodents (Yin et al., 1995; Mayr and Montminy, 2001).

Although most neuronal mRNA translation occurs in the cell body, dendrites can also synthesize proteins. Several mRNAs relevant to LTP are found in the dendrites, such as those encoding the α-CaMKII subunit and AMPA receptor subunits (Miyashiro et al., 1994; Steward and Schuman, 2001). When dendrites are mechanically isolated from the cell bodies, protein-synthesis-dependent LTP can still occur (Kang and Schuman, 1996). Similarly, when the dendritic localization domain for α-CaMKII mRNA is deleted, LTP is reduced (Miller et al., 2002). Some studies have indicated that LTP-inducing stimulation can increase the rate of dendritic protein synthesis. For example, α-CaMKII mRNA associates with dendritic polyribosomes following synaptic stimulation (Steward and Schuman, 2001). Schuman and co-workers used a transcript coding for a fluorescent reporter protein incorporating the 3' UTR and 5' UTR for CaMKII was translated in that isolated dendrites following BDNF stimulation. The regulation of dendritic translation by synaptic stimulation may remodel the LTP signaling network locally and contribute to the synapse-specificity of persistent potentiation.

12.1.2 CaMKII

CaMKII performs multiple roles in the induction and expression of LTP. It is activated in LTP, regulating the behavior of AMPA receptors and other components of the LTP signaling network through phosphorylation. However, the abundance of CaMKII in dendrites, and particularly in the postsynaptic density, would also be consistent with a noncatalytic, structural role of this molecule in synaptic function. Indeed, it has been proposed that the insertion of new AMPA receptors into the synapse depends on the establishment of new postsynaptic "slots," in which CaMKII serves an essentially structural role (Lisman and Zhabotinsky, 2001).

CaMKII activity is necessary, and probably sufficient, for the expression of an early stage of LTP that does not require de novo protein synthesis (Malenka and Nicoll, 1999). Constitutively active forms of CaMKII lead to an increase in synaptic conductance that occludes further potentiation by LTP-inducing stimuli, suggesting that the synapses have already been maximally potentiated by a common mechanism (Lledo et al., 1995). Inhibition of CaMKII activity (through pharmacological blockade or using mice containing mutations in CaMKII) prevents LTP induction (Silva et al., 1992; Otmakhov et al., 1997; Hinds et al., 1998). Inhibition of CaMKII after LTP has already been induced, however, does not prevent the maintenance of LTP, suggesting that CaMKII activity is not involved in LTP maintenance (Malinow et al., 1989; Chen et al., 2001) Conversely, it could indicate that once CaMKII is activated by LTP-inducing stimuli and positioned within the postsynaptic density, it becomes difficult to deactivate.

CaMKII Activation. CaMKII contains an associative domain, a catalytic domain containing the site for substrate and ATP binding, and an auto-inhibitory domain. It forms a holoenzyme of 8–12 subunits in a unique double-ring structure. The subunits are linked by their associative domains, while the catalytic domains protrude from the center on stalk-like structures, presumably allowing ready access to their substrates (Woodgett et al., 1983; Kanaseki et al., 1991). Under basal Ca^{2+} conditions, CaMKII is inactivated by the binding of its auto-inhibitory domain to its catalytic domain (Colbran et al., 1989). CaM binding releases this inhibition. The differing CaM binding affinities between CaMKII and the CaM-dependent phosphatase calcineurin (also known as PP2b) enable the enzymes to function as discriminators of Ca^{2+} levels. Under basal conditions, CaMKII has a very weak affinity for Ca^{2+}/CaM; therefore at low to moderate concentrations, Ca^{2+}/CaM is more likely to bind to other molecules with higher affinities, such as calcineurin (Persechini and Cronk, 1999). Calcineurin dephosphorylates, and thereby inactivates, an inhibitor that represses another protein phosphatase, protein phosphatase 1 (PP1) (Fukunaga et al., 1996). PP1 has been identified as the major phosphatase directed at CaMKII in the PSD, while another phosphatase, PP2a, predominates in the cytosol (Strack et al., 1997). Interestingly, in the cytosol, active CaMKII phosphorylates PP2a, inhibiting it for a feed-forward effect (Fukunaga et al., 2000). When CaM does bind to CaMKII in low Ca^{2+}, it quickly dissociates and only transiently activates CaMKII, which is dephosphorylated by PP2a (Hudmon and Schulman, 2002; Lisman et al., 2002) (Figure 12.1A). At higher Ca^{2+} levels, CaM-bound subunits within the holoenzyme phosphorylate their neighboring subunits at Thr286 (Hudmon and Schulman, 2002). The result of this phosphorylation is to increase the subunit's affinity for CaM 1000-fold, above that of calcineurin, primarily by increasing the dissociation time of CaM 20,000-fold (Hudmon and Schulman, 2002). This is known as CaM "trapping," and it creates a CaMKII molecule that can remain active for prolonged periods following the return of Ca^{2+} to

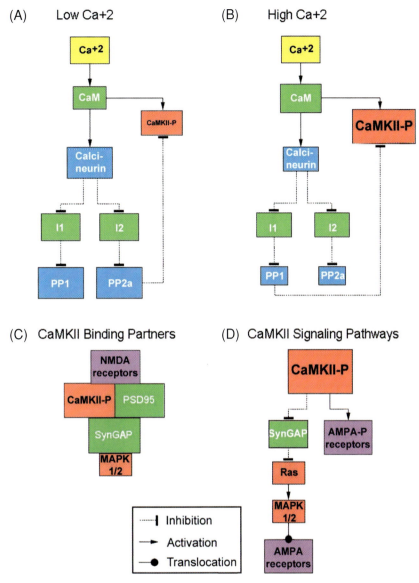

(A) Low Ca+2

(B) High Ca+2

(C) CaMKII Binding Partners

(D) CaMKII Signaling Pathways

Figure 12.1   Activation, Binding Partners and Signaling Pathways of CaMKII. (**A**) When Ca^{+2} levels are low, the phosphatase calcineurin (CaN) is primarily activated, which allows only transient activation of CaMKII. (**B**) When Ca^{+2} levels are high, coincident binding of CaM to adjacent CaMKII subunits leads to CaMKII autophosphorylation and the trapping of CaM, which results in persistent CaMKII activation. (**C**) Activated CaMKII translocates to the PSD where it binds to NMDA receptors, PSD95, and SynGap, which is in turn bound to MAPK1/2. The direct apposition of components indicates that they are binding partners. (**D**) Activation of CaMKII leads to phosphorylation of AMPA receptors resulting in increased AMPA channel conductance, as well as initiation of the MAPK signaling pathway, and the translocation of AMPA receptors to potentiated synapses. Color key (for all figures): Yellow refers to extracellular ion influx or release from intracellular stores, pink to receptors, green to proteins that bind and modulate the activity of other proteins, white to proteins that bind for compartmentalization without altering activity, red to kinases, blue to phosphatases, and magenta to transcription factors. The figures do not represent every known interaction between the signal components depicted.

basal levels (Hudmon and Schulman, 2002; Lisman et al., 2002) (Figure 12.1B).

The ring structure of the CaMKII holoenzyme not only localizes the substrate to the effector (in this case neighboring subunits within the holoenzyme), but also creates a coincidence detection device (Hudmon and Schulman, 2002; Lisman et al.,

2002). Each CaMKII subunit can only become phosphorylated if both it and its neighbor have bound CaM. Under low Ca^{2+} conditions, with CaMKII surrounded by phosphatases with higher CaM affinities, dual CaM binding leading to autophosphorylation is a low probability event; if it does occur, CaMKII is quickly de-phosphorylated by the active phosphatases. When Ca^{2+}

increases due to coincidence detection by the NMDA receptor, the probability of CaM binding to any particular subunit increases, favoring autophosphorylation (Hudmon and Schulman, 2002; Lisman et al., 2002). The activated CaMKII then translocates to the synapse, where it can bind directly to NMDA receptors and, via both NMDA receptors and the GTPase-activating protein SynGAP, to postsynaptic density protein 95 (PSD-95) (Strack et al., 2000; Bayer et al., 2001; Grant and O'Dell, 2001; Gleason et al., 2003) (Figure 12.1C). The binding to NMDA receptors has been shown to act as a "wedge" to keep CaMKII active by preventing the auto-inhibitory domain from binding the substrate domain, even if CaM dissociates (Bayer et al., 2001; Lisman et al., 2002). Also, CaMKII bound to NMDA receptors does not undergo a burst of autophosphorylation at Thr305, which normally acts as a negative feedback site by preventing CaM from re-binding to CaMKII (Shen et al., 2000; Lisman et al., 2002). The continued activity of CaMKII is dependent on saturation of phosphatases for which CaMKII is a substrate, and this occurs due to compartmentalization: The locally high concentration of CaMKII in the PSD saturates PP1, which is the major phosphatase that acts on CaMKII in that compartment (Strack et al., 1997; Lisman and Zhabotinsky, 2001).

At the synapse, CaMKII holoenzymes have been shown to be bistable: They remain in either an inactive or active state after Ca^{2+} signals return to normal (Lisman and McIntyre, 2001; Lisman and Zhabotinsky, 2001; Lisman et al., 2002). Both states are thermodynamically stable due to low CaM/CaMKII affinity and high phosphatase activity in the inactive state or high CaM/phospho-CaMKII affinity and saturated phosphatase activity in the active state. The ability to shift from a stable inactive state to a stable active state constitutes a molecular switch (Lisman and Zhabotinsky, 2001; Lisman et al., 2002). Most of the CaMKII in the spine at a potentiated synapse will be active, which may serve as a tag for the synapse to capture new proteins or mRNA for protein-dependent long-term changes in synaptic efficacy.

CaMKII Substrates. Once CaMKII is active (which occurs within 5 minutes of stimulation), it has multiple substrates available within the spine compartment (Barria et al., 1997; Soderling and Derkach, 2000) (Fig 12.1D). Among the first substrates after autophosphorylation are the local AMPA receptors of the GluR1 subtype, which CaMKII phosphorylates at Ser831 (Barria et al., 1997; Lisman et al., 2002). AMPA phosphorylation first appears as early as 15 minutes after stimulation and continues to rise for up to 60 minutes (Barria et al., 1997; Lee et al., 2000; Lisman et al., 2002). This phosphorylation site increases the conductance of the AMPA channel, potentiating synaptic transmission (Lledo et al., 1995; Soderling and Derkach, 2000). CaMKII translocates to the PSD by binding NMDA receptors, which are also substrates (Bayer et al., 2001). CaMKII also inactivates SynGAP, a PSD-95-bound GTPase that inhibits Ras by hydrolyzing GTP to GDP (Chen et al., 1998). Once Ras is released from inhibition, it sets off a cascade of interactions: Ras → Raf-1 → mitogen and extracellular regulated kinase (MEK) → MAPK (specifically the extracellular regulated protein kinase (Erk) isoforms also known as MAPK1/2 (Chen et al., 1998; Chang and Karin, 2001). Activation of MAPK1/2 plays a role in post-translational modifications at the synapse, such as the translocation of extrasynaptic AMPA receptors to the synapse.

12.1.3 Protein Kinase A (PKA)

PKA is an important modulator of early LTP and is essential for the protein synthesis-dependent form (Bernabeu et al., 1997; Frey and Morris, 1998; Otmakhova et al., 2000; Abel and Lattal, 2001; Dineley et al., 2001). PKA-dependent LTP can be induced by some forms of synaptic stimulation (Thomas et al., 1996; Blitzer et al., 1998; Brown et al., 2000). In these experiments, PKA inhibitors interfered with the expression of LTP as early as 5 minutes after stimulation, suggesting a local role for PKA. Similarly, mutant mice with reduced PKA activity also have deficits in the protein synthesis-dependent form of LTP (Abel et al., 1997). PKA is activated in response to the stimulation of adenylyl cyclase-coupled receptors. An additional important source of stimulation in LTP is likely to be the Ca^{2+}/CaM-dependent activation of types 1 and 8 adenylyl cyclase (Hanoune and Defer, 2001) (Figure 12.2A). LTP can be induced by activators of the cAMP pathway, which bypasses the early stage of LTP to induce only protein synthesis-dependent LTP (Chavez-Noriega and Stevens, 1992, 1994; Frey et al., 1993). This potentiation is blocked by CaMKII inhibitors, and it may be mediated through presynaptic mechanisms (Blitzer et al., 1995; Bear, 1997; Lisman et al., 2002). Interestingly, constitutively active forms of cAMP response element binding protein (CREB) mimic the effects of increased PKA activity on synaptic transmission (Mayr and Montminy, 2001). CREB is a transcription factor for numerous early genes thought to be involved with LTP, including zif268 and AMPA receptors (Fukunaga et al., 1996; Banke et al., 2000; Mayr and Montminy, 2001).

PKA is a holoenzyme consisting of a dimeric regulatory subunit and two monomeric catalytic subunits (Mayr and Montminy, 2001). It is kept in an inactive state through auto-inhibition by its regulatory subunit. When adenylyl cyclase is stimulated (by many mechanisms, including increased Ca^{2+} and mGluR activity), it increases cAMP formation. cAMP binds to two sites in the regulatory subunit of PKA, cleaving it from the catalytic subunit, which is then free to phosphorylate PKA substrates (Mayr and Montminy, 2001; Michel and Scott, 2002).

PKA compartmentalization and function depend on cAMP-dependent kinase anchoring proteins (AKAPs), which bind the inactive PKA regulatory subunit and, upon phosphorylation, translocate to the membrane (Fukunaga et al., 1996; Dorn and Mochly-Rosen, 2002; Michel and Scott, 2002). Frequently, the substrates of interest are also anchored by the same AKAP (Michel and Scott, 2002) (Figure 12.2B). In the PSD, PKA is bound through Yotiao to NMDA receptors and through AKAP79 to PSD-95, which also binds NMDA receptors (as well as SynGAP and through that, CaMKII) (Grant and O'Dell, 2001; Michel and Scott, 2002). NMDA receptors are enhanced by PKA activity in a Yotiao-dependent fashion indicating the important of scaffolding proteins localizing PKA for its function (Grant and O'Dell, 2001; Michel and Scott, 2002). PP1, which is indirectly inhibited by PKA, is also bound to Yotiao (Michel and Scott, 2002). Thus, the initial post-translational effects of PKA are dependent on its compartmentalization. High levels of cAMP are required for enough activated PKA to saturate local substrates, allowing it to leave the dendritic compartment and translocate to the nucleus, where it regulates gene expression (Dorn and Mochly-Rosen, 2002). Thus, similar to CaMKII and CaM, the intensity of PKA activation determines the ultimate result. The

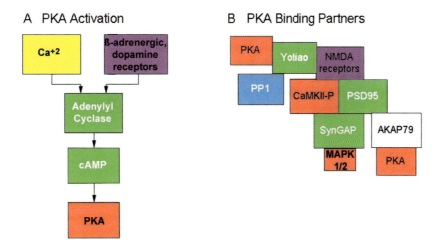

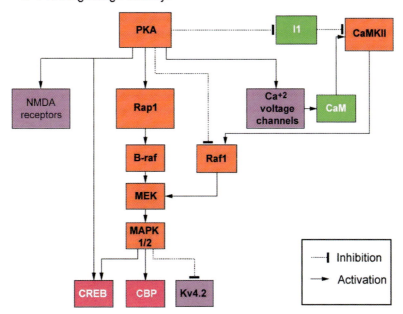

Figure 12.2   Activation, Binding Partners, and Signaling Pathways of PKA A) PKA is directly activated by cAMP, which is formed by adenylyl cyclase. Adenylyl cyclase can be activated by G-protein-coupled receptors, or by Ca^{+2} influx. B) PKA translocates to the PSD through AKAP79 or yotiao. Yotiao is a scaffolding protein that binds NMDA receptors and PP1. AKAP79 binds to PSD95 and through that to NMDA receptors. NMDA receptors are also bound to CaMKII, which associates with SynGAP and MAPK1/2, a substrate for both PKA and CaMKII. C) PKA enhances NMDA receptor activity. It also phosphorylates I1, which inhibits PP1 and thereby protects CaMKII from dephosphorylation. Activated CaMKII can then initiate the MAPK1/2 pathway through Raf-1. PKA inhibits this CaMKII-activated MAPK1/2 pathway, while initiating the MAPK1/2 pathway through Rap1 and B-raf. MAPK1/2 inhibits the potassium channel Kv4.2, which contributes to continued membrane depolarization. The PKA and MAPK1/2 pathways also stimulate the transcription of genes associated with LTP. Color key as in Fig 1.

presence of PKA in the synapse is thought to be important for its effects on early LTP, while its ability to translocate to the nucleus is essential for the protein synthesis-dependent phase (Dorn and Mochly-Rosen, 2002). In this regard, the link between PKA and PSD-95 has been shown to be important for PKA-activated transcription, similar to Yotiao's importance for synaptic effects, suggesting different pools of PKA with local versus nuclear

effects (Grant and O'Dell, 2001). Thus, the role of PKA in LTP is dictated by its compartmentalization.

PKA Substrates. Since PKA can be activated in response to many different forms of synaptic activity, its specific role in LTP is likely to depend on its integration with another signal (Figure 12.2). The mechanism by which PKA modulates early

LTP provides a nice example of such integration (Figure 12.2C). Inhibitor-1 (I-1) is a protein that, upon phosphorylation, inhibits PP1, the phosphatase that is primarily responsible for the inactivation of CaMKII in the synapse (Soderling and Derkach, 2000). I-1 is a substrate for PKA, which therefore serves to preserve CaMKII in its activated state. However, phospho-I-1 is inactivated by the Ca^{2+}-dependent phosphatase calcineurin, which is stimulated by some forms of LTP-inducing stimulation. Under these conditions, postsynaptic PKA activity is required for CaMKII activation and for LTP (Blitzer et al., 1998; Brown et al., 2000). Thus, PKA operates a phosphatase gate that modulates signaling through the CaMKII pathway.

PKA interacts with the MAPK pathway, by stimulating a GTP exchange factor (GEF) for Rap-1 (Sweatt, 2001). GEFs activate Rap-1 by exchanging GDP for GTP. PKA-activated Rap-1 activates B-raf, which then stimulates MEK, the specific activator of MAPK1/2 (Grewal et al., 2000; Sweatt, 2001; Dhillon and Kolch, 2002). This mechanism of MAPK regulation is distinct from the inhibition of SynGAP by CaMKII, suggesting that PKA and CaMKII may act synergistically to promote MAPK activation. Interestingly, PKA also inhibits Raf-1, which is the MEK kinase activated by CaMKII (Houslay and Kolch, 2000; Sweatt, 2001; Dhillon and Kolch, 2002). This suggests that PKA can inhibit the CaMKII-activated MAPK1/2 pathway where Raf-1 activity predominates. One possible outcome of this antagonism is to prevent CaMKII-activated-MAPK1/2 from translocating to the nucleus, where it could act on CREB in a PKA-independent manner. The PKA inhibition of CaMKII-activated-MAPK1/2 would therefore increase the requirement for the CREB-dependent, protein-synthesis dependent form of LTP beyond CaMKII activation, preventing stimuli that only activated CaMKII from maintaining increased synaptic efficacy. Through several distinct feed-forward mechanisms, PKA amplifies its own activation and that of CaMKII. In part, this is accomplished by increasing Ca^{2+} entry following PKA-mediated phosphorylation of NMDA receptors and voltage-dependent Ca^{2+} channels (Dineley et al., 2001; Grant and O'Dell, 2001). Furthermore, PKA-activated MAPK1/2 reduces the conductance of Kv4.2 potassium channels, resulting in a more pronounced membrane depolarization during stimulation and greater Ca^{2+} influx (Sweatt, 2001; Adams and Sweatt, 2002). However, PKA can also interfere with AMPA receptor insertion by phosphorylating the PDZ domain of stargazin, preventing its interaction with PSD-95, which suggests a negative regulatory role as well (Chetkovich et al., 2002).

Finally, PKA can translocate to the nucleus, where it can phosphorylate CREB (Bernabeu et al., 1997). Interestingly, CREB activation may require MAPK1/2 activity as well as PKA, suggesting that MAPK1/2 integrates CaMKII and PKA signals (Poser and Storm, 2001; Sweatt, 2001). Another possibility is that CREB may act as an integrator of MAPK1/2 and PKA activity in the nucleus (Impey et al., 1998). MAPK1/2 can also activate CREB binding protein (CBP), independent from CREB activation (Sweatt, 2001). CBP may act synergistically with CREB phosphorylated by PKA and/or MAPK1/2. In this regard, in other experimental paradigms, it has been shown that different genes are activated depending on the direct phosphorylation of CREB by PKA (without CBP phosphorylation) compared to the MAPK1/2 pathway (Mayr and Montminy, 2001).

12.1.4 Protein Kinase C (PKC)

While PKC activity has been shown to be involved in LTP induction and maintenance, its functions within the LTP signaling network are less well understood than those of CaMKII and PKA (Fukunaga et al., 1996; Malenka and Nicoll, 1999; Kudoh et al., 2001; Ling et al., 2002). Increased PKC activity has been shown to increase basal synaptic transmission levels, suggesting a more generalized effect on the synapse (Malenka and Nicoll, 1999). Inhibitors present prior to induction prevent LTP, although a short-term potentiation can still occur (Angenstein and Staak, 1997; Ramakers et al., 1997). Some groups have found that protein-synthesis-dependent LTP can still occur in the presence of PKC inhibitors, albeit at an attenuated level, while others have found a complete prevention (Kudoh et al., 2001). The continuing uncertainty may stem from the existence of multiple, distinctly regulated PKC isoforms. All of these isoforms are inhibited by the general PKC inhibitors used in LTP experiments, although only some may participate in LTP. $PKM\zeta$ (the catalytic domain of atypical PKC isoform $PKC\zeta$) is found in the PSD and has been implicated in both the induction and maintenance of LTP (Ling et al., 2002). Neurons transfected with dominant negative forms of $PKM\zeta$ fail to show LTP, and the postsynaptic application of $PKM\zeta$ inhibitors reverses even established LTP—something inhibitors of CaMKII and other PKC isoforms do not do (Denny et al., 1990; Otmakhov et al., 1997; Ling et al., 2002). A conventional PKC isoform, $PKC\gamma$, has also been implicated in LTP through the use of transgenic mice with a deletion of the $PKC\gamma$ gene (Saito and Shirai, 2002). PKC substrates include AMPA receptors and MAPK1/2. Since these substrates are also regulated by CaMKII and PKA, they represent likely points of integration between the PKC, CaMKII, and PKA pathways (Lee et al., 2000; Sweatt, 2001; Ling et al., 2002; Yuan et al., 2002).

PKC contains a regulatory and catalytic domain, and it also contains (as with many protein kinases) a pseudosubstrate domain that keeps the protein kinase in an inactive state (Angenstein and Staak, 1997). Conventional PKCs such as $PKC\gamma$ are activated by the binding of second messengers diacylglycerol (DAG) and Ca^{2+}, which prevent the auto-inhibitory domain from binding to the catalytic subunit (Angenstein and Staak, 1997) (Figure 12.3A). One potential source of the DAG that initially activates PKC is the mGluR stimulation in response to synaptic glutamate release. DAG is associated with lipids in the membrane, so that the activation of PKC is limited by its location in a manner similar to that of PKA (Jaken and Parker, 2000). Atypical PKC isoforms such as $PKC\zeta$ are activated by the cleavage (and therefore activation) of the catalytic domain from the auto-inhibitory domain by Ca^{2+}-dependent proteases, or by protein binding to the auto-inhibitory domain (Fukunaga et al., 1996; Angenstein and Staak, 1997; Jaken and Parker, 2000).

The PKC pathway has a complex compartmentalization system (Figure 12.3B). STICKs (substrates that interact with protein C-kinase) include localization signals that, upon phosphorylation, translocate the PKC-STICK complex to the membrane (Jaken and Parker, 2000). STICKs usually have a decreased affinity for PKC after activation, thereby releasing it, which would direct PKC activity to substrates in locations near that region of the membrane (Jaken and Parker, 2000). RACKs (receptors for activated protein C-kinase), such as RACK-1, also bind to

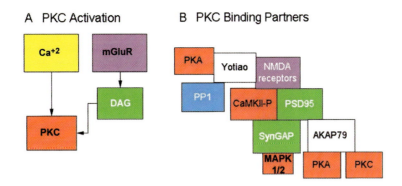

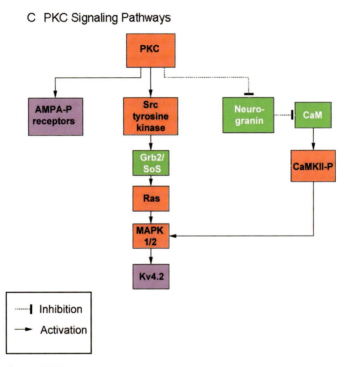

Figure 12.3 Activation, Binding Partners, and Signaling Pathways of PKC A) PKC is activated by the mGluR-mediated formation of DAG and Ca^{+2}. B) PKC binds to AKAP79 in the PSD, which positions it with PKA and CaMKII. C) PKC phosphorylates AMPA receptors at the same site as CaMKII, leading to increased channel conductance. PKC also phosphorylates neurogranin, liberating stored CaM, which can then bind and activate CaMKII. PKC can activate tyrosine kinases (such as Src), which stimulates the MAPK1/2 pathway through Ras. Color key as in Fig 1.

activated PKC, and they localize it to regions where substrates are present (Mochly-Rosen et al., 1991; Ron et al., 1994; Jaken and Parker, 2000). Thus, both STICKs and RACKs are essential for the localization of active PKC. AKAP79 is a STICK that, upon phosphorylation by PKC, translocates to the PSD membrane, bringing PKC along. Since AKAP79 also binds PKA and is required for PKA activity in the PSD, the initial phosphorylation of AKAP79 by PKC may be important for the compartmentalization of many of the components of the LTP signal transduction pathway. When CaM binds to AKAP79 in the PSD, it releases PKC, which becomes available to activate other substrates in that

area, such as neurogranin, tyrosine protein kinases, and AMPA receptors (Angenstein and Staak, 1997; Faux and Scott, 1997).

Most of the work on PKC involvement in LTP has not differentiated between PKC isoforms, although PKCζ and PKCγ have been implicated. PKC is first activated by mGluR through DAG. This is followed by a second wave of activation 30–90 minutes after stimulation due to cleavage of the auto-inhibitory domain by a Ca^{2+}-dependent protease, which leads to a long-lasting form of activation (Fukunaga et al., 1996). PKC appears to contribute to LTP during both the initial induction and during conversion to the protein synthesis-dependent form (Figure 12.3C). PKC activation by mGluRs occurs early and can lead to the phosphorylation of NMDA channels (to which mGluRs are linked via scaffolding proteins) (Ramakers et al., 1997; Soderling and Derkach, 2000; Chen et al., 2001; Dineley et al., 2001). Phosphorylation of NMDA channels by tyrosine protein kinases activated by PKC decreases the Mg^{2+} block, lowering the threshold for LTP (Soderling and Derkach, 2000; Chen et al., 2001).

Phosphorylation of GluR1 AMPA receptors at Ser831 by CaMKII or PKC leads to an increase in channel conductance, which is part of the initial phase of LTP (Lee et al., 2000). The dual activation of CaMKII and PKC may thus act in an additive fashion to potentiate AMPA receptor-mediated synaptic transmission. Conversely, scaffolding proteins may sequester AMPA receptors into subpopulations, one that is preferentially phosphorylated by CaMKII and another by PKC.

Neurogranin, another PKC substrate that has been linked to LTP induction, acts as a CaM sink in its nonphosphorylated state, binding CaM and preventing it from interacting with other proteins (Fukunaga et al., 1996; Ramakers et al., 1997). The phosphorylation of neurogranin by PKC releases CaM, which is then available for activation by Ca^{2+}. Consistent with the importance of CaM and CaMKII in LTP, prevention of neurogranin phosphorylation and its subsequent release of CaM prevents LTP induction (Ramakers et al., 1997). It is interesting to note that neurogranin can be dephosphorylated by calcineurin and PP1 (which calcineurin releases from inhibition), both of which would be active at moderate Ca^{2+} levels, suggesting that the phosphorylation state of neurogranin is highly regulated (Ramakers et al., 1997). Thus, at the high Ca^{2+} levels that attend NMDA receptor stimulation, PKC-mediated processes can promote the activation of CaMKII. The initial activation of PKC by mGluRs may promote the auto-phosphorylation of CaMKII. The second wave of PKC activity could lead to further CaMKII activation. Alternatively, it could initiate negative feedback because Ca^{2+} will have returned to basal levels at that time, favoring the activation of protein phosphatases over kinases.

Finally, PKC has been shown to increase the activation of MAPK by stimulating tyrosine protein kinases (Chen et al., 1998; Soderling and Derkach, 2000; Dineley et al., 2001; Sweatt, 2001). These protein kinases phosphorylate Grb2, which increases its ability to bind the GEF Son of Sevenless (SoS), leading to Ras activation and subsequent MAPK activation (Chen et al., 1998). PKC may also bind directly to Ras and modulate its activity (Dineley et al., 2001; Sweatt, 2001). PKC-activated MAPK1/2 has been shown to decrease the conductance of Kv4.2 channels, just as PKA-activated MAPK1/2 does (Sweatt, 2001; Yuan et al., 2002). MAPK1/2, therefore, is a master integrator for PKC, PKA, and CaMKII activity.

12.1.5 Mitogen-Activated Protein Kinase (MAPK)

MAPK comprises three families of three serine/threonine protein kinases: MAPK1/2 (Erk 1/2), c-Jun N terminal kinase (Jnk), and p38 MAPK. The numerous subtypes are differentially localized and regulated, with distinct substrate profiles, leading to a wide range of cellular effects. Among the MAPK families, MAPK1/2 has convincingly been implicated in early LTP, as well as late, protein synthesis-dependent LTP (Dineley et al., 2001; Sweatt, 2001; Adams and Sweatt, 2002), while roles for p38 and JNK

have not been established. The duration of the MAPK1/2 signal is influenced by PKA activity and Ca^{2+} levels, which may determine the ultimate targets and cellular location of MAPK1/2 activation (Poser and Storm, 2001).

MAPK activation is a classic case of specification and amplification, accomplished by a cascade of protein kinases (Figure 12.4A). Each MAPK has a specific activator (MAPKK, such as MEK for MAPK1/2) that can activate multiple molecules of its corresponding MAPK, which are more abundant (Chang and Karin, 2001). A similar amplification also occurs upstream upon the activation of the MAPKKs, which are substrates for selective protein kinases (MAPKKKs). In this way, serial activation of the MAPK cascade dramatically amplifies the signal while maintaining specificity. Such an arrangement endows the MAPK pathway with flexibility in its interactions with other signals, since convergence is possible at multiple levels of the cascade, with differential effects on the gain or duration of MAPK pathway activity. It has been shown that MAPK activation can also act as a temporal regulator, increasing steadily as stimuli repeat (Selcher et al., 2003).

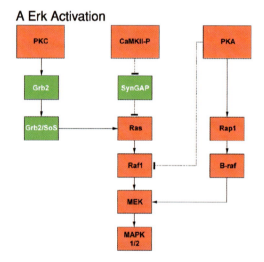

A Erk Activation

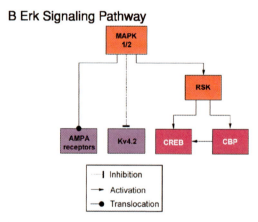

B Erk Signaling Pathway

- | Inhibition
- → Activation
- ●— Translocation

Figure 12.4 Activation and Signaling Pathways of MAPK1/2 A) MAPK1/2 is activated by PKC, CaMKII and PKA (which antagonizes the CaMKII-mediated activation of the MAPK1/2 pathway). B) MAPK1/2 induces the translocation of AMPA receptors to potentiated synapses, phosphorylates the potassium channel Kv4.2, and activates transcription factors. Color key same as in Fig 1.

Upstream activators of MAPK include receptor or nonreceptor tyrosine protein kinases that recruit adapter proteins such as Grb2, which links to the guanine nucleotide exchange factor (GEF), which triggers the release of SoS. SoS can then hydrolyze GDP from Ras, allowing Ras to bind GTP and become activated (Chen et al., 1998). Ras binds and activates Raf, the first protein kinase in the serial MAPK pathway (Dhillon and Kolch, 2002). Upon activation, Raf phosphorylates the next protein kinase in the MAPK pathway, MEK. Finally, MEK activates MAPK by phosphorylation on tyrosine and threonine.

During LTP, the MAPK1/2 response may include a transient, CaMKII-mediated component and a more persistent, PKA-mediated component. Each phase is dependent on the activation of different MEK protein kinases (Figure 12.4). The CaMKII-mediated activation of MAPK begins with the translocation of phospho-CaMKII to the postsynaptic density, where it binds and inactivates SynGAP. As a result, Ras and then Raf are activated (Chen et al., 1998; Sweatt, 2001). MAPK1/2 binds to SynGAP as well, which is tethered to NMDA channels through PSD-95. Thus, a large fraction of the CaMKII–MAPK interaction module has been localized at the synapse (Liu et al., 1999; Soderling and Derkach, 2000; Grant and O'Dell, 2001; Hardingham et al., 2001). Mutations that prevent MAPK1/2 binding to SynGAP reduce LTP (Grant and O'Dell, 2001). MAPK1/2 activation by CaMKII normally is transient and local because Raf-1 is quickly inactivated through phosphorylation by PKA (Morice et al., 1999). However, PKA can also activate the small G-protein Rap1, which then activates B-Raf followed by MEK activation (Morice et al., 1999). This pathway initiates a long-lasting phase of MAPK1/2 activation that leads to nuclear translocation (Impey et al., 1998). The temporal aspects of MAPK1/2 activation define its contribution to LTP, based on both local, post-translational effects and transcriptional regulation. Interestingly, the transient activation of MAPK1/2 by CaMKII may be restricted to MAPK1/2 that is associated with SynGAP-CaMKII, which in turn is bound through the NMDA receptor to PKA (Grant and O'Dell, 2001; Michel and Scott, 2002). MAPK1/2 can also bind to PSD-95, to which PKA is bound via AKAP79. This suggests that the dual activation of MAPK1/2 may be related to different pools of MAPK1/2 bound to upstream protein kinases via scaffolding proteins. Alternatively, PKA may inhibit Raf-1 due to proximity, but when cAMP levels are high enough, active PKA may diffuse from its initial location in the PSD and activate an unbound pool of MAPK1/2, which would then translocate to the nucleus (Impey et al., 1998). Finally, PKC enhances tyrosine protein kinase activity, which can activate Ras and initiate the MAPK cascade (Chen et al., 1998; Dineley et al., 2001; Sweatt, 2001). This effect of PKC on the MAPK pathway might be more modulatory than stimulatory.

Once MAPK1/2 has become activated, it might contribute to LTP in at least three ways: (1) translocate extrasynaptic AMPA receptors to the stimulated synapse, which will increase synaptic efficacy; (2) phosphorylate a potassium channel in the synapse (Kv4.2), decreasing channel conductance so the postsynaptic neuron remains depolarized longer following synaptic activity; and (3) translocate to the nucleus where it can activate p90 RSK, which phosphorylates CREB and its binding partner, CBP (Sweatt, 2001; Adams and Sweatt, 2002; Zhu et al., 2002) (Figure 12.4B). A Ras/MAPK1/2 pathway has been shown to drive AMPA receptors into synapses after stimulation (Zhu et al.,

2002). The phosphorylation of Kv4.2 by PKC and PKA is mediated by MAPK1/2, suggesting that there may be an additive or synergistic effect of the activation of both pathways (Adams and Sweatt, 2002). The phosphorylation of CREB and CBP may act as a molecular switch for the maintenance of LTP (Sweatt, 2001; Barco et al., 2002). CREB and CPB regulate the transcription of certain genes containing the cAMP response element (CRE) in their untranslated regulatory region (Mayr and Montminy, 2001). Some genes require both CBP and CREB for transcriptional activation, while others require only CREB (Mayr and Montminy, 2001).

12.2 COMPARTMENTALIZATION

LTP requires compartmentalization of effectors and their substrates. It is achieved through physical and functional barriers to movement, such as a narrow spine neck, or by scaffolding proteins that bind multiple proteins together in one local microenvironment, such as the PSD scaffolding protein PSD-95 (Kim and Lisman, 1999; Weng et al., 1999; Dorn and Mochly-Rosen, 2002). In either case, the effective concentration of signaling components is elevated within a select region, bringing together protein kinases and phosphatases with their substrates (Weng et al., 1999). In this regard, the specificity of scaffolding protein binding is essential for determining which signaling pathways will be activated. Signaling gradients can also be achieved by sequestering certain substrates close to a protein kinase so they are readily phosphorylated upon activation. If the protein kinase remains activated (or becomes re-activated) after local substrates have been phosphorylated, the protein kinase may be able to diffuse out to phosphorylate substrates further away, creating tiered levels of protein kinase activity and zones of regulated substrates (Dorn and Mochly-Rosen, 2002). This requires a protein kinase that is initially bound to a scaffolding protein but is released upon activation, such as PKA. In this manner, continuous or cyclical activation of PKA could lead to an integration of stimulus strength or time course that begins in the synapse and then branches out to other cellular regions, including the nucleus (Bernabeu et al., 1997; Wallenstein et al., 2002). Conversely, CaMKII translocates to the PSD after autophosphorylation where it remains, suggesting that its substrates may be more specifically confined (Grant and O'Dell, 2001). The bound state of active CaMKII and the unbound state of the PKA catalytic domain may therefore be important in mapping the intensity of system stimulation to different forms of postsynaptic potentiation, including the multiple phases of LTP. The translocation of PKC acts as a form of negative feedback initially, but also places AKAP79 in the membrane, where it binds PKA and PSD-95. The transient activation of MAPK1/2 by CaMKII prior to its prolonged activation by PKA may be dependent upon MAPK1/2 binding through SynGAP to CaMKII, while MAPK1/2 binds more indirectly to PKA via SynGAP to PSD-95 to PKA (Malenka and Nicoll, 1999; Lisman and Zhabotinsky, 2001; Soderling et al., 2001). This arrangement would facilitate the initial CaMKII-mediated activation of the MAPK pathway through Raf-1, before PKA can inhibit Raf-1 in favor of the B-raf route. As such, all of these protein kinases require strict localization for their interactions with the appropriate substrates and, in the case of CaMKII, for its continued activity.

CaMKII is localized to the PSD by direct binding to the NR1 subunit of NMDA channels, as well as through SynGAP bound to PSD-95 (which also binds NMDA channels) (Grant and O'Dell, 2001; Gleason et al., 2003). CaMKII has the strictest requirement for compartmentalization. To maintain its high level of auto-phosphorylation and activation after Ca^{2+} levels return to normal, CaMKII must be present at a high enough concentration to saturate local phosphatase activity (Lisman and Zhabotinsky, 2001). In this regard, the exclusion of all CaMKII phosphatases other than PP1 from the PSD is very important (Lisman and Zhabotinsky, 2001). If other phosphatases were to gain entry, or if PP1 could freely diffuse into and out of the PSD, phosphorylated CaMKII might not saturate the phosphatase population and would be prematurely dephosphorylated. Also, the presence of PKA sequestered in the PSD allows PKA to activate the PP1 inhibitor I-1, thereby protecting activated CaMKII (Soderling and Derkach, 2000). Thus, the requirement for sustained CaMKII activity provides a striking illustration of the importance of compartmentalization in organizing the LTP signaling network.

PKA is localized to the PSD by the AKAP Yotiao binding to NMDA receptors, and by AKAP79 binding to PSD-95, potentially creating two pools of PKA (Michel and Scott, 2002). One PKA pool would be important for CaMKII activation by repressing PP1 activity, but also inhibit CaMKII-activated MAPK1/2 after a transient activation. The other pool would be responsible for activating Rap-1 and a different pool of MAPK1/2 (Malenka and Nicoll, 1999; Lisman and Zhabotinsky, 2001; Soderling et al., 2001). Conversely, an increase in cAMP could lead to a gradient effect of PKA activity, where MAPK1/2 that is closely

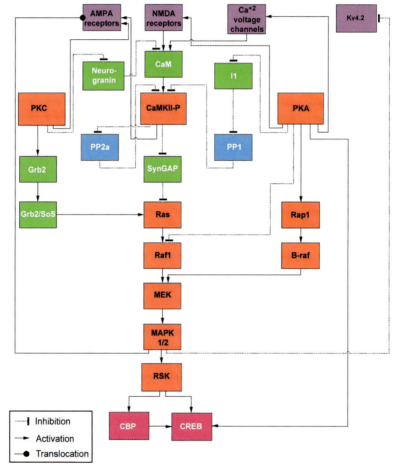

Figure 12.5 The Signaling Network for LTP Integrates Kinase Signaling Pathways Activation of NMDA receptors leads to Ca^{+2} influx. Ca^{+2} activates PKC, PKA and CaM. CaM binds and activates CaMKII, and PKC phosphorylates neurogranin, leading to the release of additional CaM. PKC also increases Ras activity, which activates the MAPK1/2 pathway. CaMKII inhibits SynGAP, which inhibits Ras. Thus the MAPK1/2 pathway is a point of integration of PKC and CaMKII. Another point of integration is the AMPA receptor, which is phosphorylated at Ser831 by CaMKII and PKC, increasing channel conductance in early LTP. CaMKII-stimulated MAPK activity leads to AMPA receptor translocation. PKA stimulates MAPK1/2 by an alternate pathway, while inhibiting CaMKII-mediated MAPK1/2 activity. The MAPK1/2 pathway, activated by both PKA and PKC, reduces the activity of Kv4.2. Together, the MAPK1/2 and PKA pathways activate transcription factors that contribute to the maintenance of LTP. Color key as in Fig 1.

bound to the PSD (and therefore activated by bound CaMKII) would be inactivated by PKA quickly, but with a high enough level of cAMP, PKA would diffuse further from its initial site and could activate a nonbound pool of MAPK1/2 (Impey et al., 1998; Giovannini et al., 2001; Grant and O'Dell, 2001; Poser and Storm, 2001). In this regard, the fact that MAPK1/2 translocates to the nucleus after PKA activation indicates that it is, in fact, not in a bound state.

PKC is directed to the membrane by AKAP79, which is dependent on previous PKC activity for phosphorylation and translocation. Upon CaM binding to AKAP79, PKC is released to phosphorylate local substrates, such as NMDA and AMPA receptors. The initial translocation of AKAP79 (which binds to PKA and through PSD-95 the NMDA receptors) due to PKC phosphorylation suggests that PKC activity may be part of the initial process by which the important molecules for LTP are compartmentalized in the PSD.

MAPK1/2 is localized to the PSD by SynGAP, which also binds CaMKII and PSD-95. The localization of MAPK1/2 allows transient activation by CaMKII and more prolonged activation by PKA. Its localization also influences the involvement of MAPK1/2 in early and more local events such as AMPA-R delivery to the PSD, and later events located distal to the potentiated region, such as nuclear phosphorylation of CREB.

12.3 CONCLUSIONS

In recent years, some of the organizing principles of the postsynaptic LTP signaling network have been established. As described in this review, they include serial amplification, multiple phases for pathway activation, the regulated clustering of components within microdomains, multi-pathway integration, and positive and negative feedback loops (Figure 12.5). Experimental strategies such as proteomics-based identification of scaffolding proteins and their binding partners, along with increasingly sophisticated modeling efforts, will improve our appreciation for the richness of this highly regulated network (Bhalla and Iyengar, 1999; Bhalla, 2003).

Empirically, the analysis of the LTP signaling network is complicated by practical limits on the spatial and temporal specificity of our experimental manipulations. For example, the application of an exogenous protein kinase activator will generally circumvent the spatial constraints of the physiologically induced signal and will likely outlast the normal time course as well. In the literature, there are many examples of treatments that increase synaptic efficiency and, often, occlude LTP. Such findings can give the erroneous impression that the LTP signaling network is redundant, when in fact it is highly integrative.

LTP represents a substantial investment on the part of a neuron. The maintenance of LTP requires macromolecular synthesis, which is energetically expensive and commits a subset of synapses to heightened efficiency for extended periods, perhaps months or years. At the level of the organism, synaptic plasticity provides an adaptive advantage, but there are also risks associated with the acquisition of inappropriate behaviors that are equivalent to false positives. Accordingly, the organization of the LTP signaling network reflects the need to avoid the unnecessary use of cellular resources and to minimize the likelihood of false positives, while permitting highly stable and localized

changes. This set of requirements may explain an often-noted property of LTP: its vulnerability to a wide range of experimental perturbations, at least during the induction period. Such behavior is to be expected from a network that features what might be called "high-order" signals, those that are generated by the integration of activity in multiple pathways. In LTP, the complexity of the integration is further increased tremendously by the spatially and temporally dynamic nature of the signaling network. A signal might be generated within a microdomain, and then either remain there or translocate elsewhere, depending on such factors as the duration of pathway activation and the phosphorylation state of scaffolding proteins. Thus, LTP is an example of a highly integrated system of complex signaling pathways interacting within microdomains to alter both the local and the nuclear environments, ultimately resulting in the persistent enhancement of specific synaptic connections.

ACKNOWLEDGMENTS

Research in our laboratories is supported by NIH grant GM-54508 to RI and NIH grant DA-15863 to EML. EAG is a postdoctoral trainee supported by NIDA training grant DA-07135.

References

Abel, T., and Lattal, K.M. (2001). Molecular mechanisms of memory acquisition, consolidation and retrieval. *Curr. Opin. Neurobiol.* **11**(2), 180–187.

Abel, T., Nguyen, P.V., Barad, M., Deuel, T.A., Kandel, E.R., and Bourtchouladze, R. (1997). Genetic demonstration of a role for PKA in the late phase of LTP and in hippocampus-based long-term memory. *Cell* **88**(5), 615–626.

Adams, J.P., and Sweatt, J.D. (2002). Molecular psychology: Roles for the ERK MAP kinase cascade in memory. *Annu. Rev. Pharmacol. Toxicol.* **42**, 135–163.

Angenstein, F., and Staak, S. (1997). Receptor-mediated activation of protein kinase C in hippocampal long-term potentiation: Facts, problems and implications. *Prog. Neuropsychopharmacol. Biol. Psychiatry* **21**(3), 427–454.

Banke, T.G., Bowie, D., Lee, H., Huganir, R.L., Schousboe, A., and Traynelis, S.F. (2000). Control of GluR1 AMPA receptor function by cAMP-dependent protein kinase. *J. Neurosci.* **20**(1), 89–102.

Barco, A., Alarcon, J.M., and Kandel, E.R. (2002). Expression of constitutively active CREB protein facilitates the late phase of long-term potentiation by enhancing synaptic capture. *Cell* **108**(5), 689–703.

Barria, A., Muller, D., Derkach, V., Griffith, L.C., and Soderling, T.R. (1997). Regulatory phosphorylation of AMPA-type glutamate receptors by CaM-KII during long-term potentiation. *Science* **276**, 2042–2045.

Bayer, K.U., De Koninck, P., Leonard, A.S., Hell, J.W., and Schulman, H. (2001). Interaction with the NMDA receptor locks CaMKII in an active conformation. *Nature (London)* **411**, 801–805.

Bear, M.F. (1997). How do memories leave their mark? *Nature (London)* **385**, 481–482.

Bernabeu, R., Bevilaqua, L., Ardenghi, P., Bromberg, E., Schmitz, P., Bianchin, M., Izquierdo, I., and Medina, J.H. (1997). Involvement of hippocampal cAMP/cAMP-dependent protein kinase signaling pathways in a late memory consolidation phase of aversively motivated learning in rats. *Proc. Natl. Acad. Sci. U.S.A.* **94**(13), 7041–7046.

Bhalla, U.S. (2003). Understanding complex signaling networks through models and metaphors. *Prog. Biophys. Mol. Biol.* **81**(1), 45–65.

Bhalla, U., and Iyengar, R. (1999). Emergent properties of networks of biological signaling pathways. *Science* **283**, 381–387.

Bliss, T.V., and Lomo, T. (1973). Long-lasting potentiation of synaptic transmission in the dentate area of the anaesthetized rabbit following stimulation of the perforant path. *J. Physiol. (London)* **232**(2), 331–356.

Blitzer, R.D., Wong, T., Nouranifar, R., Iyengar, R., and Landau, E.M. (1995). Postsynaptic cAMP pathway gates early LTP in hippocampal CA1 region. *Neuron* **15**(6), 1403–1414.

Blitzer, R.D., Connor, J.H., Brown, G.P., Wong, T., Shenolikar, S., Iyengar, R., and Landau, E.M. (1998). Gating of CaMKII by cAMP-regulated protein phosphatase activity during LTP. *Science* **280**, 1940–1942.

Brown, G.P., Blitzer, R.D., Connor, J.H., Wong, T., Shenolikar, S., Iyengar, R., and Landau, E.M. (2000). Long-term potentiation induced by theta frequency stimulation is regulated by a protein phosphatase-1-operated gate. *J. Neurosci.* **20**(21), 7880–7887.

Chang, L., and Karin, M. (2001). Mammalian MAP kinase signalling cascades. *Nature (London)* **410**, 37–40.

Chavez-Noriega, L.E., and Stevens, C.F. (1992). Modulation of synaptic efficacy in field CA1 of the rat hippocampus by forskolin. *Brain Res.* **574**(1–2), 85–92.

Chavez-Noriega, L.E., and Stevens, C.F. (1994). Increased transmitter release at excitatory synapses produced by direct activation of adenylate cyclase in rat hippocampal slices. *J. Neurosci.* **14**(1), 310–317.

Chen, H.J., Rojas-Soto, M., Oguni, A., and Kennedy, M.B. (1998). A synaptic Ras-GTPase activating protein (p135 SynGAP) inhibited by CaM kinase II. *Neuron* **20**(5), 895–904.

Chen, H.X., Otmakhov, N., Strack, S., Colbran, R.J., and Lisman, J.E. (2001). Is persistent activity of calcium/calmodulin-dependent kinase required for the maintenance of LTP? *J. Neurophysiol.* **85**(4), 1368–1376.

Chetkovich, D.M., Chen, L., Stocker, T.J., Nicoll, R.A., and Bredt, D.S. (2002). Phosphorylation of the postsynaptic density-95 (PSD-95)/discs large/zona occludens-1 binding site of stargazin regulates binding to PSD-95 and synaptic targeting of AMPA receptors. *J. Neurosci.* **22**(14), 5791–5796.

Chin, D., and Means, A.R. (2000). Calmodulin: A prototypical calcium sensor. *Trends Cell Biol.* **10**(8), 322–328.

Colbran, R.J., Smith, M.K., Schworer, C.M., Fong, Y.L., and Soderling, T.R. (1989). Regulatory domain of calcium/calmodulin-dependent protein kinase II. Mechanism of inhibition and regulation by phosphorylation. *J. Biol. Chem.* **264**(9), 4800–4804.

Denny, J.B., Polan-Curtain, J., Rodriguez, S., Wayner, M.J., and Armstrong, D.L. (1990). Evidence that protein kinase M does not maintain long-term potentiation. *Brain Res.* **534**(1–2), 201–208.

Derkach, V., Barria, A., and Soderling, T.R. (1999). Ca2+/calmodulin-kinase II enhances channel conductance of alpha-amino-3-hydroxy-5-methyl-4-isoxazolepropionate type glutamate receptors. *Proc. Natl. Acad. Sci. U.S.A.* **96**(6), 3269–3274.

Dhillon, A.S., and Kolch, W. (2002). Untying the regulation of the Raf-1 kinase. *Arch. Biochem. Biophys.* **404**(1), 3–9.

Dineley, K.T., Weeber, E.J., Atkins, C., Adams, J.P., Anderson, A.E., and Sweatt, J.D. (2001). Leitmotifs in the biochemistry of LTP induction: Amplification, integration and coordination. *J. Neurochem.* **77**(4), 961–971.

Dorn, G.W., 2nd, and Mochly-Rosen, D. (2002). Intracellular transport mechanisms of signal transducers. *Annu. Rev. Physiol.* **64**, 407–429.

Faux, M.C., and Scott, J.D. (1997). Regulation of the AKAP79-protein kinase C interaction by Ca2+/Calmodulin. *J. Biol. Chem.* **272**(27), 17038–17044.

Frey, U., and Morris, R.G. (1998). Synaptic tagging: Implications for late

maintenance of hippocampal long-term potentiation. *Trends Neurosci.* **21**(5), 181–188.

Frey, U., Krug, M., Reymann, K.G., and Matthies, H. (1988). Anisomycin, an inhibitor of protein synthesis, blocks late phases of LTP phenomena in the hippocampal CA1 region in vitro. *Brain Res.* **452**(1–2), 57–65.

Frey, U., Huang, Y.Y., and Kandel, E.R. (1993). Effects of cAMP simulate a late stage of LTP in hippocampal CA1 neurons. *Science* **260**, 1661–1664.

Fukunaga, K., Muller, D., and Miyamoto, E. (1996). CaM kinase II in long-term potentiation. *Neurochem Int.* **28**(4), 343–358.

Fukunaga, K., Muller, D., Ohmitsu, M., Bako, E., DePaoli-Roach, A.A., and Miyamoto, E. (2000). Decreased protein phosphatase 2A activity in hippocampal long-term potentiation. *J. Neurochem.* **74**(2), 807–817.

Giovannini, M.G., Blitzer, R.D., Wong, T., Asoma, K., Tsokas, P., Morrison, J.H., Iyengar, R., and Landau, E.M. (2001). Mitogen-activated protein kinase regulates early phosphorylation and delayed expression of Ca2+/calmodulin-dependent protein kinase II in long-term potentiation. *J. Neurosci.* **21**(18), 7053–7062.

Gleason, M.R., Higashijima, S., Dallman, J., Liu, K., Mandel, G., and Fetcho, J.R. (2003). Translocation of CaM kinase II to synaptic sites in vivo. *Nat. Neurosci.* **6**(3), 217–218.

Gottschalk, W., Pozzo-Miller, L.D., Figurov, A., and Lu, B. (1998). Presynaptic modulation of synaptic transmission and plasticity by brain-derived neurotrophic factor in the developing hippocampus. *J. Neurosci.* **18**(17), 6830–6839.

Grant, S.G., and O'Dell, T.J. (2001). Multiprotein complex signaling and the plasticity problem. *Curr. Opin. Neurobiol.* **11**(3), 363–368.

Grewal, S.S., Fass, D.M., Yao, H., Ellig, C.L., Goodman, R.H., and Stork, P.J. (2000). Calcium and cAMP signals differentially regulate cAMP-responsive element-binding protein function via a Rap1-extracellular signal-regulated kinase pathway. *J. Biol. Chem.* **275**(44), 34433–34441.

Hanoune, J., and Defer, N. (2001). Regulation and role of adenylyl cyclase isoforms. *Annu. Rev. Pharmacol. Toxicol.* **41**, 145–174.

Hardingham, G.E., Arnold, F.J., and Bading, H. (2001). A calcium microdomain near NMDA receptors: On switch for ERK-dependent synapse-to-nucleus communication. *Nat. Neurosci.* **4**(6), 565–566.

Hayashi, Y., Shi, S.H., Esteban, J.A., Piccini, A., Poncer, J.C. and Malinow, R. (2000). Driving AMPA receptors into synapses by LTP and CaMKII: Requirement for GluR1 and PDZ domain interaction. *Science* **287**, 2262–2267.

Hebb, D. (1949). *The Organization of Behavior: A Neuropsychological Theory.* Wiley, New York.

Hinds, H.L., Tonegawa, S., and Malinow, R. (1998). CA1 long-term potentiation is diminished but present in hippocampal slices from alpha-CaMKII mutant mice. *Learn. Mem.* **5**(4–5), 344–354.

Houslay, M.D., and Kolch, W. (2000). Cell-type specific integration of cross-talk between extracellular signal-regulated kinase and cAMP signaling. *Mol. Pharmacol.* **58**(4), 659–668.

Hudmon, A., and Schulman, H. (2002). Neuronal CA2+/calmodulin-dependent protein kinase II: the role of structure and autoregulation in cellular function. *Annu. Rev. Biochem.* **71**, 473–510.

Impey, S., Obrietan, K., Wong, S.T., Poser, S., Yano, S., Wayman, G., Deloulme, J.C., Chan, G. and Storm, D.R. (1998). Cross talk between ERK and PKA is required for Ca2+ stimulation of CREB-dependent transcription and ERK nuclear translocation. *Neuron* **21**(4), 869–883.

Jaken, S., and Parker, P.J. (2000). Protein kinase C binding partners. *BioEssays* **22**(3), 245–254.

Kanaseki, T., Ikeuchi, Y., Sugiura, H., and Yamauchi, T. (1991). Structural Features of Ca2+/calmodulin-dependent protein kinase II revealed by electron microscopy. *J. Cell Biol.* **115**(4), 1049–1060.

Kang, H., and Schuman, E.M. (1996). A requirement for local protein synthesis in neurotrophin-induced hippocampal synaptic plasticity. *Science* **273**, 1402–1406.

Kim, C.H., and Lisman, J.E. (1999). A role of actin filament in synaptic transmission and long-term potentiation. *J. Neurosci.* **19**(11), 4314–4324.

Kirkwood, A., Dudek, S.M., Gold, J.T., Aizenman, C.D. and Bear, M.F. (1993). Common forms of synaptic plasticity in the hippocampus and neocortex in vitro. *Science* **260**, 1518–1521.

Kudoh, S.N., Nagai, R., Kiyosue, K., and Taguchi, T. (2001). PKC and CaMKII dependent synaptic potentiation in cultured cerebral neurons. *Brain Res.* **915**(1), 79–87.

Lee, H.K., Barbarosie, M., Kameyama, K., Bear, M.F., and Huganir, R.L. (2000). Regulation of distinct AMPA receptor phosphorylation sites during bidirectional synaptic plasticity. *Nature (London)* **405**, 955–959.

Ling, D.S., Benardo, L.S., Serrano, P.A., Blace, N., Kelly, M.T., Crary, J.F., and Sacktor, T.C. (2002). Protein kinase Mzeta is necessary and sufficient for LTP maintenance. *Nat. Neurosci.* **5**(4), 295–296.

Lisman, J., Schulman, H., and Cline, H. (2002). The molecular basis of CaMKII function in synaptic and behavioral memory. *Nat. Rev. Neurosci.* **3**(3), 175–190.

Lisman, J.E., and McIntyre, C.C. (2001). Synaptic plasticity: A molecular memory switch. *Curr. Biol.* **11**(19), R788–R791.

Lisman, J.E. and Zhabotinsky, A.M. (2001). A model of synaptic memory: a CaMKII/PP1 switch that potentiates transmission by organizing an AMPA receptor anchoring assembly. *Neuron* **31**(2), 191–201.

Liu, J., Fukunaga, K., Yamamoto, H., Nishi, K., and Miyamoto, E. (1999). Differential roles of Ca(2+)/calmodulin-dependent protein kinase II and mitogen-activated protein kinase activation in hippocampal long-term potentiation. *J. Neurosci.* **19**(19), 8292–8299.

Lledo, P.M., Hjelmstad, G.O., Mukherji, S., Soderling, T.R., Malenka, R.C., and Nicoll, R.A. (1995). Calcium/calmodulin-dependent kinase II and long-term potentiation enhance synaptic transmission by the same mechanism. *Proc. Natl. Acad. Sci. U.S.A.* **92**(24), 11175–11179.

Luscher, C., Nicoll, R.A., Malenka, R.C., and Muller, D. (2000). Synaptic plasticity and dynamic modulation of the postsynaptic membrane. *Nat. Neurosci.* **3**(6), 545–550.

Malenka, R.C., and Nicoll, R.A. (1999). Long-term potentiation—a decade of progress? *Science* **285**, 1870–1874.

Malinow, R., Schulman, H., and Tsien, R.W. (1989). Inhibition of postsynaptic PKC or CaMKII blocks induction but not expression of LTP. *Science* **245**, 862–866.

Mayr, B., and Montminy, M. (2001). Transcriptional regulation by the phosphorylation-dependent factor CREB. *Nat. Rev. Mol. Cell. Biol.* **2**(8), 599–609.

McLennan, H. (1983). Receptors for the excitatory amino acids in the mammalian central nervous system. *Prog. Neurobiol.* **20**(3–4), 251–271.

Michel, J.J., and Scott, J.D. (2002). AKAP mediated signal transduction. *Annu. Rev. Pharmacol. Toxicol.* **42**, 235–257.

Miller, S., Yasuda, M., Coats, J.K., Jones, Y., Martone, M.E., and Mayford, M. (2002). Disruption of dendritic translation of CaMKIIalpha impairs stabilization of synaptic plasticity and memory consolidation. *Neuron* **36**(3), 507–519.

Miyashiro, K., Dichter, M., and Eberwine, J. (1994). On the nature and differential distribution of mRNAs in hippocampal neurites: Implications for neuronal functioning. *Proc. Natl. Acad. Sci. U.S.A.* **91**(23), 10800–10804.

Mochly-Rosen, D., Khaner, H., and Lopez, J. (1991). Identification of intracellular receptor proteins for activated protein kinase C. *Proc. Natl. Acad. Sci. U.S.A.* **88**(9), 3997–4000.

Morice, C., Nothias, F., Konig, S., Vernier, P., Baccarini, M., Vincent, J.D., and Barnier, J.V. (1999). Raf-1 and B-Raf proteins have similar regional distributions but differential subcellular localization in adult rat brain. *Eur. J. Neurosci.* **11**(6), 1995–2006.

Otmakhov, N., Griffith, L.C., and Lisman, J.E. (1997). Postsynaptic inhibitors of calcium/calmodulin-dependent protein kinase type II block induction but not maintenance of pairing-induced long-term potentiation. *J. Neurosci.* **17**(14), 5357–5365.

Otmakhova, N.A., Otmakhov, N., Mortenson, L.H., and Lisman, J.E. (2000). Inhibition of the cAMP pathway decreases early long-term potentiation at CA1 hippocampal synapses. *J. Neurosci.* **20**(12), 4446–4451.

Pak, J.H., Huang, F.L., Li, J., Balschun, D., Reymann, K.G., Chiang, C., Westphal, H., and Huang, K.P. (2000). Involvement of neurogranin in the modulation of calcium/calmodulin-dependent protein kinase II, synaptic plasticity, and spatial learning: A study with knockout mice. *Proc. Natl. Acad. Sci. U.S.A.* **97**(21), 11232–11237.

Persechini, A., and Cronk, B. (1999). The relationship between the free concentrations of Ca^{2+} and Ca2+-calmodulin in intact cells. *J. Biol. Chem.* **274**(11), 6827–6830.

Poser, S., and Storm, D.R. (2001). Role of Ca^{2+}-stimulated adenylyl cyclases in LTP and memory formation. *Int. J. Dev. Neurosci.* **19**(4), 387–394.

Ramakers, G.M., Pasinelli, P., Hens, J.J., Gispen, W.H., and De Graan, P.N. (1997). Protein kinase C in synaptic plasticity: Changes in the in situ phosphorylation state of identified pre- and postsynaptic substrates. *Prog. Neuropsychopharmacol. Biol. Psychiatry.* **21**(3), 455–486.

Ron, D., Chen, C.H., Caldwell, J., Jamieson, L., Orr, E., and Mochly-Rosen, D. (1994). Cloning of an intracellular receptor for protein kinase C: A homolog of the beta subunit of G proteins. *Proc. Natl. Acad. Sci. U.S.A.* **91**(3), 839–843.

Rongo, C. (2002). A fresh look at the role of CaMKII in hippocampal synaptic plasticity and memory. *Bioessays* **24**(3), 223–233.

Saito, N., and Shirai, Y. (2002). Protein kinase Cgamma (PKCgamma): Function of neuron specific isotype. *J. Biochem. (Tokyo)* **132**(5), 683–687.

Schnell, E., Sizemore, M., Karimzadegan, S., Chen, L., Bredt, D.S. and Nicoll, R.A. (2002). Direct interactions between PSD-95 and stargazin control synaptic AMPA receptor number. *Proc. Natl. Acad. Sci. U.S.A.* **99**(21), 13902–13907.

Selcher, J.C., Weeber, E.J., Christian, J., Nekrasova, T., Landreth, G.E. and Sweatt, J.D. (2003). A role for ERK MAP kinase in physiologic temporal integration in hippocampal area CA1. *Learn. Mem.* **10**(1), 26–39.

Shen, K., Teruel, M.N., Connor, J.H., Shenolikar, S., and Meyer, T. (2000). Molecular memory by reversible translocation of calcium/calmodulin-dependent protein kinase II. *Nat. Neurosci.* **3**(9), 881–886.

Shi, S., Hayashi, Y., Esteban, J.A., and Malinow, R. (2001). Subunit-specific rules governing AMPA receptor trafficking to synapses in hippocampal pyramidal neurons. *Cell* **105**(3), 331–343.

Silva, A.J., Stevens, C.F., Tonegawa, S., and Wang, Y. (1992). Deficient hippocampal long-term potentiation in alpha-calcium-calmodulin kinase II mutant mice. *Science* **257**, 201–206.

Soderling, T.R., and Derkach, V.A. (2000). Postsynaptic protein phosphorylation and LTP. *Trends Neurosci.* **23**(2), 75–80.

Soderling, T.R., Chang, B., and Brickey, D. (2001). Cellular signaling through multifunctional Ca2+/calmodulin-dependent protein kinase II. *J. Biol. Chem.* **276**(6), 3719–3722.

Steward, O., and Schuman, E.M. (2001). Protein synthesis at synaptic sites on dendrites. *Annu. Rev. Neurosci.* **24**, 299–325.

Steward, O., and Worley, P. (2002). Local synthesis of proteins at synaptic sites on dendrites: Role in synaptic plasticity and memory consolidation? *Neurobiol. Learn. Mem.* **78**(3), 508–527.

Strack, S., Barban, M.A., Wadzinski, B.E., and Colbran, R.J. (1997). Differential inactivation of postsynaptic density-associated and soluble Ca2+/calmodulin-dependent protein kinase II by protein phosphatases 1 and 2A. *J. Neurochem.* **68**(5), 2119–2128.

Strack, S., McNeill, R.B., and Colbran, R.J. (2000). Mechanism and regulation of calcium/calmodulin-dependent protein kinase II targeting to the NR2B subunit of the N-methyl-D-aspartate receptor. *J. Biol. Chem.* **275**(31), 23798–23806.

Sweatt, J.D. (1999). Toward a molecular explanation for long-term potentiation. *Learn. Mem.* **6**(5), 399–416.

Sweatt, J.D. (2001). The neuronal MAP kinase cascade: A biochemical signal integration system subserving synaptic plasticity and memory. *J. Neurochem.* **76**(1), 1–10.

Thomas, M.J., Moody, T.D., Makhinson, M., and O'Dell, T.J. (1996). Activity-dependent beta-adrenergic modulation of low frequency stimulation induced LTP in the hippocampal CA1 region. *Neuron* **17**(3), 475–482.

Wallenstein, G.V., Vago, D.R., and Walberer, A.M. (2002). Time-dependent involvement of PKA/PKC in contextual memory consolidation. *Behav. Brain Res.* **133**(2), 159–164.

Weng, G., Bhalla, U.S., and Iyengar, R. (1999). Complexity in biological signaling systems. *Science* **284**, 92–96.

Whitmarsh, A.J., Cavanagh, J., Tournier, C., Yasuda, J., and Davis, R.J. (1998). A mammalian scaffold complex that selectively mediates MAP kinase activation. *Science* **281**, 1671–1674.

Woodgett, J.R., Davison, M.T., and Cohen, P. (1983). The calmodulin-dependent glycogen synthase kinase from rabbit skeletal muscle. Purification, subunit structure and substrate specificity. *Eur. J. Biochem.* **136**(3), 481–487.

Xia, Z., Choi, E.J., Wang, F., Blazynski, C. and Storm, D.R. (1993). Type I calmodulin-sensitive adenylyl cyclase is neural specific. *J. Neurochem.* **60**(1), 305–311.

Yin, J.C., Del Vecchio, M., Zhou, H., and Tully, T. (1995). CREB as a memory modulator: Induced expression of a dCREB2 activator isoform enhances long-term memory in Drosophila. *Cell* **81**(1), 107–115.

Yuan, L.L., Adams, J.P., Swank, M., Sweatt, J.D., and Johnston, D. (2002). Protein kinase modulation of dendritic K^+ channels in hippocampus involves a mitogen-activated protein kinase pathway. *J. Neurosci.* **22**(12), 4860–4868.

Zakharenko, S.S., Zablow, L. and Siegelbaum, S.A. (2001). Visualization of changes in presynaptic function during long-term synaptic plasticity. *Nat. Neurosci.* **4**(7), 711–717.

Zhu, J.J., Qin, Y., Zhao, M., Van Aelst, L., and Malinow, R. (2002). Ras and Rap control AMPA receptor trafficking during synaptic plasticity. *Cell* **110**(4), 443–455.

Genetic Analyses of Functional Connectivity in the Nervous System

Amit Etkin, M.D., Ph.D., Gleb Shumyatsky, Ph.D.,
Christopher J. Pittenger, M.D., Ph.D., and Eric R. Kandel, M.D.

CONTENTS

Databasing the Brain. Edited by Stephen H. Koslow and Shankar Subramaniam
ISBN 0-471-30921-4 © 2005 John Wiley & Sons, Inc.

13.1 INTRODUCTION

Since the realization in the middle of the nineteenth century that specific cognitive functions can be localized to discrete brain areas, many approaches have been taken to find out how brain systems are organized and how they relate to behavior. In this chapter, we will focus on one such approach, developed only in the past decade, employing genetic manipulations in mice.

Initial attempts to define functional circuits and their modes of interaction relied on brain lesion studies. In an early example, Carl Wernicke combined his findings that a region in the superior temporal cortex is critical for the comprehension of language, with Paul Broca's earlier findings that the inferior frontal cortex is critical for the expression of language, to predict the existence of the fibers of the arcuate fasciculus, which connects these regions. Pathological lesions in humans, and more directedly placed lesions in experimental animals, can often demonstrate quite clearly the roles of a particular brain region. Generally, however, because lesions involve permanent tissue damage, they have a limited utility in (a) teasing apart specific functions for different cell types or circuit components or (b) defining temporally dynamic roles for that brain region. These limitations of lesion experiments led to the development of a variety of pharmacological approaches designed to produce specific and transient "functional lesions" that can be reversed without permanent consequences.

Using pharmacological approaches, for example, one can discriminate between the roles of various neurotransmitter systems in behavior. The effects of identified neurotransmitters can be further characterized with respect to which postsynaptic receptors mediate which aspects of the behavior. The downstream consequences of receptor activation can then be examined by exploring intracellular signaling cascades, using inhibitors of specific steps in signal transduction. Likewise, more general functions of neurons, like the ability to fire action potentials and release neurotransmitter or to recruit protein synthesis, may be transiently inhibited. The rapidity and reversibility of pharmacological approaches can supplement lesion-based methods, but they suffer from some of the same limitations as lesion studies. Intracerebrally injected drugs exert their effects on all cells in the zone of perfusion. Inhibition of protein synthesis, a commonly used approach, will target excitatory and inhibitory neurons, as well as all glial cells. Inhibition of action potential firing will block the activity of both excitatory and inhibitory neurons. In addition, pharmacological methods can only be applied to targets for which drugs exist, which are mostly neurotransmitter receptors, membrane ion channels, and some intracellular signaling molecules like protein kinases. Therefore, the vast majority of proteins expressed by neurons cannot be affected (Drews, 2000). Lastly, the lack of complete drug target specificity limits the utility of some drugs.

In recent years there has emerged a third approach to the study of brain function by perturbing it: genetic modifications in mice. When we refer to genetic methods, we mean the disruption of endogenous genes (knockouts), the alteration of endogenous genes (knockins), and the introduction of exogenous genes (transgenics), collectively referred to as "reverse genetic" methods. Genetic perturbations can allow a relatively high degree of temporal, spatial, and cell-type specificity of manipulations, particularly when compared to pharmacological and lesion-based

approaches. Some of the clearest examples of the utility of reverse genetic methods for understanding functional connectivity come from work in the field of learning and memory, which will be the primary focus of this chapter. It may be expected that the use of genetic techniques for elucidating functional circuitry will rapidly grow in the near future and contribute greatly to our understanding of many brain circuits.

13.2 THE METHODOLOGIES OF REVERSE GENETICS

13.2.1 Transgenics and the tTA System

Transgenic gene expression involves two components. First, a promoter must be selected that will effectively drive transgene expression in a particular brain region or cell type and at a particular time. Second, the cDNA being expressed must be engineered for a specific purpose, often to act as a dominant-negative or constitutively active form of a protein or simply to overexpress the wild-type protein.

Only a limited number of useful promoters that drive transgene expression in subregions of the brain have been described. Nonetheless, the clear success of transgenic approaches in the past decade is driving an active search for promoters with different spatial and temporal patterns of expression. One widely used promoter, which expresses at high levels in the postnatal forebrain (providing spatial restriction and avoiding alterations in brain development), is that of the CaMKIIα gene, introduced in 1995 by Mark Mayford and Eric Kandel (Mayford et al., 1995). Transgenic approaches using this promoter lead to gene expression in the hippocampus, cortex, and striatum.

The power of transgenic systems was further extended by Mayford in the Kandel laboratory when he used the CaMKIIα promoter to drive expression of the tTA gene (Mayford et al., 1996). tTA is a fusion protein consisting of the DNA binding domain of the bacterial tetracycline repressor and the activation domain of the strong viral transcriptional activator VP16. This chimeric transcription factor recognizes the tetracycline operator site (tetO site) in artificially constructed promoters. When expressed, tTA leads to the transcription of genes driven by tetO-containing promoters (see Figure 13.1). In the presence of the tetracycline analog doxycycline, however, tTA dissociates from its DNA targets and transcription is turned off, as tetO-driven promoters are comprised of elements from bacterial genomes and are normally silent in mammalian cells (Gossen and Bujard, 1992). Inhibiting tTA function by administration of doxycycline, thus turning the tetO transgene off, allows the investigator to determine whether the observed phenotype can be reversed. Reversibility shows that there are no permanent developmental or other consequences of transgene expression.

Because tTA- and tetO-driven transgenes are generated separately, the same tTA line can be used combinatorially to drive expression of any number of tetO-containing constructs. Alternatively, the same tetO construct may be driven by tTAs expressed from promoters targeting different brain areas. In addition, it is possible to obtain patterns of tetO transgene expression more restricted than the activating tTA's expression, an effect that seems to be dependent on the tetO construct's site of integration into the genome. For example, transgenic lines of mice may be found

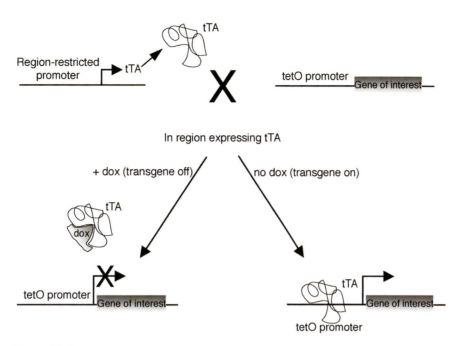

Figure 13.1   Schematic overview of the tTA–tetO transgenic system. Two independent lines of mice are made carrying a tTA transgene (*left*) expressing in a region or cell type of interest, and a tetO-driven transgene of the gene of interest (*right*). In cells expressing tTA, the tetO-driven construct will be transcribed at the basal state (no doxycycline). Administration of the tetracycline analog doxycycline inactivates tTA and turns off expression of the tetO transgene.

expressing tetO constructs selectively in the striatum or in particular cell fields of the hippocampus, even though all are driven by the same CamKIIα–tTA line. Importantly, the patterns of transgene expression obtained by breeding any tTA line to any tetO line is consistent between all of their offspring, without variation between individuals.

The tTA–tetO system allows transgene expression to be suppressed. A purely inducible system would allow transgene expression to be rapidly turned on just before an experiment, remaining silent in the absence of a pharmacological inducer. A mutant form of the tetracycline repressor has been found that activates transcription only in the presence of doxycycline, and it has been termed reverse tTA, or rtTA (Gossen et al., 1995). Isabelle Mansuy in the Kandel laboratory has successfully used this system in combination with the forebrain-specific CamKIIα promoter to drive several tetO constructs (Genoux et al., 2002; Mansuy et al., 1998).

13.2.2 Knockouts: Traditional and Cre-loxP Systems

Central to any knockout approach is the disruption of a gene's function by the deletion of portions of that gene. In traditional knockouts, the gene is deleted from the genome of every cell in the body. If a gene's expression is normally restricted to particular brain regions or cell types, then a whole-body deletion approach would not be expected to result in side effects in regions of the brain or other tissues not normally expressing the gene. An example of a traditional knockout of a gene with a relatively restricted pattern of expression is the gastrin-releasing peptide receptor study described later in this chapter. If, however,

a gene is widely expressed in the brain or is also expressed in other tissues of the body, then a more region-restricted or cell-type-restricted method of deletion is necessary. The utility and importance of regional restriction of gene deletion is well illustrated by genetic studies of the NMDA-type glutamate receptor in learning, described later in this chapter.

Brain-region or temporally restricted gene deletion can be achieved using conditional knockout approaches with the Cre-loxP system. With Cre-loxP (see Figure 13.2), loxP DNA sequences are inserted flanking the gene region to be deleted. The presence of the loxP sites, however, does not disrupt the gene. In cells in which the P1 bacteriophage Cre recombinase gene is expressed from a separate transgene, deletion of the loxP-flanked region will occur. The temporal and spatial pattern of Cre-mediated deletion depends on the promoter used to drive the Cre transgene, an issue discussed in the previous section. By expressing Cre transgenes in different brain regions, one can delete the same target gene in any number of brain regions in parallel studies. This approach has been used, for example, to study a regional restricted knockout of the NMDA receptor in the hippocampus, an area important for explicit memory.

13.2.3 Hebb's Postulate and LTP: Genetic Studies of NMDA Receptor Function in Hippocampus-Based Memory

Function and Organization of Circuits in the Hippocampus

In their classic 1957 study, William Scoville and Brenda Milner provided a dramatic demonstration of the importance of the

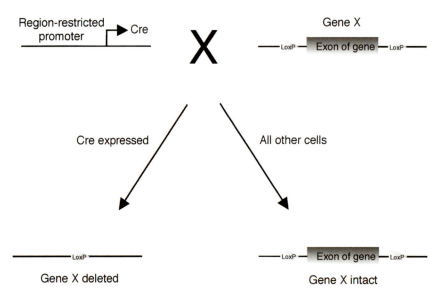

Figure 13.2   Schematic overview of the Cre-loxP conditional knockout system. Two independent lines of mice are made carrying a Cre transgene (*left*) expressing in a region or cell type of interest, along with the gene of interest flanked by loxP sites (*right*). In cells in offspring of these mice that express Cre recombinase, the loxP-flanked gene will be deleted, while in all other cells it will be left intact.

human hippocampus in learning and memory. They described a patient, H.M., who was not able to form new long-term memories after undergoing bilateral hippocampal removal for intractable epilepsy. Subsequent studies of individuals with bilateral loss of specific subfields within the hippocampus have confirmed and extended these findings (Rempel-Clower et al., 1996; Zola-Morgan et al., 1986).

The hippocampus in rodents is organized along both transverse and longitudinal axes. We will begin by discussing circuitry within its transverse axis, as genetic studies have made the greatest inroads into understanding how this aspect of hippocampal organization contributes to cognitive functions. A simplified diagram of circuits in the hippocampal formation is illustrated in Figure 13.3. Neurons in layer II of the entorhinal cortex project to the dentate gyrus by means of the perforant path. The dentate gyrus then gives rise to a mossy fiber projection that terminates on CA3 pyramidal neurons. These CA3 cells, in turn, give rise to two major projections. The Schaffer collateral pathway terminates on neurons in the CA1 region, while recurrent collaterals terminate on other cells in the CA3 region. CA1 cells then give

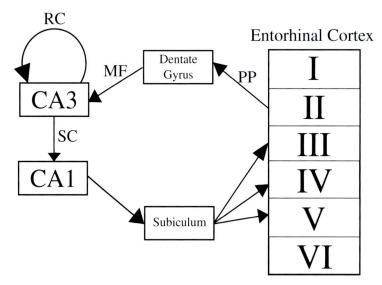

Figure 13.3   Simplified circuit diagram of the hippocampus. Shown are the entorhinal cortex, dentate gyrus, hippocampus subfields CA1 and CA3, and the subiculum. Abbreviations are: PP (perforant path), MF (mossy fiber), RC (recurrent collaterals) and SC (Schaffer collaterals). Not shown are projections from the entorhinal cortex to CA1 and CA3, which have been omitted for simplicity and are not discussed in this chapter.

rise to axons that project to the subiculum. The subiculum then projects to the deep layers of the entorhinal cortex. Finally, the entorhinal cortex projects to many of the cortical areas that had provided its initial input.

In both humans and rodents, the hippocampus has an essential role in spatial learning and memory, as well as other tasks in which associations with context, object identity, or more complex stimulus contingencies need to be made (Eichenbaum, 1996; Gerlai, 2001). Two common behavioral tasks that probe these functions of the hippocampus in rodents, the Morris water maze and contextual fear conditioning (Eichenbaum, 1996; Gerlai, 2001), are discussed in this chapter.

In the Morris water maze, the animal is put into a pool of opaque water, from which it is motivated to escape. The only way to get out of the water is to find a platform, hidden from view just beneath the level of the water, that can only be reliably found by determining its spatial location in relation to distal cues on the wall of the room (see Figure 13.4). Procedural aspects of this task can be controlled for by marking the goal platform with a visible cue, obviating any need for a spatial strategy (known as the visible platform task). Animals with hippocampal lesions perform well on the visible platform task, but fail to use spatial information to master the hidden platform version.

Contextual fear conditioning exploits the role of the hippocampus in encoding characteristics of an animal's environment and involves pairing exposure to a context (the conditioning chamber) with an aversive electric shock. When an animal is later placed in the training context, it initiates defensive freezing behavior because it has come to associate the previously neutral context (conditioned stimulus) with the aversive shock (unconditioned stimulus). The role of the hippocampus in this task is complex, possibly differentially involving the structure along its long axis, and will be discussed later in this chapter.

When an animal explores any space, it forms a neural representation of that space in the hippocampus. This spatial code can be seen in single-cell recordings in hippocampi of behaving animals. Hippocampal pyramidal neurons are normally silent, but can show dramatically elevated firing rates in a particular location as an animal is exploring their environment (see Figure 13.5). This ensemble of "place cells" has been proposed to represent the animal's position in a particular space, forming a cognitive map that can be reinstated in its entirety when the animal is placed back in an environment it had previously explored (O'Keefe and Nadel, 1978). A cell usually fires in the same spatial coordinate in distinct environments, and an ensemble of 60 simultaneously recorded place cells is sufficient to predict the animals location within 1 cm (Wilson and McNaughton, 1993). The tuning of place cell fields and the ability to activate the proper ensemble of place cells in the correct environment upon recall may be essential to the animal's recognition and spatial navigation in its environment.

Hebb's Postulate, LTP, and the NMDA Receptor

In 1949, Donald Hebb postulated that one way information may be stored in the brain is through the strengthening of synaptic connections between two neurons which occurs when the presynaptic and postsynaptic neurons were active at the same time (Hebb, 1949). Thus, if a presynaptic neuron represented information about one stimulus, and another neuron terminating on the same postsynaptic target represented information about another stimulus, the simultaneous firing of the first neuron with the postsynaptic action of the second neuron (its depolarization of its target) would lead to the "binding" of these two stimuli. This view has provided some of the theoretical framework behind the search for the molecular and cellular basis of learning and memory over the past 50 years (Bi and Poo, 2001).

The search for activity-dependent forms of synaptic plasticity in the hippocampus, which might underlie its role in learning, began with the description of a long-lasting enhancement of synaptic activity in the dentate gyrus after repeated in vivo stimulation of the perforant path (Bliss and Lømo, 1973). Long-term synaptic potentiation (LTP) has since been described in all other synapses in the hippocampal formation, and it has been most intensively studied using in vitro slice preparations (Malenka and Nicoll, 1999). LTP, like learning, is characterized by two phases. The first, a short phase lasting minutes requires covalent modification of existing proteins, but not the synthesis of new proteins. In contrast, the second phase, lasting hours to a lifetime, requires the transcription and translation of new genes (Malenka and Nicoll, 1999).

Regulating both short- and long-term process, however, must be a Hebbian "detector" of coincidence. That is, there must be a molecular mechanism underlying the ability of the postsynaptic cell to simultaneously sense its depolarization and its activation by another neuron. This detector must also be able to trigger events leading to short- and long-term forms of synaptic

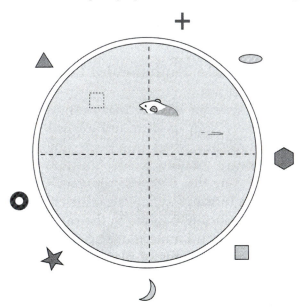

Figure 13.4 Diagram of the Morris water maze. Mice are put into an pool of opaque water, from which they will want to escape. The only way to get out of the water is to find a small platform, just below the level of the water. In the hidden platform version of the task, the animal learns to find the platform by spatial navigation based on distal cues around the pool. In the visible platform version the platform is marked with a cue, obviating any need for a spatial strategy. Learning is plotted as a decreasing latency over the course of training to finding the platform. The animal's memory of the platform location is tested by removing the platform and examining the time the animal spends searching the quadrant of the pool that had formerly contained the platform, with respect to the three other quadrants. Alternatively, crossings of the platform's former location or path length may be measured. This test session is termed a probe or transfer test.

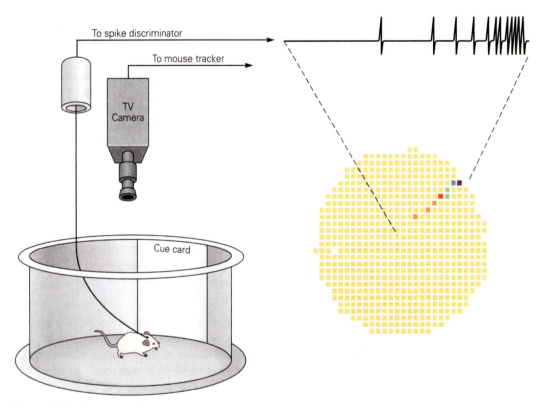

Figure 13.5   Overview of hippocampal place cell recordings. Animals are implanted with electrodes targeting the CA1 region of the hippocampus. Animals are allowed to explore a space while their position is tracked by a video camera and single unit activity is recorded from pyramidal neurons in CA1. These neurons are normally silent (denoted by a yellow color), but they fire at an increased rate as the animal approaches particular locations in the space, termed a cell's "place field."

plasticity. For many synapses, this detector appears to be the N-methyl-D-aspartate (NMDA)-type glutamate receptor. Unlike most other receptors for glutamate, the major excitatory transmitter in the brain, the NMDA receptor is capable of passing a large calcium current. Calcium is a potent activator of intracellular signaling pathways and leads to the activation of protein kinases, which modify existing proteins—a mechanism of short-term plasticity or memory. More robust or prolonged activation of protein kinase signaling pathways then results in the activation of transcription factors, which stabilize the association through the transcription of new genes. The capacity of the NMDA receptor to pass ions, including calcium, depends on release of an extracellular magnesium block, which is achieved by elevating the membrane potential (postsynaptic depolarization). In this way, glutamate released by the firing of the presynaptic neuron results in postsynaptic calcium influx only when the postsynaptic cell is already depolarized, leading to activation of downstream signaling cascades. These properties of the NMDA receptor satisfy Hebb's rule at a molecular level.

Normal function of the NMDA receptor has been shown to be essential for hippocampus-based learning as well as LTP at a number of synapses within the hippocampal formation (Morris et al., 1986,1991; Rawlins, 1996). These experiments involve disruption of NMDA function throughout the hippocampus. Since NMDA-dependent synaptic plasticity has been found at many synapses within the hippocampal formation, disrupting NMDA function in the entire hippocampus cannot elucidate which circuit components contribute to which aspects of

hippocampus-dependent behavior. Understanding this sort of functional circuitry requires more advanced techniques for restricting NMDA receptor manipulations to distinct hippocampal subfields, particularly in light of the fact that the NMDA receptor is expressed widely in the brain and its complete deletion results in perinatal lethality (Forrest et al., 1994).

The NMDA Receptor in CA1

Genetic studies of hippocampal NMDA function in learning and LTP are guided by human lesion studies of subjects with relatively selective lesions of distinct hippocampal subfields (Rempel-Clower et al., 1996; Zola-Morgan et al., 1986). These lesion studies found a severe memory impairment in human subjects with bilateral damage to the CA1 subfield. One of the major pathways providing input to CA1, as discussed earlier, is the Schaffer collateral pathway. Since LTP in this pathway has been found to be NMDA-dependent, a logical first step in a genetic dissection of hippocampal function would be the disruption of NMDA-mediated Hebbian coincidence detection in the CA1 region.

To selectively disrupt the NMDA receptor in CA1, Tsien et al. (1996a) screened mice carrying Cre transgenes driven by the CamKIIα promoter to find one expressing preferentially in the CA1 region. Using this line in combination with mice carrying loxP-flanked NMDA receptor NR1 subunits, the authors were able to delete NR1 specifically in CA1 beginning after the second postnatal week (Tsien et al., 1996a) (see Figure 13.6A, B). Postnatal Cre expression possible using the CamKIIα

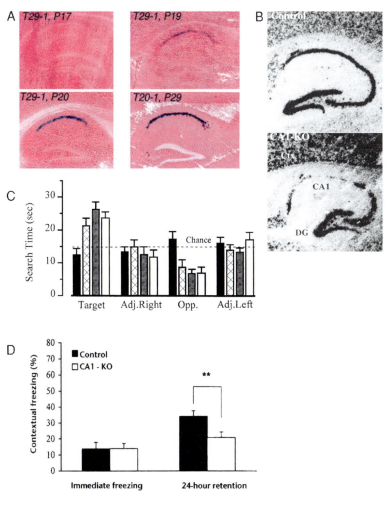

Figure 13.6 CA1 deletion of the NR1 subunit of the NMDA receptor and its behavioral consequences. (A) Cre recombinase reporter activity in the hippocampus in a line of mice showing specific deletion in the CA1 region of the hippocampus after the second postnatal week (reprinted from Tsien et al., 1996a). (B) In situ detection of NR1 transcripts in the hippocampus in control and CA1-KO mice. (C) Performance of NR1-CA1KO mice in a transfer test, reflected as search time in target versus nontarget quadrants, after training in the hidden platform version of the water maze (closed bars), compared to Cre only (hatched bars), lox-P flanked non-Cre expressing (shaded bars), and wild-type animals (open bars) (reprinted from Tsien et al., 1996b). (D) Freezing of NR1-CA1KO mice tested immediately after contextual fear conditioning, and again 24 hours later (reprinted from Rampon et al., 2000).

promoter avoids the majority of developmental effects that might confound the study of adult NMDA receptor function.

To extend findings about the importance of the CA1 region in human memory, Tsien et al. examined the behavioral consequences of restricted loss of NMDA receptor function in CA1. The authors found that NR1-CA1 knockout animals were unable to learn the location of the escape platform in the hidden platform version of the water maze (see Figure 13.6C), but performed normally in the visible platform control task. Subsequent studies found profound impairment of NR1-CA1KO animals in other hippocampus-dependent tasks, specifically contextual fear conditioning, novel object recognition, and social transmission of food preference (Rampon et al., 2000) (see Figure 13.6D).

Tsien et al. then examined LTP elicited at the Schaffer collateral pathway (CA3 to CA1) and the perforant path (entorhinal cortex to dentate gyrus). They found that CA1 LTP was completely absent in knockout mice, whereas both knockout and control animals had similar levels of LTP in the dentate gyrus after perforant path stimulation. They also tested NMDA-receptor-dependent long-term depression (LTD) in the CA1 region, and found it disturbed in knockout animals, while NMDA-receptor independent LTP in this region was normal (Tsien et al., 1996b).

This work provided the first demonstration, using a genetic lesion method (which specifically disrupted a Hebbian detector in postsynaptic neurons of the CA1 region), of how careful molecular genetic methods can illuminate functional circuitry. Furthermore, these results allowed a cleaner analysis of the dynamic role of CA1 in memory than was previously possible, without lesioning the structure. However, it is important to note that Tsien et al. stumbled across a CA1-enriched Cre line by chance, and even then in older animals recombination was not fully restricted to the CA1 region of the hippocampus (S. Tonegawa, Society for Neuroscience presentation, 2002).

Having found a striking phenotype in learning and LTP in the NR1-CA1KO animals, Tonegawa and colleagues then went on to examine the neural representation of space in the hippocampus using place cell recordings from the CA1 region (McHugh et al., 1996). Surprisingly, the authors found that place fields exist in knockout animals and are as stable as those in control animals. In comparing the spread of place fields in both groups, however, McHugh et al. found that NR1-CA1KO place fields were broader, sometimes showing multiple peaks rather than the single peak observed in control place fields. These abnormal characteristics of place fields, when viewed from an ensemble level in simultaneous recordings from large numbers of neurons, lead to a degraded representation of the space.

The NMDA Receptor in CA3

As described above, most of the hippocampal circuitry along its transverse plane is thought to be involved in a unidirectional transfer of information. It is difficult to see how this architecture could allow for whole scenes to be brought together into a unified image or point in time. The exception to this general picture is the network comprised by CA3 recurrent collateral projections onto other CA3 neurons. There is an order of magnitude more CA3–CA1 recurrent collateral projections than CA3–CA1 Schaffer collateral projections, such that the CA3 system becomes its own network, with each CA3 neuron connected to roughly 4% of all other CA3 neurons in the rat (Amaral and Witter, 1989; Ishizuka et al., 1990). Thus, information in one CA3 cell can reach most other CA3 cells within two to three synapses. Recurrent collateral excitatory input is likely to dominate the effect of other excitatory afferents, such as those provided by the mossy fiber pathway. A prominent computational theory of hippocampal function, proposed by Edmund Rolls and Alessandro Treves, holds that the network comprised of CA3 recurrent collaterals acts as an autoassociative network, allowing information originating from different cortical areas to be composited into a single "snapshot" (Treves and Rolls, 1994). This information could consist of sensory stimuli in various modalities or information on rewards, punishments, and emotional states. Likewise, during recall, the CA3 autoassociative network would be able to retrieve multiple components of a memory given only a few presented cues, allowing for pattern recognition based on relatively little information.

Autoassociative network models of CA3 function predict an important role for plasticity at CA3 recurrent collateral projections, involving a Hebbian detector such as the NMDA receptor, in spatial navigation based on few available cues. It is only when few cues are available that pattern completion would seem most important. Tonegawa and his colleagues therefore sought to identify a promoter capable of driving Cre recombinase expression specifically in CA3 cells, in order to eliminate NMDA function there by disruption of the NR1 subunit. Nakazawa et al. report a line of mice with a Cre transgene driven by the promoter of the KA-1 gene (a kainic acid-type glutamate receptor subunit), which results in Cre expression in CA3 pyramidal neurons after the fourth postnatal week (Nakazawa et al., 2002). Cre activity was found in 100% of CA3 neurons by 8 weeks of age, as well as a much lower percentage of cells in the dentate gyrus, cerebellar

granule neurons, and the nucleus of the facial nerve (see Figure 13.7A). Like CamKIIα-driven transgenes, KA1-driven Cre does not result in recombination in interneurons in the hippocampus (Sik et al., 1998). Disruption of NR1 expression can be clearly seen in the CA3 region of the hippocampus (see Figure 13.7B).

To examine the effect of CA3-NR1 deletion on spatial memory, the authors first trained NR1-CA3KO and control animals on the hidden platform version of the water maze with the full complement of spatial cues surrounding the pool. Recall of platform location, measured as the time spent occupying each quadrant of the pool with the platform removed, should lead the animal to spend more time in the quadrant that had previously contained the platform during training. When all distal cues were used, no deficit was detected in the CA3-NR1KO animals. However, when only some cues were available, a significant retrieval deficit was seen in these animals, resulting in a level of recall similar to that in normal animals in the absence of any cues (see Figure 13.7C–E). Thus, retrieval of the complete memory given partial cues (pattern completion) is defective in mice lacking Hebbian plasticity in the CA3 autoassociative network, consistent with theoretical predictions from computational theories (Treves and Rolls, 1994).

The authors then examined LTP at a number of synapses within the hippocampal formation to determine the specificity of their manipulation on plasticity at relevant synapses. Normal LTP was found in NR1-CA3KO mice at the dentate gyrus to CA3 mossy fiber pathway and the CA3 to CA1 Schaffer collateral pathway. In contrast, there was a profound deficit in the CA3 to CA3 recurrent collateral pathway shortly after tetanization. LTP at the mossy fiber pathway has not been found to depend on NMDA receptor function (Harris and Cotman, 1986; Williams and Johnston, 1988; Zalutsky and Nicoll, 1990), showing the selectivity of effect of CA3-NR1 deletion on LTP of CA3 recurrent collaterals.

Given the specific deficits in spatial memory and LTP described in the CA3-NR1KO animals, the authors then sought to examine the representation of space by CA1 place cells under full cue and partial cue retrieval conditions. CA3-NR1KO and control mice were allowed a 30-minute exploration of a novel space with a full complement of distal cues, during which time place cell recordings were acquired. After a 2-hour delay, mice were replaced in the space with either all distal cues, or only one distal cue. As seen in Figure 13.7F, place fields are stable under either retrieval condition in control mice. In contrast, place fields are consistent and stable under full cue conditions in CA3-NR1KO mice, but extremely degraded when only one cue is available.

The CA3- and CA1-NR1 knockout studies illustrate several important lessons about genetic studies of circuitry underlying different aspects of cognitive function. First, different components within a network can be analyzed separately using genetic methodologies. Physical lesions may inform behavioral investigations, but cannot be used for neurophysiological studies of the lesioned site. Genetic lesions, on the other hand, can be used at many parallel levels of analysis. Second, genetic tools can allow finer dissections of the interplay between function and anatomy than possible with pharmacological methods alone. The results of the CA1-restricted NR1 knockout were not surprising since they were predicted by previous pharmacological experiments. Results from the CA3-restricted NR1 knockout, however, came as more of a surprise, since it would be very difficult to achieve

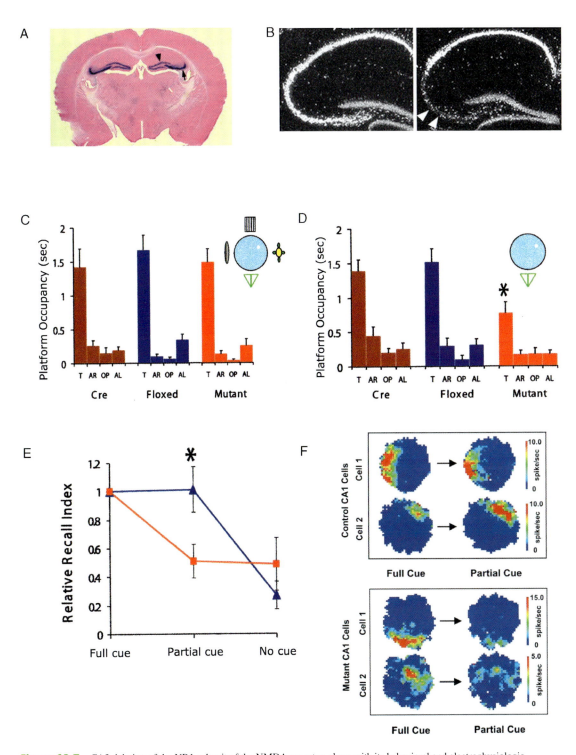

Figure 13.7 CA3 deletion of the NR1 subunit of the NMDA receptor, along with its behavioral and electrophysiologic consequences. (**A**) Cre reporter activity, seen highest in CA3, with much less recombination in the dentate gyrus. (**B**) In situ detection of NR1 transcripts in the hippocampus of NR1-CA3KO mice. (**C**) Performance of NR1-CA3KO mice ("mutant") on a transfer test with all distal cues, reflected as search time in target versus nontarget quadrants, after training in the hidden platform version of the water maze, compared to mice carrying lox-P flanked non-recombined NR1 ("floxed") or Cre-transgenic mice with intact NR1 ("Cre"). (**D**) Same as C, but recall is performed with only a partial complement of distal cues. (**E**) Quantitation of recall under full cue, partial cue, and no cue conditions, compared to the full cue condition from C and D, in NR1-CA3KO (square) or control (triangle) animals. (**F**) Representative place fields of control and NR1-CA3KO mice under full and partial cue encoding and retrieval conditions. (Reprinted from Nakazawa et al., 2002.)

such a high degree of specificity with existing pharmacological methods. Third, these studies illustrate how a number of Cre transgenic lines can be used with only one key loxP-flanked gene to lead to a more systematic understanding of (a) the role of that gene and (b) the circuitry in which it is deleted. Along these lines, Tonegawa and colleagues are now engineering lines of mice that will allow NR1 deletion in other subfields within the hippocampal formation (S. Tonegawa, Society for Neuroscience presentation, 2002). Lastly, the availability of more Cre lines with varied expression patterns will allow others to answer further functional circuitry-driven questions, of which there are many.

13.3 ARE THERE DISSOCIABLE ROLES FOR THE DORSAL AND VENTRAL HIPPOCAMPUS IN LEARNING?

The hippocampal cell fields described above extend fairly homogenously through the longitudinal extent of the hippocampus, which has an elongated structure; but the cortical regions from which they receive their projections differ. Recent results demonstrate that the molecular tools that have proven so useful in examining the different cell fields of the hippocampus can also be used to probe functional specialization along the longitudinal axis.

The rodent dorsal or septal hippocampus, which corresponds to the primate posterior hippocampus, receives input from lateral entorhinal cortex, which in turn receives the majority of its input from sensory and associational neocortex. The dorsal hippocampus is thus well positioned to process exteroceptive sensory information (paradigmatically, spatial information). The ventral or temporal hippocampus (corresponding to primate anterior hippocampus), in contrast, receives its input from medial entorhinal cortex, which collects inputs from allocortex and other limbic structures such as the amygdala. This suggests that it may have a more important role in learning situations in which interoceptive stimuli are critical parameters (Amaral and Witter, 1995). The connectivity of the anterior and posterior primate hippocampus recapitulates this dissociation (Insausti et al., 1987; Suzuki and Amaral, 1994; Witter and Amaral, 1991). Longitudinal collaterals interconnect different septotemporal levels in the dentate hilus and in region CA3. Tract tracing studies show that the dorsal two-thirds of the hippocampus is densely interconnected by these projections but is poorly connected to the ventral one-third, suggesting that these may be functionally distinct zones (Moser and Moser, 1998).

Behavioral and in vivo electrophysiological studies have supported these anatomically derived predictions. There are more place cells in the dorsal hippocampus, supporting a more important role for the dorsal hippocampus in processing spatial information; and place cells there tend to have tighter and better-defined place fields (Jung et al., 1994). Restricted lesions of the dorsal hippocampus impair learning in the Morris water maze, while lesions restricted to the ventral hippocampus do not (Moser and Moser, 1998; Moser et al., 1995). A specific role of the ventral hippocampus in learning has not been as clearly demonstrated. Contextual fear conditioning, which depends on intact hippocampal function, is spared by excitotoxic lesions of the dorsal hippocampus (Frankland et al., 1998; Maren et al., 1997), leading to the idea that this task might depend on the ventral hip-

pocampus. However, lesions of the ventral hippocampus reduce animals' basal freezing tendency, rendering quantification of fear conditioning problematic (Richmond et al., 1999); and a recent study indicates that lesions of the ventral pole of the hippocampus impair the expression of unconditioned fear, rather than the acquisition of conditioned fear (Kjelstrup et al., 2002). Thus, to date, no learning task has been convincingly demonstrated to depend specifically on intact ventral hippocampal function.

These functional dissociations along the dorsal–ventral hippocampal axis have not been extensively studied in genetically modified animals, largely due to a lack of genetic tools. No genes have yet been described to be specific to the dorsal or ventral hippocampus, or to show a dorsal–ventral gradient in their expression. Such genes would provide tools for the kind of approaches described earlier in this chapter. However, the CaMKIIα promoter can, in some circumstances, yield differential transgene expression along the dorsoventral hippocampal axis.

In a recent study, Kandel and colleagues were able to interfere with the transcription factor CREB specifically in the CA1 subregion of the dorsal hippocampus (Pittenger et al., 2002). CREB has been shown in a number of different model systems to be a critical regulator of the transition from transient to long-lasting synaptic plasticity and memory (Pittenger and Kandel, 1998; Barco et al., 2003). Short-term memory (and short-term synaptic plasticity), lasting up to 3 hours, is independent of new gene regulation, while long-term memory requires gene induction and macromolecular synthesis. In the marine snail *Aplysia*, CREB is a central regulator of this transition (Bartsch et al., 1998). Likewise, in *Drosophila*, the transition from short-term to long-term memory is facilitated by increased expression of CREB (Yin et al., 1995) and is impaired by inhibition of CREB (Yin et al., 1994). In rodents, early studies suggested a similar role (Bourtchuladze et al., 1994); but as in so many cases, clear interpretation has been complicated by, among other factors, a lack of clear restriction of molecular manipulations to meaningful subsets of functional circuits in the brain.

Pittenger et al. (2002) expressed KCREB, a dominant negative mutant of the transcription factor CREB, using the CaMKIIα promoter and the tTA system, as developed by Mayford et al. (1996) and described earlier. They found that a combination of a CaMKIIα-tTA line and a particular tetO-KCREB line resulted in KCREB expression strongly in CA1, with no detectable expression in CA3, dentate gyrus, or subiculum. Examination of coronal sections throughout the brain showed that this expression was largely restricted to the dorsal hippocampus, with expression only in rare scattered cells of the ventral hippocampus and no expression detectable at the ventral pole (see Figure 13.8). This finding indicates that, under some conditions, existing tools may already allow us to probe the long axis of the hippocampus as well as the transverse axis. These dorsal CA1-KCREB mice showed a the pattern of behavioral phenotypes consistent with a functional lesion of the dorsal hippocampus: a deficit in the Morris Water maze, but intact contextual fear conditioning. The study of these animals thus supports a dissociation already made in both rats and mice by lesion methods (Frankland et al., 1998; Maren et al., 1997); but it does so with more subtle molecular tools, in an anatomically intact hippocampus. The lack of a ventral hippocampus-specific learning task, however, prevents a clean double dissociation between the functions of the dorsal and ventral hippocampi.

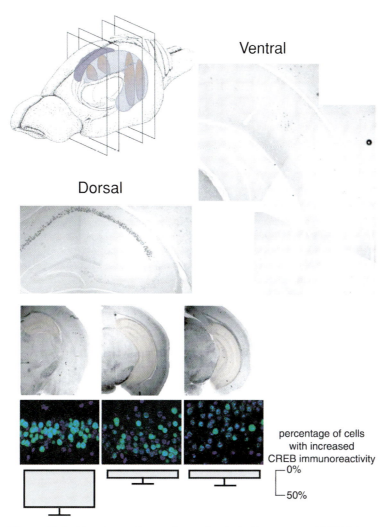

Figure 13.8 Dorsal–ventral expression gradient in the hippocampus of KCREB in transgenic animals. (**A**) Diagram of the rodent hippocampus and representative sections from the dorsal and ventral hippocampi showing differential KCREB expression. (**B**) Quantitation of the percentage of CA1 pyramidal neurons expressing KCREB from sections throughout the hippocampus. (Reprinted from Pittenger et al., 2002.)

What portion of the hippocampus, then, mediates contextual fear conditioning? Ventral hippocampal lesions impair the expression of unconditioned fear but not the acquisition of conditioned fear (Kjelstrup et al., 2002). Since dorsal hippocampal lesions also do not produce anterograde amnesia for conditioned fear, but total hippocampal lesions do (Anagnostaras et al., 1999; Kim and Fanselow, 1992; Phillips and LeDoux, 1992; Richmond et al., 1999), it seems most probable that any adequate subset of the hippocampus can mediate contextual fear. In order to probe this possibility, specifically with regard to the involvement of CREB in contextual fear conditioning, it would be desirable to examine a line of transgenic mice in the KCREB transgene expressed throughout CA1, without a dorsal–ventral gradient.

Some previous studies have also shown a dissociation between spatial learning and contextual fear conditioning similar to that observed in the dorsal CA1-KCREB transgenic mice. Mayford et al. (1995) described transgenic mice in which a constitutively active CaMKII mutant was expressed under its own promoter. In one line of mice, these investigators found impairment in spatial learning as assayed in the Barnes maze, but intact contextual fear conditioning (Bach et al., 1995). It is possible that these animals, and others in which this dissociation has been observed, had an undetected dorsal–ventral hippocampal gradient in transgene expression.

Why does the CaMKIIα promoter in conjunction with the tTA system produce this gradient of transgene expression? Our knowledge of promoter regulatory elements that might give variable expression in pyramidal cells along the long axis of the hippocampus remains too incomplete to attempt an answer to this question at this time. Since the original observation of Pittenger et al., such a gradient has been seen in other transgenic animals using the same expression system (L. Wang and E. R. Kandel unpublished observations; M. Mayford, personal communication). However, not all lines of tetO-driven transgenic mice show this expression gradient. Thus, the dorsal–ventral expression gradient observed by Pittenger et al. seems also to involve specific properties of the integration site of the tetO-KCREB transgene in this particular mouse line.

Better genetic tools to probe the longitudinal axis of the hippocampus would be valuable in addressing several questions. Do the mechanisms of hippocampal synaptic plasticity vary along this axis? Will identical molecular manipulations produce different behavioral effects if targeted to different hippocampal regions? Which behavioral tasks depend on each hippocampal region, and can these be rationalized in terms of each region's interconnectivity with entorhinal and neocortical areas? Can molecular markers confirm the two-third – one-third division suggested by tracer studies (Moser and Moser, 1998)? Or is the hippocampus functionally divided into more, or different, zones than this? It is to be hoped that, as greater attention is given by molecular geneticists to functional specialization along the long axis of the hippocampus, these and other questions will become more accessible to the behavioral biologist.

13.4 MOLECULAR AND GENETIC MAPPING OF THE NEURAL CIRCUITRY OF LEARNED FEAR IN THE AMYGDALA

The amygdala is a key component of the neural circuitry for fear, both instinctive and learned. It is found in many organisms, from reptiles to humans (Davis and Whalen, 2001; LeDoux, 2000; Maren and Fanselow, 1996). Fear has been studied in a variety of experimental organisms including snails, flies, fish, mice, and humans. Both innate and learned fear trigger a range of defensive internal and external body responses for adapting to threatening events. In this chapter we will deal mostly with learned fear.

13.4.1 Learned Fear and Amygdala Circuitry

Learned fear can be assessed quantitatively using methods that produce Pavlovian fear conditioning: quickly learned, robust and long-lasting memory for fear (Kapp et al., 1992). In this task, an initially neutral conditioned stimulus (CS) acquires significance following pairing with an aversive unconditioned stimulus (US), often after a single training event (LeDoux, 2000). After learning this association, an animal responds to the previously neutral CS with a set of defensive behavioral responses normally appropriate for the US, including freezing, increased heart rate, and startle. Thus, the CS becomes a signal in the animal's brain of impending danger. The CS may be unimodal, involving only a single cue or modality such as an auditory tone, light, smell, taste or touch, or it may be multimodal, involving several sensory modalities that are perceived together in a complex relationship of space and time, making up a context. Unimodal (cued) fear conditioning requires the amygdala but not the hippocampus. In contrast, multimodal (contextual) fear conditioning depends on both the hippocampus and the amygdala (Maren, 2001).

There are 13 nuclei in the rodent amygdala, interconnected by both excitatory and inhibitory cells (Pitkanen et al., 1997). We will consider the neural pathways for auditory and contextual fear conditioning because their circuitry is best understood. The lateral nucleus (LA) is the input region within the amygdala for auditory fear conditioning, as well as the site where association of information about the CS and US occurs. Sensory information about the CS (an auditory tone) reaches the lateral nucleus by way of two neural pathways, both of which

are essential for learned fear (Romanski and LeDoux, 1992). One pathway, the direct thalamo-amygdala pathway, originates in the medial geniculate nucleus (MGm) and the posterior intralaminar nucleus (PIN) of the thalamus. The second pathway, the indirect cortico-amygdala pathway, extends from the auditory thalamus to the auditory cortex (TE3 area), which relays the processed auditory information to the lateral amygdala (LeDoux, 2000; Maren, 2001). After information from these two inputs is associated with the US in the lateral nucleus, the output is distributed to other amygdaloid nuclei (LeDoux, 2000; Pitkanen et al., 1997), including the central nucleus (CeA). This nucleus projects, in turn, to areas in the brainstem that control autonomic (heart rate) and somatic (freezing) motor centers involved in the expression of fear (Maren, 2001). In contextual fear conditioning, sensory information about the context CS originates in the ventral hippocampus, mostly from the area CA1 and the subiculum. The basolateral (BLA) and accessory basal (AB) nuclei of the amygdala act as the input regions and the site of association of context with US information. Output from the lateral nucleus is processed by the CeA in a way similar to cued conditioning (see Figure 13.9A).

Anatomical tracing and lesion studies first demonstrated the importance of the lateral nucleus for fear conditioning (LeDoux, 2000; Pitkanen et al., 1997; Romanski and LeDoux, 1992). Subsequent physiological experiments showed that learning produces prolonged synaptic modifications similar to LTP in both of the inputs to the lateral nucleus: the thalamo-amygdala pathway (McKernan and Shinnick-Gallagher, 1997; Rogan et al., 1997) and the cortico-amygdala pathway (Tsvetkov et al., 2002). These synaptic modifications, which accompany behavioral expression of learned fear, are mechanistically similar to LTP induced artificially by electrical stimulation in tissue slices of the amygdala. For example, both types of changes involve a presynaptic enhancement of transmitter release (Tsvetkov et al., 2002). These findings and others provide the direct evidence that at least some of the mechanisms for LTP, studied in slices, are recruited behaviorally and are involved in the storage of the memory. These studies establish the amygdala as an excellent model system in the mammalian brain for analyzing the cellular and molecular mechanisms of memory storage (Malenka and Nicoll, 1999; Malleret et al., 2001; Martin et al., 2000; Stevens, 1998; Tang et al., 1999).

Here we will focus on the lateral nucleus of the amygdala because it is the most studied nucleus in the amygdala. In the lateral nucleus, like in the hippocampus and cortex, the excitatory drive of pyramidal cells is regulated by the inhibitory action of GABAergic interneurons. The details of these interactions have recently begun to emerge. Pyramidal cells in the LA constitute approximately 85% of the neuronal population and are spiny projection neurons that use glutamate as an excitatory neurotransmitter. Their excitation is controlled by fast-spiking interneurons, which innervate other interneurons in addition to synapsing onto principal cells (Muller et al., 2003). Inhibition of interneurons in the amygdala would be expected to result in an overall increase in the firing rate of pyramidal cells, as well as an increase in the excitability of their dendrites, possibly resulting in enhanced LTP of excitatory inputs from the thalamus and cortex. Interestingly, thalamic inputs themselves innervate both pyramidal cells and interneurons, suggesting the importance of a balance of connections between projection neurons

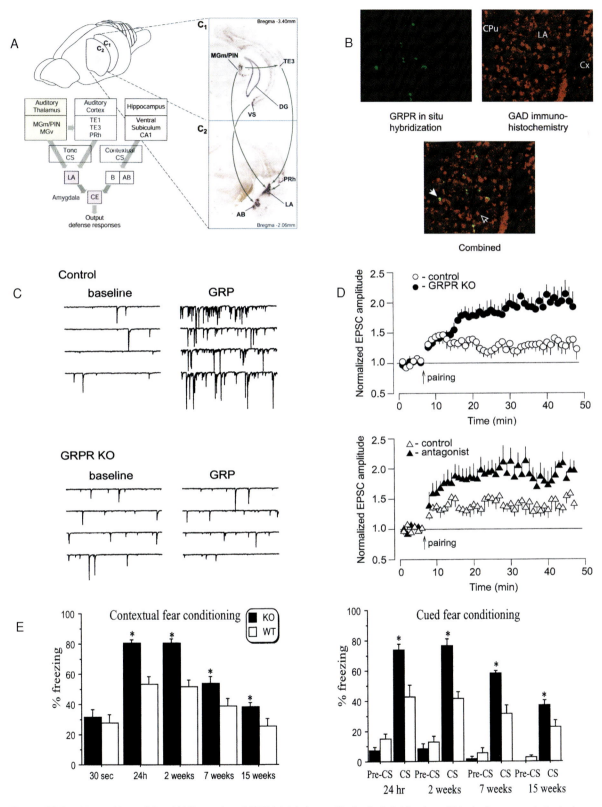

Figure 13.9   GRP and learned fear. (**A**) Expression of GRP (*right*), detected by in situ hybridization, within the fear system in the brain (*left*). (**B**) GRPR expression in a subset of amygdalar interneurons as demonstrated by co-in situ and immunohistochemistry for GRPR and the interneuron marker GAD. (**C**) Application of GRP onto amygdala slices increases spontaneous IPSCs recorded in principal neurons, an effect that is absent in GRPR KO mice. (**D**) LTP induced by pairing principal cell depolarization with stimulation of the external capsule input to the amygdala in slices from GRPR KO mice and control mice, as well as control slices treated with a GRPR antagonist. (**E**) Enhanced memory for learned fear in GRPR KO mice after contextual and cued fear conditioning over 15 weeks. (Reprinted from Shumyatsky et al., 2002.)

and interneurons (Woodson et al., 2000). The afferent projection neurons from the thalamus that innervate interneurons may act to suppress background neural noise, thus enhancing the response of LA principal cells to an incoming auditory CS. Finally, interconnections between interneurons in the amygdala may combine to form extensive inhibitory networks. These conditions together would permit the generation of synchronous oscillations in excitatory pyramidal cells, which may in turn be important for synaptic plasticity.

13.4.2 Genes and the Fear System

In contrast to the cellular physiological information that is becoming available, the specific molecular machinery that underlies synaptic plasticity in amygdala-dependent learned fear is largely unknown. Some of the common components important for synaptic plasticity and learned behavior in the hippocampus and other brain regions have also been implicated in learned fear. These include the NMDA subtype of glutamate receptors (Gewirtz and Davis, 1997; Maren et al., 1996; Miserendino et al., 1990; Tsvetkov et al., 2002), L-type Ca^{2+} channels (Tsvetkov et al., 2002; Weisskopf et al., 1999), Ras-GRF and PI-3 kinase (Brambilla et al., 1997; Lin et al., 2001), protein kinase A (Huang and Kandel, 1998; Huang et al., 2000), protein kinase C (PKC) (Weeber et al., 2000), and mitogen-activated protein kinase (MAPK) (Huang et al., 2000; Lu et al., 2001). However, these genes are widely expressed throughout the brain. An intriguing possibility is that other genes may exist which are more specifically expressed by fear-relevant circuits or parts thereof and whose functions more directly relate to learned fear.

In an attempt to tackle this question, Zirlinger et al. (2001) used microarray technology to identify genes enriched in the amygdala. In situ hybridization showed that several of these genes could be differentially localized to specific subnuclei in the amygdala. Nonetheless, these genes were also expressed elsewhere in the brain or body. Together with the results of Shumyatsky et al. (2002, described below), these studies provide molecular evidence for similarities between genetic and anatomical definitions of the amygdala and its many divisions.

Using cDNA libraries produced from single cells in the lateral amygdala, Shumyatsky et al. (2002) searched for genes whose expression was enriched in the lateral nucleus of the amygdala, resulting in the identification of the gastrin-releasing peptide (GRP). In the brain, GRP is expressed in the LA and AB and in regions relevant for both auditory and context CS processing in Pavlovian fear conditioning that send projections to these nuclei (LeDoux, 2000) (see Figure 13.9A). GRP is a 29-amino-acid mammalian homologue of the amphibian peptide bombesin (Anastasi et al., 1971; Kroog et al., 1995) and serves as a cotransmitter at some synapses. The receptor for GRP (GRPR) is a seven-transmembrane domain receptor that is coupled to $G_{\alpha q}$ protein. Downstream targets include protein kinase C and phospholipase C (Hellmich et al., 1997, 1999). GRPR activation by GRP binding ultimately leads to intracellular release of Ca^{2+} and eventually to the activation of the MAPK pathway (Sharif et al., 1997). Shumyatsky et al. found that GRP is expressed in a group of glutamatergic principal neurons enriched in zinc. Interestingly, zinc-containing glutamatergic neurons constitute a network that includes the lateral nucleus of the amygdala and

other components of the limbic system (Frederickson et al., 2000). The authors next found that, within the amygdala, the receptor for GRP, GRPR, is expressed in a subpopulation of GABAergic interneurons in the lateral nucleus. They showed that interneurons expressing GRPR constitute only a small fraction (10%) of all GABAergic interneurons in the LA (see Figure 13.9B).

13.4.3 Excitatory–Inhibitory Balance in the Amygdala, LTP, and Learned Fear

Shumyatsky et al. found that GRPR activation significantly enhances the level of tonic GABA-mediated inhibition in the lateral nucleus. Indeed, whole-cell recordings of pyramidal neurons showed that in wild-type mouse slices, adding a GRPR agonist increases the frequency of spontaneous inhibitory postsynaptic currents (sIPSCs, see Figure 13.9C). An antagonist of GRPR blocked this effect, as did picrotoxin, a blocker of $GABA_A$ receptors. Tetrodotoxin, a blocker of sodium channels, also blocked the GRP-induced increase in frequency of sIPSCs, indicating that the observed effect was coming from increased firing of action potentials in interneurons. When the authors used slices from mice that lack GRPR, application of a GRPR agonist did not evoke an increase in frequency of sIPSCs in principal cells (see Figure 13.9C), confirming that interneurons of GRPR knockout mice indeed lack GRPR-mediated inhibition of pyramidal neurons. Recent pharmacological and genetic studies have shown that the establishment of GABA-dependent balance between excitatory and inhibitory functions is critical for information processing in the amygdala (Bast et al., 2001; Krezel et al., 2001; Thielen and Shekhar, 2002). During excitation, principal cells may release GRP, which, through the binding to GRPR on interneurons, leads to GABA release. This may provide a tonic, feedforward, or feedback inhibitory control of CS processing by principal cells.

Using whole-cell recordings from principal cells, Shumyatsky et al. next found that pyramidal neurons in the lateral nucleus of the amygdala in GRPR knockout animals display enhanced LTP of the cortico-amygdala pathway (see Figure 13.9D). To rule out a possible developmental or compensatory effect of loss of the GRPR gene on the LTP phenotype, the authors showed that a GRPR antagonist led to an enhancement of LTP in wild-type slices similar to that observed in GRPR knockout animals. In agreement with this study, previous work has demonstrated that modulation of the level of GABA-mediated inhibition of the principal cells by interneurons in the amygdala may determine how easily LTP is induced at synapses in the amygdala (Krezel et al., 2001; Rammes et al., 2000).

Because LTP in the amygdala correlates with the memory of fear and use the same mechanisms (Tsvetkov et al., 2002), the authors next went on to analyze the memory of learned fear in GRPR knockout mice. Testing of knockout animals in both cued and contextual fear conditioning revealed enhanced memory for fear in both tasks when animals were tested for long-term memory at 24 hours after training (see Figure 13.9E). Cued fear conditioning depends on the amygdala but not the hippocampus. Even when tested up to 15 weeks after training, GRPR knockout mice displayed enhanced memory for fear. Interestingly, the authors showed that GRPR knockout mice had normal memory for fear when tested for short- and intermediate-term memory at 30 minutes and 4 hours.

To determine the specificity of the phenotype, the authors examined spatial learning in the water maze task, in which they found that the knockout animals performed similarly to controls. Knockout mice also display no alteration in innate fear or anxiety, assessed using naturally anxiogenic tasks like the elevated plus maze and light-dark box test, further demonstrating the specificity of this manipulation within the fear system in the brain. Since both learned fear and other anxiety phenotypes are GABA-dependent, this also suggests that principal cells and interneurons may form independent circuits within the amygdala for the processing of fear and anxiety.

13.5 CORTICAL EXCITATORY–INHIBITORY BALANCE AND VISUAL PLASTICITY AND FUNCTION

Experience-dependent plasticity during development is critical for the correct wiring and function of many cortical areas. Efforts at understanding how these processes happen have increasingly pointed to mechanistic overlap with cellular and molecular processes involved in adult plasticity in structures like the hippocampus and amygdala. A paradigmatic example of cortical developmental plasticity is provided by study of primary visual cortex. Findings here have guided much of the thought about the development of other neocortical areas. During in utero development, the cortical distribution of each eye's input begins to segregate into ocular dominance columns, an organizational scheme that does not fully mature until a few weeks after birth (Katz and Crowley, 2002). The driving force behind this segregation is thought to be the result of correlated, spontaneously generated "retinal waves" of activity. Competition established for thalamic and cortical targets between inputs from either eye eventually results in one source innervating a particular cortical column and excluding input from the either eye (Katz and Crowley, 2002).

The interocular competition continues after eye opening and is driven by visual stimulation. At some point, several weeks after birth in rodents and longer in humans, the ocular dominance columns stabilize, and the ongoing competition between eyes for representational territory in the cortex ceases. The period of time when these columns are still susceptible to alterations in input is referred to as the "critical period." Critical periods have been described in a large number of organisms and sensory modalities (Berardi et al., 2000; Katz and Crowley, 2002). Very little, however, is understood about (a) the cellular and molecular basis for critical period length and (b) the capacity of the cortex for plasticity, such as that illustrated by the expansion of one eye's representation after a period of monocular deprivation (MD). Critical periods also relate to the maturation of sensory functions, such as visual acuity; the end of the critical period for MD roughly coincides with the complete development of visual acuity (Berardi et al., 2000). Until recently, the association between critical period plasticity and sensory maturation has been merely correlative, in the absence of manipulations that have cleanly shown an effect on both functions.

An attractive model for the plasticity underlying the critical period and maturation of visual acuity is that eye-specific inputs compete for a limiting amount of target-derived neurotrophins, which serve as "survival" signals that maintain the recipient inputs over other inputs (Thoenen, 1995). Neurotrophins have also been shown to have an effect on the regulation of GABAergic interneurons, whose inhibitory activity may "sculpting" the activity and capacity for plasticity of excitatory neurons in ocular dominance columns. Brain-derived neurotrophic factor (BDNF) has been shown to regulate the strength of synaptic inhibition (Rutherford et al., 1997, 1998) as well as to enhance interneuron development in culture (Collazo et al., 1992; Ip et al., 1993; Marty et al., 1996; Nawa et al., 1994; Widmer and Hefti, 1994). A study altering the function of the GABAergic system has supported a role for interneurons in critical period plasticity (Hensch et al., 1998). Hensch et al. studied mice lacking the 65-kD isoform of the GABA-synthesizing enzyme glutamic acid decarboxylase (GAD65), which is found mostly in synaptic terminals. Mice lacking the other isoform (GAD67), which is mostly in cell somas and dendrites, die at birth. The authors found that GAD65 knockout mice, after a brief period of MD, failed to shift their visual cortical responsiveness toward input from the open eye. They then confirmed that this effect was due to blunted GABAergic interneuron function by rescuing the capacity to shift ocular dominance after MD by intracerebral infusion of diazepam, a benzodiazepine that enhances currents through activated GABA receptors.

Drawing on the these findings, Huang et al. (1999) sought to determine whether the visual cortical critical period, sensory maturation, and interneuron–excitatory neuronal balance can be shifted by genetically supplying excess BDNF. They used the CamKIIα promoter to express excess BDNF, which they found to be most dramatically increased by the third postnatal week. If BDNF availability regulates experience-dependent plasticity, and because BDNF levels normally rise to a peak in the visual cortex by two weeks after birth, then the extra BDNF should enhance cortical development. The authors found a more rapid development of interneurons within the visual cortex, consistent with the effects of BDNF on cultured interneurons (see Figure 13.10A). Accelerated development of interneurons led to larger IPSCs in cortical slices, with EPSC amplitudes remaining unaffected (Figure 13.10B).

If the activity of interneurons in the cortex regulates the capacity of the cortex for experience-dependent plasticity, then visual cortical LTP should be dampened at an earlier age. Indeed, Huang et al. were able to elicit cortical LTP at young ages, but were no longer able to do so in slices from mice by the 34th postnatal day. This LTP critical period has been shown to depend on the maturation of intracortical inhibitory circuitry (Kirkwood and Bear, 1994). In BDNF transgenic animals, LTP amplitudes decline one week earlier than in controls (see Figure 13.10C). The decreased LTP in transgenic animals was rescued by application of picrotoxin, which blocks GABA-A receptors, and would be expected to dampen the effect of interneurons on the induction and maintenance of LTP.

Huang et al. next explored the effects of shortening the critical period for intracortical synaptic plasticity on the critical period for response to MD, as well as the maturation of visual acuity in their mice. Visual-evoked potentials (VEPs) recorded in the binocular visual cortex are normally larger for stimulation of the contralateral eye than for the ipsilateral eye, reflected as a greater VEP ratio. This bias is eliminated after MD. However, MD after

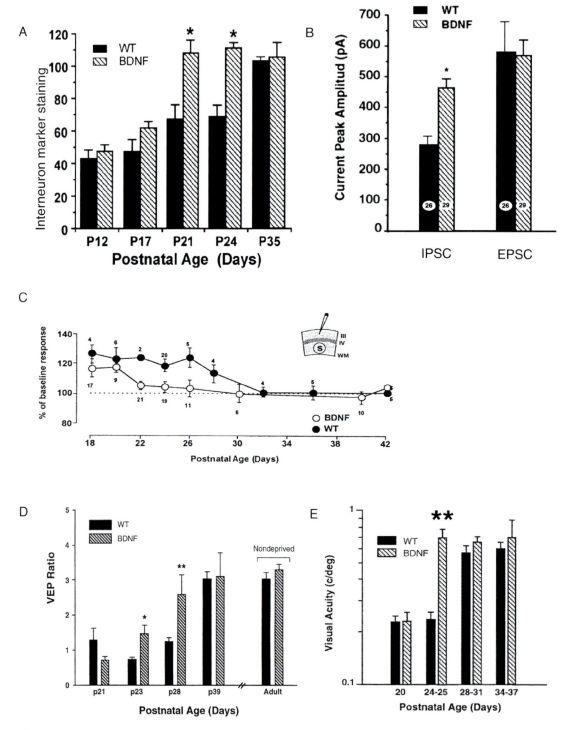

Figure 13.10   Cortical BDNF regulates interneuron development, ocular dominance plasticity, and visual acuity maturation. (**A**) Quantitation of interneuron development by detection of an immunohistochemical marker for interneurons in BDNF transgenic and control mice. (**B**) Increased amplitudes of IPSCs in BDNF transgenic mice, but no change in EPSC amplitudes recorded in cortical slices from 23- to 26-day-old animals. (**C**) Amplitudes of LTP elicited in cortical slices from BDNF transgenic animals between 18 and 24 days of age. (**D**) Visual evoked potential (VEP) ratios reflecting capacity for ocular dominance plasticity (low VEP) during and after the critical period for MD interocular representational shift (high VEP after critical period). (**E**) Precocious maturation of visual acuity in BDNF transgenic animals. (Reprinted from Huang et al., 1999.)

the critical period has ended does not result in a lowering of the VEP ratio. As shown in Figure 13.10D, Huang et al. found that the capacity to shift ocular dominance to the ipsilateral eye (a decrease in VEP ratio) was equally possible in transgenics and controls after MD starting at 21 days postnatal, but by 28 days was no longer possible in transgenics, while controls were still capable of shifting. This shortened critical period for ocular dominance coincides with the precocious ending of the LTP critical period and the point of enhanced maturation of interneurons in the BDNF transgenic animals.

If mechanisms underlying critical period plasticity relate to those underlying sensory function maturation, then a similar precocious development of visual acuity would be expected in transgenic animals. Indeed, Huang et al. found that these animals completed maturation of their visual acuity by 25 days postnatal, while wild-type animals did not do so until 31 days (see Figure 13.10E). This study, using a simple transgenic scheme, was able to provide the most compelling evidence at that time of how the development of circuitry within the visual cortex leads to higher-level functions like ocular dominance column critical periods and visual acuity maturation. In addition, their study lent weight to the neurotrophin hypothesis of cortical circuit development, demonstrated the importance of intracortical inhibitory processes, and mechanistically tied the processes underlying critical period plasticity with those underlying sensory function maturation. Similar experiments employing the tTA–tetO transgenic system may be able to extend these findings by identifying, for example, the period in which excess BDNF can elicit the effects observed, by providing temporal regulation of BDNF expression.

13.6 FUTURE DIRECTIONS

The studies described in this chapter explore the contribution of different subregions of the hippocampus to memory, as well as the contribution of an excitatory–inhibitory balance in the amygdala and visual cortex to learned fear and visual sensory function, respectively. Together they represent the clearest inroads genetic approaches have made so far into the understanding of how brain circuitry orchestrates cognitive functions. As these approaches build interest, we can no doubt expect accelerated progress. To ensure success in this area, several technical issues must receive continued attention.

First, there are currently very few promoters driving expression of transgenes in restricted areas of the nervous system. Better molecular parcellation of brain regions—for example, by molecularly distinguishing between nuclei or subnuclei in the amygdala or between different cortical areas—will provide reagents for future transgenic studies. Likewise, finding promoters that differentially target various neuronal subtypes will further define the effects of genetic manipulations. New promoters will advance both transgene systems like the tTA–tetO system and conditional knockout schemes that rely on the specificity of Cre recombinase expression. By understanding how a combination of specific transcriptional elements can drive transgene expression in specific cellular populations, synthetic promoters may be designed to result in expression patterns not found in endogenous promoters. An important question is whether genes can be found, like GRP, that express throughout a distributed functional system in the brain. If so, gene expression may identify brain regions that are interconnected for a particular common function and may allow manipulations to be made throughout the whole system, rather than in single components.

Second, much greater effort must go into improving temporal control of the effects of genetic manipulations. As mentioned near the beginning of this chapter, molecular and systems-level compensations can happen within days of when a transgene turns on or when a gene is deleted. The great strength of genetic methods is the specificity of their effect with respect to cell type, brain region, and molecular system. Pharmacological methods have their strength in the rapidity of drug effect. Coupling genetic manipulations with pharmacologic regulation of their activities may be one avenue to capitalize on the strengths of both approaches, while minimizing their weaknesses. An example is a transgene system where the protein expressed is normally sequestered in the cytoplasm and is only translocated to the nucleus in the presence of a specific pharmacological agent. Such a system was used in a recent study employing a forebrain-expressed dominant-negative form of the transcription factor CREB, similar to that described earlier in this chapter (Kida et al., 2002). While this scheme is useful only for transgenes that exert their effects in the nuclear compartment, a similar approach could be employed for transgenes with roles in the cytoplasm and not the nucleus.

Lastly, genetic tracers of circuitry or neuronal activity represent a future direction that has only very recently begun to be explored. One example is a study by Zou et al. (2001), who used a genetically expressed tracer system where an olfactory-specific promoter was used to drive expression of barley lectin (Horowitz et al., 1999). Barley lectin is anterogradely and trans-synaptically transported within expressing neurons. With this marker, these authors were able to show the stereotyped pattern of projection of olfactory neurons within the olfactory bulb and into the olfactory cortex. Another study used a fluorescently detected protein tracer that can be passed trans-synaptically (Maskos et al., 2002). It may be possible to design tracers that will only pass through activated synapses, allowing determination of dynamic changes in synaptic communication using genetic tools.

In sum, many of the weaknesses of lesion and pharmacological methods of circuit analysis can be overcome by reverse genetic techniques, allowing many parallel analyses in the absence of tissue damage and with a high degree of molecular target specificity. While many of the successes of genetic methodologies have come with relative serendipity, these tools are attracting broader interest. In addition, the possibility of coupling the strengths of pharmacological and genetic methodologies promises an even greater degree of flexibility in the kinds of scientific problems that may be addressed. Were this chapter written five years from now, no doubt many of the topics discussed will seem grossly out of date. We predict that in the coming years, highly specific manipulations will be published that will explore subtle and diverse aspects of how particular circuitry contributes to particular aspects of behavior, and we are very excited at these prospects.

References

Amaral, D. G., and Witter, M. P. (1989). The three-dimensional organization of the hippocampal formation: A review of anatomical data. *Neuroscience* **31**, 571–591.

Amaral, G., and Witter, M. P. (1995). Hippocampal Formation, In *The Rat Nervous System*, 2nd edition, (G. Paxinos, ed.). Academic Press, San Diego, CA. pp. 443–493.

Anagnostaras, S. G., Maren, S., and Fanselow, M. S. (1999). Temporally graded retrograde amnesia of contextual fear after hippocampal damage in rats: Within-subjects examination. *J. Neurosci.* **19**, 1106–1114.

Anastasi, A., Erspamer, V., and Bucci, M. (1971). Isolation and structure of bombesin and alytesin, 2 analogous active peptides from the skin of the European amphibians Bombina and Alytes. *Experientia* **27**, 166–167.

Bach, M. E., Hawkins, R. D., Osman, M., Kandel, E. R., and Mayford, M. (1995). Impairment of spatial but not contextual memory in CaMKII mutant mice with a selective loss of hippocampal LTP in the range of the theta frequency. *Cell* **81**, 905–915.

Barco, A., Pittenger, C., and Kandel, E. R. (2003). CREB, memory enhancement and the treatment of memory disorders: Promises, pitfalls and prospects. *Expert Opin. Ther. Targets* **7**, 101–114.

Bartsch, D., Casadio, A., Karl, K. A., Serodio, P., and Kandel, E. R. (1998). CREB1 encodes a nuclear activator, a repressor, and a cytoplasmic modulator that form a regulatory unit critical for long-term facilitation. *Cell* **95**, 211–223.

Bast, T., Zhang, W. N., and Feldon, J. (2001). The ventral hippocampus and fear conditioning in rats. Different anterograde amnesias of fear after tetrodotoxin inactivation and infusion of the GABA(A) agonist muscimol. *Exp. Brain Res.* **139**, 39–52.

Berardi, N., Pizzorusso, T., and Maffei, L. (2000). Critical periods during sensory development. *Curr. Opin. Neurobiol.* **10**, 138–145.

Bi, G., and Poo, M. (2001). Synaptic modification by correlated activity: Hebb's postulate revisited. *Annu. Rev. Neurosci.* **24**, 139–166.

Bliss, T. V. P., and Lømo, T. (1973). Long-lasting potentiation of synaptic transmission in the dentate area of the unanaesthetized rabbit following stimulation of the perforant path. *J. Physiol. (London)* **232**, 331–356.

Bourtchuladze, R., Frenguelli, B., Blendy, J., Cioffi, D., Schutz, G., and Silva, A. J. (1994). Deficient long-term memory in mice with a targeted mutation of the cAMP-responsive element-binding protein. *Cell* **79**, 59–68.

Brambilla, R., Gnesutta, N., Minichiello, L., White, G., Roylance, A. J., Herron, C. E., Ramsey, M., Wolfer, D. P., Cestari, V., Rossi-Arnaud, C., Grant, S. G., Chapman, P. F., Lipp, H. P., Sturani, E., and Klein, R. (1997). A role for the Ras signalling pathway in synaptic transmission and long-term memory. *Nature (London)* **390**, 281–286.

Collazo, D., Takahashi, H., and McKay, R. D. (1992). Cellular targets and trophic functions of neurotrophin-3 in the developing rat hippocampus. *Neuron* **9**, 643–656.

Davis, M., and Whalen, P. J. (2001). The amygdala: Vigilance and emotion. *Mol. Psychiatry* **6**, 13–34.

Drews, J. (2000). Drug discovery: A historical perspective. *Science* **287**, 1960–1964.

Eichenbaum, H. (1996). Is the rodent hippocampus just for 'place'? *Curr. Opin. Neurobiol.* **6**, 187–195.

Forrest, D., Yuzaki, M., Soares, H. D., Ng, L., Luk, D. C., Sheng, M., Stewart, C. L., Morgan, J. I., Connor, J. A., and Curran, T. (1994). Targeted disruption of NMDA receptor 1 gene abolishes NMDA response and results in neonatal death. *Neuron* **13**, 325–338.

Frankland, P. W., Cestari, V., Filipkowski, R. K., McDonald, R. J., and Silva, A. J. (1998). The dorsal hippocampus is essential for context discrimination but not for contextual conditioning. *Behav. Neurosci.* **112**, 863–874.

Frederickson, C. J., Suh, S. W., Silva, D., and Thompson, R. B. (2000). Importance of zinc in the central nervous system: The zinc-containing neuron. *J. Nutr.* **130**, 1471S–1483S.

Genoux, D., Haditsch, U., Knobloch, M., Michalon, A., Storm, D., and Mansuy, I. M. (2002). Protein phosphatase 1 is a molecular constraint on learning and memory. *Nature (London)* **418**, 970–975.

Gerlai, R. (2001). Behavioral tests of hippocampal function: Simple paradigms complex problems. *Behav. Brain Res.* **125**, 269–277.

Gewirtz, J. C., and Davis, M. (1997). Second-order fear conditioning prevented by blocking NMDA receptors in amygdala. *Nature (London)* **388**, 471–474.

Gossen, M., and Bujard, H. (1992). Tight control of gene expression in mammalian cells by tetracycline-responsive promoters. *Proc. Natl. Acad. Sci. U. S. A.* **89**, 5547–5551.

Gossen, M., Freundlieb, S., Bender, G., Muller, G., Hillen, W., and Bujard, H. (1995). Transcriptional activation by tetracyclines in mammalian cells. *Science* **268**, 1766–1769.

Harris, E. W., and Cotman, C. W. (1986). Long-term potentiation of guinea pig mossy fiber responses is not blocked by *N*-methyl D-aspartate antagonists. *Neurosci. Lett.* **70**, 132–137.

Hebb, D. O. (1949). *The Organization of Behavior*. Wiley, New York.

Hellmich, M. R., Battey, J. F., and Northup, J. K. (1997). Selective reconstitution of gastrin-releasing peptide receptor with G alpha q. *Proc. Natl. Acad. Sci. U. S. A.* **94**, 751–756.

Hellmich, M. R., Ives, K. L., Udupi, V., Soloff, M. S., Greeley, G. H., Jr., Christensen, B. N., and Townsend, C. M., Jr. (1999). Multiple protein kinase pathways are involved in gastrin-releasing peptide receptor-regulated secretion. *J. Biol. Chem.* **274**, 23901–23909.

Hensch, T. K., Fagiolini, M., Mataga, N., Stryker, M. P., Baekkeskov, S., and Kash, S. F. (1998). Local GABA circuit control of experience-dependent plasticity in developing visual cortex. *Science* **282**, 1504–1508.

Horowitz, L. F., Montmayeur, J. P., Echelard, Y., and Buck, L. B. (1999). A genetic approach to trace neural circuits. *Proc. Natl. Acad. Sci. U. S. A.* **96**, 3194–3199.

Huang, Y. Y., and Kandel, E. R. (1998). Postsynaptic induction and PKA-dependent expression of LTP in the lateral amygdala. *Neuron* **21**, 169–178.

Huang, Y. Y., Martin, K. C., and Kandel, E. R. (2000). Both protein kinase A and mitogen-activated protein kinase are required in the amygdala for the macromolecular synthesis-dependent late phase of long-term potentiation. *J. Neurosci.* **20**, 6317–6325.

Huang, Z. J., Kirkwood, A., Pizzorusso, T., Porciatti, V., Morales, B., Bear, M. F., Maffei, L., and Tonegawa, S. (1999). BDNF regulates the maturation of inhibition and the critical period of plasticity in mouse visual cortex. *Cell* **98**, 739–755.

Insausti, R., Amaral, D. G., and Cowan, W. M. (1987). The entorhinal cortex of the monkey: II. Cortical afferents. *J. Comp. Neurol.* **264**, 356–395.

Ip, N. Y., Li, Y., Yancopoulos, G. D., and Lindsay, R. M. (1993). Cultured hippocampal neurons show responses to BDNF, NT-3, and NT-4, but not NGF. *J. Neurosci.* **13**, 3394–3405.

Ishizuka, N., Weber, J., and Amaral, D. G. (1990). Organization of intrahippocampal projections originating from CA3 pyramidal cells in the rat. *J. Comp. Neurol.* **295**, 580–623.

Jung, M. W., Wiener, S. I., and McNaughton, B. L. (1994). Comparison of spatial firing characteristics of units in dorsal and ventral hippocampus of the rat. *J. Neurosci.* **14**, 7347–7356.

Kapp, B. S., Whalen, P. G., Supple, W. F., and Pascoe, J. P. (1992). Amygdaloid contributions to conditioned arousal and sensory information processing. In *The Amygdala: Neurobiological Aspects of Emotion, Memory, and Mental Dysfunction* (J. P. Aggleton, ed.), pp. 229–254. Wiley, New York.

Katz, L. C., and Crowley, J. C. (2002). Development of cortical circuits: Lessons from ocular dominance columns. *Nat. Rev. Neurosci.* **3**, 34–42.

Kida, S., Josselyn, S. A., de Ortiz, S. P., Kogan, J. H., Chevere, I., Masushige, S., and Silva, A. J. (2002). CREB required for the stability of new and reactivated fear memories. *Nat. Neurosci.* **5**, 348–355.

Kim, J. J., and Fanselow, M. S. (1992). Modality-specific retrograde amnesia of fear. *Science* **256**, 675–677.

Kirkwood, A., and Bear, M. F. (1994). Hebbian synapses in visual cortex. *J. Neurosci.* **14**, 1634–1645.

Kjelstrup, K. G., Tuvnes, F. A., Steffenach, H. A., Murison, R., Moser, E. I., and Moser, M. B. (2002). Reduced fear expression after lesions of the ventral hippocampus. *Proc. Natl. Acad. Sci. U. S. A.* **99**, 10825–10830.

Krezel, W., Dupont, S., Krust, A., Chambon, P., and Chapman, P. F. (2001). Increased anxiety and synaptic plasticity in estrogen receptor beta-deficient mice. *Proc. Natl. Acad. Sci. U. S. A.* **98**, 12278–12282.

Kroog, G. S., Jensen, R. T., and Battey, J. F. (1995). Mammalian bombesin receptors. *Med. Res. Rev.* **15**, 389–417.

LeDoux, J. E. (2000). Emotion circuits in the brain. *Annu. Rev. Neurosci.* **23**, 155–184.

Lin, C. H., Yeh, S. H., Lu, K. T., Leu, T. H., Chang, W. C., and Gean, P. W. (2001). A role for the PI-3 kinase signaling pathway in fear conditioning and synaptic plasticity in the amygdala. *Neuron* **31**, 841–851.

Lu, K. T., Walker, D. L., and Davis, M. (2001). Mitogen-activated protein kinase cascade in the basolateral nucleus of amygdala is involved in extinction of fear-potentiated startle. *J. Neurosci.* **21**, RC162.

Malenka, R. C., and Nicoll, R. A. (1999). Long-term potentiation—a decade of progress? *Science* **285**, 1870–1874.

Malleret, G., Haditsch, U., Genoux, D., Jones, M. W., Bliss, T. V., Vanhoose, A. M., Weitlauf, C., Kandel, E. R., Winder, D. G., and Mansuy, I. M. (2001). Inducible and reversible enhancement of learning, memory, and long-term potentiation by genetic inhibition of calcineurin. *Cell* **104**, 675–686.

Mansuy, I. M., Winder, D. G., Moallem, T. M., Osman, M., Mayford, M., Hawkins, R. D., and Kandel, E. R. (1998). Inducible and reversible gene expression with the rtTA system for the study of memory. *Neuron* **21**, 257–265.

Maren, S. (2001). Neurobiology of Pavlovian fear conditioning. *Annu. Rev. Neurosci.* **24**, 897–931.

Maren, S., and Fanselow, M. S. (1996). The amygdala and fear conditioning: Has the nut been cracked? *Neuron* **16**, 237–240.

Maren, S., Aharonov, G., Stote, D. L., and Fanselow, M. S. (1996). N-methyl-D-aspartate receptors in the basolateral amygdala are required for both acquisition and expression of conditional fear in rats. *Behav. Neurosci.* **110**, 1365–1374.

Maren, S., Aharonov, G., and Fanselow, M. S. (1997). Neurotoxic lesions of the dorsal hippocampus and Pavlovian fear conditioning in rats. *Behav. Brain. Res.* **88**, 261–274.

Martin, S. J., Grimwood, P. D., and Morris, R. G. (2000). Synaptic plasticity and memory: An evaluation of the hypothesis. *Annu. Rev. Neurosci.* **23**, 649–711.

Marty, S., Carroll, P., Cellerino, A., Castren, E., Staiger, V., Thoenen, H., and Lindholm, D. (1996). Brain-derived neurotrophic factor promotes the differentiation of various hippocampal nonpyramidal neurons, including Cajal–Retzius cells, in organotypic slice cultures. *J. Neurosci.* **16**, 675–687.

Maskos, U., Kissa, K., St. Cloment, C., and Brulet, P. (2002). Retrograde trans-synaptic transfer of green fluorescent protein allows the genetic mapping of neuronal circuits in transgenic mice. *Proc. Natl. Acad. Sci. U. S. A.* **99**, 10120–10125.

Mayford, M., Wang, J., Kandel, E. R., and O'Dell, T. J. (1995). CaMKII regulates the frequency-response function of hippocampal synapses for the production of both LTD and LTP. *Cell* **81**, 891–904.

Mayford, M., Bach, M. E., Huang, Y. Y., Wang, L., Hawkins, R. D., and Kandel, E. R. (1996). Control of memory formation through regulated expression of a CaMKII transgene. *Science* **274**, 1678–1683.

McHugh, T. J., Blum, K. I., Tsien, J. Z., Tonegawa, S., and Wilson, M. A. (1996). Impaired hippocampal representation of space in CA1-specific NMDAR1 knockout mice. *Cell* **87**, 1339–1349.

McKernan, M. G., and Shinnick-Gallagher, P. (1997). Fear conditioning induces a lasting potentiation of synaptic currents in vitro. *Nature (London)* **390**, 607–611.

Miserendino, M. J., Sananes, C. B., Melia, K. R., and Davis, M. (1990). Blocking of acquisition but not expression of conditioned fear-potentiated startle by NMDA antagonists in the amygdala. *Nature (London)* **345**, 716–718.

Morris, R. G., Anderson, E., Lynch, G. S., and Baudry, M. (1986). Selective impairment of learning and blockade of long-term potentiation by an N-methyl-D-aspartate receptor antagonist, AP5. *Nature (London)* **319**, 774–776.

Morris, R. G., Davis, S., and Butcher, S. P. (1991). Hippocampal synaptic plasticity and NMDA receptors: A role in information storage? In *Long-term Potentiation: A Debate of Current Issues* (M. Baudry and J. Davis, eds.), pp. 267–300. MIT Press, Cambridge, MA.

Moser, M. B., and Moser, E. I. (1998). Functional differentiation in the hippocampus. *Hippocampus* **8**, 608–619.

Moser, M. B., Moser, E. I., Forrest, E., Andersen, P., and Morris, R. G. (1995). Spatial learning with a minislab in the dorsal hippocampus. *Proc. Natl. Acad. Sci. U. S. A.* **92**, 9697–9701.

Muller, J. F., Mascagni, F., and McDonald, A. J. (2003). Synaptic connections of distinct interneuronal subpopulations in the rat basolateral amygdalar nucleus. *J. Comp. Neurol.* **456**, 217–236.

Nakazawa, K., Quirk, M. C., Chitwood, R. A., Watanabe, M., Yeckel, M. F., Sun, L. D., Kato, A., Carr, C. A., Johnston, D., Wilson, M. A., and Tonegawa, S. (2002). Requirement for hippocampal CA3 NMDA receptors in associative memory recall. *Science* **297**, 211–218.

Nawa, H., Pelleymounter, M. A., and Carnahan, J. (1994). Intraventricular administration of BDNF increases neuropeptide expression in newborn rat brain. *J. Neurosci.* **14**, 3751–3765.

O'Keefe, J., and Nadel, L. (1978). *The Hippocampus as a Cognitive Map.* Oxford University Press, New York.

Phillips, R. G., and LeDoux, J. E. (1992). Differential contribution of amygdala and hippocampus to cued and contextual fear conditioning. *Behav. Neurosci.* **106**, 274–285.

Pitkanen, A., Savander, V., and LeDoux, J. E. (1997). Organization of intra-amygdaloid circuitries in the rat: An emerging framework for understanding functions of the amygdala. *Trends Neurosci.* **20**, 517–523.

Pittenger, C., and Kandel, E. (1998). A genetic switch for long-term memory. *C. R. Acad. Sci. III* **321**, 91–96.

Pittenger, C., Huang, Y. Y., Paletzki, R. F., Bourtchouladze, R., Scanlin, H., Vronskaya, S., and Kandel, E. R. (2002). Reversible inhibition of CREB/ATF transcription factors in region CA1 of the dorsal hippocampus disrupts hippocampus-dependent spatial memory. *Neuron* **34**, 447–462.

Rammes, G., Steckler, T., Kresse, A., Schutz, G., Zieglgansberger, W., and Lutz, B. (2000). Synaptic plasticity in the basolateral amygdala in transgenic mice expressing dominant-negative cAMP response element-binding protein (CREB) in forebrain. *Eur. J. Neurosci.* **12**, 2534–2546.

Rampon, C., Tang, Y. P., Goodhouse, J., Shimizu, E., Kyin, M., and Tsien, J. Z. (2000). Enrichment induces structural changes and recovery from nonspatial memory deficits in CA1 NMDAR1-knockout mice. *Nat. Neurosci.* **3**, 238–244.

Rawlins, J. N. P. (1996). NMDA receptors, synaptic plasticity, and learning and memory. In *Excitatory Amino Acids and the Cerebral Cortex*, (F. Conti and T. P. Hichks, eds.), pp. 275–284. MIT Press, Cambridge, MA.

Rempel-Clower, N. L., Zola, S. M., Squire, L. R., and Amaral, D. G. (1996). Three cases of enduring memory impairment after bilateral damage limited to the hippocampal formation. *J. Neurosci.* **16**, 5233–5255.

Richmond, M. A., Yee, B. K., Pouzet, B., Veenman, L., Rawlins, J. N., Feldon, J., and Bannerman, D. M. (1999). Dissociating context and space within the hippocampus: Effects of complete, dorsal, and ventral excitotoxic hippocampal lesions on conditioned freezing and spatial learning. *Behav. Neurosci.* **113**, 1189–1203.

Rogan, M. T., Staubli, U. V., and LeDoux, J. E. (1997). Fear conditioning induces associative long-term potentiation in the amygdala. *Nature (London)* **390**, 604–607.

Romanski, L. M., and LeDoux, J. E. (1992). Equipotentiality of thalamo-amygdala and thalamo-cortico-amygdala circuits in auditory fear conditioning. *J. Neurosci.* **12**, 4501–4509.

Rutherford, L. C., DeWan, A., Lauer, H. M., and Turrigiano, G. G. (1997). Brain-derived neurotrophic factor mediates the activity-dependent regulation of inhibition in neocortical cultures. *J. Neurosci.* **17**, 4527–4535.

Rutherford, L. C., Nelson, S. B., and Turrigiano, G. G. (1998). BDNF has opposite effects on the quantal amplitude of pyramidal neuron and interneuron excitatory synapses. *Neuron* **21**, 521–530.

Scoville, W. B., and Milner, W. B. (1957). Loss of recent memory after bilateral hippocampal lesions. *J. Neurol. Neurosurg. Psychiatry* **20**, 11–21.

Sharif, T. R., Luo, W., and Sharif, M. (1997). Functional expression of bombesin receptor in most adult and pediatric human glioblastoma cell lines; role in mitogenesis and in stimulating the mitogen-activated protein kinase pathway. *Mol. Cell. Endocrinol.* **130**, 119–130.

Shumyatsky, G. P., Tsvetkov, E., Malleret, G., Vronskaya, S., Hatton, M., Hampton, L., Battey, J. F., Dulac, C., Kandel, E. R., and Bolshakov, V. Y. (2002). Identification of a signaling network in lateral nucleus of amygdala important for inhibiting memory specifically related to learned fear. *Cell* **111**, 905–918.

Sik, A., Hajos, N., Gulacsi, A., Mody, I., and Freund, T. F. (1998). The absence of a major Ca^{2+} signaling pathway in GABAergic neurons of the hippocampus. *Proc. Natl. Acad. Sci. U. S. A.* **95**, 3245–3250.

Stevens, C. F. (1998). A million dollar question: Does LTP = memory? *Neuron* **20**, 1–2.

Suzuki, W. A., and Amaral, D. G. (1994). Perirhinal and parahippocampal cortices of the macaque monkey: Cortical afferents. *J. Comp. Neurol.* **350**, 497–533.

Tang, Y. P., Shimizu, E., Dube, G. R., Rampon, C., Kerchner, G. A., Zhuo, M., Liu, G., and Tsien, J. Z. (1999). Genetic enhancement of learning and memory in mice. *Nature (London)* **401**, 63–69.

Thielen, S. K., and Shekhar, A. (2002). Amygdala priming results in conditioned place avoidance. *Pharmacol. Biochem. Behav.* **71**, 401–406.

Thoenen, H. (1995). Neurotrophins and neuronal plasticity. *Science* **270**, 593–598.

Treves, A., and Rolls, E. T. (1994). Computational analysis of the role of the hippocampus in memory. *Hippocampus* **4**, 374–391.

Tsien, J. Z., Chen, D. F., Gerber, D., Tom, C., Mercer, E. H., Anderson, D. J., Mayford, M., Kandel, E. R., and Tonegawa, S. (1996a). Subregion- and cell type-restricted gene knockout in mouse brain. *Cell* **87**, 1317–1326.

Tsien, J. Z., Huerta, P. T., and Tonegawa, S. (1996b). The essential role of hippocampal CA1 NMDA receptor-dependent synaptic plasticity in spatial memory. *Cell* **87**, 1327–1338.

Tsvetkov, E., Carlezon, W. A., Benes, F. M., Kandel, E. R., and Bolshakov, V. Y. (2002). Fear conditioning occludes LTP-induced presynaptic enhancement of synaptic transmission in the cortical pathway to the lateral amygdala. *Neuron* **34**, 289–300.

Weeber, E. J., Atkins, C. M., Selcher, J. C., Varga, A. W., Mirnikjoo, B., Paylor, R., Leitges, M., and Sweatt, J. D. (2000). A role for the beta isoform of protein kinase C in fear conditioning. *J. Neurosci.* **20**, 5906–5914.

Weisskopf, M. G., Bauer, E. P., and LeDoux, J. E. (1999). L-type voltage-gated calcium channels mediate NMDA-independent associative long-term potentiation at thalamic input synapses to the amygdala. *J. Neurosci.* **19**, 10512–10519.

Widmer, H. R., and Hefti, F. (1994). Stimulation of GABAergic neuron differentiation by NT-4/5 in cultures of rat cerebral cortex. *Brain Res. Dev. Brain Res.* **80**, 279–284.

Williams, S., and Johnston, D. (1988). Muscarinic depression of long-term potentiation in CA3 hippocampal neurons. *Science* **242**, 84–87.

Wilson, M. A., and McNaughton, B. L. (1993). Dynamics of the hippocampal ensemble code for space. *Science* **261**, 1055–1058.

Witter, M. P., and Amaral, D. G. (1991). Entorhinal cortex of the monkey: V. Projections to the dentate gyrus, hippocampus, and subicular complex. *J. Comp. Neurol.* **307**, 437–459.

Woodson, W., Farb, C. R., and Ledoux, J. E. (2000). Afferents from the auditory thalamus synapse on inhibitory interneurons in the lateral nucleus of the amygdala. *Synapse* **38**, 124–137.

Yin, J. C., Wallach, J. S., Del Vecchio, M., Wilder, E. L., Zhou, H., Quinn, W. G., and Tully, T. (1994). Induction of a dominant negative CREB transgene specifically blocks long-term memory in Drosophila. *Cell* **79**, 49–58.

Yin, J. C., Del Vecchio, M., Zhou, H., and Tully, T. (1995). CREB as a memory modulator: Induced expression of a dCREB2 activator isoform enhances long-term memory in *Drosophila*. *Cell* **81**, 107–115.

Zalutsky, R. A., and Nicoll, R. A. (1990). Comparison of two forms of long-term potentiation in single hippocampal neurons. *Science* **248**, 1619–1624.

Zirlinger, M., Kreiman, G., and Anderson, D. J. (2001). Amygdala-enriched genes identified by microarray technology are restricted to specific amygdaloid subnuclei. *Proc. Natl. Acad. Sci. U. S. A.* **98**, 5270–5275.

Zola-Morgan, S., Squire, L. R., and Amaral, D. G. (1986). Human amnesia and the medial temporal region: Enduring memory impairment following a bilateral lesion limited to field CA1 of the hippocampus. *J. Neurosci.* **6**, 2950–2967.

Zou, Z., Horowitz, L. F., Montmayeur, J. P., Snapper, S., and Buck, L. B. (2001). Genetic tracing reveals a stereotyped sensory map in the olfactory cortex. *Nature (London)* **414**, 173–179.

GeneWays: A System for Extracting, Analyzing, Visualizing, and Integrating Molecular Pathway Data

Andrey Rzhetsky, Ph.D., Ivan Iossifov, Tomohiro Koike, Michael Krauthammer, M.D., Ph.D., Pauline Kra, Ph.D., Mitzi Morris, Hong Yu, Ph.D., Pablo Ariel Duboue, Wubin Weng, John W. Wilbur, M.D., Ph.D., Vasileios Hatzivassiloglou, Ph.D., and Carol Friedman, Ph.D.

CONTENTS

Databasing the Brain. Edited by Stephen H. Koslow and Shankar Subramaniam
ISBN 0-471-30921-4 © 2005 John Wiley & Sons, Inc.

14.1 INTRODUCTION

Imagine a tribe of bright, but ignorant, cavepeople trying to understand the operation of a modern car by analyzing a collection of damaged cars produced by various makers. After many hours of hard manual labor, the cavepeople disassemble the cars into myriad small parts. Some are damaged, whereas some are intact. Some pairs of pieces interact with each other, whereas others do not interact. Some pieces are different in different cars, yet apparently have the same function. The leap to understanding the whole from knowing the parts requires reduction of redundant or conflicting pieces of information to a consistent consensus model that can be used for dynamics analysis. Researchers in the field of molecular biology of the post-genome era are in a situation similar to that of the junkyard cavepeople, except that they are contemplating a collection of diverse pieces of cellular machinery. Complicating the researchers' horizon, the identical piece of cellular machinery may play different roles in different cells of the same organism, or even within the same cell but under different environmental conditions, analogous to a Swiss Army knife in the car glove compartment. The number of nodes in human molecular networks is measured in hundreds of thousands when all substances (genes, RNAs, proteins, and other molecules) are considered together. These numerous substances can, in turn, be present or absent in dozens of cell types in humans; clearly, the complexity is too great to yield to manual analysis. Thus, with the hope of relieving the information overload currently assaulting scientists, we are developing GeneWays, a computer system that integrates a battery of tools for automatic gathering and processing of knowledge on molecular pathways.

14.2 BACKGROUND

It would be impossible to give a complete review of the vast area spanning text analysis and molecular-interactions databases, even were we to allow this review to consume the page limit of this chapter. Nevertheless, it is important to give at least a cursory overview of recent accomplishments and key research areas related to the work described in the current chapter. These key research topics correspond to the major computational problems encountered by a researcher on her long and winding road from a collection of plain-English texts to a useful database of molecular interactions.

First, given a large database of abstracts of journal articles, such as PubMed (http://www3.ncbi.nlm.nih.gov/Entrez/index.html), the researcher needs to distinguish articles relevant to her interests from millions of nonrelevant ones. For example, she might be interested in articles having "cell cycle" in the title or abstract, and less interested in articles talking about supercolliders or fur export. This task is *document sorting*; it can be viewed as a classical task of machine learning, the problem of how to do automated classification of objects into two or more classes—"relevant" and "nonrelevant" in this case. Such classification can start with a set of examples provided with known class assignment (*supervised* machine-learning methods) or without such a training dataset (*unsupervised* learning) (Shatkay et al., 2000, Shatkay and Wilbur, 2000a,b). The implemented unsupervised approaches to document sorting include clustering of article abstracts (Iliopoulos et al., 2001), assuming that relevant and nonrelevant clusters are likely to form separate groups. Supervised methods applied to this problem include naïve Bayes classifier (Craven and Kumlien, 1999; Marcotte et al., 2001) and support-vector machines (Joachims, 1999, 2001).

Second, given a set of documents that she believes is relevant to her interests, the researcher needs to *identify terms* (Jacquemin, 2001), such as names of genes, proteins, diseases, and tissues. Term identification is a critical text preprocessing stage required by many natural-language processing engines, including GENIES (Friedman et al., 2001). Researchers attempted to attack this problem by inferring morphological rules that guide generation of a term (Fukuda et al., 1998; Tanabe and Wilbur, 2002), by using parts-of-speech tagging engines that can help the downstream applications to identify multiword noun phrases (Fukuda et al., 1998; Proux et al., 1998), grammar rules (Gaizauskas et al., 2000), combinations of rule-based and dictionary-based methods (Rindflesch et al., 1999), support-vector machines (Kazama et al., 2002), hidden Markov models (Collier et al., 2000), and naïve Bayes and decision-trees classifiers (Nobota et al., 1999). It appears that the problem of tagging biological terms is a difficult one and that we may achieve better results by combining several of these approaches. Early approaches (Fukuda et al., 1998; Proux et al., 1998) were tested on small test sets and reported excellent results where the precision and recall were over 90%. However, more recent results reported for larger test sets achieved results that ranged in the mid 70s to 80s (Gaizauskas et al., 2000; Krauthammer et al., 2000) for precision and recall.

Third, having identified terms, our researcher realizes that the problem of term identification is confounded when a term has multiple meanings (*term ambiguity*) and when multiple terms correspond to the same concept (*term synonymy*). For example, the name *p21* can refer to a gene, a protein, or a messenger RNA, depending on the sentence context. Deducing the right meaning is known as *sense disambiguation*, a problem that can be tackled with machine-learning approaches, such as those using naïve Bayes, decision trees, or inductive-learning classifiers (Hatzivassiloglou et al., 2001). The most common examples of the synonymous names are pairs of abbreviated and complete protein names (e.g., *il2* stands for *interleukin-2*; both terms often occur in biological texts). The problem of synonyms can be alleviated with automatically generated dictionaries (Liu et al., 2001; Pustejovsky et al., 2001; Yu et al., 2002). See Liu 2001 (Liu et al., 2001) for an overview of word sense disambiguation applied to the biomedical domain.

Fourth, once she has identified and disambiguated such terms, our researcher wants to do *information extraction* (remember that our researcher wants to extract information about molecular interactions). She has her choice of methods that vary in complexity and success. The first group of approaches are "correlation methods" that exploit information about co-occurrence of terms in articles or abstracts (Jenssen et al., 2001; Shatkay et al., 2000; Stapley and Benoit, 2000; Stephens et al., 2001). In a more sophisticated form, such methods are based on a hidden Markov model (Leek, 1997) that requires no dictionary of terms. Methods of the second group target information extraction via *template matching*: They identify regular expressions in the text using a term dictionary and a collection of hand-crafted patterns (Blaschke et al., 1999; Ng and Wong, 1999; Ono et al., 2001; Pustejovsky et al., 2002b). Methods of the third group explicitly use *formal grammars* that can identify nested structures in a sentence. In a nutshell, a grammar is a set of allowed symbols (usually divided into terminal words that we observe in a sentence and nonterminal, invisible symbols that serve as intermediates in the imaginary process of generating a sentence) and a set of production rules that have the capability of expressing not only regular expressions but also nested structures. Production rules are used to generate a sentence by stepwise substitution, starting with a single top-level nonterminal symbol and ending with a sentence that contains only terminal words. Given a grammar and a valid sentence, we can reconstruct the sequence of substitution events (usually expressed as a *parse tree*) leading to generation of the sentence by the grammar; this process is called *parsing*. The GeneWays project as well as two other molecular-biology-related linguistic projects (Friedman et al., 2001; Park et al., 2001; Yakushiji et al., 2001) use full grammar parsers. Since different projects have different foci and are typically tested on small datasets, it is currently impossible to tell with confidence what is the relative performance of these methods, although we expect that grammar-based methods have a higher precision. A grammar-based method, however, requires access to a dictionary listing properties of the words it recognizes (the *lexicon*) and information about allowable combinations or patterns of words that are encoded in its rules. Such information is currently supplied by manual analysis of sample texts in consultation with domain experts.

Fifth, imagine that our researcher has struggled through a multiplicity of research articles and has managed to extract a large number of statements; she now needs to store this information in a database. Therefore, she requires a knowledge model on which to build a database schema. Various knowledge models for molecular-biology data have been suggested over the past few years, and many of them have been implemented in databases; all these databases (except the GeneWays database) were created manually (see Stevens et al., 2000, for a review). The most famous projects of this type include the EcoCyc/MetaCyc knowledge base and ontology, the primary emphasis of which is bacterial pathways (Karp et al., 2002a,b), the Gene Ontology of sequence/structure conservation across eukaryotes (Ashburner et al., 2000), the Tambis Ontology (Baker et al., 1998), the Ontology of Molecular Biology (Schulze-Kremer, 1998), the ontology for conceptual modeling of biological information (Paton et al., 2000), and RiboWeb: the ontology/database of structural models of the ribosome (Altman et al., 1999). There are also various databases of molecular interactions that have an implicit ontology: KEGG, LIGAND (Goto et al., 2000; Kanehisa and Goto, 2000) (databases of diverse molecular interactions and protein–ligand interactions, respectively); BIND (Salama et al., 2001), DIP (Xenarios et al., 2002), and MINT (Zanzoni et al., 2002) (three databases of protein–protein interactions); BindingDB (Chen et al., 2001) (a knowledge base of diverse molecular interactions and associated affinity information); and COMPEL (Kel-Margoulis et al., 2000) (a compendium of protein–DNA interactions). The GeneWays project is also provided with a knowledge model (Rzhetsky et al., 2000) that is fine-tuned for analysis of signal-transduction pathways in eukaryotes, but can be used for representing bacterial data as well.

Sixth, thinking of summary of molecular interactions as a blueprint of a computer chip (a real computer chip is usually less complex than a living cell), our researcher certainly needs to visualize fragments of the map to get insights into mechanisms of the "chip's" work; graph drawing is a large field in its own right. An excellent review of available methods related to molecular biology is provided by Uetz et al. (2002); a general treatment of graph drawing problems can be found in book by Di Battista et al. (1999).

Seventh, rather than straggling with individual tools every time she needs to process a new batch of a few thousand articles, the researcher may decide to integrate the previous six computational steps in a single system. The GeneWays system described in this chapter is just such an integrated system. Similar systems include the PIES system in Singapore, developed for analysis of protein–protein interactions described in journal abstracts (Ng and Wong, 1999; Wong, 2000, 2001a,b), the GENIA system in Japan (Collier et al., 1999), which uses knowledge extraction from both article abstracts and full articles to cross-index those articles with Internet-based databases, and the United States–developed MEDSTRACT system (Pustejovsky et al., 2002a,b), which extracts relationships of the form "A inhibits B" from journal-article abstracts.

We have set the context for our own project, GeneWays, by covering briefly the work that other groups have done on molecular pathways and on automated analysis of research articles. We have been developing the GeneWays system for 5 years at Columbia University; we recently used it for analysis of nearly 100,000 full-text articles and were able to populate a prototype database with 1.5 million statements. Without in any way denigrating the importance of competing systems, we believe that GeneWays is a state-of-art system that can be considerably extended and enhanced and that can be used as a tool for exciting research projects.

14.3 GENEWAYS: MOTIVATION AND ANATOMY

The word "GeneWays" probably emerged from an aberrant fusion of words "genes" and "pathways." The system was designed with the ambitious goal of automating extraction of information on molecular interactions locked in the text of journal articles.

Since the potential scope of the term "information on molecular interactions" is immense, at the first phase of the system development, we decided to focus on molecular interactions pertinent to signal-transduction pathways. Although the division of molecular pathways into "metabolic" and "signal transduction" is probably just a convenient way of looking at the elements of an interconnected unified system, there are distinct differences between these two types of pathways. Metabolic pathways mostly deal with tremendously diverse chemical alterations of relatively small molecules, whereas signal-transduction pathways are relatively poor in chemical mechanisms and predominantly involve "switch-on" and "switch-off" interactions among large molecules, such as genes and proteins. In an article describing a signal-transduction pathway, statements that "protein A binds protein B," "protein C phosphorylates protein D," and "protein E activates gene F" are seen frequently, although gene and protein names (A to F in the current example) can be drawn from a sizable list of nearly a hundred thousand names. Signal-transduction pathways, therefore, seem to be an easier target for information extraction from free text, although soon after starting the project we realized that even this "easier" task is extremely difficult to perform correctly.

GeneWays is designed to extract relations (or actions as we call them in our ontology; see Rzhetsky et al., 2000) between substances or processes. If we think about pathways as oriented graphs, we can divide relations into two groups: direct and indirect. Direct relations, which usually are physical interactions between substances, correspond to a single edge in the graph; indirect relations link two nodes (substances or processes) with a series of two or more edges. Direct relations in the current version of GeneWays include *N-acylate*, *N-glycosylate*, *O-glycosylate*, *acetylate*, *attach* (= *bind*), *createbond*, *degrade*, *demethylate*, *dephosphorylate*, *breakbond*, *methylate*, *overexpress*, *phosphorylate*, *express*, *contain*, *transcribe*, *release*, *interact*, and *substitute*. Indirect relations (which occasionally can also correspond to direct relations) include *activate*, *actupon*, *cause*, *generate*, *inactivate*, *limit*, *promote*, and *signal*. The GeneWays database currently maintains the following concept types: complex, disease, domain, gene, geneorprotein, process, protein, species, and smallmolecule.

Only a subset of these concepts (gene, geneorprotein, process, protein, and smallmolecule) can serve as vertices of a pathway graph; GeneWays uses the remainder to capture additional information about defined vertices and edges of the oriented graph. (We are currently implementing the additional concepts described in Rzhetsky et al., 2000).

Here, we describe two views of GeneWays: from the perspectives of a system developer and of a user.

From the point of view of a developer, GeneWays looks as in Figure 14.1 (asterisks identify modules of the system that are developed but not yet integrated). We can think of a system as an engine that processes raw data to create a structured product. The "raw data" that come into the system are represented by electronic copies of research articles coming from the World Wide Web, such as from the websites of scientific journals for which developers have a legitimate subscription. The task of collection and local accommodation of numerous research articles (we have approximately 85,000 full-text articles in the current system) is done by a GeneWays module called the "Download Agent," which saves retrieved text into a local database, as shown in Figure 14.1. The heart of the system comprises the modules shown inside the big arrow in Figure 14.1. First, the Term Identifier module (Krauthammer et al., 2000) identifies biologically important concepts in the text, such as names of genes, proteins, processes, small molecules, and diseases. Many such terms have synonyms and homonyms, so the Synonym/Homonym Resolver module clarifies the meaning by assigning a "canonical" name to each concept multiple aliases. Furthermore, there are other kinds of ambiguity associated with terms. For example, term "interleukin-2" can identify the corresponding gene, a messenger RNA, or the protein, depending on context. The Term Classifier module (Hatzivassiloglou et al., 2001) resolves sense ambiguity of this type. GENIES is a natural-language processing parser (Friedman et al., 2001) that takes as input plain text with identified and tagged concepts (for example, term "interleukin-2" can be tagged as "<substance='p'>interleukin-2 <\substance>", where 'p' stands for "protein"). The output of GENIES is represented with semantic trees that are not intended to be directly comprehended by humans, because they represent complex nested relationships captured from text in a machine-readable form.

An example of a GENIES parsing is shown in Figure 14.2.

The Simplifier module takes these complex output trees and unwinds them into simple binary statements (an example of a simple binary statement is "interleukin-2 binds interleukin-2 receptor"; the statement links two substances, interleukin-2 and interleukin-2 receptor, with action "bind"). The resulting simplified statements are saved into the Interaction Knowledge Base,

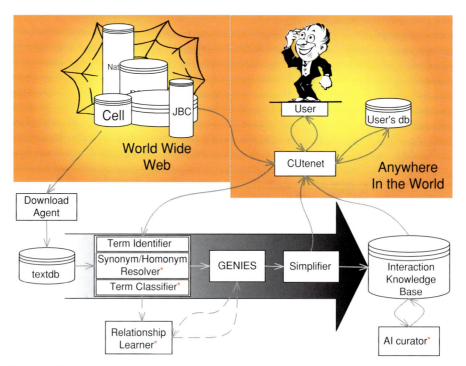

Figure 14.1   A simplified view of the GeneWays system.

```
action, promote,
                [geneorprotein,mdm2],
                [action, degrade,
                                [process, ubiquitin proteolytic pathway],
                                [geneorprotein, p53]
                ],
]
```

Figure 14.2 The GENIES parsing of the sentence "Recent studies have reported that mdm2 promotes the rapid degradation of p53 through the ubiquitin proteolytic pathway."

which is the main resource associated with the GeneWays system. The Interaction Knowledge Base is implemented on the basis of a commercial relational database (Oracle 9i), and it is built on GeneWays ontology (Rzhetsky et al., 2000).

Note that the automatically generated knowledge base is of necessity noisy: the GeneWays system extracts some percentage of statements incorrectly, and, even among correctly extracted statements, we should expect redundancy and contradictions. Therefore, the database requires curation, a process in which the original statements are annotated with statements regarding confidence in the corresponding information. The traditional way to perform such curation is through manual labor of human experts, a monumental task even for the database at its current size of roughly 2 million statements extracted from 100,000 articles. To reduce the manual work, we are implementing a Curator module that would allow GeneWays to compute the estimates of reliability automatically. We recently suggested a plausible approach to the curation and annotation problem, and we are in the process of implementing it (Krauthammer et al., 2002).

The two remaining modules are the CUtenet and Relationship Learner. We describe the first, CUtenet, later when explaining the user's perspective. The Relationship Learner module has a unique role within GeneWays because its relationships with other modules (shown by dashed lines in Figure 14.1) is different from the other relationships in the system. Most of the relationships in the figure (shown by solid arrows) depict flow of information during the data processing that leads to populating the Knowledge Base. The Relation Learner module works with the output of Term Identification/Disambiguation module to identify new semantic patterns that developers can use later to improve GENIES; therefore, the arrows connecting the Relationship Learner module with the rest of the system depict information flow during system-improvement cycles, rather than during data-processing cycles.

From the point of view of a user, the system is represented by its portal, CUtenet (pronounced "See-u- tenet," which stands for "Columbia University tenet," or "cute net," whichever you prefer; see Koike and Rzhetsky, 2000), a standalone program that accesses both the Knowledge Base and the GeneWays pipeline, as directed by a user. The primary function of CUtenet is visualization of user-defined pathways. Recently we augmented the program to access the GeneWays Interactions Knowledge Base, to (a) retrieve various interactions defined by a query formulated by a user and (b) visualize these interactions on the monitor. Moreover, the user can request information about the sentences corresponding to individual interactions and even can see the full articles from which the sentences were extracted. As an illustration of how the system works, let us consider the following example. Imagine that you are interested in a substance, the

protein called collagen. You are formulating a query equivalent to a question "Show me all interactions for collagen." The total number of interactions for a single substance stored in the Knowledge Base can be overwhelming (for collagen it is more than a thousand) and you need some mechanisms for reducing complexity of the output. Since each relation is frequently captured by GeneWays more than once from different sentences in the same as well as in distinct articles, the simplest filter for reducing the complexity of CUtenet figures is the number of times that each relation is entered into the Knowledge Base from independent sentences. In the case of collagen, the requirement that the interaction with collagen occur in the database at least 15 times retrieves collection of only 12 interactions (Figure 14.3A). Reduction of threshold to 10, 5, and 0 repetitions brings about 25, 74, and 1335 interactions, respectively (see Figures 14.3B, 14.3C, and 14.3D, respectively). Clearly, it is not very useful to show all 1335 interactions available for collagen at once. We certainly realize that this simple filter is imperfect because the statements repeated more frequently are not necessarily more important or more reliable than those repeated less frequently; nevertheless, this simple filter is better than no filter at all. We are developing a set of sophisticated filters that will allow users to select intuitive concepts for choosing among statements, such as the probability of a statement being true (see Krauthammer et al., 2002). In the current version of GeneWays a user can "walk" through the database by requesting that she visualize interactions for substances that are already shown on the screen (Figure 14.3E). Furthermore, by clicking on a graph edge in CUtenet window the user can retrieve original sentences corresponding to the interaction and full articles containing these sentences (see Figure 14.4).

An alternative way to access GeneWays system through CUtenet is shown in Figure 14.2; here, the user submits a request for processing of her favorite journal article through the GeneWays pipeline; such processing is culminated with a visualization of the extracted relationships. For example, by processing a plain-text version of a Cell article (Yao et al., 2002), GeneWays produced Figure 14.5, which shows 46 interactions. This is a relatively high number: The average number of interactions extracted by GeneWays pipeline from an average Cell article is half this number. To give you a more objective view of the number of interactions per article, we computed distributions of the number of statements per article extracted by GeneWays from three journals: *Cell*, *Journal of Molecular Biology*, and *Science*, from collection of articles spanning the past 5 years (Figure 14.6). The average number of statements extracted from a single article in these three journals was 18.06, 20.72, and 4.33, respectively. These numbers may appear at first to be somewhat low, especially for *Science* magazine; recall, however, that we analyzed all articles from each journal, and that *Science* publishes articles in all fields of science, rather than only in biology. It is natural, therefore, that an article on theoretical physics typically contains little information about interactions between genes and proteins.

14.4 EVALUATION OF EXTRACTION PRECISION

One of the important properties of a system is precision, defined as the ratio of the statements extracted correctly to the total number of extracted statements.

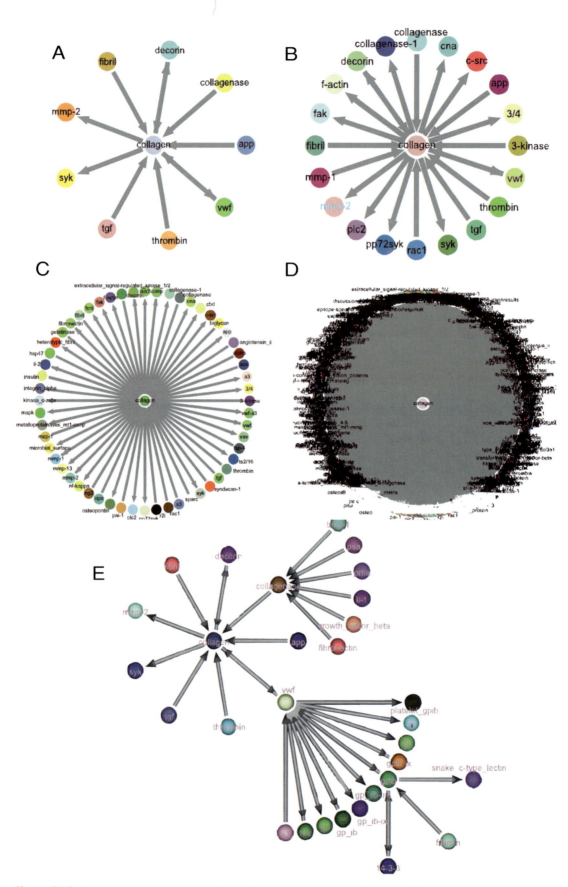

Figure 14.3   Examples of output of queries of Interaction Knowledge Base visualized with CUtenet module.

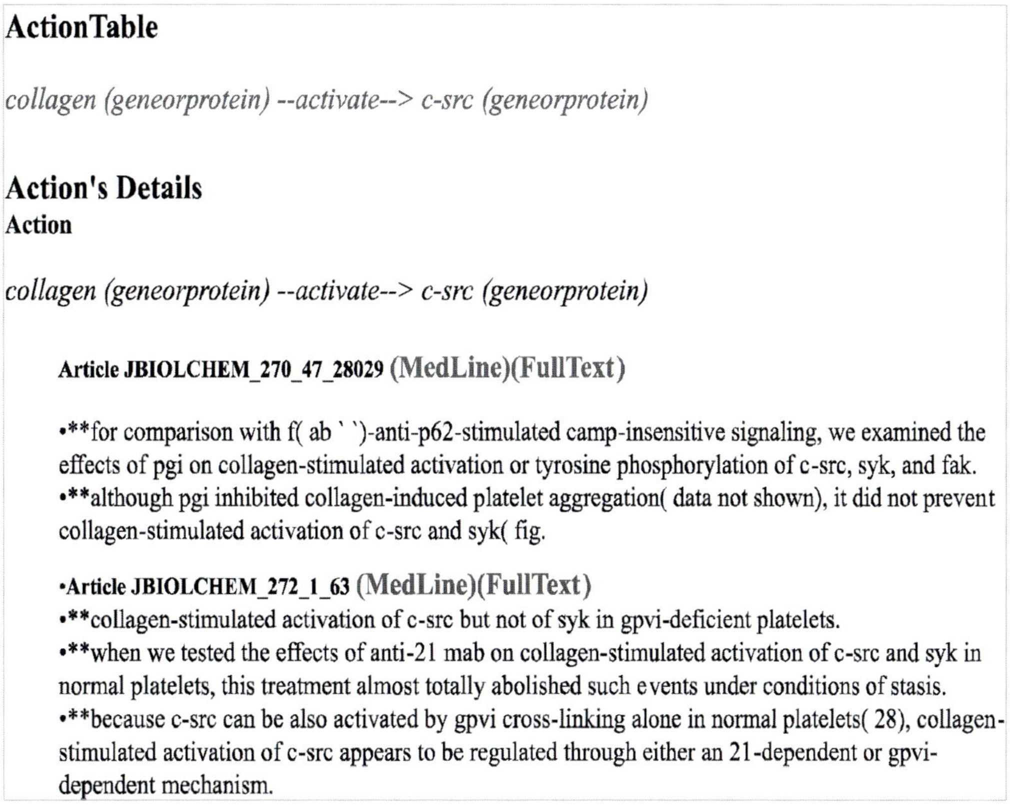

ActionTable

collagen (geneorprotein) --activate--> c-src (geneorprotein)

Action's Details
Action

collagen (geneorprotein) --activate--> c-src (geneorprotein)

Article JBIOLCHEM_270_47_28029 (MedLine)(FullText)

•**for comparison with f(ab ` `)-anti-p62-stimulated camp-insensitive signaling, we examined the effects of pgi on collagen-stimulated activation or tyrosine phosphorylation of c-src, syk, and fak.
•**although pgi inhibited collagen-induced platelet aggregation(data not shown), it did not prevent collagen-stimulated activation of c-src and syk(fig.

•**Article JBIOLCHEM_272_1_63** (MedLine)(FullText)
•**collagen-stimulated activation of c-src but not of syk in gpvi-deficient platelets.
•**when we tested the effects of anti-21 mab on collagen-stimulated activation of c-src and syk in normal platelets, this treatment almost totally abolished such events under conditions of stasis.
•**because c-src can be also activated by gpvi cross-linking alone in normal platelets(28), collagen-stimulated activation of c-src appears to be regulated through either an 21-dependent or gpvi-dependent mechanism.

Figure 14.4 A simplified view of information regarding interaction "collagen activates c-src" provided by the GeneWays Knowledge Base. For each interaction visualized by CUtenet, a user can obtain a list of sentences containing corresponding piece of information and complete articles containing these sentences.

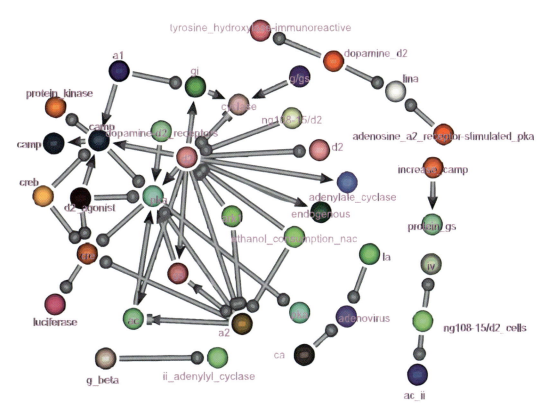

Figure 14.5 Results of GeneWays analysis of a single Cell article.

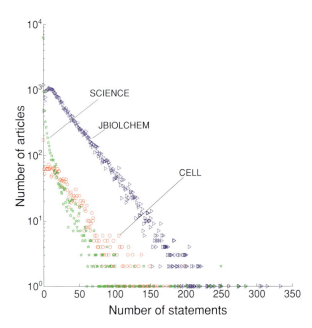

Figure 14.6   Distribution of the number of statements extracted by GeneWays from a single journal article for three journals: *Science* (SCIENCE in the figure), *Cell* (CELL), and *Journal of Biological Chemistry* (JBIOLCHEM).

can express extremely complex statements about the underlying biological systems. In contrast, the GeneWays Knowledge Base is designed to capture a "stochastic" view of the field, where statements tend to repeat, to conflict, and to have complex temporal distribution. The GeneWays Knowledge Base is likely to include a larger number of errors than do the manual databases (note that, in general, rigorous evaluations of the precision of the manual databases are not undertaken), and the number of types of relationships extracted automatically is smaller than can be extracted by a human expert. However, automatic systems can populate quickly an extremely large database (much larger than our current database of 1.5 million statements), and repetitive conflicting statements extracted automatically can be treated essentially as experimental data (see Krauthammer et al., 2002). Since the volume of text data currently available is tremendous, statistical approaches to analysis of statements extracted from the literature appear both promising and requisite.

14.5.2 Nobody (or System) Is Perfect

GeneWays in its current form has limitations. As follows from our evaluation of system precision, the noisiest part of the system is associated with term tagging. In general, it is difficult to identify a name of substance or a process in a text: Our favorite examples of difficult gene names include forever young (in plant Arabidopsis thaliana) and mothers against decapentaplegic (in *Drosophila melanogaster*). Improved term tagging is therefore likely to lead to a significant reduction in the error rate. Another plague crippling the system is associated with term *synonymy*. Although we compiled a database of gene/protein name synonyms, the dictionary approach alone appears to be insufficient. For example, "p53" and "p53 tumor suppressor" are currently stored in the GeneWays knowledge base as separate substances; these expressions have the same meaning but are difficult to recognize automatically as synonyms.

A totally different difficulty is associated with "translations" between sublanguages in scientific community. The same chemical event may be expressed in several strikingly different ways (different sublanguages) in different subdisciplines' research literature. For example, in the language of molecular biology, the statement "protein kinase A activates and phosphorylates protein B" means the same thing as expression

$$B + ATP \underset{\longleftarrow}{\overset{A}{\longrightarrow}} B \bullet P + ADP$$

To evaluate the precision of GeneWays, we selected 2500 of the most frequent unique statements (out of several hundred thousand unique statements that are currently stored in GeneWays Knowledge Base). We then had an expert in molecular biology go through the whole 2500 list, checking correctness of extraction; the whole endeavor took a few weeks. According to this expert evaluation, 125 statements of the 2500 either were extracted with errors or corresponded to "phantom statements" generated by the GeneWays system. We then traced all stages of processing for each of these 125 statements, and we found out that 100 of them were incorrect due to errors in term identification, 12 due to GENIES errors, and 5 due to Simplifier errors; 8 were actually correct (expert's error, as judged by the developers team). Therefore, according to this evaluation, GeneWays' precision was 95%; its recall was previously evaluated to be about 65% (Friedman et al., 2001). We conclude that the quality of data on molecular interactions produced by GeneWays system appears to be comparable to quality of data produced by high-throughput experiments, such as yeast two-hybrid experiments.

14.5 DISCUSSION

14.5.1 Hand-Made Databases and Automatically Produced Ones

There are a few popular molecular interaction databases, such as EcoCyc (Karp et al., 2000) and KEGG (Kanehisa and Goto, 2000), that are populated by groups of careful experts. Such "manual" databases are designed to provide a consensus view of the evolving field of molecular biology (devoid of redundancy and inconsistencies), usually have a low error rate, and

in the language of biochemistry (where ATP and ADP stand for adenosine triphosphate and adenosine diphosphate, respectively, and *P denotes a phosphate residue). Note that, in the biochemical description, kinase A is not part of the equation, but rather is merely a catalyst facilitating the reaction. A hard-core biochemist may argue that what molecular biologists say is incorrect; however, since both communities are able to understand their own statements correctly, we are dealing with two sublanguages requiring translation from one to another. If the articles analyzed by GeneWays are written in the language of molecular biology, but potential users of the resulting database speak in biochemical language (which is probably more precise), then automated "translation" of statements may become necessary.

14.5.3 Werewolves of Biological Terminology

There is a difficulty of recognizing terms "p53" and "p53 tumor suppressor" as synonyms; here the major problem is in deciding where protein name ends and a description of its function starts.

There are more extreme cases when a single term can be correctly interpreted in multiple ways. Our favorite example is protein name "MAPKKK," which stands for "mitogen-activated protein kinase kinase kinase."

Consider a hypothetical sentence "Mitogen-activated protein kinase kinase kinase phosphorylates protein Y." Term recognition here is a real problem because "mitogen," "mitogen-activated protein kinase," and "mitogen-activated protein kinase kinase" are valid substance names that are important for capturing pathway information contained in the sentence; the sentence contains four interactions, namely, "mitogen activates mitogen-activated protein kinase kinase kinase," "mitogen activated protein kinase kinase kinase activates and phosphorylates mitogen activated protein kinase kinase," "mitogen activated protein kinase kinase activates and phosphorylates mitogen activated protein kinase," and "mitogen activated protein kinase kinase kinase phosphorylates protein Y."

An ideal text processing engine should be capable of extracting all these interactions; and the GeneWays system, which currently cannot perform a nested tagging of substance names, has plenty of room for improvement.

References

Altman, R.B., Bada, M., Chai, X.J., Whirl Carillo, M., Chen, R.O., and Abernethy, N.F. (1999). RiboWeb: An ontology-based system for collaborative molecular biology. *IEEE Intell. Syst.* **14**, 68–76.

Ashburner, M., Ball, C.A., Blake, J.A., Botstein, D., Butler, H., Cherry, J.M., Davis, A.P., Dolinski, K., Dwight, S.S., Eppig, J.T., Harris, M.A., Hill, D.P., Issel-Tarver, L., Kasarskis, A., Lewis, S., Matese, J.C., Richardson, J.E., Ringwald, M., Rubin, G.M., and Sherlock, G. (2000). Gene ontology: Tool for the unification of biology. The Gene Ontology Consortium. *Nat. Genet.* **25**, 25–29.

Baker, P.G., Brass, A., Bechhofer, S., Goble, C., Paton, N., and Stevens, R. (1998). TAMBIS—Transparent Access to Multiple Bioinformatics Information Sources. *ISMB* **6**, 25–34.

Blaschke, C., Andrade, M.A., Ouzounis, C., and Valencia, A. (1999). Automatic extraction of biological information from scientific text: Protein–protein interactions. *ISMB* **7**, 60–67.

Chen, X., Liu, M., and Gilson, M.K. (2001). BindingDB: A web-accessible molecular recognition database. *Comb. Chem. High Throughput Screen* **4**, 719–725.

Collier, N., Park, H.S., Ogata, N., Tateisi, Y., Nobota, C., Ohta, T., Sekimizu, T., Imai, H., Ibushi, K., and Tsujii, J. (1999). *The GENIA Project: Corpus-Based Knowledge Acquisition and Information Extraction from Genome Research Papers,* EACL, pp. 271–271.

Collier, N., Nobata, C., and Tsujii, J. (2000). Extracting the Names of Genes and Gene Products with a Hidden Markov Model, In: *Proc. COLING 2000,* pp. 201–207. Saarbruken, Germany.

Craven, M., and Kumlien, J. (1999). Constructing biological knowledge bases by extracting information from text sources. *Proc. Int. Conf. Intell. Syst. Mol. Biol.,* pp. 77–86.

Di Battista, G., Eades, P., Tamassia, R., and Tollis, I.G. (1999). *Graph Drawing. Algorithms for the Visualization of Graphs.* Prentice Hall, Upper Saddle River, NJ.

Friedman, C., Kra, P., Yu, H., Krauthammer, M., and Rzhetsky, A. (2001). GENIES: A natural-language processing system for the extrac-

tion of molecular pathways from journal articles. *Bioinformatics* **17** (Suppl. 1), S74–82.

Fukuda, K., Tamura, A., Tsunoda, T., and Takagi, T. (1998). Toward information extraction: Identifying protein names from biological papers. *Pac. Symp. Biocomput.,* pp. 707–718.

Gaizauskas, R., Demetriou, G., and Humphreys, K. (2000). Term recognition and classification in biological science journal articles. *2nd Int. Conf. Nat. Lang. Process. (NLP-2000),* Patras, Greece, pp. 37–44.

Goto, S., Nishioka, T., and Kanehisa, M. (2000). LIGAND: Chemical database of enzyme reactions. *Nucleic Acids Res.* **28**, 380–382.

Hatzivassiloglou, V., Duboue, P.A., and Rzhetsky, A. (2001). Disambiguating proteins, genes, and RNA in text: A machine learning approach. *Bioinformatics* **17** (Suppl. 1), S97–S106.

Iliopoulos, I., Enright, A.J., and Ouzounis, C.A. (2001). Textquest: Document clustering of Medline abstracts for concept discovery in molecular biology. *Pac. Symp. Biocomput.,* pp. 384–395.

Jacquemin, C. (2001). *Spotting and Discovering Terms through Natural Language Processing.* MIT Press, Cambridge, MA.

Jenssen, T.K., Laegreid, A., Komorowski, J., and Hovig, E. (2001). A literature network of human genes for high-throughput analysis of gene expression. *Nat. Genet.* **28**, 21–28.

Joachims, T. (1999). Transductive inference for text classifcation using support vector machines. *Int. Conf. Mach. Learn. (ICML)* pp. 200–209.

Joachims, T. (2001). A statistical learning model of text classification with support vector machines. *SIGIR-01, 24th ACM International Conference on Research and Development in Information Retrieval* (W.B. Croft, D.J. Harper, D.H. Kraft, and J. Zobel, eds.), pp. 128–136. ACM Press, New York.

Kanehisa, M., and Goto, S. (2000). KEGG: Kyoto encyclopedia of genes and genomes. *Nucleic Acids Res.* **28**, 27–30.

Karp, P.D., Riley, M., Saier, M., Paulsen, I.T., Paley, S.M., and Pellegrini-Toole, A. (2000). The EcoCyc and MetaCyc databases. *Nucleic Acids Res.* **28**, 56–59.

Karp, P.D., Riley, M., Saier, M., Paulsen, I.T., Collado-Vides, J., Paley, S.M., Pellegrini-Toole, A., Bonavides, C., and Gama-Castro, S. (2002a). The EcoCyc Database. *Nucleic Acids Res.* **30**, 56–58.

Karp, P.D., Riley, M., Paley, S.M., and Pellegrini-Toole, A. (2002b). The MetaCyc Database. *Nucleic Acids Res.* **30**, 59–61.

Kazama, J., Makino, T., Ohta, Y., and Tsujii, J. (2002). Tuning support vector machines for biomedical named entity recognition. In *ACL-02,* Philadelphia, on-line.

Kel-Margoulis, O.V., Romashchenko, A.G., Kolchanov, N.A., Wingender, E., and Kel, A.E. (2000). COMPEL: A database on composite regulatory elements providing combinatorial transcriptional regulation. *Nucleic Acids Res.* **28**, 311–315.

Koike, T., and Rzhetsky, A. (2000). A graphic editor for analyzing signal-transduction pathways. *Gene* **259**, 235–244.

Krauthammer, M., Rzhetsky, A., Morozov, P., and Friedman, C. (2000). Using BLAST for identifying gene and protein names in journal articles. *Gene* **259**, 245–252.

Krauthammer, M., Kra, P., Iossifov, I., Gomez, S.M., Hripcsak, G., Hatzivassiloglou, V., Friedman, C., and Rzhetsky, A. (2002). Of truth and pathways: Chasing bits of information through myriads of articles. *Bioinformatics* **18** (Suppl. 1), S249–S257.

Leek, R.L. (1997). Information extraction using hidden markov models. In: *Computer Science,* p. 40. University of California, San Diego.

Liu, H., Lussier, Y.A., and Friedman, C. (2001). Disambiguating ambiguous biomedical terms in biomedical narrative text: An unsupervised method. *J. Biomed. Inf.* **34**, 249–261.

Marcotte, E.M., Xenarios, I., and Eisenberg, D. (2001). Mining literature for protein–protein interactions. *Bioinformatics* **17**, 359–363.

Ng, S.K., and Wong, M. (1999). Toward routine automatic pathway discovery from on-line scientific text abstracts. *Genome Inf.,* pp. 104–112.

Nobota, C., Collier, N., and Tsujii, J. (1999). Automatic term identification and classification in biological texts. *Proc. Nat. Lang. Pac. Rim Symp.*, pp. 369–374.

Ono, T., Hishigaki, H., Tanigami, A., and Takagi, T. (2001). Automated extraction of information on protein–protein interactions from the biological literature. *Bioinformatics* **17**, 155–161.

Park, J.C., Kim, H.S., and Kim, J.J. (2001). Bidirectional incremental parsing for automatic pathway identification with combinatory categorial grammar. *Pac. Symp. Biocomput.*, pp. 396–407.

Paton, N.W., Khan, S.A., Hayes, A., Moussouni, F., Brass, A., Eilbeck, K., Goble, C.A., Hubbard, S.J., and Oliver, S.G. (2000). Conceptual modelling of genomic information. *Bioinformatics* **16**, 548–557.

Proux, D., Rechenmann, F., Julliard, L., Pillet, V.V., and Jacq, B. (1998). Detecting gene symbols and names in biological texts: A first step toward pertinent information extraction. *Genome Infa. Ser. Workshop Genome Inf.* **9**, 72–80.

Pustejovsky, J., Castano, J., Cochran, B., Kotecki, M., and Morrell, M. (2001). Automatic extraction of acronym-meaning pairs from MEDLINE databases. *Medinfo* **10**, 371–375.

Pustejovsky, J., Castano, J., Sauri, R., Rumshisky, A., Zhang, J., and Luo, W. (2002a). Medstract: Creating large-scale information servers for biomedical libraries. *ACL-02*, Philadelphia.

Pustejovsky, J., Castano, J., Zhang, J., Kotecki, M., and Cochran, B. (2002b). Robust relational parsing over biomedical literature: Extracting inhibit relations. *Pac. Symp. Biocomput.*, pp. 362–373.

Rindflesch, T.C., Hunter, L., and Aronson, A.R. (1999). Mining molecular binding terminology from biomedical text. *Proc AMIA Symp.*, pp. 127–131.

Rzhetsky, A., Koike, T., Kalachikov, S., Gomez, S.M., Krauthammer, M., Kaplan, S.H., Kra, P., Russo, J.J., and Friedman, C. (2000). A knowledge model for analysis and simulation of regulatory networks. *Bioinformatics* **16**, 1120–1128.

Salama, J.J., Donaldson, I., and Hogue, C.W. (2001). Automatic annotation of BIND molecular interactions from three-dimensional structures. *Biopolymers* **61**, 111–120.

Schulze-Kremer, S. (1998). Ontologies for molecular biology. *Pac. Symp. Biocomput.*, pp. 695–706.

Shatkay, H., and Wilbur, J. (2000a). Finding themes in Medline documents: probabilistic similarity search. *ISMB*.

Shatkay, H., and Wilbur, W.J. (2000b). Finding themes in medline abstracts. *Advances in Digital Libraries. IEEE*, pp. 183–192.

Shatkay, H., Edwards, S., Wilbur, W.J., and Boguski, M. (2000). Genes, themes and microarrays: Using information retrieval for large-scale gene analysis. *Proc. Int. Conf. Intell. Syst. Mol. Biol.* **8**, 317–328.

Stapley, B.J., and Benoit, G. (2000). Biobibliometrics: Information retrieval and visualization from co-occurrences of gene names in Medline abstracts. *Pac. Symp. Biocomput.*, pp. 529–540.

Stephens, M., Palakal, M., Mukhopadhyay, S., Raje, R., and Mostafa, J. (2001). Detecting gene relations from Medline abstracts. *Pac. Symp. Biocomput.*, pp. 483–495.

Stevens, R., Goble, C.A., and Bechofer, S. (2000). Ontology-based knowledge representation for bioinformatics. *Brief. BioInf.* **1**, 398–414(399).

Tanabe, L. and Wilbur, W.J. (2002). Tagging gene and protein names in biomedical text. *Bioinformatics* **18**, 1124–1132.

Uetz, P., Ideker, T., and Schwikowski, B. (2002). Visualization and integration of protein-protein interactions. In *Protein–Protein Interactions — A Molecular Cloning Manual* (E. Golemis, ed.), pp. 623–646. Cold Spring Harbor Lab. Press, Cold Spring, NY.

Wong, L. (2000). Kleisli, a functional query system. *J. Functi. Programm.* **10**, 19–56.

Wong, L. (2001a). Bioinformatics integration simplified: The Kleisli way. In *Frontiers in Human Genetics: Diseases and Technologies* (P.S. Lai and E. Yap, eds.), pp. 79–90. World Scientific, Singapore.

Wong, L. (2001b). PIES, a protein interaction extraction system. *Pac. Symp. Biocomput.*, pp. 520–531.

Xenarios, I., Salwinski, L., Duan, X.J., Higney, P., Kim, S.M., and Eisenberg, D. (2002). DIP, the Database of Interacting Proteins: A research tool for studying cellular networks of protein interactions. *Nucleic Acids Res.* **30**, 303–305.

Yakushiji, A., Tateisi, Y., Miyao, Y., and Tsujii, J. (2001). Event extraction from biomedical papers using a full parser. *Pac. Symp. Biocomput.*, pp. 408–419.

Yao, L., Arolfo, M.P., Dohrman, D.P., Jiang, Z., Fan, P., Fuchs, S., Janak, P.H., Gordon, A.S., and Diamond, I. (2002). Betagamma dimers mediate synergy of dopamine D2 and adenosine A2 receptor-stimulated PKA signaling and regulate ethanol consumption. *Cell* **109**, 733–743.

Yu, H., Hripcsak, G., and Friedman, C. (2002). Mapping abbreviations to full forms in biomedical articles. *J. Am. Med. Inf. Assoc.* **9**, 262–272.

Zanzoni, A., Montecchi-Palazzi, L., Quondam, M., Ausiello, G., Helmer-Citterich, M., and Cesareni, G. (2002). MINT: A Molecular INTeraction database. *FEBS Lett.* **513**, 135–140.

Cellular Morphology at the Cellular Level

Christopher J. Beaver, Ph.D., Robert C. Cannon, Ph.D., and
Dennis A. Turner, M.A., M.D.

CONTENTS

Databasing the Brain. Edited by Stephen H. Koslow and Shankar Subramaniam
ISBN 0-471-30921-4 © 2005 John Wiley & Sons, Inc.

15.1 INTRODUCTION

Neurons possess an elegant and elaborate shape, with highly specialized axonal and dendritic processes for information shaping and information transfer. However, the relationship of neuron appearance and structure to dendritic function remains a complex field, requiring various types of mathematical models (Ascoli et al., 2001a; Turner et al., 1991). Computer-assisted reconstruction techniques are now widely available and in common use, facilitating a quantitative approach to single neuron function and to brain models, using measured neurons as the primary structural platform. A rough estimate would suggest that worldwide there is a pool of at least 25,000 measured neurons, in digital form (such as Neurolucida format). However, only a fraction of these cells (less than a few hundred) are available in the public domain, in a few websites. Though synthesized neurons largely recreate most of the complex dendritic structure (Ascoli et al., 2001a), the variance between neurons within the same class is hard to replicate, until further measurements can indicate the range of variation.

Since the function of the brain depends upon the complex interactions between neurons, it would seem reasonable to assemble regional and whole-brain maps from a collection of neurons, with their specialized shape as the basis for interactions. Representative neurons within various structures could be replicated (with stochastic variance), allowing complex spatial overlap and interactions to be modeled on a large, regional scale, such as assembling a hippocampus with realistic elements. However, all current brain maps rely upon either (a) histological stains demonstrating only the soma of neurons (such as Nissl stain) or (b) acetylcholinesterase stains, rather than neuronal structure per se. The purpose of these maps has been to distinguish regions within the brain, rather than to model the function of regions. Thus, the availability of larger sets of real, measured neurons could allow the assembly of collections of neurons into a regional structure, including neuropil, accompanied by neuronal function and interconnections. For this analysis situation, the variation between neurons as well as the distinction of classes of neurons are critical elements of a large dataset.

The analysis of neuronal structure and dendrites has followed several different paths, depending on the analytic need and the level of resolution required. For example, dendrites have been compared between cell classes, during development, and during pathological alterations in neuronal structure (Cannon et al., 1999). Due to the time and effort required for both individual neuron visualization and computer-aided reconstruction, such measured neurons are valuable and have been placed in neuronal databases for open dissemination. The actual acquisition and measurement of neurons from experimental preparations has proven itself to be a significant limitation. If better methods were available for selectively and completely demonstrating neuronal structure, then more neurons could become available as well. Neuronal databases and enhancing potential uses of available neurons will promote optimal use of existing neurons and will facilitate building neuronal, circuitry, and brain models. The visualizations of real neurons can then also be compared critically to synthetic neurons, particularly examining the degree of differences between neurons of varying classes (Ascoli, 1999).

These levels of analysis of neuronal structure and function, and regional brain function, underlay the strong need for enhancing access to the large numbers of neurons already measured. Because of frequent requests from the modeling community for more access, we established one of the first open, web databases of neuron structure in 1998 (Cannon et al., 1998), with over a hundred neurons from the hippocampus. Though further cell contributions were solicited and encouraged, none occurred over several years. Thus, it became clear that a better method was required for neuron interchange. The consistent problem was that sharing of neuronal structures takes effort and an altruistic sense of trust, in that the originators of the cells deserved some credit for their work, beyond the publication of their original data. This issue has been faced in many networks of scientists, beginning with protein, DNA, and RNA, databases, but in which a shared sense of effort and contribution could be rewarding to both the contributors and users, who overlapped. In the case of neuronal structures, the experimentalists generating the data and performing most of the work largely do not engage in complex modeling schemes. Therefore, they may not necessarily understand the importance and utility of sharing, nor the modeling, which uses these structures as a base. Likewise, the modelers using the data often do not appreciate the effort, which is required for obtaining and measuring neurons. Since there are two separate groups involved, rather than overlapping users and contributors, as in the case of many databases, additional methods for interchange have been sought. (Turner et al., 2002).

These issues have led us to newer approaches, using a different informatics base than before, particularly as Web-based sharing has become more sophisticated, which we will discuss further. Use of measured neurons also requires some background as to how the neurons were obtained, reconstructed, and presented. This chapter outlines current approaches to labeling neurons in vivo and in vitro, as well as reconstruction, techniques and limitations. The manipulation and analysis of these neurons will then be considered, together with issues concerning Web-based storage and dissemination.

15.2 TYPES OF CELLULAR DATASETS

Measured neurons are now most frequently obtained using physiological injection of a dye (particularly biocytin) during sharp or patch intracellular recordings (Figure 15.1). A large amount of dye can be rapidly injected, and dendritic diffusion is enhanced in live neurons. These procedures require a physiological preparation, such as in vivo cerebral cortex, or slices (400–500 μm) maintained in vitro. The tissue is then fixed and processed for biocytin, usually resulting in an isolated neuron visualized within the tissue. This isolation facilitates the reconstruction, since there is no ambiguity about the identity of processes in all sections (i.e., there is only one cell stained). Subsequent measurements are altered, however, by whatever processes of fixation and clearing are used. Most neurons available in the several Web databases are derived from this type of visualization, usually from fixed tissue (Cannon et al., 1998; Pyapali and Turner, 1996). Appropriate correction factors may be applied to decrease possible histological or measurement distortions, if specific information is available (Horcholle-Bossavit et al., 2000).

Neuronal reconstruction methods have arisen from classic neuroanatomy, to portray the outline of neuronal structure. Traditional techniques include camera lucida tracings and, more

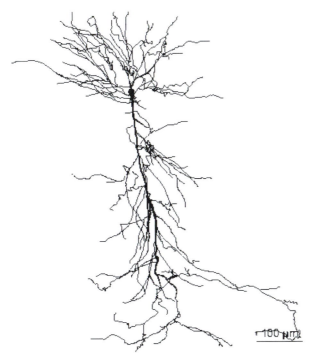

Figure 15.1   An example of a hippocampal CA1 cell reconstructed using the Neurolucida program. Cell n401t can be found in the www .compneuro.org copy of the Duke/Southampton database of neuronal morphology.

recently, computer-generated linear representations of structure. Preparing camera lucida tracings involves a compound microscope and a drawing tube, which allows the projection of an image of the microscopic specimen onto a sheet of paper. The systems currently in use allow the visualization of both the specimen and the paper through the microscope, so that a direct comparison can be made between the structure being traced and the drawing. Using a high-magnification lens (typically a 100× planapochromat), fine details of dendritic structure can be achieved, to the level of resolution of light microscopy (∼0.4 μm). The resulting planar representation of a neuron may subsequently be pieced together from multiple sheets.

The most common reconstruction system currently available, Neurolucida (Microbrightfield, Manchester, VT), uses a tracking apparatus for both planar and vertical measurements. This system consists of a motorized stage and focus mechanism, with a readout that can be transferred to a computerized database. The system keeps track of where the investigator is pointing/focusing in all three dimensions, and it saves a log with length and orientation coordinates for all specified nodes. Dendritic diameters are measured using a variable spot cursor, which can be matched to the diameter at each node. The resulting neuronal structure contains (a) a series of connected nodes, with calibrated, three-dimensional locations, and (b) cylinders representing dendrites between the nodes. As with any approximation system, a proportionally large number of cylinders may be used to represent, for example, a tapering dendrite or a neuronal cell body, depending on the degree of fidelity desired. However, dendritic spines are rarely included or counted, which may result in significant loss of the actual surface contour, but can be added in later analysis (Turner, 1984).

Improvements to visualizing live neuronal structure include gene-based approaches, such as live transfection with green fluorescent protein, or some variant (Lo et al., 1994). The advantage of using live cells is that the dye or protein label is then actively diffused throughout all processes of the cell. This is in comparison to fixed tissue approaches, such as (a) Lucifer yellow injections into lightly fixed tissue or (b) Golgi stains. In these approaches, the diffusion of the dye or the staining of the cell from outside is much less complete, as estimated by total dendritic length measurements. For example, Lucifer-yellow-filled dentate granule cells averaged approximately 1 mm in total dendritic length, similar to Golgi, but far less than is commonly measured using intracellular injections into live cells and reconstructions, averaging 4–5 mm. As new techniques become available which offer both selectivity and complete cell staining, these may replace intracellular injections as the method of choice, provided that cellular measurements confirm the completeness of the staining.

15.2.1 Availability and Types of Neuronal Archives

Several laboratories do share anatomical neuron files with the rest of the scientific community, but most data are available only for the hippocampal formation. At present, three organized hippocampal cell archives are accessible through the Web: Duke/Southampton (Turner Lab; Cannon et al, 1998), Claiborne Lab, and Gulyas' data. The Duke/Southampton archive currently contains 124 complete cells (all from the hippocampus), and it constitutes the largest public resource to date (www.cns.soton.ac.uk or www.compneuro.org/CDROM/ nmorph). For each cell, the archive contains the anatomical file (in the standard SWC or Stockley–Wheal–Cannon format; Cannon et al., 1998; Ascoli et al., 2001a), a gif or jpg image of the neuron with calibration bar, a set of metadata (including cell type, authors and publication reference, species and age of the animal, type and date of preparation, shrinkage factor, and soma area), and a text file including additional comments and information by the authors (e.g., type of electrode used, input resistance and time constant, etc.).

Existing neuronal databases remain limited in accessibility, scope, and usefulness, in spite of available tools. Despite various attempts to develop repositories of cell architecture data, so far attempts to distribute morphological data have been unsuccessful. What types of solutions exist, and do they meet the requirement for data dispersion? Three general categories of solutions currently exist, to meet this demand (Table 13.1): (1) local websites, or FTP sites; (2) central databases; or (3) federated, common format databases. We will discuss each of these in turn, along with their relative advantages and disadvantages.

Local Websites or FTP Sites on a Lab or University Machine

Most morphological data currently available can be found on individual lab machines or websites. Such a solution is primarily advantageous for the lab itself for the following reasons: It allows easy access to the data and makes sharing data with collaborators straightforward; it allows the individual lab to maintain control

TABLE 13.1 A comparison of different database formats

Solution Property	Centralized Databases[1]	Lab Websites[2]	Federated Data Sites[3]
Integrates across datasets	Yes	No	Yes
Investigator retains control	No	Yes	Yes
Investigator gets credit	No	Yes	Yes
Integrates with information management tools	Partial	No	Yes
Handles heterogeneous data	No	Yes	Yes
Handles evolving data	No	Partial	Yes
Can be set up without IT expertise	No	No	Yes
Can be used without IT expertise	Partial	Partial	Yes
Handles complex queries	Yes	No	Yes
Useable without curators	Partial	Yes	Yes
Supports curation	Yes	No	Yes
Supports rapidly changing data	Yes	No	No
Cost < 5% grant total	No	Yes	Yes

Suitable for: (1) well-funded communities with agreed data standards; (2) Individuals with unique data formats; (3) Communities with heterogeneous and evolving datasets and data types.

of the intellectual property (to ensure that appropriate credit is given); and important annotations and insights can be placed with the data to aid in interpretation. The disadvantage to this arrangement is that the data are confined to remote locations and is relatively inaccessible to other researchers who do not have direct contact with the lab providing the data. While individual labs may be willing to share the information, the collections are only discovered through extensive searches of the Internet, using search engines that are not well suited for tracking scientific data, or through interactions at scientific meetings. The development of a website-based interface may also present a hurdle for some labs in terms of the expertise or time required. Idiosyncratic methods for storage of data and accompanying annotations can also engender data transfer/interpretation problems unless solutions are provided.

Several examples of the local cell morphology databases are available over the Internet; and there are, undoubtedly, other yet unpublished caches of data scattered throughout institutions. The archive of Brenda Claiborne's lab at the University of Texas, San Antonio (www.utsa.edu/claibornelab), for example, contains over 50 examples of cells reconstructed from intracellular fills of hippocampal neurons. The data archive includes a thumbnail image with scale bar, the index number linked to the raw data, the age of the animal in days (ranging from 14 to 368), the sex, and the total dendritic length in microns. The archive is enriched by an extensive usage note, which specifies important metadata information (reconstruction system, accuracy, reference to publications). Similarly, Gulyas' archive at the Institute of Experimental Medicine in Budapest, Hungary (www.koki.hu/~gulyas/ca1cells) contains 18 CA1 pyramidal cells and 67 CA1 interneurons (further divided in three groups on the basis of their immunoreactivity). For each cell, the archive includes an index entry, a jpg image, and the anatomical files in virtual reality format, segment format, tree format (both of which are extensively described in the archive), and the aforementioned SWC format. Metadata, including publication references, and additional information, such as synapse density from electron microscopy, are also linked to the archive. In addition to these databases, which are available to any individual, there are other collections, which are distributed by their owners upon request. But, there is no common access, or method for keeping up to date, because these demand-only sites are not publicized on the Web or at meetings.

Centralized Databases

Centralized databases depend upon soliciting submissions of data to a central location, assembling the information into a standard format, and maintaining a Web interface for uploading data and performing queries. Implementation of this type of solution has been successful in generating interactive databases that collect and manage information on such items as protein (http://www.rcsb.org/pdb/) or gene sequences (http://www.ncbi.nlm.nih.gov/Genbank/; http://genome.ucsc.edu/) or FMRI data (http://www.fmridc.org). Other centralized databases are more heavily managed resources that do not require the participation of the original researchers. They are populated from published work, require substantial management, and are usually specific to a particular community of interest. Examples of such repositories include the Sense lab project and Coco Mac (collations of connectivity data on the macaque brain). The former is a 10-year effort by an international network of researchers to develop a comprehensive approach to building integrated, multidisciplinary models of neurons and neural systems, using the olfactory pathway as a model. The latter is an approach to produce a systematic record of the known wiring of the primate brain, containing details of hundreds of tracing studies in their original descriptions, to which further data are continuously added.

The advantages of this type of data management are that it ensures that the data are maintained in a standardized format, and that a user interface facilitates the addition of new information in a timely manner and in maintaining the information at a single point where all users have equal access. Successful centralized databases require two key components, however. First, because these databases involve enormous effort, an individual or organization must be willing to expend the capital and/or time to develop and maintain the database and user interface. This has, in the past, meant that a governmental agency or academic institution be willing to subsidize the development of the database. The second and perhaps more important component is the enlistment of individual data contributions to the database. This is, to a great degree, dependent upon ensuring an easy transfer of information to the database, but also upon there being an incentive to contribute. For example, if one contributes, then one may have free access to all of the data, which could be useful for enhancing the contributor's work.

Development of a centralized database of neuronal morphology has encountered difficulty on at least two levels. First, the number of potential contributors and end-users for the information in such a database is much smaller than the demand for the information in the databases of protein or genetic sequences, for example. This means that the pressure to ensure the precedence of one's data in a database is much lower. Second, the data are complicated to obtain and manage, in comparison to more straightforward data (i.e., genetic sequences, amino acid names). Consequently, in comparison with the effort asked of the curator(s) of the database and the end-user, the onus is placed on the data providers, who do the work to analyze the samples, generate meaningful data files, and send them to the repository. The combination of these two factors has created reluctance on the part of researchers to contribute to such a database.

The apparent benefits of a centralized database, however, have encouraged attempts to generate a repository of neuronal morphological data and elicit contributions. The Duke/Southampton cell archive contains a collection of 124 cells with accompanying metadata, organized with thumbnail images of the neurons and data in the SWC format primarily from a single source. A viewer (CVAPP), supplied at the archive site, can be used for reading files in its own (swc) or Neurolucida format, saving structures for use in the neuron or genesis modeling packages, tracing branches out from any given point to check connections, checking for loops, editing joining points, or shifting whole branches. Examples of the user interface and the viewer program are shown in (Figures 15.2 and 15.3).

This database met with only a moderate degree of success, however. The database was used extensively for the retrieval of data and software, but no contributions were received.

Federated Data Sites

The basis of this management solution is to combine the best components of the two alternatives, and make contributing as undemanding a task as possible. According to this plan, each lab maintains a collection of data files on a local source disk, which can be shared, with some rules. The individual data sources would use software provided by a centralized service to produce a relatively standard interface. The centralized service also acts as a network hub, whereby other researchers can perform a single search across many sites and browse through the data samples of information. Obviously, this approach depends upon easy identification and notification of available data, requiring some form of common signaling or software format, which is installed at every site. The system should be comprised of a user-friendly data-publishing tool and a server, linking data from remote sites. Data providers are able to document or annotate their data as they see fit before exporting the data to a website both as browsable Web pages and as XML. Once data have been annotated and transferred to the provider's own website, their involvement in the process is complete. It is the task of the data users and database experts to work out how best to federate the disparate sources of data so they can be exploited as a coherent entity. Creating common entry points, describing datasets as accurately as possible, and providing sufficient metadata to distinguish data are all critical for this process to function effectively.

The federated database solution, therefore, addresses two of the biggest hurdles to data dissemination via centralized database: the obligation to document data according to formats imposed by the database constructor and the perceived loss of ownership. At the same time it federates data from scattered individual websites. Using recent, flexible technology with the Web as a form of distributed database, it eliminates the hurdle of the effort placed on individual users to the data at a centralized site, while, through the central server, permitting efficient browsing by other users not available with the local website approach. Contributors are able to maintain both intellectual and physical control over the data. Issues of data conversion could be easily handled by making conversion tools available at the central server. No judgment about the quality of data is made; individual end-users can evaluate the quality of the data.

The federated solution will require the standardization of certain amount of information from the data providers, but less than a centralized approach that demands that data providers standardize their metadata. A good proportion of this data, such as area definitions and cell types, will already be in a usable format, and a standard format will evolve through the interaction of users with the database. The mechanism involves creation of new metadata linking the metadata on one site to that on another. This might, for example, be a synonym table to indicate that the field "area" in one set of records should be treated as equivalent to "region" in another. Current tools provide a small set of such functions for data users to improve the performance of a federating server. This is one of the most active areas of ongoing development, because the success of a federated distributed approach to databasing depends to a large extent on the breadth and consistency achievable by a federated cataloguing and search mechanism, in addition to mechanisms that contributors may have for identifying and differentiating data using a common format scheme.

Annotation "as one sees fit" could involve defining and then filling in forms, or adopting forms used elsewhere for similar data. In most cases, the resulting structure will be complex, since ideally it should contain the same information as is required for the methods section of a publication. However, unlike a centralized database, the decision of what to include rests entirely with the data provider. The utility of the data to others is, of course, heavily dependent on the associated annotation, so providing relevant information is crucial. We have investigated the start of a set of software tools to provide this type of functionality, which we now describe further.

15.2.2 Axiope

Axiope (www.axiope.org) is an initial attempt to provide a framework and software tools for constructing and establishing federated data sharing. Although the site and tools are still under development, they are available in a preliminary version. This preliminary version includes examples of federated cell morphology archives, including data originally located in the Duke/Southampton archive (Cannon et al, 1998). The current server software in the Axiope system accepts messages from the client tool whenever a new website (data site) is created. It then visits the site and extracts all the metadata with standard http requests (Figure 15.4). The server software can then construct

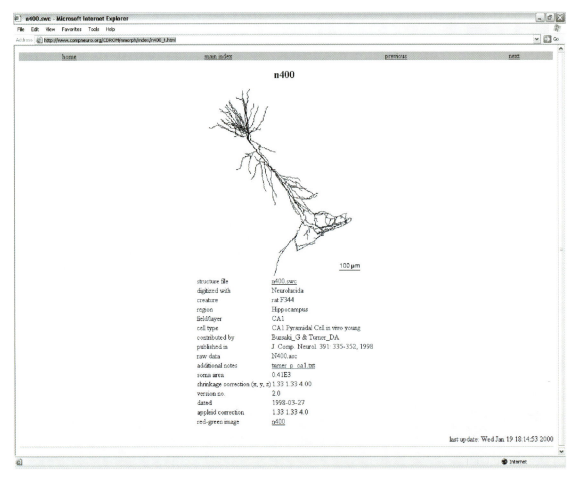

Figure 15.2   A screen shot of a cell and the associated data in the www.compneuro.org database. Accompanying data include the animal species, brain region, publication source and date, raw data, metadata, and correction factors.

a searchable catalog of the contents of one site and can attempt to merge metadata from different sites, which contain the same type of data. At its simplest, this involves spotting terms shared by the metadata on two sites (for example, "keywords" is likely to appear in a great many records). In general, however, effective federation is not possible without either more standardization on the individual sites or the provision of further metadata information to be used by the server to associate datasets. At this time there are few "standards" in terms of how to (a) provide or modify these metadata to allow federation and (b) search for common types of cells.

Although still in its infancy, using the Axiope system "off the shelf" already improves upon the results achieved in the Duke/Southampton archive, which involved considerable custom software development. However, the archive also included valuable tools for editing and modifying cell structures, including CellViewer, which extended considerably the capabilities of the proprietary Neurolucida software tools (Microbrightfield, Inc). These tools, such as used in editing or converting files, are not immediately available. Almost all the cost of using the system involves metadata entry, which is the one inescapable component that can only be done by data providers. Further developments will add computational modules, providing enhanced visualization options or even complex searching based on derived properties, building on the original metadata. The major dilemma is that

the providers of the data get little out of constructing a database and doing the work of providing the metadata, other than a sense of altruism and contribution to the neuroscience community at large.

Axiope Catalyzer is a combined data management, data-sharing, and data publication tool. It can be used on its own by a single user or a small group, but it also interoperates with other software and services to provide a complete data management system for large groups of users spread over different sites. The first use of the program is to create and edit local catalogs containing descriptive information about resources. In many cases the resources will be files of binary data (images, documents, data from digital acquisition systems), but equally catalogs can hold information about other things (books, CDs, people, etc.).

Unlike a text document, the information that a catalog can hold is highly structured. For example, a simple entry about a file might have fields for its name, size, and creation date. Each entry must correspond to a form that the system knows about. For those who are familiar with relational databases, the use of "form" here corresponds roughly to a "table" in database terminology. But unlike a database, the user is free to change the forms at will. That is, fields can be added and removed, and their names or types can be changed without problems. This is the most important feature of Catalyzer compared with other systems. It recognizes that the user may not, a priori, know all the pieces of information

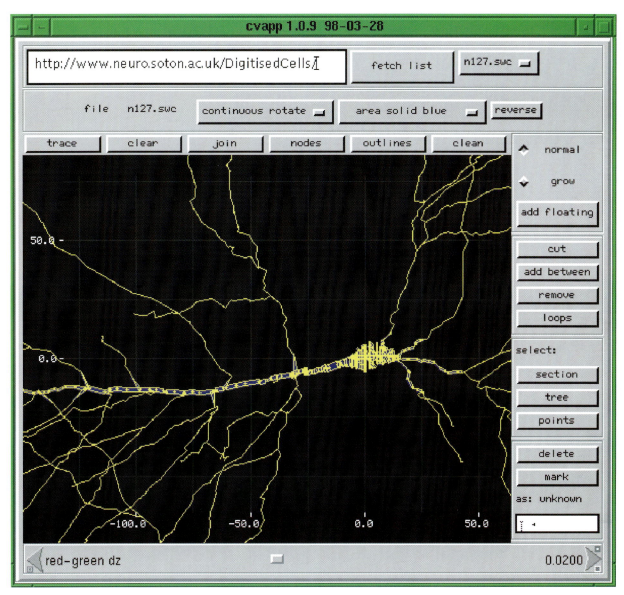

Figure 15.3   Screen shot of the CVAPP program interface.

that should be entered. It therefore makes it simple to modify the underlying form while data are being entered.

Once one or more catalogs have been created, Catalyzer provides several ways to mine and share the data. There is a built-in search system to relocate information rapidly. There are also facilities for creating websites from the catalog and for uploading them to a Web server. Such sites can be used to share information either within a group on a private server, or globally on the World Wide Web.

The Data Model

The data model adopted for Catalyzer is closely related to the object model used in the combination of XML with XML schema. Information stored in XML files is organized as a hierarchy or tree of elements. Elements can be of a variety of types, and for each type an XML schema can be used to specify what it is allowed to contain in terms of attributes (name–value pairs) and other elements.

In the same way, the Axiope data model is based around "catalogs" (analogous to whole XML documents), each of which can contain a variety of "records" (analogous to the elements in an XML document). Each record must correspond to a "form" (analogous to an XML schema) that specifies what pieces of information it should contain. The model is distinct from XML in being highly restricted. In effect, it presents a selected subset of what is possible with XML in a way that is accessible to a non-expert user. Using XML directly gives more expressive power, but also requires much more technical expertise.

The terminology is derived from familiar applications such as library catalogs. Such a catalog may contain a large number of records for different types of items: books, journals, electronic resources, etc. The different types of items could all be described using subsets of the same format (e.g., leaving the ISBN field blank for items that do not have an ISBN) or there might be different types of records for the different types of object. Axiope catalogs use the latter approach, and much of the user interface

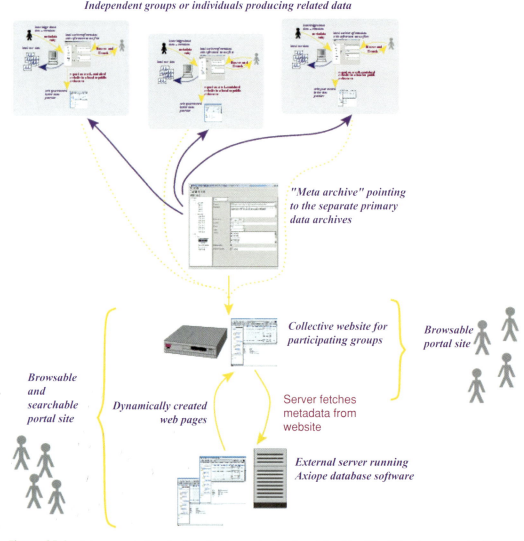

Independent groups or individuals producing related data

"Meta archive" pointing to the separate primary data archives

Collective website for participating groups

Browsable portal site

Browsable and searchable portal site

Dynamically created web pages

Server fetches metadata from website

External server running Axiope database software

Figure 15.4 Axiope organization. A schematic diagram showing the relationship of the different components of the Axiope system.

is involved with creating and editing the forms that are used to specify what information is required for particular types of records.

Forms are lists of field specifications, each of which governs a single item of information in a corresponding record. The specification includes the name of the item (roughly like the column heading in a spreadsheet) and its type, and it can hold additional descriptive information about the item that can later be presented to the user as help text. There are a number of possible types supported including single-line sections of text, extended areas of text, numbers, dates, times, URLs, references to local files, and Boolean (true/false) values. For the short text elements there are further options to specify how they should be presented including lists and menus where the possible values can be specified as part of the form. Items in a form can also be extracted for reuse elsewhere as new data types. One particular application is in the use of taxonomies. In this case—for example, in labeling a brain region—a menu can be built up containing the possible valid terms. Once created, the field specification comprising the data type and possible values can be stored and reused in different contexts. In this way it allows storing and sharing of a taxonomy for the particular domain.

The tree-based data model used in Catalyzer lies in the space between spreadsheets and relational databases. It can be used to create the equivalent of a single table by setting up a form for the column headings and a series of records for the rows. But it can also be used to set up a large number of different forms, each like a table in a relational database. The forms can specify that that certain fields are required to make references between records of different types, thereby covering much of the "foreign key" functionality of a database.

Inevitably, as a hybrid technology, there also have to be disadvantages or payoffs compared with the alternatives. The Axiope solution deliberately abandons several key features of relational databases. Most notably, there is no support for transactions (the ability for several users to edit the same data at the same time without problems) and there is no independent query language. The system is also much more computer-intensive; but in view of the dramatic increases in hardware technology, this is no longer important for most laboratory applications. The user interface is slightly less convenient and flexible than a spreadsheet, but this seems to be a necessary price in order that the resulting documents have sufficient structure to be readily processed later on.

Importing and Generating

Any data management system that claims to be useful in a lab environment must solve the problem of the costs associated with metadata entry compared with the advantages of having data in the system. Minimizing the effort involved in getting data in is therefore a primary concern. Catalyzer currently supports two ways of creating catalogs from external metadata sources. The first is a catalog generator that uses properties from the file system to describe a folder tree of files. The second is an importer from tables of comma-separated values (CSV) as can be exported from spreadsheets and database programs. This provides a simple way to read in data from multiple spreadsheets into a catalog.

Folder Tree Description

The generator can be run over a folder tree and creates a catalog that exactly matches the structure of the original folder hierarchy. Folders are mapped to records of type Folder within the catalog, and files are mapped to records that conform to an internally generated "file" form. They are populated with a link to the resources (the file itself), the file name, size, file extension, and the names of its parent directories. The generate function can be used as a convenient starting point for adding metadata about files. Once the initial catalog has been generated, forms can be extended to include more information about the cited resources. For example, a catalog created from a directory of images will contain records using the standard image form. This can be extended, or copied and modified, to include fields particular to describing the image, such as acquisition parameters or content description.

Importing from Spreadsheets in CSV Format

Most spreadsheet applications (e.g., Excel, OpenOffice) have a "Save as CSV" or "Export to CSV" facility. Once a spreadsheet has been exported as a CSV file, Catalyzer can import it into an existing catalog. This creates a new form with fields named according to the column headings in the first line of the spreadsheet, as well as one new record for each row of the spreadsheet.

User Interface Examples

Figures 15.5 and 15.6 show the two main views provided for creating, editing, and browsing catalogs. The first view shows a single record at a time and provides facilities for modifying the forms in the system and entering data into records. The second view looks more like a spreadsheet. Not all the form-editing functions are available in this mode but it does facilitate rapid data entry and manipulating multiple records at a time. Figure 15.7 shows a view of the website that is created by default from the morphology archive example.

Potential Problems and Limitations

There are currently a number of limitations of Axiope. For example, it is not yet sufficiently complete to set up an entire website at an independent location. Although the tools available so far are likely to make the process of cataloguing data easier, particularly for internal use, these do not substitute for website generation.

One of the most difficult aspects of the metadata, which requires specification, is the assessment of the insecurity of the data, in terms of accuracy compared to a standard. For example, to what percent or degree of accuracy are the dendritic diameter measurements? Specification of accuracy and experimental errors is a critical concern for modeling, so these errors can be included in the analysis and bracketed. Thus, these items are as critical as the data itself in any subsequent use.

Using the tools available on the Axiope website, we were able to establish a remote website (on the Axiope server) and to set up criteria for metadata entry and manipulation. However, it is not clear how this process would be generalized and standardized, to allow various types of data and to generate additional websites through a set of access rules. This would decrease the burden of work for the provider, not to have to maintain a website and worry about security or marketing, since this would be provided by the main website. This would be a valuable option.

Even with easier website generation and local control of the data, there still remains a significant mismatch between providers and users in this case of morphometric neuronal databases. This neuron example remains unlike gene databases and protein databases, where a mutual interest and rationale for communication coexist, because providers get additional data with which to compare to their own data. If the availability was high and there were sufficient data providers using such a common platform, then experimentalists may eventually access the federated database, to be able to easily compare others' cells with their own.

Additionally, federated databases still require that the provider generate high-quality data, which are uniform in the sense that the data can be specified into sets and preferably have already been published, and thus subjected to peer review, confirming the quality of the data.

15.2.3 Utility of Measured Neurons

Measurements from Digitized Neurons

One of the advantages of storing reconstructed neurons in digital format is the possibility to extract any morphological information in a consistent, reliable, and automated way. The software accompanying the Neurolucida reconstruction system provides limited morphological analysis. Recently, a more general, powerful, flexible, portable, and nonproprietary software tool to measure morphological parameters, called L-Measure, was publicly released (http://www.krasnow.gmu.edu/L-Neuron; Scorcioni and Ascoli, 2001). L-Measure consists of a C++ engine controlled by a user-friendly Java graphical interface (tested for Windows and Linux systems). Users select the cells to be analyzed, define the parameters of the analysis, and specify the parts of the dendrites that should be considered or excluded. The engine parses the digital files and returns the requested information as raw data columns (which can be imported in spreadsheet program or statistical packages) or as simple statistical summaries. L-Measure can read all known digital morphological formats (including Neurolucida files and the format adopted in the Duke/Southampton archive). Any combination of more than 40 morphological parameters can be measured, including diameter, length, surface, volume, distance (Euclidian or along the path), angles, overall spread and extend of the trees, topological

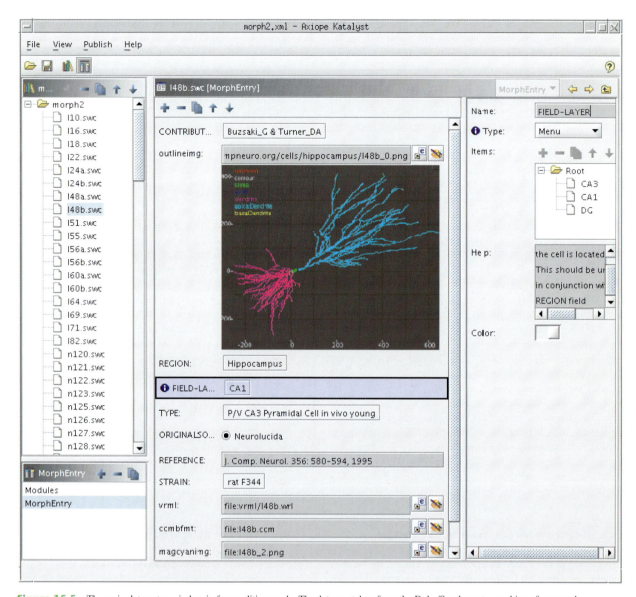

Figure 15.5   The main data entry window in form-editing mode. The data are taken from the Duke/Southampton archive of neuronal morphology and includes structures and images of digitized cells. On the left is a tree view of all the records in the archive. The large middle area shows a single record, and the panel on the right shows the highlighted field from the form. In the original archive, all the metadata were sections of text. On importing to Catalyzer, this can be easily converted to menus, list selections, or "radio buttons." The right panel shows the results for the item originally labeled "FIELD-LAYER." This has been converted into a menu with three items corresponding to the values present in original archive. The use of a menu makes subsequent data entry (for adding new cells) much less time-consuming and reduces errors.

symmetry, number of termination and bifurcation events, and so on. A typical L-Measure query, corresponding to an "advanced" Sholl analysis, could be: "Plot the number of branches of diameter greater than 1 μm versus their Euclidian distance from the soma." A comprehensive analysis of hippocampal neurons using these types of analysis revealed large differences between hippocampal granule cells and pyramidal cells in particular (Cannon et al., 1999).

Simulations of Dendritic Growth and Branching
Growth and branching characteristics of neurons determine the three-dimensional appearance of dendrites, within externally imposed constraints. Quantitative analysis of such growth includes definition of likely branch angles, likelihood of branching at

particular distances, branching complexity, and whether dendrites behave as fractal entities in terms of organizational patterns (Cannon et al., 1999). A critical question in cellular and developmental neuroanatomy is how much of the structural complexity is captured by a given morphological analysis. Computational modeling may provide an empirical answer. If synthetic neurons can be generated on the basis of the measured statistics, real and synthetic neurons of a given class can be quantitatively compared to search for any discrepancy, thus guiding the investigator toward a more complete description of the natural structures. L-Neuron (Ascoli and Krichmar, 2000) is one of the software tools that generates synthetic neurons in a format compatible with CVAPP, L-Measure, and common electrophysiological simulators.

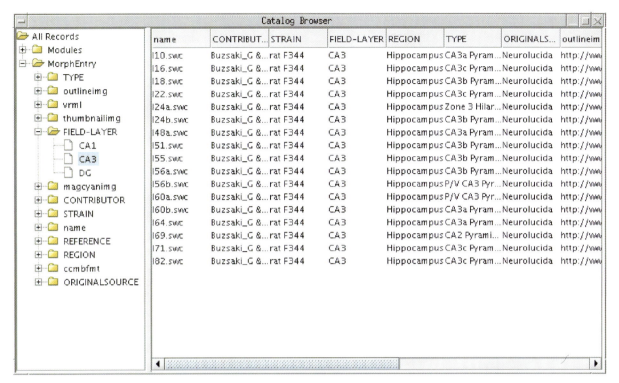

Figure 15.6 The "browse" view of the morphology archive. Whereas the main window shows the records as they are organized in the catalog, the browse view shows the logical structure of the records. The tree is organized by record type, then by the fields present in each record, and finally by the values the field takes on. In this example, the display area shows all records that have the value "CA3" for the field named "FIELD-LAYER" in records of type "MorphEntry."

L-Neuron (publicly available in Windows and Linux alpha versions) uses an iterative approach to defining synthetic neurons based on the local branch diameter (stochastically determining dendritic elongation, bifurcation, and termination) and three-dimensional directional biases (constraining dendritic orientation). Simulated neurons can be rendered and stored as digital files like the real, reconstructed neurons and can thus be subjected to the same morphometric analysis (Ascoli et al., 2001b; Ascoli and Samsonovich, 2002). They can also be used as substrates for electrophysiological and biophysical simulations. Examples of virtual neurons generated with this approach can be found in the Virtual Neuromorphology Electronic Database at the L-Neuron website (http://www.krasnow.gmu.edu/L-Neuron/). In addition to those implemented in L-Neuron, other algorithms have been developed by several laboratories to simulate dendritic morphology (Ascoli, 2002).

Modeling of Dendritic Function

An extensive discussion of the practical aspects of electrophysiological models and their computational simulations is obviously beyond the scope of the present chapter. Extended texts have been devoted to this topic, including excellent recent books (Bower and Beeman, 1998; Koch and Segev, 1998; De Schutter, 2000) and reviews (Hines and Carnevale, 1997). Practical aspects of biophysical simulations based on anatomical reconstructions such as those described in this chapter have also been recently discussed (Lazarewicz et al., 2002a). The following sections are kept at a general level to provide the non-expert reader with a functional context for Web-based archives of neuronal morphology.

Detailed comparison of neuronal shape and function necessarily involves the use of detailed morphometric data. Since in most cases dendrites are roughly assumed to be cylindrical, cable models represent a first approximation. Dendritic dimensions and lengths are transformed into an electrical (electrotonic) image of a neuron. The electrotonic length values can be calculated for each dendritic path and for the whole neuron (Turner, 1984). Description of synaptic potential conduction may also be simulated onto discrete dendritic sites, for estimation of synaptic efficacy. The severe limitations of cable methods are the slow computation times, the limited number of input, and output sites that may be included, and the lack of a format to include voltage-dependent properties (Turner et al., 1991, 1996).

The most common approach to estimating neuronal function is the use of compartmental models, which divide the neuronal structure into connected equipotential regions. These models can represent a wide variety of dendritic morphologies and spatially varying membrane properties (Migliore et al., 1995; Turner et al., 1991, 1996). If there are many small compartments, the model can be regarded as a discrete numerical solution to the cable equations described above. Within each compartment the membrane and intracellular parameters can be lumped and the model then reduces to a system of (possibly nonlinear) differential equations. The inclusion of appropriate neuronal parameters is critical, but often difficult to specify quantitatively. There are many parameters for which good quantitative estimates are not available, thus increasing the need for extrapolation of data. These techniques offer the clear advantage of incorporating a number of nonlinear properties that cannot be included in the cable model format and also the inclusion of multiple, concurrent processes as well as

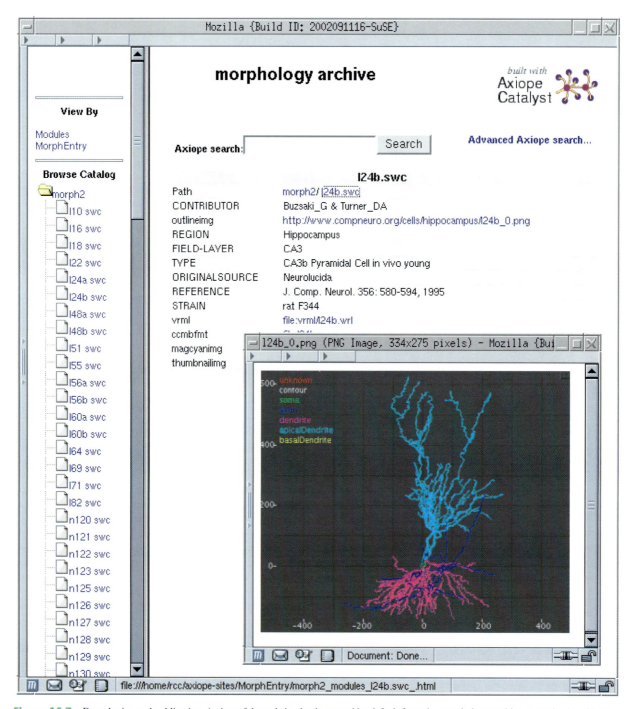

Figure 15.7 Data sharing and publication. A view of the website that is created by default from the morphology archive example. As with the main window, the catalog structure is shown as an expandable tree on the left. The center area shows a single-page record view. The site also contains paginated lists of items. The inset panel shows an image of one of the cells opened in a pop-up window.

multiple recording sites. The interaction between the distribution of nonlinear properties and the details of dendritic structure can result in extremely complex subcellular mechanisms of spike initiation and propagation (Lazarewicz et al., 2002b).

A very important (and often overlooked) aspect of morphologically realistic biophysical models of neuronal electrophysiology is the spatial resolution necessary to achieve numerical stability. The basic assumption of compartmental models is that membrane voltage is constant within each compartment. This assumption is used in the numerical approximation of the differential equations describing passive and active membrane properties. In reality, however, the voltage gradient is spatially continuous. The error introduced in the approximate numerical solution depends on the size of the model compartments: The smaller the compartment, the better the approximation. This relationship heavily depends on the type and complexity of the biophysical model, the specific kinetic equations, and the activity pattern under consideration. On the other hand, it is also important not to use *too small* a compartment size, which would slow down the simulation time as well as introduce the risk of numerical instability. While several methods are available to estimate the appropriate compartment size (Lazarewicz et al., 2002a), it is important to verify the stability of detailed biophysical models.

Structure–Activity Correlations

The parallel simulation of morphological and electrophysiological properties also allows a complete and direct investigation of the structure–activity relationship in single cells. One example is the simulation of synaptic potentials in CA1 hippocampal pyramidal neurons by cable modeling (Turner, 1984). Using this format of cable modeling, a reasonable fit was obtained between model predictions and physiological estimates of dendritic EPSPs (Turner, 1988). In a recent study, Krichmar and coworkers loaded the same complex model of passive and active membrane properties onto all reconstructed CA3 pyramidal cell morphologies available at the Duke/Southampton archive. Upon identical stimulation protocols, each individual cell produced a distinct simulated response. Overall, the electrophysiological behavior of these cells under simulated somatic current clamp conditions varied from quiescence to regular spiking, complex spiking, and irregular bursting (Krichmar et al., 2002). This broad range of firing characteristics (which is known for real CA3 pyramidal cells) was completely due in this modeling study to the morphological variability, since the biophysical and stimulation parameters were held constant for all cells (see also Migliore et al., 1995). The results demonstrated that dendritic size has a nontrivial effect on neuronal excitability (e.g., larger trees result in lower bursting rates but higher spiking rates within bursts) and that the specific dendritic branching pattern may influence burst duration. A similar study was carried out on a different subset of neurons from the Duke/Southampton archive, to determine the possible electrophysiological effects of cortical lesions (Nasuto et al., 2001).

Real neurons are both morphologically and biophysically heterogeneous. Thus, the direct experimental investigation of the influence of dendritic morphology on the electrophysiological behavior of neurons may require an exceedingly large number of recordings to isolate statistically the most significant correlation parameters. The use of computational modeling, together with archived digital reconstructions of real neurons, makes this type of study possible. In addition, simulated neurons may be synthesized with extreme morphological characteristics and run through the biophysical simulations to verify the correlation trends between structural and electrophysiological properties (Ascoli et al, 2001b). Finally, this approach may be extended to the study of the structure–activity relationship under more realistic conditions of synaptic stimulations, which are practically impossible to fully control in experimental preparations.

Extension to Network Modeling

Once validated in comparison to cohorts of measured neurons, large numbers of anatomically realistic synthetic neurons can also be assembled into real-scale networks. An essential step in this process is to define the appropriate surfaces and volumes in space that delimit the anatomical boundaries for the spatial distribution of neurons in the tissue. Real or synthetic neurons can then be virtually packed in space based on their preferential axes of orientation (e.g., Senft, and Ascoli, 1999). These constructs can provide a bridge between single-cell morphology and system-level stereology, as well as between subcellular bio-

physical properties and emergent network activity. There is the potential to build circuitry and regional brain models based on both a combination of measured neurons and synthetic neurons, with stochastic variability added to enhance the resemblance of simulated neurons to a large population of cells within the brain (Senft and Ascoli, 1999). Defining the innate characteristics of neuronal growth and associated physiological functioning should enhance the simulation of a large population of synthetic neurons, as well as enhance their application toward realistic network and regional simulation.

15.3 DISCUSSION

As the number of laboratories using Neurolucida grows, there may become stronger pressure for sharing, as experimentalists and users want direct comparisons of their data with that in the published literature. Thus, there may be a shift toward more contribution to the world's literature and databases, as contributors in effect also become users, for the purpose of comparing their own data to that of others. This change would mirror that of other databases, as contributors perceive a need to use data from the entire database for their own purposes, rather than just an altruistic donation of data. This effect may stem from a need for experimentalists to have direct comparison of other's results, rather than just indirect comparison from literature values. This is a separate need from using available data for modeling of single cells, networks of neurons, or building a global brain atlas based on cellular structures, rather than nuclear density profiles.

15.3.1 Usefulness of Cell Archives in General

The use of multiple, individual neurons for modeling incorporates inherent variability across neurons, and it provides a realistic structural base for various types of structure–function analyses (Turner, 1984). The trend has clearly progressed toward the use of reconstructed neurons in modeling, as cell reconstructions have become more widespread, more neurons are available in archives, and computer-modeling programs can now easily accommodate complex neurons. Since central archiving has not proven feasible or popular, considerable hope may be placed in the use of federated data base structures, such as Axiope described above, wherein data are not stored centrally, but rather accessed from multiple servers. This type of system may overcome many of the objections of groups generating neuronal structures by the maintenance of authorship and control, and it possibly could lead to thousands of neurons being available for public dissemination and subsequent analysis. The widespread availability of neurons may then lead to a cellular analysis of brain circuitry, rather than a systems-level approach commonly in use to approximate brain function.

15.3.2 Usefulness of Axiope Specifically

Axiope is clearly in its very beginning in terms of capabilities, usefulness, and features. Unlike the previous generation of CellViewer in the Duke/Southampton archive, all of the tools are directed toward informatics and databasing, rather than tools for the users to view, access, and edit neuronal structures. These tools are not sufficient to establish an independent database, but provide a base for specifying metadata and for organization of data. There need to be better methods of publicizing the availability and use, since currently only the modeling community and users are aware of this capability. One potential method for enhancing contributions is to (a) point out the potential usefulness to even contributors and (b) advertise through the various companies involved, such as Microbrightfield. Thus, further improvement in the tools available to easily construct an independent website, structure metadata, and enhance visibility and availability on the Web will be critical, as well as tools for editing and using the structures, beyond those commercially available. Contributors will need to be enticed to develop such independent, federated websites, possibly through funding to provide some on-site support for establishing the sites and generating the appropriate metadata.

15.3.3 Issues of Databases in General

Quality of the data is critical, including estimates of accuracy and limitations. Should included data be published? Publishing data allows the contributing laboratories to have primary credit and to establish a peer-reviewed base, as well as to advertise the database in which the data exist for public access. Data should be in the open domain after publication, since the contributors have already benefited from use and access to the data. How to optimally encourage open databasing? A symmetrical interface would be optimal, wherein both parties get something out of the contribution, in addition to providing a benefit to the general scientific community. Also, if there are a really a limited number of users, the provider may not sense that there is a demand, which may be the case for the vast majority of Neurolucida users. Popularizing the creation of locally controlled, federated websites for enhancing access to the data will be critical, confirming that access can be controlled and that credit for the original work can be maintained (particularly through prior publication), while still allowing considerable altruistic benefit. These are all issues for future funding, which should include allowance to assist users in setting up databases with an on-site expert, coaxing to encourage contribution by peer-reviewed journal and grant funding sources, and enlarging the domain of contributors and users so that a mutual benefit can be achieved on both sides.

References

Ascoli, G.A. (1999). Progress and perspectives in computational neuroanatomy. *Anat. Rec.* **257**, 195–207.

Ascoli, G.A. (2002). Neuroanatomical algorithms for dendritic modeling. *Network—Comput. Neural Syst.* **13**, 247–260.

Ascoli, G.A., and Krichmar. J.L. (2000). L-Neuron: A modeling tool for the efficient generation and parsimonious description of dendritic morphology. *Neurocomputing*, **33**, 1003.

Ascoli, G.A., and Samsonovich, A. (2002). Bayesian morphometry of hippocampal cells suggests same-cell somatodendritic repulsion. *Adv. Neural Inf. Process. Syst.* **14**, 133–139.

Ascoli, G.A., Krichmar, J.L., Nasuto, S.J., and Senft, S.L. (2001a). Generation, description and storage of dendritic morphology data. *Philos. Trans. R. Soc. London, Ser. B* **356**, 1131–1145.

Ascoli, G.A., Krichmar, J.L., Scorcioni, R., Nasuto, S.J., and Senft, S.L. (2001b). Computer generation and quantitative morphometric analysis of virtual neurons. *Anat. Embryol.* **204**, 283–301.

Bower, J.M., and Beeman, D. (1998). *The Book of Genesis. Exploring Realistic Neural Models with the GEneral NEural SImulation System*, 2nd ed. Springer-Verlag, New York.

Cannon, R.C., Turner, D.A., Pyapali, G.K., and Wheal, H.V. (1998). Online archive of reconstructed hippocampal neurons using CellViewer. *J. Neurosci. Methods* **84**, 49–54.

Cannon, R.C., Wheal, H.V., and Turner, D.A. (1999). Dendrites of classes of hippocampal neurons differ in structural complexity and branching patterns. *J. Comp. Neurol.* **413**, 619–633.

De Schutter, E. (2000). *Computational Neuroscience: Realistic Modeling for Experimentalists*. CRC Press, Boca Raton, FL.

Hines, M.L., and Carnevale, N.T. (1997). The NEURON simulation environment. *Neural Comput.* **9**, 1179–1209.

Horcholle-Bossavit, G., Gogan, P., Ivanov, Y., Korogod, S., and Tyc-Dumont, S. (2000). The problem of morphological noise in reconstructed dendritic arborizations. *J. Neuroci. Methods* **95**, 83–93.

Koch, C., and Segev, I. (1998). *Methods in Neural Modeling: From Ions to Networks*. MIT Press, Boston, MA.

Krichmar, J.L., Nasuto, S.J., Scorcioni, R., Washington, S.D., and Ascoli, G.A. (2002). Effects of dendritic morphology on CA3 pyramidal cell electrophysiology: A simulation study. *Brain Res.* **941**, 11–28.

Lazarewicz, M.T., Boer-Iwema, S., and Ascoli, G.A. (2002a). Practical aspects in anatomically accurate simulations of neuronal electrophysiology. In *Computational Neuroanatomy: Principles and Methods* (G. Ascoli, ed.). Humana Press, Totowa, NJ.

Lazarewicz, M.T., Migliore, M., and Ascoli, G.A. (2002b). A new bursting model of CA3 pyramidal cell physiology suggests multiple locations for spike initiation. *BioSystems* **67**, 129–137.

Lo, D.C., McAllister, A.K., and Katz, L.C. (1994). Neuronal transfection in brain slices using particle-mediated gene transfer. *Neuron* **13**, 1263–1268.

Migliore, M., Cook, E.P., Jaffe, D.B., Turner, D.A., and Johnston, D. (1995). Computer simulations of morphologically reconstructed CA3 hippocampal neurons. *J. Neurophysiol.* **73**, 1157–1168.

Nasuto, S.J., Krichmar, J.L., and Ascoli., G.A. (2001). A computational study of the relationship between neuronal morphology and electrophysiology in an Alzheimer's Disease model. *Neurocomputing* **38–40**, 1477–1487.

Pyapali, G.K., and Turner, D.A. (1996). Increased dendritic extent in CA1 hippocampal pyramidal cells from aged F344 rats. *Neurobiol. Aging* **17**, 601–611.

Scorcioni, R., and Ascoli, G.A. (2001). Algorithmic extraction of morphological statistics from electronic archives of neuroanatomy. *Lect. Notes Comput. Sci.* **2084**, 30–37.

Senft, S.L., and Ascoli, G.A. (1999). Reconstruction of brain networks by algorithmic amplification of morphometry data. *Lect. Notes Comput. Sci.* **1606**, 25–33.

Turner, D.A. (1984). Conductance transients onto dendritic spines in a segmental cable model of CA1 and dentate hippocampal neurons. *Biophys. J.* **46**, 85–96.

Turner, D. A. (1988). Waveform and amplitude characteristics of evoked responses to dendritic stimulation of CA guinea-pig pyramidal cells. *J. Physiol.* (London) **395**, 419–439.

Turner, D.A., Wheal, H.V., Cole, H., and Stockley, E. (1991). Three-dimensional reconstructions and analysis of the cable properties of neurones. In *Cellular Neurobiology* (J. Chad and H.V. Wheal, eds.), p. 225. Oxford University Press, Oxford.

Turner, D.A., Isaac, J., Chen, Y., Stockley, E.W., and Wheal, H.V. (1996). Analysis of dendritic synaptic sites in hippocampal CA1 pyramidal cells—assessment of variability and nonuniformity. In *Excitatory Synaptic Transmission* (H.V. Wheal, ed.), p. 171. Academic Press, San Diego, CA.

Turner, D.A., Cannon, R.C., and Ascoli, G.A. (2002). Web-based neuronal archives: Neuronal morphometric and electrotonic analysis. In *Neuroscience Databases* (R. Kotter, ed.), p. 81. Kluwer Academic Press, Boston.

Cellular Models

Raimond L. Winslow, Ph.D.

CONTENTS

Databasing the Brain. Edited by Stephen H. Koslow and Shankar Subramaniam
ISBN 0-471-30921-4 © 2005 John Wiley & Sons, Inc.

16.1 INTRODUCTION

Cardiac electrophysiology is a field with a rich history of integrative modeling extending back to the early 1960s and the first models of the cardiac action potential developed by Noble (1960, 1962) shortly after the Hodgkin and Huxley model of the squid action potential. Indeed, some of the most fundamental advances in modeling of the cell, such as the formulation of dynamic models of regulation of intracellular ion concentration, were first developed in this field (DiFrancesco and Noble, 1985). In this chapter, we will review the state of the art in computational modeling of the cardiac myocyte. Without doubt, there are clear morphological differences between cardiac myocytes and neurons— specifically, that myocytes lack the complex dendritic architecture that many neurons use as a substrate for signal processing and integration. However, there are many issues that must be faced by both modelers of neuronal and myocyte function. Chief among these is how models of discrete processes (ion channels, membrane transporters, signaling pathways) may be formulated and integrated to obtain a more complete picture of the molecular basis of cell function. This theme will be addressed in this chapter.

16.2 BACKGROUND INFORMATION

16.2.1 Structure of the Myocyte

Cell Morphology
The heart may be divided into several different regions, with muscle cells in each region serving a distinct role. Rhythmic electrical activity of the heart is initiated within the sinoatrial (SA) node, located in the right atrium near the base of the superior vena cava. Ion channels present in the SA node cell membrane give rise to oscillating electrical activity. Frequency entrainment of these oscillations is ensured within the SA node since adjacent cells are electrically coupled by gap junctions. Waves of electrical depolarization generated in the SA node propagate into and through the right and left atria, initiating contraction of these muscle masses and hence filling of the right and left ventricles. The depolarization wave is then conducted through cells of the atrioventricular (AV) node and the Purkinje fiber system to initiate electrical depolarization and contraction of the ventricles.

Figures 16.1A and 16.1B show the morphology of different cardiac muscle cells. Morphology is diverse, ranging from small spindle-shaped SA node cells (Figure 16.1A—1,2,3) to longer cylindrical atrial cells (Figure 16.1A—4) and ventricular muscle fibers (Figure 16.1B). This figure makes clear a significant difference between neuronal and cardiac myocyte morphology: Myocytes lack the complex, branching dendritic architecture common to many neurons. Since the vast majority of muscle cells of the heart are ventricular muscle cells, and since abnormality of function of these cells is of such importance to human disease, the remainder of this chapter will focus on properties of the ventricular myocyte.

Intracellular Organization
Figure 16.1C shows a schematic illustration of the sarcomere, the fundamental unit of contraction of the cardiac muscle fiber, and its relationship to other organelles within the myocyte. The sarcomere is the contractile apparatus between adjacent Z-disks. The H-zone contains the thick (myosin) filaments and is the region within which there is no overlap with thin (actin) filaments. The A-band is the region spanned by the length of the thick filaments. The shaded region in Figure 16.1C represents the region of overlap of thick and thin filaments. Muscle contraction is accomplished by the sliding motion of the thick and thin filaments relative to one another in this region in response to elevated levels of intracellular calcium (Ca^{2+}) and ATP hydrolysis.

Three additional structures shown in Figure 16.1C have particular importance. These are the T-tubules, the mitochondria, and the network and junctional sarcoplasmic reticulum (NSR and JSR, respectively). T-tubules are cylindrical invaginations of the sarcolemma that extend deep into the cell and approach the sarcoplasmic reticulum (SR). L-type Ca^{2+} channels (LCCs) are localized predominantly to this region of the T-tubule membrane. The SR is a luminal organelle located throughout the interior of the cell and is involved in the uptake, sequestration, and release of Ca^{2+}. SR can be divided into two main components known as junctional SR (JSR) and network SR (NSR). The JSR is that portion of the SR most closely approximating the T-tubules, and the distance between these two structures is being $\sim$12–15 nm. The close proximity of these two structures forms a restricted subspace in which Ca^{2+} levels are thought to rise to very high concentrations—a region also known as the diadic space. The majority of the SR Ca^{2+} release channels, known as the ryanodine receptors (RyRs), are found in the JSR directly apposed to the LCCs. Finally, Figure 16.1C shows that mitochondria are located predominantly in the A-band region and near the z-lines, so that ATP production is located near the major sources of ATP utilization (e.g., muscle contraction and pumping of Ca^{2+} into the NSR).

16.2.2 Ion Channels and Electrical Excitability

Overview of the Cardiac Action Potential and Calcium Transients
While neurons and myocytes differ with respect to cellular morphology, they share one stfrong similarity. In both cell types, a complement of voltage-gated ion channels, membrane pumps, and exchangers interact to give rise to complex electrical behavior. A major focus of molecular cardiobiology has been to identify these transporter systems, characterize them both experimentally and in terms of biophysically detailed models, and understand how their interactions support the information processing needs of the myocyte. Here, we summarize the major ion transporters that have been modeled in the cardiac myocyte.

The complement of voltage-gated ion channels expressed in mammalian cardiac ventricular myocytes isolated from different species differs significantly. The main factor producing this variation is difference in heart rate, ranging from as high as $\sim$ 600 beats per minute (bpm) in murine to $\sim$ 60 bpm in larger mammalian myocytes. These dramatic differences in heart rate are supported by differences in expression of ion channels regulating action potential (AP) duration. Here, we focus on sarcolemmal ion channels expressed in large mammalian species such as canine and human.

(A)

(B)

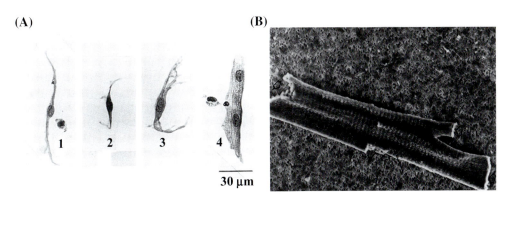

(C)

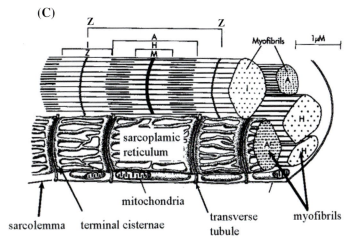

Figure 16.1 (**A**) Micrographs of sino-atrial node cells illustrating morphology. Cells numbered 1–3 are examples of elongated spindle cells (1), spindle cells (2) and spider cells (3) located in the sinus node. Cell number 4 is an atrial cell. Reprinted from Verheijck et al. (1998), with permission of the American Heart Association. (**B**) Scanning electron micrograph of a ventricular myocyte Reprinted from Fozzard et al. (1991), with permission of Raven Press. (**C**) Structure of the cardiac sarcomere. Reprinted from Roos (1997), with permission of Academic Press.

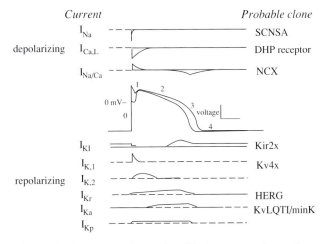

Figure 16.2 Schematic illustration of the large mammalian cardiac ventricular myocyte action potential (membrane potential in millivolts as a function of time) illustrating depolarizing and repolarizing current (*left*) and alias gene names encoding each of these currents. Reprinted from Tomaselli and Marban (1999), with permission of the American Heart Association.

Figure 16.2 shows a schematic illustration of the large mammalian cardiac action potential. The currents mediating the AP upstroke (Phase 0) are the fast inward sodium (Na) current (denoted as I_{Na}; for review, see Marban et al., 1998) and, to a lesser extent, the L-Type Ca^{2+} current (denoted $I_{Ca,L}$; for review, see Kemp and Hell, 2000). The Phase 1 notch, which is apparent in ventricular myocytes isolated from epi- and mid-myocardial regions but which is largely absent in those isolated from the endocardium, is produced by activation of the voltage-dependent transient outward potassium (K) current (denoted $I_{to,1}$). In the canine, a transient voltage-independent Ca^{2+}-modulated Cl^- current contributes to the Phase 1 notch (denoted $I_{to,2}$); however, this current is not known to be expressed in human. The Phase 2 plateau is a time during which membrane conductance is very low, with potential being determined by a delicate balance between small inward and outward currents. The major inward plateau current is $I_{Ca,L}$, and major outward plateau currents are generated by the rapid and slow-activating delayed outward rectifier currents, denoted as I_{Kr} and I_{Ks}, respectively, and the plateau K current I_{Kp}. Finally, repolarization Phase 3 is produced by the hyperpolarizing activated inward rectifier K current I_{K1}.

Three major ion transporters and exchangers play a critically important role in shaping properties of the cardiac AP and the Ca^{2+} transient, as well as in long-term regulation of intracellular ion concentrations. These are the sarcolemmal Na^+–K^+ pump, the sarcolemmal Na^+–Ca^{2+} exchanger, and the SR Ca^{2+}-ATPase. The sarcolemmal Na^+–K^+ pump, present in virtually all mammalian cell membranes, extrudes 3 Na^+ ions while importing 2 K^+ ions on each cycle. This pump functions to keep intracellular Na^+ concentration low, thereby maintaining the external versus internal gradient of Na^+ by extruding Na^+ that enters during each AP. Cycling of this pump requires hydrolysis of 1 ATP molecule, and generates a net outward movement of 1 positive charge, thus contributing to outward membrane current and influencing resting membrane potential.

The sarcolemmal Na^+–Ca^{2+} exchanger imports three Na^+ ions for every Ca^{2+} ion extruded, yielding a net charge movement. It is driven by both transmembrane voltage and intra- and extracellular Na^+ and Ca^{2+} ion concentrations. It functions in forward mode during diastole, in which case it extrudes Ca^{2+} and imports Na^+, thus generating a net inward current. It is the principal means by which Ca^{2+} is extruded from the myocyte following each AP, particularly during the diastolic interval. Due to the voltage- and Ca^{2+}-sensitivity of the exchanger, experimental evidence indicates that it can function in reverse mode during the plateau phase of the AP, in which case it extrudes Na^+ and imports Ca^{2+}, thus generating a net outward current (see $I_{Na/Ca}$ current in Figure 16.2).

A second major cytoplasmic Ca^{2+} extrusion mechanism is the SR Ca^{2+}-ATPase. This ATPase pumps Ca^{2+} from the cytosol into the NSR. The SR Ca^{2+}-ATPase has both forward and reverse components (Shannon et al., 1997), with the reverse component serving to prevent overloading of the SR with Ca^{2+} at rest. An additional Ca^{2+} extrusion mechanism is the sarcolemmal Ca^{2+}-ATPase. This Ca^{2+} pump hydrolyzes ATP to transport Ca^{2+} out of the cell. However, it contributes a sarcolemmal current that is small relative to that of the Na^+–Ca^{2+} exchanger, with estimates indicating perhaps that as little as 3% of Ca^{2+} extrusion from the myocyte is mediated by this pump.

Excitation–Contraction Coupling

Excitation–contraction (EC) coupling is the process by which opening of LCCs leads to Ca^{2+} release from the JSR, triggering muscle contraction (for an excellent review, see Bers, 2002). EC coupling involves a close interplay between LCCs and RyRs within the diadic space. During the initial stages of the action potential, voltage-gated LCCs in the sarcolemmal membrane open, allowing the entry of Ca^{2+} into the diadic space. As Ca^{2+} concentration in the region near the T-tubules increases, Ca^{2+} binds to the RyR, increasing their open probability and leading to Ca^{2+} release from the JSR in a process known as Ca^{2+}-induced Ca^{2+} release (CICR). The amount of Ca^{2+} released from the JSR is significantly more than the amount of trigger Ca^{2+} entering via LCCs. One reason for this is that there are more RyRs than LCCs in mammalian cardiac ventricular myocytes. Estimates of this RyR:LCC ratio vary from 8:1 in rat, 5.6:1 in humans, and 4:1 in guinea pig (Bers and Stiffel, 1993). Freeze-fracture electron micrographs show that the RyR are arranged in a regular lattice while the LCCs are located randomly in the T-tubular membrane (Franzini-Armstrong and Protasi, 1997).

The phenomenon of CICR has been studied extensively using both experiments and models. Experiments have show that there are three major properties of CICR: (1) low RyR open probability at rest; (2) high gain; and (3) graded Ca^{2+} release. CICR occurs in response to the opening of the RyR due to Ca^{2+} entry through LCCs (Fabiato, 1983). At rest, the RyR have a very low open probability, yet when exposed to Ca^{2+}, open probability increases, causing a rapid increase in myoplasmic Ca^{2+}. The result is that in response to the influx of "trigger" Ca^{2+} via LCCs, a much greater amount of Ca^{2+} is released from the JSR via the RyR. The ratio of Ca^{2+} released from JSR to the amount of trigger Ca^{2+} entering the myocyte is referred to as the *EC coupling gain*. *Graded release* refers to the phenomenon, originally observed by Fabiato and co-workers (Fabiato, 1985), that Ca^{2+} release from JSR is graded according to the amount of trigger Ca^{2+} entering the cell via LCCs.

16.3 TECHNICAL DETAILS AND METHODOLOGY

16.3.1 Computational Models of the Myocyte

Development of myocyte models began in the early 1960s with the publication of Purkinje fiber action potential models based on the Hodgkin–Huxley model of the squid action potential (Fitzhugh, 1960; Noble, 1960). Subsequent elaboration of these and other models led to development of the first biophysically based cell model describing interactions between voltage-gated membrane currents, membrane pumps and exchangers that regulate Ca^{2+}, Na^+, and K^+ levels, and additional intracellular Ca^{2+} cycling processes in the cardiac myocyte—the so-called DiFrancesco-Noble model of the Purkinje fiber (DiFrancesco and Noble, 1985). This landmark model established the conceptual framework from which all subsequent models of the myocyte have been derived, and it has served as a prototype for neuronal models incorporating regulation of intracellular ion concentrations. Derivative myocyte models include the cardiac ventricular myocyte (Irvine et al., 1999; Jafri et al., 1998; Luo and Rudy, 1991, 1994; Noble et al., 1991), SA node cells (Demir et al., 1994; Dokos et al., 1996; Noble and Noble, 1984; Wilders et al., 1991) and atrial myocytes (Courtemanche et al., 1998; Nygren et al., 1998). Biophysical and biochemical processes incorporated in these models now include: (a) voltage-dependent membrane currents, in some instances based on detailed Markov chain models of ion channel function (Clancy and Rudy, 1999; Greenstein et al., 2000; Jafri et al., 1998); (b) membrane pumps and transporters; (c) intracellular calcium cycling (Jafri et al., 1998; Puglisi and Bers, 2001); (d) excitation–contraction coupling and isometric force generation (Rice et al., 2000); and (e) energy production via the tri-carboxylic acid cycle and oxidative phosphorylation (Cortassa et al., 2002, 2003). These models have proven reproductive and predictive properties and have been applied to advance our understanding of myocyte function in both health and disease (Clancy and Rudy, 1999; Irvine et al., 1999; Mazhari et al., 2001; Shaw and Rudy, 1997). In the following sections we describe how each of these components of myocyte models are formulated.

16.3.2 Myocyte Model Components—Ion Channels

For many years Hodgkin–Huxley models have been the standard for describing membrane current kinetics (Hodgkin and Huxley, 1952). However, data obtained using new experimental approaches, in particular those for producing recombinant channels by co-expression of genes encoding pore-forming and accessory channel subunits in host cells, have shown these models have significant limitations. First, while these models can be expanded to an equivalent Markov chain representation having multiple closed and inactivated states (Hille), many single-channel behaviors such as mean open time, first latency, and a broad range of other kinetics behaviors cannot be described using these equivalent Markov models (Chay, 1991; Horn and Vandenberg, 1984). Second, where it has been studied in detail, as is the case for cardiac Na channels, Hodgkin–Huxley models are insufficient for reproducing behaviors that may be critically state-dependent, such as how ionic channels interact with drugs and toxins (Irvine and Winslow, 1996; Liu and Rasmusson, 1997). Accordingly, much recent effort in modeling of cardiac ionic currents has focused on development of biophysically detailed Markov chain models of channel gating.

Both of these themes—the quantitative characterization of the interactions between pore-forming channel and modulatory subunits and drug-channel interactions—are of general importance to the characterization and modeling of all excitable cells. In the following sections, we therefore describe important aspects of ion channel/current modeling that have emerged in the field of computational cardiology. As a case study, we describe a Markov chain model of the cardiac Na channel to illustrate fundamental principles of channel/membrane current modeling and how the parameters of such models may be constrained using a diversity of experimental data. In subsequent sections, we will show how this model may be extended to the quantitative description and prediction of drug–channel interactions. We will also illustrate the development of a model of the I_{Kr} current based on co-expression studies investigating interactions between the pore-forming channel subunit and an accessory protein, characterize these interaction in terms of a model, and demonstrate how such models are of value in making specific predictions regarding effects of loss of function and point mutations.

16.3.3 Modeling of the Cardiac Na Current

As in many excitable cells, the cardiac Na channel is responsible for the rapid upstroke phase of the AP. We have recently developed a Markov chain model of the cardiac Na channel that is able to reproduce and predict a wide range of single-channel and whole-cell current properties (Irvine et al., 1999). Structure of this model is shown in Figure 16.3A. The channel can occupy any of 13 states. The top row of states corresponds to zero to four voltage sensors being activated (C_0 through C_4) plus an additional conformational change required for opening ($C_4 \rightarrow O_1$ and $C_4 \rightarrow O_2$). The bottom row of states corresponds to channel inactivation. Affinity of the inactivation particle binding site is hypothesized to increase by a scaling factor (a) as the channel activates and to decrease by the same factor as the channel deactivates. Closed-closed and closed-open transitions (horizontal transitions) are voltage-dependent, and closed-inactivated transitions (vertical transitions) are voltage independent. Transition rates are of a form given by Eyring rate theory (Hille, 1992), and include explicit temperature dependence:

$$\lambda = \frac{kT}{h} \exp\left(\frac{-\Delta H_\lambda}{RT} + \frac{\Delta S_\lambda}{R} + \frac{z_\lambda FV}{RT} \right), \qquad (16.1)$$

where k is Boltzmann's constants, T is the absolute temperature, h is Planck's constant, R is the gas constant, F is Faraday's constant, ΔH_λ is the change in enthalpy, ΔS_λ is the change in entropy, z_λ is the effective valence (i.e., the charge moved times the fractional distance the charge is moved through the membrane), and V is the membrane potential in volts.

The probability of occupying any particular channel state is described by a set of ordinary differential equations, written in matrix notation as

$$\frac{\partial \mathbf{P}(t)}{\partial t} = \mathbf{W}\mathbf{P}(t), \qquad (16.2)$$

where $\mathbf{P}(t)$ is a vector state occupancy probabilities and $\mathbf{W}$ is the state transition matrix. $\mathbf{W}$ is in general a function of voltage and time. For the voltage-clamp conditions generally used to constrain ion current models, $\mathbf{W}$ is piece wise time-independent; thus Eq. (16.2) has the analytic solution

$$\mathbf{P}(t) = \exp\left(\mathbf{W}t\right)\mathbf{P}(0). \qquad (16.3)$$

Current through an ensemble of Na channels is calculated as

$$I_{Na}(t) = N G_{Na} P_{open}(t)(V(t) - E_{Na}(t)), \qquad (16.4)$$

where $I_{Na}(t)$ is Na current, N is the number of Na channels, G_{Na} is single channel conductance, $P_{open}(t)$ is the probability of occupying the open states ($O_1 + O_2$), $V(t)$ is membrane potential, and $E_{Na}(t)$ is the reversal potential for Na given by the Nernst equation.

The number of coupled differential equations, and hence the number of parameters that need to be constrained for the model, may be reduced through application of two fundamental principles. First, the state occupancy probabilities for a Markov chain model must sum to one. Hence, through application of this principle, the number of dynamic equations may be reduced by one.

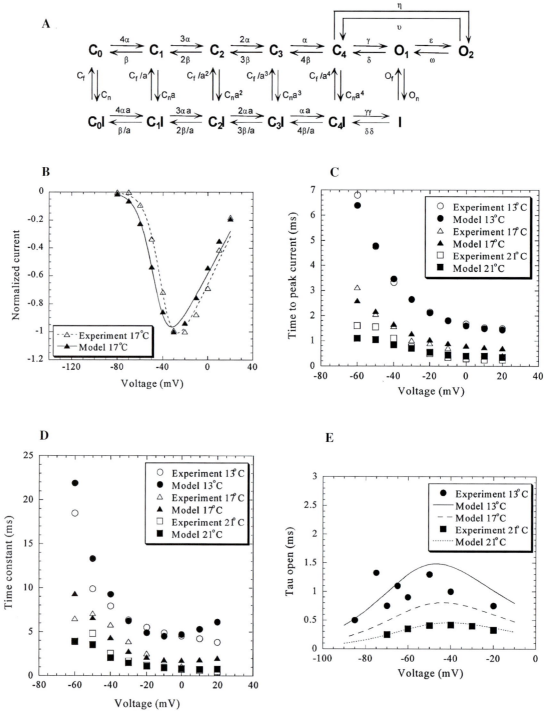

Figure 16.3 (**A**) Markov chain model of the human cardiac Na channel. States C_{0-4} are closed states, states $O_{1,2}$ are open, conducting states, and states C_{1-4I} and I are inactivated states. (**B**) Normalized peak Na current (ordinate) as a function of membrane potential (millivolts; abscissa). Open and filled symbols are experimental and model data at 17°C, respectively. (**C**) Time to peak Na current (milliseconds; ordinate) as a function of membrane potential (millivolts; abscissa). Experimental and model data are compared at 13°C, 17°C and 21°C. Data at 13°C, and 21°C are model fits, and data at 17°C is a model prediction. (**D, E**) Model predictions (solid, dashed, and dotted lines) at 13°C, 17°C, and 21°C, respectively, of single channel open time (msec; ordinate) as a function of membrane potential (mV; abscissa). Filled symbols are experimental data.

In addition, there are several loops in the model that must satisfy the principle of microscopic reversibility. Microscopic reversibility is derived from the law of conservation of energy and states that the product of rate constants when traversing a loop clockwise must be equal to the product when traversing the same loop counterclockwise (Hille, 1992). For the closed–closed–inactivated loops, satisfying microscopic reversibility requires that the transitions among the closed–inactivated states be scaled by a, the same factor used to scale the transitions between rows. Microscopic reversibility is preserved around the closed–open–inactivated loop by isolating the ΔH, ΔS, and z terms in the product and satisfying each term separately using the following equations:

$$\Delta H_{\gamma\gamma} = \Delta H_{\gamma} + \Delta H_{on} + \Delta H_{\delta\delta} + \Delta H_{cf} \\ + 8RT \ln a - \Delta H_{\delta} - \Delta H_{cn} - \Delta H_{of}, \quad (16.5)$$

$$\Delta S_{\gamma\gamma} = \Delta S_{\gamma} + \Delta S_{on} + \Delta S_{\delta\delta} + \Delta S_{cf} - \Delta S_{\delta} \\ - \Delta S_{cn} - \Delta S_{of}, \quad (16.6)$$

$$z_{\gamma\gamma} = z_{\gamma} + z_{on} + z_{\delta} + z_{of} - z_{\delta\delta}. \quad (16.7)$$

Similarly, microscopic reversibility is preserved around the closed–open–open loop using the following equations for ΔH_{η}, ΔS_{η}, and z_{η}:

$$\Delta H_{\eta} = \Delta H_{\gamma} + \Delta H_{\varepsilon} + \Delta H_{\nu} - \Delta H_{\delta} - \Delta H_{\omega}, \quad (16.8)$$

$$\Delta S_{\eta} = \Delta S_{\gamma} + \Delta S_{\varepsilon} + \Delta S_{\nu} - \Delta S_{\delta} - \Delta S_{\omega}, \quad (16.9)$$

$$z_{\eta} = z_{\gamma} + z_{\delta} - z_{\nu}. \quad (16.10)$$

These microscopic reversibility constraints thus reduce the dimension of the parameter estimation problem, as transition rates $\gamma\gamma$ and η are fully constrained.

The model of Figure 16.3A may also be viewed as a Markov chain description of single-channel behavior. Single-channel gating may be simulated using the method of Clay and DeFelice (1983). In this method, the length of time a channel stays in its current state (i.e., its dwell time denoted as T_j) is calculated according to the formula

$$T_j = -(\ln r) \left/ \sum_{k=1}^{x} \lambda_{jk} \right., \quad (16.11)$$

where r is a random variable drawn from a uniform distribution on the interval [0,1] and λ_{jk} is the transition rate from state j to state k. The sum is over the x pathways out of state j. The resulting dwell time T_j is an exponentially distributed random variable with parameter $\lambda = \sum_{k=1}^{x} \lambda_{jk}$. At the end of the dwell time, the new state of the channel is determined by assigning random numbers to a portion of the interval [0,1] based on the probabilities of changing to neighboring states. These probabilities are equal to the rate constant for a particular transition divided by the sum of the rate constants for all possible transitions. Once the new state is determined, another random number is used to calculate the dwell time in the new state. At an instantaneous voltage step, channels remain in their current state, but the dwell times are recalculated.

Extensive experimental data were required to fully determine the model parameters. The majority of data were taken from human SCN5A-encoded Na channels. Experimental data obtained at temperatures of 13°C and 21°C were used to constrain the model, and the ability of the model to predict data collected at 17°C was tested. Constraining data included (a) ionic currents in response to voltage-clamp; (b) gating charge accumulation; (c) steady-state inactivation curve; (d) rate of tail current relaxation; (e) time course of recovery from inactivation; and (f) single-channel open times. A cost function defined as the squared error between simulated and experimental data (including both whole-cell current and single-channel data) was minimized to determine an optimal model parameter set. A simulated annealing algorithm (Corana, 1987) was needed to perform this minimization, because the cost function exhibited many local minima. The resulting model was able to reproduce a broad range of membrane current data (Irvine et al., 1999), and Figures 16.3B–E demonstrates the ability of the model to predict channel/current properties at 17°C as well as single-channel data not included in the fitting process.

16.3.4 Myocyte Model Components—Intracellular Ion Concentration Changes

The majority of integrative models of the myocyte available today are of a type known as "common pool" models (Stern, 1992), the structure of which is shown in Figure 16.4A. In such models, Ca^{2+} flux through both LCCs and ryanodine-sensitive Ca^{2+} release channels in the JSR membrane is directed into a single common Ca^{2+} compartment referred to as the subspace (as in Figure 16.3). The subspace represents the total volume of the ~5000 diadic spaces present in the ventricular myocyte. In a modeling *tour de force*, Stern demonstrated that common pool models are structurally unstable, exhibiting all-or-none Ca^{2+} release except (possibly) over some narrow range of model parameters. This instability occurs because Ca^{2+} release from JSR produces a large, rapid increase of Ca^{2+} concentration in the subspace. This, in turn, results in a very strong positive feedback effect in which increased binding of Ca^{2+} to RyR induces further RyR channel opening and release of Ca^{2+}. Despite this inability to reproduce experimentally measured properties of graded JSR Ca^{2+} release, common pool models have been very successful in reproducing and predicting a range of myocyte behaviors. This includes properties of interval–force relationships that depend heavily on proper dynamic modeling of intracellular Ca^{2+} uptake and release mechanisms (Rice et al., 2000).

We illustrate the process of modeling time-varying changes of intracellular ion concentration with reference to the common pool model architecture shown in Figure 16.4A. In this model, there are four distinct Ca^{2+} compartments (the cytosol, subspace, NSR, and JSR) and one Na and K compartment (the cytosol). Note that in present myocyte models, the cytosolic concentrations of both Na^+, K^+, and Ca^{2+} are assumed to be uniform. The time rate of change of concentration C_i of the ith ionic species in a given compartment is given by

$$\frac{dC_i(t)}{dt} = \frac{I_i(t)}{z_i FV}, \quad (16.12)$$

where $C_i(t)$ is concentration (typically mM) of species i, t is time (typically milliseconds), $I_i(t)$ is net current into the compartment carried by species i (typically in pA), z_i is valence of the ith

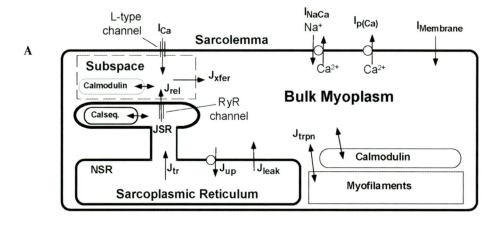

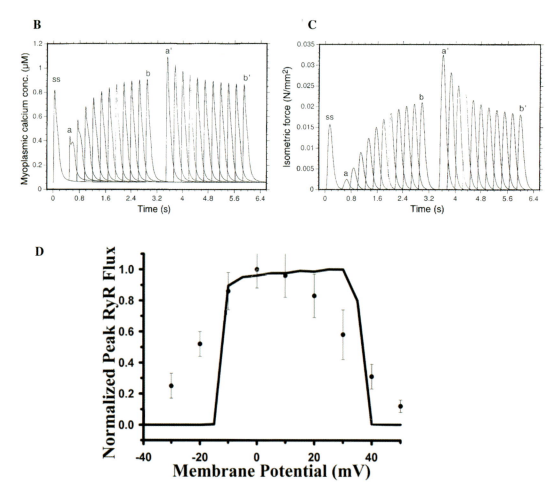

Figure 16.4 (**A**) Schematic illustration of the structure of common pool ventricular myocyte models. (**B**) Extrasystolic restitution and post-extrasystolic potentiation responses showing cytosolic Ca^{2+} concentration (micromolar; ordinate) as a function of time (seconds; abscissa) predicted by the common pool ventricular myocyte using the stimulus protocols described in the text. (**C**) As in part B, but with isometric predicted (N/mm^2) from the Ca^{2+} transients in part B; force plotted on the ordinate. (**D**) Normalized peak RyR release flux (ordinate) as a function of membrane potential (millivolts; abscissa) for rat ventricular myocytes (symbols) and for the common pool model (solid line).

species, F is Faradays constant, and V is compartment volume (typically in units of pL). One such equation may be defined for the concentration of each ionic species in each model compartment. Ion flux between compartments, related to the term $I_i(t)$ in Eq. (16.12), is produced either by (a) diffusion due to differences in ion species concentration between adjacent compartments (as is the case for the flux term J_{xfer} in Figure 16.4A representing Ca^{2+} diffusion from the subspace to the cytosol), (b) gating of ion channels in the sarcolemmal or JSR membrane (as is the case for Ca^{2+} flux J_{rel} in Figure 16.4A from the JSR into the subspace through RyR channels), or (c) the action of membrane transporters and exchangers (for example, Ca^{2+} flux through the SR Ca^{2+}-ATPase, labeled J_{up} in Figure 16.4A). The form of the algebraic equations describing function of membrane transporters and exchangers, including their concentration-, voltage-, and in some instances ATP-dependence, may be found in the published equations for a number of myocyte models. In addition, buffering of Ca^{2+} by negatively charged phospholipids head groups in the sarcolemmal and JSR subspace membrane, by cytosolic myofilaments (troponin) and calsequestrin in the JSR, is modeled. Buffering due to mechanisms other than myofilaments is described using the rapid buffer approximation of Wagner and Keizer (1994).

16.3.5 Myocyte Model Components—Composite Equations for Common Pool Models

Common pool models of the cardiac myocyte consist of systems of nonlinear ordinary differential-algebraic equations describing the time evolution of model state variables. These state variables are (a) probability of occupancy of ion channel states [Eq. (16.2)] and current flux through open channels [Eq. (16.4)], (b) concentrations of ion species in model compartments [Eq. (16.12)], and (c) time evolution of membrane potential. Currently, all biophysically detailed model of the myocyte assume that since these cells are spatially compact, they are isopotential, with time rate of change of membrane potential given by

$$\frac{dv(t)}{dt} = -\left\{ \sum_i I_i^{ion}[v(t)] + \sum_i I_i^{pump}[v(t), c(t)] \right\},$$
(16.13)

where $v(t)$ is membrane potential, $I_i^{ion}[v(t)]$ is current carried by the ith membrane current, and $I_i^{pump}[v(t), c(t)]$ is current through the ith membrane pump/exchanger, which can depend on both membrane potential $v(t)$ and the relevant ion concentration $c(t)$.

Figure 16.8 shows examples of simulated APs and Ca^{2+} transients compared with those measured from isolated canine ventricular myocytes. These data illustrate the point that common pool models have been quite successful in reconstruction of AP properties. Results of a more challenging simulation, obtained using the Jafri–Rice–Winslow model of the guinea pig ventricular myocyte (Jafri et al., 1998), are shown in Figures 16.4B and 16.4C. In these simulations, the model cell was paced at an interstimulus interval of 1.5 s until a steady-state was reached. Panels B and C of Figure 16.4 show model Ca^{2+} (Figure 16.4B) and isometric force (Figure 16.4C) transients in response to the pacing stimuli. Isometric force was computed using a model developed by Rice et al., 2000; (Jafri et al., 1998), that relates cytosolic Ca^{2+} level to isometric force. Ca^{2+} and force transients labeled

SS in Figures 16.4B and 16.4c are responses to the last stimulus of this periodic pulse train. Following cessation of the periodic stimulus train, a second stimulus (denoted S2) is delivered at a variable interval. Responses labeled a and b in Figures 16.4B and 16.4C are Ca^{2+} and force transients in response to short or long S2 intervals, respectively, following cessation of the periodic pulse train. A third stimulus (denoted S3) is delivered 3 s after the corresponding S2 stimulus. Responses to the S3 stimuli delivered 3 s after the corresponding S2 stimuli a and b are labeled in Figures 16.3B and 16.3C as a' and b', respectively.

S2 responses are seen to increase in amplitude as a function of the S2 interval. This property is known as extrasystolic restitution and is determined primarily by the rate of recovery of the RyR from adaptation following opening and release of Ca^{2+} from the JSR (Rice et al., 2000). S3 responses are seen to decline in amplitude depending on the corresponding S2 interval. This property is known as post-extrasystolic potentiation. The mechanism can be understood by contrasting the magnitude of Ca^{2+} release events for S2 responses a and b in Figure 16.4B. Since the S2 interval is long for response b, this provides ample time for SR Ca^{2+} levels to be restored by the SR Ca^{2+}-ATPase, and Ca^{2+} release amplitude is large. This large release event depletes SR Ca^{2+} levels substantially so that the release event b' initiated 3 s later by the S3 stimulus is small. The converse is true for release events a and a'.

The responses shown in Figures 16.4B and 16.3C agree very well with experimental data (Yue et al., 1985), and they show that common pool models are very capable of describing complex intracellular Ca^{2+} cycling processes determined by the rate of SR Ca^{2+} loading by the SR Ca^{2+}-ATPase. However, as we have noted above, common pool models are not able to reproduce a very fundamental behavior of cardiac myocytes: SR Ca^{2+} release that is smoothly and continuously graded with influx of trigger Ca^{2+} through sarcolemmal LCCs. This failure is demonstrated in Figure 16.4D. This figure shows normalized peak Ca^{2+} flux through RyR channels (ordinate) as a function of membrane potential (mV; abscissa). Filled circles are experimental measurements from the work of Wier et al. (1994), showing that release flux increases smoothly to a maximum flux at about 0 mV and then decreases to near zero at more depolarized potentials. Release flux increases from −40 to 0 mV since over this potential range, open probability of LCCs increases very steeply reaching a maximum value. Release flux decreases over the potential range greater than 0 mV because electrical driving force on Ca^{2+} decreases monotonically. The solid line shows release flux for the Jafri–Rice–Winslow guinea pig ventricular myocyte model. Release is all-or-none, with (a) regenerative release initiated at a membrane potential causing opening of a sufficient number of LCCs (~ -15 mV) and (b) release terminating at potential for which electrical driving force is reduced to a critical level ($\sim +40$ mV).

This all-or-none behavior of Ca^{2+} release in common pool models has very important implications for model dynamics. LCCs undergo not only voltage- but also Ca^{2+}-dependent inactivation (Bers and Perez-Reyes, 1999; Peterson et al., 1999). Inactivation depends on local subspace Ca^{2+} concentration and occurs as Ca^{2+} binding to calmodulin (Peterson et al., 1999), which is tethered to the LCC, induces the channel to switch from a normal mode of gating to a mode in which transitions to open states are extremely rare. Recent experimental data have demonstrated

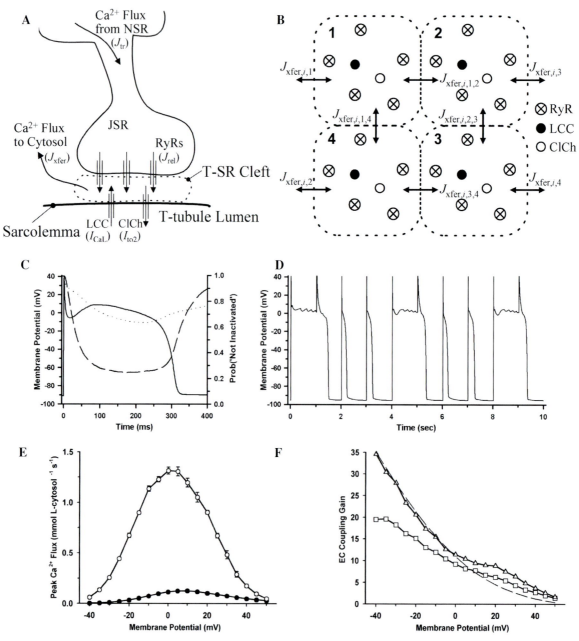

Figure 16.5 (**A**) Structure of the LCC–RyR complex, referred to as the functional unit (FU). A single LCC in the sarcolemmal membrane is associated with 5 RyR in the closely apposed JSR membrane. ClCh denotes a single Ca^{2+}-modulated Cl^- channel that is thought to be co-located in the dyadic space. (**B**) Structure of the Ca^{2+} release unit (CaRU). Each CaRU consists of 4 FUs, with Ca^{2+} diffusion between adjacent FUs and into the surrounding cytosolic space. (**C**) Solid line is an action potential (membrane potential in millivolts, left ordinate; time in milliseconds, abscissa) predicted by the local-control myocyte model. Dotted line is the fraction of channels (right ordinate) not voltage-inactivated, and dashed line is the fraction not Ca^{2+}-inactivated during the action potential shown by the solid line. (**D**) Behavior of the common pool myocyte model when the balance between voltage- and Ca^{2+}-inactivation is as shown in panel C. Note instability of action potentials. (**E**) Peak Ca^{2+} flux (ordinate) through RyRs (open symbols) and LCCs (filled symbols) as a function of membrane potential (millivolts; abscissa). (**F**) EC coupling gain (ordinate: ratio of peak RyR to LCC flux) as a function of membrane potential (millivolts; abscissa).

that voltage-dependent inactivation of LCCs is a slow and weak process, whereas Ca^{2+}-dependent inactivation is relatively fast and strong (Linz and Meyer, 1998; Peterson et al., 1999). This implies in turn that there is a very strong coupling between (a) Ca^{2+} release from JSR into the local subspace and (b) regulation of inactivation of the LCC. When this newly revealed

balance between voltage- and Ca^{2+}-dependent inactivation is incorporated into common pool models, the models become unstable, exhibiting alternating short- and long- duration APs (Greenstein and Winslow, 2002; Mazhari et al., 2001) (an example of this is shown in Figure 16.5D). The reason for this is intuitively clear: Since JSR Ca^{2+} release is all-or-none in these models, Ca^{2+}-dependent inactivation of LCCs is all-or-none, depending on whether release has or has not occurred. Since L-type Ca^{2+} current is a major contributor to inward current during the plateau phase of the AP (see Figure 16.2), its biphasic inactivation leads to instability of AP duration. This, unfortunately, constitutes a fatal weakness of biophysically based common pool models.

16.3.6 Local Control Models of the Myocyte

The fundamental failure of common pool models described above suggests that more biophysically based models of EC coupling must be developed and investigated. Understanding of the mechanisms by which Ca^{2+} influx via LCCs triggers Ca^{2+} release from the JSR has advanced tremendously with the development of experimental techniques for simultaneous measurement of LCC currents and Ca^{2+} transients and detection of local Ca^{2+} transients, and this has given rise to the local control hypothesis of EC coupling (Bers and Stiffel, 1993; Sham, 1997; Stern, 1992; Wier et al., 1994). As illustrated schematically in Figure 16.5A, this hypothesis asserts that opening of an *individual* LCC in the T-tubular membrane triggers Ca^{2+} release from the *small cluster* of RyRs located in the closely apposed ($\sim$12 nm) JSR membrane. Thus, the local control hypothesis asserts that release is all-or-none at the level of these individual groupings of LCCs and RyRs (referred to as the functional unit). However, LCC:RyR clusters are physically separated at the ends of the sarcomeres (Franzini-Armstrong et al., 1999). These clusters therefore function in an approximately independent fashion. The local control hypothesis asserts that graded control of SR Ca^{2+} release, in which Ca^{2+} release from JSR is a smooth, continuous function of Ca^{2+} influx, is achieved by the statistical recruitment of elementary Ca^{2+} release events in these independent diadic spaces. Thus, at the heart of the local-control hypothesis is the assertion that the co-localization of LCCs and RyRs is a structural component that is fundamental to the property of graded Ca^{2+} release and force generation at the level of the cell. This concept of channel co-localization contributing in fundamental ways to cell behavior is another general theme of biophysical signal processing in excitable cells.

We have recently implemented a local-control model of myocyte function (Greenstein and Winslow, 2002). As a compromise between structural and biophysical detail versus tractability, a "minimal model" of local control of Ca^{2+} release, referred to as the Ca^{2+} release unit (CaRU) model, was developed. Figure 16.5B shows a schematic of the CaRU model. This model is intended to mimic the properties of Ca^{2+} sparks in the T-tubule/SR (T-SR) junction (Ca^{2+} sparks are elementary SR Ca^{2+} release events arising from opening of a cluster of RyRs) (Cheng et al., 1993). Figure 16.5B shows a cross section of the model T-SR cleft, which is divided into four individual diadic subspace compartments arranged on a 2×2 grid. Each subspace (SS) compartment contains a single LCC and 5 RyRs in its JSR and sarcolemmal membranes, respectively. All 20 RyRs in the CaRU communicate with a single local JSR volume. The 5:1 RyR to LCC stoichiometry is chosen to be consistent with recent estimates indicating that a single LCC typically triggers the opening of 4–6 RyRs (Wang et al., 2001). Each subspace is

treated as a single compartment in which Ca^{2+} concentration is uniform; however, Ca^{2+} may diffuse passively to neighboring subspaces within the same CaRU. The division of the CaRU into four subunits allows for the possibility that an LCC may trigger Ca^{2+} release in adjacent subspaces (i.e., RyR recruitment) under conditions where unitary LCC currents are large. Since LCC:RyR clusters are physically separated (Franzini-Armstrong et al., 1999), each model CaRU is assumed to function independently of other CaRUs. Upon activation of RyRs, subspace Ca^{2+} concentration will become elevated. This Ca^{2+} will freely diffuse to either adjacent subspace compartments (J_{iss}) or into the cytosol (J_{xfer}) along its concentration gradient. The local JSR compartment is refilled via passive diffusion of Ca^{2+} from the network SR (NSR) compartment (J_{tr}).

The local control simulation algorithm is described in detail in the Appendix of Greenstein and Winslow (2002). Simulation of the dynamics of each CaRU requires both (a) numerical integration of the ordinary differential equations describing local subspace and JSR Ca^{2+} balance and (b) Monte Carlo simulation of LCC and RyR channel gating in the approximately $\sim$12,500 CaRUs in the cell (there are $\sim$50,000 LCCs per ventricular myocyte). The state of each channel is described by a set of discrete-valued random variables that evolve in time as described by Markov processes. Time steps for CaRU simulations are adaptive and are chosen to be sufficiently small based on channel transition rates. The CaRU simulations occur within the (larger) time step used for the numerical integration of the system of ordinary differential equations describing the time evolution of global state variables. As a result of the embedded Monte Carlo simulation, all model state variables and ionic currents/fluxes will contain a component of stochastic noise. These fluctuations introduce a degree of variability to simulation output.

Figures 16.5C–F show macroscopic properties of APs and SR Ca^{2+} release in this hybrid stochastic/ODE model. Figure 16.5C shows the relative balance between the fraction of LCCs *not* voltage-inactivated (dotted line) and *not* Ca^{2+}-inactivated (dashed line) during the AP. These fractions were designed to fit the experimental data of Linz and Meyer (1998). The solid line shows a local-control model AP. This AP should be contrasted with those produced by the common pool model when the same relationship between LCC voltage- and Ca^{2+}-dependent inactivation as shown in Figure 16.5C is used. Clearly, the local-control model exhibits stable APs, whereas the common pool model does not. Figure 16.5E shows the voltage dependence of peak LCC Ca^{2+} influx ($F_{LCC(max)}$—filled circles, ordinate) and peak RyR Ca^{2+} release flux ($F_{RyR(max)}$—open circles, ordinate) in response to voltage-clamp steps to the indicated potentials (mV, abscissa). Ca^{2+} release flux is a smooth and continuous function of membrane potential, and hence trigger Ca^{2+}, as shown by the experimental data in Figure 16.4D. EC coupling gain may be defined as by Wier et al. (1994), as the ratio $F_{RyR(max)}/F_{LCC(max)}$, and is plotted as a function of voltage in Figure 16.5F (triangles). EC coupling gain is monotonically decreasing with increasing membrane potential, and it agrees with corresponding experimental measurements made by Wier (Wier et al., 1994). The role of inter-subspace coupling on gain is demonstrated in Figure 16.5F, by comparison of control simulations (triangles) to those in the absence of inter-subspace coupling (squares). With inter-subspace coupling intact, EC coupling gain is greater at all potentials, but the increase in gain is most dramatic at more negative potentials. In this negative voltage range, LCC open probability is submaximal, leading to sparse LCC openings. However, unitary current magnitude is relatively high, so that in the presence of Ca^{2+} diffusion within the CaRU, the rise in local Ca^{2+} due to the triggering action of a single LCC can recruit and activate RyRs in adjacent subspace compartments within the same T-SR junction. The net effect of inter-subspace coupling is therefore to increase the magnitude and slope of the gain function preferentially in the negative voltage range. These simulations therefore offer an intriguing glimpse of how the co-localization and stochastic gating of individual channel complexes can have a profound effect on the overall, integrative behavior of the cell.

16.4 CURRENT APPLICATIONS

16.4.1 Models of Drug–Channel Interactions

A major motivation for our development of the cardiac Na channel model was to test the ability of such a model to predict drug–channel interactions. Several models have been proposed to explain block of cardiac sodium channels by class I antiarrhythmic drugs such as lidocaine and its analogs. The modulated receptor (MR) model of drug–channel interaction, advanced simultaneously by Hondeghem and Katzung (1977) and Hille (1977), is perhaps the most widely known theory of drug–channel interactions. It is a very general theory that asserts that (a) a drug can bind to any channel state, (b) the binding affinity to the channel is a function of channel state, and (c) drug-bound channels are nonconducting. The allosteric effector (AE) model is particular to lidocaine action on the Na channel, and it asserts that the action of lidocaine is to "stabilize" the inactivated states by increasing transition rates into, and decreasing transition rates out of, these states. The AE model does not incorporate drug-bound states. In previous work (Irvine, 1998), we examined the ability of these two models to quantitatively reproduce effects of lidocaine on Na current properties. Analyses demonstrated that the Hondeghem–Katzung MR model was able to reproduce data on which fitting of model parameters was based (use-dependence, onset of block, recovery from block, and dose–response data). However, the resulting model did not yield reasonable drug affinities for each state, did not preserve microscopic reversibility, and could not predict experimental data not included in the fitting process such as experimentally measured changes in the charge–voltage curve due to drug binding (Irvine, 1998). The AE model was able to reproduce the dose–response curve and drug-induced shift of the steady-state inactivation function measured experimentally, but could not reproduce or predict any time-dependent actions of lidocaine (Irvine, 1998).

We subsequently demonstrated that the failure of the AE model to reproduce time-dependent drug action was due to its lack of drug-bound states; indeed the addition of a single drug-bound state dramatically enhanced the ability of this model to explain kinetic data. We further hypothesized that the inadequacy of the MR model of lidocaine action was not due to any fundamental flaw in the theoretical underpinnings of the model, but, rather, was due to implementation of the model based on inadequate characterization of the Na current itself.

We therefore investigated the ability of an MR model of lidocaine action based on the Irvine model of the cardiac Na channel (Figure 16.3A) to reproduce and predict data on lidocaine binding. The structure of this model is shown in Figure 16.6A. We assumed, for simplicity, that each open and inactivated state has a corresponding drug-bound state. Transition rates into and out of these drug-bound states are controlled by the eight parameters $L_r, h, K_r, g, K_I, L_I, L_A,$ and K_A. Scaling factors h and g are such that drug affinity increases with transitions toward the inactivated state I. Application of the principle of microscopic reversibility to loops involving drug-bound states reduces the number of free parameters to be estimated to constrain a model of drug action to seven—precisely the same number of parameters specifying the original Hondeghem–Katzung model (Hondeghem and Katzung, 1977). Results are shown in Figures 16.6B–16.6G. The model very accurately reproduces data on which model parameters were fit (use-dependent block in Figure 16.6B; rate of recovery from block in Figure 16.6C; dose–response relationship in Figure 16.6D; and effects of lidocaine on steady-state inactivation in Figure 16.6E). The model also predicts experimental data not included in the fitting process (frequency dependence of drug block, to be compared with experimental data of Figure 16.6B in Furukawa et al.; current–voltage relationships in Figure 16.6G).

These results indicate that modulated-receptor theory provides a solid conceptual framework from which to develop quantitative and predictive models of drug-channel interactions. Substantial experimental data are required to constrain such models. However, in our view, the collection of such data, the development of drug–channel interaction models, and their integration into whole-cell models to predict effects of drugs on cell function is entirely feasible and necessary in the final stages of the antiarrhythmic drug development process.

16.4.2 Integrative Modeling of Channel Subunit Interactions, Loss of Function, and Point Mutations

The cardiac delayed rectifier potassium current $I_{K,r}$ mediates repolarization of the action potential. The human *ether-à-go-go related gene* (HERG) (Sanguinetti et al., 1995) encodes the pore-forming subunit of $I_{K,r}$ (Sanguinetti et al., 1995; Sanguinetti and Jurkiewicz, 1990). However, channels formed by co-expression of KCNE2 (encoding mink-related peptide 1, MiRP1) and the pore-forming subunit HERG resemble native cardiac $I_{K,r}$ channels more closely in their gating and unitary conductance, modulation by extracellular potassium, and inhibition by class III antiarrhythmic drugs (e.g., E-4031) than do HERG-encoded channels (Abbott et al., 1999). Mutations in either HERG or KCNE2 are linked to heritable or acquired long QT syndrome—a syndrome in which AP duration is prolonged, thus contributing to increased likelihood of arrhythmia.

Mazhari et al. (2001) characterized the functional interactions between HERG and KCNE2 in order to define underlying mechanisms for action potential prolongation and long QT syndrome. To do this, HERG channels and green fluorescent protein (GFP) were transfected in HEK 293 cells. When studying effects of HERG/KCNE2 co-expression, these same cells were also transfected with hKCNE2 and red fluorescent protein (RFP). GFP- and RFP-positive cells were studied within

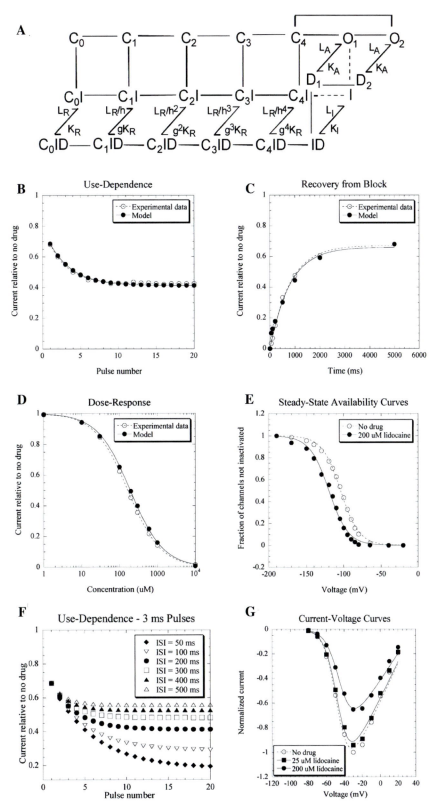

Figure 16.6 (**A**) Model of lidocaine action on cardiac Na channel. States shown in the background are those of the cardiac Na channel model, as described in Figure 16.3. States in foreground labeled $C_{0-4}ID$, ID, and $D_{1,2}$ are drug-bound states. Transition rates into and out of the drug-bound states are as described in the text. (**B**) Relative Na current amplitude (ordinate) as a function of stimulus number (abscissa) in the presence of 200 μM lidocaine. Open symbols are experimental, and filled symbols are model data. These data are referred to as use-dependent data in the text. (**C**) Recovery of Na current amplitude from lidocaine block ate (ordinate) as a function of time (milliseconds; abscissa). Open symbols are experimental, and filled symbols are model data. (**D**) Steady-state current relative to control (ordinate) as a function of lidocaine concentration (micromolar; abscissa). (**E**) Steady-state inactivation (ordinate) as a function of membrane potential (millivolts; abscissa) with and without 200 μM lidocaine. (**F**) Use-dependence data over a range of inerpulse intervals. (**G**) Peak current–voltage relationships with and without 25 and 200 μM lidocaine.

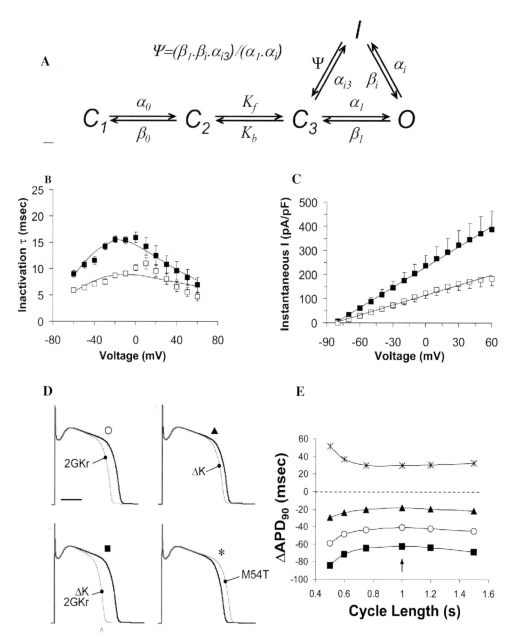

Figure 16.7 (**A**) Structure of the HERG-KCNE2 model. Transition rate Ψ is constrained by applying the principle of microscopic reversibility to the $C_3 \rightleftarrows O \rightleftarrows I$ loop. (**B**) Inactivation time constant (milliseconds; ordinate) as a function of membrane potential (millivolts; abscissa) for HERG- (open symbols) and HERG-KCNE2-encoded channels. Solid lines are model fits to the data. (**C**) Current density (pA/pF; ordinate) as a function of membrane potential (millivolts; abscissa) for HERG- (open symbols) and HERG-KCNE2-encoded channels. Solid lines are model fits to the data. (**D**) Effects of alterations in kinetics and/or conductivity on action potential repolarization; o, Effects of increase in conductivity alone (GKr→2GKr); ▲ effect of changes in kinetics (ΔK) alone (from HERG-KCNE2 to HERG model; ■ effects of changes in conductivity (GKr→2GKr) and kinetics (ΔK); ∗, effects of M54T-hKCNE2 mutation on action potential repolarization. (**E**) Changes in AP duration relative to "wild-type" models (ΔAPD_{90}, milliseconds; ordinate) as function of cycle length (seconds; abscissa) for each case described in above panels (symbols match the symbols shown above each action potential trace). Here negative and positive values indicate a decrease and an increase in APD_{90} compared to "wild-type" model, respectively. Cycle length of 1 s in panels A–D (arrow in panel E).

24 to 36 hours of transfection. Figure 16.7 shows the results. Figure 16.7A shows a Markov chain model of the HERG- and HERG/KCNE2-encoded channel accounting for experimental data on voltage-dependent steady-state and time-varying properties of activation, deactivation, and inactivation. Model formulation, constraints, and parameter fitting were as described above for the Na channel model, with the exception that parameters optimizing model fit were estimated using the Nelder–Mead simplex method. Co-expression of HERG/KCNE2 was shown to have no significant effect on steady-state activation or inactivation properties. Co-expression produced a modest acceleration of activation and deactivation kinetics (not shown), a significant reduction of inactivation time constants (Figure 16.7B), and an approximate factor-of-two reduction of current density (Figure 16.7C). The most significant change in model rate constants was an increase in the α_{i3} transition rate between C_3 and I, consistent with the accelerated inactivation kinetics.

To interpret the functional significance of these changes, the HERG/KCNE2 model was incorporated into an integrative model of the canine ventricular myocyte (Irvine et al., 1999), and the significance of *loss of function* mutations to KCNE2 was simulated under the assumption that such mutations ablate the HERG/KCNE2 interactions described above (Figure 16.7D). Simulation results showed that loss of function mutations would result in AP duration shortening, not lengthening and thus cannot explain the phenotype of heritable or acquired long QT syndrome.

The HERG/KCNE2 model was then used to predict effects of a point mutation associated with KCNE2 (referred to as the M54T mutation) (Abbott et al., 1999). This mutation results in a two- to threefold increase in deactivation rates and a modest reduction in activation slope factor, compared with wild-type HERG/KCNE2 co-expression (Abbott et al., 1999). To make this prediction, rate constants of the HERG/KCNE2 Markov model were adjusted, based on the data of Abbott et al. (1999), to include kinetic changes associated with the M54T mutation and incorporated in the action potential model. These changes result in an increased AP duration (Figure 16.7D, panel M54T). More importantly, the prolongation of action potential was larger at shorter pacing cycles (Figure 16.7E). This is consistent with previous clinical observations in which M54T mutation resulted in prolongation of the QT interval in the electrocardiogram of the affected patient during exercise (Abbott et al., 1999).

These results illustrate the importance of biophysically detailed channel modeling and the incorporation of these models into integrative models of excitable cells. Continued development of such models and their application to prediction of drug–channel interactions, loss of function, and point mutations remains a critically important, emerging aspect of computational cardiology.

16.4.3 Modeling the Molecular Basis of Disease—Heart Failure

Heart failure (HF), the most common cardiovascular disorder, is characterized by ventricular dilatation, decreased myocardial contractility, and reduced cardiac output. Prevalence in the general population is over 4.5 million, and it increases with age to levels as high as 10%. New cases number approximately 400,000 per year. Patient prognosis is poor, with mortality roughly 15% at one year, increasing to 80% at 6 years subsequent to diagnosis. It is now the leading cause of sudden cardiac death (SCD) in the United States. An increased understanding of the molecular basis of this disease therefore offers the possibility of improved treatments that can reduce the risk of SCD.

Experimental studies have now identified two critical aspect of the cellular phenotype of heart failure. First, ventricular myocytes isolated from failing human (Beuckelmann et al., 1992) and canine (Kaab et al., 1996; O'Rourke et al., 1999) hearts exhibit significant AP prolongation. An example is shown in Figure 16.8. APs shown by solid and dotted lines in Figure 16.8A were recorded from normal and failing canine midmyocardial ventricular myocytes, respectively. Duration of the failing AP ($\sim$660 ms) is roughly twice that of the normal ($\sim$330 ms). AP duration is controlled by the balance between inward and outward membrane currents, primarily during the plateau phase of the AP. Possible explanations for this prolongation are therefore HF-induced up-regulation of inward currents and/or down-regulation of outward currents. Second, failing ventricular myocytes exhibit altered Ca^{2+} transients. An example of normal and failing Ca^{2+} transients obtained simultaneously with the AP recordings of Figure 16.8A, and measured using Indo-1 fluorescence ratio, is shown in Figure 16.8B. Differences between normal (solid line) and failing (dotted line) Ca^{2+} transients include (a) reduced amplitude and (b) reduced rate of decline of the Ca^{2+} transient subsequent to repolarization of the AP.

There is little evidence to support the idea that an up-regulation of inward currents is responsible for prolongation of AP duration in HF, because the majority of measurements of whole-cell Na^+ and Ca^{2+} current density show no change in the density of these currents (O'Rourke et al., 1999). However, down-regulation of voltage-gated K currents is known to occur in HF. Measurements of whole-cell inward rectifier current I_{K1} show that current density at hyperpolarized membrane potentials is reduced in HF by $\sim$50% in humans (Beuckelmann et al., 1993) and by $\sim$40% in dogs (Kaab et al., 1996). Measurements of I_{to1} show that in end-stage HF human and canine tachycardia pacing-induced HF indicate that current density is reduced by up to 70% in HF (Näbauer et al., 1996). Human and canine Ca^{2+}-independent transient outward current I_{to1} is a combination of currents encoded by the KCND3 and KCNA4 genes (Dixon et al., 1996; Näbauer et al., 1996), and KCND3 expression has been shown to be reduced in HF (Kaab et al., 1998). There appears to be no change in expression of the HERG or KCNQ1 gene encoding α-subunits of the I_{Kr} and I_{Ks} channels, respectively, in HF.

Expression of diverse proteins involved in the processes of EC coupling have also been measured in normal and failing myocytes. These proteins include (a) the SR Ca^{2+}-ATPase encoded by the SERCA2 gene, (b) the phospholamban protein encoded by the PLN gene, and (c) the sodium–calcium (Na^+–Ca^{2+}) exchanger protein encoded by the NCX1 gene. Measurements indicate that there is an approximate 50% reduction of SERCA2 mRNA (O'Rourke et al., 1999; Studer et al., 1994; Takahashi et al., 1992), expressed SR Ca^{2+}-ATPase protein level and direct SR Ca^{2+}-ATPase uptake rate (O'Rourke et al., 1999) during HF. There is a 55% increase in NCX1 mRNA levels, along with an approximate factor-of-two increase in Na^+–Ca^{2+} exchange activity in human (Reinecke et al., 1996; Studer et al., 1994)

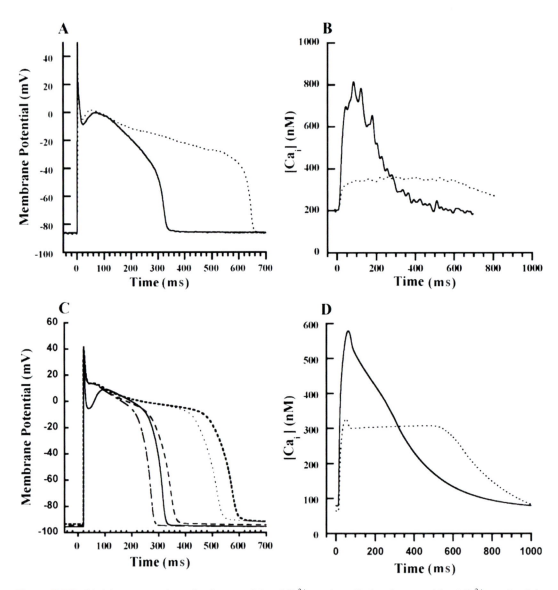

Figure 16.8 Model versus experimental action potentials and Ca^{2+} transients. Each action potential and Ca^{2+} transient is in response to a 1-Hz pulse train, with responses measured in the steady state. (**A**) Experimentally measured membrane potential (millivolts; ordinate) as a function of time (milliseconds; abscissa) in normal (solid) and failing (dotted) canine myocytes. (**B**) Experimentally measured cytosolic Ca^{2+} concentration (nanomoles/liter; ordinate) as a function of time (milliseconds; abscissa) for normal (solid) and failing (dotted) canine ventricular myocytes. (**C**). Membrane potential (millivolts; ordinate) as a function of time (milliseconds; abscissa) simulated using the normal canine myocyte model (solid) and with the successive down-regulation of I_{to1} (dot–dashed, 66% down-regulation), I_{K1} (long-dashed, down-regulation by 32%), SERCA2 (rightmost short-dashed, down-regulation by 62%) and NCX1 (dotted, up-egulation by 75%). (**D**) Cytosoli Ca^{2+} concentration (nanomoles/liter; ordinate) as a function of time (milliseconds; abscissa) simulated using the normal (solid) and heart failure (dotted) model. Reprinted from Winslow et al. (2001), with permission of the Royal Society of London.

and canine HF (O'Rourke et al., 1999). There is uncertainty as to whether mRNA and expressed protein level of phospholamban is decreased (Arai et al., 1993) or unchanged (Meyer et al., 1995) in human HF, and there is evidence that expressed protein level is decreased by a percentage amount equal to that of the SR Ca^{2+}-ATPase in failing canine heart (O'Rourke et al., 1999).

An important question is, "How do the experimentally measured changes in gene expression, protein levels, and current densities impact on the morphology of the AP and Ca^{2+} transient?" In particular, it is key to know which of these changes

has the greatest functional effect. To answer this question, we have developed a computational model of the failing ventricular myocyte (Irvine et al., 1999). The above data suggest the following *minimal* model of altered repolarization and Ca^{2+} handling in ventricular cells from the failing canine heart: (a) reduced expression of I_{K1} and I_{to1}, (b) down-regulation of the SR Ca^{2+}-ATPase, and (c) up-regulation of the electrogenic Na$^+$–Ca^{2+} exchanger. Since the density, but not the kinetic behavior, of each of the four transporters and ion currents comprising the minimal model appears altered in HF, we incorporated information on this altered gene and protein expression in the canine cell model by varying the density of these four membrane transporters (I_{to1}, I_{K1}, SR Ca^{2+}-ATPase, and Na^+–Ca^{2+} exchanger) within experimentally derived limits (O'Rourke et al., 1999).

The model has been used to test the hypothesis that this minimal set of heart-failure-induced changes can account for prolongation of AP duration, as well as decreased peak amplitude and decay rate of the Ca^{2+} transient observed in failing myocytes. Figures 16.8C and 16.8D demonstrate the ability of the model to reconstruct APs and Ca^{2+} transients measured in both normal and failing canine midmyocardial ventricular myocytes. Figure 16.8C shows a normal model AP (solid line) and also shows model APs corresponding to the additive effects of sequential down-regulation of I_{to1} (by 62%; dot–dashed line), I_{K1} (by 32%; long-dashed line), and SR Ca^{2+}-ATPase (by 62%; rightmost short-dashed bold line), followed by up-regulation of Na^+–Ca^{2+} exchanger (by 75%; dotted line). Changes of transporter amplitude were based on average values derived from experiments using midmyocardial failing canine ventricular myocytes. Model simulations indicate that down-regulation of I_{to1} produces a modest *shortening*, not lengthening, of AP duration. On first consideration, this seems to be an anomalous effect since I_{to1} is an outward K current, but is one that agrees with the experimental results of Zygmunt et al. (1997) in canine myocytes (see their Figure 2). The mechanism of this AP duration shortening has been investigated in detail using computational models (Greenstein et al., 2000), and results show that reduction of the Phase 1 notch depth through down-regulation of I_{to1} reduces the electrical driving force on inward Ca^{2+} current and hence shortens AP duration. The additional down-regulation of I_{K1} (long-dashed line) produces modest AP prolongation, consistent with the fact that outward current through I_{K1} is activated primarily at potentials which are hyperpolarized relative to the plateau potential. The most striking result is shown by the short-dashed line in Figure 16.8C: Significant AP prolongation occurs following down-regulation of SR Ca^{2+}-ATPase. This down-regulation results in a near doubling of AP duration that is similar to that observed experimentally (Figure 16.8A). Finally, the model predicts that up-regulation of Na^+–Ca^{2+} exchanger, when superimposed on these other changes, contributes to modest APD *shortening* due to reverse mode Na^+–Ca^{2+} exchange and generation of a net outward current during the plateau phase of the AP.

This modeling has provided important insights into the mechanism of AP prolongation and altered Ca^{2+} transients in heart failure. Prior to this work, the consensus was that down-regulation of the genes encoding the I_{to1} and I_{K1} outward K currents was responsible for AP prolongation—a very intuitive and reasonable hypothesis. The model indicates that this is not likely to be the case. Rather, the main contributor to AP prolongation involves down-regulation of the gene encoding the SR Ca^{2+}-ATPase. Subsequent model simulations have shown that down-regulation of this transport process alone has a severe effect on prolongation of the AP—a prediction confirmed by experiments in which cyclopiazonic acid is used to block SR Ca^{2+}-ATPase transport (Mazhari et al., 2001).

The modeling described above illustrates the value of using quantitative models to interpret the consequences of changes in gene and protein expression on cell function. It also serves as a reminder of how intuition, or a "mental-model," regarding the ways in which highly nonlinear interactions between various subcellular processes affect cell behavior may lead us in the wrong direction. Prediction of cellular phenotype using knowledge of underlying molecular changes *must* be based on interpretations derived from quantitative experimentally based models.

16.5 LIMITATIONS

The complexity of biological models, including those of the cardiac myocyte, is increasing rapidly. This complexity makes the reliable publication and exchange of models difficult. It is literally the case that very few biophysically detailed models of the myocyte make it into publication free of errors. It is therefore critically important that methods for the error-free dissemination of computational and mathematical models be developed. XML-based markup languages such as CellML and the Systems Biology Markup Language (SBML) are being developed to support the error-free exchange of models independently of the hardware and software architectures on which these models will run. An application programming interface for CellML is being developed, and several groups are developing software for automated source code generation from CellML files. However, the extent to which the CellML project will continue to develop is at present unclear. A second approach to the dissemination of models involves the use of web services. Web services (Coyle, 2002), a technology building on the ability of Simple Object Access Protocol (SOAP) to support distributed network communication, has great potential as a tool for making both data and computational algorithms transparently available to other software applications, thus facilitating the machine discovery, communication, and analyses of biological data and models.

16.6 PROSPECTS FOR THE FUTURE

16.6.1 New Myocyte Model Components—Mitochondrial Energy Production

Approximately 2% of cellular ATP is consumed on each heartbeat. The major processes consuming ATP in the myocyte are muscle contraction, activity of the SR Ca^{2+}-ATPase, and Na–K pumping. In addition, cellular ATP levels influences ion channel function including the sarcolemmal ATP-modulated K channel (Nichols and Lederer, 1990). Thus an emerging area of interest in computational cardiology is to develop biochemically based models of energy production and utilization in the myocytes. Such models are critically important to the quantitative understanding of the effects of ischemia on heart function.

Recently, we have formulated an integrated thermokinetic model of cardiac mitochondrial energetics comprising the tricarboxylic acid (TCA) cycle, oxidative phosphorylation, and mitochondrial Ca^{2+} handling (Cortassa et al., 2003). This model describes dynamics of key regulatory effectors of TCA cycle enzymes and the production of NADH and $FADH_2$. These molecules are used by the electron transport chain to establish a proton motive force ($\Delta\mu_H$) which then drives the F_1F_0-ATPase. Mitochondrial matrix Ca^{2+} is also a model state variable. Mitochondrial Ca^{2+} concentration is determined by the Ca^{2+} uniporter and Na^+/Ca^{2+} exchanger activities, and it regulates activity of the TCA cycle enzymes isocitrate dehydrogenase (IDH) and α-ketoglutarate dehydrogenase (KGDH). The model is described by 12 ordinary differential equations that represent $\Delta \Psi_m$ (mitochondrial membrane potential) and matrix concentrations of Ca^{2+}, NADH, ADP, and TCA cycle intermediates. The model is able to reproduce experimental data concerning mitochondrial bioenergetics, Ca^{2+} dynamics, and respiratory control, relying only on the fundamental properties of the system. The time-dependent behavior of the model, under conditions simulating an increase in workload, closely reproduces the experimentally observed mitochondrial NADH dynamics in heart trabeculae subjected to changes in pacing frequency. The steady-state and time-dependent behavior of the model support the role of mitochondrial matrix Ca^{2+} in matching energy supply with demand in cardiac cells. Further development and testing of this model, its integration into models of the myocyte, and the use of these models to investigate myocyte responses to ischemia are required.

16.6.2 New Myocyte Model Components—Signal Transduction Pathways

Virtually no quantitative modeling has addressed the functioning of signal transduction pathways in the cardiac myocyte. Perhaps the most important of these pathways are the $\beta 1$- and $\beta 2$-adrenergic signaling pathway. An excellent review of these pathways and of their action on cardiac LCCs is provided in Kemp and Hell (2000). Briefly, stimulation of the $\beta 1$-adrenergic receptor ($\beta 1AR$) or $\beta 2AR$ leads to a G-protein-mediated activation of adenylyl cyclase (AC) and increased production of cAMP. This, in turn, activates protein kinase A (PKA). Evidence suggests that the $\beta 2AR$, LCC, PKA, AC, and cAMP phosphodiesterases may assemble to form a highly localized signaling complex (Davare et al., 2001) regulating phosphorylation of LCCs. Actions of the $\beta 2ARs$ are more global, producing phosphorylation of not only LCCs, but also of phospholamban (which in its phosphorylated form dissociates from the SR Ca^{2+}-ATPase, relieving inhibition of this transporter), HERG-encoded channels (I_{Kr} current), and KCNQ1-encoded channels (I_{Ks} current). There are possible effects on the RyR, but these reports are controversial (Li et al., 2002; Marx et al., 2000). Since this signaling pathway is critical to regulation of heart rate and force of contraction, quantitative modeling of this signaling cascade and its action on ion channels and transporters is of fundamental significance in computational cardiology.

The above are only two important examples of directions for the future development of cardiac models. Each of these (and other directions) involves the modeling of processes (e.g., mitochondrial energy production), or development of conceptual approaches (e.g., in the case of b2AR signaling, modeling of local protein assemblies, whose dynamics may be mediated by interactions between very small numbers of molecules), that are of general importance in the field of computational biology.

References

Abbott, G.W., Sesti, F., Splawski, I., Buck, M.E., Lehmann, M.H., Timothy, K.W., Keating, M.T., and Goldstein, S.A. (1999). MiRP1 forms IKr potassium channels with HERG and is associated with cardiac arrhythmia. *Cell* **97**(2), 175–187.

Arai, M., Alpert, N.R., MacLennan, D.H., Barton, P., and Periasamy, M. (1993). Alterations in sarcoplasmic reticulum gene expression in human heart failure: A possible mechanism for alterations in systolic and diastolic properties of the failing myocardium. *Circ. Res.* **72**, 463–469.

Bers, D., and Stiffel, V. (1993). Ratio of ryanodine to dihidropyridine receptors in cardiac and skeletal muscle and implications for E–C coupling. *Am. J. Physiol.* **264**(6 Pt 1), C1587–C1593.

Bers, D.M. (2002). Cardiac excitation-contraction coupling. *Nature (London)* **415**, 198–205.

Bers, D.M., and Perez-Reyes, E. (1999). Ca channels in cardiac myocytes: Structure and function in Ca influx and intracellular Ca release. *Cardiovasc. Res.* **42**(2), 339–360.

Beuckelmann, D.J., Nabauer, M., and Erdmann, E. (1992). Intracellular calcium handling in isolated ventricular myocytes from patients with terminal heart failure. *Circulation* **85**, 1046–1055.

Beuckelmann, D.J., Nabauer, M., and Erdmann, E. (1993). Alterations of K^+ currents in isolated human ventricular myocytes from patients with terminal heart failure. *Circ. Res.* **73**, 379–385.

Chay, T.R. (1991). The Hodgkin–Huxley Na^+ channel model versus the five-state Markovian model. *Biopolymers* **31**(13), 1483–1502.

Cheng, H., Lederer, W.J., and Cannell, M.B. (1993). Calcium sparks: Elementary events underlying excitation–contraction coupling in heart muscle. *Science* **262**, 740–744.

Clancy, C., and Rudy, Y. (1999). Linking a genetic defect to its cellular phenotype in a cardiac arrhythmia. *Nature (London)* **400**, 566–569.

Clay, J.R., and DeFelice, L.J. (1983). Relationship between membrane excitability and single channel open-close kinetics. *Biophys. J.* **42**, 151–157.

Corana, E.A. (1987). Minimizing multimodal functions of continuous variables with the simulated annealing algorithm. *ACM Trans. Math. Software* **13**(3), 262–280.

Cortassa, S., Aon, M., Winslow, R.L., Marban, E., and O'Rourke, B. (2002). Modeling mitochondrial Ca2+ dynamics abd energy metabolism. *Biophys. J.* **82**(1), 115a.

Cortassa, S., Aon, M., Marban, E., Winslow, R., and O'Rourke, B. (2003). An integrated model of cardiac mitochondrial energy metabolism and calcium dynamics. *Biophys. J.* **84**, 2734–2755.

Courtemanche, M., Ramirez, R.J., and Nattel, S. (1998). Ionic mechanisms underlying human atrial action potential properties: Insights from a mathematical model. *Am. J. Physiol. Heart Circ. Physiol.* **275**(1), H301–H321.

Coyle, F.P. (2002). *XML, Web Services, and the Data Revolution.* Addison-Wesley, Boston, MA.

Davare, M.A., Avdonin, V., Hall, D.D., Peden, E.M., Burette, A., Weinberg, R.J., Horne, M.C., and Hoshi, T., and Hell, J.W. (2001). A beta2 adrenergic receptor signaling complex assembled with the Ca^{2+} channel Cav1.2. *Science* **293**, 98–101.

Demir, S.S., Clark, J.W., Murphey, C.R., and Giles, W.R. (1994). A mathematical model of a rabbit sinoatrial node cell. *Am. J. Physiol.* **266**(3 Pt 1), C832–C852.

DiFrancesco, D., and Noble, D. (1985). A model of cardiac electrical activity incorporating ionic pumps and concentration changes. *Philos. Trans. R. Soc. London, Ser. B* **307**, 353–398.

Dixon, J.E., Shi, W., Wang, H.-S., McDonald, C., Yu, H., Wymore, R.S., Cohen, I.S., and McKinnon, D. (1996). Role of the Kv4.3 K^+ channel in ventricular muscle: A molecular correlate for the transient outward current. *Circ. Res.* **79**, 659–668.

Dokos, S., Celler, B., and Lovell, N. (1996). Ion currents underlying sinoatrial node pacemaker activity: A new single cell mathematical model. *J. Theor. Biol.* **181**(3), 245–272.

Fabiato, A. (1983). Calcium-induced release of calcium from the cardiac sarcoplasmic reticulum. *Am. J. Physiol.* **245**(1), C1–C14.

Fabiato, A. (1985). Time and calcium dependence of activation and inactivation of calcium-induced release of calcium from the sarcoplasmic reticulum of a skinned canine cardiac Purkinje cell. *J. Gen. Physiol.* **85**, 247–289.

Fitzhugh, R. (1960). Thresholds and plateaus in the Hodgkin–Huxley nerve equations. *J. Gen. Physiol.* **43**, 867–896.

Fozzard et al. (1991). *The Heart and Cardiovascular System: Scientific Foundations.* Raven Press, New York.

Franzini-Armstrong, C., and Protasi, F., (1997). Ryanodine receptors of striated muscles: A complex channel capable of multiple interactions. *Physiol. Rev.* **77**(3), 699–729.

Franzini-Armstrong, C., Protasi, F., and Ramesh, V. (1999). Shape, size, and distribution of Ca(2+) release units and couplons in skeletal and cardiac muscles. *Biophys. J.* **77**(3), 1528–1539.

Greenstein, J., and Winslow, R.L. (2002). An integrative model of the cardiac ventricular myocyte incorporating local control of Ca^{2+} release. *Biophys. J.* **83**, 2918–2945.

Greenstein, J., Po, S., Wu, R., Tomaselli, G., and Winslow, R.L. (2000). Role of the calcium-independent transient outward current Ito1 in action potential morphology and duration. *Circ. Res.* **87**, 1026.

Hille, B. (1977). Local anaesthetics: Hydrophilic and hydrophobic pathways for the drug-receptor reactions. *J. Gen. Physiol.* **69**, 497–515.

Hille, B. (1992). *Ionic Channels of Excitable Membranes,* 2nd ed., pp. 341–345. Sinauer, Sunderland, MA.

Hodgkin, A.L., and Huxley, A.F. (1952). Currents carried by sodium and potassium ions through the membrane of the giant axon of Loligo. *J. Physiol. (London)* **116**, 449–472.

Hondeghem, L.M., and Katzung, B.G. (1977). Time- and voltage-dependent interactions of antiarrhythmic drugs with cardiac sodium channels. *Biochim. Biophys. Acta* **472**, 373–398.

Horn, R., and Vandenberg, C.A. (1984). Statistical properties of single sodium channels. *J. Gen. Physiol.* **84**(4), 505–534.

Irvine, L. (1998). *Models of the Cardiac Na Channel and the Action of Lidocaine.* Johns Hopkins University School of Medicine, Baltimore, MD.

Irvine, L., and Winslow, R. (1996). Numerical studies of use-dependent block of cardiac sodium channels by quinidine on spiral wave reentry.; 1996; pp. 613–616. *Computers in Cardiology,* IEEE Press. Indianapolis, IN.

Irvine, L., Jafri, M.S., and Winslow, R.L. (1999). Cardiac sodium channel Markov model with temperature dependence and recovery from inactivation. *Biophys. J.* **76**, 1868–1885.

Jafri, S., Rice, J.J., and Winslow, R.L. (1998). Cardiac Ca^{2+} dynamics: The roles of ryanodine receptor adaptation and sarcoplasmic reticulum load. *Biophys. J.* **74**, 1149–1168.

Kaab, S., Nuss, H.B., Chiamvimonvat, N., O'Rourke, B., Pak, P.H., Kass, D.A., Marban, E., and Tomaselli, G.F. (1996). Ionic mechanism of action potential prolongation in ventricular myocytes from dogs with pacing-induced heart failure. *Circ. Res.* **78**(2), 262–273.

Kaab, S., Dixon, J., Duc, J., Ashen, D., Näbauer, M., Beuckelmann, D.J., Steinbeck, D., McKinnon, D., and Tomaselli, G.F. (1998). Molecular basis of transient outward potassium current downregulation in human heart failure: A decrease in Kv4.3 mRNA correlates with a reduction in current density. *Circulation* **98**, 1383–1393.

Kemp, T.J., and Hell, J.W. (2000). Regulation of cardiac L-type calcium channels by protein kinase A and protein kinase C. *Circ. Res.* **87**, 1095.

Li, Y., Kranias, E.G., Mignery, G.A., and Bers, D.M. (2002). Protein kinase A phosphorylation of the ryanodine receptor does not affect calcium sparks in mouse ventricular myocytes. *Circ. Res.* **90**, 309–316.

Linz, K.W., and Meyer, R. (1998). Control of L-type calcium current during the action potential of guinea-pig ventricular myocytes. *J. Physiol. (London)* **513**(Pt 2), 425–442.

Liu, S., and Rasmusson, R.L. (1997). Hodgkin-Huxley and partially coupled inactivation models yield different voltage dependence of block. *Am. J. Physiol.* **272**(4 Pt 2), H2013–H2022.

Luo, C., and Rudy, Y. (1991). A model of the ventricular cardiac action potential. Depolarization, repolarization and their interaction. *Circ. Res.* **68**, 1501–1526.

Luo, C.H., and Rudy, Y. (1994). A dynamic model of the cardiac ventricular action potential: I. Simulations of ionic currents and concentration changes. *Circ. Res.* **74**, 1071–1096.

Marban, E., Yamagishi, T., and Tomaselli, G.F. (1998). Structure and function of voltage-gated sodium channels. *J. Physiol. (London)* **508**(Pt 3), 647–657.

Marx, S.O., Reiken, S., Hisamatsu. Y., Jayaraman, T., Burkhoff, D., Rosemblit, N., and Marks, A.R. (2000). PKA phosphorylation dissociates FKBP12.6 from the calcium release channel (ryanodine receptor): Defective regulation in failing hearts. *Cell* **101**(4), 365–376.

Mazhari, R., Greenstein, J.L., Winslow, R.L., Marban, E., and Nuss, H.B. (2001). Molecular interactions between two long-QT syndrome gene products, HERG and KCNE2, rationalized by in vitro and in silico analysis. *Circ. Res.* **89**(1), 33–38.

Meyer, M., Schillinger, W., Pieske, B., Holubarsch, C., Heilmann, C., Posival, H., Kuwajima, G., Mikoshiba, K., Just, H., and Hasenfuss, G. (1995). Alterations of sarcoplasmic reticulum proteins in failing human dilated cardiomyopathy. *Circulation* **92**, 778-784.

Näbauer, M., Beuckelmann, D.J., Überfuhr, P., and Steinbeck, G. (1996). Regional differences in current density and rate-dependent properties of the transient outward current in subepicardial and subendocardial myocytes of human left ventricle. *Circulation* **93**, 168–177.

Nichols, C.G., and Lederer, W.J. (1990). The regulation of ATP-sensitive K channel activity in intact and permeabilized rat ventricular cells. *J. Physiol. (London)* **423**, 91–110.

Noble, D. (1960). Cardiac action and pace maker potentials based on the Hodgkin–Huxley equations. *Nature (London)* **188**, 495.

Noble, D. (1962). A modification of the Hodgin–Huxley equations applicable to Purkinje fiber action and pacemaker potentials. *J. Physiol. (London)* **160**, 317.

Noble, D., and Noble, S.J. (1984). A model of sino-atrial node electrical activity based on a modification of the DiFrancesco–Noble (1984) equations. *Proc. R. Soc. London, Ser. B* **222**, 295–304.

Noble, D.S., Noble, S.J., Bett, C.L., Earm, Y.E., Ko, W.K., and So, I.K. (1991). The role of sodium–calcium exchange during the cadiac action potential. *Ann. N.Y. Acad. Sci.* **639**, 334–354.

Nygren, A., Fiset, C., Firek, L., Clark, J.W., Lindblad, D.S., Clark, R.B., and Giles, W.R. (1998). Mathematical model of an adult human atrial cell : The role of K^+ currents in repolarization. *Circ. Res.* **82**(1), 63–81.

O'Rourke, B., Peng, L.F., Kaab, S., Tunin, R., Tomaselli, G.F., Kass, D.A., and Marban, E. (1999). Mechanisms of altered excitation-contraction coupling in canine tachycardia-induced heart: Experimental studies. *Circ. Res.* **84**, 562–570.

Peterson, B., DeMaria, C., Adelman, J., and Yue, D. (1999). Calmodulin is the Ca^{2+} sensor for Ca2+-dependent inactivation of L-type calcium channels. *Neuron,* pp. 549–558.

Puglisi, J.L., and Bers, D.M. (2001). LabHEART: An interactive computer model of rabbit ventricular myocyte ion channels and Ca transport. *Am. J. Physiol.* **281**(6), C2049–C2060.

Reinecke, H., Studer, R., Vetter, R., Holtz, J., and Drexler, H. (1996). Cardiac Na/Ca exchange activity in patients with end-stage heart failure. *Cardiovasc. Res.* **31**, 48–54.

Rice, J.J., Jafri, M.S., and Winslow, R.L. (2000). Modeling short-term interval–force relations in cardiac muscle. *Am. J. Physiol.* **278**, H913–H931.

Roos (1997) Mechanics and force production. In *The Myocardium.* Academic Press, San Diego, CA.

Sanguinetti, M.C., and Jurkiewicz, N.K. (1990). Two components of cardiac delayed rectifier K^+ current: Differential sensitivity to block by Class III antiarrhythmic agents. *J. Gen. Physiol.* **96**, 192–215.

Sanguinetti, M.C., Jiang, C., Curran, M.E., and Keating, M.T. (1995). A mechanistic link between an inherited and an acquired cardiac arrhythmia: HERG encodes the IKr potassium channel. *Cell* **81**(2), 299–307.

Sham, J.S.K. (1997). Ca^{2+} release-induced inactivation of Ca2+ current in rat ventricular myocytes: Evidence for local Ca^{2+} signalling. *J. Physiol. (London)* **500**(2), 285–295.

Shannon, T.R., Ginsberg, K.S., and Bers, D.M. (1997). SR Ca uptake rate in permeabilized ventricular myocytes is limited by reverse rate of the SR Ca pump. *Biophys. J.* **72**(2), A167.

Shaw, R.M., and Rudy, Y. (1997). Electrophysiologic effects of acute myocardial ischemia. A mechanistic investigation of action potential conduction and conduction failure. *Circ. Res.* **80**(1), 124–138.

Stern, M. (1992). Theory of excitation-contraction coupling in cardiac muscle. *Biophys. J.* **63**, 497–517.

Studer, R., Reinecke, H., Bilger, J., Eschenhagen, T., Bohm, M., Hasenfuss, G., Just, H., Holtz, J., and Drexler, H. (1994). Gene expression of the Na^+–Ca^{2+} exchanger in end-stage human heart failure. *Circ. Res.* **75**, 443–453.

Takahashi, T., Allen, P.D., Lacro, R.V., Marks, A.R., Dennis, A.R., Schoen, F.J., Grossman, W., Marsh, J.D., and Izumo, S. (1992). Expression of dihydropyridine receptor (Ca^{2+} channel) and calsequestrin genes in the myocardium of patients with end-stage heart failure. *J. Clin. Invest.* **90**, 927–935.

Tomaselli and Marban (1999). Electrophysiological remodeling in hypertrophy and heart failure. *Circ. Res.* **42**, 270–283.

Verheijck et al. (1998). Distribution of atrial and nodal cells within the rabbit sinoatrial node: Models of sinoatrial transition. *Circulation* **97**, 1623–1631.

Wagner, J., and Keizer, J. (1994). Effects of rapid buffers on Ca^{2+} diffusion and Ca^{2+} oscillations. *Biophys. J.* **67**(1).

Wang, S.Q., Song, L.S., Lakatta, E.G., and Cheng, H. (2001). Ca^{2+} signalling between single L-type Ca^{2+} channels and ryanodine receptors in heart cells. *Nature (London)* **410**, 592–596.

Wier, W.G., Egan, T.M., Lopez-Lopez, J.R., and Balke, C.W. (1994). Local control of excitation–contraction coupling in rat heart cells. *J. Physiol. (London)* **474**(3), 463–471.

Wilders, R., Jongsma, H.J., and van Ginneken, A.C. (1991). Pacemaker activity of the rabbit sinoatrial node. A comparison of mathematical models. *Biophys. J.* **60**(5), 1202–1216.

Winslow, R., Greenstein, J., Tomaselli, G., and O'Rourke, B. (2001). Computational model of the failing myocyte: Relating altered gene expression to cellular function. *Philos. Trans. R. Soc. London, Ser. A* **359**, 1187–1200.

Yue, D.T., Burkhoff, D., Franz, M.R., Hunter, W.C., and Sagawa, K. (1985). Postextrasystolic potentiation of the isolated canine left ventricle: Relationship to mechanical restitution. *Circ. Res.* **56**, 340–350.

Zygmunt, A.C., Robitelle, D.C., and Eddlestone, G.T. (1997). Ito1 dictates behavior of ICl(Ca) during early repolarization of canine ventricle. *Am. J. Physiol.* **273**, H1096–H1106.

Electrophysiological Models

Mark E. Nelson, Ph.D.

CONTENTS

17

Databasing the Brain. Edited by Stephen H. Koslow and Shankar Subramaniam
ISBN 0-471-30921-4 © 2005 John Wiley & Sons, Inc.

17.1 INTRODUCTION

Understanding the electrophysiological basis of neural coding, communication, and information processing is central to modern neuroscience research. Mathematical modeling and computer simulation have become an integral part of the neuroscientist's toolbox for exploring these phenomena at a variety of levels of organization, from the biophysical basis of current flow through individual ion channels, to the modeling of aspects of cognitive function arising from the distributed activity of large populations of neurons. Neural models can be constructed at many levels of abstraction. Some types of scientific questions can be addressed using highly reduced models that treat neurons as simple threshold devices, while other questions require detailed models of membrane biophysics and intracellular signaling networks.

This chapter will focus on the tools and techniques for constructing *biophysically detailed compartmental models* of individual neurons and local networks (Koch and Segev, 1998; De Schutter 2001). Such models are well positioned to take advantage of emerging neuroinformatics approaches that can potentially link such models with a wealth of empirical data currently being compiled and organized into large neuroscientific databases (Huerta et al., 1993; Koslow and Huerta, 1997; Shepherd et al., 1998). In contrast, highly abstracted models lack sufficient biological detail to establish meaningful links to these databases, while large systems-level models involving multiple brain regions are generally too diverse in structure and function for neuroinformatics approaches to be productive. In the intermediated term over the next several years, biophysically detailed models of single neurons and local networks will likely provide the most fruitful level of analysis for uncovering new functional relationships, dynamical principles, and information processing strategies.

How do electrophysiological models fit into an informatics approach to neuroscience? This question is perhaps best answered in the context of a bottom-up view of the problem. Starting at the molecular level, sequence-based informatics approaches are being used to reveal information about structural and evolutionary relationships among ion channels and receptor proteins. Molecular dynamics simulations can help establish links from the structural level to the functional properties of individual ion channel and receptor complexes. Electrophysiological models come into play at the next level of organization where information-processing properties emerge from the dynamic interactions of multiple channel and receptor types at the single-neuron level and interactions of multiple neurons at the network level.

Neurons typically contain numerous types of ion channels and membrane receptors. Different types of neurons express different combinations of these proteins, with varying densities and varying spatial distributions. There are, for example, dozens of different types of voltage-gated K^+ channels, but an individual neuron may only express a few of these, and the expression might be restricted to the soma or to particular regions of the dendrites (Rudy, 1988; Coetzee et al., 1999). Different types of K^+ channels vary in their electrophysiological properties, such as activation and inactivation voltages, time constants, and conductances. This heterogeneity suggests that K^+ channels may be differentially expressed and distributed in order to shape the electrophysiological response properties of individual neurons for carrying out particular types of information processing tasks.

The contributions of different ion channels to the information processing capabilities of the system cannot be deduced from the properties of individual ion channels alone. Rather, functional properties at the single-neuron level must be evaluated in the presence of an appropriate mix of channel types, densities, and distributions and in the context of physiologically relevant spatiotemporal patterns of input. For example, certain types of K^+ channels from the Kv3 gene family are known to be activated only at rather depolarized membrane potentials and tend to have fast activation and inactivation time constants (Rudy et al., 1999). Using biophysically detailed compartmental models, neuroscientists have been able to achieve a detailed understanding of how Kv3 channel properties contribute to temporal signal processing in the electric sense of weakly electric fish (Rashid et al., 2001a,b; Doiron et al., 2001) and in the mammalian auditory system (Wang et al., 1998). If appropriate databases were available and suitable neuroinformatics tools existed, one could imagine undertaking a variety of interesting comparative investigations regarding the functional role of Kv3 channels in other species, other sensory systems, and other neural information processing contexts.

17.2 BACKGROUND

This chapter assumes a general familiarity with the neurophysiological and biophysical mechanisms associated with electrical signaling in neurons. Good introductory material in this area can be found in numerous undergraduate textbooks (e.g., Delcomyn, 1998; Shepherd, 1994). Recent advanced texts are available that provide detailed, up-to-date coverage in areas such as the cellular and molecular biology of nerve cells (e.g., Levitan and Kaczmarek, 2002) and the biophysical properties of ion channels (e.g., Hille, 2001). A brief glossary is provided here as a convenient reference for some of the key terminology and functional concepts.

17.2.1 Glossary

Action Potential: a transient electrical impulse that propagates along an axon and serves as the most common form of electrical signaling between neurons. The duration is typically on the order of a millisecond, and the amplitude is on the order of 100 mV. Action potentials are also called nerve impulses or spikes.

Axon: an output branch of a neuron that conducts action potentials away from the site of initiation and conveys electrical signals to other neurons or effectors. The axon usually starts off as a single long branch but may terminate in a complex branched arbor that distributes outputs to large numbers of target neurons.

Compartmental Model: a single neuron model that divides the cell into multiple spatial compartments. Each compartment can have different properties (length, diameter, membrane voltage, ion channel densities, etc.). The model produces a coupled set of differential equations that are solved using numerical integration techniques.

Conductance: a measure of the ease with which electric current flows through a material; the reciprocal of resistance; conductance units are siemens; conductance (siemens) is a measure of current (amperes) divided by voltage (volts).

Dendrite: an input branch of a neuron that typically receives synaptic contacts from other neurons and conveys graded electrical potentials to other parts of the neuron.

Equilibrium Potential: For an individual type of ion, the membrane potential at which the effects of the electrical potential difference and the concentration gradient across the membrane are balanced so as to produce no net ion flux. Also called the Nernst potential. Calculated from the Nernst equation: $E_{ion} = (RT/zF) \ln([ion]_{out}/[ion]_{in})$, where R is the universal gas constant, T is the temperature in degrees Kelvin, z is the ionic charge, F is Faraday's constant, and $[ion]_{in}$ and $[ion]_{out}$ are ionic concentrations.

Gating: the process by which ion channels open and close so as to regulate the flow of ions. Voltage-gated ion channels change their gating state based on the local electrical potential difference across the cell membrane. Ligand-gated channels change their gating state based on the binding of signaling molecules (neurotransmitters). Gating usually involves a change in the conformation of the channel protein.

Hodgkin–Huxley Model: originally, a specific model of the ionic basis of the action potential in squid giant axon developed by Hodgkin and Huxley and published in 1952. More generally, any model that uses a Hodgkin–Huxley-type formalism to describe macroscopic ionic currents in nerve cells based on voltage-dependent gating properties.

Macroscopic Conductance: the electrical conductance arising from a population of single channel conductances; often quoted in terms of conductance per unit area of membrane; typically in the range of millisiemens (mS) per square centimeters.

Markov Model: a formalism for modeling stochastic processes that treats the behavior of a system as a series of transitions between distinct states. In the context of ion channels, these distinct states represent different conformational states. Some conformations will correspond to closed states, some to open states, and some to inactivated states.

Membrane Potential: The electrical potential in the intracellular space. If the potential outside of the cell is used as a reference (0 mV), then a typical membrane potential for a neuron at rest would be on the order of -70 mV. The choice of reference point, however, is a matter of convention. In some cases the inside of the cell is used as a reference, in which case the resting membrane potential would be 0 mV and the external potential would be $+70$ mV.

Monte Carlo Method: a numerical method for simulating the behavior of a stochastic system by using random numbers to generate possible outcomes based on a model of the underlying probability distribution.

Single-Channel Conductance: the electrical conductance of a single ion channel in the open state; typically between 1 and 150 pS (picosiemens).

Soma: the cell body of the neuron; contains the nucleus and much of the metabolic machinery of the cell.

Reversal Potential: the membrane potential at which no net current flows through an open ion channel or activated synapse. If the channel or synapse is permeable to a single ionic species, then the reversal potential is equal to the equilibrium potential for that ion. Otherwise, the reversal potential reflects a weighted sum of the equilibrium potentials for all permeant ionic species.

Voltage-Clamp: a technique for recording the ionic currents across the cell membrane during controlled changes in the membrane potential. Fast feedback circuitry is used to maintain the membrane potential at the desired "command" level.

17.3 TECHNICAL DETAILS AND METHODOLOGY

17.3.1 Hodgkin–Huxley Models

The core mathematical framework for modern biophysically based neural modeling was developed half a century ago by Sir Alan Hodgkin and Sir Andrew Huxley. They carried out an elegant series of electrophysiological experiments on the squid giant axon in the late 1940s and early 1950s. The squid giant axon is notable for its extraordinarily large diameter ($\sim$0.5 mm). Most axons in the squid nervous system and in other nervous systems are typically at least 100 times thinner. The large size of the squid giant axon is a specialization for rapid conduction of action potentials that trigger the contraction of the squid's mantle when escaping from a predator. In addition to being beneficial for the squid, the large diameter of the giant axon was beneficial for Hodgkin and Huxley because it permitted manipulations that were not technically feasible in smaller axons that had been used in biophysical studies up to that point. In a well-designed series of experiments, Hodgkin and Huxley systematically demonstrated how the macroscopic ionic currents in the squid giant axon could be understood in terms of changes in Na^+ and K^+ conductances in the axon membrane. Based on a series of voltage-clamp experiments, they developed a detailed mathematical model of the voltage-dependent and time-dependent properties of the Na^+ and K^+ conductances. The empirical work led to the development of a coupled set of differential equations describing the ionic basis of the action potential (Hodgkin and Huxley, 1952), which became known as the Hodgkin–Huxley (HH) model. The real predictive power of the model became evident when Hodgkin and Huxley demonstrated that numerical integration of these differential equations (using a hand-cranked mechanical calculator!) could accurately reproduce all the key biophysical properties of the action potential. For this outstanding achievement, Hodgkin and Huxley were awarded the 1963 Nobel Prize in Physiology and Medicine (shared with Sir John Eccles for his work on the biophysical basis of synaptic transmission).

Electrical Equivalent Circuits

In biophysically based neural modeling, the electrical properties of a neuron are represented in terms of an electrical equivalent circuit. Capacitors are used to model the charge storage capacity of the cell membrane, resistors are used to model the various types of ion channels embedded in membrane, and batteries are used to represent the electrochemical potentials established by differing intra- and extracellular ion concentrations. In their seminal paper on the biophysical basis of the action potential, Hodgkin and Huxley (1952) modeled a segment of squid giant axon using an equivalent circuit similar to that shown in Figure 17.1. In the equivalent circuit, the current across the membrane has two major components, one associated with the membrane capacitance and one associated with the flow of ions through resistive membrane channels. The capacitive current I_c

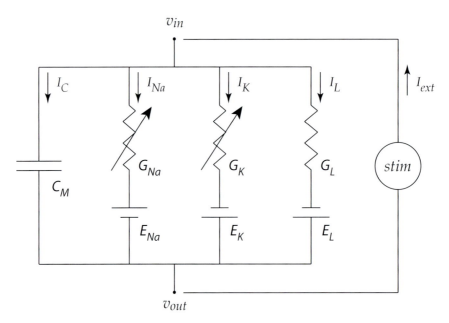

Figure 17.1 Electrical equivalent circuit for a short segment of squid giant axon. The capacitor represents the capacitance of the cell membrane; the two variable resistors represent voltage-dependent Na$^+$ and K$^+$ conductances, the fixed resistor represents a voltage-independent leakage conductance, and the three batteries represent reversal potentials for the corresponding conductances. The pathway labeled "*stim*" represents an externally applied current, such as might be introduced via an intracellular electrode. The sign conventions for the various currents are indicated by the directions of the corresponding arrows. Note that the arrow for the external stimulus current I_{ext} is directed from outside to inside (i.e., inward stimulus current is positive), whereas arrows for the ionic currents I_{Na}, I_K, and I_L are directed from inside to outside (i.e., outward ionic currents are positive). After Hodgkin and Huxley (1952).

is defined by the rate of change of charge q at the membrane surface: $I_c = dq/dt$. The charge $q(t)$ is related to the instantaneous membrane voltage $V_m(t)$ and membrane capacitance C_m by the relationship $q = C_m V_m$. Thus the capacitive current can be rewritten as $I_c = C_m\, dV_m/dt$. In the Hodgkin–Huxley model of the squid axon, the ionic current I_{ion} is subdivided into three distinct components: a sodium current I_{Na}, a potassium current I_K, and a small leakage current I_L that is primarily carried by chloride ions. The behavior of an electrical circuit of the type shown in Figure 17.1 can be described by a differential equation of the general form

$$C_m \frac{dV_m}{dt} + I_{ion} = I_{ext}, \qquad (17.1)$$

where I_{ext} is an externally applied current, such as might be introduced through an intracellular electrode. Equation (17.1) is the fundamental equation relating the change in membrane potential to the currents flowing across the membrane.

Macroscopic Ionic Currents

The individual ionic currents I_{Na}, I_K, and I_L shown in Figure 17.1 represent the macroscopic currents flowing through a large population of individual ion channels. In HH-style models, the macroscopic current is assumed to be related to the membrane voltage through an Ohm's law relationship of the form $V = IR$. In many cases it is more convenient to express this relationship in terms of conductance rather than resistance, in which case Ohm's law becomes $I = GV$, where the conductance G is the

inverse of resistance, $G = 1/R$. In applying this relationship to ion channels, the equilibrium potential E_k for each ion type also needs to be taken into account. This is the potential at which the net ionic current flowing across the membrane would be zero. The equilibrium potentials are represented by the batteries in Figure 17.1. The current is proportional to the conductance times the difference between the membrane potential V_m and the equilibrium potential E_k. The total ionic current I_{ion} is the algebraic sum of the individual contributions from all participating channel types found in the cell membrane:

$$I_{ion} = \sum_k I_k = \sum_k G_k(V_m - E_k), \qquad (17.2)$$

which expands to the following expression for the Hodgkin–Huxley model of the squid axon:

$$I_{ion} = G_{Na}(V_m - E_{Na}) + G_K(V_m - E_K) + G_L(V_m - E_L). \qquad (17.3)$$

Note that individual ionic currents can be positive or negative depending on whether or not the membrane voltage is above or below the equilibrium potential. This raises the question of sign conventions. Is a positive ionic current flowing into or out of the cell? The most commonly used sign convention in neural modeling is that ionic current flowing *out* of the cell is positive and ionic current flowing into the cell is negative (see subsection entitled "Sign Conventions" for more details).

In general, the conductances are not constant values, but can depend on other factors like the membrane voltage or the intracellular calcium concentration. In order to explain their experimental data, Hodgkin and Huxley postulated that G_{Na} and G_K were voltage-dependent quantities, whereas the leakage current G_L was taken to be constant. Thus the resistor symbols in Figure 17.1 are shown as variable resistors for G_{Na} and G_K, and as a fixed resistor for G_L. Today, we know that the voltage-dependence of G_{Na} and G_K can be related to the biophysical properties of the individual ion channels that contribute to the macroscopic conductances. Although Hodgkin and Huxley did not know about the properties of individual membrane channels when they developed their model, it will be convenient for us to describe the voltage-dependent aspects of their model in those terms.

Gates

The macroscopic conductances of the HH model can be considered to arise from the combined effects of a large number of microscopic ion channels embedded in the membrane. Each individual ion channel can be thought of as containing one or more physical *gates* that regulate the flow of ions through the channel. An individual gate can be in one of two states, *permissive* or *nonpermissive*. When *all* of the gates for a particular channel are in the permissive state, ions can pass through the channel and the channel is *open*. If any of the gates are in the nonpermissive state, ions cannot flow and the channel is *closed*. Although it might seem more natural to speak of *gates* as being *open* or *closed*, a great deal of confusion can be avoided by consistently using the terminology *permissive* and *nonpermissive* for gates while reserving the terms *open* and *closed* for channels.

The voltage-dependence of ionic conductances is incorporated into the HH model by assuming that the probability for an individual gate to be in the permissive or nonpermissive state depends on the value of the membrane voltage. If we consider gates of a particular type i, we can define a probability p_i, ranging between 0 and 1, which represents the *probability* of an individual gate being in the permissive state. If we consider a large number of channels, rather than an individual channel, we can also interpret p_i as the fraction of gates in that population that are in the permissive state. At some point in time t, let $p_i(t)$ represent the fraction of gates that are in the permissive state. Consequently, $1 - p_i(t)$ must be in the nonpermissive state.

$$
\begin{array}{ccc}
\text{fraction in} & \xrightarrow{\alpha_i(V)} & \text{fraction in} \\
\text{nonpermissive} & & \text{permissive} \\
\text{state, } 1 - p_i(t) & \xleftarrow{\beta_i(V)} & \text{state, } p_i(t)
\end{array}
$$

The rate at which gates transition from the nonpermissive state to the permissive state is denoted by a variable $\alpha_i(V)$, which has units of s^{-1}. Note that this "rate constant" is not really constant, but depends on membrane voltage V. Similarly, there is a second rate constant, $\beta_i(V)$, describing the transition rate from the permissive to the nonpermissive state. Transitions between permissive and nonpermissive states in the HH model are assumed to obey first-order kinetics:

$$
\frac{dp_i}{dt} = \alpha_i(V)(1 - p_i) - \beta_i(V)p_i, \tag{17.4}
$$

where $\alpha_i(V)$ and $\beta_i(V)$ are voltage-dependent. If the membrane voltage V_m is clamped at some fixed value V, then the fraction of gates in the permissive state will eventually reach a steady-state

value (i.e., $dp_i/dt = 0$) as $t \to \infty$ given by

$$
p_{i,t\to\infty} = \frac{\alpha_i(V)}{\alpha_i(V) + \beta_i(V)}. \tag{17.5}
$$

The time course for approaching this equilibrium value is described by a simple exponential with time constant $\tau_i(V)$ given by

$$
\tau_i(V) = \frac{1}{\alpha_i(V) + \beta_i(V)}. \tag{17.6}
$$

When an individual channel is open, it contributes some small, fixed value to the total conductance and zero otherwise. The macroscopic conductance for a large population of channels is thus proportional to the number of channels in the open state, which is, in turn, proportional to the probability that the associated gates are in their permissive state. Thus the macroscopic conductance G_k due to channels of type k, with constituent gates of type i, is proportional to the *product* of the individual gate probabilities p_i:

$$
G_k = \bar{g}_k \prod_i p_i, \tag{17.7}
$$

where $\bar{g}_k$ is a normalization constant that determines the maximum possible conductance when all the channels are open (i.e., all gates are in the permissive state).

We have presented Eqs. (17.4)–(17.7) using a generalized notation that can be applied to a wide variety of conductances beyond those found in the squid axon. To conform to the standard notation of the HH model, the probability variable p_i in Eqs. (17.4)–(17.7) is replaced by a variable that represents the gate type. For example, Hodgkin and Huxley modeled the sodium conductance using three gates of a type labeled "m" and one gate of type "h". Applying Eq. (17.7) to the sodium channel using both the generalized notation and the standard notation yields

$$
G_{Na} = \bar{g}_{Na} p_m^3 p_h = \bar{g}_{Na} m^3 h. \tag{17.8}
$$

Similarly, the potassium conductance is modeled with four identical "n" gates:

$$
G_K = \bar{g}_K p_n^4 = \bar{g}_{Na} n^4. \tag{17.9}
$$

Summarizing the ionic currents in the HH model in standard notation, we have

$$
I_{ion} = \bar{g}_{Na} m^3 h(V_m - E_{Na}) + \bar{g}_K n^4(V_m - E_K) + g_L(V_m - E_L), \tag{17.10}
$$

$$
\frac{dm}{dt} = \alpha_m(V)(1 - m) - \beta_m(V)m, \tag{17.11}
$$

$$
\frac{dh}{dt} = \alpha_h(V)(1 - h) - \beta_h(V)h, \tag{17.12}
$$

$$
\frac{dn}{dt} = \alpha_n(V)(1 - n) - \beta_n(V)n. \tag{17.13}
$$

To completely specify the model, the one task that remains is to specify how the six rate constants in Eqs. (17.11)–(17.13) depend on the membrane voltage. Then Eqs. (17.10)–(17.13), together with Eq. (17.1), completely specify the behavior of the membrane potential V_m in the HH model of the squid giant axon.

Sign Conventions

Note that the appearance of I_{ion} on the left-hand side of Eq. (17.1) and I_{ext} on the right indicates that they have opposite *sign conventions*. As the equation is written, a positive external current I_{ext} will tend to depolarize the cell (i.e., make V_m more positive) while a positive ionic current I_{ion} will tend to hyperpolarize the cell (i.e., make V_m more negative). This sign convention for ionic currents is sometimes referred to as the neurophysiological or physiologists' convention. This convention is conveniently summarized by the phrase "inward negative," meaning that an inward flow of positive ions into the cell is considered a negative current. This convention perhaps arose from the fact that when one studies an ionic current in a voltage-clamp experiment, rather than measuring the ionic current directly, one actually measures the clamp current that is necessary to counterbalance it. Thus an inward flow of positive ions is observed as a negative-going clamp current, hence explaining the "inward negative" convention. Some neural simulation software packages, such as GEN-ESIS, use the opposite sign convention (inward positive), since that allows all currents to be treated consistently. In the figures shown in this chapter, membrane currents are plotted using the neurophysiological convention (inward negative).

Voltage Conventions

While we're on the topic of *conventions*, there are two more issues that should be discussed here. The first concerns the *value* of the membrane potential V_m. Recall that potentials are relative; only potential differences can be measured directly. Thus when defining the intracellular potential V_m, one is free to choose a convention that defines the resting intracellular potential to be zero (the convention used by Hodgkin and Huxley), or one could choose a convention that defines the extracellular potential to be zero, in which case the resting intracellular potential would be

around -70 mV. In either case the potential *difference* across the membrane is the same, it's simply a matter of how "zero" is defined. Most simulation software packages allow the user to select a voltage reference convention they like.

The second convention we need to discuss concerns the *sign* of the membrane potential. The modern convention is that depolarization makes the membrane potential V_m more positive. However, Hodgkin and Huxley (1952) used the opposite sign convention (depolarization negative) in their article. In the figures in this chapter, we use the modern convention that depolarization is positive.

At a conceptual level, the choice of conventions for currents and voltages is inconsequential; however, at the implementation level it matters a great deal, since inconsistencies will cause the model to behave incorrectly. The most important thing in choosing conventions is to ensure that the choices are internally consistent. One must pay careful attention to these issues when implementing a simulation using equations from a published model, since it may be necessary to convert the empirical results reported using one set of conventions into a form that is consistent with one's own model conventions.

Rate Constants

How did Hodgkin and Huxley go about determining the voltage-dependence of the rate constants α and β that appear in Eqs. (17.11)–(17.13)? How did they determine that the potassium conductance should be modeled with four n gates, but that the sodium conductance required three m gates and one h gate? In order to answer these questions, we need to look in more detail at the type of data that can be obtained from voltage-clamp experiments.

Figure 17.2 shows simulated voltage-clamp data, similar to those obtained by Hodgkin and Huxley in their studies of squid

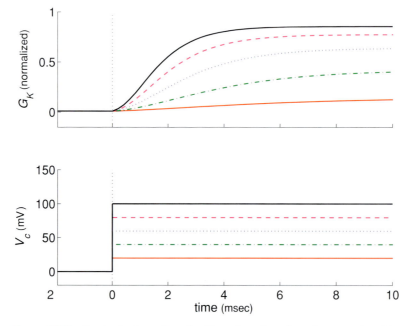

Figure 17.2 Simulated voltage-clamp data illustrating voltage-dependent properties of the K$^+$ conductance in squid giant axon. The command voltage V_c (mV) is shown in the lower panel, and the K$^+$ current is shown in the upper panel. Simulation parameters are from the Hodgkin and Huxley model (1952).

giant axon. In these experiments, Hodgkin and Huxley used voltage-clamp circuitry to step the membrane potential from the resting level (0 mV) to a steady depolarized level. The figure shows the time course of the change in normalized K$^+$ conductance for several different voltage steps. Three qualitative effects are apparent in the data. First, the steady-state conductance level increases with increasing membrane depolarization. Second, the onset of the conductance change becomes faster with increasing depolarization. Third, there is a slight temporal delay between the start of the voltage step and the change in conductance.

In the simulated voltage-clamp experiments illustrated in Figure 17.2, the membrane potential starts in the resting state ($V_m = 0$, using the HH voltage convention) and is then instantaneously stepped to a new clamp voltage V_c. What is the time course of the state variable n, which controls gating of the K$^+$ channel, under these circumstances? Recall that the differential equation governing the state variable n is given by

$$\frac{dn}{dt} = \alpha_n(V)(1 - n) - \beta_n(V)n. \tag{17.14}$$

Initially, with $V_m = 0$, the state variable n has a steady-state value (i.e., when $dn/dt = 0$) given by Eq. (17.5):

$$n_\infty(0) = \frac{\alpha_n(0)}{\alpha_n(0) + \beta_n(0)}. \tag{17.15}$$

When V_m is clamped to a new level V_c, the gating variable n will eventually reach a new steady-state value given by

$$n_\infty(V_c) = \frac{\alpha_n(V_c)}{\alpha_n(V_c) + \beta_n(V_c)}. \tag{17.16}$$

The solution to Eq. (17.14) that satisfies these boundary conditions is a simple exponential of the form

$$n(t) = n_\infty(V_c) - (n_\infty(V_c) - n_0(0))e^{-t/\tau_n(V_c)}. \tag{17.17}$$

Given Eq. (17.17), which describes the time course of n in response to a step change in command voltage, one could try fitting curves of this form to the conductance data shown in Figure 17.2 by finding values of $n_\infty(V_c)$, $n_\infty(0)$, and $\tau_n(V_c)$ that give the best fit to the data for each value of V_c. Figure 17.3 illustrates this process, using some simulated conductance data generated by the Hodgkin–Huxley model. Recall that n takes on values between 0 and 1, so in order to fit the conductance data, n must be multiplied by a normalization constant $\bar{g}_K$ that has units of conductance. For simplicity, the normalized conductance $G_K / \bar{g}_K$ is plotted. The dotted line in Figure 17.3 shows the best-fit results for a simple exponential curve of the form given in Eq. (17.17). While this simple form does a reasonable job of capturing the general time course of the conductance change, it fails to reproduce the sigmoidal shape and the temporal delay in onset. This discrepancy is most apparent near the onset of the conductance change, shown in the inset of Figure 17.3. Hodgkin and Huxley realized that a better fit could be obtained if they considered the conductance to be proportional to a higher power of n. Figure 17.3 shows the results of fitting the conductance data using a form $G_K = \bar{g}_k n^j$ with powers of j ranging from 1 to 4. Using this sort of fitting procedure, Hodgkin and Huxley determined that a reasonable fit to the K$^+$ conductance data could be obtained using an exponent of $j = 4$. Thus they arrived at a description for the K$^+$ conductance under voltage-clamp

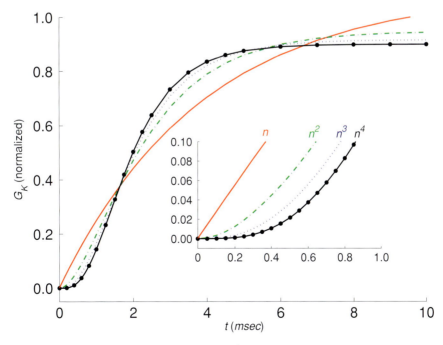

Figure 17.3   Best-fit curves of the form $G_k = \bar{g}_K n^j$ ($j = 1$–4) for simulated conductance versus time data. The inset shows an enlargement of the first millisecond of the response. The initial inflection in the curve cannot be well-fit by a simple exponential (dotted line) which rises linearly from zero. Successively higher powers of j ($j = 2$: dot–dashed; $j = 3$: dashed line) result in a better fit to the initial inflection. In this case, $j = 4$ (solid line) gives the best fit.

conditions given by

$$G_K = \bar{g}_K n^4 = \bar{g}_K \left[n_\infty(V_c) - (n_\infty(V_c) - n_\infty(0))e^{-t/\tau_n} \right]^4 \tag{17.18}$$

Activation and Inactivation Gates

The strategy that Hodgkin and Huxley used for modeling the sodium conductance is similar to that described above for the potassium conductance, except that the sodium conductance shows a more complex behavior. In response to a step change in clamp voltage, the sodium conductance exhibits a transient response (Figure 17.4), whereas the potassium conductance exhibits a sustained response (Figure 17.2). Sodium channels inactivate whereas the potassium channels do not. To model this process, Hodgkin and Huxley postulated that the sodium channels had two types of gates, an activation gate, which they labeled m, and an inactivation gate, which they labeled h. Again, boundary conditions dictated that m and h must follow a time course given by

$$m(t) = m_\infty(V_c) - (m_\infty(V_c) - m_\infty(0))e^{-t/\tau_m(V_c)}, \tag{17.19}$$

$$h(t) = h_\infty(V_c) - (h_\infty(V_c) - h_\infty(0))e^{-t/\tau_h(V_c)}. \tag{17.20}$$

Hodgkin and Huxley made some further simplifications by observing that the sodium conductance in the resting state is small compared to the value obtained during a large depolarization hence they were able to neglect $m_\infty(0)$ in their fitting procedure. Likewise, steady-state inactivation is nearly complete for large depolarizations, so $h_\infty(V_c)$ could also be eliminated from the fitting procedure. With these simplifications, Hodgkin and Huxley were able to fit the remaining parameters from the voltage-clamp data. The sodium conductance G_{Na} was thus modeled by an expression of the form $G_{Na} = \bar{g}_{Na} m^3 h$.

Parameterizing the Rate Constants

By fitting voltage-clamp data as discussed above, steady-state conductance values and time constants can be empirically determined as a function of command voltage for each of the gating variables associated with a particular channel. Using Eqs. (17.5) and (17.6), the steady-state conductance values and time constants can be transformed into expressions for the forward and backward rate constants α and β. For example, for the potassium channel n gate we have

$$\alpha_n(V) = \frac{n_\infty(V)}{\tau_n(V)}, \tag{17.21}$$

$$\beta_n(V) = \frac{1 - n_\infty(V)}{\tau_n(V)}. \tag{17.22}$$

Thus there are two equivalent representations for the voltage-dependence of a channel. One representation specifies the voltage-dependence of the rate constants, which we'll call the α/β representation. The other representation specifies the voltage-dependence of the steady state conductance and the time constant, which we'll call the n_∞/τ representation. These two representations are interchangeable, and one can easily convert between them using the algebraic relationships in Eqs. (17.5) and (17.6) (for transforming from α/β to n_∞/τ) and Eqs. (17.21) and (17.22) (for transforming from n_∞/τ to α/β). In general, experimentalists tend to use the n_∞/τ representation because it maps more directly onto the results of voltage-clamp experiments. Modelers, on the other hand, tend to express voltage-dependences using the α/β representation, because it maps more directly onto the gating equations Eqs. (17.11)–(17.13) in the standard formulation of the Hodgkin–Huxley model.

Voltage-clamp experiments yield estimates of n_∞/τ or α/β only at the discrete clamp voltages V_c used in the experiment. Numerical integration of the HH model, however, requires that n_∞/τ or α/β values be specified over a continuous range of membrane voltages, since the membrane potential varies

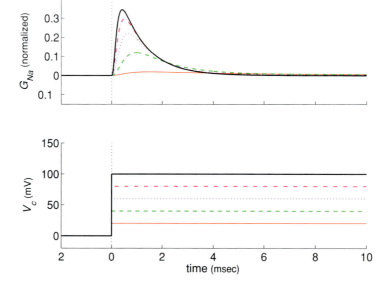

Figure 17.4   Simulated voltage-clamp data illustrating activation and inactivation properties of the Na$^+$ conductance in squid giant axon. The command voltage V_c is shown in the lower panel, and the Na$^+$ current is shown in the upper panel. Simulation parameters are from the Hodgkin and Huxley model (1952).

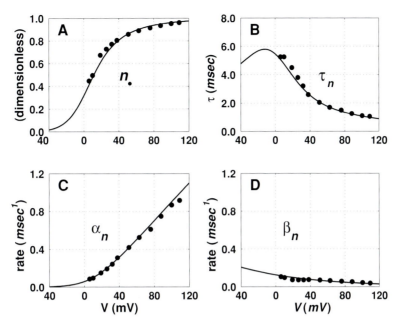

Figure 17.5 Parametric fits to voltage-dependence of the K$^+$ conductance in the HH model. (**A**) Steady-state value n_∞; (**B**) time constant τ_n (**C**) forward rate constant α_n; and (**D**) backward rate constant β_n. Data points are from Table 1 of Hodgkin and Huxley (1952). Solid lines in panels C and D are parametric fits to the rate data. The best-fit curves correspond to Eqs. (17.23) and (17.24), respectively. Solid lines in panels A and B are the transformations of the α/β functions into the n_∞/τ representation using Eqs. (17.5) and (17.6).

continuously in the model. Typically, voltage-dependences are expressed as a continuous function of voltage, and the task for the modeler becomes one of determining the parameter values that best fit the data. As an illustration, the closed circles in Figure 17.5A,B represent empirical data on $n_\infty(V_c)$ and $\tau_n(V_c)$ obtained by Hodgkin and Huxley (Table 1, Hodgkin and Huxley, 1952). The data points in Figure 17.5C,D show the same data set transformed into the α/β representation. Hodgkin and Huxley used the following functional forms to parameterize their K$^+$ conductance results (shown as solid lines in Figure 17.5):

$$\alpha_n(V) = \frac{0.01(10 - V)}{\exp\left(\frac{10 - V}{10}\right) - 1},$$ (17.23)

$$\beta_n(V) = 0.125 \exp(-V/80).$$ (17.24)

If Eqs. (17.23) and (17.24) above are compared with Eqs. (17.12) and (17.13) from the original article (Hodgkin and Huxley, 1952), you will note that the sign of the membrane voltage has been changed to correspond to the modern convention (see subsection entitled "Voltage Conventions" above). Hodgkin and Huxley used similar functional forms to describe the voltage-dependence of the m and h gates of the sodium channel:

$$\alpha_m(V) = \frac{0.1(25 - V)}{\exp\left(\frac{25 - V}{10}\right) - 1},$$ (17.25)

$$\beta_m(V) = 4 \exp(-V/18),$$ (17.26)

$$\alpha_h(V) = 0.07 \exp(-V/20),$$ (17.27)

$$\beta_h(V) = \frac{1}{\exp\left(\frac{30 - V}{10}\right) + 1}.$$ (17.28)

In neural simulation software packages, the rate constants in HH-style models are often parameterized using a generic functional form:

$$\alpha(V) = \frac{A + BV}{C + H \exp\left(\frac{V + D}{F}\right)}.$$ (17.29)

In general, this functional form may require up to six parameters (A, B, C, D, F, H) to fully specify the rate equation. However, in many cases adequate fits to the data can be obtained using far fewer parameters. Fortunately, Eq. (17.29) is flexible enough that it can be transformed into simpler functional forms by setting certain parameters to either 0 or 1. For example, if the voltage-clamp data can be adequately fit by an exponential function over the relevant range of voltages, then setting $B = 0, C = 0, D = 0$, and $H = 1$ in Eq. (17.29), results in a simple exponential form, $a(V) = A \exp(-V/F)$, with just two free parameters $(A$ and $F)$ to be fit to the data. Similarly, setting $B = 0, C = 1$ and $H = 1$ gives a sigmoidal function with three free parameters $(A, D$, and $F)$.

One other technical note is that certain function forms can become indeterminate at certain voltage values. For example, the expression for $\alpha_n(V)$ in Eq. (17.23) evaluates to the indeterminate form 0/0 at $V = 10$. The solution to this problem is to apply L'Hôpital's rule, which states that if $f(x)$ and $g(x)$ approach 0 as x approaches a, and $f'(x)/g'(x)$ approaches L as x approaches a, then the ratio $f(x)/g(x)$ approaches L as well. Using this rule, it can be shown that $\alpha_n(10) = 0.1$. When implementing HH-style rate functions in computer code, care must be taken to handle such cases appropriately.

Calcium-Dependent Channels

Certain types of ion channels are influenced by both membrane voltage and intracellular calcium concentration. Although calcium-dependence was not part of the original HH model, it is straightforward to extend the HH framework to handle this case. Calcium-dependence is typically implemented by modifying the α/β rate equations to include an additional state variable representing the intracellular calcium concentration. For example, Traub (1982) proposed a model of intrinsic bursting in hippocampal neurons that included a slow calcium-dependent potassium conductance G_s. This conductance was modeled using an HH-style rate equation that depends on both membrane voltage V and intracellular calcium concentration χ. Traub (1982) modeled the slow potassium conductance as $G_s = \bar{g}_s q$, where q is a standard HH gating variable with first-order kinetics:

$$\frac{dq}{dt} = \alpha_q(1 - q) - \beta_q q. \tag{17.30}$$

The voltage- and calcium-dependence were incorporated into the rate equations as follows:

$$\alpha_q(\chi, V) = \exp(V/27)\frac{0.005(200 - \chi)}{\exp\left(\frac{200-\chi}{20}\right) - 1}, \tag{17.31}$$

$$\beta_q = 0.002. \tag{17.32}$$

Conductances that depend on both membrane voltage and calcium concentration are rarely as well characterized experimentally as are ordinary voltage-dependent channels. In part this is due to the technical challenges in trying to achieve a "calcium clamp" to precisely quantify the calcium-dependence. Furthermore, voltage-clamp experiments on these conductances are more difficult to interpret because even though the membrane voltage is held fixed by the clamp circuitry, the intracellular calcium concentration is varying during the clamp. Consequently, modelers must often devise rate equations for such channels based on more qualitative criteria than are used for regular voltage-dependent channels. To simplify this task, it is common to take one of the α/β rate equations as a constant [as was done for β_q in Eq. (17.32) above] and to put all of the voltage- and calcium-dependence into the other rate equation. This reduces the number of unknown parameters in the model, and it simplifies searching the parameter space.

For understanding the effects on channel gating, the region of space in which the calcium concentration must be known is a thin shell just inside the membrane surface. The calcium concentration in this region can be significantly different from the bulk concentration in the interior of the cell. Calcium enters this shell region primarily through the influx of Ca^{2+} ions through membrane calcium channels. Calcium leaves the shell region due to diffusion and buffering. A simple model of intracellular calcium dynamics describes this process by a differential equation of the form (Traub, 1982)

$$\frac{d\chi}{dt} = AI_{Ca} - B\chi, \tag{17.33}$$

where A is a constant related to the volume of the shell and the conversion of coulombs to moles of ions, while B is a rate constant representing the effects of diffusion and buffering. As a technical note, recall that ionic currents are typically defined as "inward negative" (see subsection entitled "Sign Conventions" above). Using this convention, the constant A in Eq. (17.33)

will be a negative number, such that inward (negative) calcium current will cause a positive change in calcium concentration χ. For a discussion of more advanced techniques for modeling calcium dynamics, see Yamada et al. (1998).

17.3.2 Markov Models of Individual Channels

The HH framework has been extremely successful for developing quantitative models of macroscopic currents observed in single neurons. However, a different approach must be used if one is interested in modeling the currents flowing through individual channels. At the microscopic level, gating of individual ion channels is a stochastic process. Transitions between permissive and nonpermissive gating states take place by probabilistic transitions between different conformational states of the ion channel complex. Certain conformational states allow ions to move through the channel, while others do not. When monitored experimentally in single-channel patch-clamp recordings, for example, individual channels are observed to fluctuate randomly between open and closed states.

Markov models provide a framework for describing the microscopic currents through individual ion channels (Destexhe and Huguenard, 2001). The basic assumption underlying the Markov model formalism is that the opening and closing of ion channels can be described as a series of transitions between distinct conformational states. Certain states may correspond to the channel being open, closed, inactivated, and so on. Transitions between different states occur according to a set of transition probabilities. Figure 17.6 shows a generic Markov model consisting of 5 states S_i and 10 transition probabilities p_{ij}. Note that the number of transition probabilities will depend on the topology of the Markov model. For example, a fully connected 5-state model, in which any state could transition to any other state, would have 20 transition probabilities. Part of the task of designing a Markov model involves determining how many states are involved, which transitions are allowed, and which are forbidden. The forbidden transitions don't appear in the diagram. The modeler's task then becomes one of determining values for the remaining allowed transition probabilities.

The probability to find the system in state S_i at some time t is defined as $P_i(t)$. The transition probability p_{ij} is the conditional probability of finding the system in a new state j if it has recently been in state i. The time evolution of $P_i(t)$ can be written as

$$\frac{dP_i(t)}{dt} = \sum_{j=1}^{n} P_j(t)p_{ji} - \sum_{j=1}^{n} P_i(t)p_{ij}. \tag{17.34}$$

The first term on the right-hand side of this equation represents the increase in probability of finding the system in state S_i due to transitions entering this state from other states. The second term represents the decrease in probability due to transitions out of state S_i into other states. If there is a large population of identical channels, then $P_i(t)$ can be interpreted as the fraction of channels in state S_i and the transition probabilities p_{ij} can be interpreted as rate constants. Thus Markov models provide a convenient formalism for linking the gating properties of individual channels to the behavior of macroscopic currents as described by the HH model.

Figure 17.7A shows a five-state Markov model that corresponds to the n^4 gating kinetics of the HH K^+ channel model.

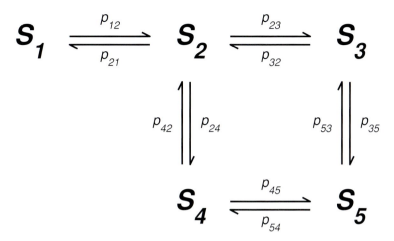

Figure 17.6 A representative Markov model diagram. This particular model has five distinct states $S_1 - S_5$ and 10 transition probabilities p_{ij}. In Markov models of ion channels, each state represents a putative conformational state of the ion channel complex. Some conformations will correspond to closed states, some to open states, and some to inactivated states.

The Markov model has five distinct states, $n_0 - n_4$, where the subscript represents the number of HH gates in the permissive state. When the channel is in state n_1, for example, one of the gates is in the permissive configuration and three of the gates are nonpermissive. Ions can flow through the channel only when all gates are in the permissive state (state n_4); all other states correspond to closed states. The transition probabilities between states can be calculated from the forward (α_n) and reverse (β_n) rate constants of the HH K$^+$ channel model and the assumption that each

gate behaves independently. There are four possible ways that the n_0 state can transition to the n_1 state, so the corresponding transition rate is $4\alpha_n$. The full set of transition probabilities that correspond to the HH model kinetics are shown in the figure.

The sequence of openings and closings of an individual channel can be simulated using Monte Carlo techniques to randomly generate state transitions with the specified probabilities. Recall that the HH rate constants are voltage-dependent, so a change in membrane voltage (Figure 17.7B) will result in a shift of all

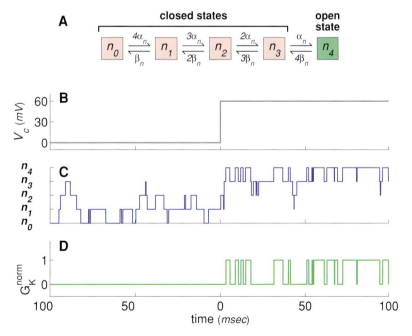

Figure 17.7 A Markov model of the HH K$^+$ conductance. **(A)** The Markov model has four closed states $n_0 - n_3$ and one open state n_4. The subscript corresponds to the number of n gates in the permissive state. **(B)** Command voltage in a simulated voltage-clamp experiment. **(C)** Monte Carlo simulation of state transitions of the Markov model in response to a step change in command voltage. **(D)** Normalized conductance of the K$^+$ channel. The channel is open ($G_K^{\text{norm}} = 1$) whenever the system is in state n_4, otherwise the channel is closed ($G_K^{\text{norm}} = 0$).

(A) HH Na channel

inactivated states

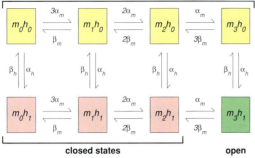

closed states   **open**

(B) Na channel (Patlak, 1991)

inactivated states

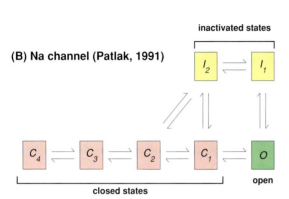

open

closed states

Figure 17.8 Two different Markov models of the Na^+ conductance. (**A**) The Markov model corresponding to independent activation and inactivation gating of the HH model has eight distinct states. The subscripts on the states represent the number of gates of each type in the permissive state. The channel is open only when all three m gates and the h gate are in the permissive state (m_3h_1). (**B**) A model proposed by Patlak (1991) which includes interactions between the activation and inactivation gates. This model provides a better description of actual voltage-clamp data from squid axon than does the HH model.

the transition probabilities and hence a shift in the probability distribution of states. Figure 17.7C shows a Monte Carlo simulation of the time history of state transitions before and after a step change in clamp voltage. When the membrane is clamped to the resting voltage ($V_C = 0$), the system spends most of its time in states $n_0 - n_2$, which are all closed states. When the membrane is clamped to a depolarized voltage ($V_C = 60$), the system spends most of its time in states $n_2 - n_4$. The channel is open whenever the model is in state n_4, as reflected in the conductance record shown in Fig. 17.7D.

Figure 17.8A shows an 8-state Markov model that corresponds to the m^3h gating kinetics of the HH Na^+ channel. The model is in the open state only when all gates are permissive (state m_3h_1). Any state in which the inactivation gate is nonpermissive (h_0) corresponds to an inactivated state of the channel. According to the HH model, the behavior of the inactivation gate (h) is independent of the three activation gates (m). This is reflected in the Markov model by the fact that transitions to an inactivated state can potentially occur from any open or closed state. However, careful experimental studies of Na^+ channel gating kinetics have revealed that activation and inactivation processes

are not completely independent. Figure 17.8B shows a more recent Markov model of Na^+ channel gating (Patlak, 1991) that provides a better description of the data.

17.3.3 Synaptic Models

Thus far the techniques in this chapter have focused primarily on modeling voltage-dependent channels. Equally important from a functional perspective are the ligand-gated channels that mediate chemical synaptic transmission. When an action potential arrives at the presynaptic terminal of a chemical synapse, neurotransmitter is released into the synaptic cleft. Neurotransmitter molecules subsequently bind to ligand-gated receptors in the postsynaptic membrane, causing changes in ionic current flow across the membrane. In an equivalent electrical circuit model (Figure 17.1), ligand-gated channels are represented by additional resistive pathways across the membrane.

For simulating synaptic activation in neural models, the details of synaptic release, diffusion, and receptor binding are often abstracted into a simpler form that describes the postsynaptic conductance as a time-dependent function. The arrival of an action potential at a synapse at time t_{spike} gives rise to a transient change in a postsynaptic conductance that is often modeled using the alpha function (Rall, 1967):

$$G_{syn}(t) = g_{peak}e/\tau_{syn}(t - t_{spike})e^{-(t-t_{spike})/\tau_{syn}} \quad \text{for } t \geq t_{spike}. \tag{17.35}$$

The peak of the conductance change occurs at time $t = t_{spike} + \tau_{syn}$, and the conductance value at this time is g_{peak}. The synaptic current I_{syn} associated with the synapse is modeled by $I_{syn}(t) = G_{syn}(t)(V - E_{syn})$, where E_{syn} is the reversal potential of the synapse.

When a synapse is activated by a sequence of action potentials, the net change in conductance is often modeled as a linear summation of the contributions from each individual action potential. A straightforward implementation based on Eq. (17.35) would require keeping a time history of spike activity and summating over all previous spike times. However, this approach is computationally inefficient and rarely used in large-scale simulations. There are more efficient methods involving either the reformulation of the conductance change as a second-order differential equation (Wilson and Bower, 1989) or reorganization of the computation to require the storage of only two running sums per synapse, rather than a complete time history of activation (Srinivasan and Chiel, 1993).

Another technique for modeling synaptic conductances utilizes a Markov model approach (Destexhe et al., 1998). The simplest form of such models involves only a single open state and a single closed state. Such two-state models can be represented by

$$C + T \underset{\beta}{\overset{\alpha}{\rightleftarrows}} O \tag{17.36}$$

where C is a closed state, O is an open state, T represents neurotransmitter, and α and β are forward and backward rate constants, respectively. Unlike the Hodgkin and Huxley model, the rate constants, α and β, are independent of membrane voltage. Let the fraction of receptors in the open state be represented by r, and let the neurotransmitter concentration be denoted by $[T]$.

Then the first-order kinetic equation for this system is

$$\frac{dr}{dt} = \alpha[T](1 - r) - \beta r. \qquad (17.37)$$

One simple way to model the neurotransmitter concentration is to assume that a constant amplitude pulse of transmitter is released when the action potential arrives at the presynaptic terminal, in which case Eq. (17.37) can be solved analytically for $r(t)$ (Destexhe et al. 1994). The synaptic current is then modeled by

$$I_{syn}(t) = \bar{g}_{syn} r(t)(V - E_{syn}). \qquad (17.38)$$

In general, Markov models of this type can be much more sophisticated than the two-state model presented above. These more detailed models can have multiple states representing various open, closed, and desensitized configurations. Such biophysically rich Markov models may be particularly useful when using a neuroinformatics approach to investigate how receptor properties are altered by variations in the molecular structure and subunit composition of particular ligand-gated receptors.

Metabotropic Receptors

Up to this point, we have been discussing ionotropic receptors for which neurotransmitter binding causes direct and immediate gating of an associated ion channel. Metabotropic receptors, on the other hand, exert their influence indirectly by acting through an intracellular second messenger system. For metabotropic receptors, neurotransmitter binding leads to the activation of intracellular biochemical pathways, which may ultimately link to the opening or closing of second messenger gated ion channels. The cascade of reactions that take place in such systems can be modeled using a combination of Markov models for the components that have discrete states and standard biochemical reaction kinetics for describing chemical concentrations that vary continuously (Destexhe et al., 1994). For example, the binding of transmitter T to a metabotropic receptor R, leading to the formation of an activated receptor state R^*, might be described by a two-state Markov model:

$$R + T \;\xrightleftharpoons{}\; R^* \qquad (17.39)$$

Following receptor activation, there could be several intermediate biochemical reactions of the general form

$$A + B \;\underset{\beta}{\overset{\alpha}{\rightleftharpoons}}\; X + Y \qquad (17.40)$$

which can be modeled using standard reaction kinetics (Bhalla, 2001). In Eq. (17.40), α and β are forward and backward rate constants for the reaction. The chemical concentrations are governed by a rate equation of the form

$$d[A]/dt = -\alpha[A][B] + \beta[X][Y] \qquad (17.41)$$

and a set of relationships that reflect the stoichiometry of the reaction

$$d[A]/dt = d[B]/dt = -d[X]/dt = -d[Y]/dt. \qquad (17.42)$$

In a second messenger cascade, one of the reactants appearing on the left-hand side of one of the biochemical reactions would be the activated receptor R^*, and one of the products appearing on the right-hand side would be a second messenger Z that could serve as a ligand for a postsynaptic ion channel. The gating of

this second messenger gated channel could then be described by a Markov model, such as the following two-state model,

$$C + Z \;\xrightleftharpoons{}\; O \qquad (17.43)$$

or by a more complex multi-state model. For example, Destexhe et al. (1994) found that a four-state Markov model was needed to adequately fit both the rising and decaying phases of a G-protein-activated GABA$_B$ receptor current.

17.3.4 Multicompartment Models

A simple electrical equivalent circuit, such as that shown in Figure 17.1, can be used to model a localized region of nerve cell membrane. In general, however, neurons have spatially extended axons and dendrites with heterogeneous properties. Different regions of the cell will have different diameters and varying types and densities of ion channels and receptors. Furthermore, quantities such as the local membrane potential and the local intracellular calcium concentration can vary significantly across the spatial extent of a neuron. Multicompartment models provide a means for handling the spatial complexity of neuron morphology and the heterogeneity of physical properties. Figure 17.9 illustrates the compartmental modeling approach for a segment of dendritic membrane. The multicompartment modeling approach divides the neuron into a number of smaller spatial compartments, each of which can be modeled with an electrical equivalent circuit similar to Figure 17.1. The components of the equivalent circuit and their numerical values can vary from compartment to compartment, depending on the particular types of conductances found in different regions of the cell. Neighboring compartments are coupled by axial currents that flow between compartments in the intracellular space. The membrane potential for compartment i, V_i, is related to the membrane potentials in neighboring compartments, V_{i-1} and V_{i+1}, by

$$C_m \frac{dV_i}{dt} + I_{ion} = \frac{(V_{i-1} - V_i)}{r_{i-1,i}} + \frac{(V_{i+1} - V_i)}{r_{i+1,i}}, \qquad (17.44)$$

where C_m and I_{ion} are based on the equivalent circuit for compartment i. The terms $r_{i \pm 1,i}$ represent the axial resistances between neighboring compartments, and the terms $(V_{i \pm 1,i} - V_i)/r_{i \pm 1,i}$ represent the axial currents. Similar relationships exist for branch points where an axonal or dendritic segment splits into two or more subsegments. Using these techniques, multicompartment models can describe arbitrarily complex cell morphologies. Detailed advice on how to construct, parameterize, and test multicompartment models can be found in Segev and Burke (1998) and De Schutter and Steuber (2001).

17.3.5 Network Models

Previous sections have covered techniques for modeling single neurons, ion channels, and individual synapses. Using these techniques, it is relatively straightforward to create network models, in which the spike outputs from certain model neurons provide synaptic inputs to other neurons in the network. There are two main issues to consider in constructing network-level models.

(A) dendrite

(B) compartmentalization

$i - 1$ i $i + 1$

(C) equivalent circuit

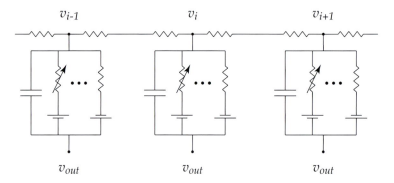

v_{i-1} v_i v_{i+1}

v_{out} v_{out} v_{out}

Figure 17.9 Compartmental approach for single-neuron modeling. The dendrites (A) are divided into distinct regions that are represented by cylindrical compartments (B). Each compartment can have different physical characteristics (membrane potential, length, diameter, channel types, channel densities, etc.). The physical properties are modeled by an electrical equivalent circuit (C). In the circuit model, neighboring compartments are coupled by resistors representing the axial resistance of the intracellular space. Branch points are handled in a similar manner (not shown).

One involves choosing an appropriate mathematical representation for the propagation of action potentials between neurons. The other issue has to do with techniques for specifying the synaptic connectivity within the network.

In principle, the propagation action potentials between neurons could be handled using Hodgkin–Huxley conductances and a multicompartmental description of the axon and its terminal arbor. This approach is sometimes used when the scientific questions being addressed pertain explicitly to mechanisms of action potential propagation (Manor et al., 1991). However, it is computationally expensive to use a full multicompartment model to describe every axon and terminal arbor in a large network. Because of the all-or-none nature of the action potential, it is often possible to use a more efficient technique in which action potentials are represented as discrete temporal events. In this event-based approach, an action potential generated by neuron i at time t_i is represented as a time-stamped event that is used to trigger synaptic input to a target neuron j after some time delay Δt_{ij}. Propagation along the axon is not modeled explicitly; rather it is implicit in the axonal propagation delay Δt_{ij}. Recall that a single axon typically makes synaptic contacts with multiple target neurons. In general, the propagation delay Δt_{ij} can have different numerical values for each of the possible postsynaptic targets.

The second issue in network modeling involves specification of the connectivity between neurons. For small network models, this is often handled on a case-by-case basis, whereas large network models usually require a rule-based approach. For example, a model of an invertebrate central pattern generator might involve 10 neurons with an average of five synapses per neuron, resulting in approximately 50 synaptic connections. Specification of the synaptic properties (receptor type, reversal potential, peak conductance, propagation delay, etc.) could easily be handled on a synapse-by-synapse basis. In contrast, a network model of a local region of mammalian visual cortex might involve on the order of 10,000 neurons with an average of 100 synapses per neuron, resulting in one million synaptic connections. In this case, a synapse-by-synapse specification would be unfeasible and a rule-based approach would be utilized. For example, a connection rule might specify that all neurons of type A (e.g., inhibitory interneurons) make a particular type of synaptic connection (e.g., GABAergic) with all neurons of type B (e.g., pyramidal cells) that lie within a fixed radius. The rule might also specify how the peak conductance and axonal propagation delay vary with target distance.

17.3.6 Software Tools

Fortunately, sophisticated software packages are available to facilitate the development, implementation, and dissemination of biophysically detailed neural models. Two of the most widely used tools are GENESIS (GEneral NEural SImulation System) (Bower and Beeman, 1998; Bower et al., 2002) and NEURON (Hines and Carnevale, 2002). Both of these modeling environments are designed for constructing biophysically detailed multicompartment models of single neurons, and they also provide modeling tools that span from the molecular level to the network level. Both GENESIS and NEURON provide high-level languages for model specification, predefined sets of neural building blocks, and graphical user interface elements for simulation control and visualization. To construct a specific neural model, the model specification language is used to define and link appropriate sets of predefined building blocks to create a functional model. The basic building blocks include such things as compartments or cable segments for modeling neuron morphology,

voltage-gated and ligand-gated conductances, components for intracellular diffusion and buffering of ions, chemical and electrical synapses, and various forms of synaptic plasticity. Other building blocks provide the model with external inputs and outputs, including file I/O and graphical displays. Some building blocks provide models of electrophysiological instrumentation like stimulus generators and voltage-clamp circuits, which allow users to closely model the experimental setups that are used in empirical studies. Custom user-defined elements can be created if the required modeling component is not already part of the predefined set of building blocks. More information on these modeling environments, including documentation, tutorials, users groups, and workshop announcements, can be found on the Web by following the links provided in the Web Resources section.

17.4 CURRENT APPLICATIONS

Hundreds of biophysically detailed neural models have been developed using GENESIS, NEURON, and similar modeling tools. The scientific issues addressed in these models span a broad range of topics, including intracellular signaling, dendritic processing, neural oscillations, central pattern generation, motor control, sensory coding, feature extraction, learning, and memory. See the subsection entitled "Web Resources" for links to research publications that have been generated using GENESIS and NEURON. An illustrative example of this type of biophysically detailed modeling approach is provided by the cerebellar Purkinje cell model developed by De Schutter and Bower (1994a,b) using GENESIS. The dendritic morphology shown in Figure 17.10 contains approximately 1600 distinct compartments with lengths and diameters based on detailed anatomical

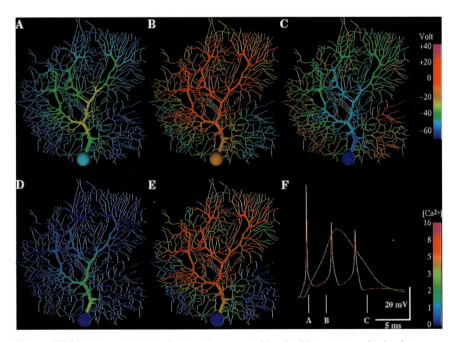

Figure 17.10   Representations of the membrane potential and calcium concentration in a large compartmental model of a cerebellar Purkinje cell following synaptic activation. **(A–C)** Membrane potential 1.4, 4.0, and 10.0 ms after synaptic activation. **(D, E)** Intracellular Ca^{2+} concentration 1.4 and 4.0 ms after activation. **(F)** Membrane potential (red trace) and Ca^{2+} concentration (green trace) in the cell body following activation. The vertical white bars indicate the times at which the false color images in panels **A–E** were generated. From De Schutter and Bower (1994b, with permission).

reconstructions of an actual Purkinje cell (Rapp et al., 1992). The model includes 10 different types of voltage-dependent channels: two Na^+ channels (fast and persistent), two Ca^{2+} channels (T-type and P-type), three voltage-dependent K^+ channels, and two Ca^{2+}-dependent K^+ channels. The channel properties were modeled using Hodgkin–Huxley equations, and the modeling parameters were constrained by empirical voltage-clamp data where available. The channels were distributed differentially over three zones of the Purkinje cell. Synaptic inputs were modeled using a dual exponential version of the alpha function [Eq. (17.35)] that allows for different time constants for the rising and falling phases of the synaptic waveform (Wilson and Bower, 1989). Figure 17.10 shows the response of the model to a large synchronous synaptic activation over a large portion of the dendritic tree. This pattern of synaptic input represents activation of the Purkinje cell by a climbing fiber input. The so-called "complex spike" response of a Purkinje cell to climbing fiber stimulation has been well studied experimentally. The ability of the model to reproduce known membrane voltage and intracellular calcium characteristics of a complex spike was one of the benchmarks for tuning certain model parameters and for evaluating the underlying modeling assumptions. The simulation results summarized in Figure 17.10 represent only one of several studies carried out using the Purkinje cell model (De Schutter and Bower, 1994a,b). After tuning the model to reproduce a range of in vitro firing behaviors, the model was used to make predictions about the in vivo firing patterns of Purkinje cells. The model has been particularly useful in elucidating the role of dendritic inhibition in shaping neural response properties.

17.5 LIMITATIONS

There are several limitations to keep in mind when developing biophysically detailed neural models. Perhaps one of the most important is that such models are actually highly impoverished relative to the true richness and complexity of the underlying biology. Even though these models are described as "biophysically detailed," many aspects of cell and membrane physiology have been stripped away in the modeling process. The art of creating a good model involves knowing which details are important and which details can be safely disregarded. However, details that are unimportant in one functional context may become pivotal in a different context. Thus, one should avoid thinking of any particular model, such as the Purkinje cell model described above, as a full and complete description of the underlying biological system.

It is better to think of a neural model as an extended hypothesis that is designed to address a restricted range of neurobiological function. As an extended hypothesis, each model embodies a large number of assumptions. Certain assumptions will be well supported by empirical data, while others will be largely speculative. For the purpose of hypothesis testing, it is important to keep track of all the underlying assumptions and the corresponding empirical constraints on those assumptions. This is one area where neuroinformatics tools can play a key role in helping modelers establish and document links between each assumption and the set of empirical results that impact that particular assumption. In terms of hypothesis testing, an important limitation to keep in mind is that even if a neural model successfully reproduces certain empirical results, it does not imply that all the underlying assumptions in the model are true. Likewise, if a model fails to agree with some piece of empirical data, the fact that the "extended hypothesis" is falsified does not directly indicate which of the underlying assumptions might be responsible for the disagreement. Therefore, it is not particularly useful to simply label a neural model as "right" or "wrong." Instead, neural modeling should be viewed as an integral component of the scientific method, in which progress is made through multiple iterations of experimental observation, hypothesis generation (model building), prediction (model simulation), and testing (comparison with empirical data).

17.6 OUTLOOK

Based on research trends over the past decade, it is clear that both neuroinformatics and electrophysiological modeling are becoming increasingly important tools for exploring the functional properties of neural systems. Several ongoing research and development efforts are leading toward a convergence and integration of neuroinformatics and modeling tools that will greatly enhance the ability of neuroscientists to make use of these powerful approaches. Much of this development effort is taking place in the context of the Human Brain Project (Huerta et al., 1993; Koslow and Huerta, 1997; Shepherd et al., 1998). Several research groups are actively developing large electrophysiological databases, common data representations to facilitate information sharing, software tools for electrophysiological data analysis and visualization, neuroinformatics tools for search and retrieval, and neuroinformatics-based extensions to neural modeling software packages. Overviews of several of these projects are available in Koslow and Huerta (1997), and more information on the current status of these various efforts can be found on the Human Brain Project website (see subsection entitled "Web Resources").

References

Bhalla, U.S. (2001). Modeling networks of signaling pathways. In *Computational Neuroscience: Realistic Modeling for Experimentalists* (E. De Schutter, ed.), pp. 25–48. CRC Press, Boca Raton, FL.

Bower, J.M., and Beeman, D., eds. (1998). *The Book of GENESIS: Exploring Realistic Neural Models with the GEneral NEural SImulation System*. Springer-Verlag, New York.

Bower, J.M., Beeman, D., and Hucka, M. (2002). GENESIS simulation system. In *The Handbook of Brain Theory and Neural Networks*. (M.A. Arbib, ed.), 2nd ed., pp. 475–478. MIT Press, Cambridge, MA.

Coetzee, W.A., Amarillo, Y., Chiu, J., Chow, A., Lau, D., McCormack, T., Moreno, H., Nadal, M.S., Ozaita, A., Pountney, D., Saganich, M., Vega-Saenz de Miera, E., and Rudy, B. (1999). Molecular diversity of K^+ channels. *Ann. N.Y. Acad. Sci.* **868**, 233–285.

Delcomyn, F. (1998). *Foundations of Neurobiology*. Freeman, New York.

De Schutter, E., ed. (2001). *Computational Neuroscience: Realistic Modeling for Experimentalists*. CRC Press, Boca Raton, FL.

De Schutter, E., and Bower, J.M. (1994a). An active membrane model of the cerebellar Purkinje cell I. Simulation of current clamps in slice. *J. Neurophysiol.* **71**, 375–400.

De Schutter, E., and Bower, J.M. (1994b). An active membrane model of the cerebellar Purkinje cell: II. Simulation of synaptic responses. *J. Neurophysiol.* **71**, 401–419.

De Schutter, E., and Steuber, V. (2001). Modeling simple and complex active neurons. In *Computational Neuroscience: Realistic Modeling for Experimentalists* (E. De Schutter, ed.), pp. 233–257, CRC Press, Boca Raton, FL.

Destexhe, A., and Huguenard, J. (2001). Which formalism to use for modeling voltage-dependent conductances? In *Computational Neuroscience: Realistic Modeling for Experimentalists*. (E. De Schutter, ed.), pp. 129–157. CRC Press, Boca Raton, FL.

Destexhe, A., Mainen, Z.F., and Sejnowski, T.J. (1994). Synthesis of models for excitable membranes, synaptic transmission and neuromodulation using a common kinetic formalism. *J. Comput. Neurosci.* **1**, 195–230.

Destexhe, A., Mainen, Z.F., and Sejnowski, T.J. (1998). Kinetic models of synaptic transmission. In *Methods in Neuronal Modeling: From Ions to Networks*. C. Koch and I. Segev, (eds.), 2nd ed., pp. 1–26. MIT Press, Cambridge, MA.

Doiron, B., Longtin, A., Turner, R.W., and Maler, L. (2001). Model of gamma frequency burst discharge generated by conditional backpropagation. *J. Neurophysiol.* **86**, 1523–1545.

Hille, B. (2001). *Ion Channels of Excitable Membranes*, 3rd ed. Sinauer, Sunderland, MA.

Hines, M.L., and Carnevale, N.T. (2002). NEURON simulation environment. In *The Handbook of Brain Theory and Neural Networks*. (M.A. Arbib, ed.), 2nd ed., pp. 769–773, MIT Press, Cambridge, MA.

Hodgkin, A.L., and Huxley, A.F. (1952). A quantitative description of membrane current and its application to conduction and excitation in nerve. *J. Physiol. (London)* **117**, 500–544.

Huerta, M.F., Koslow, S.H., and Leshner, A.I. (1993). The Human Brain Project: An international resource. *Trends Neurosci.* **16**, 436–438.

Koch, C., and Segev, I., eds. (1998). *Methods in Neuronal Modeling: From Ions to Networks*, 2nd ed. MIT Press, Cambridge, MA.

Koslow, S., and Huerta, M., eds. (1997). *Neuroinformatics: An Overview of the Human Brain Project*. Erlbaum, Mahwah, NJ.

Levitan, I.B., and Kaczmarek, L.K. (2002). *The Neuron: Cell and Molecular Biology*, 3rd ed. Oxford University Press, New York.

Manor, Y., Gonczarowski, J., and Segev, I. (1991). Propagation of action potentials along complex axonal trees. Model and implementation. *Biophys. J.* **60**, 1411–1423.

Patlak, J.B. (1991). Molecular kinetics of voltage-dependent Na$^+$ channels. *Physiol. Rev.* **71**, 1047–1080.

Rall, W. (1967). Distinguishing theoretical synaptic potentials computed for different soma-dendritic distributions of synaptic input. *J. Neurophysiol.* **30**, 1138–1168.

Rapp, M., Yarom, Y., and Segev, I. (1992). The impact of parallel fiber background activity on the cable properties of cerebellar Purkinje cells. *Neural Comput.* **4**, 518–533.

Rashid, A.J., Dunn, R.J., and Turner, R.W. (2001a). A prominent soma-dendritic distribution of Kv3.3 K$^+$ channels in electrosensory and cerebellar neurons. *J. Comp. Neurol.* **441** (3), 234–247.

Rashid, A.J., Morales, E., Turner, R.W., and Dunn, R.J. (2001b). The contribution of dendritic Kv3 K$^+$ channels to burst discharge in a sensory neuron. *J. Neurosci.* **21** (1), 125–135.

Rudy, B. (1988). Diversity and ubiquity of K channels. *Neuroscience* **25**, 729–749.

Rudy, B., Chow, A., Lau, D., Amarillo, Y., Ozaita, A., Saganich, M., Moreno, H., Nadal, M.S., Hernandez-Pineda, R., Hernandez-Cruz, A., Erisir, A., Leonard, C., and Vega-Saenz de Miera, E. (1999). Contributions of Kv3 channels to neuronal excitability. *Ann. N. Y. Acad. Sci.* **868**, 304–343.

Segev, I. (1998). Cable and compartmental models of dendritic trees. In *The Book of GENESIS: Exploring Realistic Neural Models with the GEneral NEural SImulation System*. (J.M. Bower and D. Beeman, eds.), pp. 51–77. Springer-Verlag, New York.

Segev, I., and Burke, R.E. (1998). Compartmental models of complex neurons. In *Methods in Neuronal Modeling: From Ions to Networks*. (C. Koch and I. Segev, eds.), 2nd ed., pp. 63–96. MIT Press, Cambridge, MA.

Shepherd, G.M. (1994). *Neurobiology*, 3rd ed. Oxford University Press, New York.

Shepherd, G.M., Mirsky, J.S., Healy, M.D., Singer, M.S., Skoufos, E., Hines, M.S., Nadkarni, P.M., and Miller, P.L. (1998). The Human Brain Project: Neuroinformatics tools for integrating, searching, and modeling multidisciplinary neuroscience data. *Trends Neurosci.* **21**, 460–468.

Srinivasan, R., and Chiel, H.J. (1993). Fast calculation of synaptic conductances. *Neural Comput.* **5**, 200–204.

Traub, R.D. (1982). Simulation of intrinsic bursting in CA3 hippocampal neurons. *Neuroscience* **7**, 1233–1242.

Wang, L.-Y., Gan, L., Forsythe, I.D., and Kaczmarek, L.K. (1998). Contribution of the Kv3.1 potassium channel to high-frequency firing in mouse auditory neurons. *J. Physiol (London)* **509**, 183–194.

Wilson, M.A., and Bower, J.M. (1989). The simulation of large-scale neuronal networks. In *Methods in Neuronal Modeling: From Synapses to Networks* (C. Koch and I. Segev, eds), pp. 291–334, MIT Press, Cambridge, MA.

Yamada, W., Koch, C., and Adams, P. (1998). Multiple channels and calcium dynamics. In *Methods in Neuronal Modeling: From Synapses to Networks* (C. Koch, and I. Segev, ed.), 2nd ed. pp. 137–170. MIT Press, Cambridge, MA.

Web Resources

CATACOMB: a simulation system for biologically based network models (http://www.compneuro.org/catacomb/).

GENESIS: a general-purpose simulation system for single neuron and network models (http://www.genesis-sim.org/GENESIS/). For a list of research publication using the GENESIS simulator, see http://www.genesis-sim.org/GENESIS/pubs.html.

Human Brain Project: a multi-agency program supporting neuroinformatics research and development (http://www.nimh.nih.gov/neuroinformatics/index.cfm).

NEOSIM: a prototype for the next generation of neural simulators with plug-in support for models developed in other simulation environments, such as GENESIS and NEURON (http://www.neosim.org/).

NEURON simulator: a general-purpose simulation system for single neuron and network models (http://www.neuron.yale.edu/). For a list of research publication using the NEURON simulator, see http://www.neuron.yale.edu/neuron/bib/usednrn.html.

NeuroML: a prototype markup language for describing neuroscience simulation models (http://www.neuroml.org/).

NeuroSys: a prototype database system providing informatics and modeling components (http://cns.montana.edu/research/neurosys/).

NTSA Workbench: a prototype database system for neuronal time-series data (http://soma.npa.uiuc.edu/isnpa/isnpa.html).

SenseLab: prototype databases of cell properties, membrane properties, and neural models (http://senselab.med.yale.edu/senselab/).

USC Brain Project: a prototype neuroinformatics tools for linking multiple databases and neural models (http://www-hbp.usc.edu/).

Models of Neuronal Outgrowth

Duncan E. Donohue, B.A. and Giorgio A. Ascoli, Ph.D.

CONTENTS

Databasing the Brain. Edited by Stephen H. Koslow and Shankar Subramaniam
ISBN 0-471-30921-4 © 2005 John Wiley & Sons, Inc.

18.1 INTRODUCTION

The focus of this chapter is on the anatomy of single neurons with particular attention to the issues of the description and generation of dendritic morphology and neurite navigation. Neuronal structure has been the subject of experimental investigation since the early days of neuroscience. Yet the complexity and variability of both dendritic and axonal trees constitute formidable challenges to the comprehensive statistical description of their geometry. Neurons are typically grouped in morphological classes based on the prominent geometrical features of their arborizations. Dendrites, for example, can have a relatively uniform distribution in space (e.g., stellate morphology), a planar organization (e.g., retinal ganglion cells), or a polarized structure (e.g., basal and apical trees in pyramidal cells). Dendritic morphologies can be specific with respect to the location in the brain. For example, pyramidal cells have different anatomical characteristics in the hippocampus and in the neocortex and, within the cortex, in different lobes (e.g., occipital or prefrontal) and layers (e.g., layer II or V). At an extreme, peculiar morphologies are only known to be present in specific regions (e.g., Purkinje cells in the cerebellum).

It is important to realize that considerable variability is also present within a given morphological class. For example, within hippocampal pyramidal cells, CA3 and CA1 neurons have different morphologies. Within CA3 cells, the structure of dendritic trees varies depending on sublocation (CA3a, CA3b, CA3c). Similarly, granule cells in the dentate gyrus have different geometrical features depending on their position in the blades as well as their depth in the somatic layer. Aside from important but subtle issues of subclassification, even a uniform morphological subclass contains cells that are not identical to each other. Thus, a complete anatomical characterization based on cellular populations is, by necessity, statistical in nature. Nevertheless, the morphological differences among classes are greater than those among subclasses and individual neurons, and this is the basis upon which classification is performed (Figure 18.1).

Axons are typically divided in long and local projections. A given neuronal class can be further characterized on the basis of the classes of neurons it synapses upon, the neurotransmitter(s) it releases, and a variety of chemical markers typically revealed immunohistochemically. Locally, axons also display prominently recognizable anatomical features, such as climbing, parallel, and mossy fibers. Both dendritic and axonal morphologies contribute to determine the local and regional connectivity of the network. Neurites also affect signal dynamics by introducing delays, possible transmission failures, electrotonic attenuation, spatial and temporal summation, and nonlinear active integration.

It is striking that the unparalleled variety of neuronal shapes develops from undifferentiated cell progenitors. It is now apparent that both intrinsic and extrinsic factors are responsible for the adult morphology (Lasek and Black, 1988). Intrinsic determinants include gene regulation and expression as well as internal dynamics of enzymatic networks. External determinants include chemical signals released by other neurons, glia, and endocrine system. Since extrinsic messages are usually transduced intracellularly, the interplay between nature and nurture is extremely difficult to tease out.

Recent research suggests that a computational approach using stochastic models is a viable strategy to define, search, and represent the essential and complete sets of morphological parameters of neuronal shape. In this approach, synthetic neurons are generated within a virtual environment based on experimental measures collected from real traced cells. The goal is to design a model resulting in simulated neurons that are indistinguishable (by exhaustive statistical analysis or expert visual inspection) from the real cells of the corresponding morphological class. The parameters of the model may also reflect specific observed or hypothesized mechanisms of real neuronal development.

In this chapter we briefly summarize the current biological knowledge of neuronal development. From those premises, we review the most successful and/or biologically plausible models of neuronal structure and outgrowth. In addition, we discuss the technical and computational aspects of experimental neuronal tracing and their mathematical representation. We conclude with an overview of current and future applications, limitations, and potential for neuroscience research.

18.2 RELEVANT BACKGROUND ON DENDRITIC DEVELOPMENT

Our knowledge of neuronal development is increasing rapidly. Advances in many relevant fields such as genetics and imaging are steadily elucidating pieces of the puzzle (for examples see Luo, 2002). However, a complete story of how neuronal morphologies mature has yet to develop. For every proposed "rule" of neuronal development, exceptions are quickly discovered. Factors promoting neurite extension in one context have been found to have the opposite effect in others (Song et al., 1998). Chemicals may have different effects in different cells (Luo, 2002), and even in different parts of the same cell (Kryl et al., 1999). Thus, the account of neuronal development that follows must be taken as general and incomplete.

There is evidence that neurons go through distinct stages of development on their way to maturation. Each of these stages contributes to the myriad of different shapes and dizzying selection of branching patterns observed in the adult nervous system. Many of the proteins that play leading roles in neuronal development are expressed differently in young cells than in more mature ones (Leclerc et al., 1996; Yacoubian and Lo, 2000). Growing neurons can start out with one morphology only to switch to a completely different shape later in development (Vercelli et al., 1992). These morphological changes are ultimately the result of finely controlled cellular and molecular mechanisms that are beginning to be uncovered.

The complex branching structures demonstrated by neurons are generated by the behavior of growth cones. Growth cones are the polymorphic mobile ends of neurites that guide their development. The dynamics of growth cones can be summarized by a set of basic movements (Figure 18.2). After stemming from the soma of a neuron (initiation), a growth cone can extend, retract, turn, or bifurcate. Bifurcating can occur as the splitting of an extending growth cone (terminal branching or bifurcating proper), or as the result of the formation of a new growth cone from the side of an existing branch (interstitial branching). These five behaviors are, in turn, mediated by the cytoarchitecture

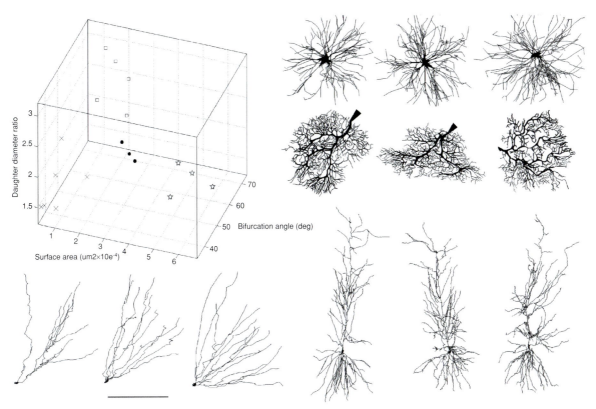

Figure 18.1   Three-dimensional scatter plot of morphological parameters measured from neurons belonging to various morphological classes. Measured parameters are the average diameter ratio between the larger and the smaller branches at bifurcations (height of the box), total surface area (width), and average amplitude angle at bifurcations (depth). Three exemplars of each neuronal class are illustrated: (clockwise) motoneurons (stars, 4 neurons measured), Purkinje cells (filled circles, 3 neurons measured), CA1 pyramidal cells (empty squares, 5 neurons measured), and granule cells (crosses, 6 neurons measured). Surface area of motoneurons has been reduced 10-fold in the 3D scatterplot. Scale bar (microns): Granule cells and Purkinje cells, 200; motoneurons, 1000; pyramidal cells: 300. Reproduced with permission from Ascoli (2002b).

of growth cones, specifically microtubules and actin filaments. Microtubules stabilize the long-term structure of neurites and act in organelle transport, while actin filaments are involved in the dynamic behavior of growth cones.

Microtubules are polymers of tubulin heterodimers. Their polar construction gives them a quickly growing end called the plus-end and a slowly growing end called the minus-end. Axons contain microtubules with plus-ends distal to the soma, while dendrites start out with only plus-end distal microtubules, but later add minus-end distal microtubules as well. Minus-end distal microtubules are found in higher concentrations proximal to the soma and may allow selective organelle transport to dendrites (Baas et al., 1989). In addition to molecular motors for transport such as kinesins and dyneins, microtubules can bind microtubule-associated proteins (MAPs). Two main functions of MAPs are to bind microtubules together into relatively stable bundles and to allow them to associate with other molecules (Kobayashi and Mundel, 1998). Different MAPs are found in dendrites and axons, and several MAPs are differentially expressed during the various stages of cell maturation (Leclerc et al., 1996). Tau is a major axonal MAP, while MAP2 is important in dendritic development (Kobayashi and Mundel, 1998).

Actin filaments form the growth cone at the tips of developing neurites. The growth cone has two major components: (a) long finger-like filopodia composed of filaments of actin with plus-ends distal and (b) webbing-like lamellipodia in the recesses between filopodia (Figure 18.2). Lamellipodia contain a less organized matrix of actin (Gallo and Letourneau, 2000). The actin is polymerized by the addition of monomers along the leading edge of both lamellipodia and filopodia. At the same time, filaments are being transported back toward the bundled microtubules by myosin motor proteins and depolymerized from their minus-ends (Lin et al., 1996). As outlined below, the relative rates of retrograde transport, polymerization, and depolymerization control the extension and retraction of filopodia, and ultimately the direction of neurite growth (Lin and Forscher, 1995). It should be stressed that the interactions and mutual control between actin filaments and microtubules are complex and not thoroughly understood (e.g., Schaefer et al., 2002).

Neurite initiation is one of the least understood events in neuronal development. It involves the polymerization of microtubules and actin filaments in the formation of an extending growth cone on the cell's surface. The extension of the initiated growth cone is better characterized. For extension it appears that visible actin filopodia are not required (Zheng et al., 1996). In normally developing cells, however, where growth cones are present, actin-rich lamellipodia and filopodia are invaded by a few stray microtubules (Dent et al., 1999). As the

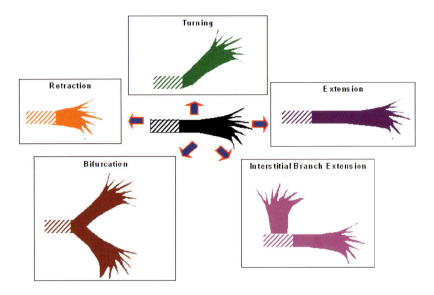

Neurite action:	Microtubules	Actin filaments
Initiate	Polymerization from cell body	Polymerization Growth cone formation
Extend	Polymerization at tips	Polymerization required
Retract	Depolymerization at tips	Polymerization required
Bifurcate	Polymerization, Stabilization of 2 filopodia	Depolymerization/retrograde transport increased in center, decreased on outside edges
Interstitial Branching	Polymer transport, Filopodia stabilization	Local destabilization and polymerization Filopodia formation
Turning	Stabilized on side of turn, destabilization on opposite side	Depolymerization/retrograde transport decreased on side of turn increased on opposite side

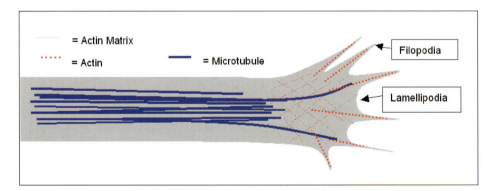

Figure 18.2   The five "elementary" movements of growth cones after initiation (*top*), with the corresponding behaviors of microtubules and actin filaments (*center*) and a schematic representation of a growth cone detail (*bottom*).

growth cone moves forward, it is invaded and stabilized by the extending microtubules (Goldberg and Burmeister, 1986; Aletta and Grene, 1988). The question of whether the microtubules that act in extending the neurite are grown from small monomer- to oligomer-sized pieces, or whether larger polymers are being transported to cause extension, is still under debate (for polymer transport viewpoint, see Baas, 1997; for smaller subunit transport viewpoint, see Kobayashi and Mundel, 1998). Minus-end microtubules appear to also be initiated in conjunction with the centrosome and then transported minus-end first down the

dendrites, where they can be further extended by polymerization (Kobayashi and Mundel, 1998).

Retraction of neuronal processes essentially follows the opposite mechanism of their extension, involving microtubule depolymerization and growth cone collapse. It is interesting, however, that actin polymerization has been shown to be required for (Solomon and Magendantz, 1981) and causative of (Jalink et al., 1993; Fukushima et al., 1998, 2000, 2002a) neurite retraction. Myosin motors are also required for retraction (Ahmad et al., 2000).

Filopodia are thought to control neurite turning, and perhaps elongation, by sampling the extracellular environment for chemoattractants and chemorepellants (Gallo and Letourneau, 2000; Zhou et al., 2002). While microtubules stabilize neuronal outgrowths, and probably play a leading role in the overall extension and retraction of neurites (Bently and Toroian-Raymond, 1986), turning in response to extracellular cues requires actin filaments (Zheng et al., 1996). During a turn, the retrograde transport of the actin filaments is slowed on the side of chemoattractants, leading to filopodia extension in the same direction (Lin and Forscher, 1993; Suter et al., 1998). Conversely, chemorepellants cause growth cone collapse on the side of their detection, producing a turn in the opposite direction (Fan and Raper, 1995). There is also evidence, however, that direct perturbation of microtubules can lead to growth cone turning (Buck and Zheng, 2002).

Bifurcation of a growth cone could be thought of in terms of turning in two directions at once. In appropriate circumstances, the chemoattractive forces on opposite sides of the growth cone are sufficient to support filopodia extension in two different directions. For example, Zhou et al. (2002) showed that application of ML-7, a chemorepellant normally causing growth cone turning and actin bundle loss, could also cause growth cone splitting. Specifically, application of ML-7 to the center of the extending growth cone caused neurite bifurcation (Zhou et al., 2002; also reviewed in Luo, 2002). Another important mechanism for branch formation in both axons (e.g., Dent et al., 1999) and dendrites (e.g, Dailey and Smith, 1996) is the formation of interstitial branches. These branches start off as single filopodia extending from the side of the neurite following actin and microtubule destabilization. Both axons and dendrites appear to develop areas of high filopodial activity along their shafts where these interstitial branches are likely to occur. If these extending filopodia are invaded by microtubules, they can be stabilized and form new side branches (Luo, 2002). The microtubules in side branches are different from those stabilizing primary neurite extension in that they appear to be transported primarily as polymers and not individual units (reviewed in Gallo and Letourneau, 2000). Dendrite interstitial branches have also been found to be affected by genetic manipulation differently from primary dendrites (Gao et al., 1999).

The different behaviors that growth cones demonstrate during neurite development are controlled by a complex cascade of cellular signaling pathways. In addition to MAPs and molecular motors, several proteins are thought to act directly on microtubules and actin filaments to influence neuronal growth. One such family of proteins is that of the actin depolymerizing factor (ADF, also known as cofilin; for review, see Bamburg, 1999). Phosphorylation of ADF inhibits its depolymerizing activity on actin filaments (Sumi et al, 2001). Another important player is the adenomatous polyposis coli protein (APC), which preferentially binds to (Nathke et al., 1996) and stabilizes (Zumbrunn et al., 2001) the plus-end of microtubules. The exact series of events leading to activation or inactivation of these and other cytoskeletal proteins is unclear. Some general principles, however, do emerge from piecing together the parts of various known pathways. These principles are discussed below together with examples of some of the better understood pathways.

A fundamental element of cytoskeletal development is the control of actin and microtubule binding proteins, either directly or through intermediary phosphatases and kinases, by members of the Rho family of GTPases (for reviews see Whitford et al., 2002; Bishop and Hall, 2000; Redmond and Ghosh, 2002). The three main members of the Rho family are Rho, Rac, and Cdc42. As GTPases, they are active when bound to GTP and inactive when bound to GDP. When active, they generally affect cytoskeletal development by releasing downstream kinases from autoinhibition. These kinases, directly or through further intermediaries, affect actin filaments, microtubules, and molecular motors (see also Tang, 2003). The Rho family of GTPases is regulated by a wide variety of guanosine nucleotide exchange factors (GEFs) and GTPase-activating proteins (GAPs). GEFs activate Rho GTPases by facilitating their exchange of GDP for GTP, while GAPs inactivate Rhos by speeding the catalysis of GTP to GDP. Rho GTPases, GAPs, and GEFs are, in turn, controlled by a variety of extrinsic pathways, initiated by the binding of a cell surface receptor to its extracellular ligand. Several of these pathways are summarized in Figure 18.3 and Table 18.1.

A much studied transduction pathway begins with the binding of semaphorin III A (Sema3A, previously known as collapsin-1) to its cell membrane receptor complex composed of neuropilin-1 and plexin-1 (reviewed in Tang, 2003, and Whitford et al., 2002). Sema3A is released from cells in the marginal zone and is capable of attracting cortical pyramidal cell apical dendrites toward the marginal zone, while simultaneously repelling their axons. This dual effect is controlled by guanosine $3',5'$-monophosphate (cGMP) levels, which are higher in apical dendrites than in axons due to asymmetric location of the cGMP's producing enzyme, soluble guanylate cyclase (Song et al., 1998; Polleux et al., 2000). Sema3A acts as chemoattractant in the presence of high cGMP levels and as chemorepellants with low cGMP levels, possibly through a direct interaction between the plexin-1 portion of its receptor complex and a Rho GTPase (Tang, 2003).

A very important set of pathways affecting Rho GTPases involves the binding of extracellular neurotrophins to their tropomyosin-related kinase (Trk) family and p75 receptors. Among the several neurotrophins involved in neuronal development, the three best characterized are nerve growth factor (NGF), brain-derived growth factor (BDNF), neurotrophin-4 (NT-4/5) and neurotrophin-3 (NT3). These neurotrophins bind to TrkA, TrkB, TrkC, and p75 receptors with varying specificity, as shown in Figure 18.3. The ligand specificity of TrkA and TrkB can be limited to NGF and BDNF, respectively, if those receptors bind cooperatively with the p75 receptor (Luo, 2002). The effects of neurotrophin binding vary, and they can result in specific outcomes in different cortical layers (Whitford et al., 2002; see Table 1 for examples) and cell classes (Kryl et al., 1999). TrkB and TrkC are expressed in a shortened form in more mature neurons (Kryl et al., 1999). Visual cortex pyramidal neurons react to activation of full-length TrkB receptors, found in young neurons, by increased proximal dendritic branching, and to activation of the shortened splice variant, found in older neurons, with extension of distal dendrites (Polleux et al., 2000).

Another relatively well-characterized pathway that affects Rho GTPase activity and neuronal development requires the binding of lysophosphatidic acid (LPA) to LPA1 or LPA2 receptors (Berezovska et al., 1999; Fukushima et al., 2002b). LPA receptors are up-regulated in cells in the ventricular zone (Fukushima et al., 2000). Activation of this pathway causes neurite retraction through Rho activity (Fukushima et al., 1998).

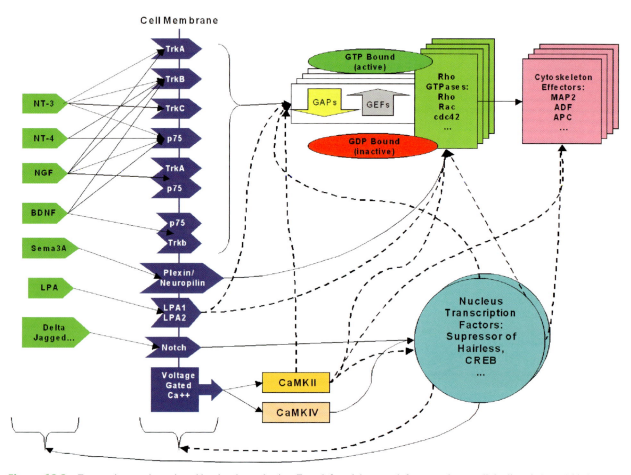

Figure 18.3 Enzymatic cascades activated by signal transduction. From left to right, growth factors and extracellular ligands (green) bind to membrane receptors (blue). These affect enzymes such as CaM kinases (yellow, *bottom*) and Rho GTPases (light green, *top*), either directly or through the action of activators (GEFs, gray) and deactivators (GAPs, yellow). Enzymes can activate nuclear transcription factors (light blue) and cytoskeleton effectors (pink). Transcription factors regulate the production of practically all players in this scheme. Solid arrows: commonly accepted effects. Dotted arrows: putative effects.

With a different mechanism, the Notch receptor pathway is activated when the growing cell contacts membrane-bound ligands on neighboring cells. Binding of these ligands (including Delta,

TABLE 18.1 Effects of membrane receptor activation on dendritic growth

Membrane Receptor	Effects
Neuropilin-1/Plexin-1	Chemoattraction (high cGMP) Chemorepulsion (low cGMP)
TrkA	Enhanced dendritic growth via MAP2 activation
TrkB	Increased proximal dendrite branching (full length) Distal branches elongation (T-1 truncated)
TrkC	Dendrite destabilization and increased branching in layer 4, but not layer 5, cortical neurons
p75	Neurite outgrowth via RhoA deactivation
LPA1, LPA2	Actin depolymerization/growth cone collapse Actin polymerization/retraction
Notch	Contact-dependent inhibition of growth Decreased length, increased branching

Serrate, and Jagged) to the extracellular domain of Notch releases the intracellular domain, which enters the nucleus and binds to Suppressor of Hairless transcription factor (for review, see Redmond and Ghosh, 2002). The eventual effect of this binding is also neurite retraction (Fukushima et al., 1998, 2002b; Sestan et al., 1999).

Synaptic activity also has strong effects on neuronal development. While some aspects of neuronal development, including initial synapse formation, proceed regularly in the absence of synaptic neurotransmitter release, other aspects, such as synapse maintenance, do not (Verhage et al., 2000). Tectal neurons require glutamate binding at both NMDA and AMPA receptors to reach their normal size (Rajan and Cline, 1998). Pyramidal neuron (Maletic-Savatic et al., 1999) and motoneuron (Kalb, 1994) dendrites also respond to NMDA activity by enhancing growth. More generally, electrical activity in the form of increased intracellular calcium has powerful effects on the development of neurons. Like Notch signaling, calcium can exert its developmental effects through activation of transcription factors. Intracellular calcium raised by the activation of voltage-gated calcium channels can activate calcium/calmodulin-dependent protein kinase IV (CaMKIV). CaMKIV can, in turn, activate cyclic-AMP-responsive-element binding protein (CREB), which affects transcription and potentiates dendritic growth (Redmond et al., 2002). Calcium activation of CaMKII has also been shown to have an effect on dendritic morphology by stabilizing and limiting growth (Wu and Cline, 1998).

These pathways emphasize the importance of ligand-initiated signaling pathways in controlling neuronal development. It is important to remember, however, that, in addition to responding to external chemical cues, growth cones movement is also a function of cell membrane tension (Condron and Zinn, 1997) and extracellular electrical fields (e.g., Jaffe and Poo, 1979; Macias et al., 2000).

18.3 A SURVEY OF COMPUTATIONAL MODELS OF NEURITE OUTGROWTH

From the section above, it may be concluded that underlying the shape of individual neurons, and eventually the connectivity of the entire brain, is the behavior of individual growth cones. Specifically, the mechanics of microtubules, actin filaments, and their associated proteins allow the growth cone to extend, retract, bifurcate, and turn to create the branching structures of neurons. The fundamental importance of these molecular interactions is highlighted by the complex cascade of enzymatic pathways providing finely tuned regulation and control. Many of the above mechanisms are being investigated through the use of computational models. Biologically driven models of neuronal development can be divided into three main categories. At a low level, neuron development can be modeled in terms of growth cone behavior. Models of this type deal with the details of growth cone extension, retraction, and navigation. These models range from the specifics of actin polymerization and depolymerization in the lamellipodia (Li et al., 1994) to the long-range navigation of axons through complex environments of chemical gradients (Segev and Ben-Jacob, 2000). At an intermediate level, neurons can be modeled in terms of their overall branching structure. These models include those based on one-dimensional stochastic processes (Nowakowski et al., 1992) and those incorporating more complex elements such as realistic time bins (van Pelt et al., 1997) or two- and three-dimensional interactions (Samsonovich and Ascoli, 2003). At a third, higher level, neuronal development can be modeled on the network scale (for review see van Ooyen, 2001). Here we concentrate on models at the first two levels (see summary in Table 18.2).

Li et al. (1994) attempt to account for fundamental aspects of growth cone behavior by modeling the polymerization and depolymerization of the actin filaments constituting the lamellipodia. In this model, the polymerization at the leading edge of the growth cone pushes the lamellipodia forward, while the depolymerization in the internal part of the growth cone near the microtubule core acts to pull the middle of the growth cone forward. Also included in this model are parameters describing the neck of the growth cone and the adhesion of the growth cone to its substrate. The growth cones resulting from this model reproduce several of the behaviors of real growth cones, including variable rates of extension and retraction (Li et al., 1994).

At a similar level of detail, van Veen and van Pelt (1994a, b) deal with microtubules rather than with actin dynamics. This model is based on microtubule polymerization and depolymerization and the resulting cell membrane tension. Though the model does not produce branched tree structures, it includes a mechanism of competition of multiple growth cones for tubulin diffusing from the soma. Growth rate is predicted to depend only minimally on diameter (due to tubulin diffusion rate) and not at all on neuritic length. In contrast, growth cone extension is found to be very sensitive to even small changes in the relative rates of polymerization and depolymerization of microtubules (van Veen and van Pelt, 1994b).

At a slightly higher level than that of actin filaments and microtubules, the models focus on the behavior of filopodia. Buettner and colleagues stochastically populate a virtual growth cone with filopodia in order to investigate path-finding behavior (Buettner et al., 1994; Buettner, 1995). The lengths, life spans, and numbers of extending filopodia are shown to be important in determining the target detecting abilities of the growth cone. A related model by Goodhill and Urbach (1999) computes the gradient sensing ability of growth cones given biologically driven assumptions, such as growth cone sampling radii and receptor number. The main predictions are that growth cones use a spatial rather than temporal mechanism for sensing gradients and that higher gradients of bound factor than of diffusible factor are required for detection by the growth cone (Goodhill and Urbach, 1999).

A different modeling approach to neuronal development aims at the description of the branching structures of neurites (usually dendrites). While these models do not generally include mechanistic aspects of growth cone dynamics, they may be based on underlying developmental principles or may attempt to realistically follow the spatiotemporal course of shape change in growing neurons. These models of dendritic branching typically aim to recreate the structure of real cells based on stochastic processes (Table 18.2). In these processes, branching and termination probabilities may depend on parameters with a direct link to growth mechanisms (e.g., via diffusion of chemicals), such as distance from the soma, branch order, or distance since last bifurcation. Alternatively, branching and termination probabilities may be correlated to local or global parameters that are less directly reducible to the biophysics of growth cone. These parameters include the total number of terminations and local diameter. A growth cone does not have access to the information of how many tips are being (or have been) formed. Similarly, the diameter of a dendritic branch continues to increase well after the extension of that branch (as it supports an increasing number of dendrites downstream). Nevertheless, both total number of terminations and (adult) diameters are indirectly linked to developmental mechanisms such as total metabolic resource and dendritic maintenance (so-called "hidden parameters").

Models of dendritic branching can address the issue of bifurcation patterns in one, two, or three spatial dimensions. One-dimensional models ignore all angle information and generally represent branching in terms of distances from the soma and other branch points. One-dimensional models still contain considerable information regarding tree topology (e.g., bifurcation partition asymmetry, branch points, path length, and, in principle, diameters). Two-dimensional models include angle information, but only on one plane. These models could be thought of as representing cultured neurons, which are restricted to movement in the two-dimensional surface provided by the plate. In addition to branch angle information, two-dimensional models can include

TABLE 18.2 Summary of computational models of neuronal outgrowth

Model	Type	Mechanisms/Elements	Behavior/Results	Cell Class
Li et al., 1994	Growth Cone/Molecular	Actin behavior in the lamellipodia.	Variable growth rates, retraction.	Generic
van Veen and van Pelt, 1994a	Growth Cone/Molecular	Microtubule polymerization and diffusion, membrane tension.	Strong dependence of growth rates on polymerization/ dempolyerization rates. Constant elongation rates.	Generic
Buettner et al., 1994	Growth Cone	Filopodia behavior in growth cone chemoreception.	Filopodia initiation, extension, and retraction modeled. Growth cone able to cross over nonpermissive media in realistic way.	Generic
Goodhill and Urbach, 1999	Growth Cone	Filopodia behavior in growth cone chemoreception.	The chemodetection ability is computed. Spatial, not temporal gradient detection supported by model as likely mode of chemonavigation.	Generic
Nowakowski et al., 1992	1D Branching	Branching probability based on current branch length.	Recreates branch length distributions of real cells.	Cat dorsal horn cells.
Ireland et al., 1985; Carriquiry et al., 1991	1D Branching	Branching probability based on branch order and number of branches at previous order.	Predicts time of initiation and speed of growth using data from cells of different ages.	Apical dendrites of rat entorhinal cortex.
Burke et al., 1992	1D Branching	Branching and termination probabilities based on branch diameter and additional parameters.	Recreates dendrograms properties.	Cat alpha-motoneurons
Ascoli et al, 2001a,b; Donohue et al, 2002	1D/3D Branching	Diameter threshold determines branching and terminations (Hillman's description).	Dendrogram properties and (with the addition of tropism) spatial distribution (3D branching).	Cat motoneurons, rat CA1 pyramidal cells, guinea pig Purkinje cells
van Pelt and Verwer, 1986 (QS); Dityatev, 1995	1D Branching	Branching probability allowed in interstitial branches and controlled by branch order.	Branching generally restricted to terminal segments in cells used.	Cat, rat, and frog motoneurons. Cultured rat striatal neurons.
van Pelt et al., 1997 (BEST)	1D Branching	Branching probability controlled by tree size and branch order. Realistic time bins.	Time course of branching probabilities is predicted.	Cat, rat, and frog motoneurons. Rat Purkinje cells and pyramidal basal dendrites. Human granule cells.
Hely et al., 2001	1D Branching/Molecular	Branching probability controlled by MAP2 phosphorylation and diffusion.	Variation of model parameters able to produce a wide variety of morphologies.	Rat pyramidal basal dendrites.
Li et al., 1992; Li and Qin, 1996	2D Branching	Branching probability controlled by membrane tension, lateral inhibition, and contact inhibition.	Recreates branch angles and allows biologically plausible contact-initiated mechanisms.	Generic
Senft and Ascoli, 1999	3D Branching, navigation	Growth modeled by subsequent stemming, extending, and branching phases. Includes axonal navigation.	Created a large-scale cellular model of region CA1 of the rat hippocampus.	Rat CA1 pyramidal cells

TABLE 18.2 *(Continued)*

Model	Type	Mechanisms/Elements	Behavior/Results	Cell Class
Robert and Sweeney, 1997	2D Navigation	Growth cone response to electric fields.	Recreates the turning of growth cones toward negative electrode through filopodia behavior.	Generic
Segev and Ben-Jacob, 2000	2D Navigation	Growth cone chemonavigation with obstacles and multiple targets.	Axons are able to follow chemical gradients to find targets. Growth cone switches from chemorepellation to chemoattraction.	Generic
Samsonovich and Ascoli, 2003	3D Orientating	Control of 3D structure by directional, straight, and radial growth.	Determination made that radial growth from soma is more important than directional growth.	Rat hippocampal principal neurons.
Senft, 2002	3D Navigation	Stochastic navigation and branching through realistic 3D space.	Realistic 3D pathways could prove useful for developmental and physiological models.	Mouse thalamocortical afferents.

dendrite–dendrite interaction properties and trophic influences. Three-dimensional models include the full range of angle information required to describe a branching structure in three dimensions. These models are able to fully represent the general shape of actual dendrites. Complicating the issue of model dimensionality is the fact that both one- and three-dimensional model results are displayed in print in the two dimensions provided by paper media. In addition, many models that include information in only one dimension (length) are referred to as two-dimensional (e.g., Kliemann, 1987). All of the models discussed below are one-dimensional unless otherwise noted.

One of the simplest branching models is that by Nowakowski et al. (1992), in which the branching probability of a growing dendrite depends on the distance from the last bifurcation point. Immediately after a branching event, the probability of subsequent growth cone branching is temporarily suppressed. The probability rises toward an asymptotic value as a function of distance from the last branch point (or soma if no branching has yet occurred). This model accounts for almost 90% of the sample variance when fitted to the segment length data from 17 reconstructed cat spinal dorsal horn neurons, demonstrating the usefulness of even simple stochastic processes (Nowakowski et al., 1992).

A model developed at Iowa State University (Ireland et al., 1985; Kliemann, 1987) varies bifurcation probability by branch order. Branch order is defined for each point in a dendritic tree as the number of bifurcations separating that point from the soma, plus one. In this model, bifurcation probability may also depend on the number of branches at previous orders (i.e., on the branching history). Realistic time bins are obtained by measuring parameters from cells of different ages, allowing the extrapolation of branch statistics back to the initial dendritic sprouting (Ireland et al., 1985). Interestingly, the model can fit branch distributions from basal and young apical dendritic data, but not from older apical trees (Carriquiry et al., 1991). The authors attribute this discrepancy to the ability of apical trees to branch interstitially proximal to the soma fairly late in development. The

introduction of branching inhibition following previous branching in a later version of this model allows for the generation of more complex morphologies, including (a) the sparse stem with side-branch structures typical of apical dendrites and (b) bushier trees typical of basal dendrites (Uemura et al., 1995).

A different family of models relates the probability of a dendrite to bifurcate or terminate to branch diameter. The rationale for this approach is based on (a) Hillman's early electron microscopy studies, suggesting that a minimum number of microtubules is necessary to sustain a dendritic bifurcation, and (b) linking the number of microtubules to branch diameter (Hillman, 1979). According to this view, when a dendrite bifurcates, the set of microtubules is divided between the two daughters. A first quantitative model related to these ideas is that developed by Burke et al. (1992) to describe the branching statistics of dendritic trees in spinal motoneurons. Bifurcation, elongation, and termination probabilities are quantitatively extracted from experimental data and used to stochastically generate simulated branching structures. This model also fits the dependence of the diameter of both daughters on the parent diameter. This description captures several morphological features of motoneurons, including the distribution of terminations and of the average diameter value with branch order. However, in order to appropriately fit the maximum branch order and the distribution of bifurcations with path distance from the soma, the model has to include additional parameters describing the influence of previous branching behavior and somatic distance on the probability of bifurcating.

Ascoli and Krichmar (2000) propose a more direct implementation of Hillman's ideas. Here, dendrites elongate according to measured statistics and then bifurcate and terminate directly based on a diameter threshold, also measured from real data (Figure 18.4). The daughter diameters at bifurcations is assigned based on the parent diameter using Rall's "3/2" rule (Rall et al., 1992). Again it is important to emphasize that this algorithmic description does not represent a direct developmental mechanism, since the model parameters are measured from adult cells, and dendritic branches are thinner during growth than in the final, mature shape. Nonetheless, the model constitutes a complete statistical description of dendrograms properties based on biophysically plausible constraints. This model can reproduce a variety of morphological features of hippocampal pyramidal cells (Figure 18.5), cerebellar Purkinje cells, and spinal motoneurons (Ascoli et al., 2001a,b; Donohue et al., 2002).

All of the above models only allow growth and bifurcation at terminal tips. However, as mentioned in the previous section, interstitial branching is commonplace in several neuronal classes (e.g., Daily and Smith, 1996). The QS model by van Pelt and Verwer (1986) allows branching at both terminal and nonterminal segments, resulting in a range of behaviors from all interstitial to all terminal branching. In addition, branching probability decreases with branch order. This model is used, for example, by Dityatev et al. (1995) in the comparative analysis of motoneuron morphologies. In later models, van Pelt and colleagues again assume all branching to be terminal (van Pelt et al., 1997). A simple constant branching probability (B) is made dependent on the size of the whole tree (E) and on branch order (S). The addition of time bins based on real data (T) result in the "BEST" model

for dendritic development. In effect, this is a two-dimensional model, with one dimension of space and one of time.

Attempting to close the cross-scale gap between biochemical processes and the formation of branching structures during development are models including control of bifurcation and elongation by molecular interactions. In Hely et al. (2001), dendritic growth depends on the calcium-mediated phosphorylation state of microtubule-associated protein 2 (MAP2). In this model, calcium influx is a function of the surface-to-volume ratio (which is related to dendritic diameter), while MAP2 is controlled (through diffusion) by the distance from the soma. The phosphorylation state of MAP2 in response to calcium is complex, due to the concomitant action of several enzymes, such as CaMKII and calcineurin. When dephosphorylated, MAP2 binds to and stabilizes microtubules, thus promoting elongation. Phosphorylation causes MAP2 to dissociate from microtubules and promotes branching. A variety of different dendritic morphologies (including basal dendritic trees of pyramidal cells) can be created by varying the rates of phosphorylation and dephosphorylation.

Two-dimensional models include the description of branch angles and spatial interactions of neurites. In Li et al. (1992),

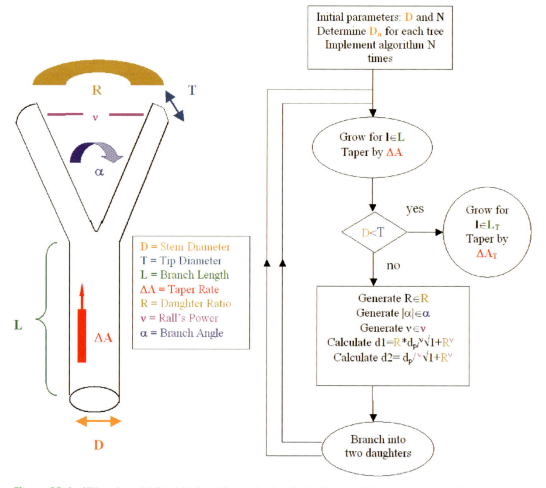

Figure 18.4 Hillman's model. Dendritic branching can be described with a set of diameter-dependent rules. A branch grows for a certain length (L) while tapering (ΔA), and then it bifurcates or terminates depending on the final diameter and a diameter threshold (T). If the branch bifurcates, it stems two daughter branches at an angle α, with diameters determined by a ratio (R) and Rall's power ν. The two new stems behave like the original branch (recursion). Reproduced with permission from Ascoli (1999).

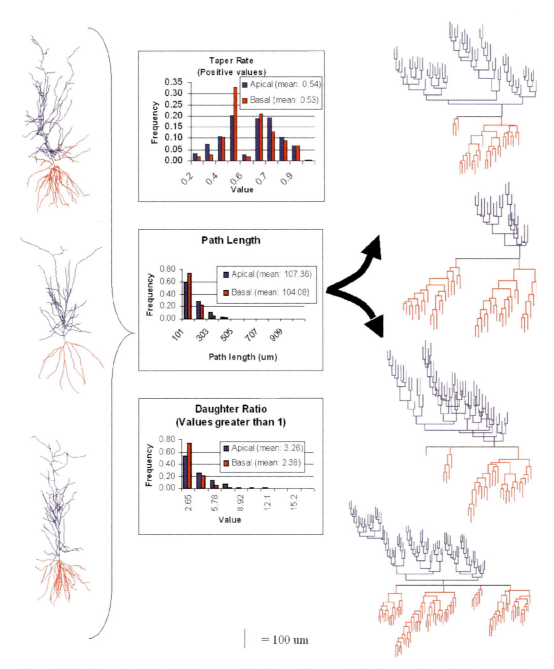

Figure 18.5   Generation of virtual dendrograms with L-Neuron. From a pool of real neurons (exemplified on the left, top to bottom, by Turner's cells n404, n406, n402) a set of statistics for the algorithmic parameters is extracted. The examples in the center show the distribution of the taper rate, path length, and daughter ratio (top to bottom). The algorithm can then produce an arbitrary number of non-identical virtual dendrograms (four examples are shown on the right). Scale bar for both real and modeled dendrograms is 100 μm.

growth cone extension in one part of the cell plays an inhibitory effects on the advance of growth cones in other parts of the cell. This distal branching–branching inhibition, similar to previously mentioned mechanisms (Uemura et al., 1995), allows differential growth rates in different regions of the neuron. In addition, this model includes a form of contact inhibition: If a growing neurite hits another neurite, it stops and possibly retracts (Li et al., 1992). As discussed earlier, Notch is thought to mediate analogous contact inhibition in real cells. Within this framework, virtual cells extend and branch in gradients of growth

factor, resulting in meaningful bifurcation angle data. A later two-dimensional model, by Li and Qin (1996), controls branching through the tension created by the cell membrane as the growth cone extends, much like the van Veen and van Pelt (1994a) model discussed above. This model allows unequal growth rates for different growth cones, and it also uses membrane tension to compute realistic bifurcation angles.

Computational modeling has also been used to address specific issues related to neurite navigation, independent of branching behavior. For example, Robert and Sweeney (1997)

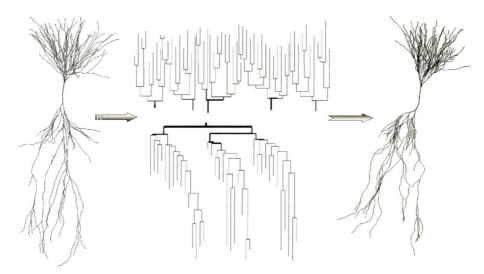

Figure 18.6   Remodeling dendritic orientation. After removing all angle information from the experimental data (*left*), dendrograms are reoriented solely based on stem direction, repulsion from the soma, and noise. Arrow length: 100 μm.

quantitatively captured growth cones response to electric fields in vitro (turning toward a negative and away from a positive electrode) with a model based on filopodia dynamics. In a similar model by Segev and Ben-Jacob (2000), growth cone navigate through relatively complex environments, including obstacles and multiple targets, using chemoattractant and chemorepellant cues.

Only few models attempt to create complete three-dimensional virtual cells that can be compared to real neurons. Samsonovich and Ascoli (2003) propose a simple stochastic description of the 3D dendritic orientation of hippocampal principal cells. According to this model, dendrites select their relative growth direction based on the radial vector from the soma (which depends on the local position), on a fixed "external" directional bias, and on (constant) random noise. These factors can be quantitatively measured from real adult neurons and used to "re-grow" virtual cells based on the one-dimensional branching structure of real cells (dendrograms). Two important results emerge from this study. First, for all morphological classes (basal and apical dendrites of CA3 and CA1 pyramidal cells, and dentate granule cells) only the repulsion from the soma (and not the fixed external bias) is shown to be statistical significant. Somatic repulsion is also quantitatively assessed for each morphological class relative to the unitary tendency of straight growth. Second, the virtual neurons generated by this model are strikingly similar to the original ones (see, e.g., Figure 18.6), as indicated both by visual comparison and statistical morphological analysis. Thus, most information of dendritic spatial orientation can be effectively captured with only two parameters (somatic repulsion and noise). This study also raises important questions about the biophysical mechanisms that could underlie such processes during development (Samsonovich and Ascoli, 2003).

In the ArborVitae model, dendritic growth is described as a sequence of discrete phases: stemming, extending, bifurcating, and again extending, with several parameters determining the statistics of branch length and diameter (Senft and Ascoli, 1999). Arbitrarily complex structures can be obtained by appending successive layers of more and more distal dendrites

starting from the soma. This model also provides a description of 3D navigation of neurites based on attraction (or repulsion) to specific targets. In the case of extending axons growing toward dendritic fields, a provision is also included to establish synaptic contacts (Figure 18.7). The explanatory link between axonal navigation and network formation may be too complex to allow the creation of a model based on lower-level mechanistic principles (e.g., growth cone dynamics). However, several parameters in ArborVitae have important biological correlates. For example, the attraction of an axon to a dendrite is inhibited once the distance between the two is below a given threshold ("Too Close?" in Figure 18.7). Varying the value of this threshold gives rise to different fiber anatomies, such as perforant (high threshold), sprouting (intermediate), and climbing (low threshold). Interestingly, the same model of navigation can be used (after appropriate parameter-tuning) to describe dendritic orientation (Ascoli et al., 2001a,b).

Given the importance of the spatial location of chemical and cellular cues in neurite growth, an important recent development consists of embedding the ArborVitae model in a virtual 3D environment that reproduces the 3D distributions of neurons from real anatomical data. Senft demonstrates this approach by "seeding" thalamic and cortical cells in the appropriate anatomical coordinates (as reconstructed from digitized mouse atlases) and allowing axons to project toward their targets as described above (Senft, 2002). The resulting simulations represent an invaluable computational benchmark to develop and test mechanistic hypotheses of axonal navigation (Figure 18.8).

18.4 TECHNICAL DETAILS, METHODOLOGY, AND CURRENT APPLICATIONS

A fundamental issue of morphological modeling regards the representation, reproducibility, and quality control of primary data. The most common source of experimental data regarding dendritic structure is the intracellular injection of dyes, which allows

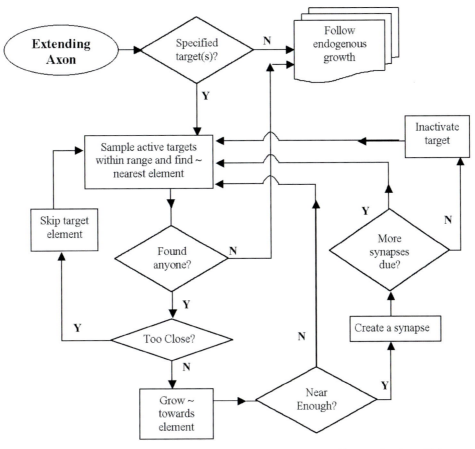

Figure 18.7   Axonal navigation in the ArborVitae model. Axons grow toward (or away from) specified targets (somata, dendrites, or other axons of specific cell classes and layers), based on a set of parameters determining, for example, intensity of trophic effect and duration of the refractory period after synaptic contacts are made.

the selective staining of single cells. This preparation, however, is highly sensitive to subtle variations in the experimental protocol, especially because of shrinkage and tissue deformation due to dehydration and sectioning. While shrinkage can be measured and corrected by post-processing, a particularly daunting problem is the resulting amplification of noise, particularly perpendicular to the surface of the section (for a complete discussion of these issues, see Ascoli et al., 2001a).

Once the tissue sections containing the stained neurons are mounted and coverslipped on a microscope slide, the morphological structure is digitized by a human operator with the aid of tracing software and computerized control of stage movements. This step in data acquisition is absolutely necessary for reliable quantitative analyses and any neuroinformatics application, because it transforms analogic images in a digital, machine readable format. However, the reconstruction process is long, difficult, subjective, and error-prone. As a result, digitized morphologies may contain system- and operator-specific idiosyncratic biases, and the same intracellularly injected, stained, and mounted neuron can yield two fairly different digital reconstructions by two different researchers (Jaeger, 2001).

The reconstruction of a single neuron, from the experimental preparation to data digitization, can take several weeks of work of skillful researchers. This, along with the above issues of data reproducibility, make data sharing an incredibly important practice in this field. Several archives of hippocampal neurons are

now publicly available through the internet (see web resources; for a review, also see Ascoli, 2002a). A project for cataloging neocortical cells is also currently underway (see Chapter 19). In addition, tools such as those included in the Axiope project are being developed which will hopefully add greatly to the availability of digitized neurons (see Chapter 15; Goddard et al, 2003).

Important aspects regarding the social and technical advantages and disadvantages of alternative database designs are currently being discussed in the neuroinformatics literature (see, e.g., Cannon et al., 2002; Turner et al., 2002).

In addition to contributing to our understanding of the biological mechanisms underlying dendritic outgrowth and axonal navigation (as outlined in the previous section), a foremost application of morphological models is that of efficient databasing. The statistical distributions of model parameters constitute a compact, quantitative description of complex the data of neuronal structure. For example, while the digital reconstructions of 100 cells from a single morphological class typically represent a ~1-MB (compressed) file, the model specifications, together with the entire parameter statistics for 10 morphological classes only amount to ~100 kB of information. In addition, these stochastic models can generate *any* number of nonidentical synthetic neurons from a given (finite) number of experimental reconstructions. These means of data compression and amplification are useful features for the representation, description, and simulation of the anatomical complexity and variability

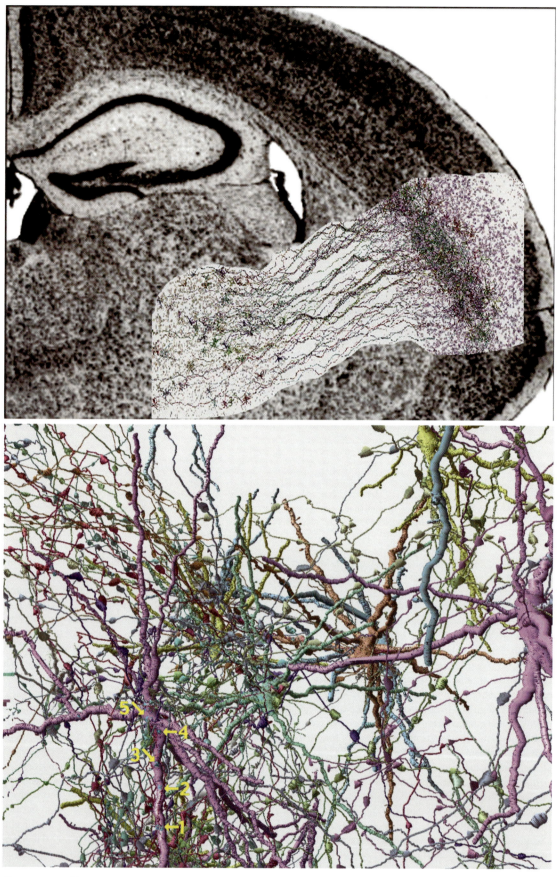

Figure 18.8 **Top panel:** Simulated thalamocortical pathways in ArborVitae overlaid on Nissl-stained coronal sections from the Sidman mouse Atlas. In the underlying grayscale structure (real data), the hippocampus, thalamus, and portions of the target cortex are prominent. Some of the simulated cells (each given a separate hue) have elaborated dendrites and send axons running obliquely up to the cerebral cortex (*upper-right*). **Bottom panel**: High magnification of a simulated cortical region. The tiny spots on the target cells' somata and dendrites are ArborVitae's representations for synapses. An axon entering the field in the *lower left* walks up the nearby dendrite, making contacts (1–5) at regular intervals. Adapted with permission from Senft (2002).

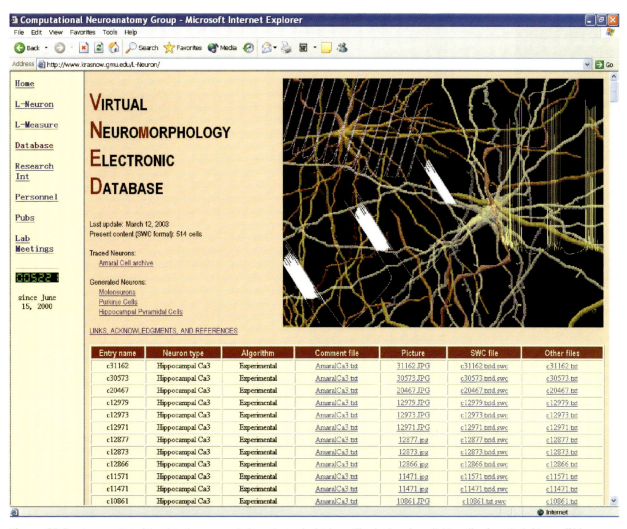

Figure 18.9   A screenshot of the virtual neuromorphology electronic database. The database is available online from the L-Neuron Web page, and it contains pictures (JPEG and bitmap) and digital files of neuronal structures generated with a variety of models, and of real reconstructed neurons. The database also includes comment and algorithm files for the different cell groups, links to conversion routines, and references to the original experimental data.

of the central nervous system, and they constitute an important added value to neuroinformatics. An image from an example of electronic morphological database, containing both real and simulated data (Ascoli et al., 2001a), is shown in Figure 18.9.

Neuroinformatics tools are now available for the visualization, exploration, editing, conversion, and measurement of digital morphological data from experimental preparations or public archives. Two free, user-friendly software programs in particular constitute indispensable utilities for computational neuroanatomy studies: L-Measure and Cvapp. L-Measure reads all known digital formats of neuronal morphology and outputs a variety of measurements, including lengths, surfaces, distances, angles, fractal dimension, and any arbitrary combination thereof (Scorcioni and Ascoli, 2001). L-Measure, available in Windows and Linux versions, can be operated by scripts or through its graphical user interface (Figure 18.10). This program can also use any of the measured parameters to specify further analyses, making it possible for example to obtain the distribution of terminations "with diameter greater than 1 micron" as a function of the path distance from the soma.

Cvapp graphically displays digital morphological files (Figure 18.11) and allows users to zoom, pan, rotate, and color-code reconstructed or simulated neurons (Cannon et al., 1998). This tool also provides a means of editing the structure by (dis) connecting dendrites, adding or removing points, changing diameters, and converting morphological files into formats suitable for electrophysiological simulations with popular modeling packages such as Genesis and Neuron. A variety of other available routines, while not as carefully curated as Cvapp and L-Measure, are also useful for computational neuroanatomists. These tools include Neuron_morpho, an ImageJ plugin that allows digitization of image stacks, Swc2hoc, which converts morphological files into Neuron simulation files with optimized spatial compartmental meshing (Lazarewicz et al., 2002), and Dendro1, which extracts and saves dendrogram properties from morphological files (Donohue et al., 2002). These and other routines are publicly available through the Internet (see Web Resources).

In addition to the above-mentioned tools for the visualization and analysis of digitized cells, projects are also under

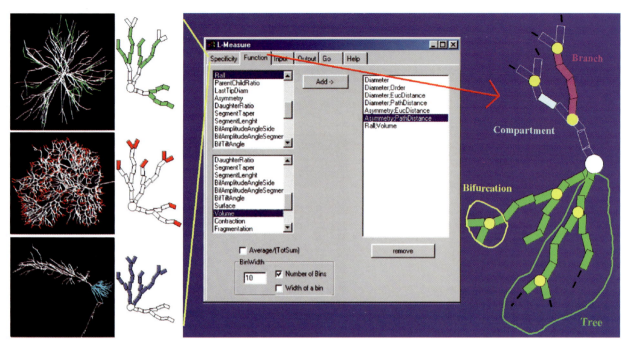

Figure 18.10 The L-Measure graphical user interface (*middle*) allows to extract a variety of measurements defined by specific functions at the compartment, bifurcation, branch, or tree level (*right scheme*). Parameters can be specifically measured by selecting sub-groups of dendrites with the specificity function. The examples on the *left*, showing pairs of real and schematized neurons, represent dendritic regions of branch order two (*top*, a motoneuron), terminal tips (*middle*, a Purkinje cell), and basal trees (*bottom*, a pyramidal cell).

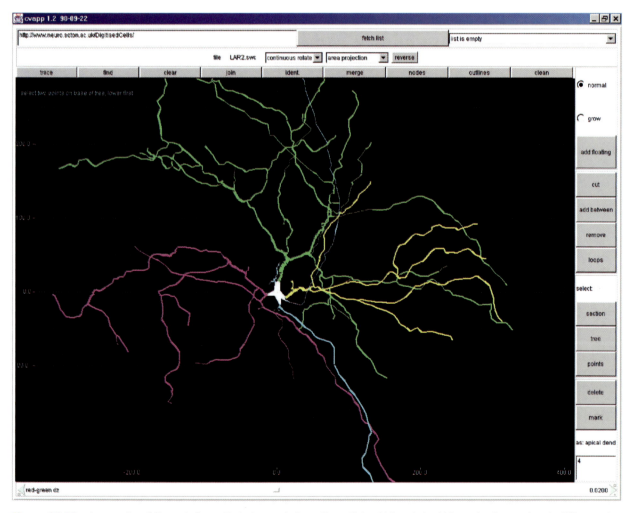

Figure 18.11 A screenshot of Cannon's Cvapp displaying a retinal ganglion cells in which each dendritic tree has been assigned a different color.

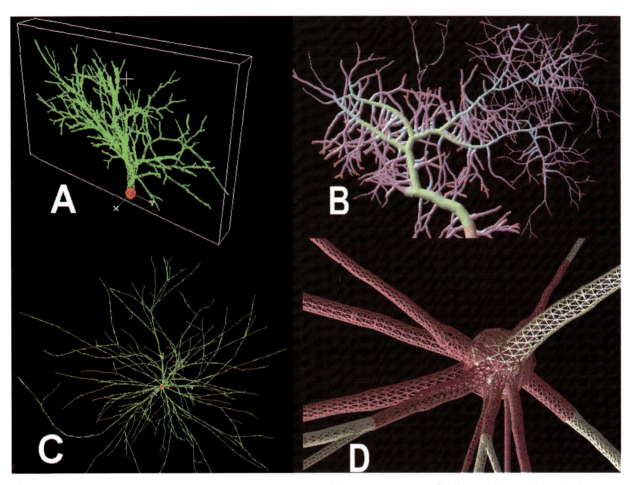

Figure 18.12 Examples of virtual Purkinje cells (**A**, L-Neuron; **B**, ArborVitae) and motoneurons (**C**, L-Neuron; **D**, ArborVitae). Panels are not in scale: Full views are presented in Parts A and C, while details are shown in Parts B (dendritic branches) and D (soma and dendritic stems).

development that aim at facilitating the creation, management, and sharing of neuronal databases. The Axiope suite (described in detail in Chapter 15) includes software that automates many of the steps needed to (a) create flexible databases of digitized neurons and (b) share those databases through a federated system. In principle, this approach provides for many cells, real and generated, to be accessed from a single source while allowing the labs that generate the data to retain ownership and control of it.

Another important resource for researchers in need of experimental data on which to base their morphological models is constituted by the NMDB project (see Chapter 19). The goal of the Neocortical Microcircuit Data Base is to catalogue and share data, including digitized tracings, from neocortical cells. This database is being populated with additional information about the cells it contains, including genetic and detailed electrophysiological information. When these data become publicly available and accessible, they will provide a rich repertoire for complex searches and data mining.

Tools such as L-Measure are important to extract model parameters from experimental data. Not all models listed in Table 18.2, however, constitute a complete description of dendritic morphology. In addition to the distinction between one- two-, and three-dimensional models (with only the latter ones capturing the full extent of spatial orientation), it should be noted that models in which branching depends on path distance, previous

bifurcations, and the total number of terminations seldom include a description of diameter. Two software programs have been described that allow the simulation of complete neuronal structures: ArborVitae (Senft and Ascoli, 1999; Senft, 2002), which was discussed in the previous section, and L-Neuron (Ascoli and Krichmar, 2000). L-Neuron implements a variety of the diameter-based models described in this chapter (e.g., Hillman, 1979; Burke et al., 1992). Examples of virtual cells generated by L-Neuron and ArborVitae are shown in Figure 18.12 (see also Ascoli et al., 2001b).

Two important aspects should be mentioned regarding the preparation, execution, and analysis of these simulations. The first one concerns the consistency of the statistical description and the stability of the algorithm. The statistical description of the parameters used in the model, which are measured from real cells, must be reproduced in the simulated cells. The second aspect regards the completeness of the algorithm, and it tackles the question of the amount of information effectively captured by the model and its parameters. To probe this aspect, the statistical distribution of parameters not explicitly used by the model is measured from real and simulated cells and is compared. If the two groups are statistically indistinguishable, these tested "emergent" parameters represent redundancies, which do not need to be specified in addition to the model parameters. Analysis of emergent parameters is also useful in the comparison between

Parameter	Exp. (N=1×6)	Hillman/PK (N=10×6)	Burke (N=10×6)
Total Length (μm)	102,799	94,609 (10213)	87,876 (7230)
Total Area (μm²)	562,363	401,511 (46,572)	464,713 (38,049)
Asymmetry	0.462	0.48 (0.01)	0.42 (0.01)
Number of Bifurcations	158.3	159.7 (18.28)	122.4 (9.91)
Maximum Order	10.3	11.8 (0.74)	9.2 (0.31)
Average path to tips (μm)	1137.1	1153.4 (31.43)	1199.2 (27.73)
Average distance to tips (μm)	932.7	953.3 (23.7)	1055.4 (22.3)
Height (95%)	1600	1565.4 (78.70)	1939.8 (70.82)
Width (95%)	1882	1493.99 (149.77)	1956.30 (107.48)
Depth (95%)	1728	1790.34 (92.60)	2256.51 (70.22)

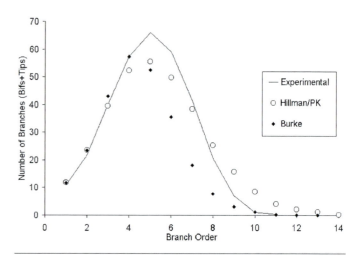

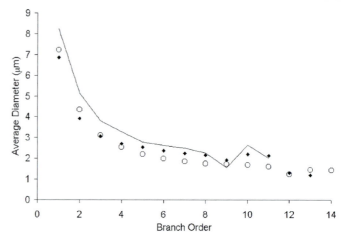

Figure 18.13   Examples of comparative analysis of emergent morphological parameters from real and synthetic motoneurons. Ten groups of six virtual neurons each were generated. Tabulated values (*left*) refer to the average over the six cells, for the real neurons, and to the mean and (standard deviation) of the averages over the 10 groups, for the virtual neurons. Plots on the right represent distributions of dendritic segments (*top*) and diameters (*bottom*) versus branch order.

alternative algorithms and in the development of more powerful models (Figure 18.13).

A final application of dendritic morphology modeling is the generation of anatomically accurate substrates for biophysically detailed simulations of electrophysiological activity (Ascoli, 1999). These types of studies are now widely adopted in the investigation of the structure–function relationship, but are limited by the small number of available neuroanatomical reconstructions (e.g., Krichmar et al., 2002). Stochastic models can provide both (a) the necessary amount of data and (b) specific "extreme" cases that may be useful to test or further develop mechanistic hypotheses.

18.5 LIMITATIONS AND FUTURE IN NEUROSCIENCE

"All models are wrong. Some models are useful." It is important to remember that anatomical models are not the same as experimentally acquired data. The very gist of the modeling approach is to select a subset of the information present in the experimental data and to discard the rest (assumed as "redundant").

The verification of the model is empirical (Are synthetic neurons statistically distinguishable from the real ones?). Thus, the results are not "mathematically proven," but are just working hypotheses that may be confuted by further analysis and/or experiments.

In addition to addressing particular questions about development such as the role of actin dynamics or the response to electrical fields, developmental models can offer other insights into the mechanisms by which neurons come by their shape. Even the selection of model parameters can shed light onto development. If a model is not sufficient to capture all aspects of neuronal growth and maturation, it may provide a good starting point to look for additional important concepts yet to be discovered. A successful model can yield insights into what may be happening in real cells, and an unsuccessful one can point out shortcomings in our understanding.

One area in which computational modeling should prove useful is in the elucidation of the complex interaction of microtubules, actin filaments, and other cytoskeletal proteins in the development of neurons. The debate over the relative importance of microtubule and actin dynamics in the behavior of growth cones is not yet settled. While most low-level models have so far only incorporated either actin filaments or microtubules, both are required for normal growth cone behavior, and the complex interaction between the two is just beginning to be investigated.

In comparing various models of dendritic branching, it is important to underline that most of these models were built and tested using different real cell data. It may be hard to determine if a model is selected to fit the data, or vice versa. After the first modeling studies suggesting proximal interstitial branching of apical dendrites late in development (Carriquiry et al., 1991), experimental verification confirmed that interstitial branching may be as prominent (if not more) than terminal branching (Dailey and Smith, 1996). Few of the subsequent models based on "terminal-only" branching have been validated (or criticized) using apical data. Other subtle differences or selections in available or analyzed data of real neurons, such as terminal versus interbifurcation segment length or growth, could also affect the particular model design (Nowakowski et al., 1992).

In addition, compared to adult morphology, developmental data are too sparse to presently allow the design of a "complete" growth model. The number of biological parameters constraining outgrowth is huge. They include not only genetic determinants (which can be implemented, at least in principle, in algorithms), but also tightly controlled environmental influences. For example, dendritic and axonal growth are mutually dependent and also are affected by network activity, cellular metabolism, physical boundaries, and so on. Thus, a truly complete model of neuronal outgrowth would have to include molecular details and system level developmental dynamics.

Despite these limitations, computational modeling of neuronal morphology holds the potential of a deeper understanding of neuronal structure and how it relates to both cellular activity and macroscopic anatomy. While a restricted number of experimentally reconstructed neurons for each morphological class can be included in a comprehensive archive of brain data, stochastic models (and the corresponding statistical distributions) will be essential for the establishment of data-based knowledge of cellular morphology. Once the computational power becomes available, stochastic models will also constitute a practical means to generate real-scale, morphologically complete, cellular-level representations of the brain. This will likely lead to the generation of anatomically realistic interconnected neuronal networks for the study of system-level dynamic behavior. Notwithstanding the difficulty to include structural plasticity at multiple temporal and spatial scales in anatomical models, such an approach could fill an important explanatory gap between basic cellular processes and larger-scale brain activity (e.g., EEG and functional imaging).

The computational and graphical representation of neuronal structure already plays an important role in the education of new generations of neuroscientists. The type of data representation embodied in this approach is also suitable for applications in the search and analysis of pharmacologically active substances. The success of this approach will critically depend on the integration of cellular anatomy with other related neuroinformatics efforts, across modality and scale, including the characterization of four-dimensional gene expression, metabolic networks, and the biophysical determinants of electrophysiological activity, synaptic plasticity, system-level rhythms, and behavioral and cognitive attributes.

ACKNOWLEDGMENT

This work was supported by Human Brain Project grant R01-NS39600, jointly funded by NINDS and NIMH (National Institutes of Health).

References

Ahmad, F.J., Hughey, J., Wittmann, T., Hyman, A., Greaser, M., and Baas, P.W. (2000). Motor proteins regulate force interactions between microtubules and microfilaments in the axon. *Nat. Cell. Biol.* **2**(5), 276–280.

Aletta, J.M., and Greene, L.A. (1988). Growth cone configuration and advance: A time-lapse study using video-enhanced differential interference contrast microscopy. *J. Neurosci.* **8**(4), 1425–1435.

Ascoli, G.A. (1999). Progress and perspectives in computational neuroanatomy. *Anat. Rec.* **257**(6), 195–207.

Ascoli, G.A. (2002a). Neuroanatomical algorithms for dendritic modelling. *Network* **13**(3), 247–260.

Ascoli, G.A., ed. (2002b). *Computational Neuroanatomy: Principles and Methods.* Humana Press, Totowa, NJ.

Ascoli, G.A., and Krichmar, J.L. (2000). L-Neuron: A modeling tool for the efficient generation and parsimonious description of dendritic morphology. *Neurocomputing* **32–33**, 1003–1011.

Senft and Ascoli, 1999.

Ascoli, G.A., Krichmar, J.L., Nasuto, S.J., and Senft, S.L. (2001a). Generation, description and storage of dendritic morphology data. *Philos. Trans. R. Soc. London, Ser. B* **356**, 1131–1145.

Ascoli, G.A., Krichmar, J.L., Scorcioni, R., Nasuto, S.J., and Senft, S.L. (2001b). Computer generation and quantitative morphometric analysis of virtual neurons. *Anat. Embryol.* **204**(4), 283–301.

Baas, P.W. (1997). Microtubules and axonal growth. *Curr. Opin. Cell Biol.* **9**(1), 29–36.

Baas, P.W., Black, M.M., and Banker, G.A. (1989). Changes in microtubule polarity orientation during the development of hippocampal neurons in culture. *J. Cell Biol.* **109**(6 Pt 1), 3085–3094.

Bamburg, J.R. (1999). Proteins of the ADF/cofilin family: Essential regulators of actin dynamics. *Annu. Rev. Cell Dev. Biol.* **15**, 185–230.

Bentley, D., and Toroian-Raymond, A. (1986). Disoriented pathfinding by pioneer neurone growth cones deprived of filopodia by cytochalasin treatment. *Nature (London)* **323**, 712–715.

Berezovska, O., McLean, P., Knowles, R., Frosh, M., Lu, F.M., Lux, S.E., and Hyman, B.T. (1999). Notch1 inhibits neurite outgrowth in postmitotic primary neurons. *Neuroscience* **93**(2), 433–439.

Bishop, A.L., and Hall, A. (2000). Rho GTPases and their effector proteins. *Biochem. J.* **348**(Pt 2), 241–255.

Buck, K.B., and Zheng, J.Q. (2002). Growth cone turning induced by direct local modification of microtubule dynamics. *J. Neurosci.* **22**(21), 9358–9367.

Buettner, H.M. (1995). Computer simulation of nerve growth cone filopodial dynamics for visualization and analysis. *Cell Motil. Cytoskel.* **32**(3), 187–204.

Buettner, H.M., Pittman, R.N., and Ivins, J.K. (1994). A model of neurite extension across regions of nonpermissive substrate: Simulations based on experimental measurement of growth cone motility and filopodial dynamics. *Dev. Biol.* **163**(2), 407–422.

Burke, R.E., Marks, W.B., and Ulfhake, B. (1992). A parsimonious description of motoneuron dendritic morphology using computer simulation. *J. Neurosci.* **12**(6), 2403–2416.

Cannon, R.C., Turner, D.A., Pyapali, G.K., and Wheal, H.V. (1998). An on-line archive of reconstructed hippocampal neurons. *J. Neurosci. Methods* **84**(1–2), 49–54.

Cannon, R.C., Howell, F.W., Goddard, N.H., and De Schutter, E. (2002). Non-curated distributed databases for experimental data and models in neuroscience. *Network* **13**(3), 415–428.

Carriquiry, A.L., Ireland, W.P., Kliemann, W., and Uemura, E. (1991). Statistical evaluation of dendritic growth models. *Bull. Math. Biol.* **53**(4), 579–589.

Condron, B.G., and Zinn, K. (1997). Regulated neurite tension as a mechanism for determination of neuronal arbor geometries in vivo. *Curr. Biol.* **7**(10), 813–816.

Dailey, M.E., and Smith, S.J. (1996). The dynamics of dendritic structure in developing hippocampal slices. *J. Neurosci.* **16**(9), 2983–2994.

Dent, E.W., Callaway, J.L., Szebenyi, G., Baas, P.W., and Kalil, K. (1999). Reorganization and movement of microtubules in axonal growth cones and developing interstitial branches. *J. Neurosci.* **19**(20), 8894–8908.

Dityatev, A.E., Chmykhova, N.M., Studer, L., Karamian, O.A., Kozhanov, V.M., and Clamann, H.P. (1995). Comparison of the topology and growth rules of motoneuronal dendrites. *J. Comp. Neurol.* **363**(3), 505–516.

Donohue, D.E., Scorcioni, R., and Ascoli, G.A. (2002). Generation and description of neuronal morphology using L-Neuron: A case study. In *Computational Neuroanatomy: Principles and Methods*, (G. Ascoli, ed.), pp. 49–70. Humana Press, Totowa, NJ.

Fan, J., and Raper, J.A. (1995). Localized collapsing cues can steer growth cones without inducing their full collapse. *Neuron* **14**(2), 263–274.

Fukushima, N., Kimura, Y., and Chun, J. (1998). A single receptor encoded by vzg-1/lpA1/edg-2 couples to G proteins and mediates multiple cellular responses to lysophosphatidic acid. *Proc. Natl. Acad. Sci. U.S.A.* **95**(11), 6151–6156.

Fukushima, N., Weiner, J.A., and Chun, J. (2000). Lysophosphatidic acid (LPA) is a novel extracellular regulator of cortical neuroblast morphology. *Dev. Biol.* **228**(1), 6–18.

Fukushima, N., Ishii, I., Habara, Y., Allen, C.B., and Chun, J. (2002a). Dual regulation of actin rearrangement through lysophosphatidic acid receptor in neuroblast cell lines: Actin depolymerization by Ca(2+)-alpha-actinin and polymerization by rho. *Mol. Biol. Cell.* **13**(8), 2692–2705.

Fukushima, N., Weiner, J.A., Kaushal, D., Contos, J.J., Rehen, S.K., Kingsbury, M.A., Kim, K.Y., and Chun, J. (2002b). Lysophosphatidic acid influences the morphology and motility of young, postmitotic cortical neurons. *Mol. Cell. Neurosci.* **20**(2), 271–282.

Gallo, G., and Letourneau, P.C. (2000). Neurotrophins and the dynamic regulation of the neuronal cytoskeleton. *J. Neurobiol.* **44**(2), 159–173.

Gao, F.B., Brenman, J.E., Jan, L.Y., and Jan, Y.N. (1999). Genes regulating dendritic outgrowth, branching, and routing in *Drosophila. Genes Dev.* **13**(19), 2549–2561.

Goddard, N.H., Cannon, R.C., and Howell, F.W. (2003). Axiope tools for data management and data sharing. *Neuroinformatics* **1**, 271–284.

Goldberg, D.J., and Burmeister, D.W. (1986). Stages in axon formation: Observations of growth of *Aplysia* axons in culture using video-enhanced contrast-differential interference contrast microscopy. *J. Cell Biol.* **103**(5), 1921–1931.

Goodhill, G.J., and Urbach, J.S. (1999). Theoretical analysis of gradient detection by growth cones. *J. Neurobiol.* **41**(2), 230–241.

Hely, T.A., Graham, B., and Ooyen, A.V. (2001). A computational model of dendrite elongation and branching based on MAP2 phosphorylation. *J. Theor. Biol.* **210**(3), 375–384.

Hillman, D.E. (1979). Neuronal shape parameters and substructures as a basis of neuronal form. In *The Neurosciences: Fourth Study Program* (F. Schmitt, ed.), pp. 477–498. MIT Press, Cambridge, MA.

Ireland, W., Heidel, J., and Uemura, E. (1985). A mathematical model for the growth of dendritic trees. *Neurosci. Lett.* **54**(2–3), 243–249.

Jaeger, D. (2001). Accurate reconstruction of neuronal morphology. In *Computational Neuroscience: Realistic Modeling for Experimentalists* (E. De Schutter, ed.) CRC Press, Boca Raton, FL, pp. 159–178.

Jaffe, L.F., and Poo, M.M. (1979). Neurites grow faster towards the cathode than the anode in a steady field. *J. Exp. Zool.* **209**(1), 115–128.

Jalink, K., Eichholtz, T., Postma, F.R., van Corven, E.J., and Moolenaar, W.H. (1993). Lysophosphatidic acid induces neuronal shape changes via a novel, receptor-mediated signaling pathway: Similarity to thrombin action. *Cell Growth Differ.* **4**(4), 247–255.

Kalb, R.G. (1994). Regulation of motor neuron dendrite growth by NMDA receptor activation. *Development* **120**(11), 3063–3071.

Kliemann, W. (1987). A stochastic dynamical model for the characterization of the geometrical structure of dendritic processes. *Bull. Math. Biol.* **49**(2), 135–152.

Kobayashi, N., and Mundel, P. (1998). A role of microtubules during the formation of cell processes in neuronal and non-neuronal cells. *Cell Tissue Res.* **291**(2), 163–174.

Krichmar, J.L., Nasuto, S.J., Scorcioni, R., Washington, S.D., and Ascoli, G.A. (2002). Effects of dendritic morphology on CA3 pyramidal cell electrophysiology: A simulation study. *Brain Res.* **941**(1–2), 11–28.

Kryl, D., Yacoubian, T., Haapasalo, A., Castren, E., Lo, D., and Barker, P.A. (1999). Subcellular localization of full-length and truncated Trk receptor isoforms in polarized neurons and epithelial cells. *J. Neurosci.*, **19**(14), 5823–5833.

Lasek, R.J., and Black, M.M. (1988). *Intrinsic Determinants of Neuronal Form and Function.* Liss., New York.

Lazarewicz, M.T., Boer-Iwema, S., and Ascoli, G.A. (2002). Practical aspects in anatomically accurate simulations of neuronal electrophysiology. In *Computational Neuroanatomy: Principles and Methods* (G. Ascoli, ed.), pp. 127–148. Humana Press, Totowa, NJ.

Leclerc, N., Baas, P.W., Garner, C.C., and Kosik, K.S. (1996). Juvenile and mature MAP2 isoforms induce distinct patterns of process outgrowth. *Mol. Biol. Cell.* **7**(3), 443–455.

Li, G.H., and Qin, C.D. (1996). A model for neurite growth and neuronal morphogenesis. *Math. Biosci.* **132**(1), 97–110.

Li, G.H., Qin, C.D., and Wang, Z.S. (1992). Neurite branching pattern formation: modeling and computer simulation. *J. Theor. Biol.* **157**(4), 463–486.

Li, G.H., Qin, C.D., and Li, M.H. (1994). On the mechanisms of growth cone locomotion: Modeling and computer simulation. *J. Theor. Biol.* **169**(4), 355–362.

Lin, C.H., and Forscher, P. (1993). Cytoskeletal remodeling during growth cone-target interactions. *J. Cell Biol.* **121**(6), 1369–1383.

Lin, C.H., and Forscher, P. (1995). Growth cone advance is inversely proportional to retrograde F-actin flow. *Neuron* **14**(4), 763–771.

Lin, C.H., Espreafico, E.M., Mooseker, M.S., and Forscher, P. (1996). Myosin drives retrograde F-actin flow in neuronal growth cones. *Neuron* **16**(4), 769–782.

Luo, L. (2002). Actin cytoskeleton regulation in neuronal morphogenesis and structural plasticity. *Annu. Rev. Cell Dev. Biol.* **18**, 601–635.

Macias, M.Y., Battocletti, J.H., Sutton, C.H., Pintar, F.A., and Maiman, D.J. (2000). Directed and enhanced neurite growth with pulsed magnetic field stimulation. *Bioelectromagnetics* **21**(4), 272–286.

Maletic-Savatic, M., Malinow, R., and Svoboda, K. (1999). Rapid dendritic morphogenesis in CA1 hippocampal dendrites induced by synaptic activity. *Science* **283**, 1923–1927.

Nathke, I.S., Adams, C.L., Polakis, P., Sellin, J.H., and Nelson, W.J. (1996). The adenomatous polyposis coli tumor suppressor protein localizes to plasma membrane sites involved in active cell migration. *J. Cell Biol.* **134**(1), 165–179.

Nowakowski, R.S., Hayes, N.L., and Egger, M.D. (1992). Competitive interactions during dendritic growth: A simple stochastic growth algorithm. *Brain Res.* **576**(1), 152–156.

Polleux, F., Morrow, T., and Ghosh, A. (2000). Semaphorin 3A is a chemoattractant for cortical apical dendrites. *Nature (London)* **404**, 567–573.

Rajan, I., and Cline, H.T. (1998). Glutamate receptor activity is required for normal development of tectal cell dendrites in vivo. *J. Neurosci.* **18**(19), 7836–7846.

Rall, W., Burke, R.E., Holmes, W.R., Jack, J.J, Redman, S.J., and Segev, I. (1992). Matching dendritic neuron models to experimental data. *Physiol. Rev.* **72**(4 Suppl.), S159–S186.

Redmond, L., and Ghosh, A. (2002). The role of Notch and Rho GTPase signaling in the control of dendritic development. *Curr. Opin. Neurobiol.* **11**(1), 111–117.

Redmond, L., Kashani, A.H., and Ghosh, A. (2002). Calcium regulation of dendritic growth via CaM kinase IV and CREB-mediated transcription. *Neuron* **34**(6), 999–1010.

Robert, M.E., and Sweeney, J.D. (1997). Computer model: Investigating role of filopodia-based steering in experimental neurite galvanotropism. *J. Theor. Biol.* **188**(3), 277–288.

Samsonovich, A.V., and Ascoli, G.A. (2003). Statistical morphological analysis of hippocampal principal neurons indicates cell-specific repulsion of dendrites from their own cell. *J. Neurosci. Res.* **71**(2), 173–187.

Schaefer, A.W., Kabir, N., and Forscher, P. (2002). Filopodia and actin arcs guide the assembly and transport of two populations of microtubules with unique dynamic parameters in neuronal growth cones. *J. Cell Biol.* **158**(1), 139–152.

Scorcioni, R., and Ascoli, G.A. (2001). Algorithmic extraction of morphological statistics from electronic archives of neuroanatomy. *Lect. Notes Comput. Sci.* **2084**, 30–37.

Segev, R., and Ben-Jacob, E. (2000). Generic modeling of chemotactic based self-wiring of neural networks. *Neural Networks.* **13**(2), 185–199.

Senft, S.L. (2002). Axonal navigation through voxel substrates. A strategy for reconstructing brain circuitry. In *Computational Neuroanatomy: Principles and Methods* G.A. (Ascoli, ed.). Humana Press, Totowa, NJ, pp. 245–270.

Senft, S.L., and Ascoli, G.A. (1999). Reconstruction of brain networks by algorithmic amplification of morphometry data. *Lect. Notes Comput. Sci.* **1606**, 25–33.

Sestan, N., Artavanis-Tsakonas, S., and Rakic, P. (1999). Contact-dependent inhibition of cortical neurite growth mediated by notch signaling. *Science* **286**, 741–746.

Solomon, F., and Magendantz, M. (1981). Cytochalasin separates microtubule disassembly from loss of asymmetric morphology. *J. Cell Biol.* **89**(1), 157–161.

Song, H., Ming, G., He, Z., Lehmann, M., McKerracher, L., Tessier-Lavigne, M., and Poo, M. (1998). Conversion of neuronal growth cone responses from repulsion to attraction by cyclic nucleotides. *Science* **281**, 1515–1518.

Sumi, T., Matsumoto, K., Shibuya, A., and Nakamura, T. (2001). Activation of LIM kinases by myotonic dystrophy kinase-related Cdc42-binding kinase alpha. *J. Biol. Chem.* **276**(25), 23092–23096.

Suter, D.M., Errante, L.D., Belotserkovsky, V., and Forscher, P. (1998). The Ig superfamily cell adhesion molecule, apCAM, mediates growth cone steering by substrate-cytoskeletal coupling. *J. Cell Biol.* **141**(1), 227–240.

Tang, B.L. (2003). Inhibitors of neuronal regeneration: Mediators and signaling mechanisms. *Neurochem. Int.* **42**(3), 189–203.

Turner, D.A., Cannon, R.C., and Ascoli, G.A. (2002). Web-based neuronal archives: Neuronal morphometric and electrotonic analysis. In *Neuroscience Databases—A Practical Guide,* (R. Kotter, ed.), pp. 81–98. Elsevier, Amsterdam.

Uemura, E., Carriquiry, A., Kliemann, W., and Goodwin, J. (1995). Mathematical modeling of dendritic growth in vitro. *Brain Res.* **671**(2), 187–194.

van Ooyen, A. (2001). Competition in the development of nerve connections: A review of models. *Network* **12**(1), R1–R47.

van Pelt, J., and Verwer, R.W. (1986). Topological properties of binary trees grown with order-dependent branching probabilities. *Bull. Math. Biol.* **48**(2), 197–211.

van Pelt, J., Dityatev, A.E., and Uylings, H.B. (1997). Natural variability in the number of dendritic segments: Model-based inferences about branching during neurite outgrowth. *J. Comp. Neurol.* **387**(3), 325–340.

van Veen, M.P., and van Pelt, J. (1994a). Dynamic mechanisms of neuronal outgrowth. *Prog. Brain Res.* **102**, 95–108.

van Veen, M.P., and van Pelt, J. (1994b). Neuritic growth rate described by modeling microtubule dynamics. *Bull. Math. Biol.* **56**(2), 249–273.

Vercelli, A., Assal, F., and Innocenti, G.M. (1992). Emergence of callosally projecting neurons with stellate morphology in the visual cortex of the kitten. *Exp. Brain Res.* **90**(2), 346–358.

Verhage, M., Maia, A.S., Plomp, J.J., Brussaard, A.B., Heeroma, J.H., Vermeer, H., Toonen, R.F., Hammer, R.E., van den Berg, T.K., Missler, M., Geuze, H.J., and Sudhof, T.C. (2000). Synaptic assembly of the brain in the absence of neurotransmitter secretion. *Science* **287**, 864–869.

Whitford, K.L., Dijkhuizen, P., Polleux, F., and Ghosh, A. (2002). Molecular control of cortical dendrite development. *Annu. Rev. Neurosci.* **25**, 127–149.

Wu, G.Y., and Cline, H.T. (1998). Stabilization of dendritic arbor structure in vivo by CaMKII. *Science* **279**, 222–226.

Yacoubian, T.A., and Lo, D.C. (2000). Truncated and full-length TrkB receptors regulate distinct modes of dendritic growth. *Nat. Neurosci.* **3**(4), 342–349.

Zheng, J.Q., Wan, J.J., and Poo, M.M. (1996). Essential role of filopodia in chemotropic turning of nerve growth cone induced by a glutamate gradient. *J. Neurosci.* **16**(3), 1140–1149.

Zhou, F.Q., Waterman-Storer, C.M., and Cohan, C.S. (2002). Focal loss of actin bundles causes microtubule redistribution and growth cone turning. *J. Cell Biol.* **157**(5), 839–849.

Zumbrunn, J., Kinoshita, K., Hyman, A.A., and Nathke, I.S. (2001). Binding of the adenomatous polyposis coli protein to microtubules increases microtubule stability and is regulated by GSK3 beta phosphorylation. *Curr. Biol.* **11**(1), 44–49.

Web Resources

L-Neuron: http://www.krasnow.gmu.edu/L-Neuron (case-sensitive). Includes the Virtual Neuromorphology Electronic Database, a collection of real and simulated neurons, the L-Neuron and L-Measure software programs, to generate and measure neuronal morphology, and other neuroinformatics tools.

Compneuro: http://www.compneuro.org/CDROM/nmorph/cellArchive.html. Includes the Duke/Southampton Archive of Cellular Morphology, as well as the Cvapp visualization and editing tool.

Claiborne's archive: http://www.utsa.edu/claibornelab. A collection of reconstructed principal cells from the hippocampus.

Gulyas' archive: http://www.koki.hu/~gulyas/ca1cells/index.htm. A collection of reconstructed CA1 pyramidal cells and interneurons.

Neuron_morpho: http://www.maths.soton.ac.uk/staff/D'Alessandro/morpho. A Java plugin for ImageJ to digitally reconstruct neurites from image stacks.

DATABASE APPLICATIONS

PART

III

The Neocortical Microcircuit Database (NMDB)

Henry Markram, Ph.D., Xiaozhong Luo, Ph.D., Gilad Silberberg, Ph.D., Maria Toledo-Rodriguez, Ph.D., and Anirudh Gupta, Ph.D.

CONTENTS

Databasing the Brain. Edited by Stephen H. Koslow and Shankar Subramaniam
ISBN 0-471-30921-4 © 2005 John Wiley & Sons, Inc.

19.1 INTRODUCTION

The neocortex is a major part of the mammalian brain and is made up of a continuous microcircuit sheet of stereotypic neuron types that are also intricately interconnected in a stereotypical manner (Ramón y Cajál, 1911; Lorente de Nó, 1938; Peters and Jones, 1984; Peters, 1987; White, 1989; DeFelipe, 1997; Thomson and Deuchars, 1997; Somogyi et al., 1998; Gupta et al., 2000; Thomson and Bannister, 2003; Toledo-Rodriguez et al., 2003). During sensory input, assemblies of neurons dynamically form to parcellate the neocortex into functionally specialized modules allowing sensory processing (Hubel and Wiesel, 1977; Szentagothai, 1978; Eccles, 1984; Purves et al., 1992; Shmuel and Grivald, 1996; Mountcastle, 1998). The microcircuitry of these modules lies at the heart of the information processing capability of the neocortex enabling perception, attention, memory, and higher cognitive functions. The *Neocortical Microcircuit Database* (NMDB; http://microcircuit.epfl.ch) has been constructed in order to organize the anatomical, physiological, and molecular properties of neocortical microcircuits.

The NMDB is based on numerically breaking down key properties of neurons and connections in order to produce "profiles" which form the foundation of the database. Neurons are characterized in terms of their morphological, physiological, and gene expression profiles. Synaptic connections are characterized in terms of their morphological and physiological profiles. Neuron morphology profiles are obtained from a detailed morphometric breakdown of 3D-reconstructed neurons *(m-profiles)*, neuron physiology profiles are obtained from a detailed measurement of the electrophysiological responses to a series of stimulus protocols *(e-profiles)*, and neuron gene expression profiles are obtained from single-cell RT-PCR data *(g-profiles)* and in the near future will be obtained from gene-chips. Synaptic connections are characterized by the identity of the pre- and postsynaptic neurons *(sn-profile)*, the anatomy of synaptic connections are characterized by the axonal and dendritic location of light microscopically identified putative synapses *(sm-profile)*, and the physiology of synaptic connections are characterized by a profile of electrophysiological parameters obtained from recordings of postsynaptic potential and current responses (PSPs and PSCs) to a series of stimulation protocols applied to the presynaptic neuron *(se-profile)*.

The development of the database will follow a series of phases. In the first phase, the database is focused on indexing neuronal and synaptic profiles. In the second phase, canonical neuron types and synaptic connections will be defined. In the third phase, the distribution and relative numbers of neurons will be defined for the different layers of the neocortex, and in the final phase the microcircuit will be fully reconstructed for visual guided exploration of its structure and physiology. In parallel to the NMDB, a database of mathematical models of the different types of neurons, synaptic connections, and various-sized microcircuits is being constructed to eventually allow visualization of the activated neocortical microcircuit. In later phases, various long-range input and output pathways into neocortical microcircuits will be incorporated into the database to enable linking of microcircuits between different brain areas and regions. The NMDB is currently populated with microcircuit data primarily of juvenile rat somatosensory cortex, but data from any species, area, and age can be entered.

19.2 NEOCORTICAL MICROCIRCUIT DATA

The data for the neocortical microcircuit is organized into the following profiles:

19.2.1 Experiment Profile

The experiment profile *(x-profile)* of a neuron provides general information about the cell. It serves as a general I.D. of each cell, including:

Experiment Date.

Cell Name. This is denoted by experimenter and used as the main I.D. in the database.

Layer. Laminar location of the soma.

Anatomical Type. Subjective anatomical classification.

Electrophysiological Type. Subjective electrophysiological classification.

Genetic Type. Description of main molecular properties.

Connections. This indicates whether there are any synaptic connections with other cells.

Reconstructed. This tells whether or not the cell was 3D-computer reconstructed.

Experimenter. Full name and affiliation of the person who performed the experiment.

Animal Species, Neocortical Area, Age, Weight, and Sex.

Comments. Notes and other information that describes the cell and its characteristics.

19.2.2 Neuronal Profiles

A functional module in many species can take the form of a column of the neocortex that is around 300–500 μm in diameter and 1–2 mm in height (Mountcastle, 1957, 1998; Hubel and Wiesel, 1962; Jones et al., 1975; Goldman and Nauta, 1977). This diameter corresponds well with the dimensions of the basal axonal and dendritic local clusters of pyramidal neurons. In the rat there are about 10,000 neurons in such a column (Ren et al., 1992; Beaulieu, 1993). Pyramidal cells (PCs) can display some heterogeneity in morphology and electrophysiology, especially across layers (DeFelipe and Farinas, 1992; Callaway, 1998), but their most important differences are in terms of their input and output pathways (Jones, 1981; White, 1989; Hubener et al., 1990). Interneurons are a highly heterogeneous neuronal population (DeFelipe, 1993, 2002; Cauli et al., 1997; Kawaguchi and Kubota, 1997; Gupta et al., 2000; Thomson and Bannister, 2003; Toledo-Rodriguez et al., 2003 with nine major anatomical classes, at least 15 electrophysiological classes, and more than 15 different molecular classes based on (co-) expression of neuropeptides and calcium binding proteins. While the NMDB provides a subjective classification scheme only for operational purposes, such as conveniently searching for cells with a particular general behavior and naming pre- and postsynaptic neurons involved in connections, the database is mostly based on objective numerical profiles of various neuronal and synaptic properties allowing independent analyses. Neurons are characterized in terms of their *m-profiles, e-profiles, and g-profiles*.

m-Profile

To obtain the *m-profile* for a neuron, the neuron must be fully 3D-computer reconstructed and converted into Neurolucida format (Glaser and Glaser, 1990). This 3D-model must be uploaded into the database, and a database tool automatically performs an extensive morphometric analysis on the model neuron and enters the *m-profile* into the database. This method ensures a standardized *m-profile* for all neurons.

Conventions. In the morphometric analysis, the axonal and dendritic arbors are described in terms of trees, segments, branch orders, nodes, and terminals (see Figure 19.1).

- *Trees.* The axon and dendrites (including apical dendrite) of a neuron are structurally described as trees, which emerge from the somata (Figure 19.1a). Three types of trees are defined: axonal, dendritic, and apical dendritic. Normally, there is only one axonal tree, one or more dendritic trees, and one apical dendrite tree, but some cells can have more than one axon tree or lack any one of the three kinds because of an incomplete reconstruction.
- *Segment and Branch Order.* A segment indicates a section of a branch, either end of which is a root, a node, or a terminal (see next paragraph). No branching occurs between the starting and ending point of a segment. Segments are labeled with order numbers that represent how far they are from the root, not measured by actual distance but by how many nodes they have passed (Figure 19.1a). For example, a root segment has an order number of 1. Segments branched from the root segment share the order number of 2. Segments that go further have the order number of 3, and so on. An order number does not, therefore, uniquely identify a *single* segment; instead, it identifies a *group* or a *layer* of segments, and there could be many segments in a tree or across several trees sharing the same order number (Figure 19.1a).
- *Node and Terminal.* A node is the ending point of a segment and starting point of linked segments (not the ending point of a branch), whereas a terminal denotes not only the ending point of a segment, but also that of the branch of a tree (i.e., there will be no segment next to it; see Figure 19.1a).

The Morphometric Analysis. The neuron analysis tool actually performs eight kinds of analyses according to different points of view on each of the three tree types.

- *Neuron Summary.* Provides a summary of the analysis for all trees of each type. For example, the dendrite has four trees and 10 nodes, and the total length of all of these four trees is

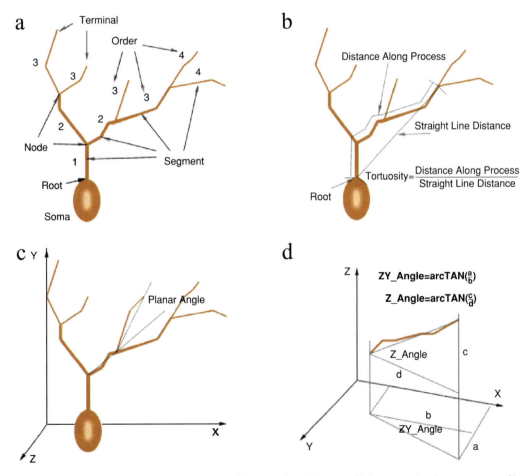

Figure 19.1   Morphometric analysis. Definitions of the (**a**) analyzed elements, (**b**) the tortuousity of a tree or a part of it, (**c**) the planar angle of a segment, and (**d**) the angular projections on the Cartesian axes.

1152.449 μm, the mean length of these four trees is 288.112 μm, and so on.

- **Segment Analysis.** This generates data on every single segment which are identified by a combination of tree type, tree number, and order number. Here, end-type "Branch" means that the ending point of the segment is a node, therefore there will be another segment next to it, while all others end types are terminals. For example, a segment that is dendritic tree #1 and order #1 has a length of 10.289 μm, an XY angle of 76.686 degrees, and the end is a "Branch."

- **Terminal Analysis.** This focuses on nodes and terminals of every single tree. The data show how many nodes and terminals there are in every tree of a certain type, and the length and mean length and other parameters are applied to these nodes or terminals. Here, the node length is the total length of segments that have no terminal ends, and the terminal length is for those segments that have terminal ends.

- **Node Analysis.** This obtains the distance along the process from every node to the tree root, the straight-line distance from the node to the root, and the tortuousity of this node, which is the former distance divided by the latter one, showing how much the process curves (see Figrue 19.1b).

- **Tree Analysis.** This provides information about every group of segments that have the same order number within each tree.

- **Individual Analysis.** This produces data about all segments in every tree, such as the total length and the mean segment length.

- **Tree Totals.** This outputs data about every group of segments that have the same order number in all trees of each type, *not* in each tree. The length is total length of all segments with the same order number on all trees of a certain type.

- **Sholl Analysis.** This is a useful tool for quantifying structural properties such as the degree and areas of arborization of dendritic and axonal trees. The analysis is carried out by means of dividing the space occupied by the neuron into finely spaced concentric spheres, the center of which is the soma (Sholl, 1956). Data from the Sholl analysis include the structural analysis of each type of trees, providing the total length, number of intersections, nodes, ends, and so on, in each intersphere space ("Sholl shell").

The detailed morphometric analysis extracts more than 70 values:

- **EachTreeLengthMean.** Mean of lengths of segments with same orders in each tree.
- **EachTreeLengthVar.** Standard deviation of lengths of segments with same orders in each tree.
- **EachTreeOrderLength.** Sum of lengths of segments with same orders in each tree.
- **EachTreeOrderNodes.** Number of nodes with same orders in a tree.
- **EachTreeOrderQty.** Number of segments with same orders in each tree.
- **EachTreeOrderSurface.** Sum of surfaces of segments with same orders in each tree.
- **EachTreeOrderVolume.** Sum of volumes of segments with same orders in each tree.

- **EachTreeSurfaceMean.** Mean of surfaces of segments with same orders in each tree.
- **EachTreeSurfaceVar.** Standard deviation of surfaces of segments with same orders in each tree.
- **EachTreeVolumeMean.** Mean of volumes of segments with same orders in each tree.
- **EachTreeVolumeVar.** Standard deviation of volumes of segments with same orders in each tree.
- **EndingQty.** Number of ends on all trees.
- **IndivTreeLengthMean.** Mean of segment lengths in a tree.
- **IndivTreeLengthVar.** Standard deviation of segment lengths in a tree.
- **IndivTreeOrderNodes.** Number of nodes in a tree.
- **IndivTreeOrderSurface.** Total surfaces of segments in a tree.
- **IndivTreeOrderVolume.** Total volumes of segments in a tree.
- **IndivTreeSurfaceMean.** Mean of segment surfaces in a tree.
- **IndivTreeSurfaceVar.** Standard deviation of segment surfaces in a tree.
- **IndivTreeTotalLength.** Total length of segments in a tree.
- **IndivTreeVolumeMean.** Mean of segment volumes in a tree.
- **IndivTreeVolumeVar.** Standard deviation of segment volumes in a tree.
- **LengthMeanOfNode.** Mean of lengths from the farthest node of every branch to the root of each tree, not including segments with endings.
- **LengthMeanOfTerm.** Mean of lengths from every ending to the root of each tree.
- **LengthOfNode.** Sum of all lengths from the farthest node of every branch to the root of each tree, not including segments with endings.
- **LengthOfTerminal.** Sum of all lengths from every ending to the root of each tree.
- **LengthVarOfNode.** Standard deviation of lengths from the farthest node of every branch to the root of each tree, not including segments with endings.
- **LengthVarOfTerm.** Standard deviation of lengths from every ending to the root of each tree.
- **MaxOrderOfTrees.** Maximal segment order in all trees.
- **MeanLengthOfTrees.** Mean length of all segments of trees.
- **MeanOrderOfTrees.** Mean of segment orders of trees.
- **NodeDistanceOfProcess.** Distance from each node to the root of a tree.
- **NodeProcessTortuosity.** Tortuosity of process from each node to the root of a tree.
- **NodeQty.** Number of nodes on all trees.
- **NodeStraightDistance.** Straight line distance from each node to the start of the root of a tree.
- **PlanarAngle.** Angle between a segment and extension of its parent segment (Figure 19.1c).
- **SegmentBaseDiameter.** Diameter at the start point of a segment.
- **SegmentLength.** Segment length of a tree.
- **SegmentOrder.** Segment order within a tree.
- **SegmentSurface.** Segment surface of a tree.

- *SegmentTortuosity.* Segment tortuosity of a tree.
- *SegmentVolume.* Segment volume of a tree.
- *SurfaceMeanOfNode.* Mean of surfaces from the farthest node of every branch to the root of each tree, not including segments with endings.
- *SurfaceMeanOfTerm.* Mean of surfaces from every ending to the root of each tree.
- *SurfaceMeanOfTrees.* Mean surface of segments of trees.
- *SurfaceOfNode.* Sum of all surfaces from the farthest node of every branch to the root of each tree, not including segments with endings.
- *SurfaceOfTerminal.* Sum of all surfaces from every ending to the root of each tree.
- *SurfaceVarOfNode.* Standard deviation of surfaces from the farthest node of every branch to the root of each tree, not including segments with endings.
- *SurfaceVarOfTerm.* Standard deviation of surfaces from every ending to the root of each tree.
- *TotalLengthOfTrees.* Total length of all segments of trees.
- *TotalSurfaceOfTrees.* Total surface of all segments of trees.
- *TotalVolumeOfTrees.* Total volume of all segments of trees.
- *TreeQty.* Number of trees.
- *TreeTotaLengthVar.* Standard deviation of segment lengths with same orders of all trees.
- *TreeTotalLengthMean.* Mean of segment lengths with same orders of all trees.
- *TreeTotalOrderLength.* Total length of segments with same orders of all trees.
- *TreeTotalOrderNodes.* Number of nodes with same orders of all trees.
- *TreeTotalOrderQty.* Number of segments with same orders of all trees.
- *TreeTotalOrderSurface.* Total surface of segments with same orders of all trees.
- *TreeTotalOrderVolume.* Total volume of segments with same orders of all trees.
- *TreeTotalSurfaceMean.* Mean of segment surfaces with same orders of all trees.
- *TreeTotalSurfaceVar.* Standard deviation of segment surfaces with same orders of all trees.
- *TreeTotalVolumeMean.* Mean of segment volumes with same orders of all trees.
- *TreeTotalVolumeVar.* Standard deviation of segment volumes with same orders of all trees.
- *VolumeMeanOfNode.* Mean of volumes from the farthest node of every branch to the root of each tree, not including segments with endings.
- *VolumeMeanOfTerm.* Mean of volumes from every ending to the root of each tree.
- *VolumeMeanOfTrees.* Mean volume of segments of trees.
- *VolumeOfNode.* Sum of all volumes from the farthest node of every branch to the root of each tree, not including segments with endings.
- *VolumeOfTerminal.* Sum of all volumes from every ending to the root of each tree.

- *VolumeVarOfNode.* Standard deviation of volumes from the farthest node of every branch to the root of each tree, not including segments with endings.
- *VolumeVarOfTerm.* Standard deviation of volumes from every ending to the root of each tree.
- *XY_Angle.* Angle between projection of a segment on XY plane and X axis (Figure 19.1d).
- *Z_Angle.* Angle between a segment and its projection on XY plane (Figure 19.1d).

e-Profile

The electrophysiological profile of neurons is obtained by applying a series of different stimulation pulses to a neuron during intracellular or whole-cell patch-clamp recordings. The responses to these pulses are measured to obtain a spectrum of electrophysiological parameters (EPs). The database currently provides entry of as many as 140 such EPs. This array of EPs is referred to as the *e-profile*. Due to inter-lab differences in the methods used to record from cells, the EPs are not analyzed automatically by a resident tool in the database. At a later stage a tool will be available for automatic analysis of a series of text files that contain responses to prescribed pulse patterns. Nevertheless, the database provides a template table into which the data can be entered for as many EPs as possible. In addition to filling out the *e-profile* table, images of electrical responses and files containing raw traces can be uploaded into the database for online viewing or for download. In addition to the *e-profile*, a subjective classification of the overall behavior must be entered by the user. The general behavior is defined according to the neuronal discharge response to step current injections (Gupta et al., 2000). The employed definitions of neuronal discharge responses extend and refine previous classification schemes; see Toledo-Rodriguez et al. (2003) for comparison of different classification schemes.

The electrophysiological profile of neocortical neurons was obtained from their responses to various current injections. We aimed to examine the electrophysiology in sufficient detail to capture most of the active and passive biophysical properties of a neuron. For this purpose a set of stimulation protocols for the extraction of more than 140 EPs was created (Figure 19.2).

- *ADP.* After depolarization generated by a short burst of APs.
- *AP_Drop.* Change in the AP amplitude upon injection of a strong current pulse (Figure 19.2c). When analyzed this stimulus set gives rise to the following electrical parameters.
  - Drop in first to second AP: the decrease in the amplitude of the second AP evoked by a step current pulse that is five times the threshold current.
  - Change to steady state: the change in the AP amplitude during a burst of APs evoked by a step current pulse that is five times threshold current.
  - Change to steady state after second AP: the change in the AP amplitude from the second AP during a burst of APs evoked by a step current pulse that is five times threshold current.
  - Maximum rate of AP change: maximum rate of change in the AP amplitude during a burst evoked by five times threshold current.

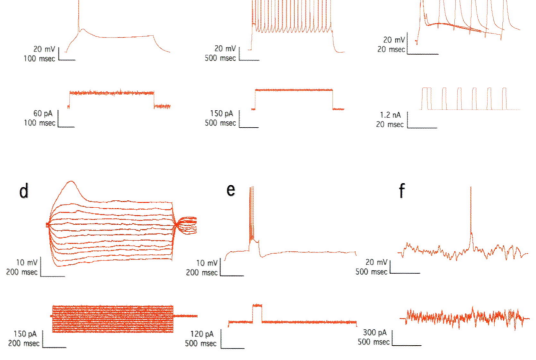

Figure 19.2   The *e-profile*. Some of the different electrophysiological stimulation protocols employed for the extraction of the *e-profile*. Voltage traces (*upper panels*) and current traces (*lower panels*) of (**a**) threshold step, (**b**) supra-threshold step, (**c**) AP recovery test, (**d**) *IV* curves, (**e**) s-AHP, and (**f**) noise filtering.

- **AP_Threshhold.** Injection of a ramp current to determine the threshold of AP generation. When analyzed, this stimulus set gives rise to the following electrical parameters:
  - AP threshold: threshold for discharge APs during a ramp depolarization.
  - AHP after first AP in the ramp: amplitude of the AHP after the first AP generated by a ramp current pulse.
- **AP_Wave Form.** Shape of the first two APs generated just above threshold. When analyzed, this stimulus set gives rise to the following electrical parameters:
  - (1st or 2nd) AP amplitude: average amplitude of the first or second AP.
  - AP duration: average time for the first or second AP from onset to offset. AP onset is defined as the time near the voltage inflection when the voltage changes more than 0.5 mV within 50 μs while the offset is defined as the return of V_m to the onset value.
  - (1st or 2nd) AP duration half-width; average time for the first or second AP from half amp to the same voltage occurring during AP fall phase.
  - (1st or 2nd) AP rise time: first or second AP duration from onset to the peak.
  - (1st or 2nd) AP fall time: first or second AP duration from peak to the offset.
  - (1st or 2nd) AP rise rate: first or second AP_amp/rise time.
  - (1st or 2nd) AP fall rate: first or second AP_amp/fall time.
  - Fast AHP: amplitude from first or second AP onset to minimum voltage.

- **AP_Change.** From the analysis of the change between first and second APs the following electrical parameters are obtained:
  - Change in AP amplitude: percent change in AP amplitude between the first and second AP.
  - Change in AP duration: percent change in AP duration between the first and second AP.
  - Change in AP duration half-width: percent change in AP duration half width between the first and second AP.
  - Change in AP rise rate: percent change in AP rise rate between the first and second AP.
  - Change in AP fall rate: percent change in AP fall rate between the first and second AP.
  - Change in AP fast AHP: percent change in AHP amplitude between the first and second AP.
- **Delta.** Membrane time constant for brief hyperpolarizing current pulses, which, when analyzed, gives rise to the electrical parameter delta average decay time constant, the time constant for the membrane to depolarize after a brief hyperpolarizing current injection.
- **Discharge.** Long depolarizing pulses to determine AP discharge patterns (Figures 19.2a, 19.2b). When analyzed, this stimulus set gives rise to the following electrical parameters:
  - Slope of ID threshold: the slope of the current–discharge relationship from rest.
  - Average delay to first AP: time from current pulse onset to first AP.
  - SD of delay to first AP: SD of delays to first AP.

- Average delay to second AP: the average delay for the cell to generate a second AP.
- SD of delay to second AP: SD of delays to second AP; Average initial burst interval, average interspike interval for the first three APs.
- SD of average initial burst interval: SD of the average interspike interval for the first three APs.
- Average initial accommodation: the initial change in the interspike interval during a burst.
- Average steady-state accommodation: the steady-state degree to which the interspike interval changes during a burst.
- Rate of accommodation to steady state.
- Average accommodation at steady state.
- Average rate of accommodation during steady state.
- CV: the average coefficient of variation of the AP discharge.
- Average skew discharge: reflects the distribution of the CV of the discharge.
- Average discharge stuttering: a measure of the degree to which the AP discharge "stutters" or changes its frequency.
- *Discharge_Depol.* Depolarizing currents to determine discharge during depolarization.
- *Discharge_Hyperpol.* Depolarizing currents to determine discharge during hyperpolarization.
- *Discharge_Threshold.* Depolarizing currents near AP threshold. When analyzed this stimulus set gave rise to the following electrical parameters;
  - Slope of ID threshold: the minimum current required to drive the cell to threshold.
  - Average discharge at threshold: the average discharge at depolarization currents near threshold.
  - Average delay to first AP: time from current pulse onset to first AP at depolarization currents near threshold.
  - SD of delay to first AP: SD of delays to first AP at depolarization currents near threshold.
  - Average delay to second AP: the average delay for the cell to generate a second AP at depolarization currents near threshold.
  - SD of delay to second AP: SD of delays to second AP at depolarization currents near threshold.
  - Average initial burst interval: average interspikes interval for the first three APs at depolarization currents near threshold.
  - SD of average initial burst interval: SD of the average interspike interval for the first three APs at depolarization currents near threshold.
- *IV.* Subthreshold current–voltage relationship (Figure 19.2d). When analyzed, this stimulus set gave rise to the following electrical parameters:
  - Input resistance for peak: maximum input resistance (V at peak of voltage response to I current injection).
  - Input resistance for steady state: input resistance at steady state (V at steady state of voltage response to I current injection).
  - Rectification index for peak IV: change in input resistance at peak voltage.
- Rectification index for steady state IV: change in input resistance at steady-state voltage.
- Maximum sag; difference between exponentially extrapolated voltage and steady-state voltage.
- *MTC.* Hyperpolarizing test current for input resistance during priming either hyperpolarization, depolarization or neither.
- *Noise_Filtering.* Subthreshold white noise current injection to determine filtering properties at different MPs.
- *Noise Spiking.* Noisy current injection to determine response reliability (Figure 19.2f).
- *Resting Membrane Potential.* The membrane voltage at the onset of whole-cell recording.
- *Sag.* Hyperpolarizing currents to examine hyperpolarizing-activated currents.
- *SAHP.* Hyperpolarization after a burst of APs (Figure 19.2e).
  - AHP amplitude 1: maximal amplitude of the AHP recorded after a burst of APs.
  - AHP amplitude 2: amplitude of the AHP at 100 ms after the end of a burst of APs.
  - Time to maximal AHP: time to the maximal AHP since the end of the burst.
- *Sine_Spectrum.* Testing Neuronal response to a sine spectrum.
- *AP_Recovery.* Brief pairs of depolarizing pulses to evoked APs and test for recovery of AP amplitudes.
- *Spontaneous.* No current injection at resting membrane potential to examine spontaneous activity.
- *True_Noise.* Voltage response to white noise current injection.

g-Profile

The genetic profiles are obtained from single-cell RT-PCR studies (Monyer and Jonas, 1995; Cauli et al., 1997, 2000; Wang et al., 2002; Baranauskas et al., 2003; Toledo-Rodriguez et al., 2003). The gel strip image must be uploaded, and the *g-profile* template table must be filled out. The *g-profile* is divided into functionally characterized groups of genes such as calcium binding proteins, neuropeptides, neurotransmitter enzymes, structural proteins, and so on.

Our PCR protocol contains up to 50 mRNA species (Figure 19.3), including more than 30 channel subunits (Kv1.1, Kv1.2, Kv1.4, Kv1.6, Kvβ1, Kvβ2, Kv2.1, Kv2.2, Kv3.1, Kv3.2, Kv3.3, Kv3.4, Kv4.2, Kv4.3, KChIP1, KChIP2, KChIP3, HCN1, HCN2, HCN3, HCN4, SK1, SK2, SK3, Caα1A, Caα1B, Caα1E Caα1G, Caα1H, Caα1I, Caβ1, Caβ3, and Caβ4), three calcium-binding proteins (CB, PV, and CR), 10 neuropeptides (NPY, VIP, SOM, CCK, POMC, pENK, Dyn, SP, CRH, and CGRP), and five enzymes (Gad65, Gad67, ChAT, nNOS, and GAPDH), see Figure 19.3.

19.2.3 Synaptic Profiles

Multineuron recording allows simultaneous characterization of synaptic connections in terms of the pre- and postsynaptic neurons as well as the anatomical and physiological properties of the connection.

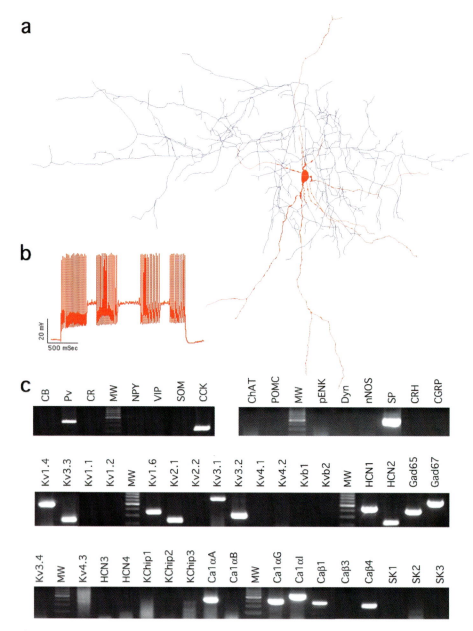

Figure 19.3 The *g-profile*. A representative sample of the morphological, electrophysiological, and genetic properties of a neocortical interneuron using single-cell multiplex RT-PCR combined with patch-clamp recordings. (**a**) 3D-computer reconstruction of the layer IV LBC: Soma and dendrites are in red, and axons are in blue. (**b**) Discharge response of the same neuron to suprathreshold current injection is a classical stuttering response (c-STUT). (**c**) PCR examination of the layer IV LBC: agarose gel showing the expression of mRNAs encoding for voltage-activated K^+ channels (Kv1.1/2/4/6, Kv1.2/1/2, Kv2.1/2, Kv3.1/2/3/4, Kv4.2/3, KchIP1/2/3); K^+/Na^+-permeable hyperpolarization-activated channels (HCN1/2/3/4); calcium-activated K^+ channels (SK1/2/3); voltage-activated Ca^{2+} channels (Ca1A/B/E/G/E/I, Ca1/3/4); calcium-binding proteins [calbindin (CB), parvalbumin (PV), calretinin (CR)]; neuropeptides [neuropeptide Y (NPY), vasoactive intestinal peptide (VIP), somatostatin (SOM), cholecystokinin (CCK), proenkephalin (pENK), proopiomelanocortin (POMC), dynorphin (Dyn), substance P (SP), corticotropin releasing hormone (CRH), and calcitonin gene-related peptide (CGRP)]; and enzymes [glutamic acid decarboxylase (Gad 65 and Gad 67), choline acetyltransferase (ChAT), nitric oxide synthase (nNOS), and glyceraldehyde-3-phosphate dehydogenase (GAPDH)].

sn-Profile

The *sn-profile* is a set of two names that describe the identity of the pre- and postsynaptic neurons involved in the connection. The *sn-profile* contains the anatomical and electrophysiological neuron types according to subjective criteria for classifying their morphological and electrical diversity (e.g., Layer IV, LBC, d-NAC to Layer IV, PC, RS).

sm-Profile

The anatomy of a synaptic connection is described by the *sm-profile*. This profile contains information about the numbers of putative synapses, their location on the axonal and dendritic arbors of the pre- and postsynaptic neuron, respectively (also referred to as synaptic innervation patterns), the axonal and dendritic geometric and steady-state electrotonic distances of each of the putative synapses, and so on (Figure 19.4).

Synaptic Connections

- **Synapse #.** This denotes the particular synapse under investigation.
- **Presynaptic Innervation Pattern (pre-SIP).** This describes the numbers, localization, and distributions of the synapses employed in a connection from the perspective of the source neuron.
- **Primary Axonal Collateral (PAC).** Axon collaterals (with all their sequential branches) that emerge *directly* from the axonal main stem. The axonal main stem is defined as originating directly from the neuron's soma, ultimately leaving the cortical sheet and entering the white matter (WM) for projection neurons (pyramidal cells) or as the axon that extends beyond the local axonal clusters continuing to give rise to secondary clusters and/or extends furthest from the source cell soma for local circuit neurons (interneurons).
- **Axonal Branch Order (ABO).** Branching frequency of a neuronal axon tree. The branch order increases stepwise at every node and reflects the number of branch points between the bouton forming the synapse and the soma of the source neuron. It may impact the axonal delay as well as invasion of action potentials (APs) (e.g., branch point failures) to the respective boutons.
- **Geometric Distance (L).** Geometric distance of the synapse-forming bouton from the soma *along* the axonal tree.
- **Postsynaptic Innervation Pattern (post-SIP).** This describes the numbers and localization of the synapses employed in a connection from the perspective of the target neuron. It resembles previous descriptions of target-domain specific innervation, but includes additional useful parameters.
- **Dendritic Branch Order (DBO).** This denotes the location of the synapse *along* the dendritic arbor according to the branching frequency of the dendritic tree. The branch order increases stepwise at every node and reflects the number of branchpoints between the synapse and the soma of the target neuron. Depending on the type of dendrite targeted, one may distinguish between:
  - Basal dendritic branch order (bDO).
  - Apical dendritic branch order (aDO).
  - Apical oblique dendritic branch order (aODO).

- Tuft dendritic branch order (tDO) for pyramidal neurons. Interneurons display a main, vertical "apical-like" dendrite (often inverted and directed toward the WM), and the targeted dendrites may be equally distinguished on this basis.
- Geometric distance (L): geometric distance of the synapse-forming bouton from the soma *along* the dendritic tree.
- Steady-state electrotonic distance (X): electrotonic distance of synapse under steady-state conditions assuming a passive compartmental model with homogenous distribution of non-voltage-activated/resting conductances. Obtaining X values requires determining geometric lengths of dendritic segments, their radii (r), their internal/axial resistivity (R_i) and their membrane resistance (R_m) (for details, see Markram et al., 1998).

- **Innervated Dendritic Fraction.** The fraction of the dendritic trees of the postsynaptic neuron receiving contacts from the presynaptic neuron.
- **Innervating PAC Fraction.** The fraction of the primary axonal collaterals used to innervate the postsynaptic neuron.
- **Ratio of Synapses Sharing Same Axonal Segments.** Of particular interest when considering divergent connections with different temporal dynamics, since it stresses synaptic independence.
- **Ratio of Neighboring Synapses Located Within 10 μm.** Of particular interest when considering synaptic interactions (such as saturation, etc.).
- **Rough Post-SIP.** In addition to obtaining histograms of synaptic distributions in terms of PAC, ABO, DBO, L, and X, it is favorable to determine a rough post-SIP in a connection.
  - Total# synapses.
  - # and % of somatic synapses: numbers and percentages of synapses in a particular connection that are formed onto the soma of the target cell.
  - # and % of bD synapses: numbers and percentages of synapses in a particular connection that are formed onto the basal dendrites of the target cell.
  - # and % of aD synapses: numbers and percentages of synapses in a particular connection that are formed onto the apical dendrite of the target cell.
  - # and % of aOD synapses: numbers and percentages of synapses in a particular connection that are formed onto the apical oblique dendrites of the target cell.
  - # and % of tuft synapses: numbers and percentages of synapses in a particular connection that are formed onto the tuft dendrites of the target cell.
  - Geometric distances (L): mean $\pm$ standard deviation (s.d.).
  - Electrotonic distances (X): mean $\pm$ standard deviation (s.d.).

Determination of such a rough post-SIP is especially useful for implementation in reduced, multicompartmental neuron models to study the effects of differential synaptic distributions.

se-Profile

The electrophysiological properties of synapses are characterized in terms of the biophysical, quantal, and dynamic properties (Figure 19.5). The *biophysical* properties focus on the

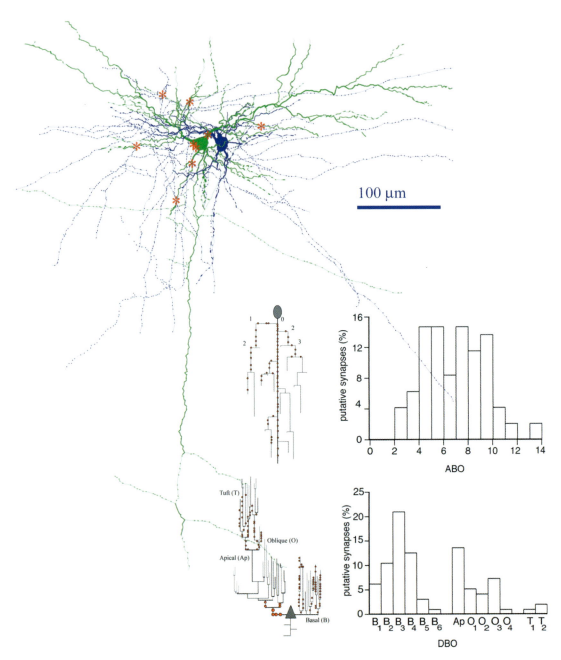

Figure 19.4   The *sm-profile*. Anatomical properties of neocortical connections. (*top*) 3D reconstruction of a specific type of interneuron (blue) innervating a pyramidal neuron (green) in neocortical layers II/III. Nine putative synapses (red asterisks) were formed in this connection (2 axo-somatic, 7 axo-dendritic). These putative axo-dendritic synapses were formed onto 3/4 dendritic trees (innervated dendritic fraction, 0.75; see text for details) and originated from 5/7 primary axonal collaterals (innervating PAC fraction, 0.71; see text for details). (*middle*) Schematic representation of presynaptic innervation patterns. (*middle left*) Boutons utilized in the formation of synaptic contacts in distinct connections are distributed along the axonal arbor of the source neuron (illustrated for an interneuron), which can be described in terms of (i) axonal branch orders (ABOs, subset indicated by numbers for illustrative purposes; see text for details) and/or primary axonal collaterals (PACs; see text for details). (*middle right*) Bouton distributions characteristic of a particular neocortical connection (specific type of interneuron to pyramidal cell (PC)) are plotted as a histogram in relation to the utilized axonal branch orders (ABO). (*bottom*) Schematic representation of postsynaptic innervation patterns. (*bottom left*) Locations of synaptic contacts employed in distinct connections are distributed along the somato-dendritic domain of the postsynaptic neuron (illustrated for a PC), which may be functionally compartmentalized into soma and distinct dendritic regions [tuft (T), apical dendritic (AP), apical oblique (O), and basal dendritic (B)]. A more detailed characterization of synaptic locations includes the dendritic branch orders (for each distinct dendritic compartment and layers), as well as geometric and electrotonic distances (*L* and *X*, respectively) (see text for details). (*bottom right*) Synaptic distributions characteristic of a particular neocortical interneuron to PC connection are plotted as a histogram in relation to the utilized dendritic branch orders (DBO). Modified and reproduced with permission from Wang et al. (2002).

amplitudes; latencies; rise and decay times of PSPs and/or PSCs; synaptic conductances; synaptic charge transfer; and so on. The *quantal* parameters include estimates of quantal size, probability of release, and number of release sites. The *dynamic* properties include the time constants governing the rates of recovery from synaptic depression and facilitation as well as the absolute and utilization of synaptic efficacy parameters (Ase and Use, respectively; see below).

Description of Physiological Parameters

- *Synapse Type.* This denotes the nature (excitatory or inhibitory) and type of the connection [according to the underlying time constant of recovery from synaptic facilitation (F) and depression (D), respectively; see below].
  - GABAergic F_1 type: inhibitory connection with F outlasting D (F:D ratio $>>10$).
  - GABAergic F_2 type: inhibitory connection with D outlasting F (D:F ratio $>>10$).
  - GABAergic F_3 type: inhibitory connection with similar D and F (F:D/D:F ratio $<<5$).
  - Glutamatergic F type: excitatory connection with F outlasting D.
  - Glutamatergic D type: excitatory connection with D outlasting F.

- *Physiological Properties.*
  - Reversal potential: determines the nature (excitatory or inhibitory) of the connection by determining the target neuron's membrane potential at which synaptic current flow reverses (Figure 19.5a).
  - Pharmacological block: determines the subtype of receptors engaged (e.g., CNQX—AMPA: APV—NMDA; Bicuculline—GABAa; particular antagonists for mGluRs and GABA-B) (Figure 19.5b).

- *Unitary Responses (Figure 19.5c).*
  - Failures: Defined when no synaptic response is detectable after eliciting a presynaptic AP, in particular defined as "events" with synaptic potential and/or current response amplitudes (PSP/PSC) below the r.m.s (root mean square) of baseline/resting membrane voltage or current noise; a minimum of 50–100 evoked APs is required to obtain estimates of failures.
  - Release probability (Pr): Probability of synaptic PSP/PSC responses (amplitudes > r.m.s of baseline noise) to presynaptically eliciting APs—that is, defined as (1-Failures); a minimum of 50–100 evoked APs is required to obtain estimates of failures.
  - Amplitudes: Average ± standard deviation (s.d.) of PSPs/PSCs; needs specification of holding potential at which synaptic responses were acquired.
  - Quantal size (q): response obtained by the release of a single synaptic vesicle provided that no postsynaptic receptor saturation occurs.
  - Binomial n: binomially defined numbers of synapses within a particular connection; binomial $n = (1 - Pr)/CV^2 * Pr$.
  - CV of amplitudes: degree of variability of PSPs/PSCs; defined as s.d./mean of synaptic response amplitudes.

- RT of PSPs/PSCs: average ± s.d. of voltage and/or current rise times (RT) of synaptic responses; RTs determined as time to rise from 20–80% peak amplitude (at least 30–50 sweeps excluding failures).
- DTC of PSPs/PSCs: average ± s.d. of voltage and/or current decay time constants (DTC) of synaptic responses estimated as single exponentials because of multiple distributed release sites typical of neocortical connections.
- Latency of PSPs/PSCs: average ± s.d. of synaptic response latencies; latency of PSPs/PSCs defined as the time from the peak of the presynaptic AP to 10% of the PSP/PSC amplitude; this measure includes axonal, synaptic and dendritic delays.
- Charge: integral of mean synaptic current response yields the charge transferred from the synapse to the site of recording (usually somatic).
- Conductance (G_1)/connection: chord conductances (G_1) determined as the slope of a line fit (linear regression) through amplitude of unitary responses (or first responses if protocols with trains of stimuli are employed) and intersection at the reversal potential (provided that rectification is negligible).
- G_{max} of connection: maximal conductances (G_{max}) calculated as G_1/Pr or G_1/Use (see below); denotes the conductance given that Pr is 1 at all synapses—that is, synaptic release is occurring at all release sites within a particular connection.
- Anatomical G_{syn}: per synapse conductance estimated according to anatomically defined numbers of synapses within a particular connection (G_{max}/anatomical n).
- Binomial G_{syn}: per synapse conductance estimated according to binomially defined numbers of synapses within a particular connection (G_{max}/binomial n).

- *Synaptic Dynamics (Figure 19.5d).* The synaptic parameters governing the dynamic response behavior of synaptic transmission (Ase, Use, D, and F; see below) are obtained by a model of frequency-dependent synaptic transmission (Markram et al., 1998) and iteratively fitting average synaptic potential and/or current responses to trains of presynaptic APs of different frequencies (ranging from 0.1 to 100 Hz). The model is based on earlier concepts of the release process (Betz, 1970; Byrne, 1982), which can be rephrased by stating that the fraction (Use) of the synaptic efficacy utilized by an AP becomes instantaneously unavailable for subsequent use and recovers with a time constant of D. Synaptic facilitation was modeled by including a facilitating mechanisms in the model as a pulsed increase in Use by each AP. The running value of Use is referred to as u, while Use remains a parameter that applies to the first AP in a train. The value u decays with a single exponential F to the resting value Use (Markram et al., 1998). Whereas others (Magleby, 1987) have modeled facilitation and depression employing several time constants, the single time constant of recovery from synaptic depression and facilitation obtained using this phenomenological model is sufficient to fully capture the main features of synaptic transmission at any given neocortical connection. In particular, the parameters are not constrained by assumptions about the underlying biophysical mechanisms of synaptic transmission.

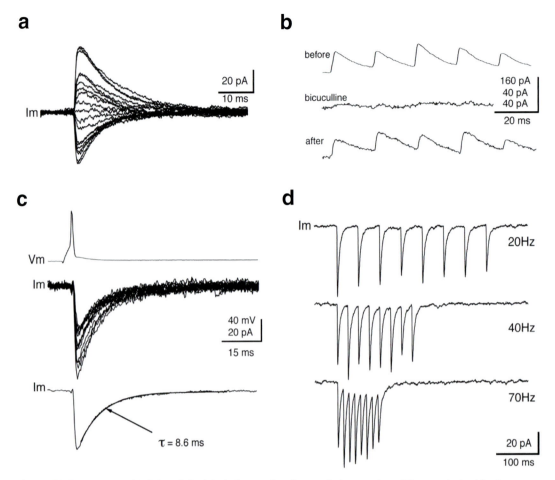

Figure 19.5 The *se-Profile*. Selected physiological properties of neocortical connections. All traces obtained for the same specific type of interneuron to PC connections illustrated in Figure 19.4. **(a)** Synaptic reversal potential. Average synaptic currents (20–40 trials for each trace) obtained for a wide range of postsynaptic holding potentials (V_{hold} ranging from −85 to +15 mV) are overlaid. Synaptic currents reversing near −40 mV confirm GABAergic nature of synaptic transmission. **(b)** Synaptic pharmacology. Synaptic current responses in postsynaptic PC ($V_{hold} = 0$ mV) to high-frequency stimulation of presynaptic interneuron (50-Hz train, 30 APs; only the first five PSCs shown) were completely and reversibly blocked by bath application of bicuculline (10 μM), confirming that the synaptic responses were purely GABA-A receptor mediated. **(c)** Consecutive trials of unitary current responses in postsynaptic PC (*middle panel*, 15 traces overlaid) to action potentials (APs) in the presynaptic neuron (*upper trace*). *Lower trace* represents average of traces shown in middle panel. A single-exponential fit shows characteristic slow decay of GABAergic currents. Obtained traces may be further utilized to obtain a more complete description of the transmission properties at this particular connection, including latencies, rise times (RT), amplitude distributions (including CV of amplitudes, failures, release probabilities, quantal sizes, and binomial n), and decay time constants (DTC) (see text for details). **(d)** Dynamic properties of synaptic transmission. Average synaptic responses (30–50 trials for each trace) to trains of regularly spaced APs of different frequencies. Note the seeming absence of short duration facilitation at low frequencies (*upper panel*, 20 Hz), which is gradually unmasked at higher frequencies (*middle trace* 40 Hz, *lower trace* 70 Hz). The obtained average traces can be implemented in models of synaptic transmission to extract the relevant dynamic parameters of synaptic transmission at each particular connection (i.e., Ase, Use, *F, D, F:D-*, and/or *D:F* ratios; see text for details). Modified and reproduced with permission from Wang et al. (2002).

- Absolute synaptic efficacy (Ase): maximal voltage and/or current response of a connection. Ase is equivalent to the quantal size multiplied by the number of release sites and an electrotonic attenuation factor. Ase thus represents the maximal response and may also be obtained given that Pr or Use equals 1; that is, release occurs at every synapse within a particular connection.

- Utilization of synaptic efficacy (Use): equivalent to Pr provided that it can be established that the mechanism of

frequency-dependence of synaptic transmission is located purely presynaptically.

- Time constant for recovery from synaptic depression (*D*): the time constant by which the connection recovers from depression that may be due to several different biophysical mechanisms such as vesicle depletion, a functional refractory period of release due to sequential decrease of Ca^{2+} influx (Klein et al., 1980; Zucker, 1989), or desensitization of the Ca^{2+}-induced machinery.

- Time constant for recovery from synaptic facilitation (F): the time constant by which the connection recovers from facilitation that may be due to several different biophysical mechanisms mainly involving intraterminal Ca homeostasis (Zucker, 1989; Fisher et al., 1997).
- Ratio of synaptic recovery time constants ($D{:}F$ ratio and/or $F{:}D$ ratio): indicates the impact of the different dynamic processes underlying frequency-dependence of synaptic transmission.

19.3 PLATFORM OF THE NMDB

At this stage of the development we chose a testing platform that will be as nonspecialized as possible. Once the database nears completion, we will be able to determine better the state-of-art platform that will best suit this database. The structure is such that transferring the database to a selected system will be straightforward.

- *Database Management System.* Microsoft Access
- *Operating System.* Microsoft Windows 2000 Server
- *Web Server.* Microsoft IIS 5.0
- *Programming Languages and Techniques.*
  - MATLAB 6—Online Analysis Tool creation, application, and implementation.
  - VBScript, JavaScript, ASP, HTML, CSS—Web page building, server side paging, data presentation.
  - SQL, ADO—data retrieval and update, database manipulation.

19.4 STRUCTURE OF THE NMDB

Tables
The database is actually a set of tables, and each table contains records with a group of fields that are closely related. These tables are used to store data effectively, not to present data. The tables are divided into the four different categories, not by physical grouping but by their relations to the *index* table. If one opens an mdb file in MS-ACCESS and viewed the tables, you would not be able to tell which is which, unless you saw their relations. The tables are organized around cells and synaptic connections.

Single Cells
Each of the profiles (categories m, e, g) has one or more tables containing specific types of properties and corresponding data. Above the specific profiles is a group of general neuronal properties and their values, which form a cell *index* table. This could also be called "category 0."

Each category of properties has a *sub-index* table and one or more *regular* tables. Each table contains a "CellName" field, among others. In the *index* table and *sub-index* tables, this field is defined as the *primary key* that identifies the record stored in the table, and it acts as a foreign key in *sub-index* tables and *regular* tables. Exceptions appear in tables in the category of morphology, where "CellName" and "TreeType" together are defined as co-primary key or foreign key, in which case only the information about a certain type of tree (e.g. axon) is needed.

Synaptic Connections
There is only an *index* table and several regular tables about connections at this stage of the NMDB. Similar to cells, "ConnectionName" is the primary key in the *index* table and a foreign key in *regular* tables, utilized to identify a record in a table.

Relationships
Single cell and connection data are related and appear in different tables. There is need to ensure correct and simultaneous updating of the relevant tables when any changes of the data are performed. The database therefore includes relationships between tables (cells and connections): *Index* table and *sub-index* table have one-to-one or one-to-many relationships; likewise, *sub-index* tables and *regular* tables have relationships of one-to-one or one-to-many (Figure 19.6). Cells serve as properties of a connection (pre- and postsynaptic cells, respectively), and synaptic connections are properties of a cell. Relationships are therefore defined between tables *across* the two entities. Hence, in the first case, "CellName" is the primary key in a cell table and is a foreign key in a connection *index* table, and the relationship is one-to-many. *Index property* tables and relationships are defined in order to enable correct and simultaneous updating of data across all the relevant tables.

19.5 ORGANIZATION OF THE NMDB

The database is organized into four main components: database, online tools, image gallery, and download section.

Database
This section is for users to retrieve data and information from the NMDB and has four parts:

- *Cells.* General information and descriptions of single cells and their m-, e-, and g-profiles, as well as other data related to the cells together with images, graphs, and figures.
- *Connections.* Information and data of synaptic connections between cells, the sn-, sm-, and se-profiles, and the relevant figures.
- *DB Statistics.* Statistics of the NMDB with numbers of single cells in the database, numbers of connections, the statistics of every category, and so on.
- *DB Definitions.* Definitions and descriptions of the terms that are used in the NMDB.

Online Tools
This section provides different tools for users of the NMDB.

- *Neuron Analysis Tool.* Fully analyzes the morphological features of a cell and automatically saves the data it generates as well as the m-profile into the NMDB.
- *File Upload Tool.* Upload cells (Neurolucida files) that Neuron Analysis Tool needs as an input; upload cells for presentation that users want to share with others.

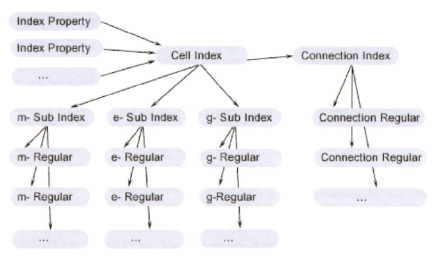

Figure 19.6   Relationships between tables. Database tables are interlinked in a vectorial manner, enabling simultaneous and accurate updating of all relevant tables after making changes in the source table. Illustrated is the scheme of relationships between the different *(index, sub-index, and regular)* table types.

- *Profiles Analysis Tool.* Processes a certain group of selected profiles; obtains the mean, max, min, standard deviations, and so on, for every parameter in the profile.
- *DB Data Input Tool.* Allows manual entering of data into the NMDB.
- *DB Table Edit Tool.* Allows editing/modifying tables, adding new/deleting fields, and so on.

Gallery

This section contains searchable database pictures as well as others.

- *Cell Images.* Displays a certain group of cell images selected from all of the cell images in the NMDB according to specified search criteria.
- *3D-Rotation Demo.* Provides two examples that roughly show a cell rotating in three dimensions.

Other kinds of images such as *e-profile* figures and *g-profile* figures will also be added to the section.

Downloads

This area hosts different types of files available for users to download.

- *Presentations.* PowerPoint files, Excel files, Word files, Movie clips, Igor Files, and so on, that had been uploaded by other users.
- *NeuroLucida Files.* Reconstructed cells in Neurolucida ascii file format.
- *Other.* Other downloads can be carried out directly during browsing the pages of the website—for example, cell image downloads, and so on.

Searching the NMDB

Retrieving data in the NMDB can be achieved in several different ways.

- *Combinatorial Search Engine.* There is a flexible combinatorial search engine. One can begin by searching cells, cell images, profiles (*m-, e-, and g-profiles*), or connections and then follow the links in the result page to isolate the data. Searching cells can be achieved by combinations of (a) all cells, (b) specific cell types, (c) cells in specific layers, or (d) various combinations. At this stage, searching is based on combinations of morphological and electrophysiological parameters. At a later stage, searching combinations will include all key parameters across the different profiles.
- *Key Word Search.* Searching can also be performed by typing in keywords.
- *Any Point Access.* At any point of the database, a cell or connection can be accessed allowing surfing within the database.
- *Cell Gallery.* All cells can be viewed in a gallery, and the database can be entered through clicking a cell of interest.

19.6 SUBMITTING DATA TO THE NMDB

Because of the diversity and complexity of the data, there are several avenues to submit data to the NMDB and we will continue seeking new ways to allow easier, faster, and more flexible entering of data.

The first avenue is to submit the Neurolucida file in ascii format with the File Upload Tool, then use the Neuron Analysis Tool with the option of "Save result to database" checked. This ensures that morphological data of the cell will be automatically saved to the NMDB. Note that, in order to count boutons using Sholl Analysis, users should use "Marker 11" or "Marker 2" defined in Neurolucida to represent a bouton in computer reconstruction; other markers are not recognized at this stage.

The second will be to use an automatic analysis tool to save *e-profiles* to the NMDB. This tool will be available at a later stage. The third avenue is to use the *DB Data Edit Tool* to fill out template tables that the NMDB provides for different kinds of data entering. This is used to submit the general information about the experiment, cells, and connections, along with those profile data that have not yet been covered by any other submitting tools.

19.7 USING THE ONLINE TOOLS

Neuron Analysis Tool

This tool is used to comprehensively analyze morphological features of a cell. It produces a large amount of data, as well as a cell image and *m-profile* (data and graph). All these are then automatically saved into the database as morphological features, which can be selectively retrieved later. The tool can also be used for private analysis without submitting data to the NMDB. Neurolucida files with ".asc" extension are used as the input of the tool. Using the Neuron Analysis Tool is simple and user-friendly: The user simply selects the cell name in the submitted list and then clicks the analyze button. However, it is important to know that general information about the cell should be entered as extensively as possible to allow the search engine to retrieve all relevant cells.

File Upload Tool

This tool is for users to submit their Neurolucida files or to upload their presentations to the website. Note that "Cell Name" is a crucial item in the database and it uniquely identifies a cell. Once the cell is submitted, its name normally will not be changed from then on.

Profile Processing Tool

This tool is provided to analyze a group of profiles (*m-*, *e-*, or *g-profiles*), obtaining the mean, standard deviation, maximum, and minimum values of all the columns of the profiles that have been selected according to certain searching criteria (which is the same as when searching for a certain group of cells).

19.8 EDITING THE NMDB

Modification of data or tables can be made to almost all the tables in the database by registered members. However, editing a table or adding new or deleting fields of tables requires higher privileges (see administering the NMDB).

Adding New or Updating Records

For the morphological data, this is carried out by using the Neuron Analysis Tool. When it prompts that there already exists a cell with the same name in the database, the user is required to confirm whether the cell is to be reanalyzed and updated before continuing the process. When manually adding a new record, it is important to respect and follow the relationships between tables. Data should be entered in the order of the *index* table, then *sub-index* table, then regular table, which means that when adding a cell, one has to add at least the cell name to the cell *index* table, then to the *sub-index* tables, and only then the rest of the tables. In addition, one can only add a cell name to a regular table or sub-index table by selecting from a list of cell names

available to the table. If the cell name was not in the list, one has to go back to add it first in the index table.

Editing Tables

The database supports limited and restricted editing on most of the tables, thereby deleting existing fields from and adding new fields to a table. Once a field was deleted, the data in the field are removed permanently and cannot be retrieved any further.

19.9 ADMINISTRATION OF THE NMDB

The NMDB is open for free access by anyone, and all data in the database are available for free download. Attached to each dataset are the experimenters name and affiliation. We ask that users that download data notify experimenters and add their names to the acknowledgments of any publications resulting from the use of the data. In addition, users should indicate that the source of the data is from the NMDB.

Users that intend to submit data must first register. A review and confirmation of membership will be e-mailed together with a username and password. At the time of registration, users can also apply for various privileges that allow modifications of the database. Only the super-user can alter table structures.

The Private NMDB

The NMDB is being constructed in two formats. The one format currently available is the generally accessible format for all user submissions and access. The second format that we aim to have completed soon is a private lab format (NMDB$_p$). This will allow users to privately organize their microcircuit data before publication. After publication, users will be able to simply click a "Publish Data" bouton to send their data to the public NMDB.

19.10 SUMMARY

The *Neocortical Microcircuit Database* (NMDB) is the first microcircuit database that allows the systematic storage and retrieval of microcircuit data. At this stage it provides a method to enter the basic anatomical, physiological, and molecular properties of neurons and synaptic connections. There are many other properties that still need to be added, such as synaptic molecular profiles as well as biophysical properties and distribution of ion channels and receptors. We are also further developing the NMDB analysis tools that will allow relational analysis of the data for the purpose of research and for the purpose of constructing programs that can extract data from the database for automated reconstruction of complete microcircuits according to precise biological parameters. Even further developments include the use of such reconstruction algorithms to feed virtual reality platforms to allow 3D visualization and interactive surfing of the microcircuit. A parallel database will allow simulation of microcircuit dynamics. Finally, the database will be linked to other relevant databases such as the Neuronal Development Database (NDDB), Molecular Networks Database (MNDB), Biochemical Dynamics Database (BDDB), Neuron Dynamics Database (NDDB), Synaptic Plasticity Database (SPDB), and a Network Dynamics Database (NDDB).

References

Baranauskas, G., Tkatch, T., Nagata, K., Yeh, J.Z., and Surmeier, D.J. (2003). Kv3.4 subunits enhance the repolarizing efficiency of Kv3.1 channels in fast-spiking neurons. *Nat. Neurosci.* **6**, 258–266.

Beaulieu, C. (1993). Numerical data on neocortical neurons in adult rat, with special reference to the GABA population. *Brain Res.* **609**, 284–292.

Betz, W.J. (1970). Depression of transmitter release at the neuromuscular junction of the frog. *J. Physiol. (London)* **206**, 629–644.

Byrne, J.H. (1982). Analysis of synaptic depression contributing to habituation of gill-withdrawal reflex in *Aplysia californica. J. Neurophysiol.* **48**, 431–438.

Callaway, E.M. (1998). Local circuits in primary visual cortex of the macaque monkey. *Annu. Rev. Neurosci.* **21**, 47–74.

Cauli, B., Audinat, E., Lambolez, B., Angulo, M.C., Ropert, N., Tsuzuki, K., Hestrin, S., and Rossier, J. (1997). Molecular and physiological diversity of cortical non pyramidal cells. *J. Neurosci.* **17**, 3894–3906.

Cauli, B., Porter, J.T., Tsuzuki, K., Lambolez, B., Rossier, J., Quenet, B., and Audinat, E. (2000). Classification of fusiform neocortical interneurons based on unsupervised clustering. *Proc. Natl. Acad. Sci. U.S.A.* **97**, 6144–6149.

DeFelipe, J. (1993). Neocortical neuronal diversity: Chemical heterogeneity revealed by colocalization studies of classic neurotransmitters, neuropeptides, calcium-binding proteins, and cell surface molecules. *Cereb. Cortex.* **3**, 273–289.

DeFelipe, J. (1997). Types of neurons, synaptic connections and chemical characteristics of cells immunoreactive for calbindin-D28K, parvalbumin and calretinin in the neocortex. *J. Chem. Neuroanat.* **14**, 1–19.

DeFelipe, J. (2002). Cortical interneurons: From Cajál to 2001. *Prog. Brain. Res.* **136**, 215–238.

DeFelipe, J., and Farinas, I. (1992). The pyramidal neuron of the cerebral cortex : Morphological and chemical characteristics of the synaptic inputs. *Prog. Neurobiol.* **39**, 563–607.

Eccles, J.C. (1984). The cerebral neocortex. A theory of its operation. In *Cerebral Cortex: Functional Properties of Cortical Cells* (E.G. Jones and A. Peters, eds.), pp. 1–36. Plenum Press, New York.

Fisher, S.A., Fischer, T.M., and Carew, T.J. (1997). Multiple overlapping processes underlying short-term synaptic enhancement. *Trends Neurosci.* **20**, 170–177.

Glaser, J.R., and Glaser, E.M. (1990). Neuron imaging with Neurolucida—a PC-based system for image combining microscopy. *Comput. Med. Imaging Graph.* **14**, 307–317.

Goldman, P.S., and Nauta, W.J. (1977). Columnar distribution of cortico-cortical fibers in the frontal association, limbic, and motor cortex of the developing rhesus monkey. *Brain Res.* **122**, 393–413.

Gupta, A., Wang, Y., and Markram, H. (2000). Organizing principles for a diversity of GABAergic interneurons and synapses in the neocortex. *Science* **287**, 273–278.

Hubel, D.H., and Wiesel, T.N. (1962). Receptive fields, binocular interaction and functional architecture in the cat's visual cortex. *J. Physiol. (London)* **160**, 106–154.

Hubel, D.H., and Wiesel. T.N. (1977). Ferrier lecture. Functional architecture of macaque monkey visual cortex. *Proc. R. Soc. London, Ser. B* **198**, 1–59.

Hubener, M., Schwarz, C., and Bolz, J. (1990). Morphological types of projection neurons in layer 5 of cat visual cortex. *J. Comp. Neurol.* **301**, 655–674.

Jones, E.G. (1981). Anatomy of cerebral cortex: Columnar input–output organisation. In *The Organization of the Cerebral Cortex* (F.O. Schmitt, F.G. Wordan, G. Adelman, and S.G. Dennis, eds.), pp. 200–235. MIT Press, Cambridge, MA.

Jones, E.G., Burton, H., and Porter, R. (1975). Commissural and cortico-cortical "columns" in the somatic sensory cortex of primates. *Science* **190**, 572–574.

Kawaguchi, Y., and Kubota, Y. (1997). GABAeric cell subtypes and their synaptic connections in rat frontal cortex. *Cereb. Cortex* **7**, 476–486.

Klein, M., Shapiro, E., and Kandel, E.R. (1980). Synaptic plasticity and the modulation of the Ca2+ current. *J. Exp. Biol.* **89**, 117–157.

Lorente de Nó, R. (1938). Cerebral cortex: Architecture, intracortical connections, motor projections. In *Physiology of the Nervous System* (J.F. Fulton, ed.), 1st ed., pp. 291–329. Oxford University Press, New York.

Magleby, K.L. (1987). Short-term changes in synaptic efficacy. In *Synaptic Function* (G.M. Edelman, W.E. Gall, and W.M., Cowan, eds.), pp. 21–56. Wiley, New York.

Markram, H., Wang, Y., and Tsodyks, M. (1998). Differential signaling via the same axon of neocortical pyramidal neurons. *Proc. Natl. Acad. Sci. U.S.A.* **95**, 5323–5328.

Monyer, H., and Jonas, P. (1995). Polymerase chain reaction analysis of ion channel expression in single neurons of brain slices. In *Single-Channel Recordings* (B. Sakmann, and E. Neher, eds.), pp. 357–373. Plenum Press, New York.

Mountcastle, V.B. (1957). Modality and topographic properties of single neurons in a cat's somatosensory cortex. *J. Neurophysiol.* **20**, 408–434.

Mountcastle, V.B. (1998). *Perceptual Neuroscience: The Cerebral Cortex.* Harvard University Press, Cambridge, MA and London.

Peters, A. (1987). Synaptic specificity in the cerebral cortex. In *Synaptic Functions* (G.M. Edelman, W.E. Gall, and W.M. Cowan, eds.), pp. 373–397. Wiley, New York

Peters, A., and E.G. Jones, eds. (1984). *Cellular Componenets of the Cerebral Cortex.* Plenum Press, New York.

Purves, D., Riddle, D.R., and LaMantia, A.S. (1992). Iterated patterns of brain circuitry (or how the cortex gets its spots). *Trends Neurosci.* **15**, 362–368.

Ramón, y Cajál, S. (1911). *Histology of the Nervous System.* Oxford University Press, New York.

Ren, J.Q., Aika, Y., Heizmann, C.W., and Kosaka, T. (1992). Quantitative analysis of neurons and glial cells in the rat somatosensory cortex, with special reference to GABAergic neurons and parvalbumin-containing neurons. *Exp. Brain Res.* **92**, 1–14.

Shmuel, A., and Grinvald, A. (1996). Functional organization for direction of motion and its relationship to orientation maps in cat area 18. *J. Neurosci.* **16**, 6945–6964.

Sholl, D.A. (1956). *The Organization of the Cerebral Cortex.* Methuen, London.

Somogyi, P., Tamas, G., Lujan, R., and Buhl, E.H. (1998). Salient features of synaptic organisation in the cerebral cortex. *Brain Res. Rev.* **26**, 113–135.

Szentagothai, J. (1978). The Ferrier Lecture, 1977. The neuron network of the cerebral cortex: A functional interpretation. *Proc. R. Soc. London, Ser. B* **201**, 219–248.

Thomson, A.M., and Bannister, A.P. (2003). Interlaminar connections in the neocortex. *Cereb. Cortex* **13**, 5–14.

Thomson, A.M., and Deuchars, J. (1997). Synaptic interactions in neocortical local circuits: Dual intracellular recordings in vitro. *Cereb. Cortex* **7**, 510–522.

Toledo-Rodriguez, M., Gupta, A., Wang, Y., Wu, C.Z., and Markram, H. (2003). Neocortex: Basic neuron types. In *The Handbook of Brain Theory and Neural Networks* (M.A. Arbib, ed.), 2nd ed., MIT Press, Cambridge, MA, pp. 719–725.

Wang, Y., Gupta, A., Toledo-Rodriguez, M., Wu, C.Z., and Markram, H. (2002). Anatomical, physiological, molecular and circuit properties of nest basket cells in the developing somatosensory cortex. *Cereb. Cortex.* **12**, 395–410.

White, E.L. (1989). *Cortical Circuits. Synaptic Organization of the Cerebral Cortex.* Birkhäuser, Boston, MA.

Zucker, R.S. (1989). Short-term synaptic plasticity. *Annu. Rev. Neurosci.* **12**, 13–31.

SenseLab: A Decade of Experience with Multilevel Multidisciplinary Neuroscience Databases

*Luis Marenco, M.D., Chiquito J. Crasto, Ph.D., Nian Liu, Ph.D.,
Michele Migliore, Ph.D., Jian Liu, Thomas M. Morse, Ph.D., Michael L.
Hines, Ph.D., Prakash M. Nadkarni, M.D., Perry L. Miller, M.D., Ph.D.
and Gordon M. Shepherd, M.D., D. Phil.*

CONTENTS

20

Databasing the Brain. Edited by Stephen H. Koslow and Shankar Subramaniam
ISBN 0-471-30921-4 © 2005 John Wiley & Sons, Inc.

20.1 INTRODUCTION

The year 2003 marks the end of the first decade of the Human Brain Project (Pechura and Martin, 1991; Koslow, 2002), an ambitious and farsighted initiative to develop the tools of informatics in the service of neuroscience research. It also marks the first decade of SenseLab as a pilot project within the program. It is therefore timely to summarize the progress that has been made thus far and to indicate some of the new directions for the future.

20.2 BACKGROUND

Internet-accessible databases for genes and proteins have played an essential role in the great progress being made in molecular biology. It therefore seems obvious that electronic databases should have a similarly salutary effect on research in specific types of organ systems. Of these, the brain is the most complex and therefore presents the greatest challenge. An increasing flood of data is being produced by an increasing range of methods at all levels of organization, from molecules to behavior. The data range from sequence data at the molecular level, through cell biochemistry, structure, and function, through circuits and brain scans, to monitoring of behavior (Shepherd et al., 1998; Miller et al., 2001).

Despite the obvious advantages, the building of databases to support this work has been slow to develop. There are several reasons. First and foremost, much of the data is in a form far more complicated than the simple sequences of genes and proteins. For example, it includes complex structures of cells and their organelles; physiological recordings of complex activity taking place over different time scales; and brain scans from different subjects, which have to undergo complex procedures for alignment in order to make comparisons between them.

Solving these problems has become the province of the new scientific discipline of neuroinformatics. It is a crossroads discipline, requiring several types of expertise. First are *experimental neuroscientists*, who generate the data and understand its strengths and weaknesses. Second are *computer modelers*, people skilled in simulations of molecules, cells, and systems, in order to integrate the data so that cells and systems can be simulated in ways that assist in the interpretation of the experimental results and guide new experiments. Third, it requires *informatics specialists*, researchers who understand how to construct databases containing complex data and data objects. Most importantly, these specialists are the ones who can devise the tools that enable one to search the data to yield insights not otherwise obtainable.

The SenseLab project is built on these three pillars. The *experimental neuroscience* subproject is aimed at analyzing the olfactory pathway as a model system. The olfactory bulb was one of the first regions of the brain in which experimental analysis was combined with computational modeling to give insight into the functional organization of dendrites and local circuits. It has continued as a favorable target for these studies. SenseLab thus had as a primary aim to pursue this combined experimental–theoretical approach. A *computational modeling* subproject supports the building of models for this purpose. But the key to bringing together these two approaches was to build databases

of the experimental data and the computational models, as well as innovative search tools for extracting complex multilevel, multidisciplinary data from them. For this it was essential to have *informatics specialists* dedicated full time to the collaboration.

The Human Brain Project has provided the means for the SenseLab project to build these pillars and work toward these goals. The strategy has been to respond to immediate needs for carrying out experiments and computer modeling of the properties of cells and circuits in the olfactory pathway, in order to understand the neural basis of behavior in this system. This has required databases at successive levels of organization, from genes and proteins, through synapses and microcircuits, to cells and systems. It has required building new types of search tools that permit these critical data to be extracted at the different levels and combined into working models. And finally it has required extending the database to other well-analyzed types of neurons in other brain regions in order to compare functional organization in different cells and systems for general principles.

This chapter briefly describes our efforts to create a suite of databases aimed at supporting our experimental studies. Our studies have two main aims. The first is to understand better the neural basis of this sensory modality—the sense of smell. Three databases—of olfactory receptor proteins (ORDB), olfactory receptor ligands (OdorDB), and odor maps (OdorMapDB)—are devoted to this task. The second main aim is to understand better the roles of dendrites and neuronal microcircuits in processing information in this pathway. Three more databases—of membrane properties expressed in cells (CellPRopDB) and in parts of cells (NeuronDB), as well as of models that incorporate these functional properties (ModelDB)—support this task. Our analysis includes close comparisons with these mechanisms in other systems of the brain. The interrelated databases are found at senselab.med.yale.edu.

20.3 OLFACTORY DATABASES

The sense of smell starts with cells in the nose that detect different odor molecules. The detective work is done by a large family of olfactory receptor proteins in the sensory cells (Buck and Axel, 1991). This is the largest family of G-protein-coupled receptors and the largest gene family in the genome (Pilpel et al., 1998; Mombaerts, 1999). A first set of three databases supports research on these receptors.

Olfactory Receptor Database (ORDB) contains more than 5000 olfactory and olfactory-related receptor genes and proteins, including the recently identified complete olfactory genomes of the human and the mouse. The database is officially supported by the National Institute for Deafness and Other Communicative Disorders (NIDCD), with an scientific advisory board that reports to the institute. The database was formed initially in response to requests from laboratories engaged in the early cloning and sequencing of the olfactory receptors. Currently, it is supporting the effort to arrive at a rational nomenclature for these large gene families. It is also supporting data mining of the receptors to identify motifs critical in receptor functions (Skoufos, 1999). Effective dissemination of information to users necessitates populating the database in a timely and efficient manner. The interoperative tool AUTOPOP populates ORDB from sources of gene and protein sequences (Crasto et al., 2002).

It also supports molecular modeling of the receptors to gain insight into their binding pockets and differential interactions with different odor molecules (Singer, 2000).

Odor Database (OdorDB) contains the odor molecules that have been shown to interact with the receptors, based on experiments on receptors expressed in heterologous systems or overexpressed in native cells. One of the main conclusions from these studies is that the receptors and odor molecules interact with each other in complex combinatorial ways. This presents a problem in analyzing receptor–ligand specificity that is far more demanding than in traditional pharmacological studies. Close links between OdorDB and ORDB allow the user to carry out the two essential types of operations: Query all odor molecules shown to act with a given receptor, and query all receptors shown to interact with a given odor.

The activity aroused in the sensory cells is transmitted to the next relay station, the olfactory bulb, where it sets up spatial activity patterns—called odor images—equivalent to visual images in the visual pathway (Xu et al., 2000).

Odor Map Database (OdorMapDB) contains maps of these activity patterns, as they have been recorded in animals with 2-deoxyglucose labelling and high-resolution fMRI in response to odor stimulation. Tools have been developed to allow the three-dimensional activity patterns in specific layers of the olfactory bulb to be traced and projected onto two-dimensional flat maps, in order to facilitate the quantitative analysis of the activity patterns (Liu et al., 2003). Tools have also been developed to allow a given map to be superimposed onto one or more other maps, in order to characterize quantitatively and qualitatively the differences between the maps, which are believed to underlie the ability to discriminate different odors (Liu et al., 2002).

These three databases thus support studies on how odor space is represented as neural images in the brain, one of the most intriguing and challenging problems in contemporary sensory neuroscience.

20.4 NEURONAL DATABASES

The activity maps are then processed by microcircuits in the olfactory bulb, for export as activity patterns to the next higher stages underlying the perception of smell. A second suite of three databases supports experimental and computational analysis of membrane properties (receptors, channels, transmitters) of neurons involved in processing the neural information by these microcircuits. By construction of sophisticated search tools, these databases allow comparisons to be made between these and analogous microcircuits in other brain regions, as a first step toward extracting the general principles underlying the neural basis of perception.

Cell Properties Database (CellPropDB) supports a proteomics approach to the membrane properties found in different neuronal cell types in different brain regions. When molecular biologists first identify a particular gene or protein in a given neuronal cell type in a given region, it is deposited here. The Web page for each cell type displays an inventory of receptors, channels, and transmitters, with those found in that type highlighted. A unique *multiple properties search tool* allows the user to search all neuron types for any arbitrary combination of properties. This *complex search tool* is a specific example of a more general *metadata-driven ad hoc query tool* utilizing the flexibility of the EAV/CR framework (Marenco et al., 2003). This carries to the cellular level the ability in gene and protein databases to search for sequence motifs, a search capacity that has been crucial to identifying families of genes and proteins. The same ability is now possible at the cellular level, to identify motifs of expressed properties that will define families of cells on a molecular basis.

At a more detailed level, *Neuron Database (NeuronDB)* represents those membrane properties as they are found in different combinations in different neuron compartments (axon, soma, dendrites). This database also provides a further enhancement of the *multiple properties search tool* that enables the investigator to search across neurons for arbitrary "expression" motifs within the different neuronal compartments. This is a first step toward enabling neuroscientists to identify complex motifs of properties that are shared by different neurons, which is central to identifying their functional phenotype. It is thus the type of informatics tool needed for developing the field of functional proteomics.

The last step in studying these properties is to integrate them in computational models. This job is facilitated by Model Database (ModelDB) (Davison et al., 2002; Migliore et al., 2003). ModelDB and NeuronDB share the ontology of channels, receptors, and neurotransmitters, which makes it convenient to search for models with particular properties or to search for experimental data in NeuronDB relevant to particular models. Other relationships among models are described in ModelDB with a concept ontology and through citation relationships. ModelDB currrently contains over 100 computational models of different types of neurons and their properties, with uniform formatting and curating. The models are fully documented by being linked to peer-reviewed publications. Many of these models run in NEURON (Hines and Carnevale, 2001). All models can be downloaded and run as they were published, or the files may be changed as the user wishes in order to develop new insights into a model of a particular property or neuron type.

The functionality of each of these databases, and the ability to navigate smoothly between them and to other related databases on the Web, is explained on the websites.

20.5 CRITERIA OF PROGRESS

Usefulness. SenseLab started with ORDB for supporting cloning and sequencing of the olfactory receptor gene family by laboratories in the field. The next was NeuronDB, to support electrophysiological studies of the active properties of distal dendrites, together with ModelDB, to support the parallel exploration of dendritic compartmental models. The other databases have been added in response to new needs over the intervening years, most recently OdorMapDB. The usefulness of the databases is indicated by the total of over 348,000 hits of SenseLab over the past year, with the highest number going to ORDB (98,000) and ModelDB (71,000).

Evaluation. In order to go beyond the totals and obtain better insight into how the databases are being used, we instituted a Web page that pops up when a user hits one of the databases more than 100 times in a day. The page simply asks the user to indicate who they are, how they are using the website, and any suggestions for improving the database to serve their needs. This has given rise to a number of responses that have shown the broad uses that are being found for the databases, along with valuable suggestions for corrections or improvements.

20.6 CHALLENGES FOR THE FUTURE

The obvious challenges for the future will focus on continually improving the functionality of the databases, as well as building in new properties and new search tools. The flexible EAV/CR database architecture (Nadkarni et al., 1999) will be critical in facilitating these improvements and expansions, as detailed in the companion chapter (Chapter 5) in this volume.

Here we identify several challenges that will be important not only for SenseLab but for all databases serving the wider neuroscience community.

Standards. The standards for accepting data into a database was one of the first faced by the original Institute of Medicine committee that recommended setting up the Human Brain Project, and one that almost stopped the project before it could get started. Who determines what data go into a database? After much discussion, it was decided that rather than try to set up a central mechanisms for deciding on all data, it would be better to encourage individual pilot projects to deal with this problem in the different areas of neuroscience. As the Web has exploded, this has been a wise policy, in line with the openness that is a hallmark of the Web.

In the case of SenseLab, our databases are mainly based on published data. Thus the standards for the data are the standards of the journals that accepted the articles that reported the data.

Unpublished Data. SenseLab has also included unpublished data. This was initially in response to requests from sequencing laboratories to place fragmentary sequences obtained but not published in an anonymous database section. Anyone doing a blast search and getting a hit for that sequence could contact the producing laboratory to initiate a dialogue on whether one or the other laboratory had an interest in completing the sequencing.

There are, of course, mountains of unpublished data of many types: fine structural images, stained cell images, physiological recordings, brain scans, and so on. There are many who feel that one of the aims of neuroscience databases should be to make those data openly available. We see no problem with individuals mounting those data on their own websites, although there are potential copyright problems if authors wish to publish those data in journals at some later date. For centrally curated databases such as those in SenseLab, a more serious problem with including unpublished data is the potentially overwhelming quantity and variable quality of that data. At present we feel that the highest priority needs to be put on populating the databases with high-quality published data.

Populating Databases. A database is of little use unless it is adequately filled with data, adequately "populated" as one says. Many an ambitious database project has foundered on this problem, when, for example, there is a great idea for a database, but without data no one shows up. A partially populated database may be worse, because a search can actually be misleading if it is incomplete.

To deal with this problem, SenseLab has in the first instance relied on data entry by its senior investigators and colleagues. It has employed students to enter data, some working over the Internet at distant sites, constituting a virtual "global laboratory." It has actively explored methods for automated data entry through natural language algorithms. The last approach has been hampered by a number of factors: the diverse ways that authors report their data, the large numbers of alternate terms for the same cell or property, the journal formats that make it difficult for automated searches, and the limited value of key words for automated recovery. It is clear that there needs to be a reassessment of these aspects of traditional journal publishing in order to facilitate better the ability to extract complex data automatically from electronic journal pages.

Ontologies and Interoperability. The problem of alternate terms for the same cell or property is looming larger and larger as the scope of databases grows. The science (or perhaps art) of terminology has been referred to as *ontology*. Standardization is important because different terms may mean the same thing (pyramidal neuron; Betz cell), or the same term may have multiple meanings in different contexts (Purkinje cell in heart or cerebellum). Reconciling these different terms and meanings in machine understandable ways is the province of the science (or perhaps art) of interoperability.

At present, databases are developed by different laboratories for different purposes. It is obvious that the need to coordinate them will make interoperability one of the major challenges for neuroinformatics in the foreseeable future. This problem is addressed by other authors in the present volume.

SenseLab has had to confront this problem in the multiple terms that have been applied to different olfactory receptor (OR) genes and proteins. For example, for the human ORs, at least four different nomenclatures have been proposed. Since there are 1000 or so OR genes, the problem is not trivial. At present, ORDB is addressing this problem through an internal interoperability by aligning the different genes with their different terms so that the user can decide for her/himself which term to use for a given OR.

The informatics infrastructure that facilitates interoperability between SenseLab databases and for expansion to implement new databases is described in a companion chapter (Chapter 5) in this volume.

Sharing. A final problem, perhaps the most important and difficult for neuroinformatics, is that this new discipline requires an ethic of sharing data (Koslow, 2002). This seems obvious enough; it has been the key to the success of the gene and protein databases, which owe their growth and utility to a simple rule: When a new gene or protein is identified, a condition of publication is that it is deposited in the appropriate database so that all can access it for further study and comparison with other sequences. Without this rule, there is no sharing, and there are no data for analysis and comparison.

In the field of neuroscience, by comparison, there is not yet this ethic of data sharing. Although the identification of a new property of a neuron or glia cell or neural circuit may be published electronically, the data sit on the electronic page and can only be extracted by laborious searches on key words. What is needed is placing the data at the time of publication into databases where sophisticated search tools can enable colleagues to carry out analyses and comparisons that go beyond the initial analysis in the primary publication. We thus have the two principles that must guide the development of effective databases to support neuroscience research: sharing and searching.

The first steps toward this goal are being taken in the field. The fMRI Data Center (http://www.fmridc.org) is beginning to receive fMRI brain scan data that include not only the published results but also the datasets that produced them. Others are building databases of nerve cell morphology and electrophysiological recordings of nerve cell activity (see the website for the Human Brain Project (http://www.nimh.nih.gov/neuroinformatics/index.cfm). Links are being forged at an increasing rate between the different databases.

20.7 TOWARD A BRAIN DATABASE

From this review of SenseLab and of other databases that are well along in their development, it appears that considerable progress has been made in neuroinformatics in the past 10 years, from tentative promises to robust and useful informatics tools supporting neuroscience research. These pilot studies can be viewed as having established the feasibility of using neuroinformatics to provide critical support for neuroscience research. In addition, they have constructed databases that provide systematic inventories of data on brain structure and function.

The next challenge is to organize these mulitplying databases into a metadatabase of neuroscience databases. As a first step toward the this goal, a Human Brain Project Database (http://senselab.med.yale.edu/hbpdb) site has been constructed, leveraging the EAV/CR approach described above. HBPDB provides search tools for organizing and accessing the data contained in the 35 projects of the Human Brain Project, in terms of multiple levels of organization and multiple types of data. It is a step toward the original vision of "Mapping the Brain and its Functions" (Pechura and Martin, 1991). The ability to realize this vision—to map the brain as well as the Human Genome Project has mapped the genome—will be limited only by our willingness to share our data, link our databases, and provide the search tools to make sense of it all (Marenco et al., 2003).

ACKNOWLEDGMENTS

Our research has been supported by National Institutes of Health grants P20LM07253 and T15LM07056 from the National Library of Medicine; the Human Brain Project/Neuroinformatics funded jointly by the National Institute on Deafness and Other Communication Disorders, National Institutes of Mental Health, National Institute of Neurological Disorders and Stroke, National Institute on Aging, and National Science Foundation; and the Department of Defense under a Multiple University Research Initiative (MURI).

References

Buck, L., and Axel, R.A. (1991). A novel multigene family may encode odorant receptors: A molecular basis for odor recognition. *Cell* **65**, 175–187.

Crasto, C., Marenco, L., Miller, P., Shepherd, G. (2002). Olfactory Receptor Database: A metadata–driven automated population from sources of gene and protein sequences. *Nucleic Acids Research* **30**, 354–360.

Davison, A.P., Morse, T.M., Migliore, M., Marenco, L., Shepherd. G.M., and Hines, M.L. (2002). ModelDB: A resource for neuronal and network modeling. In *Neuroscience Databases: A Practical Guide* (R. Kotter, ed.), pp. 99–109. Kluwer Academic Publishers, Dordrecht, The Netherlands.

Hines, M.L., and Carnevale, N.T. (2001). NEURON: A tool for neuroscientists. *The Neuroscientist* **7**, 123–135.

Koslow, S.H. (2002). Opinion: Sharing primary data: A threat or asset to discovery? *Nat. Rev. Neurosci.* **3**, 311–313.

Liu, N., Marenco, L., Nadkarni, P., Miller, P., and Shepherd, G.M. (2002). Senselab odor map database (OdorMapDB) and tools for online regional brain image analysis. The Human Brain Project Annual Meeting, Bethesda, MD, May.

Liu, N., Xu, F., Marenco, L., Miller, P., and Shepherd, G.M. (2003). Mapping the rodent olfactory bulb: OdorMapBuilder, a generic software program to unfold 3D spatial activity patterns. The Human Brain Project Annual Conference, Bethesda, MD, May.

Marenco, L., Tosches, N., Crasto, C.J., Shepherd, G.M., Miller, P.L, and Nadkarni, P.M. (2003). Achieving evolvable web-database bioscience applications using the EAV/CR framework. *J. Am. Med. Inf. Assoc.* **10** (5), 444–453.

Migliore, M., Morse, T.M., Davison, A.P., Marenco, L., Shepherd, G.M., and Hines, M.L. (2003). ModelDB. Making models publicly accessible to support computational neuroscience. *Neuroinformatics* **1**, 131–134.

Miller, P.L., Nadkarni, P., Singer, M., Marenco, L., Hines, M., and Shepherd, G. (2001). Integration of multidisciplinary sensory data: A pilot model of the human brain project approach. *J. Am. Med. Inf. Assoc.* **8**, 34–48.

Mombaerts, P. (1999). Seven-transmembrane proteins are odorant and chemosensory receptors. *Science* **286**, 707–711.

Nadkarni, P.M., Marenco, L., Chen, R., Skoufos, E., Shepherd, G., and Miller, P. (1999). Organization of heterogeneous scientific data using the EAV/CR representation. *J. Am. Med. Inf. Assoc.* **6**, 478–493.

Pechura, C.M., and Martin, J.B., eds. (1991). *Mapping the Brain and its Functions. Integrating Enabling Technologies into Neuroscience Research.* National Academy Press, Washington, DC.

Pilpel, Y., Sosinsky, A., and Lancet, D. (1998). Molecular biology of olfactory receptors. *Essays Biochem.* **18**. 243–250.

Shepherd, G., Mirsky, J.S., Healy, M.P., Singer, M.S., Skoufos, E., Hines, M.S., Nadkarni, P., and Miller, P.L. (1998). The Human Brain Project: Neuroinformatics tools for integrating, searching and modeling multidisciplinary neuroscience data. *Trends Neurosci.* **21**, 460–468.

Singer, M.S. (2000). Analysis of the molecular basis for octanal interactions in the expressed rat 17 olfactory receptor. *Chem. Senses* **25**, 155–165.

Skoufos, E. (1999). Conserved sequence motifs of olfactory receptor-like proteins may participate in upstream and downstream signal transduction. *Receptors & Channels* **6**, 401–413.

Xu, F., Greer, C.A., and Shepherd, G.M. (2000). Odor maps in the olfactory bulb. *J. Comp. Neurol.* **422**, 489–495.

Three-Dimensional Visualization and Analysis of Wiring Patterns in the Brain: Experiments, Tools, Models, and Databases

Jan G. Bjaalie, M.D., Ph.D. and *Trygve B. Leergaard, M.D., Ph.D.*

CONTENTS

21

Databasing the Brain. Edited by Stephen H. Koslow and Shankar Subramaniam
ISBN 0-471-30921-4 © 2005 John Wiley & Sons, Inc.

21.1 INTRODUCTION

The central nervous system is essentially a conglomerate of nerve cells, arborizations, and supportive tissue. Aggregates of such basic components constitute a multitude of units, each with a characteristic intrinsic structural organization and a unique pattern of connectivity with other units. Through complex but highly organized wiring patterns, different units form distributed functional systems. The study of these wiring patterns of the brain is a classical neuroscience discipline, providing important contributions toward understanding of brain function. Indeed, the function of any given region of the brain is influenced by the particular connections of that region with other parts of the nervous system. The scope of the present chapter is to review and evaluate commonly used as well as recently introduced strategies and approaches for exploring the wiring patterns in the brain, as well as to discuss new trends and possibilities in the field.

Connectivity in the brain may be studied at variable resolutions with several anatomical and physiological techniques, as well as with tomographic imaging tools. Knowledge about connections in the brain builds profoundly on data collected from lesion and axonal tracing experiments (Brodal, 1981; Heimer and Robards, 1981; Heimer and Zaborszky, 1989; Swanson, 2000; Martin, 2002). Axonal tracers are suitable not only for demonstrating the presence of connections but also for characterizing detailed architectonical features of neuronal cell populations and their projections. Currently available axonal tracer substances allow investigation of neural connections at levels of resolution unmatched by other techniques (Van Haeften and Wouterlood, 2000; Vercelli et al., 2000; Wouterlood et al., 2002b). Many existing new techniques for investigating brain connectivity have been developed recently (see, e.g., Conturo et al., 1999; Mori, 2002; Senft, 2002; Saleem et al., 2002). Nevertheless, axonal tracers represent the gold standard for mapping wiring patterns in the brain.

Analysis of brain circuitry with axonal tracers may be conducted at different levels. One is the system level, dealing with the organization of neuronal populations and projection systems in the brain (Gerfen, 1992; Knierim and Van Essen, 1992; Brodal and Bjaalie, 1997; Ruigrok, 1997; Rouiller and Welker, 2000; Leergaard, 2003). Other, more detailed levels cover studies of the morphology and properties of individual neurons and their local connections (Martin, 2002; Wang et al., 2002). Here we will focus on the system level. At this level, axonal tracers can be used to (1) detect connections between different parts of the brain, at the level of cortical regions and areas, and nuclei of deeper structures, (2) identify distribution and number of neurons projecting to a defined target, or extent and densities of axonal projections originating from a specific site in the brain,

and (3) outline overall principles of topography and structural design of brain connections.

Over the past decades, many investigators have used a variety of tracer methods and contributed huge amounts of data and insights about connections in the brain (see, e.g., Brodal, 1981; Nieuwenhuys et al., 1988; Voogd, 1995; Kamper et al., 2002). Most of the connections in the brain have now been analyzed in some detail with axonal tracing methods. There are, nevertheless, reasons to suggest that the potentials of these methods have not been fully exploited. First, many of the classical studies have relied on methods that are less sensitive, and allow less detailed morphological visualization, than more recently introduced axonal tracing methods (Brodal, 1981; Heimer and Robards, 1981; Heimer and Zaborszky, 1989; Vercelli et al., 2000). Second, most of the investigations performed to date have studied tracer labeling patterns in selected single sections only. This may have resulted in an incomplete understanding of the principles of topographical organization in several systems. Collecting information from more complete series of sections, along with adding three-dimensional (3D) reconstruction, visualization, and analyses, has turned out to provide new data and new dimensions of understanding of brain circuitry (see, e.g., Bjaalie et al., 1991; Nikundiwe et al., 1994; Malmierca et al., 1995, 1998; Leergaard et al., 1995, 2000a,b, 2004; Arecchi-Bouchhioua et al., 1996; Bajo et al., 1999; Nadasdy and Zaborszky, 2001; Parent et al., 2001; Diaz et al., 2003). Third, in neuroanatomy, as in most other neuroscience disciplines, researchers are faced with a lack of standardized formats for data presentation, along with a general lack of tools and environments for sharing and re-use of lower level data (Amari et al., 2002; Koslow, 2000, 2002; Bjaalie, 2002). The value of the information contained in the spatial patterns revealed with tracing methods will increase considerably if information is presented in formats allowing direct comparison with similar and other data types, available for the same specific brain location (Bjaalie, 2002).

In this chapter, we will provide an overview of some of the recent developments in the study of brain connections at the system level. The primary examples are taken from the rat somatosensory and cerebro-cerebellar systems. Following general considerations on data types and procedures, along with a mini-review of the most commonly used axonal tracing methods, we will consider the state of the art for data acquisition, 3D reconstruction from serial section data, and visualization and analysis of neuronal tracing data. Finally, we will consider strategies for high-throughput data recording, digital atlasing, and databasing that will be needed to accelerate progress in the field.

21.2 DATA TYPES AND BASIC PROCEDURES

Data describing brain connectivity at the system level can be of multiple categories, ranging from tables listing presence or absence of axonal tracer labeling, or numbers of tracer labeled cells in different structures of the brain, via wiring diagrams showing the individual components that are connected (Nieuwenhuys et al., 1988; Welker et al., 1988; Swanson, 2000), to maps of complex 3D distribution patterns (Nikundiwe et al., 1994; Leergaard et al., 1995, 2000b; Merchan and Berbel, 1996;

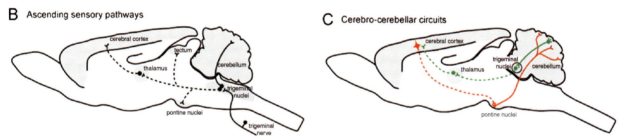

Figure 21.1   Schematic diagrams of somatosensory pathways in the rat central nervous system, showed to exemplify communication of data extracted from connectivity studies. (**A**) Diagram of the input and output relationships of the primary somatosensory cortex (modified from Welker et al., 1988, with permission). Sensory impulses from the face/head reach the contralateral primary somatosensory cortex via the trigeminal nuclei and the thalamus. The cortex, in turn, projects bilaterally to multiple cortical and subcortical targets. (**B**) Schematic overview of ascending somatosensory pathways. Somatosensory impulses arising from the head (red) enter the brain by way of the trigeminal nerve. Within the central nervous system, sensory pathways from the face are synaptically interrupted in the trigeminal nuclei and are projected to the cortex via the thalamus, with collateral projections to various other targets, such as the cerebellum, pontine nuclei, and superior colliculus. (**C**) Schematic depiction of the large cerebro-cerebellar pathways, which reciprocally links the cerebral cortex with the contralateral cerebellar cortex by way of the pontine nuclei, the deep cerebellar nuclei, and the thalamus. Thin lines in Part A indicate quantitatively minor projections. Dashed lines in Parts B and C indicate contralateral projections. Cb, cerebellum; CP, caudate-putamen; MI, primary motor cortex, PN, pontine nuclei; PO, posterior complex thalamus; PR, perirhinal cortex; RT, reticular nucleus thalamus; SI, primary somatosensory cortex, SII, secondary somatosensory cortex; SC, superior colliculus; VB, ventrobasal complex thalamus.

Malmierca et al., 1998). Here we will briefly exemplify simple wiring diagrams (Figure 21.1) and complex topographical maps (Figure 21.2), and we will outline the basic methods and procedures that are typically coupled to the production of these data types.

Simple wiring diagrams are shown in Figure 21.1. These diagrams illustrate the most basic data type related to brain connectivity, the presence of connections between units in the brain (areas and nuclei). Here, rat somatosensory-related connections are shown. It is seen that the face-related part of the somatosensory cortex has a number of input and output connections (Figure 21.1A; Welker et al., 1988). The sensory information from the face enters the brain via the trigeminal nerve and nucleus and is distributed to multiple targets in the brain, including the cerebral cortex, via the thalamus (Figure 21.1A,B). From the cerebral cortex, somatosensory information (as well as multiple other modalities) is conveyed through large re-entrant circuits

via the cerebellum (Figure 21.1C; Middleton and Strick, 2000) and basal ganglia (Alexander et al., 1986, 1990). Much of the present understanding of these and related systems is derived from axonal tracing investigations.

To accurately chart all the connections present in each wiring diagram, data from numerous experiments, published in multiple reports, are lumped together. Furthermore, all areas and nuclei included are defined using architectonic and physiological criteria. The definition of the boundaries of structures in the brain, as well as the naming of the structures, is far from trivial. Many boundaries are only vaguely defined, and different investigators tend to emphasize different criteria. Several different parcellation and naming schemes therefore exist for some parts of the brain, hampering the communication of results collected from such regions (compare, e.g., Paxinos and Watson, 1998; Swanson, 1999; also, see Bowden and Martin, 1995; Goddard et al., 2001). We will return to these problems below, in the context of data comparison

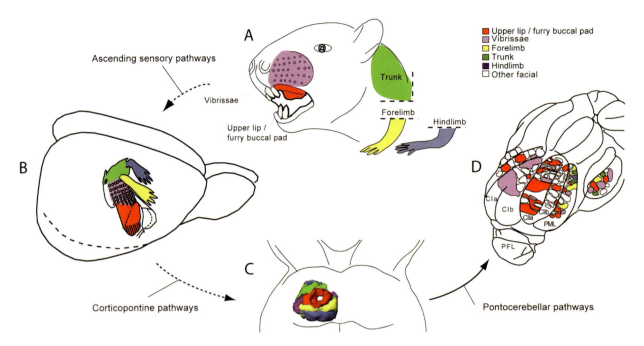

Figure 21.2 Map transformations in the rat somatosensory system. Sensory information from the the body surfaces on the left side (**A**) is orderly projected through ascending pathways and represented in a somatotopic map in the right primary somatosensory cortex (**B**). The essentially 2D cortical somatotopic brain map (**B**) is projected to the pontine nuclei by way of corticopontine projections. In the pontine nuclei, the cortical map is transformed into a more complex 3D map in the right pontine nuclei (**C**). The cerebellar cortex, in turn, contains an even more complex map referred to as a fractured map (**D**). Dashed lines indicate projection pathways contralateral to the peripheral sensory surface. Part A, modified from Voogd (1995), with permission; Part B, modified from Welker (1971) with permission; Part C, based on Leergaard et al. (2000a,b); Part D, modified from Voogd (1995) with permission and adjusted according to data from Bower et al. (1981) and Bower (1997).

and coordinate systems, and emphasize fundamental approaches that need to be pursued in order to increase the value of the data collected in different experiments.

Data describing topographical maps, and the transformations of such maps from one brain structure onto another, are much more complex than the pure wiring diagram data. Figure 21.2 exemplifies somatosensory topographical data. Here, cortical and subcortical representations of the body surface are shown in the cerebral cortex, the pontine nuclei (the group of nuclei intercalated in the pathways between the cerebral cortex and the cerebellum), and the cerebellar cortex. The cerebral cortical and cerebellar maps (Figure 21.2B,D) are generated from electrophysiological micromapping data (Woolsey, 1958; Welker, 1971; Woolsey and Van der Loos, 1970; Shambes et al., 1978; Bower and Kassel, 1990). The pontine map (Figure 21.2C) is derived from axonal tracing data with the use of tracers injected under electrophysiological guidance into various locations of the cortex (Figure 21.2B). The pontine map shown is based on a combination of data from altogether 38 axonal tracing experiments and 51 tracer injections (Leergaard et al., 2000a,b). The use of computerized 3D reconstructions and a common coordinate system was fundamental for revealing this pattern (for references and discussion, see Leergaard, 2003).

At a more detailed level, it is seen that the projections from the left body surface to the right cerebral cortex (Figure 21.2A,B, with intermediate stages shown in Figure 21.1 A,B) are orderly arranged, resulting in a fairly continuous point-to-point mapping of cutaneous representations in the primary somatosensory cortex (Figure 21.2B). The cerebral cortex projects onto

the ipsilateral pontine nuclei (corticopontine projection), and in this system the continuous and essentially two-dimensional (2D) cortical map is transformed onto a more complex 3D map (Figure 21.2C). Individual body representations are here represented in a more distributed fashion, with two or more representations for each major body part. The overall spatial relationships between each body part are, nevertheless, preserved. The pontine nuclei, in turn, project to the contralateral cerebellar cortex (pontocerebellar projection). The mossy fiber axons of pontine neurons give rise to a patchy map in the left cerebellar cortex (Figure 21.2D), where somatotopic order is apparently lost, and sensory representations are dispersed in a 2D, fractured map (Shambes et al., 1978; Bower and Kassel, 1990). Moreover, the pontine input to the cerebellum is in registration with the trigeminal input (Figure 21.1B), producing a coherent map of cutaneous representations from different sources of input (Bower et al., 1981). These map transformations are thought to reflect properties of the processing taking place in the pathways between the cerebral cortex and the cerebellum.

The map shown in Figure 21.2C is of particular interest in the context of 3D analysis of data from axonal tracing experiments, a topic that we will focus on in the present chapter. A sequence of experimental and technical manipulations leads to this type of end result. First, axonal tracers are delivered into defined locations in the brain. The animal survives and the tracer is transported along the axons to label the connections of the tracer delivery site. Following perfusion, the relevant brain tissue is sectioned serially and processed, before data acquisition in the microscope is performed. The ensuing data may be submitted to

3D reconstruction, visualization, and analysis, as well as to transfer to a common coordinate system for across animal comparisons. Finally, data averaging and simplification may lead to the construction of a single map based on data from multiple experimental animals and tracer injections (Leergaard et al., 2000a,b).

In the following, we will deal with the methodological and technical details related to the use of axonal tracers and the further manipulation of tracing data, as well as with the fundamental problems with the use of axonal tracing techniques, data analysis, and data re-use and exchange. The examples shown in Figures 21.3–21.9 will be from the system illustrated in Figure 21.2.

21.3 AXONAL TRACING METHODS

Historically, neural tracing methods involved degeneration studies as well as studies of anterograde and retrograde axonal transport of macromolecules, such as isotopically labeled materials, tritiated amino acids, and exogeneous macromolecular markers (for historical reviews, see, e.g., Cowan and Cuénod, 1975; Brodal, 1981; Mesulam, 1982; Kuypers and Huisman, 1984; Heimer and Zaborszky, 1989). More recently, a variety of more sensitive tracing methods have been developed. These include dextran amines (Glover et al., 1986; Veenman et al., 1992; Reiner et al., 2000; Van Haeften and Wouterlood, 2000), Phaseolus vulgaris leucoagglutinin (Gerfen and Sawchenko, 1984; Wouterlood and Groenewegen, 1991), carbocyanine dyes (Honig and Hume, 1986, 1989; Thanos and Bonhoeffer, 1987; Sparks et al., 2000), bacterial toxins (Sawchenko and Gerfen, 1985; Angelucci et al., 1996), and latex microspheres (Katz et al., 1984; Katz and Iarovici, 1990), and they allow high-resolution studies of anatomic connections between brain regions (for a review, see Vercelli et al., 2000).

The large majority of axonal tracers requires in vivo delivery of a tracer substance into a defined part of the brain (Figure 21.3). During a postoperative survival time (usually several days), neural tracers are transported either along axons, from cell bodies to terminal fields (anterograde transport, Figures 21.3B and 21.4A–C), or in the opposite direction (retrograde transport, Figures 21.3C and 21.4D–F). Following various steps of processing, tracer labeled fibers and cells can be visualized microscopically in sections through the brain tissue. Sections of interest are often counterstained to visualize architectonic features, at the level of cellular and myelin staining, or identification of specific molecules (see, e.g., Skirboll et al., 1989; Chronwall et al., 1989; Wouterlood et al., 2002b; also, see Paxinos et al., 1999). Such staining is often required for the identification of boundaries of areas and nuclei that contain the labeling (see, e.g., Brevik et al., 2001).

Some tracers can be used to identify detailed anatomical information such as branches and varicosities of individual axons, at the level of light and fluorescence microscopy (Figure 21.4A–C). A few can be identified also at the electron microscopic level, providing access to interpretations at the synapse level. Axonal tracers can be combined for double (Figure 21.4) and triple (Kuypers and Huisman, 1984; Skirboll et al., 1989), and in some cases quadruple, labeling (Akintunde and Buxton, 1992), to facilitate analysis of complex topographical patterns. In combination with immunocytochemistry or in situ hybridization techniques, specific molecules or gene expressions can be assigned to tracer-labeled neuronal elements (Lewis et al., 1988; Skirboll et al., 1989; Chronwall et al., 1989). Finally, the localization of tracer injection sites can be functionally characterized using electrophysiological mapping or functional imaging prior to tracer delivery (Livingstone and Hubel, 1988; Arieli and Grinvald, 2002).

Variable amounts of tracer are delivered into the brain depending on the purpose and resolution of an investigation. Large amounts—for example, deposited at multiple adjacent sites—may be used to fill up as much as possible a well-defined brain region. Such experiments can be used to identify multiple sources of input (see, e.g., Bjaalie and Brodal, 1983; Dietrichs et al., 1983; Vassbø et al., 1999), and/or all outputs from the region (Leergaard et al., 2000b). Variations of this strategy include application of tracer into transected nerves or pathways (see, e.g., Armengol and Salinas, 1991; Glover, 1995; Liss and Wiberg, 1997; Holtzer et al., 2003), or application into peripheral nervous sheets, such as the retina or olfactory epithelium (e.g., Shatz, 1983; Pautler et al., 1998; Pautler and Koretsky, 2002). Although useful for some analyses, the use of large amounts of tracers generally provides data of limited specificity, since any given brain area contains a variety of neuronal populations with different characteristics, all of which are affected by the tracer delivered into the tissue. Smaller quantities of tracers, deposited in well-defined and restricted parts of the brain, are needed for collecting information about inputs and outputs at increasing levels of granularity and to study topographical organization (e.g., Brodal, 1982; Ruigrok, 1997; Köbbert et al., 2000; Leergaard and Bjaalie, 2002). Minute deposits of tracer allow investigation of the detailed connections of individual neurons (Zaborszky and Duque, 2000; Sugihara et al., 2001). The latter approach is technically demanding, usually requiring careful microscopic analysis across many sections. By combining tracing techniques with other methods, such as electrophysiological recording during tracer delivery, or identification of molecular distribution patterns in relation to tracer labeling in tissue sections, ensuing data may be better characterized in terms of functional parameters.

The ability of selected neurotrophic viruses (herpes, rabies, and pseudorabies) to infect chains of synaptically linked neurons has opened for advanced trans-synaptic studies of the origins of axonal projections to peripheral targets (Kuypers and Ugolini, 1990; Ugolini, 1995; Aston-Jones and Card, 2000), as well as tracing across several links in complex brain circuits (Middleton and Strick, 2000; Kelly and Strick, 2000). Another recently introduced technique is tomographic imaging, providing new opportunities for in vivo mapping of brain circuitry. The magnetic resonance imaging (MRI) detectable T1 contrast agent manganese was recently employed for tracing of major efferent sensory pathways in rodents (Pautler et al., 1998; Lin et al., 2001; Watanabe et al., 2001; Pautler and Koretsky, 2002), as well as for visualization of central nervous circuits in monkeys and birds (Saleem et al., 2002; Van der Linden et al., 2002).

The in vivo application of tracer substances, as discussed above, is difficult to use in some areas of animal research, such as the study of connectivity early in development, and, of course, cannot be used in humans. Several investigators are pursuing possibilities for noninvasive pathway tracing using diffusion tensor imaging (reviewed in Mori, 2002) for tracing of major projection systems in humans as well as in animals. The level of detail

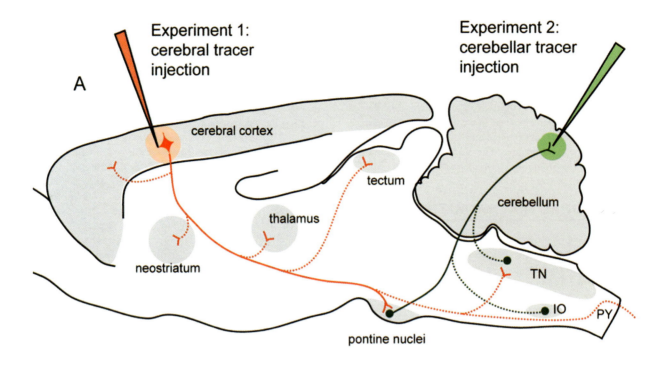

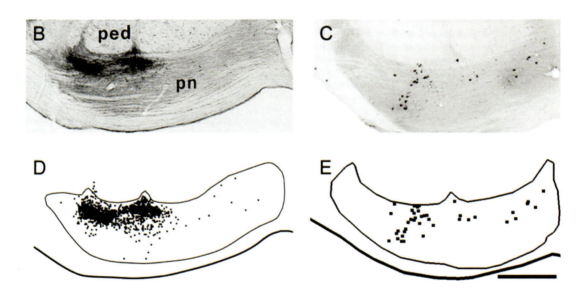

Figure 21.3 Anterograde and retrograde tracing in rat cerebrocerebellar system—that is, the corticopontine (solid red lines) and pontocerebellar (solid green lines) pathways. In Experiment 1, a neural tracer is focally placed within the cerebral cortex and anterogradely transported to label the trajectories and terminal fields in multiple cortical and subcortical target regions. Note that some subcortical projections arise from different populations of cortical cells, while other projection are collaterals with a common cortical origin. In Experiment 2, a neural tracer is placed in a restricted region of the cerebellar cortex and retrogradely transported to label neurons in multiple regions of origin that send their axons to this particular injection site. For several practical reasons, and as shown in this example, connectivity studies are limited to studying connections of only one or a few sites at a time. (**B, C**). Photomicrographs and dot maps of transverse sections through the pontine nuclei with anterogradely labeled axonal plexuses arising from the cerebral cortex (**B**), and retrogradely labeled neurons that project to the cerebellum (**C**). The labeled axonal plexuses (**C**) are semiquantitatively coded as dot densities corresponding to densities of labeled axons (**D**), while labeled neurons are recorded as one dot per labeled cell (**E**). Part B contains data published in Leergaard et al. (2000b). Scale bar, 500 μm (Parts B and C).

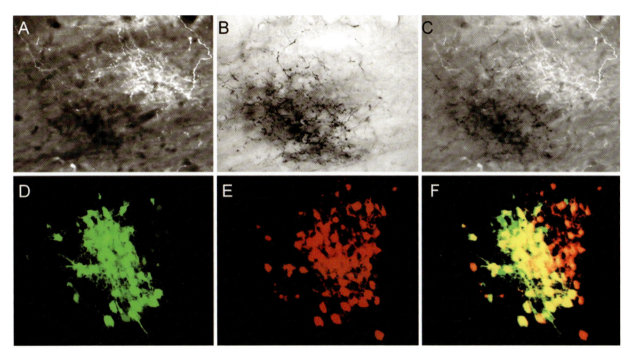

Figure 21.4 Examples showing dual anterograde tracing of axonal plexuses (**A–C**) and dual retrograde tracing of neurons (**D–F**). *Anterograde tracing* (**A–C**): Focal injection of the two tracers rhodamine-conjugated dextran amine (FluoroRuby) and biotinylated dextran amine (BDA) into individual whisker representations in the rat primary somatosensory cortex labeled axonal plexuses in multiple brain regions, here exemplified in corresponding photomicrographs of a transverse section through the pontine nuclei. Part A shows fluorescent FluoroRuby labeling in a section viewed with fluorescence microscopy, Part B shows BDA labeling in the same section viewed with bright field microscopy, and Part C shows an overlay of the two images. *Retrograde tracing* (**D–F**): Two partially overlapping populations of vestibulospinal and vestibulo-ocular neuron groups in a chicken embryo were retrogradely labeled by application of fluorescein-conjugated dextran amines into the contralateral cervical spinal cord, along with application of rhodamine-conjugated dextran amine into the ipsilateral medial longitudinal fascicle. Part D shows labeled (green) contralaterally projecting vestibulospinal neurons, Part E shows labeled (red) ipsilaterally projecting vestibuloocular neurons, and Part F shows the two populations together. Parts A–C, data from Leergaard et al. (2000a); Parts D–F, modified from Diaz et al. (2003), with permission.

provided by these techniques is limited. A method that can produce potentially more accurate data is the postmortem tracing techniques, relying on the use of carbocyanine dyes in fixed tissue (reviewed in Honig and Hume, 1989; Köbbert et al., 2000). An obvious disadvantage with the postmortem approach is the limited diffusion of the tracers that precludes tracing over longer distances. Whereas in vivo application of carbocyanine dyes in rats (for example) labels even the longest projection systems, only few examples of postmortem tracing of these dyes over longer distances have been reported (O'Leary and Terashima, 1988).

In summary, a collection of axonal tracing methods, delivering high sensitivity and detailed anatomical information, are available today. The tracers can be used in different combinations for multiple labeling experiments, as well as in combinations with other techniques. This allows the investigator to chart connectivity throughout many systems of the brain, as well as to perform sophisticated experiments to test hypotheses related to anatomical and functional organization.

21.4 DATA ACQUISITION

Anatomic studies of system level connectivity generally require microtome sectioning of the brain tissue followed by microscopic analysis. Historically, microphotographs and camera lucida drawings were the primary data acquisition methods. Today, digital images and computerized geometrical representations have largely replaced the classical approaches.

Image data are captured with a camera attached to the microscope. At the magnification necessary for the camera to acquire data at single-cell resolution (Figure 21.4), the diameter of the object field of a compound microscope is only a fraction of the dimension of most sections studied. This necessitates that multiple images are collected and stitched together to produce a composite image, or mosaic, covering the region of interest in each section (Figures 21.3B,C and 21.5B). This process is typically aided by a motorized, computer-controlled, microscope stage and can be implemented with commonly used image analysis systems (see, e.g., Lillehaug et al., 2002).

Geometry data are essentially point and line representations of features in the section images (Figure 21.3D,E), such as boundaries of areas and nuclei (coded as lines) and labeled cells or axonal varicosities (coded as points). Such geometry data can be acquired with a multitude of approaches (for a review, see Bjaalie, 1992), two of which will be dealt with here: semiautomatic and automatic methods. The most commonly used semiautomatic method is based on the principle of image-combining computerized microscopy, introduced by Glaser and Van der Loos (Glaser et al., 1979, 1983) and implemented by several investigators and companies (Capowski, 1985, 1989; Glaser and Glaser, 1990; Leergaard and Bjaalie, 1995). In brief, this method mixes a computer-generated image and the image of the specimen. A graphical overlay is manually introduced on the image of the specimen, and the user performs a manual segmentation of the structures of interest by pointing with a mouse to place graphic symbols, lines, and points, representing different data categories (Figure 21.5A). The microscope is equipped with a motorized stage; and as the position of the microscope stage is changed, the graphical overlay moves synchronously with the image of the specimen. Thus, the user views the specimen with an overlay of lines and dots, providing direct feedback during

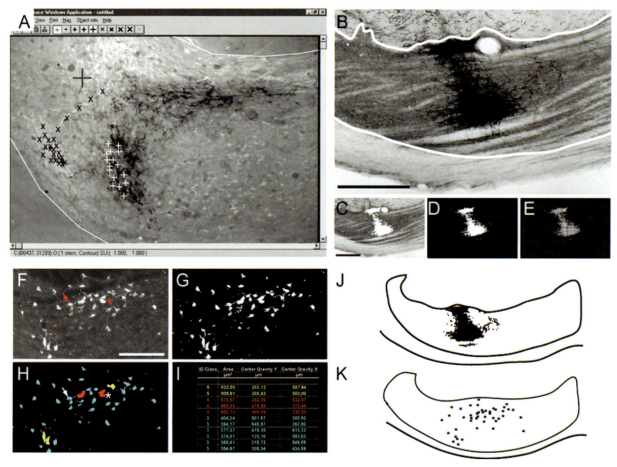

Figure 21.5 Semiautomatic (**A**) and automatic (**B–K**) methods for data acquisition from serial sections. *Semiautomatic data acquisition* (**A**): The graphical user interface of the data acquisition program MicroTrace as seen in the field of view of an image-combining computer microscope. The specimen is a transverse section through the right pontine nuclei viewed with fluorescence microscopy (through the Leitz N2.1 rhodamine filter block). Axonal plexuses were anterogradely labeled after injection of rhodamine-conjugated dextran amine and biotinylated dextran amine in electrophysiologically defined individual whisker representations in SI (data from Leergaard et al., 2000a, with permission). The borders of the pontine nuclei are digitized as lines, and two categories of symbols are placed above parts of the labeled regions. The large cross illustrates the computer screen cursor. *Automatic data acquisition of labeled axonal plexuses* (**B**): Digital photomicrograph of labeled corticopontine axonal plexuses (black), visible in dense and loose plexuses. The boundary of the pontine nuclei are indicated by a white line. Labeled fibers, appearing as black or shades of gray, were identified from the digitized gray-scale image by choosing a threshold value from minimum 0 (black) to a certain maximum level (dark gray) and represented as white (**C**). A binary image (**D**) was created from a chosen threshold value. By superimposing a grid onto the binary image (**E**), it was possible to obtain *x, y* coordinates from squares containing more than a given number of white pixels, which in turn was represented as black dots (**J**). *Automatic detection of labeled cell distributions:* Digital grey-scale image of transverse section through the pontine nuclei (**F**), showing pontocerebellar neurons labeled following injection of the fluorescent tracer rhodamine conjugated dextran amine into the cerebellar crus IIa. The digital image was converted to a binary black and white image by applying a threshold value (**G**). Binary operators were applied to remove artifacts, fill closed gaps, and separate particles. Dense clusters of overlapping cells were not separated (*asterisk*), while partially overlapping cells, connected only by a thin bridge (*arrow*), were separated (**H**). A "detect particle" function embedded in the analySIS software was used to count the resulting particles and sort them in a table according to area. A part of this table is shown in **I**. Unique colors were assigned to different area ranges, and spatial coordinates *x, y* were automatically assigned to the center of gravity of each particle and could be represented by black dots in a graphic plot (**K**). Careful comparison of the original "raw image" with the analyzed image shows that most labeled neurons are recognized correctly. Some cells, however, were not counted as separate objects due to the merging of overlapping cells. Weakly labeled cells, cellular fragments, and small artifacts (pink) that gave rise to particle sizes below the defined size (area) range for labeled cells were excluded from the output table. Scale bars, 500 μm (**B–E**) and 200 μm (**F–I**). Part A is reproduced from Leergaard and Bjaalie (2002), with permission; Parts B–K are modified from Lillehaug et al. (2002), with permission.

the data entry procedure. This type of data acquisition has been used in numerous recent publications (for extensive bibliography, see http://www.microbrightfield.com/) and allows accurate and rapid recording of structures at high magnification. An example of a system configuration is detailed in Leergaard and Bjaalie (1995). Software for image-combining microscopy include MicroTrace (Leergaard and Bjaalie, 1995) and Neurolucida (MicroBrightField, Colchester, VT, USA).

More automatic data acquisition methods have improved in recent years and now provide opportunities for reliable recording of tissue elements across large brain regions. Practical implementation is often best achieved with combined use of the semiautomatic and automatic methods. The basis for the automatic methods is image data, as discussed above. Usually, composite images covering the region of interest are used. To automatically assign coordinates to large numbers of structures—for example, populations of labeled objects (stained neuronal cell bodies, or clusters of axonal plexuses)—it is useful to first create binary images with a high signal contrast (Figure 21.5 D,G). To further distinguish signals of interest from background noise, images can be filtered with different threshold values, and objects sorted according to shape and size, before localization coordinates are assigned (Figure 21.5 H,I; see, e.g., Lillehaug et al., 2002). Image analysis tools allow recording of labeled axonal plexuses, neurons, and structural boundaries in stained sections. Confocal microscopy may add value to automatic methods, by providing high-resolution information in multiple thin optical sections through thick sections (Wouterlood et al., 2002a).

21.5 THREE-DIMENSIONAL RECONSTRUCTION FROM SERIAL SECTIONS

Following data acquisition, the next key step is the co-registration of the information collected from the sections. Stacks of 2D digital images can be aligned and used as a basis for 3D volume renderings. For the purpose of axonal tracing, however, the by far most commonly used approach is to use 2D geometric data from each section—that is, point- and line-coded data representing the objects of interest—as a basis for 3D reconstruction. The geometry data may be collected with direct semiautomatic digitization, or with image analysis approaches relying on the initial collection of digital images, as explained above.

This section will outline the basic approach for reconstruction of 3D spatial relationships from 2D point- and line-coded datasets derived from serial sections. The precision of a 3D reconstruction based on data from serial sections depends heavily on the section quality, the data acquisition method, and the procedure for section alignment. It is essential to minimize section distortions and to obtain complete series of sections. In general, 3D co-registration of sections is obtained with the use of identifiable anatomic landmarks present in the histological material. The outer boundaries of recognizable structures, vessels, or fiducial markers, recorded together with the data of interest, are often used for alignment. Block-face images, captured during microtomy, may also be used as a basis for the 3D reconstruction (Annese and Toga, 2002). Based on data acquisition as exemplified above, section images, or geometric representations

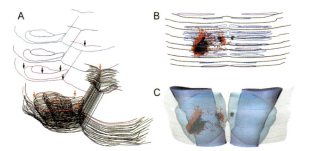

Figure 21.6   Assembly and visualization of a 3D reconstruction of the rat pontine nuclei (data from Leergaard et al., 2000a). (**A**). Series of digitized transverse sections through the pontine nuclei are aligned according to multiple anatomical landmarks (*arrows*), by careful positioning of individual (colored) sections while inspecting the growing 3D reconstruction (black sections) from multiple angles of view on the computer screen. The 3D reconstruction is shown here in an oblique rostrolateral view. (**B, C**) Ventral views of the complete reconstruction. Part B shows a line-and-point representation, with different anatomic structures displayed in different colors and with labeled axons shown as red and black dots. In Part C, the boundaries of the brain stem, the pontine nuclei, and the descending fiber tracts are shown as transparent or solid surfaces.

based on section images (consisting of point and line data) are brought in 3D registration using different approaches depending on the software solutions used. For example, with the Neurolucida software (MicroBrightField, Colchester, VT), stacks of section drawings are aligned during the data acquisition procedure by matching the positions of several landmarks in neighboring sections. With other software (e.g., Micro3D developed in our laboratory, cf. http://www.nesys.uio.no/), more dynamic alignment is performed by maneuvering individual sections and matching multiple landmarks using real-time rotation of the reconstruction during alignment (Figure 21.6). To further aid the inspection of the 3D reconstructions from various angles of view, digitized sections from control brains, cut in section planes orthogonal to the series to be reconstructed, may be used as templates for alignment (see, e.g., Brevik et al., 2001). To ensure that the reconstruction retains natural proportions, sections submitted to histological procedures are measured in the x,y plane, and linear size adjustments are introduced in the final reconstruction to maintain correct in vivo proportions.

21.6 BASIC AND ADVANCED VISUALIZATION OF 3D GEOMETRIC MODELS

The co-registered stacks of 2D point and line data make up a 3D reconstruction. The simplest graphical communication (visualization) of the information contained in the 3D reconstruction is the mere presentation of data as points and lines, as they were originally coded (Figure 21.6B). The line data, in our context referred to as contour lines, represent boundaries of regions inside the brain or the external surface of the brain. The stacks of contour lines can be used as a basis for computerized re-synthesis of biological surfaces. This is done by producing a mesh of 3D polygons to geometrically define the surfaces. Various methods and algorithms are used to perform this operation (see, e.g., West and Skytte, 1986; Toga, 1990; Bjaalie et al., 1997). Computer graphics is then used to produce visualizations of this geometry, including solid and transparent viewing, shading, and lighting of the surfaces, to explore properties of the multiple datasets present in a given reconstruction (Figures 21.6C, 21.7A,D–F, and 21.8). Point data, representing individual cell bodies or axonal plexuses, can also be submitted to geometrical modeling and mathematical analysis. For example, each point can be expanded from a single pixel presentation to multiple pixels (larger points) or to spheres. (The modeling of spheres follows the same principles as outlined for surfaces.). Different visualizations of point distributions representing tracer labeling are exemplified in Figures 21.6–21.8.

Here we will exemplify basic visualization procedures with data from our investigations of the rat cerebro-cerebellar system. In a double anterograde tracing experiment of subcortical projections from rat whisker barrel cortex (Figures 21.6 and 21.8; adopted from Leergaard et al., 2000b), two axonal tracers were injected into separate whisker barrels of the primary somatosensory cortex, and the ensuing axonal labeling in the pontine nuclei was mapped and reconstructed in 3D. The point data (scatter plots) shown in Figure 21.6B,C and in Figure 21.8 (left column) show the distribution of the labeled terminal fields from the two injection sites (red and black) inside the pontine nuclei. The ventral surface of the brain stem (gray), and the boundary surface of the pontine nuclei (blue) are both made transparent to allow inspection of the distribution patterns. An orderly arrangement of the axonal terminal fields is seen, with one tracer label located inside the other (closer to the central part of the pontine nuclei, on one side). Technical details and data interpretations are provided in Leergaard et al. (2000b).

Basic visualization of multiple structures in the rat brain stem is shown in Figure 21.7D–F. Here, a model of the rat brain stem and cerebellum is seen from different angles of view. Inside the brain stem (transparent grey surface), the boundaries of the descending fiber tract (green) and various precerebellar nuclei (other colors) are displayed as solid surfaces. The pontine nuclei (dealt with in the example above) is shown here as a solid yellow surface surrounded by a bounding box, representing a local coordinate system for this part of the brain. The 3D distribution of four different populations of pontocerebellar neurons, retrogradely labeled by dual tracer implantation into the cerebellar hemispheres in two rats, is represented as colored squares (dots) within the pontine coordinate system (Figure 21.7B).

The aim of the basic 3D visualization exemplified above is to communicate overall spatial information about brain regions and labeling patterns. More advanced manipulation and visualization is needed to reveal more detailed information contained in the datasets. Hence, solutions that allow users to interactively select and combine different tools and to integrate data calculations with visualization are important. In the following, we will briefly review four computerized procedures that in our hands have turned out to be useful for analysis of axonal tracing data: surface modeling, density gradient analysis, clipping, and overlap analysis. Datasets from our ongoing studies of the rat pontine nuclei, along with our custom software Micro3D (http://www.nesys.uio.no), will be used to exemplify the procedures (Figure 21.8).

Computerized *surface modeling* of the outer boundaries of clusters of cells or axonal terminals facilitates visualization of the size, shape, and extent of the volume occupied by the labeled objects (Figure 21.8). Surfaces demonstrate 3D shapes better than point clouds. Since the densities of tracer labeled neural objects are usually inhomogeneous, the external boundaries of the zones of labeling are often gradual and thus difficult to define. Several methods may be applied to define a geometry that surrounds a point population, including manual digitization of contour lines surrounding the labeling (Malmierca et al., 1998), density gradient analysis followed by segmentation of regions containing densities above a predefined threshold (Leergaard et al., 2000a,b), and approaches using more complex calculations (Nadasdy and Zaborszky, 2001).

Density gradient analysis is useful for characterizing inhomogeneous distributions of objects. With a 2D density analysis (Malmierca et al., 1998; Vassbø et al., 1999; Figure 21.8), the user first chooses a suitable angle of view and then superimposes a 2D grid on the 3D reconstruction. Grey levels or pseudo-colors, determined by the densities recorded, are assigned to each square in the grid. Variable degrees of averaging across squares can be used to visualize density gradients at different levels of resolution. Density gradient analysis can also be performed with a true 3D approach, by subdividing the reconstruction into cubes (voxels) and calculating the densities within each cube (Nadasdy and Zaborszky, 2001). The problem with the latter approach, at least at the level of conventional journal (paper) publications, is the technical difficulty of communicating the 3D density visualizations.

Clipping in three-dimensional space is another powerful computer graphics tool. It removes objects outside a user-defined volume and thus allows the investigator to re-slice the 3D reconstruction at any chosen section angle and thickness, independent of the original sections (Figure 21.8). This provides opportunities for a much more complete analysis of spatial distribution patterns than viewing restricted to the original section planes. Bi-directional slicing at perpendicular angles was crucial for understanding previously neglected complex organization patterns in the pontine nuclei (Leergaard et al., 2000b, their Figure 6).

Analysis of spatial *overlap* between cell groups or terminal fields connected with different sources or targets must be carefully adopted to the goal of the analysis and the underlying biological problem. The definition of overlap or segregation is not trivial, and different analytical models should be employed for different purposes (Bjaalie et al., 1991). In the context of operations performed by brain connections, it is relevant to determine to what degree projections arising from separate source regions are convergent or divergent—that is, to what extent projections from separated injection sites overlap in a target zone, or extend terminal fields to multiple separated target regions. Such analysis typically relies on data from double-labeling experiments. One approach employed by several investigators has been to first apply a grid to each section and then count the number of events (points of different categories representing labeling) in each bin in the grid, before computing statistics from the data (He et al., 1993; Alloway et al., 1999; Leergaard et al., 2000a, 2004). The mathematics involved is straightforward, whereas the biological interpretation will be influenced by assumptions about the organization of the part of the brain being analyzed, such as the distances at which functional interaction between the labeled objects can occur.

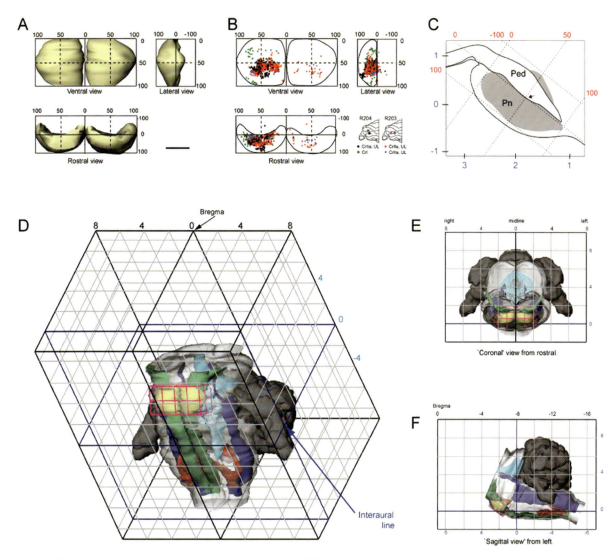

Figure 21.7   Local and global coordinate systems for describing different levels of localization in the rat brain. (**A, B**) Computer-generated 3D reconstruction of the pontine nuclei and labeled pontocerebellar neurons, from three angles of view (reproduced from Brevik et al., 2001, with permission). In Part A, the architectonically defined external boundaries of the pontine nuclei are shown as solid surfaces. In Part B, pontocerebellar projection neurons, labeled after tracer implantation into crus I and IIa of the rat cerebellum, are shown as colored dots. Each dot represents one neuron. The size and location of the implantation sites are shown in the inset line drawings in Part B. Coordinate systems of relative values from 0% to 100% are used. The halfway (50%) reference lines are shown as dotted lines. Curved solid lines represent nuclear boundaries. (**C**) Diagram for translating coordinates between the local pontine coordinate system and the skull based stereotaxic atlas of Paxinos and Watson (1998). A sagittal section is reproduced from Paxinos and Watson, with permission. Dotted lines and red numbers represent the local pontine coordinate system; the blue numbers on the surrounding frame indicate the skull-based stereotaxic coordinates of Paxinos and Watson (1998). (**D–F**) A 3D reconstruction of the brain stem, with selected brain stem nuclei in different colors and the cerebellum in gray, together with the local pontine and global stereotaxic coordinate systems in oblique lateral (**D**), coronal (**E**), and sagittal (**F**) views. Data from Brevik et al. (2001), Bar: 1 mm (Parts A and B).

21.7 COMBINING DATA FROM DIFFERENT EXPERIMENTS: GLOBAL AND LOCAL COORDINATE SYSTEMS

Axonal tracing techniques available today offer new and unique possibilities for mapping of connections in the brain. Data collected by different laboratories, however, are often difficult to compare due to variation in the resolution of images documenting the findings, the plane of sectioning, the use of different section spacing, and dissimilar techniques for data documentation.

This is common for most anatomic brain mapping investigations (Swanson, 1995, 2000; Schmahmann et al., 1999; Thompson et al., 1997, 2000; Toga and Thompson, 1999; Geyer et al., 2001; Toga et al., 2001). Furthermore, individual variability in size and shape of the brain regions investigated may pose problems even for comparisons of experimental results collected with rigid standardized procedures. The original approach for data comparison in neuroanatomy was to transfer anatomic distribution data to standard drawings of sections through the region of interest (Brodal, 1940, 1978; Jansen and Brodal, 1940; Voogd,

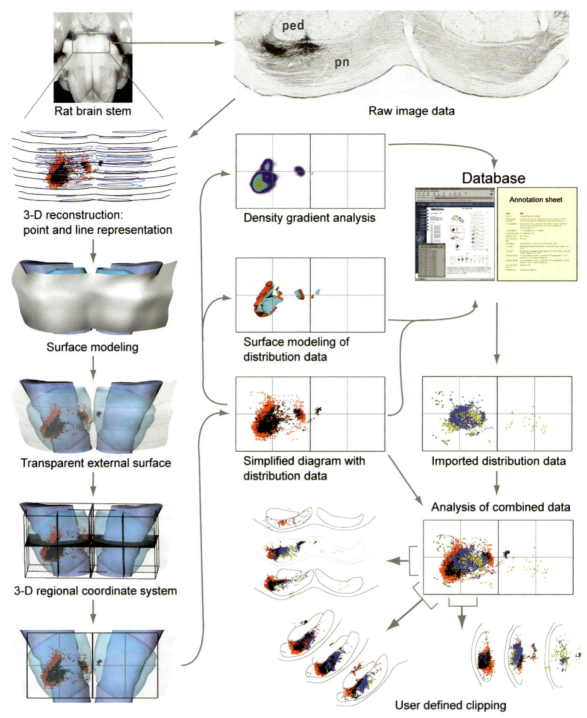

Figure 21.8   Example of computerized analysis of neuronal distribution data. The rat pons is isolated and sectioned, and data from serial sections are used to build a 3D computerized reconstruction containing information about the distribution of specific neuronal elements. In this example, two different axonal plexuses were labeled after injection of two axonal tracers into individual whisker representations in the primary somatosensory cortex (data from Leergaard et al., 2000a). In the initial analysis, the outer boundaries of the brain and the pontine nuclei are modeled as surfaces, whereas points represent distributions of anterogradely labeled axonal plexuses (red and black dots). A local pontine 3D coordinate system is introduced to define the region of interest (see also Figure 21.7). A simplified diagram is used to show the distribution of labeling within the local coordinate system. Examples of further computational analysis (surface modeling of the main clusters of points, and density gradient analysis for one of the point categories) are shown. The new experimental data are deposited in a database. Data from the same brain region and different experimental animals are superimposed in the same coordinate system. From the database, different datasets can be downloaded and combinations of data viewed and further analyzed. In this example, data from retrograde tracing of the pontocerebellar projection (blue, yellow, and green dots representing pontine neurons labeled after different tracer injections in the cerebellum, data from Brevik et al., 2001) are combined with the anterograde corticopontine tracing data (red and black dots). Dynamic subdividing (clipping) of the combined datasets into sections of chosen thickness and orientation add a further dimension to the analysis of the 3D topographic map. Ped, peduncle; pn, pontine nuclei.

1967; Brodal and Bjaalie, 1992). Each drawing or plot from an experiment was then carefully (and manually) translated to the standard drawing that provided the best match. Although usually not explicitly defined, this procedure logically implied the use of a common coordinate system. Below, we will consider the use of precisely defined coordinate systems, logically divided into two categories: global and local.

Global coordinate systems for the brain are usually defined by landmarks in the skull. Conventional atlases present line drawings of sections through the brain in skull-based coordinate systems (Paxinos and Watson, 1998; Swanson, 1999; Paxinos and Franklin, 2001). However, once the brain has been removed from the skull for further analysis, the direct reference to the global coordinate system is lost. Local coordinate systems therefore represent a valuable supplement, since they do not rely on skull landmarks but instead are based on boundaries and landmarks in the tissue (see, e.g., Brevik et al., 2001). A local coordinate system does not necessarily have to fit the exact size and shape of the analyzed region, as long as it is based on distinct and easily reproducible boundaries or landmarks. For anatomy at the light microscopic level, local coordinate systems are particularly useful for assigning exact locations to distribution data and also for comparison of data collected in different experimental animals. One example is the local 3D coordinate system for the pontine nuclei in rat. The 3D coordinate system for this brain stem region is based on a cuboid bounding box, with a specific orientation, fitted to cytoarchitectonic boundaries and landmarks in the region of interest (Figure 21.7A,B; Leergaard et al., 2000a,b; Brevik et al., 2001). The location of the bounding box is also defined in relation to standard (skull-based) atlas coordinates, allowing transfer of data to other coordinate systems (Figure 21.7C–F). The local pontine nuclei coordinate system has been used for presentation of data originating in different brains. Data from so far 127 tracer injections in 74 animals (work in progress) have been superimposed in this coordinate system and analyzed with software tools for 3D analysis (as exemplified in Figure 21.8).

Three-dimensional coordinate systems offer distinct advantages for investigations of deeper structures of the brain, and they are also used for analysis of cerebral cortical distribution data. Two-dimensional coordinate systems, however, provide an alternative for the cerebral cortex. The cortex is a laminar structure, topologically equivalent to a 2D sheet. Advanced software for flattening of the complicated folded primate cortex has been developed and used in multiple anatomic and physiological investigations based on light microscopic section data (Drury et al., 1996). Several new tools aimed at assisting in the analysis of fMRI data are developed to further facilitate cortical distribution analysis. These include optimized flattening software, tools for transfer of data from the cortex of different brains onto an average representation, and novel surface-based coordinate systems (Fischl et al., 2002). The neuroimaging community, which relies on powerful tools for analysis of a rapidly increasing population of experimental subjects, is a driving force in this endeavor.

A major concern for all brain distribution analysis is individual variability. In the rat brain stem (Figures 21.6–21.8), this variability is limited. An accurate analysis can be performed by directly superimposing data from different animals in the same internal nuclear coordinate system, adjusting only for size differences (linear warping or affine transformations). For example, repeated, identical experimental manipulation in a series of

rats may produce almost identical distribution patterns of data superimposed in the same local coordinate system (Leergaard et al., 2000b). Neural systems of higher mammals are known to contain larger individual variability, requiring various measures to allow comparison of data from different animals. The use of advanced warping methods and probabilistic atlases is particularly important for primate and human brain studies (Roland and Zilles, 1994; Mazziotta et al., 1995, 2001).

21.8 METHODOLOGICAL LIMITATIONS

In this section we will first discuss properties of axonal tracers that may restrict their use, with emphasis of the ensuing limiting consequences for the experimental design. Second, we will deal with methodological problems related to data acquisition and 3D reconstruction of axonal tracing data from serial sections.

Overall, the available modern tracing techniques and data acquisition methods are able to accurately code the distribution of labeling through serial sections. The numbers recorded are, however, less reliable, for reasons related to tracer sensitivity and specificity, as well as data acquisition procedures, as discussed below. Tracing methods should therefore primarily be used to detect relative density gradients and distributions. Care must be taken when absolute values are recorded and comparisons are made with use of different experimental protocols, tracers, and data acquisition methods.

The ideal axonal tracer has a high sensitivity and specificity. The sensitivity is determined by the uptake of the tracer into neuronal cell bodies, dendrites, or axonal terminal fields, as well as by the transport of the tracer along the axons. The most modern tracing methods are highly sensitive. Only very small amounts of such tracers need to be deposited to give rise to uptake and transport. For example, the dextran amines, introduced in the 1980s and 1990s (Glover et al., 1986; Veenman et al., 1992), are considerably more sensitive than other tracers, such as horseradish peroxidase conjugates, commonly used over several decades (Mesulam, 1982). Several currently available axonal tracers (e.g., dextran amines, *Phaseolus vulgaris*–leucoagglutinin, and Cholera-toxin; Vercelli et al., 2000) label complete axonal trajectories, with sufficient contrast between labeling and background to allow visualization of individual axons with varicosities. Individual axons and their collaterals can be traced across multiple sections. Nevertheless, it is not easy to verify that a tracer is taken up and transported by *all* neuronal cell bodies located within the boundaries of a tracer injection site, or by all axonal terminal fields, exposed to the tracer. Uptake is apparently restricted at the boundaries of the injection or implantation sites, where the concentration of tracer is lower than close to the center of the application site.

The specificity of a tracer is determined by the ability of the tracer to label only a well-defined group of structural elements. Usually, this implies that a specific tracer labels only neurons that arise from or project directly to an injection site. For example, uptake of tracer by fibers passing through the site of tracer administration may give rise to erroneous interpretations. Some tracers may give rise to confusion due to transport in the retrograde direction followed by anterograde transport into collateral axonal branches of the labeled neurons. Another complicating factor, seen with a few tracers only, is transsynaptic transport. These factors potentially reduce the specificity of the labeling but can nevertheless be exploited in relevant experimental designs. For example, while uptake in passing fibers is usually regarded as a methodological problem, it is an essential property of tracers applied to transected nerves or pathways (Arecchi-Bouchhioua et al., 1996; Holtzer et al., 2003). Complex network designs may be systematically studied by use of collateral labeling effects (Malmierca et al., 1998) or by well-controlled trans-synaptic transport of some tracers (Ugolini, 1995; Kelly and Strick, 2000, Tang et al., 1999; Saleem et al., 2002). In general, valid data about connections in the brain can only be obtained when the experimental design takes into consideration tracer properties, as discussed above.

Data acquisition methods provide further challenges, particularly with regard to identification of tracer labeled elements and determination of the correct numbers of labeled elements. Identification of labeled structures can usually be performed with high accuracy using semiautomatic data acquisition methods, as outlined above. The investigator uses a set of predetermined criteria for recording the labeled elements. Labeled cells are usually recorded with a coordinate corresponding to the center of the cell profile visible in the section. Axonal terminal fields may be recorded as a dotted pattern, and the density of labeling may be displayed as variable dot densities (Figures 21.3D,E and 21.5). Several anterograde tracers show the full length of the fibers as well as varicosities along the fibers. Plotting of individual varicosities are, however, often problematic in areas with dense labeling. Thus, in such areas, the plots do not accurately distinguish different density levels, although the distribution of the labeling can be correctly shown (Figures 21.3D and 21.5A).

Automatic methods are potentially more fast and accurate than the semiautomatic ones but have important limitations both with regard to speed and plotting accuracy. With 3D reconstruction, large numbers of sections need to be collected, and large mosaic images (covering multiple fields on view in each section) are often required. In order to handle the ensuing large amounts of data, efficient routines and data management procedures must be established (see discussion below). Problems with regard to detection of labeled cell bodies relate primarily to incompletely labeled cell bodies (giving rise to false duplication of labeled cells) and merging of overlapping labeled cell bodies (making it difficult to identify each cell). With the use of erosion and dilation binary operators and tools for particle separation available in image analysis systems, these problems are reduced but not completely removed. The identification of dense labeled axonal plexuses is readily done by automatic systems, whereas weak anterograde labeling—in particular, individual fibers—are more difficult to detect with the automatic procedures. These and related issues are discussed in detail by Lillehaug et al. (2002).

Regardless of the data acquisition method, it is important to be aware of the possibilities for double counting of neuronal cell bodies that are present in adjacent sections and to take appropriate measures when more exact estimates of numerical values are relevant. A discussion of the methods that need to be added to the repertoire presented in this chapter is provided by Coggeshall and Lekan (1996).

The limitations of the 3D reconstruction procedures are mostly determined by technical issues related to tissue handling and the ensuing section quality. The truthfulness of the shape and proportions of a 3D reconstruction is closely related to the quality of sections. Sectioning and histology need to be performed carefully to reduce the distortions and damage to sections to a minimum. Section shrinkage should be measured and taken into consideration when preparing the final 3D reconstruction (for a discussion of these and related issues, see Bookstein, 1990; Annese and Toga, 2002). Furthermore, properties inherent to the brain structure pose specific challenges for visualization, data analysis, and interpretations. One such feature is the gradual architectonic and functional transitions that are frequently observed in the brain. These make it difficult to unambiguously define the boundaries of a given brain region; thus, visualizations usually demonstrate only one of several possible parcellation schemes (Amunts et al., 2000; Amunts and Zilles, 2001; Kötter et al., 2001; Rademacher et al., 2001; Toga et al., 2001; Fischl et al., 2002; Zilles et al., 2002). A solution to this problem is the use of variable boundary definitions and ranges of tolerances of deviations from a basic parcellation scheme.

21.9 FUTURE PERSPECTIVES

The techniques and approaches discussed in the present chapter offer new possibilities for analysis of system level brain connectivity. Currently available axonal tracers are highly sensitive, and with the correct tuning of tracer properties and experimental design, high-quality data about brain connectivity can readily be collected with use of the existing repertoire of axonal tracer substances. With regard to future perspectives, we will therefore bypass further discussions of axonal tracing techniques. Our focus will be on two issues that we believe will be of utmost importance for the further development of the field: high-throughput data acquisition methods, and web-accessible digital brain atlases coupled to databases and analytical tools.

21.9.1 High-Throughput Data Acquisition

Methods for accurate and fast collection of data from large series of sections will be essential to perform large-scale computerized visualization and analysis of brain wiring patterns, as well as to collect multiple other data types from serial sections. A logical starting point for high-throughput data acquisition is the use of automated methods, discussed above. With use of motorized microscope stages, cameras for the capturing of multiple images, and the relevant software and hardware configurations, it is possible to create mosaics of images covering large regions of each section. Such complete section images can subsequently be submitted to segmentation procedures that will produce an interpretation of the labeling patterns of interest, such as the coordinates representing labeled cells and axonal terminal fields shown in Figure 21.5. Improvements of image analysis methods, as well as optimization of routines for segmentation, will be important in this endeavor. An interesting new trend in neuroscience- and biology-related bright field microscopy is the use of large microscope stages that can accommodate several slides. Images are collected at cellular resolution in a stepwise fashion and are assembled into a composite image for each section, relying on the coordinates provided by the microscope stage. Data collection and assembly of the series of composite images is thus automated and require only limited human intervention. Such multi-slide automatic microscope systems are currently used to collect nonradioactive in situ hybridization data from mice (Carson et al., 2002; Visel et al., 2002; see also http://www.genepaint.org/) but are equally well-suited for detection of other colored signals imaged with bright field microscopy. Further developments to handle also fluorescence data seem realistic to achieve. We consider the new increasingly automated microscopy methods as critical for implementing future more complete anatomical mapping of major connections in the brains of experimental animals.

21.9.2 Digital Brain Atlases and Databases

Comparisons of data collected from different experimental animals and different laboratories are of crucial importance for obtaining new knowledge in many neuroscience disciplines. This applies also to the field of brain systems organization. Earlier, we discussed how the use of local and global coordinate systems facilitates such comparison. We anticipate that spatial reference systems, digital atlases, and accompanying databases will play an increasing role for the handling of brain systems data, such as axonal tracing data from experimental animals, mapped in well-defined coordinate systems.

There are lessons to learn from the fields of human neuroimaging and human multimodality brain analysis. These disciplines have taken the lead in developing digital atlas and database capabilities (for references, see Mazziotta and Toga, 2002; Fox and Lancaster, 2002; Van Horn and Gazzaniga, 2002). The analysis of spatial distribution data from the human brain is particularly demanding since comparison of data must capture a large variability of brain structure and function, among individual brains, and across levels of investigation and imaging modalities. Digital brain atlases provide a structural framework in which datasets—collected at different scales, with different methods, and from different subjects—can be integrated. Efficient use of digital atlases requires the presence of accompanying mathematical tools for spatial translation of data from a given brain onto the atlas, as well as for translation of data from one atlas to another. A number of different strategies and solutions to accomplish this have been implemented (for a review, see Roland and

Zilles, 1994; Mazziotta et al., 1995, 2001). The basic tools and experiences from this field can readily be applied to the simpler situation provided by experimental animals (Bjaalie, 2002). Only limited variability between brains is seen when, for example, rats of the same strain, gender, and age are selected for analysis. This simplifies the transfer of data to and from digital atlases.

Axonal tracing data from experimental animals published to date have mostly been presented in selected single sections, without a clear reference to normalized 3D space. By collecting data about brain connections in standardized coordinate systems, as discussed in the present chapter, it will be possible to transfer data from many experimental animals onto standard digital atlases to facilitate across animal comparison. These atlases can incorporate multiple other data types. Recent progress in the mapping of gene expression data and molecular level data from the rodent brain provides novel opportunities for synergies with developments outlined in the present chapter. Atlases, preferably combining different data categories, will necessarily have to be linked to databases in order to provide access to search routines, to lower-level data, to detailed annotations of the data that are spatially defined in the atlas, and to tools for visualization and analyses. Figures 21.9 and 21.10 show work in progress from our laboratory on the development of web applets and digital atlases (for references, see Bjaalie, 2002). The atlas framework shown in Figure 21.9 covers a single region of the rat brain (the pontine nuclei on one side) and is based on a local coordinate system. The associated database currently contains more than 120 axonal tracing datasets. The user will be able, via the Internet, to independently choose combinations of datasets and thus test different models of input–output relationships in the pontine nuclei. Figure 21.10 shows a multimodal atlas with several different data types superimposed in the same framework to facilitate integration in 3D space.

Tools for meta-analysis and modeling of structure–function relationships will play an important role in the building of knowledge that will improve the understanding of the brain. The atlas–database concept will provide an environment for data

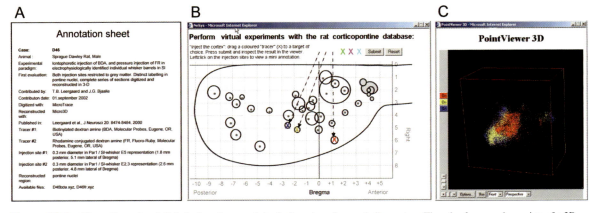

Figure 21.9 Three-dimensional digital atlas of connectivity in the rat cerebrocerebellar system. The atlas framework consists of a 3D coordinate system with corticopontine axonal distribution data from different single and multiple anterograde tracer experiments coded as points. Annotations for all experiments are included (**A**), together with an HTML-based user interface for selecting combinations of datasets (**B**), and a Java-based viewer for online visualization and rotation of selected datasets in the archive. The graphical user interface (**B**) shows the localization and size of the tracer injections included in the archive. A mouse click on an injection site gives access to the annotation sheet for the corresponding experiment (**A**). The user may select a combination of datasets by dragging colored symbols onto the injection site(s) of choice, in a sense performing "virtual" injections, before submitting the selection to the Java viewer applet (**C**). In the viewer, the selected datasets appear as 3D point populations with colors as previously defined by the user. A bounding box representing the local pontine coordinate system can be activated, and the datasets may be turned on and off in the viewer. The data archive will be accessible from http://www.nesys.uio.no/. Work in progress by S. Hajnoczi, D. Øyan, T.B. Leergaard, C. Pettersen, , S. Lillehaug, and J.G. Bjaalie.

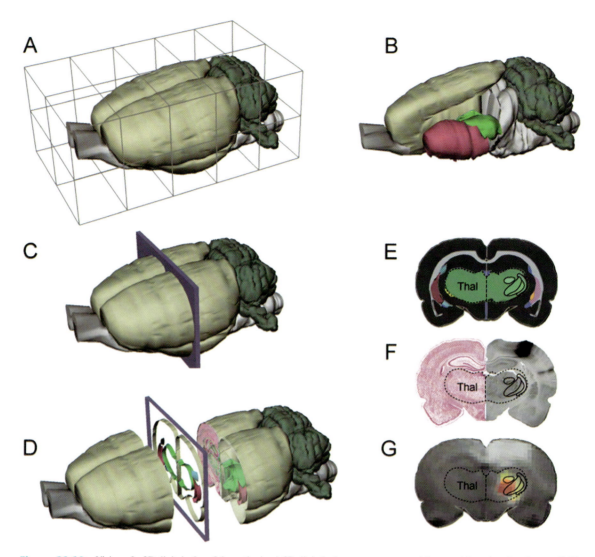

Figure 21.10 Vision of a 3D digital atlas of the rat brain. A 3D digital atlas was reconstructed from serial section drawings available in the Swanson atlas of the rat brain (Swanson, 1999) and was combined with a 3D reconstruction of a rat cerebellum and brain stem (data from Brevik et al., 2001). (**A**) The reconstructed external brain surface is shown as a solid surface. Stereotaxic atlas coordinates are shown as a bounding box with selected reference lines. (**B**) The left cortical hemisphere and underlying white matter are removed to reveal the external boundaries of the basal ganglia (pink) and thalamus (green) as solid surfaces. The solid blue box in **C** and **D** indicates the position and orientation of the coronal slice shown in (**E–G**). A user-defined slice (of arbitrary thickness and orientation) through the 3D atlas, here exemplified as a coronal slice, may be used to view different data modalities registered in the atlas. The atlas thus provides access to a segmentation of a number of structures (**E**), corresponding histological information (**F**, left side), and connectivity data obtained with conventional section-based techniques (**F**, right side) or with tomographic techniques (**G**). The thionin-stained data in part F (left side) are modified from Swanson (1999), with permission. The tracing data in part F (right side) are modified from Leergaard et al. (2000b), with permission. Part G contains unpublished manganese enhanced MRI tracing data by T.B. Leergaard, J.G. Bjaalie, A. Devor, L.L. Wald, and A.M. Dale.

integration, which in our judgment will not be possible to create with conventional approaches. From the perspective of the individual researcher, the minimum requirement will be that the digital atlas provides a framework of the major structures in the brain of the experimental animal in question, and that it contains an accompanying repertoire of tools that allow translation of data to atlas coordinates. This will imply at least the use of affine (linear) transformations to match experimental datasets to atlas templates. A number of more sophisticated approaches may be added to increase precision of data transfer and the value of results extracted from the atlas and database (Toga, 1999).

21.10 CONCLUDING REMARKS

There is little doubt that modern tract-tracing methods will continue to be useful in basic neuroanatomic research. The further study of the complexity and totality of brain circuits requires the use of accurate and rapid data acquisition methods, translation of data to well-defined coordinate systems, data management strategies that will enable investigators to more efficiently store and retrieve datasets for repeated analysis in new contexts, and a suite of tools made available to large groups of investigators that will allow such analysis and re-analysis to take place. We anticipate that digital atlases, databases, and associated tools, made available via the Internet, will be of considerable use in future neuroanatomy. In the present chapter, we have focused on the data that can be generated with high-resolution axonal tracing methods, to unravel detailed topography and principles of spatial organization of cell groups and axonal terminal fields. Numerous methods for identification of other neural characteristics, including methods for detection of neurotransmitters, enzymes, receptors, and gene products (see, e.g., Storm-Mathisen and Ottersen, 1987, 1990; Brandtzaeg, 1998; Danbolt et al., 1998; Petralia and Wenthold, 1999; Carson et al., 2002), can be used to provide spatial coordinates describing aspects of spatial distribution similar to the ones demonstrated in the present chapter. The combined mapping of accurate data about brain connections and wiring patterns, along with data of other categories including the gene and molecular level, represents a novel and potentially powerful research strategy that will benefit strongly from the use of gradually improving tools for computational data manipulation and data management.

ACKNOWLEDGMENTS

We thank all members of the NeSys laboratory for encouraging support and discussions, and we also thank all our collaborators contributing data to the database hosted by our laboratory (http://www.nesys.uio.no/). Software developments described in this chapter were performed with the valuable assistance of Christian Pettersen, Stefan Hajnoczi, and Daniel Øyan. This work was supported by grants from The Research Council of Norway and EC grant QLG3-CT 2001-002256.

References

Akintunde, A., and Buxton, D.F., (1992). Origins and collateralization of corticospinal, corticopontine, corticorubral and corticostriatal tracts: A multiple retrograde fluorescent tracing study. *Brain Res.* **586**, 208–218.

Alexander, G.E., DeLong, M.R., and Strick, P.L. (1986). Parallel organization of functionally segregated circuits linking basal ganglia and cortex. *Annu. Rev. Neurosci.* **9**, 357–381.

Alexander, G.E., Crutcher, M.D., and DeLong M.R., (1990). Basal ganglia-thalamocortical circuits: Parallel substrates for motor, oculomotor, "prefrontal" and "limbic" functions. *Prog. Brain Res.* **85**, 119–146.

Alloway, K.D., Crist, J., Mutic, J.J., and Roy, S.A. (1999). Corticostriatal projections from rat barrel cortex have an anisotropic organization that correlates with vibrissal whisking behavior. *J. Neurosci.* **19**, 10908–10922.

Amari, S.-I., Beltrame, F., Bjaalie, J.G., Dalkara, T., De Schutter, E., Egan, G.F., Goddard, N., Gonzalez, C., Grillner, S., Herz, A., Hoffmann, K.-P., Jaaskelainen, I., Koslow, S.H., Lee, S.-U., Matthiessen, L., Miller, P.L., Mira, da Silva, F., Novak, M., Ravindranath, V., Ritz, R., Ruotsalainen, U., Sebestra, V., Subramaniam, S., Tang, Y., Toga, A.W., Usui, S., Van Pelt, J., Verschure, P., Willshaw, D., and Wrobel, A. (2002). Neuroinformatics: The integration of shared databases and tools towards integrative neuroscience. *J. Int. Neurosci.* **1**, 117–128.

Amunts, K., and Zilles, K. (2001). Advances in cytoarchitectonic mapping of the human cerebral cortex. *Neuroimaging Clin. North Am.* **11**, 151–69, vii.

Amunts, K., Malikovic, A., Mohlberg, H., Schormann, T., and Zilles, K. (2000). Brodmann's areas 17 and 18 brought into stereotaxic space—where and how variable? *NeuroImage* **11**, 66–84.

Angelucci, A., Clasca, F., and Sur, M. (1996). Anterograde axonal tracing with the subunit B of cholera toxin: A highly sensitive immunohistochemical protocol for revealing fine axonal morphology in adult and neonatal brains. *J. Neurosci. Methods* **65**, 101–112.

Annese, J., and Toga, A.W. (2002). Postmortem anatomy. In *Brain Mapping: The Methods* (A.W., Toga and J.C. Mazziotta, eds.), pp 537–571. Academic Press, San Diego, CA.

Arecchi-Bouchhioua, P., Yelnik, J., Francois, C., Percheron, G., and Tande, D. (1996). 3-D tracing of biocytin-labelled pallido-thalamic axons in the monkey. *NeuroReport* **7**, 981–984.

Arieli, A., and Grinvald, A. (2002). Optical imaging combined with targeted electrical recordings, microstimulation, or tracer injections. *J. Neurosci. Methods* **116**, 15–28.

Armengol, J.A., and Salinas, P. (1991). Analysis of the ipsi- and contralateral location of the neurons of the nucleus reticularis tegmenti pontis projecting to the cerebellum and of the trajectory of their axons within the pons to the brachium pontis. An "in vivo" and "in vitro" study. *J. Hirnforsch.* **32**, 715–724.

Aston-Jones, G., and Card, J.P. (2000). Use of pseudorabies virus to delineate multisynaptic circuits in brain: Opportunities and limitations. *J. Neurosci. Methods* **103**, 51–61.

Bajo, V.M., Merchán, M.A., Malmierca, M.S., Nodal, F.R., and Bjaalie, J.G. (1999). Topographic organization of the dorsal nucleus of the lateral lemniscus in the cat. *J. Comp. Neurol.* **407**, 349–366.

Bjaalie, J.G. (1992). Three-dimensional computer reconstructions in neuroanatomy: Basic principles and methods for quantitative analysis. In *Quantitative Methods in Neuroanatomy* (M.G. Steward ed.), pp. 249–293. Wiley, Chichester.

Bjaalie, J.G. (2002). Localization in the brain: New solutions emerging. *Nat. Neurosci. Rev.* **3**, 322–325.

Bjaalie, J.G., and Brodal, P. (1983). Distribution in area 17 of neurons projecting to the pontine nuclei: A quantitative study in the cat with retrograde transport of HRP-WGA. *J. Comp Neurol.* **221**, 289–303.

Bjaalie, J.G., Diggle, P.J., Nikundiwe, A., Karagülle, T., and Brodal, P. (1991). Spatial segregation between populations of ponto-cerebellar neurons: Statistical analysis of multivariate spatial interactions. *Anat. Rec.* **231**, 510–523.

Bjaalie, J.G., Daehlen, M., and Stensby, T.V. (1997). Surface modelling from biomedical data. In *Numerical Methods and Software Tools in Industrial Mathematics*, (M. Daehlen and A. Tveito, eds.), pp. 9–26. Birkhaeuser, Boston, MA.

Bookstein, F.L. (1990). Distortion correction. In *Three-dimensional Neuroimaging* (A.W. Toga, ed.), pp. 235–249. Raven Press, New York.

Bowden, D.M., and Martin, R.F. (1995). NeuroNames brain hierarchy. *NeuroImage* **2**, 63–83.

Bower, J.M. (1997). Control of sensory data acquisition. *Int. Rev. Neurobiol.* **41**, 489–513.

Bower, J.M., and Kassel, J. (1990). Variability in tactile projection patterns to cerebellar folia crus IIA of the Norway rat. *J. Comp. Neurol.* **302**, 768–778.

Bower, J.M., Beermann, D.H., Gibson, J.M., Shambes, G.M., and Welker, W. (1981). Principles of organization of a cerebro-cerebellar circuit. Micromapping the projections from cerebral (SI) to cerebellar (granule cell layer) tactile areas of rats. *Brain Behav. Evol.* **18**, 1–18.

Brandtzaeg, P. (1998). The increasing power of immunohistochemistry and immunocytochemistry. *J. Immunol. Methods* **216**, 49–67.

Brevik, A., Leergaard, T.B., Svanevik, M., and Bjaalie, J.G. (2001). Three-dimensional computerised atlas of the rat brain stem precerebellar system: Approaches for mapping, visualization, and comparison of spatial distribution data. *Anat. Embryol.* **204**, 319–332.

Brodal, A. (1940). The cerebellum of the rabitt. A topographical atlas of the folia as revealed in transverse sections. *J. Comp. Neurol.* **72**, 63–81.

Brodal, P. (1978). The corticopontine projection in the rhesus monkey. Origin and principles of organization. *Brain* **101**, 251–283.

Brodal, A. (1981). *Neurological Anatomy in Relation to Clinical Medicine.* Oxford University Press, Oxford.

Brodal, P. (1982). The cerebropontocerebellar pathway: Salient features of its organization. *Exp. Brain Res., Suppl.* **6**, 108–132.

Brodal, P., and Bjaalie, J.G. (1992). Organization of the pontine nuclei. *Neurosci. Res.* **13**, 83–118.

Brodal, P., and Bjaalie, J.G. (1997). Salient anatomic features of the cortico-ponto-cerebellar pathway. *Prog. Brain Res.* **114**, 227–249.

Capowski, J.J. (1985). The reconstruction, display, and analysis of neuronal structures using a computer microscope. In *The Microcomputer in Cell and Neurobiology Research* (R.R. Mize, ed.), pp. 85–109. Elsevier, New York.

Capowski, J.J. (1989). *Computer Techniques in Neuroanatomy.* Plenum Press, New York.

Carson, J.P., Thaller, C., and Eichele, G. (2002). A transcriptome atlas of the mouse brain at cellular resolution. *Curr. Opin. Neurobiol.* **12**, 562–565.

Chronwall, B.M., Lewis, M.E., Schwaber, J.S., and O'Donohue, T.L. (1989). *In situ* hybridization combined with retrograde fluorescent tract tracing. In: *Neuroanatomical Tract-Tracing Method 2: Recent Progress* (L., Heimer and L. Zaborszky, eds.), pp, 265–298. Plenum Press, New York.

Coggeshall, R.E., and Lekan, H.A. (1996). Methods for determining numbers of cells and synapses: A case for more uniform standards of review. *J. Comp Neurol.* **364**, 6–15.

Conturo, T.E., Lori, N.F., Cull, T.S., Akbudak, E., Snyder, A.Z., Shimony, J.S., McKinstry, R.C., Burton, H., and Raichle, M.E. (1999). Tracking neuronal fiber pathways in the living human brain. *Proc. Natl. Acad. Sci. U.S.A.* **96**, 10422–10427.

Cowan, W.M., and Cuénod, M. (1975). *The Use of Axonal Transport for Studies of Neuronal Connectivity.* Elsevier, Amsterdam.

Danbolt, N.C., Lehre, K.P., Dehnes, Y., Chaudhry, F.A., and Levy, L.M. (1998). Localization of transporters using transporter-specific antibodies. *Methods Enzymol.* **296**, 388–407.

Diaz, C., Glover, J.C., Puelles, L., and Bjaalie, J.G. (2003). The relationship between hodological and cytoarchitectonic organization in the vestibular complex of the 11-day chicken embryo. *J. Comp Neurol.* **457**, 87–105.

Dietrichs, E., Bjaalie, J.G., and Brodal, P. (1983). Do pontocerebellar fibers send collaterals to the cerebellar nuclei? *Brain Res.* **259**, 127–131.

Drury, H.A., Van Essen, D.C., Anderson, C.H., Lee, C.W., Coogan, T.A., and Lewis, J.W. (1996). Computerized mappings of the cerebral cortex: A multiresolution flattening method and a surface-based coordinate system. *J. Cogn Neurosci.* **8**, 1–28.

Fischl, B., Salat, D.H., Busa, E., Albert, M., Dieterich, M., Haselgrove, C., van der, Kouwe, A., Killiany, R., Kennedy, D., Klaveness, S., Montillo, A., Makris, N., Rosen, B., and Dale, A.M. (2002). Whole brain segmentation: Automated labeling of neuroanatomical structures in the human brain. *Neuron* **33**, 341–355.

Fox, P.T., and Lancaster, J.L. (2002). Opinion: Mapping context and content: The BrainMap model. *Nat. Rev. Neurosci.* **3**, 319–321.

Gerfen, C.R. (1992). The neostriatal mosaic: Multiple levels of compartmental organization in the basal ganglia. *Annu. Rev. Neurosci.* **15**, 285–320.

Gerfen, C.R., and Sawchenko, P.E. (1984). An anterograde neuroanatomical tracing method that shows the detailed morphology of neurons, their axons and terminals: Immunohistochemical localization of an axonally transported plant lectin, *Phaseolus vulgaris* leucoagglutinin (PHA-L). *Brain Res.* **290**, 219–238.

Geyer, S., Schleicher, A., Schormann, T., Mohlberg, H., Bodegård, A., Roland, P. E., and Zilles, K. (2001). Integration of microstructural and functional aspects of human somatosensory areas 3a, 3b, and 1 on the basis of a computerized brain atlas. *Anat. Embryol.* **204**, 351–366.

Glaser, E. M., Gissler, M., and Van der Loos, H. (1979). An interactive camera lucida computer-microscope. *Soc. Neurosci. Abstr.* **5**, 1697.

Glaser, E.M. Tagamets, M., McMullen, N.T., and Van der Loos, H. (1983). The image-combining computer microscope—an interactive instrument for morphometry of the nervous system. *J. Neurosci. Methods* **8**, 17–32.

Glaser, J.R., and Glaser, E.M. (1990). Neuron imaging with Neurolucida—a PC-based system for image combining microscopy. *Comput. Med. Imaging Graph.* **14**, 307–317.

Glover, J. C., (1995). Retrograde and anterograde axonal tracing with fluorescent dextran-amines in the embryonic nervous system. *Neurosci. Protocols* **30**, 1–13.

Glover, J.C., Petursdottir, G., and Jansen, J.K. (1986). Fluorescent dextran-amines used as axonal tracers in the nervous system of the chicken embryo. *J. Neurosci. Methods* **18**, 243–254.

Goddard, N.H., Hucka, M., Howell, F., Cornelis, H., Shankar, K., and Beeman, D. (2001). Towards NeuroML: Model description methods for collaborative modelling in neuroscience. *Philos. Trans. R. Soc. London, Ser. B* **356**, 1209–1228.

He, S.Q., Dum, R.P., and Strick, P.L. (1993). Topographic organization of corticospinal projections from the frontal lobe: Motor areas on the lateral surface of the hemisphere. *J. Neurosci.* **13**, 952–980.

Heimer, L., and Robards, M.J. (1981). *Neuroanatomical Tract-tracing Methods.* Plenum Press, New York.

Heimer, L., and Zaborszky, L. (1989). *Neuroanatomical Tract-tracing Methods 2. Recent Progress.* Plenum Press, New York and London.

Holtzer, C.A., Marani, E., van Dijk, G.J., and Thomeer, R.T. (2003). Repair of ventral root avulsion using autologous nerve grafts in cats. *J. Peripher. Nerv. Syst.* **8**, 17–22.

Honig, M.G., and Hume, R.I. (1986). Fluorescent carbocyanine dyes allow living neurons of identified origin to be studied in long-term cultures. *J. Cell Biol.* **103**, 171–187.

Honig, M.G., and Hume, R.I. (1989). Carbocyanine dyes. Novel markers for labelling neurons. *Trends Neurosci.* **12**, 336–338.

Jansen, J., and Brodal, A. (1940). Experimental studies on the intrinsic fibers of the cerebellum. II. The cortico-nuclear projection. *J. Comp. Neurol.* **73**, 267–321.

Kamper L., Bozkurt, A., Rybacki, K., Geissler, A., Gerken, I., Stephan, K.E., and Kötter, R. (2002). An introduction to CoCOMac-Online. The online-interface of the primate connectivity database CoCoMac. In: *Neuroscience Databases. A Practical Guide,* (R. Kötter ed.), pp. 155–169. Kluwer Academic Press, Norwell, MA.

Katz, L.C., and Iarovici, D.M. (1990). Green fluorescent latex microspheres: A new retrograde tracer. *Neuroscience* **34**, 511–520.

Katz, L.C., Burkhalter, A., and Dreyer, W.J. (1984). Fluorescent latex microspheres as a retrograde neuronal marker for in vivo and in vitro studies of visual cortex. *Nature (London)* **310**, 498–500.

Kelly, R.M., and Strick, P.L. (2000). Rabies as a transneuronal tracer of circuits in the central nervous system. *J. Neurosci. Methods* **103**, 63–71.

Knierim, J.J., and van Essen, D.C. (1992). Visual cortex: Cartography, connectivity, and concurrent processing. *Curr. Opin. Neurobiol.* **2**, 150–155.

Köbbert, C., Apps, R., Bechmann, I., Lanciego, J.L., Mey, J., and Thanos, S. (2000). Current concepts in neuroanatomical tracing. *Prog. Neurobiol.* **62**, 327–351.

Koslow, S.H. (2000). Should the neuroscience community make a paradigm shift to sharing primary data? *Nat. Neurosci.* **3**, 863–865.

Koslow, S.H. (2002). OPINION: Sharing primary data: A threat or asset to discovery? *Nat. Rev. Neurosci.* **3**, 311–313.

Kötter, R., Stephan, K.E., Palomero-Gallagher, N., Geyer, S., Schleicher, A., and Zilles, K. (2001). Multimodal characterization of cortical areas by multivariate analyses of receptor binding and connectivity data. *Anat. Embryol.* **204**, 333–350.

Kuypers, H.G., and Huisman, A.M. (1984). Fluorescent tracers In: *Advances in Cellular Neurobiology,* (S. Fedoroff, ed.), pp. 307–340. Academic Press, Orlando, FL.

Kuypers, H.G., and Ugolini G. (1990). Viruses as transneuronal tracers. *Trends Neurosci.* **13**, 71–75.

Leergaard, T.B., (2003). Clustered and laminar topographic patterns in rat cerebropontine pathways. *Anat. Embryol.* **206**, 149–162.

Leergaard, T.B., and Bjaalie, J.G. (1995). Semi-automatic data acquisition for quantitative neuroanatomy. MicroTrace—computer programme for recording of the spatial distribution of neuronal populations. *Neurosci. Res.* **22**, 231–243.

Leergaard, T.B., and Bjaalie, J.G. (2002). Architecture of sensory map transformations: axonal tracing in combination with 3-D reconstruction, geometric modeling, and quantitative analyses, In: *Computational Neuroanatomy: Principles and Methods* (G. Ascoli, ed.), pp. 119–217. Humana Press, Totowa, NJ.

Leergaard, T.B., Lakke, E.A., and Bjaalie, J.G. (1995). Topographical organization in the early postnatal corticopontine projection: A carbocyanine dye and 3-D computer reconstruction study in the rat. *J. Comp. Neurol.* **361**, 77–94.

Leergaard, T.B., Alloway, K.D., Mutic, J.J., and Bjaalie, J.G. (2000a). Three-dimensional topography of corticopontine projections from rat barrel cortex: Correlations with corticostriatal organization. *J. Neurosci.* **20**, 8474–8484.

Leergaard, T.B., Lyngstad, K.A., Thompson, J.H., Taeymans, S., Vos, B.P., De Schutter, E., Bower, J.M., and Bjaalie, J.G., (2000b). Rat somatosensory cerebropontocerebellar pathways: Spatial relationships of the somatotopic map of the primary somatosensory cortex are preserved in a three-dimensional clustered pontine map. *J. Comp Neurol.* **422**, 246–266.

Leergaard, T.B., Alloway, K.D., Pham, T.A., Bolstad, I., Hoffer, Z., Pettersen, C., Bjaalie, J.G. (2004). Three-dimensional topography of corticopontine projections from rat sensorimotor cortex: Comparisons with corticostriatal projections reveal diverse integrative organization. *J. Comp. Neurol.* **478**, 306–322.

Lewis, M.E., Krause, R.G., and Roberts-Lewis, J.M. (1988). Recent developments in the use of synthetic oligonucleotides for in situ hybridization histochemistry. *Synapse* **2**, 308–316.

Lillehaug, S., Oyan, D., Leergaard, T.B., and Bjaalie, J.G. (2002). Comparison of semiautomatic and automatic data acquisition methods for studying three-dimensional distributions of large neuronal populations and axonal plexuses. *Network* **13**, 343–356.

Lin, C.P., Tseng, W.Y., Cheng, H.C., and Chen, J.H. (2001). Validation of diffusion tensor magnetic resonance axonal fiber imaging with registered manganese-enhanced optic tracts. *NeuroImage* **14**, 1035–1047.

Liss, A.G., and Wiberg, M. (1997). Loss of nerve endings in the spinal dorsal horn after a peripheral nerve injury. An anatomical study in *Macaca fascicularis* monkeys. *Eur. J. Neurosci.* **9**, 2187–2192.

Livingstone, M., and Hubel, D. (1988). Segregation of form, color, movement, and depth: Anatomy, physiology, and perception. *Science* **240**, 740–749.

Malmierca, M.S., Rees, A., Le Beau, F.E., and Bjaalie, J.G. (1995). Laminar organization of frequency-defined local axons within and between the inferior colliculi of the guinea pig. *J. Comp. Neurol.* **357**, 124–144.

Malmierca, M.S., Leergaard, T.B., Bajo, V.M., Bjaalie, J.G., and Merchan, M.A. (1998). Anatomic evidence of a three-dimensional mosaic pattern of tonotopic organization in the ventral complex of the lateral lemniscus in Cat. *J. Neurosci.* **18**, 10603–10618.

Martin, K.A. (2002). Microcircuits in visual cortex. *Curr. Opin. Neurobiol.* **12**, 418–425.

Mazziotta, J.C., and Toga, A.W. (2002). Speculations about the future, In *Brain Mapping. The Methods* (A.W., Toga and J.C. Mazziotta, eds.), pp. 829–856. Academic Press, San Diego, CA.

Mazziotta, J.C., Toga, A.W., Evans, A.C., Fox, P.T., and Lancaster, J.L. (1995). Digital brain atlases. *Trends Neurosci.* **18**, 210–211.

Mazziotta, J.C., Toga, A.W., Evans, A.C., Fox, P.T., Lancaster, J.L., Zilles, K., Woods, R., Paus, T., Simpson, G., Pike, B., Holmes, C., Collins, L., Thompson, P., MacDonald, D., Iacoboni, M., Schormann, T., Amunts, K., Palomero-Gallagher, N., Geyer, S., Parsons, L., Narr, K., Kabani, N., Le Goualher, G., Boomsma, D., Cannon, T., Kawashima, R., and Mazoyer, B. (2001). A probabilistic atlas and reference system for the human brain: International Consortium for Brain Mapping (ICBM). *Philos. Trans. R. Soc. London, Ser. B* **356**, 1293–1322.

Merchan, M.A., and Berbel, P. (1996). Anatomy of the ventral nucleus of the lateral lemniscus in rats: A nucleus with a concentric laminar organization. *J. Comp Neurol.* **372**, 245–263.

Mesulam, M.-M. (1982). Principles of horseradish peroxidase neurohistochemistry and their applications for tracing neural pathways—Axonal transport, enzyme histochemistry and light microscopic analysis. In *Tracing Neural Connections with Horseradish Peroxidase* (M.-M. Mesulam ed.), pp. 1–151. Wiley, Chichester.

Middleton, F.A., and Strick, P.L. (2000). Basal ganglia and cerebellar loops: Motor and cognitive circuits. *Brain Res. Rev.* **31**, 236–250.

Mori, S. (2002). Principles, methods, and applications of diffusion tensor imaging. In: *Brain Mapping. The Methods* (A.W., Toga and J.C. Mazziotta, eds.), pp. 379–398. Academic Press, San Diego, CA.

Nadasdy, Z., and Zaborszky, L. (2001). Visualization of density relations in large-scale neural networks. *Anat. Embryol.* **204**, 303–317.

Nieuwenhuys, R., Voogd, J., and Van Huijzen, C. (1988). *The Human Central Nervous System. A Synopsis and Atlas.* Springer-Verlag, Berlin.

Nikundiwe, A.M., Bjaalie, J.G., and Brodal, P. (1994). Lamellar organization of pontocerebellar neuronal populations. A multi- tracer and 3-D computer reconstruction study in the cat. *Eur. J. Neurosci.* **6**, 173–186.

O'Leary, D.D., and Terashima, T. (1988). Cortical axons branch to multiple subcortical targets by interstitial axon budding: Implications for target recognition and "waiting periods." *Neuron* **1**, 901–910.

Parent, M., Levesque, M., and Parent, A. (2001). Two types of projection neurons in the internal pallidum of primates: Single-axon tracing and three-dimensional reconstruction. *J. Comp Neurol.* **439**, 162–175.

Pautler, R.G., and Koretsky, A.P. (2002). Tracing odor-induced activation in the olfactory bulbs of mice using manganese-enhanced magnetic resonance imaging. *NeuroImage* **16**, 441–448.

Pautler, R.G., Silva, A.C. and Koretsky, A.P. (1998). In vivo neuronal tract tracing using manganese-enhanced magnetic resonance imaging. *Magn. Reson. Med.* **40**, 740–748.

Paxinos, G., and Franklin, K.B.J. (2001). *The Mouse Brain in Stereotaxic Coordinates.* Academic, San Diego, CA.

Paxinos, G., and Watson, C. (1998). *The Rat Brain in Stereotaxic Coordinates.* Academic Press, San Diego, CA.

Paxinos, G., Carrive, P., and Wang, H., and Wang, P.-Y. (1999). *Chemoarchitectonic Atlas of the Rat Brainstem.* Academic Press, San Diego, CA.

Petralia, R.S., and Wenthold, R.J. (1999). Immunocytochemistry of NMDA receptors. *Methods Mol. Biol.* **128**, 73–92.

Rademacher, J., Burgel, U., Geyer, S., Schormann, T., Schleicher, A., Freund, H.J., and Zilles, K., (2001). Variability and asymmetry in the human precentral motor system. A cytoarchitectonic and myeloarchitectonic brain mapping study. *Brain* **124**, 2232–2258.

Reiner, A., Veenman, C.L., Medina, L., Jiao, Y., Del Mar, and N., Honig, M.G. (2000). Pathway tracing using biotinylated dextran amines. *J. Neurosci. Methods* **103**, 23–37.

Roland, P.E., and Zilles, K. (1994). Brain atlases—a new research tool. *Trends Neurosci.* **17**, 458–467.

Rouiller, E.M., and Welker, E. (2000). A comparative analysis of the morphology of corticothalamic projections in mammals. *Brain Res. Bull.* **53**, 727–741.

Ruigrok, T.J. (1997). Cerebellar nuclei: The olivary connection. *Prog. Brain Res.* **114**, 167–192.

Saleem, K.S., Pauls, J.M., Augath, M., Trinath, T., Prause, B.A., Hashikawa, T., and Logothetis, N.K. (2002). Magnetic resonance imaging of neuronal connections in the macaque monkey. *Neuron* **34**, 685–700.

Sawchenko, P.E., and Gerfen, C.R. (1985). Plant lectins and bacterial toxins as tools for tracing neuronal connections. *Trends Neurosci.* **8**, 378–384.

Schmahmann, J.D., Doyon, J., McDonald, D., Holmes, C., Lavoie, K., Hurwitz, A.S., Kabani, N., Toga, A., Evans, A., and Petrides, M. (1999). Three-dimensional MRI atlas of the human cerebellum in proportional stereotaxic space. *NeuroImage* **10**, 233–260.

Senft, S.L. (2002) Axonal navigation through voxel substrates: A strategy for reconstructing brain circuitry. In *Computational Neuroanatomy: Principles and Methods* (G. Ascoli, ed.), pp. 245–270. Humana Press, Totowa, NJ.

Shambes, G.M., Gibson, J.M., and Welker, W. (1978). Fractured somatotopy in granule cell tactile areas of rat cerebellar hemispheres revealed by micromapping. *Brain Behav. Evol.* **15**, 94–140.

Shatz, C.J. (1983). The prenatal development of the cat's retinogeniculate pathway. *J. Neurosci.* **3**, 482–499.

Skirboll, L.R., Thor, K., Helke, C., Hökfelt, T., Robertson, B., and Long, R. (1989). Use of retrograde fluorescent tracers in combinations with immunohistochemical methods. In: *Neuroanatomical Tract-Tracing Methods 2: Recent Progress,* (L., Heimer and L. Zaborszky eds.), pp. 5–18. Plenum Press, New York.

Sparks, D.L., Lue, L.F., Martin, T.A., and Rogers J. (2000). Neural tract tracing using Di-I: A review and a new method to make fast Di-I faster in human brain. *J. Neurosci. Methods* **103**, 3–10.

Storm-Mathisen, J., and Ottersen, O.P. (1987). Tracing of neurons with glutamate or gamma-aminobutyrate as putative transmitters. *Biochem. Soc. Trans.* **15**, 210–213.

Storm-Mathisen, J., and Ottersen, O.P. (1990). Immunocytochemistry of glutamate at the synaptic level. *J. Histochem. Cytochem.* **38**, 1733–1743.

Sugihara, I., Wu, H.S., and Shinoda, Y. (2001). The entire trajectories of single olivocerebellar axons in the cerebellar cortex and their contribution to Cerebellar compartmentalization. *J. Neurosci.* **21**, 7715–7723.

Swanson, L.W. (1995). Mapping the human brain: Past, present, and future. *Trends Neurosci.* **18**, 471–474.

Swanson, L.W. (1999). *Brain Maps: Structure of the Rat Brain.* Elsevier, Amsterdam.

Swanson, L.W. (2000). A history of neuroanatomical mapping. In *Brain Mapping. The Systems,* (A.W., Toga and J.C. Mazziotta, eds.), pp. 77–109. Academic Press, San Diego, CA.

Tang, Y., Rampin, O., Giuliano, F., and Ugolini, G. (1999). Spinal and brain circuits to motoneurons of the bulbospongiosus muscle: Retrograde transneuronal tracing with rabies virus. *J. Comp. Neurol.* **414**, 167–192.

Thanos, S., and Bonhoeffer, F. (1987). Axonal arborization in the developing chick retinotectal system. *J. Comp. Neurol.* **261**, 155–164.

Thompson, P.M., MacDonald, D., Mega, M.S., Holmes, C.J., Evans, A.C., and Toga, A.W. (1997). Detection and mapping of abnormal brain structure with a probabilistic atlas of cortical surfaces. *J. Comput-Assisted Tomogr.* **21**, 567–581.

Thompson, P.M., Woods, R.P., Mega, M.S., and Toga, A.W. (2000). Mathematical/computational challenges in creating deformable and probabilistic atlases of the human brain. *Hum. Brain Mapp.* **9**, 81–92.

Toga, A.W. (1990). Three-dimensional reconstruction. In *Three-Dimensional Imaging* (A.W. Toga, ed.), pp. 251–272. Raven Press, New York.

Toga, A.W., ed. (1999). *Brain Warping.* Academic Press, San Diego, CA.

Toga, A.W., and Thompson, P. (1999). An introduction to brain warping. In *Brain Warping,* A.W. Toga ed.), pp. 1–26. Academic Press, San Diego, CA.

Toga, A.W., Thompson, P.M., Mega, M.S., Narr, K.L., and Blanton, R.E. (2001). Probabalistic approaches for atlasing normal and disease-specific brain variability. *Anat. Embryol.* **204**, 267–282.

Ugolini, G. (1995). Specificity of rabies virus as a transneuronal tracer of motor networks: Transfer from hypoglossal motoneurons to connected second- order and higher order central nervous system cell groups. *J. Comp. Neurol.* **356**, 457–480.

Van der Linden, A., Verhoye, M., Van Meir, V., Tindemans, I., Eens, M., Absil, P., and Balthazart, J. (2002). In vivo manganese-enhanced magnetic resonance imaging reveals connections and functional properties of the songbird vocal control system. *Neuroscience* **112**, 467–474.

Van Haeften, T., and Wouterlood, F.G. (2000). Neuroanatomical tracing at high resolution. *J. Neurosci. Methods* **103**, 107–116.

Van Horn, J.D., and Gazzaniga, M.S. (2002). Opinion: Databasing fMRI studies towards a 'discovery science' of brain function. *Nat. Rev. Neurosci.* **3**, 314–318.

Vassbø, K., Nicotra, G., Wiberg, M., and Bjaalie, J.G. (1999). Monkey somatosensory cerebrocerebellar pathways: Uneven densities of corticopontine neurons in different body representations of areas 3b, 1, and 2. *J. Comp. Neurol.* **406**, 109–128.

Veenman, C.L., Reiner, A., and Honig, M.G. (1992). Biotinylated dextran amine as an anterograde tracer for single- and double-labeling studies. *J. Neurosci. Methods* **41**, 239–254.

Vercelli, A., Repici, M., Garbossa, D., and Grimaldi, A. (2000). Recent techniques for tracing pathways in the central nervous system of developing and adult mammals. *Brain Res. Bull.* **51**, 11–28.

Visel, A., Ahdidan, J., and Eichele, G. (2002). A gene expression map of the mouse brain: Genepaint.org—a database of gene expression patterns. In *Neuroscience Databases: A Practical Guide* (R. Kötter, ed.), pp. 19–36. Kluwer Academic Publishers, Boston, MA.

Voogd, J. (1967). Comparative aspects of the structure and fibre connexions of the mammalian cerebellum. *Prog. Brain Res.* **25**, 94–134.

Voogd, J. (1995). Cerebellum. In *Gray's Anatomy,* L.H., Bannister, M.M., Berry, P., Collins, M., Dyson, J.E., Dussek, and M.W.J. Ferguson, eds.), pp. 1027–1065. Churchill-Livingstone, New York.

Wang, Y., Gupta, A., Toledo-Rodriguez, M., Wu, C.Z., and Markram, H. (2002). Anatomical, physiological, molecular and circuit properties of nest basket cells in the developing somatosensory cortex. *Cereb. Cortex* **12**, 395–410.

Watanabe, T., Michaelis, T., and Frahm, J. (2001). Mapping of retinal projections in the living rat using high-resolution 3D gradient-echo MRI with Mn^{2+}-induced contrast. *Magn. Reson. Med.* **46**, 424–429.

Welker, C. (1971). Microelectrode delineation of fine grain somatotopic organization of (SmI) cerebral neocortex in albino rat. *Brain Res.* **26**, 259–275.

Welker, E., Hoogland, P.V., and Van der Loos, H. (1988). Organization of feedback and feedforward projections of the barrel cortex: A PHA-L study in the mouse. *Exp. Brain Res.* **73**, 411–435.

West, M.J., and Skytte, J., (1986). Anatomical modeling with computer-aided design. *Comput. Biomed. Res.* **19**, 535–542.

Woolsey, C.N. (1958), Organization of somatic sensory and motor areas of the cerebral cortex. In *Biological and Biochemical Bases of Behavior* (H.F., Harlow and C.N. Woolsey, eds.), pp. 63–81. University of Wisconsin Press, Madison.

Woolsey, T.A., and Van der Loos, H. (1970). The structural organization of layer IV in the somatosensory region (SI) of mouse cerebral cortex. The description of a cortical field composed of discrete cytoarchitectonic units. *Brain Res.* **17**, 205–242.

Wouterlood, F.G., and Groenewegen, H.J. (1991). The Phaseolus vulgaris–leucoagglutinin tracing technique for the study of neuronal connections. *Prog. Histochem. Cytochem.* **22**, 1–78.

Wouterlood, F.G., and Van Haeften, T., Blijleven, N., Perez-Templado, P., and Perez-Templado, H. (2002a). Double-label confocal laser-scanning microscopy, image restoration, and real-time three-dimensional reconstruction to study axons in the central nervous system and their contacts with target neurons. *Appl. Immunohistochem. Mol. Morphol.* **10**, 85–95.

Wouterlood, F.G., Vinkenoog, M., and van den Oever, M. (2002b). Tracing tools to resolve neural circuits. *Network* **13**, 327–342.

Zaborszky, L., and Duque, A. (2000). Local synaptic connections of basal forebrain neurons. *Behav. Brain Res.* **115**, 143–158.

Zilles, K., Schleicher, A., Palomero-Gallagher, N., and Amunts, K. (2002). Quantitative analysis of cyto- and receptor architecture of the human brain. In *Brain Mapping: The Methods.* (A.W. Toga, ed.), pp. 573–602. Academic Press, San Diego, CA.

Surface-Based Atlases and a Database of Cortical Structure and Function

David C. Van Essen, Ph.D., John Harwell, M.S., Donna Hanlon, M.S., and James Dickson, M.S.

CONTENTS

Databasing the Brain. Edited by Stephen H. Koslow and Shankar Subramaniam
ISBN 0-471-30921-4 © 2005 John Wiley & Sons, Inc.

22.1 INTRODUCTION

The dominant structure of the mammalian brain is the cerebral cortex, which has been intensively studied for more than a century using an increasingly diverse and powerful array of techniques. Our current state of understanding about cortical organization and function presents a conundrum for neuroscientists. On the one hand, staggering amounts of experimental data have been obtained, and conventional scientific publications are seriously limited in their ability to provide access to vast amounts of valuable data. Moreover, new data are being acquired at an explosive rate, particularly with the advent of modern neuroimaging techniques. On the other hand, a firm understanding is lacking on a number of fundamental issues. Most notably, while there is widespread agreement that the cortex can be subdivided into many distinct areas ($\sim$100–200 areas in primates), there is much uncertainty and controversy regarding the precise identity and location of many of these areas, even for the most intensively studied species. In this regard, cortical cartographers are in a state comparable to that of seventeenth-century cartographers of the earth's surface, who generated many competing versions of maps of the earth's major geographical and societal subdivisions, all fragmentary and imprecise.

In attempting to decipher cortical organization, one major hurdle relates to the extensive convolutions that give cortex its distinctive appearance. These convolutions make it difficult to visualize complex spatial relationships among the various cortical nooks and crannies. Another equally vexing but intriguing problem is that of individual variability. In humans, the pattern of convolutions in any individual is as distinctive as that of one's fingerprint patterns. This "geographic" variability is further compounded by individual differences in the size and location of each cortical area.

Surface-based visualization and analysis methods offer a powerful approach for coping with the problems imposed by cortical convolutions and their variability. Surface maps of the cortex provide the same kind of visualization advantages as do two-dimensional maps of the earth's surface. In addition, they are extremely valuable for dealing with the issues of individual variability and with comparisons across species.

Our laboratory is engaged in a two-pronged thrust in computerized cortical cartography. One effort has been to develop brain-mapping software that is useful for cortical surface visualization and analysis (Van Essen et al., 2001a). Another is to develop surface-based atlases for several species that serve as repositories for an increasingly diverse range of experimental data (Van Essen and Drury, 1997; Van Essen et al., 2001b; Van Essen, 2002a, 2004). Here we provide a progress report on the current state of these atlases and the options for making effective use of them. Particular emphasis will be on an atlas of visual cortical areas, because the visual cortex has been especially in-

tensively studied and is the best understood functional modality. However, the analyses generalize to all of cerebral cortex and even to the more highly convoluted cerebellar cortex.

Section 22.2 introduces the primate and rodent atlases and illustrates multiple ways in which each of them can be visualized. Section 22.3 discusses strategies for mapping experimental data onto atlases and illustrates a variety of exemplar datasets. One set of examples are from the macaque monkey, where the data are richest and most diverse and where many important technical and conceptual issues can be clearly articulated. Additional examples are included from the human atlas maps and also from interspecies registration between macaque and human cortex. Sections 22.4 and 22.5 address the database issues involved in making the atlases and associated datasets readily available to the neuroscience community and in updating the atlases as information continues to pour in.

22.2 VISUALIZING ATLAS VOLUMES AND SURFACES

High-resolution structural MRI provides an invaluable window on many aspects of brain structure, including the complexity of cortical folds, and it circumvents the problems of physical distortions and section alignment problems that arise when dealing with tissue sections processed after conventional histological sectioning. Accordingly, we have used high-resolution structural MRI as our primary substrate for atlases of human, macaque, rat, and mouse. The macaque atlas brain, illustrated in Figure 22.1A by a coronal section through both cerebral hemispheres, has excellent contrast between gray and white matter that allows accurate visualization of the entire cortical sheet. Because cortical shape changes significantly from one slice to the next, it takes a complete sequence of images (i.e., an image volume) to accurately represent the complete pattern of convolutions.

22.2.1 Segmentation and Surface Reconstruction

Explicit surface reconstructions provide the most compact and efficient way to represent cortical shape. A key intermediate stage in cortical surface reconstruction involves generation of a volume segmentation, in which the boundary between black and white voxels represents the shape of the cortex. The atlases discussed here were generated using the SureFit method for cortical segmentation (Van Essen et al., 2001a). Figure 22.1B shows a slice through a segmentation (translucent red) that is overlaid on the same structural MRI slice (right hemisphere only) shown in Figure 22.1A. The SureFit method aims for a segmentation boundary that runs midway through the cortical thickness (i.e., cortical layer 4). This estimate is obtained by a multistage process that first generates separate probabilistic maps for the location of the cortical inner boundary (gray–white transition) and the outer boundary (pial surface). The segmentation boundary is then set midway between the most likely position of the inner and outer boundaries. The segmentation boundary is tessellated to form a geometrically defined surface, as shown in a cross section overlaid on the structural MRI (Figure 22.1C) and in a lateral view of the entire fiducial surface (Figure 22.1D).

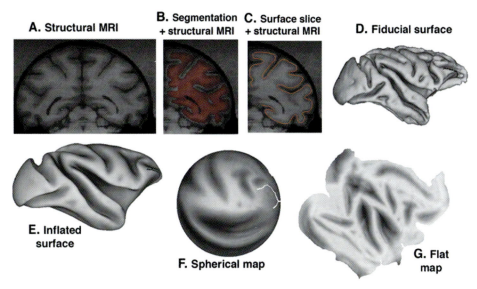

Figure 22.1 Segmentation, surface reconstruction and reconfiguration of structural MRI-based macaque brain atlas generated using SureFit and Caret software. Structural MRI data courtesy of N. Logothetis.

Besides SureFit, a number of other methods for automated or semiautomated cortical segmentation and surface reconstruction have been developed over the past decade, including Freesurfer (Fischl et al., 1999a), BrainVoyager (Goebel et al., 1998), and mrGray (Teo et al., 1997). All of these methods have similar objectives, but they differ in the algorithms used and in the fidelity with which the resultant surfaces represent the precise cortical shape. Because SureFit targets the cortical mid-thickness, the surface area of any given patch corresponds closely to the associated cortical volume. In contrast, the alternative segmentation methods noted above target the boundary between gray and white matter; the surface area measurements they yield are poorly correlated with associated cortical volume along the gyral crests and sulcal fundi. As the fidelity of fMRI and other analysis methods increases, these distinctions will become increasingly significant, making it particularly desirable that the atlases used as repositories be as accurate as possible.

22.2.2 Surface Visualization and Manipulation Using Caret

Once a surface has been generated, its shape can be manipulated in a variety of ways that facilitate visualization. The examples presented in this chapter were processed and visualized using Caret software developed in our laboratory. Figures 22.1E–G illustrate the three most commonly used configurations besides the fiducial surface. In the inflated surface (Figure 22.1E), the convolutions are smoothed out, but a representation of the original shape is preserved by displaying a "depth map" to indicate the distance from each node from the external hull of the hemisphere, so that deeper sulci appear darker. In a spherical map (Figure 22.1F), the inflated surface is projected to a geometrically defined sphere, and distortions in surface area (relative to the fiducial surface) are further reduced while preserving the spherical shape. Spherical maps serve two key roles: first, as a substrate for defining surface-based coordinates of latitude and longitude;

and second, as the preferred substrate for surface-based registration between an individual and an atlas (see below). The cortical flat map (Figure 22.1G) has a standard set of cuts used to reduce distortions, much as is done for flat maps of the earth's surface. However, alternate flat map configurations, based on a different set of cuts, are available when it is desirable to avoid cuts in a particular region of interest, such as prefrontal cortex.

22.2.3 Primate and Rodent Atlases

Figure 22.2 shows MRI slices from atlases of four major experimental species. The human (Figure 22.2A) and macaque (Figure 22.2B) atlases are illustrated by coronal slices that intersect cerebrum and cerebellum; the mouse (Figure 22.2C) and rat (Figure 22.2D) atlases are illustrated by horizontal slices that intersect cerebellum, cerebrum, and olfactory bulb. The human, macaque, and rat cerebral cortices were segmented using SureFit, yielding surface reconstructions that accurately captured the complete patterns of cortical folds (Figure 22.3). The rat cerebral cortex was reconstructed in Caret from digitized section contours.

The cerebellar cortex was more difficult to segment because it is only about one-third the thickness of cerebral cortex in each species, and because the human and macaque cerebellum contain numerous tertiary folds (folia) in addition to the primary and secondary folds (lobules and lamellae). The folia are barely visible in the macaque atlas MRI and are mostly unresolvable in the human atlas MRI (cf. Figure 22.2). A customized modification of the SureFit segmentation algorithm was used to obtain a reasonable initial segmentation of the cerebellum, which was then manually edited to generate a refined segmentation (Van Essen, 2002b). For the mouse and rat, the resultant cerebellar surface reconstructions faithfully represent nearly all of its convolutions (except for the flocculus and paraflocculus, which were poorly resolved in the MRI). The imperfections in the macaque and human reconstructions are more significant: The macaque

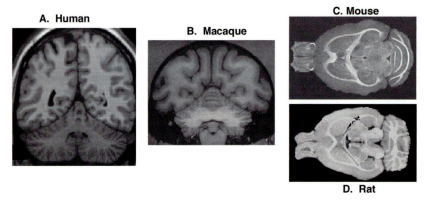

Figure 22.2   Structural MRI slides of human, macaque, mouse, and rat brains. Data courtesy of C. Holmes and A. Toga (human); N. Logothetis (macaque); R. Jacobs and A. Toga (mouse); and V. Song (rat).

reconstruction lacks an estimated one-third of the folia, and the human reconstruction lacks most of the folia and a few of the smaller lamellae.

Each of the 12 atlas surfaces was processed to obtain a standard set of inflated, spherical, and flat map configurations, like those already illustrated for the macaque right cerebral hemisphere. Figure 22.4 shows the cerebral and cerebellar fiducial surfaces and flat maps for all four species. The cerebral hemispheres are colored to indicate the different cortical lobes; the cerebellum is colored to show the different lobules; buried cortex is indicated by darker shading on all maps. All surfaces were flattened using a multi-resolution distortion–reduction algorithm in Caret. This process was fully automated for the cerebral maps

(except for manually drawing a standardized set of cuts that reduce the distortions on the flat maps). For the cerebellum, extensive interactive processing was needed to deal with the highly complex pattern of intrinsic (Gaussian) curvature of the cerebellum. The shape of the cerebellum is analogous to that of a rubber glove, insofar as both have numerous domains of high intrinsic curvature (in a glove, the highly curved fingertips and saddle points between fingers). Like a glove, the cerebellum is difficult to flatten without extensive and precisely placed cuts. The cerebral cortex, in contrast, is analogous to a deflated and crumpled beach ball that has a low and uniform degree of intrinsic curvature associated with the inflated spherical surface, but relatively little additional intrinsic curvature associated with crumpling (folding) of the surface. Consequently, even with extensive cuts, the residual distortions in surface area are much greater on the cerebellar flat maps than on the cerebral cortical maps.

Parenthetically, it is noteworthy that the differences in intrinsic geometry between adult cerebral and cerebellar cortex reflects major differences in how their convolutions arise during early development. For cerebral cortex, folding occurs long after neuronal proliferation and migration have established a smooth, embryonic cortical sheet, analogous to the inflated beach ball described above; folding into the crumpled adult configuration occurs in conjunction with, and is probably driven by, the establishment of long-distance cortico cortical connections (Van Essen, 1997). For cerebellar cortex, folding occurs coextensively with massive proliferation and migration of granule cells across the cerebellar surface. This presumably allows extensive region-specific increases in cortical surface area that result in many domains having high intrinsic curvature (Van Essen, 2002b).

The fundamental compactness of surface reconstructions and flat maps is exemplified by the fact that the entire cortical sheet for three structures in each of four species can be conveniently displayed in a single illustration (Figure 22.4). Figure 22.4 shows the surface areas (measured on the fiducial surface) for each cortical structure. The ratio of cerebellar surface area to cerebral surface area (both hemispheres combined) is an estimated 46% in the mouse and 51% in the rat. In the macaque and human, these estimates are 25% and 28%, respectively, but these are underestimates because many cerebellar folia were not included, especially in the human reconstruction. The value of 1128 cm^2

Cerebral and Cerebellar Cortex

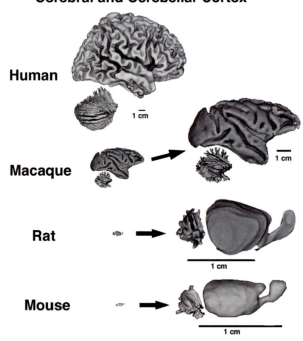

Figure 22.3   Cerebral and cerebellar surface reconstructions of human, macaque, rat, and mouse. Reconstructions are shown to scale on the left and are expanded on the right for clarity.

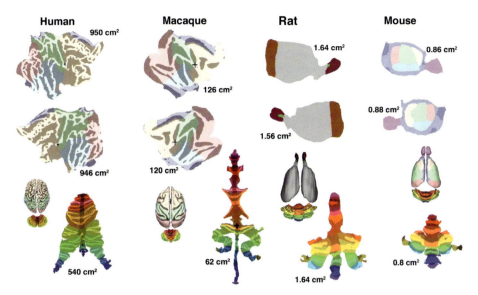

Figure 22.4   Fiducial and flat map configurations for human, macaque, and mouse cerebral and cerebellar surfaces. The surface areas adjoining each flat map represent values for the fiducial surface. For the human, the fiducial surfaces were registered to the Washington University 711-2B atlas (Ojemann et al., 1997); surface areas for the fiducial surfaces registered to the MNI-305 atlas are about 15% greater because its atlas brain dimensions are larger.

for human cerebellar cortex reported by Sultan and Braitenberg (1993) is 59% of total cerebral cortical surface area and is likely to be a more accurate estimate.

22.3 MAPPING DATA ON INDIVIDUALS AND ONTO ATLASES

Among the many types of experimental data that are appropriate for registration to a cortical atlas, the most basic involve information about the partitioning of cortex into different areas. We first consider this issue for macaque cerebral cortex, which has been intensively studied using a variety of approaches.

22.3.1 Mapping the Macaque Cortex

Architectonic Subdivisions in the Macaque
A wealth of data is available for the macaque regarding cortical areas that have been distinguished on the basis of one or more of the following criteria: (i) architecture (differences evident in histological sections), (ii) connectivity (distinctive patterns of inputs and outputs), (iii) topography (maps of sensory surfaces), (iv) single-unit neurophysiology (neuronal response characteristics), and (v) functional specializations revealed by neuroimaging, including both optical methods and fMRI. However, even after a century of effort, it can be very difficult to decide what constitutes an "area"—that is, a neurobiologically distinct domain that differs from neighboring regions in functionally important and consistent ways. Architectonics is the least directly related to function among the approaches noted above, yet it remains invaluable as an approach that can be applied systematically over large cortical expanses in multiple individuals. We demonstrate this by illustrating how a mosaic of architectonically distinct cortical subdivisions charted in a population of hemispheres can be

registered to the macaque atlas to yield a probabilistic map of cortical organization.

Figure 22.5 shows a map of cortical areas charted on an individual hemisphere (Case A), using myeloarchitecture, cytoarchitecture, and chemoarchitecture (SMI-32 immunoreactivity) to delineate anatomically distinct subdivisions (Lewis and Van Essen, 2000a). The figure illustrates several key stages in the identification and mapping process. An initial stage is to identify architectonic transitions on individual sections, as shown on an exemplar myelin-stained section in Figure 22.5A. The spatial uncertainty in identifying and localizing architectonic borders is typically 1–3 mm and sometimes more (with only rare exceptions, such as the razor-sharp V1/V2 boundary). This degree of uncertainty is comparable to the dimensions of some of the cortical areas themselves. Core regions (labels in Figure 22.5A) separated by transition regions (white bars in Figure 22.5A) were identified throughout each of the stained sections in Case A. Information from neighboring sections stained for myelin, Nissl, and SMI-32 immunoreactivity was then combined, and a composite estimate of core and transition regions was marked on digitized contours of regularly spaced sections (panel B). A surface reconstruction was then generated from these section contours (in Case A, the entire hemisphere except for frontal and occipital poles), and the identified core regions were "painted" onto the reconstructed surface (panel C). A cortical flat map generated from this surface allows all 79 architectonic subdivisions identified in Case A to be seen in a single view without distortions from foreshortening (Figure 22.5D). Figure 22.5E shows an expanded view of the intraparietal sulcus and neighboring regions with visual areas labeled. Most of these subdivisions are likely to represent genuine cortical areas because they can be distinguished on the basis of additional anatomical or functional criteria (Lewis and Van Essen, 2000a,b). However, some of them have been categorized as architectonically distinct "zones" within a larger area (e.g., zones VIPm and VIPl within area VIP); resolution

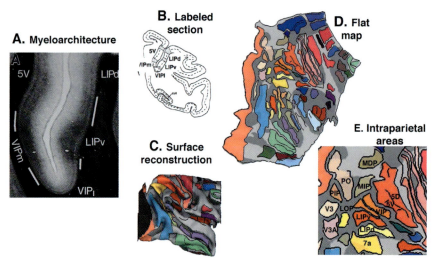

Figure 22.5   A map of architectonic subdivisions in an individual macaque hemisphere (Case A = 95D.R). (**A**) Myelin-stained coronal section through the intraparietal sulcus. (**B**) Areal boundaries and transition zones identified in panel A and transferred to a digitized contour section. (**C**) Computerized surface reconstruction of right hemisphere generated in Caret from a sequence of regularly spaced section contours. (**D**) Flat map of 79 architectonic subdivisions in Case A. (**E**) Visual areas and zones in the expanded transparietal region are labeled individually.

of whether or not these zones will eventually be regarded as genuinely distinct areas awaits additional experimental evidence using complementary analysis methods.

The same process was used to generate maps of architectonic subdivisions in four additional hemispheres (Figure 22.6) for fiducial surfaces (top row) and flat maps (middle row). In addition, the lower row shows expanded portions of flat maps for the intraparietal sulcus. In general, the same set of cortical subdivisions was found in all hemispheres (except for uncharted regions in the partial hemisphere reconstructions). Each map shows the same basic arrangement of areas, but there are many differences in detail. These differences can be grouped into three main types:

1. *Geographic Variability.* The overall pattern of cortical folding in the macaque is strikingly consistent across hemispheres: A dozen or more major sulci with consistent relationships to one another can be reliably identified. Nonetheless, these sulci vary in their precise location, dimensions (length and width), and trajectory in 3D (stereotaxic) space. A few additional sulci (e.g., the external calcarine sulcus) are prominent in some hemispheres, a mere dimple in others, and absent altogether in still others.

2. *Shape and Location.* Each cortical area varies significantly in its shape and in its location relative to geographical landmarks, such as the tip of a sulcus or the margins of a gyrus. Interestingly, most cortical areas tend to be distinctly elongated, with the long axis running approximately parallel to that of the sulcus or gyrus in which they lie. The degree of variability is difficult to quantify systematically, but for the elongated areas on the map, the variability is typically several millimeters or more along their long axis but less along their short axis (width) of these areas.

3. *Size Variability.* Well-defined cortical areas can vary by twofold or more in surface area from one individual to the next, as has been documented most thoroughly for area V1 (Filiminoff, 1932; Van Essen et al., 1984; Andrews et al., 1997; Amunts et al., 2000). Differences in areal size are of particular interest because they may contribute to individual differences in various aspects of brain function. An important issue is whether there are strong correlations in the sizes of different areas or whether they tend to vary independently. However, this is difficult to assess from the maps in Figures 22.5 and 22.6 because of the experimental uncertainties in identifying areal boundaries.

22.3.2 The Registration Problem

Atlases provide a spatial framework for assessing both the commonalities across individuals and the nature and magnitude of individual variability. Registration to an atlas involves compensating for some aspects of individual variability. Given the issues raised in the preceding section, two basic questions are: (1) What should the objectives be when registering an individual to an atlas? and (2) How can these objectives best be achieved methodologically?

Because the variability related to geography (folding) and that related to functional organization are not perfectly correlated, and because there are uncertainties associated with both types of measurement, the notion of compensating simultaneously for all aspects of variability—that is, to register one brain to another so that they are perfectly aligned in all respects—is an unattainable ideal. Instead, it is necessary to select a subset of the factors giving rise to variability as guides (constraints) for the registration process. One basic issue is whether to emphasize criteria related to geography, functional organization, or both. For the examples illustrated below, we have used mainly geographic criteria when registering an individual to an atlas of the same species, and we have used mainly functional criteria when registering between species.

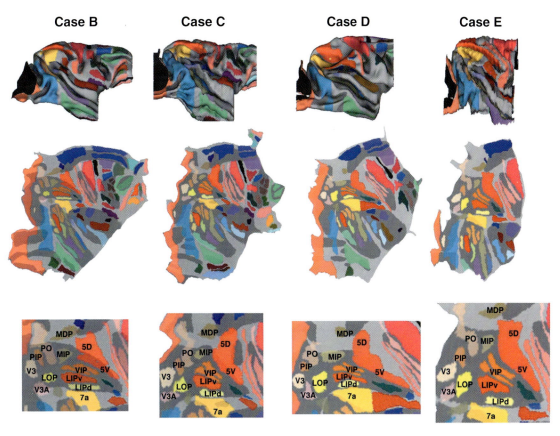

Figure 22.6 Architectonic maps in four individual macaque hemispheres. Upper panels show lateral views of fiducial surface reconstructions; middle panels show cortical flat maps of the entire reconstructed surface; lower panels show flat maps of just the intraparietal sulcus. The number of identified subdivisions is 82 in Case B (94C.R), 84 in Case C (93I.R), 64 in Case D (94I.R), and 35 in Case E (95D.R, a much less extensive reconstruction).

The many methods available for registering data to an atlas can be grouped into three general categories: *volume*-based, *slice*-based, and *surface*-based registration. Given the sheet-like structure of the cortex, surface-based registration has a fundamental advantage, because it inherently respects the topology of the cortical sheet (Van Essen et al., 1998, 2001a; Fischl et al., 1999b; see below). Hence, for data that have been mapped to surface reconstructions of individual hemispheres, such as the macaque architectonic maps in Figures 22.5, 22.6, surface-based registration is a preferred approach.

22.3.3 Generating a Probabilistic Macaque Atlas

The data from the individual maps in Figures 22.5 and 22.6 were registered to the atlas using a standard set of sulcal landmarks that were consistently identifiable in the individual and atlas hemispheres. In Figure 22.7A,B each sulcus is painted a distinct color in order to illustrate the correspondences between sulci on an individual map (Case A) and the atlas map. Landmark contours drawn along the fundus of each sulcus were used to register the individual to the atlas. The flat maps were registered directly to one another using a two-dimensional registration algorithm (Joshi, 1997; Van Essen et al., 1998). Figure 22.7C shows the resultant architectonic map of Case A after registration to the atlas. The same registration process was carried out for the four

individual architectonic maps shown in Figure 22.6. The deformed architectonic maps from all five cases were superimposed to form a probabilistic map of cortical areas (Figure 22.7D). Saturated colors signify regions where all or nearly all core areas of a given identity overlap; paler coloration indicates regions of partial overlap. In general, there is little overlap between differently labeled areas, signifying that the variability in areal location relative to geographic landmarks does not exceed the uncertainties in identifying areal boundaries. Thus, major sulcal landmarks provide a reasonably reliable basis for localizing cortical areas in the macaque.

Using the probabilistic atlas as a guide, the most likely boundaries between neighboring areas were drawn, thereby generating a "standard" map of Lewis and VE areas (Figure 22.6E). This includes 72 subdivisions, most of which are considered distinct areas but some are consistently identifiable zones within a larger area. This can also be visualized in Caret as an "areal estimation map" that has fuzzy boundaries that semiquantitatively reflect the uncertainties and probabilistic nature of the representation.

22.3.4 A Plethora of Partitioning Schemes

Besides the Lewis and Van Essen scheme illustrated in Figures 22.5–22.7, numerous other partitioning schemes for macaque cortex remain in widespread use. Some of these

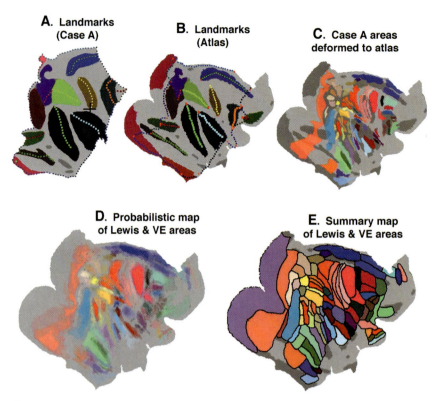

Figure 22.7 Surface-based registration and the generation of probabilistic and summary architectonic maps. (**A**) Landmark sulci and perimeter delineation in Case A, flat map. (**B**) Corresponding landmarks on the atlas flat map. (**C**) Architectonic areas from Case A registered to the atlas using a 2D surface-based registration. (**D**) Probabilistic map from Cases A–E registered to the atlas. This is an extension of a previous example of a probabilistic map of architectonically delineated visual areas that were registered to a different (non-MRI-based) macaque atlas (Van Essen et al., 2001b). (**E**) Summary architectonic map of the most likely extent of each cortical subdivision.

schemes cover only restricted regions, whereas others include the entire hemisphere. Any scheme in which areal boundaries are represented with reasonable accuracy in relation to gyral and sulcal landmarks can be brought into register with the macaque atlas using alignment and registration options available in Caret.

Using this approach, we have registered schemes from 14 published studies to the atlas, including 10 schemes for parts or all of visual cortex (see also Van Essen, 2004). Figure 22.8 illustrates two of these schemes, including that of Felleman and Van Essen in panel A and that of Ungerleider and Desimone 1986; (Desimone and Ungerleider, 1989) in panel B. Outside areas V1 and V2, the arrangement of areas differ significantly for most regions just among these three schemes (Figures 22.7E and 22.8A,B). When the other schemes are included, there is a consensus among investigators about the basic identity, location, and size of the area for only a few regions of visual cortex. Specifically, there is a strong consensus regarding areas V1, V2, MT (a.k.a. V5), and perhaps also V4 and V3A. In the case of V3 (V3d) and VP (V3v) there is agreement about the arrangement of topographic maps, but disagreement as to whether these constitute distinct areas versus separate areas that each contain partial visual representations (Lyon and Kaas, 2002; Van Essen, 2004). For most of the cortex, the selection of any given location will likely yield at least two and often more areal names that represent genuinely distinct schemes (e.g., using the node identification option in Caret). Figure 22.8C shows a composite

map, in which major clusters of areas are shown. Each cluster represents a group of functionally related areas that are consistently identified across studies even though there are marked differences regarding the arrangement of constituent areas within each cluster.

A useful way to obtain greater objectivity, specificity, and conciseness in discussions about partitioning and functional specialization is to use surface-based coordinates (latitude and longitude) to specify locations on the cortical surface (Fischl et al., 1999a; Drury et al., 1999). This is illustrated with latitude and longitude isocontours defined on the sphere (Figure 22.8D) and after projection to the flat map (Figure 22.8E).

22.3.5 Seeing the Connections

Each cortical area has a complex pattern of connections with other cortical areas and with a variety of subcortical targets. A wiring diagram of connections in macaque visual cortex reveals hundreds of pathways among several dozen visual cortical areas (Felleman and Van Essen, 1991). Such wiring diagrams are useful for a variety of purposes, including the assessment of hierarchical relationships among different areas, and they have been incorporated into CoCoMac, a text-based connectivity database for the macaque (Stephan et al., 2001). However, this type of information fails to portray the full spatial complexity

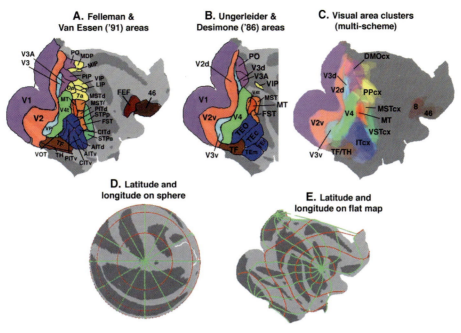

Figure 22.8 Multiple partitioning schemes and a surface-based coordinate system for macaque visual cortex. (**A**) Felleman and Van Essen (1991) visual areas. (**B**) Visual areas from the Ungerleider and Desimone scheme (Ungerleider and Desimone, 1986; Desimone and Ungerleider, 1989). (**C**) Clusters of visual areas common to the 10 partitioning schemes discussed in Van Essen (2003). (**D**) Latitude and longitude isocontours on the spherical map. (**E**) Latitude and longitude isocontours on the flat map. Latitude and longitude, along with the areal identity according to various partitioning schemes, are part of the readout provided by the node identification option in Caret.

of connection patterns and the wide range in overall strength of different pathways. Terms like "weak," "moderate," and "strong" connections are often used, but these inadequately reflect variations in connection strength that may span two or three orders of magnitude.

Figure 22.9 illustrates an alternative strategy for representing connectivity patterns on the atlas. In this example, adapted from Lewis and Van Essen (2000b), retrograde tracers were injected into two different sites, one centered on area MST (MSTdp, specifically) and the other centered on area LIPd (but encroaching into neighboring area 7a). The labeling pattern for each case was quantified in terms of connection density (labeled cells per mm^2) in the individual experimental hemispheres, and the connection density maps were registered to the atlas, where they can be compared with one another and also with

any of the cortical partitioning schemes contained in the atlas (e.g., the Lewis and Van Essen borders overlaid in Figure 22.9A,B). As shown in Figure 22.9C, additional visualization tools available in Caret facilitate comparisons of different connection patterns by showing an overlay of MST connections in red, LIPd/area-7a connections in green, and connections to both sites in yellow.

22.3.6 Slice-Based Registration

The approach illustrated in Figures 22.7–22.9 is well-suited for registration of datasets that have been charted on cortical surface maps. However, there are innumerable cases of valuable data, especially from older publications, in which the only

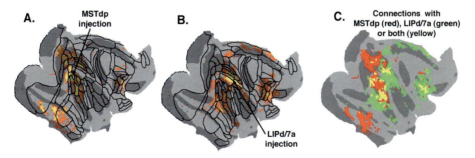

Figure 22.9 Quantitative maps of connectivity patterns in visual cortex. (Data from Lewis and Van Essen, 2000b). (**A**) Connection density from a retrograde tracer injection centered in MSTdp, with Lewis and Van Essen (2000a) areal boundaries superimposed. (**B**) Connection density from an injection centered in LIPd. (**C**) Overlap map showing regions connected with MSTdp (red), LIPd/7a (green), or both (yellow).

accessible data are displayed on photomicrographs or drawings of histological sections. In such cases, data can be mapped from individual sections onto corresponding slices of the atlas map using a slice-based registration option available in Caret. This will allow many types of data from published and unpublished studies to be incorporated into the macaque atlas with reasonable fidelity. Slice-based registration is likely to be particularly useful for analyses of cerebellar cortex, because it is impractical with current methodology to generate accurate cerebellar reconstructions for individual experimental brains in the way that can now be done routinely for cerebral cortex.

22.3.7 Mapping Function onto Surfaces

An enormous amount of information about the functional specialization of different cortical areas has been obtained by single-unit recordings in the macaque. These single-unit studies are now being complemented by functional MRI, which can be routinely carried out in macaques (Nakahara et al., 2002; Vanduffel et al., 2002; Tsao et al., 2003) just as in humans. Obviously, it is desirable to bring both single-unit and fMRI data into the atlas framework to facilitate a wide variety of comparisons.

An example of monkey fMRI data mapped to the atlas is shown in Figure 22.10. Figure 22.10A shows a map of activation by a checkerboard pattern (relative to a fixation-only baseline) on a right hemisphere flat map from an individual macaque. The activation pattern covers a large region, including cortex near the artificial cut in occipital cortex. The two-dimensional method for registering flat maps illustrated previously (Figure 22.8) encounters problems in the vicinity of cuts on either the individual or the atlas flat map. Consequently, we used instead a spherical registration algorithm, in which landmarks were drawn on the individual and atlas maps (not shown) and then projected to spherical surfaces (Figure 22.10B–E). The spherical map of the individual was then deformed to bring its landmarks into register with the atlas landmarks, using a multi-cycle registration method (Caret User Guide Part 2—online). The registered

data can be projected back to the flat maps (and to other configurations) for easier visualization. The checkerboard-specific activations were then registered to the atlas where the borders from various partitioning schemes can be overlaid. This is shown in Figure 22.10F for the Lewis and Van Essen scheme, indicating that the activations include much of area V1, V2, V3, and V4.

The options for mapping single-unit neurophysiological data depend on how the data have been encoded and represented. For example, it is increasingly common to (a) obtain structural MRI scans of monkeys used in alert recording experiments and (b) generate surface reconstructions from the MRI data. If the 3D coordinates of recording sites are recorded in conjunction with this, the recording sites can then be mapped to the individual surface and from there to the atlas.

22.3.8 Mapping Human Cortex

As with the macaque, there is great diversity in the types of anatomical and functional data that are suitable for mapping to the human atlas. Currently, the lion's share of available data for human cortex derive from noninvasive neuroimaging studies—initially PET, but now dominated by fMRI. Data from architectonic studies and other anatomical approaches are much less common, owing to the obvious difficulties associated with acquiring and processing postmortem histological data in humans. The options for registration to the human atlas fall into four major categories, each illustrated by exemplar data in Figures 22.11 and 22.12.

1. *Manual Transfer of Data.* Large amounts of experimental data, particularly from older publications, exist as illustrations that lack explicit spatial coordinates, yet are desirable to incorporate into the atlas framework. A notable example is Brodmann's cytoarchitectonic partitioning scheme for human cortex, which was published only as drawings of lateral and medial views of a single hemisphere (Brodmann, 1909). Brodmann's scheme has been mapped onto the surface-based atlas by manually

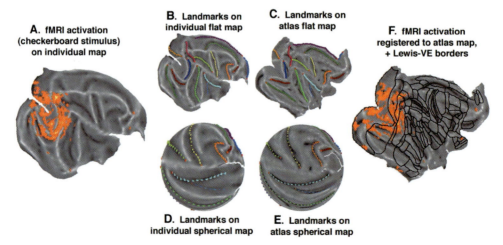

Figure 22.10   Monkey fMRI activation patterns (checkerboard stimulus, positive activations only) deformed from an individual map to the atlas map using spherical surface-based registration. (**A**) FMRI activation (checkerboard stimulus) on individual map. (**B**) Landmarks on individual flat map. (**C**) Landmarks on atlas flat map. (**D**) Landmarks on individual spherical map. (**E**) Landmarks on atlas spherical map. (**F**) fMRI activation registered to atlas map, + Lewis–VE borders. Data courtesy of D. Tsao and R. Tootell.

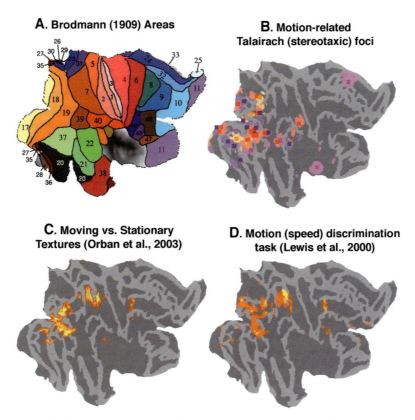

Figure 22.11   Mapping data to the human atlas surface. (**A**) Brodmann's (1909) cytoarchitectonic areas manually transferred to the atlas. (**B**) Motion-related Talairach foci projected to the atlas fiducial surface, along with uncertainty limits (10-mm radius). Coloration of each focus reflects differences in the particular type of motion-processing task (Van Essen and Drury, 1997). (**C, D**) Motion-related fMRI activations mapped to the atlas from studies by Orban et al. (2003) and Lewis et al. (2000). The volume data were mapped to the atlas surface by assigning each node in the surface an intensity value of the fMRI voxel within which the node lies in Talairach space.

estimating the location of each subdivision on the atlas surface, using common geographic landmarks as a guide (but interpolating into buried sulcal regions to provide continuity). The resultant map of Brodmann areas, shown in Figure 22.11A, is only a rough approximation to the actual pattern of architectonic

subdivisions according to Brodmann's criteria, but it nonetheless provides a useful representation for a partitioning scheme that remains in widespread use.

2. *Projection of Stereotaxic (Talairach) Foci.* Neuroimaging data are commonly registered to Talairach stereotaxic space

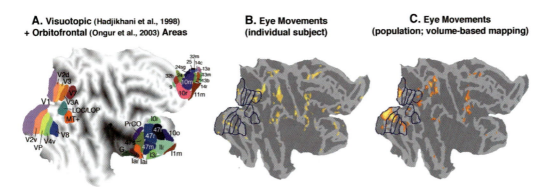

Figure 22.12   (**A**) Human visuotopic and orbitofrontal areas registered from individual flat maps (Hadjikhani et al., 1998; Öngür et al., 2003; Gusnard et al., 2003) to the atlas surface. (**B**) Eye movement fMRI activation patterns from an individual subject deformed to the atlas by surface-based registration, visuotopic areas from panel A superimposed. (**C**) Eye movement activation from a population average mapped by stereotaxic projection from Talairach space onto the atlas surface. Data for panels B and C are from Corbetta et al. (1998).

(Talairach and Tournoux, 1988) in an effort to compensate for individual variability. Once data have been registered, the spatial pattern of activation in any given experiment can be summarized by reporting the three-dimensional (x, y, z) Talairach coordinates of the center of each activation focus. This strategy has been widely adopted since it was proposed (Fox et al., 1985), and the number of Talairach foci reported in the neuroimaging literature is very large—perhaps in the tens of thousands. Because the human surface-based atlas (both the volume and the surfaces) is registered to Talairach space, activation centers of interest can be readily mapped onto the atlas fiducial surface and projected onto the atlas flat map. This is illustrated in Figure 22.11B for a set of 41 activation foci related to motion processing, reported in eight publications.

The spatial uncertainty associated with the stereotaxic projection method depends on the fidelity with which individual brains (including the atlas brain) have been registered to Talairach space using one or another of the many registration algorithms in current use. These include the original piecewise linear registration method used by Talairach and Tournoux; other low-dimensional algorithms such as rigid-body, bounding box, affine transformations, and a number of high-dimensional warping methods that compensate for some of the finer-grained aspects of brain structure (see Toga, 1999). However, even the best of these volume-based registration methods leaves considerable residual uncertainty, especially in regions where the pattern of convolutions is highly variable. Typically, the uncertainty associated with volume-based registration is in the range of 10 mm to 15 mm radius (Van Essen and Drury, 1997). The impact of this uncertainty on the mapping of stereotaxic foci can be expressed by coloring portions of the map that are within a specified range, as illustrated in Figure 22.11B by shading of all regions within a 10-mm radius of one or another motion-related focus.

3. *Stereotaxic Mapping of Volume Data.* Reporting just the centers of individual activation foci fails to represent the complexity of spatial activation patterns typically obtained using fMRI. Consequently, it is preferable to map the complete activation pattern contained in an fMRI volume onto the atlas surface. Population-average data (i.e., data from a population of subjects registered to Talairach space using one of the aforementioned volume registration methods) can be easily and quickly mapped onto the human atlas surface using an fMRI mapping tool available in Caret. Figure 22.11C,D illustrates two such examples, from motion processing studies in human cortex, one from passive viewing of moving textures (Orban et al., 2003), the other from a speed discrimination task (Lewis et al., 2000). This method has the same types of spatial uncertainties as that described for stereotaxic projection of individual foci. Hence, in comparing two or more activation patterns mapped onto the atlas surface, such as those in Figure 22.11B–D, it is important to recognize that differences may include contributions from spatial uncertainties associated with the mapping method as well as genuine differences associated with the functional activations associated with the different paradigms.

The population-average stereotaxic mapping methods illustrated in Figure 22.11 provide an efficient general way to incorporate unlimited amounts of data into the surface-based atlas framework. Currently, the primary rate-limiting step in this process is obtaining access to the vast amounts of relevant volume-based neuroimaging data that are stored in individual laboratories but not in any publicly accessible repositories (see Sections 22.4 and 22.5).

4. *Surface-Based Registration from Individuals to the Atlas.* The utility of surface-based registration for mapping data from individuals to an atlas has already been discussed and illustrated for the macaque (cf. Figures 22.7 and 22.10). The potential of this approach is even greater for human cortex, given the complexity of human cortical convolutions and the high degree of individual variability.

The reasons why surface-based registration is particularly suitable for analyses of human cortex can be illustrated by extending the previously raised analogy between cerebral cortex and a beach ball (see Section 22.2.3) to include five additional points. (i) The specification of distinct cortical areas occurs early in development, prior to cortical folding. At this stage the cortex is analogous to a fully inflated beach ball whose surface is covered by a complex mosaic of distinctly labeled (colored) patches. (ii) A collection of individual hemispheres is analogous to a collection of beach balls; individual variability in the size of each cortical area is analogous to diversity in the sizes and precise locations of the corresponding patches (areas) in different beach balls. (It is as though patches on each beach ball were painted according to the same general instructions, but with the "painters" having a limited degree of artistic freedom in their task.) (iii) Individual variability in the pattern of convolutions in humans is analogous to having each beach ball deflated and folded in a different pattern, with some commonality in the major creases but much variability in the finer-grained folds. (iv) Surface-based registration using spherical maps is analogous to inflating each individual beach ball to a sphere, then registering one sphere to another using selected geographic and/or functional criteria. This circumvents the problem of topologically incorrect registration caused by the highly variable patterns of folding. (v) Volume-based registration is analogous to deforming one crumpled beach ball to another, striving for a deformation that respects the topology of the cortical surface, but with a severe risk of topological mismatches and/or severe local distortions. Collectively, these arguments support the assertion that surface-based registration is inherently better suited than volume-based registration as a strategy for dealing with individual variability of human cortex.

Figure 22.12 illustrates two types of experimental data that have been mapped to the atlas using surface-based registration. In Figure 22.12A, maps of cortical areas were generated on individual hemispheres and then registered to the atlas. In occipital cortex, the maps show visuotopically organized areas (Van Essen, 2004, based mainly on Hadjikhani et al., 1998); in orbitofrontal cortex, the maps show architectonic subdivisions charted by Öngür et al. (2003). For comparison, Figure 22.12C shows fMRI activations from a population of subjects studied with the same eye movement paradigm, but registered volumetrically to Talairach space before mapping to the atlas. The differences are pronounced, but it is difficult to assess the relative contributions of genuine individual variability in structure and function versus technical factors related to the mapping methods (see below).

Once mapped to an atlas, results from different types of analysis can be compared objectively in a variety of ways, several of which are illustrated in Figure 22.13. Figure 22.13A shows motion-related activations of Lewis et al. (2000) and Orban et al. (2003) in red and green, respectively, with overlapping regions

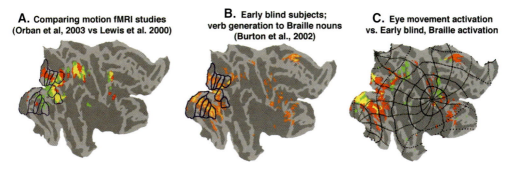

Figure 22.13 Comparison across different types of neuroimaging results. (**A**) Maps of motion-related fMRI activations from Orban et al. (2003), in red; Lewis et al. (2000), in green; and the combination in yellow, with visuotopic areal boundaries superimposed. (**B**) Extensive activation of occipital cortex in blind subjects when generating verbs in response to nouns in Braille (Burton et al., 2002). (**C**) Comparison of activations during eye movements in normal subjects in green (Figure 22.12C; cf. Corbetta et al., 1998) and the Braille activation of occipital cortex red.

in yellow, and with visuotopically organized areas superimposed as blue boundaries. The common activation regions include not only MT+, which has been implicated in motion analysis in numerous studies, but also the adjoining region LOC/LOP, area V7, the frontal eye fields (FEF), and major foci in and near the intraparietal sulcus. Comparison with the motion-related activation focus centers mapped in Figure 22.11B shows overlap with some but not all of the major clusters. The differences may reflect sensitivity and/or experimental design for the paradigms used in each study.

Figure 22.13B,C shows two additional examples of comparisons that can be facilitated using this approach: Figure 22.13B shows strikingly robust activation of occipital cortex in a population of subjects blind since early childhood during a task of generating verbs in response to words (nouns) presented as Braille stimuli (Burton et al., 2002). The overlay of visuotopic and boundaries indicates that the activation includes portions of most visuotopic areas. Figure 22.13C shows a direct comparison of the Braille activation in Figure 22.13B (red) with the eye movement-related activation in normal subjects illustrated in Figure 22.12C (green), revealing that both paradigms cause pronounced activation of a restricted portion of V1 and V2. Overlaying the latitude and longitude isocontours on this map makes the point that all portions of these complex differential activation patterns can be explicitly and concisely pinpointed using spherical coordinates.

Despite the intrinsic potential of surface-based registration for coping with individual variability in human cortex, much remains to be done to establish this as a well-validated general strategy that can be routinely applied and interpreted. The first prerequisite, obtaining high-quality surface reconstructions of individual hemispheres, is less of a bottleneck than in the past, but it nonetheless still requires a significant investment of time and effort. Once surfaces are available, there are multiple options for registering an individual to an atlas. The landmark-based approach illustrated here has the advantage of allowing selection of just those landmarks that are reliably identifiable and are likely to represent functionally corresponding regions, but it has the disadvantage of requiring user interaction and judgment applied to each individual map. An alternative approach (Fischl et al., 1999b) requires less user interaction (because the registration is constrained by automatically computed measures of hemispheric shape), but it has the disadvantage that regions of highly variable geography may distort the registration and reduce functional correspondences. Hence, it is important to validate each approach systematically and to objectively compare alternate strategies for surface-based registration, including ones based on functional subdivisions that are routinely identifiable in all normal subjects.

As with the macaque probabilistic atlas illustrated in Figure 22.7, surface-based registration offers the prospect of generating a variety of probabilistic maps of human cortex. These include maps of geographical variability (gyral and sulcal patterns) as well as functional variability associated with mapping of cortical areas (e.g., visuotopic areas) and mapping of activation patterns in response to standardized testing paradigms.

22.3.9 Interspecies Comparisons

Comparisons across species, particularly between macaques and humans, are critical for elucidating which aspects of cortical organization and function are common across species and which aspects are species-specific. Some insights on these issues can be gained from side-by-side comparisons of macaque and human cortical maps. For example, comparison of visual cortical organization in the macaque (Figures 22.7 and 22.8) versus human (Figure 22.12A) reveals several areas, such as V1, V2, and MT, that can be recognized in both species even though they differ in relative sizes and in geographic location.

Surface-based registration offers a powerful approach to making interspecies comparisons systematic and more objective and in exploring the implications of proposed homologies in relation to the organization of nearby areas. However, geographic landmarks are of limited utility for interspecies registration because of the aforementioned differences in relationship between cortical areas and geographic landmarks. For example, neither the calcarine sulcus nor the superior temporal sulcus is suitable as a geographic landmark because V1 and MT have very different locations in humans and macaques in relation to these landmarks. On the other hand, a number of functional areas that are very likely to be homologous across species can be identified in both atlases. These landmarks, shown in Figure 22.14A,B, include boundaries of areas V1, V2, MT, A1, olfactory, gustatory, and somatosensory/motor cortex (3/4), as indicated on the figure (see also Denys et al., 2004).

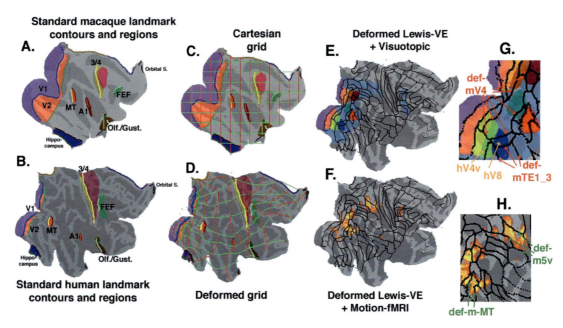

Figure 22.14   Interspecies surface-based registration. (**A, B**) Landmarks for presumed homologous regions in the macaque and human flat maps, respectively. (**C**) Cartesian grid on the macaque map. (**D**) Deformed grid on the human map. (**E, F**) Lewis and Van Essen (2000a) areal boundaries deformed to the human map and overlaid on the human visuotopic areas (panel E) and human motion-related activations from Orban et al. (2003). (**G**) Expanded portion of occipito-temporal cortex from panel E. (**H**) Expanded portion of parieto-occipital cortex from panel F.

Registration between the macaque and human maps using these landmarks results in a deformation pattern that is revealed by comparing the Cartesian grid on the macaque map (Figure 22.14C) to the deformed grid on the human map (Figure 22.14D). Overall, human cortex has 10 times the surface area of macaque cortex. Regions that are expanded much less than this 10-fold average include most of occipital cortex (e.g., human area V1 is only twice the size of macaque V1), parts of parietal cortex and motor cortex between the central sulcus, and the frontal eye fields. Regions that are expanded considerably more than average include prefrontal cortex (anterior to FEF), temporal cortex between MT and A1, and most of parietal cortex.

This standard registration can be used to assess the degree to which other less fully characterized/higher areas or regions are in register. For example, Figure 22.14E shows the deformed Lewis and Van Essen cortical areas in relation to human visuotopic areas. Deformed macaque V1, V2, MT, and adjoining areas V3 and VP are reasonably well-registered to their human counterparts, as expected given the choice of landmarks. On the other hand, deformed macaque V4 (def-mV4) has virtually no overlap with human V8 (hV4), although the latter has been proposed as a homologue of macaque V4 (Zeki, 2003). Zeki's hypothesis is not overtly disproven by the mismatch, but is made much less plausible. Figure 22.14F shows the same deformed macaque areal boundaries in relation to the motion-related fMRI activation shown in Figure 22.11. The prominent motion-related activation in human parietal cortex is centered on deformed macaque area 5, a predominantly somatosensory area. Since the actual human homologue of macaque area 5 is unlikely to be involved in visual motion processing, additional functionally based landmarks will be needed to improve the correspondence in this region (Denys et al., 2004).

A particularly attractive strategy is to compare fMRI activation patterns when similar tasks are carried out in monkeys and humans. Ideally, one might anticipate seeing activation patterns that are topologically very similar except for differential expansion that can be compensated by the type of registration illustrated in Figure 22.14. However, recent results suggest instead that the fMRI activation patterns associated with similar behavioral tasks and stimulation may show considerable species specificity (Vanduffel et al., 2002; Denys et al., 2004; Orban et al., 2004).

22.4 MANAGING AND COMMUNICATING SURFACE-RELATED DATA

22.4.1 Key Characteristics of Surface-Related Data

Databases will play an increasingly important role in allowing neuroscientists to capitalize on the explosion of surface-related experimental data becoming available for cerebral and cerebellar cortex in humans, macaques, and rodents. However, surface-related datasets pose special challenges that must be addressed in database design and implementation. The primary difficulty is not the amount of data per se, because surface reconstructions typically yield more compact datasets than the volume data from which they were derived. Rather, the challenges arise mainly from a combination of (i) the diverse nature of surface-related data and the associated file types and (ii) the complex relationships among various datasets, including relationships that result from transformations of data from individuals to an atlas and from atlases of one species to another. It is useful to elaborate on these issues before describing the specific database we have developed to address them.

22.4.2 Diversity of Data Types and File Types

The data types routinely used for surface-based analyses can be grouped into four major classes. (i) *Surface geometry* information is used to specify the precise shape, or configuration of each surface (e.g., fiducial, inflated, and flat maps). (ii) *Node attribute* information assigns each node in the surface to various categories (e.g., to a particular cortical area according to one or another partitioning scheme) and also to various real-valued measures (e.g., activation levels derived from fMRI experiments, connection densities derived from anatomical studies). (iii) *Point and contour* data represent information about arbitrary locations that may lie in between the discrete nodes of the cortical surface reconstruction and may even lie above or below the surface (e.g., boundaries between cortical areas or regions or locations of activation focus centers). (iv) *Ancillary data* includes a diversity of data types that are not explicitly spatial but are essential for viewing or manipulating surfaces (e.g., palettes used to assign colors to particular areas or boundaries; matrices used to specify affine transformations).

Another aspect of data complexity involves the growing use of various "composite" files, in which the data have a common format but derive from many independent studies. Examples from the human and macaque atlases include composite "paint" files and "border" files that may contain a dozen or more areal partitioning schemes; composite "metric" files that may contain dozens of fMRI activation patterns derived from many different studies and even different laboratories; and composite "foci" files that may contain coordinates of hundreds of Talairach activation foci related to one or another aspect of functional specialization. Keeping track of where the data originated and what scientific information is represented is a substantial challenge for such composite datasets.

The current version of Caret (v. 5.1) supports 28 distinct file types, each having a format that is useful for representing a particular type of information. Other surface-visualization software programs, (e.g., FreeSurfer, Brain Voyager) also support a diverse range of file formats, but there is little standardization of file types and file formats across different software packages. An additional complication arises from the desirability of including various types of volumetric data (e.g., structural MRI plus volumetric fMRI activations) in a surface-related database because of the intimate association between volume and surface representations (e.g., fMRI activations mapped to the cortical surface).

22.4.3 Complex Linkages Among Files

A given surface reconstruction can be associated with a large number of distinct data files. For example, the current publicly available version of the human atlas contains several dozen files just for the right cerebral hemisphere dataset. Exploration of a particular topic of interest with the aid of the atlas dataset may entail using a dozen or more of these files in a single analysis session. Several strategies are used in Caret to facilitate the efficient identification and accessing of such data (Dickson

et al., 2001). First, the data are organized around the concept of a "surface family," which includes all of the data (geometry files, "paint" files, "metric" files, etc.) that are linked to a specific surface reconstruction (e.g., a particular atlas surface). Second, a "specification file" is used to list an organized collection of files from the same surface family and to load subsets of files into Caret that are convenient to view as an ensemble. Third, each file is routinely given a name that is informative with regard to the type of file, the type of data, the surface family with which it is associated, and the date of creation or modification.

These characteristics are particularly important in relation to the growing use of surface-based registration within and across species. Any given experimental dataset can be carried through many successive transformations that are accessible for visualization and comparison. For example, the "checkerboard" fMRI activation pattern illustrated in Figure 22.10 started as a volumetric fMRI dataset acquired for an individual monkey; it was mapped onto an individual hemisphere surface (Figure 22.10A), deformed from the individual surface to the macaque atlas (Figure 22.10F), and deformed again from the macaque to the human atlas (not shown). An investigator may be interested in viewing the same dataset mapped to more than one surface (e.g., on both the macaque and human atlas, or on the individual surface as well as the macaque atlas), which puts a premium on having ready access to all of them.

A related consideration is that at each stage of analysis, a given dataset may have been processed in a variety of ways, with results that may differ either subtly or in major ways. For example, an fMRI volume dataset may have been mapped to more than one surface reconstruction (e.g., to compare SureFit versus FreeSurfer segmentation methods); the fMRI data on an individual surface may have been registered to the atlas using more than one set of landmark constraints or registration parameters; and registration between species may include an evaluation and comparison of several sets of interspecies registration landmarks. An investigator viewing results for one stage of analysis (e.g., displayed on the macaque atlas) may need to know exactly what transformations or other processing steps were associated with the particular data set being viewed. To minimize confusion and potentially erroneous interpretation of such data, it is important that the numerous steps in the analysis be encoded in a way that allows tracking of key processing steps.

22.5 THE SURFACE MANAGEMENT SYSTEM DATABASE (Sums DB)

We have developed the Surface Management System database (Sums DB), which is customized for handling surface-based neuroscience data in a way that addresses the issues raised in the preceding section. Sums DB was designed with six main criteria in mind: (i) *Ease of data entry*. The flow of data into a database depends critically on how easily investigators can submit data to the database. (ii) *Support for diverse data types*. Given the complexity and ongoing evolution of surface-related datasets, it is important that the database support a large and evolving set of data file types. (iii) *Robust search capabilities*. It is desirable to support a wide variety of searches that key on many types of information, including that contained in file names, keywords, and other metadata, as well as on spatial data (coordinates, etc.)

contained in many data types. (iv) *Flexible download options*. It is important to have flexibility that allows the user to select as few or as many files that are needed from complex collections of data files and to download these easily. (v) *Multiple security levels*. A database should permit multiple levels of sharing, allowing some data to be privately held by the individual user, other data to be shared among identified collaborators, and other data to be publicly available. (vi) *Multi-site distribution*. A single centralized database is impractical for dealing with the full spectrum of surface-related data, both published and unpublished, in current use in the neuroscience community. Hence, it is important to design a distributed database that allows users to store data locally and to allow sharing of data across multiple sites. The current design of Sums DB retains some of the features of a prototype version described previously (Dickson et al., 2001). However, it has many new features that more effectively meet the design criteria noted above.

22.5.1 Data Uploading and Processing

Data entry in Sums DB can be carried out easily using a data entry interface or using command-line entries in a unix terminal window. The most general method of data entry involves uploading an "archive" file that is a compressed collection of files, bundled so that the individual file names and directory structure are preserved. An alternative option allows a specification file and its constituent individual files to be automatically uploded in a single step.

Uploaded archive files are uncompressed, and the constituent individual files are tested to determine whether they already exist in the database (by exact file content, not just the file name). Each archive file as a whole and each of its constituent files are then assigned unique archive identifiers. Relevant file types recognized by Sums DB (currently 51 types) are further processed to extract relevant metadata from the file header, which allows for searching by keywords and comments. Metadata processing is customized for a subset of surface-related data types, but Sums DB allows any arbitrary data file to be archived. This includes volume-based datasets (structural and functional MRI) as well as text and graphics data.

Sums DB has three levels of permission control for accessing data: owner, group, and public (other), just as with Unix file systems. Data are initially accessible only to the owner, but the read and write permissions can be readily changed for individual files or for all of the files in an archive.

22.5.2 Data Searching

The current version of Sums DB allows metadata to be searched according to a number of criteria, both generic and surface-specific. The generic options include (i) any text strings contained in the file name; (ii) date of entry into the database; (iii) keywords anywhere in the data file; (iv) text string in file header comment; and (v) owner and/or permission category. The surface-specific options currently include: (i) the file type (e.g. "coordinate" file, "paint" file), (ii) archive category (e.g., tutorial data, published datasets, or atlas-fMRI data), and (iii) various surface properties (e.g., coordinate files categorized by being configuration inflated, flat, etc.). For example, selection of View

Tutorials in the main Sums DB window brings up a listing of the Caret tutorial datasets currently contained in Sums DB (Figure 22.15A). Nearly all of the illustrations in this chapter were generated from data contained in Tutorials 1–4. Next to each archive entry, a pull-down menu provides various options, such as the "show details" that is selected for TUTORIAL.3 in Figure 22.15A. This brings up a complete listing of the 130 files contained in the archive, including of all the data used to generate the interspecies registration shown in Figure 22.14. A portion of this list is shown in Figure 22.15B, which illustrates options that can be applied to individual files. For example, the "text view" option selected for the "deformation map" file listed in Figure 22.15B brings up a view of the file header (Figure 22.15C), which lists all the files used to register the macaque and human right hemispheres (including the specific landmarks used to constrain the registration).

Another mode of searching involves direct hyperlinks from published studies. A growing number of publications include hyperlinks that connect directly to the relevant Sums DB archive containing the data used for analysis and generation of figures. Figure 22.16A, for example, shows results from Astafiev et al. (2003) as seen on a web browser (their Figure 3 and the associated figure legend). Clicking on the hyperlink at the end of the legend brings up a Sums DB search result screen of the data used to generate this figure. The files (a few of which are listed in Figure 22.16B) include fMRI volume data as well as surface data, and they include a registration between macaque and human cortex that was constrained by a choice of landmarks that differs in several respects from the registration illustrated in Figure 22.14 of the present chapter.

22.5.3 Data Selection and Retrieval

Once a file listing is displayed in the Sums DB Search Results window, several options are available for reviewing or downloading the data. These include options to view metadata comments or the full text. Downloading options include immediate downloading of individual data files; transfer of a group of filenames to a clipboard that can be downloaded as a group; downloading of complete archive datasets (e.g., for the various Caret tutorials); and downloading of all of the files listed in a selected specification file.

22.6 FUTURE DIRECTIONS

Of the many additional capabilities that are desirable to incorporate into future version of Sums DB, three merit specific mention. (i) *Spatially based queries*. Spatially based query options will allow searches for combinations of node coordinates and node-based attributes. For example, this might include a search for datasets in which a user-specified location of human cortex (identified, say, by latitude and longitude or by its identity as area V1) shows fMRI activation levels above a selected criterion level in a particular type of neuroimaging study (e.g., studies involving visual attention). Analogous queries could be applied to sterotaxic data, such as neuroimaging activation foci. (ii) *Database federation*. It is impractical for any single database to contain the full set of data that may

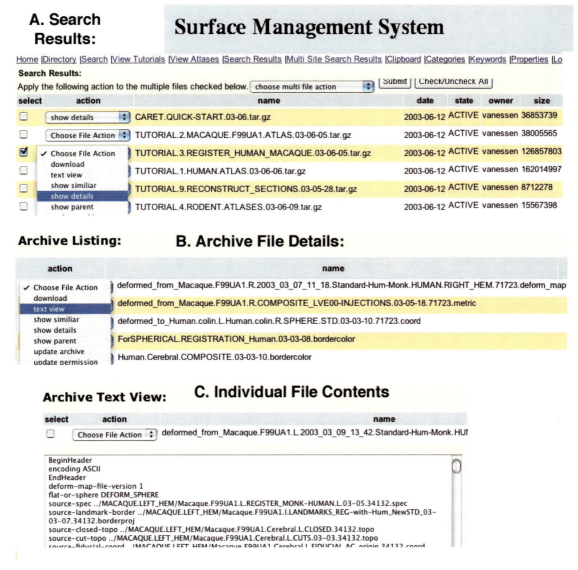

Figure 22.15 Exemplar data archives and files contained in SuMS. (**A**) Partial listing of archives obtained from selecting "View Tutorials." (**B**) Partial listing of files obtained by selecting "show details" of TUTORIAL.3. (**C**) File header obtained by selecting "text view" option for the deformation map file.

be of interest to users. Hence, it is important to achieve effective federation with other databases. For human neuroimaging data, the most important link will be to databases such as the fMRI Data Center (http://www.fmridc.org) that are customized for volume-related fMRI data, including large amounts of metadata related to how the data were acquired and analyzed (Van Horn et al., 2001; http://www.fmridc.org). Federation of Sums DB and the FMRI-DC would allow users to identify and access volume and surface datasets from the same experimental paradigm in a coordinated way even if they are stored in separate databases. For the macaque atlas, there will be natural linkages to the BrainInfo atlas (http://brainfo.rprc.washington.edu) and to CoCoMac, a database of connectivity (Stephan et al., 2001; http://www.cocomac.org). For rodent data, a natural link will be to the Cell-Centered Database (Martone et al., 2002 http://pamina2.sdsc.edu/CCDB). (iii) *Web-based data visualization.* Currently, users must have Caret or some other surface visualization software available on their own computer in order to

make use of the datasets contained in Sums DB. An important complementary capability will be to provide Web-based atlas visualization capabilities so that results of a search can be viewed, or at least previewed, without the requirement of downloading the datasets and using visualization software on one's own computer.

Figure 22.17 summarizes the integrated nature of the atlas and database approach discussed in this chapter. The investigator or student sitting at a computer terminal has access to a family of surface-based atlases, each of which provides a flexible viewing substrate for a rapidly growing body of experimental data, progressively more of which will reside in Sums DB or in databases with which it is federated, and progressively more of which will be rapidly addressable by hyperlinks contained in publications. The utility and power of this approach will hopefully counteract the sociological and other factors that are current impediments to widespread sharing of neuroscience data (Koslow, 2002; Toga, 2002).

A. Figure in publication, with hyperlink in figure legend

View larger version (47K):
[in this window]
[in a new window]

Figure 3. *A,* Macaque brain with anatomical areas (Lewis and Van Essen, 2000⊞). *B,* Deformation and mapping of macaque areas onto a human atlas brain (left hemispheres, dorsal views) using surface-based registration. Red indicates the FEF (area8), purple indicates motor areas (area4), green indicates somatosensory areas (AIP, areas 5D and 3A), and yellow indicates visual parietal areas (LIP, VIP); the lines in each area define the LIP (dorsal and ventral) within the lateral intraparietal complex and VIP subdivisions (medial and lateral) within the ventral intraparietal area. Brown indicates other parietal areas (area 7A, PO), dark red indicates the MT, and blue indicates V4. *C,* Group–average ANOVA *F(z)* map, averaged over cue direction for attention. Black borders indicate deformed macaque visual areas painted in Figure 4 A. *D, F(z)* map during saccade preparation. *E, F(z)* map during pointing preparation. Labels in italic indicate the anatomical landmarks. Abbreviations include anatomical locations in the human brain and a deformed area in the monkey (in parentheses): *sfs,* Superior frontal sulcus; *cs,* central sulcus; *PrCeG,* precentral gyrus; *PMd,* premotor dorsal. Data sets are available at http://pulvinar.wustl.edu:8081/sums/archivelist.do?archive_id = 315115.

B. Data archive in SuMS, available for downloading

Archive Detail:

Apply the following action to the multiple files checked below.   Choose Multi File Action ▾ | Submit | Check/Uncheck All |

select	action	name	state	date	username	size
☐	Choose File Action ▾	ASTAFIEV_EtAl_DATA.03-02-15.tar.gz	ACTIVE	2003-02-15	vanessen	35977261

Final data set (corrected 15feb03) Astafiev et al, J. Neurosci. 2003;23 4689-4699

Archive Listing:

select	action	name	state	date inserted	owner	size
☐	Choose File Action ▾	ASTAFIEV_ceye-att_task_time_mc+orig.BRIK.gz	ACTIVE	2003-02-15	vanessen	5692
☐	Choose File Action ▾	ASTAFIEV_ceye-att_task_time_mc+orig.HEAD	ACTIVE	2003-02-15	vanessen	1815
☐	Choose File Action ▾	ASTAFIEV_ceye_time_mc+orig.BRIK.gz	ACTIVE	2003-02-15	vanessen	24134
☐	Choose File Action ▾	ASTAFIEV_ceye_time_mc+orig.HEAD	ACTIVE	2003-02-15	vanessen	1779
☐	Choose File Action ▾	Human.colin.Cerebral.L.CLOSED.71785.topo	ACTIVE	2003-02-15	vanessen	7241280
☐	Choose File Action ▾	Human.colin.R.ASTAFIEVetAl03_Prepare_saccade_point_attend.71723.metric	ACTIVE	2003-02-15	vanessen	3652644
☐	Choose File Action ▾	Human.colin.L.ASTAFIEVetAl03_Prepare_saccade_point_attend.71785.metric	ACTIVE	2003-02-15	vanessen	3655718
☐	Choose File Action ▾	Human.colin.Cerebral.L.FIDUCIAL.TLRC.711-2B.71785.coord	ACTIVE	2003-02-15	vanessen	2670647

Figure 22.16   **(A)** Iconified Figure 3 and the associated figure legend from A stafiev et al. (2003). **(B)** A partial listing of Sums DB files in the archive identified by the hyperlink in this figure legend (http://brainmap.wustl.edu:8081/sums/archive id=315115).

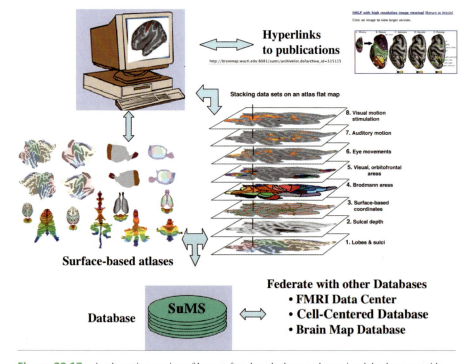

Figure 22.17   A schematic overview of how surface-based atlases and associated databases provide users with unprecedented flexibility in exploring diverse aspects of neuroscience data.

References

Amunts, K., Malikovic, A., Mohlberg, H., Schormann, T., and Zilles, K. (2000). Brodmann's areas 17 and 18 brought into stereotaxic space—where and how variable? *NeuroImage* **11**, 66–84.

Andrews, T.J., Halpern, S.D., and Purves, D. (1997). Correlated size variations in human viual cortex, lateral geniculate nucleus, and optic tract. *J. Neurosci.* **17**, 2859–2868.

Astafiev, S.V., Shulman, G.L., Stanley, C.M., Snyder, A.Z., Van Essen, D.C., and Corbetta, M. (2003). Functional organization of human intraparietal and frontal cortex for attending, looking, and pointing. *J. Neurosci.* **23**, 4689–4699.

Brodmann, K. (1909). *Vergleichende Lokalisationslehre der Grosshirnrinde.* Barth, Leipzig.

Burton, H., Snyder, A.Z., Diamond, J.B., and Raichle, M.E. (2002). Adaptive changs in early and late blind: A fMRI study of verb generation to heard nouns. *J. Neurophysiol.,* **88**, 3359–3371.

Corbetta, M., Akbudak, E., Conturo, T.E., Snyder, A.Z., Ollinger, J.M., Drury, H.A., Linenweber, M.R., Raichle, M.E., Van Essen, D.C., Petersen, S.E., and Shulman, G.L. (1998). A common network of functional areas for attention and eye movements. *Neuron* **21**, 761–773.

Denys, K., Vanduffel, W., Fize, D., Nelissen, K., Peuskens, H., Van Essen, D. and Orban, G.A. (2004). The processing of visual shape in the cerebral cortex of human and non-human primates: A functional magnetic resonance imaging study. *J. Neurosci.* **24**, 2551–2565.

Desimone, R., and Ungerleider, L. (1989). Neural mechanisms of visual processing in monkeys. In *Handbook of Neuropsychology* (F. Boller, and J. Graman, eds.), pp. 267–299. Elsevier, Amsterdam.

DeYoe, E.A., and Van Essen, D.C. (1988). Concurrent processing streams in monkey visual cortex. *Trends Neurosci.* **11**, 219–226.

Dickson, J., Drury, H., and Van Essen, D.C. (2001). Surface management system (SuMS): A surface-based database to aid cortical surface reconstruction, visualization and analysis. *Philos. Trans. R. Soc. London, Ser. B* **356**, 1277–1292.

Drury, H.A., Van Essen, D.C., Corbetta, M., and Snyder, A.Z. (1999). Surface-based analyses of the human cerebral cortex. In *Brain Warping* (A.Toga, ed.), pp. 337–363. Academic Press, San Diego, CA.

Felleman, D.J., and Van Essen, D.C. (1991). Distributed hierarchical processing in primate cerebral cortex. *Cereb. Cortex* **1**, 1–47.

Filiminov, I.N. (1932). Uber die variabilitat der grosshirnrindenstruktur. Mitteilung II. Regio occipitalis beim erwachsenen Menschen. *J. Psychol. Neurol.* **44**, 1–96.

Fischl, B., Sereno, M.I., and Dale, A.M. (1999a). Cortical surface-based analysis. I: Inflation, flattening, and a surface-based coordinate system. *NeuroImage* **9**, 195–207.

Fischl, B., Sereno, M.I., Tootell, R.B., and Dale, A.M. (1999b). High-resolution intersubject averaging and a coordinate system for the cortical surface. *Hum. Brain Mapp.* **8**, 272–284.

Fox, P.T., Perlmutter, J.S., and Raichle, M.E. (1985). A stereotactic method of anatomical localization for postron emission tomography. *J. Comput-Assisted Tomogr.* **9**, 141–153.

Goebel, R., Khorram-Sefat, D., Muckli, L., Hacker, H., and Singer, W. (1998). Functional imaging of mirror and inverse reading reveals separate coactivated networks for oculumotion and spatial transformations. *NeuroReport* **9**, 713–719.

Gusnard, D.A., Ollinger, J.M., Shulman, G.L., Cloninger, C.R., Price, J.L., Van Essen, D.C., and Raichle, M.E. (2003). Personality and brain circuitry. *Proc. Natl. Acad. Sci. U.S.A.* **100**, 3479–3484.

Hadjikhani, N., Liu, A.K., Dale, A.M., Cavanagh, P., and Tootell, R.B.H. (1998). Retinotopy and color sensitivity in human visual cortical area V8. *Nat. Neurosci.* **1**, 235–241.

Joshi, S.C. (1997). Large deformation landmark based differeomorphic for image matching. Ph.D. Thesis, Sever Institute, Washington University, St. Louis, MO.

Koslow, S.H. (2002). Should the neuroscience community make a paradigm shift to sharing primary data? *Nat. Neurosci.* **3**, 863–865 (rev.).

Lewis, J.W., and Van Essen, D.C. (2000a). Architectonic parcellation of parieto-occipital cortex and interconnected cortical regions in the Macaque monkey. *J. Comp. Neurol.* **428**, 79–111.

Lewis, J.W., and Van Essen, D.C. (2000b). Cortico-cortical connections of visual, sensorimotor, and multimodal processing areas in the parietal lobe of the Macaque monkey. *J. Comp. Neurol.* **428**, 112–137.

Lewis, J.W., Beauchamp, M.S., and DeYoe, E.A. (2000). A comparison of visual and auditory motion processing in human cerebral cortex. *Cereb. Cortex* **10**, 873–888.

Lyon, D.C., and Kaas, J.H. (2002). Evidence for a modified v3 with dorsal and ventral halves in macaque monkeys. *Neuron* **33**(3), 453–461.

Martone, M.E., Gupta, A., Wong, M., Qian, X., Sosinsky, G., Ludascher, B., and Ellisman, M.H. (2002). A cell-centered database for electron tomotgraphic data. *J. Struct. Biol.* **138**, 145–155.

Nakahara, K., Hayashi, T., Konishi, S., and Miyashita, Y. (2002). Functional MRI of macaque monkeys performing a cognitive set-shifting task. *Science* **295**, 1532–1536.

Ojemann, J.G., Akbudak, E., Snyder, A.Z., McKinstry, R.C., Raichle, M.E., and Conturo, T.E. (1997). Anatomic localization and quantitative analysis of gradient refocused echo-lanaar fMRI suceptibility artifacts. *NeuroImage* **6**, 156–167.

Öngür, D., Ferry, A.T., and Price, J.L. (2003). Architectonic subdivision of the human orbital and medial prefrontal cortex. *J. Comp. Neurol.* **460**, 425–444.

Orban, G.A., Fize, D., Peuskens, H., Denys, K., Nelissen, K., Sunaert, K., Todd, J., and Vanduffel, W. (2003). Similarities and differences in motion processing between the human and macaque brain: Evidence from fMRI. *Neuropsychologia* **41**, 1757–1768.

Orban, G.A., Van Essen, D. and Vanduffel, W. (2004). Comparative mapping of higher visual areas in monkeys and humans. *Trends Cogn. Sci.* **8**, 315–324.

Stephan, KE., Kamper, L., Bozkurt, A., Burns, G.A.P.C., Young, M.P., and Kotter, R. (2001). Advanced database methodology for the collation of connectivity data on the macaque brain (CoCoMac). *Philos. Trans. R. Soc. London, Ser. B* **356**, 1159–1186.

Sultan, F., and Braitenberg, V. (1993). Shapes and sizes of different mammalian cerebella. A study in quantitative comparative neuroanatomy. *J. Hirnforsch.* **34**, 79–92.

Talairach, J., and Tournoux, P. (1988). *Coplanar Stereotaxic Atlas of the Human Brain.* Thieme Medical, New York.

Teo, P.C., Sapiro, G., and Wandell, B.A. (1997). Creating connected representations of cortical gray matter for functiona MRI visualization. *IEEE Trans. Med. Imaging* **16**, 852–863.

Toga, A.W., ed. (1999). *Brain Warping.* Academic Press, San Diego, CA.

Toga, A.W. (2002). Neuroimage databases: The good, the bad and the ugly. *Nat. Rev. Neurosci.* **3**, 302–309.

Tsao, D.Y., Vanduffel, W., Sasaki, Y., Fize, D., Knutsen, T.A., Mandeville, J.B., Wald, L.L., Dale, A.M., Rosen, B.R., Van Essen, D.C., Livingstone, M.S., Orban, G.A., and Tootell, R.B.H. (2003). Stereopsis activates V3A and caudal intraparietal areas in macaques and humans. *Neuron* **39**, 555–568.

Ungerleider, L.G., and Desimone, R. (1986). Cortical connections of visual area MT in the macaque. *J. Comp. Neurol.* **248**, 190–222.

Vanduffel, W., Fize, D., Peuskens, H., Denys, K., Sunaert, S., Todd, J.T, and Orban, G.A. (2002). Extracting 3D from motion: Differences in human and monkey intraparietal cortex. *Science* **298**, 413–415.

Van Essen, D.C. (1997). A tension-based theory of morphogenesis and compact wiring in the central nervous sytem. *Nature (London)* **385**, 313–318.

Van Essen, D.C. (2002a). Windows on the brain: The emerging role of atlases and databases in neuroscience. *Curr. Op in. Neurobiol.* **12**, 574–579.

Van Essen, D.C. (2002b) Surface-based atlases of cerebellar cortex in human, macaque, and mouse. In *New Directions in Cerebellar Research*, S.M. Highstein, T. Thach, (eds.). *Ann. NY Acad. Sci.* **978**, 468–479.

Van Essen, D.C. (2004). Organization of visual areas in Macaque and human cerebral cortex. In *The Visual Neurosciences* (L. Chalupa and J.S. Werner, eds.). MIT Press, Cambridge, MA, pp. 501–521.

Van Essen, D.C., and Drury, H.A. (1997). Structural and functional analyses of human cerebral cortex using a surface-based atlas. *J. Neurosci.* **17**, 7079–7102.

Van Essen, D.C., Newsome, W.T., and Maunsell, J.H.R. (1984). The visual field representation in striate cortex of the macaque monkey: Asymmetries, anisotropies and individual variability. *Vision Res.* **24**, 429–448.

Van Essen, D.C., Drury, H.A., Joshi, S., and Miller, M.I. (1998). Functional and structural mapping of human cerebral cortex: Solutions are in the surfaces. *Proc. Natl. Acad. Sci. U.S.A.* **95**, 788–795.

Van Essen, D.C., Dickson, J., Harwell, J., Hanlon, D., Anderson, C.H., and Drury, H.A. (2001a). An integrated software system for surface-based analyses of cerebral cortex. *J. Am. Med. Inf. Assoc.* **8** (Special Issue on the Human Brain Project), 443–459.

Van Essen, D.C., Lewis, J.W., Drury, H.A., Hadjikhani, N., Tootell, R.B.H., Bakircioglu, M., and Miller, M.I. (2001b). Mapping visual cortex in monkeys and humans using surface-based atlases. *Vision Res.* **41**, 1359–1378.

Van Horn, J.D., Grethe, J.S., Kostelec, P., Woodward, J.B., Aslam, J.A., Rus, D., Rockmore, D., and Gazzaniga, M.S. (2001). The Functional Magnetic Resonance Imaging Data Center (fMRIDC): The challenges and rewards of large-scale databasing of neuroimaging studies. *Philos. Trans. R. Soc. London, Ser. B* **356**, 1323–1339.

Zeki, S. (2003). Improbable areas in the visual brain. *Trends Neurosci.* **26**, 23–26.

Brain Atlases of Normal Human Subjects

John Mazziotta, M.D., Ph.D.

CONTENTS

Databasing the Brain. Edited by Stephen H. Koslow and Shankar Subramaniam
ISBN 0-471-30921-4 © 2005 John Wiley & Sons, Inc.

23.1 INTRODUCTION

The quest to develop an atlas of the human brain is not new. Nearly a century ago, landmark studies by Brodmann (1909) and von Economo (von Economo and Koskinas, 1925) looked to microscopic cellular architecture as a means of developing a comprehensive atlas that would facilitate the organization of neuroscience information and a better understanding of cortical anatomy. The problem, of course, was that the labor-intensive nature of the work resulted in their ability to examine one or, at best, a few specimens. At the same time, the unknown variance of human brain structure among individuals in a population made single sample studies of limited value for atlas development.

Pan forward a hundred years and one now has access to in vivo imaging techniques that can acquire data, at 1-mm^3 resolution, about the anatomy of the entire human brain in approximately 15–20 minutes. This capability, combined with a virtual growth industry of neuroscientists, focused on understanding brain structure and function in health and disease, and the stage was set to develop modern atlases of the human brain. This is exactly the situation that currently exists.

Because the scope of this chapter is large, its focus will be directed at three types of undertakings. The first are atlas projects that seek to collect raw data from the neuroimaging community and make it available through databases, to that same community as a means to share the costly data and extend the extraction of meaningful results from studies that can be pooled in ways different from their original intent (Van Horn and Gazzaniga, 2002). The second represents the development a probabilistic atlas of the human brain where the datasets are acquired expressly for that purpose (Mazziotta et al., 1995, 2001a,b). In this strategy both raw data, from the original acquisitions, and derived data, resulting from the analysis of the original acquisitions, are linked to large sets of demographic, clinical, and behavioral information as well as DNA samples. Such a strategy allows the quantification of variance, for both structure and function, in a large human population and also provides for the recalculation of this variance in subpopulations of the main dataset. By combining these results with microscopic data of cyto- and chemoarchitecture, a comprehensive view of brain structure and function in a large population has been developed. The third strategy utilizes derived data only. It is exemplified by the BrainMap® Project (Fox and Lancaster, 1994, 1996, 2002), coordinates (based on Talaraich and Tournoux, 1988) are collected from published, peer-reviewed manuscripts in the literature and entered into a database along with the experimental variables for the functional imaging studies, thereby allowing future investigators to perform meta-analyses of these datasets in ways not originally intended by the investigators who acquired the original data. All three approaches were designed to fulfill the goal of sharing information and organizing it in a way that is manageable and benefits future neuroscientific investigations.

The unique structure and function of the human brain, relative to other organs in the body, provides a guide for how these atlases have been built. The central nervous system is divided into highly specialized regions that have unique properties in terms of cell types, connections, and organization. The functions of these units vary with time, spanning the gamut from the millennia of evolution to the millisecond choreography of neurophysiological events. This temporal and spatial specialization is well-suited to the application of informatics techniques. In fact, such methods will be required as the basis from which to begin to understand and organize the ever-increasing amount of neuroscientific information that is accumulating about this, the most complicated system known. For each brain region and for every attribute ascribed to it, data must be organized in a rational fashion that takes advantage of the brain's inherent neuroanatomy, spatial segregation, and the variance among individuals that exist for these regions. Ultimately, what is required is a multidimensional database organized with three dimensions in space and one in time along with a seemingly infinite number of attributes referable to these four physical dimensions.

As in geography, neuroscience requires accepted maps, terminologies, coordinate systems, and reference spaces to allow accurate and effective communication within the field and to allied disciplines. Geographical atlases of the earth have advantages over atlases that are anatomical in nature. Earth atlases can assume a relatively constant physical representation over thousands of years. On that single, stable construct, an infinite number of abstract representations of features can be overlaid. For earth maps, such features might include rainfall, temperature, population density, or crime rates.

Unlike geographical atlases, anatomical atlases cannot assume a single, constant physical reality. This is true despite the fact that standard atlases that utilized single subjects minimize this fundamental problem. Anatomical atlases must first deal with the fact that there are a potentially infinite number of physical realities that must be modeled to obtain an accurate, probabilistic representation of the entire population. Upon this anatomical representation, one can then overlay features in a fashion analogous to that described for earth atlases (Mazziotta, 1984). In the brain, such features might include cytoarchitecture, chemoarchitecture, blood flow distributions, metabolic rates, ligand binding, behavioral and pathologic correlates, and many others. Like earth maps, brain maps can vary in time frames ranging from milliseconds (e.g., electrophysiologic events) to minutes (e.g., skill acquisition), years (e.g., development, maturation, aging), or millennia (i.e., evolution).

Classic atlases of the human brain, or brains from other species, have been derived from a single brain, or brains from a very small number of subjects, and have employed simple scaling factors to stretch or constrict a given subject's brain to match the atlas. The result is a rigid and often inflexible system that disregards useful information about morphometric (i.e., dimensionality) and densitometric (i.e., intensity) variability among subjects.

In this chapter, descriptions of the three atlas and database strategies, described above, will be addressed along with the methods and limitations associated with attempts to develop an organizational and informatics strategy for a system as complicated as the human brain. Not intended to be exhaustive or encyclopedic, the emphasis of this chapter will be to stress strategies that have emerged to address this challenging opportunity. The associated references provide additional background and source material for the reader who would like to delve deeper into this topic.

23.2 BACKGROUND

The relationship between structure and function in the human brain, at either a macro- or microscopic level, is complex and poorly understood. Our perspective on brain function is typically equated with the methods available to measure it. For the tomographic brain imaging techniques, the results produce macroscopic estimates of where gross functional changes (typically of a hemodynamic nature) are occurring. Electromagnetic techniques can provide direct information about when these events occur, and they can supply less accurate spatial information about where. The development of an atlas for the human brain simply provides the framework in which to place these ever-accumulating datasets in a fashion that allows them to be related to one another and that begins to provide insights into the relationship between micro- and macroscopic structure and function.

23.2.1 Growth of Neuroscience and Lost Opportunities

The growth of neuroscience in the last 25 years has been extraordinary. Brain mapping and neuroimaging have witnessed a similarly exponential rise in interest, output, and productivity, although on a smaller scale. Throughout the neuroscience community, there is a general frustration with the volume of data that is generated and its relative inaccessibility in forms other than narrative text. Consider, for example, that over 13,000 Society for Neuroscience abstracts are published electronically each year. Faced with such a staggering volume of information, the individual neuroscientist typically retreats to his or her small scientific niche, resulting in ever-increasing specialization and isolation within the field.

At the same time, funding for neuroscience research has a limited return on its investment, in that only a small fraction of raw data that are collected through such funds is analyzed fully, and far less is interpreted and published. Even when published, narrative formats require arduous comparisons across experiments, methods, and species.

If there were a system that provided a logical and organized means by which to maintain data from meetings, individual experiments, or the field as a whole, referenced to the anatomy of the brain, the species and the stage in development, or duration of a pathologic process, then highly automated, content-based queries would vastly improve access, allow immediate comparisons among experiments and laboratories, provide a manageable format to assess new data at meetings or through periodic publications, provide for electronic experiments and hypothesis generation using the data of others to test theories, and greatly increase the value of every dollar spent on neuroscience research. Such an outcome requires sophisticated neuroinformatics tools, dedicated scientists committed to the successful completion of such a project in a practical fashion, and a paradigm shift in the sociology of neuroscientists with regard to information sharing (Koslow, 2000). Nevertheless, the benefits of such an approach are enormous on their own, and of even greater value if one extrapolates the current situation to even greater numbers of neuroscientists and datasets in the future.

23.2.2 Data Explosion

As the quality of neuroscientific data improves, so too does its magnitude. As spatial resolution in imaging data changes by one order of magnitude in one dimension, the volume of data points increases by a factor of 1000. In vivo imaging instruments are now routinely capable of producing 1-mm^3 resolution elements, whereas microscopic and ultrastructural studies achieve spatial resolutions 1000–100,000 times better. If one considers that 50,000–75,000 genes code for proteins of relevance to the human nervous system at some point during the lifespan, one can see the impact of assaying and storing such information across a range of spatial resolutions, as demonstrated in Figure 23.1. Current genomic technology, and future advances in it, make feasible the ability to generate vast amounts of genetic information. All of these data are in search of an organizational home referenced to the location of the sample, in neuroanatomical terms, and the time frame of the sample as a function of the development of the organism. Once again, the brain's architecture becomes the most appropriate and intuitively sensible structure in which to organize such data so as to optimize correlations between biologically related datasets. While the remainder of this chapter focuses on human brain imaging data and atlases, it is important to note the magnitude of the information management problem for neuroscience as a whole.

23.2.3 Data Integration

To demonstrate the practical uses of a sophisticated brain atlas, an example will be taken from actual experience, namely, an experiment performed by Watson et al. (1993) to identify the visual motion area of the human brain (i.e., V5 or MT) (Ungerleider and Desimone, 1986) using relative cerebral blood flow (CBF) (Mazziotta et al., 1985; Fox and Mintun, 1989) measured with PET (Figure 23.2). In the experiment, each subject had multiple PET-CBF studies in two states: The first state involved a stationary pattern of targets, while the second one involved moving targets. The significant difference between the datasets collected in these two states (Friston et al., 1991) was then superimposed on MRI data using the AIR alignment and registration algorithm (Woods et al., 1992, 1993). The result of this experiment demonstrated consistent bilateral activations of the dorsolateral, inferior occipital cortex in each subject. Furthermore, a consistent relationship between the site of increased blood flow and the frequently observed ascending limb of the inferior temporal sulcus (Ono et al., 1990) was found (Figure 23.2A). Because the investigators were knowledgeable about temporo-occipital anatomy and physiology, they recognized that this location had also been identified by Flechsig (1920) as a portion of the human cerebral cortex (Flechsig Feld 16) that is myelinated at birth (Bailey and von Bonin, 1951) (Figure 23.2B). These observations have been repeatedly confirmed by independent laboratories demonstrating the human V5 (Zeki et al., 1991) areas as a frequently detectable and robust functional landmark at the temporo-occipital junction (Dumoulin et al., 2000).

Now envision this experiment performed using advanced neuroinformatics tools and a brain atlas. Anatomical warping and segmentation tools would be used to automatically

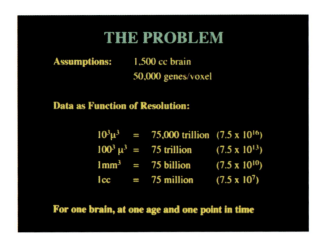

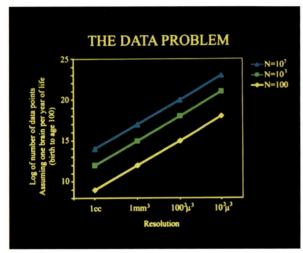

Figure 23.1 The magnitude of neuroinformatics data for the human brain. While there are a number of assumptions in this illustration, the orders of magnitude are realistic and enormous. They depict what would be involved in developing an organized data structure that combines location in the human brain with gene expression maps. (Top) This example assumes that approximately 50,000 genes may be expressed in any three-dimensional region (voxel) of the brain at any given time during development. The typical human male brain is 1500 cm³ in volume. Based on the spatial resolution used to determine gene expression (ranging from 1 cm³ to $10^3 \mu m^3$), the number of data points ranges from 75 million to 75 thousand trillion. Keep in mind that this is what is required to do just one brain at a given point in time. (Bottom) If one uses these same assumptions and the range of resolutions noted in part A, the range of data magnitudes for a series of brains collected across a population with a representative of each age from birth to age 100 results in dataset magnitudes that range from 10^9 to 10^{23}. These truly astronomical orders of magnitude will require innovative, practical neuroinformatic data structures that allow the referencing of such information as a function of both location and time.

segment and label the anatomical regions of the brain for each subject. Alignment and registration would show the effects of functional registration on macroscopic anatomy. Functional alignment and registration using the V5 activation sites would automatically demonstrate the consistent relationship between that functional region and the ascending limb of the inferior temporal gyrus. Furthermore, it would quantitate the spatial relationships between the sulcal/gyral anatomy and the functionally activated

zone across subjects. Differences in responses could be related to demographic, clinical, and genotypic (Bartley et al., 1997; Zilles et al., 1997) information, if these were collected as part of the experiment and were related to population data already available in a four-dimensional atlas. Cyto- and chemoarchitectural data would be available for automated reference with regard to this cortical zone (Clark and Miklossy, 1990; Rademacher et al., 1993). Time-series data from EEG or MEG would show the temporal relationships of this region to others (Dale et al., 1999; Ahlfors et al., 1999). Lesion data could also be accessed if such datasets had been added as an attribute (Zihl et al., 1991) (Figure 23.2C). This is in contrast to the current situation where activated cortical regions are identified, and one must laboriously search the literature to try to identify, qualitatively in experiments with different characteristics, the regions of the brain that are of experimental interest for a given neuroscientific question.

23.3 TECHNICAL DETAILS AND METHODOLOGY

23.3.1 Database Strategies

This is a chapter about human brain atlases, but such atlases are really databases organized in a graphical presentation that provide intuitive insights into the datasets that they contain. Atlases of the earth overlay information based on regional or global geography. Also, atlases of the brain contain information that is linked to brain structures by names or coordinates. There are innumerable ways to organize such data. Given the vast scope and magnitude of such an information base, the selection of the proper strategies to facilitate searches and organization is vital to success. What follows are a number of the important considerations appropriate to consider when assessing the organizational aspects of such a system. These include the development of digital libraries and warehouses, the basic dimensionality of the database structure, whether such datasets should be collected and maintained centrally or in a distributed fashion, and certain aspects of the sociology associated with data sharing and data exchange (also see Section 1.1, Chapter 1, this volume).

1. *Digital Libraries.* Raw (unprocessed) and derived data organized in the databases conceived as digital libraries and data warehouses have been and will be important to the neuroimaging community. These datasets include those with "raw" data (e.g., complete, three-dimensional, multispectral MRI structural studies of individual subjects), "scalped" (e.g., extracranial structures removed) datasets, and intensity normalized, "scalped" datasets. Access to such information may allow investigators to obtain normal control data for neuroimaging experiments or to test various methods for image analysis and display without the requirement to acquire original data on their own. Most problematic will be the distribution of "raw" datasets, because the potential for compromising subject confidentiality is an issue. Since the experimental subject's face could be reconstructed from the raw datasets, one strategy would be to alter or eliminate facial structures from the dataset prior to distribution.

2. *Four-Dimensional.* The physical world is organized into four dimensions and, thus, forms a logical and comprehensive organizational framework for any brain atlas or database. Plans

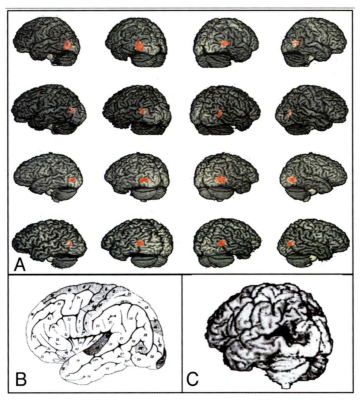

Figure 23.2 Illustrative example of human visual area V5. **(A)** Four separate subjects demonstrating bilateral CBF-PET activation of V5 superimposed on their respective structural MRI studies (Watson et al., 1993). Note the consistent relationship seen between the activated site (red) and the ascending limb of the inferior temporal sulcus that also coincides with the cortical region (seen in **B**, arrow) identified by Flechsig (1920) as being myelinated at birth. **(C)** Patient studied by Zihl et al. (1983, 1991) with damage to the V5 area resulting in a selective disturbance of visual motion perception. Source: John Watson, M.D; University of New South Wales, Sydney, Australia.

should anticipate the future inclusion of time-series data from dynamic, functional data acquisition methods such as fMRI, EEG, and MEG, requiring the fourth dimension. It is expected that spatiotemporal and purely temporal patterns of brain activity will constitute functional entities and markers of their own. These can be used for the following purposes:

a. Since function will be defined in the future by brain locations and timing of activity, the brain atlas must incorporate temporal and spatiotemporal brain activity information.

b. Spatiotemporal and temporal functional markers can be used for most of the same purposes described for the spatial functional markers (fMRI, PET): warping; correlations across subjects; an additional source of information in calculating population distributions—these can be used, for example, to inform studies with small populations, and so on.

c. Temporal and spatiotemporal information can be used to correlate brain activity across subjects in the temporal dimension.

d. They can also be used as priors for brain source estimation methods, and their probability distributions can be used for Bayesian procedures in EEG and MEG brain source localization (Schmidt et al., 1999).

With such a data structure, queries-by-content tools and strategies can be developed. These tools will allow users of the atlas to submit a query in the form of actual data (e.g., a two-dimensional image of a portion of the brain or a three-dimensional block of data) and ask the atlas to search for matches using wavelet-based or other techniques that have previously been demonstrated to be successful for two-dimensional internet searches of graphic material (Wang et al., 1997). The expansion of these approaches to three and, eventually, four dimensions will be an important neuroinformatics milestone that will find uses far beyond current applications.

Furthermore, a system organized in this fashion, along with the tools associated with it, will allow for efficient, convenient, and comprehensive access by neuroscience clients to the ever-growing data of an atlas system. The goal would not be to develop physiological models of brain function, neural connectivity, and other important neurobiological questions. But these exciting opportunities will be more easily achieved by providing a system of database interactions and structure for modelers, neuroimagers, and neuroscientists, in general. Once established and populated with data, an atlas, organized in this fashion, will allow for "electronic" hypothesis generation and experimentation using previously collected, well-described, and effectively organized data.

3. *Attributes.* It is conceptually important to understand that the database architecture of an atlas, while organized in four dimensions, to match the organization of the nervous system,

can have a very high number of attributes all referenced to these basic four dimensions. These additional attributes need not be specified at the time of establishing the original datasets included in the atlas. Some can be derived and others can be added at a later point through further examination of the original subjects (e.g., longitudinal studies, other methodologies) or by further analysis of existing data (e.g., genotyping of stored DNA samples).

The most difficult challenge to the actual organization of such an atlas is the scaling and referencing of data across major spatial or temporal domains. While most current structural brain images have a fundamental spatial unit of resolution of 1 mm^3, there is no reason why microscopic and ultrastructural information cannot appropriately populate the individual 1-mm^3 voxels of the macroscopic dataset. The same can be said of temporal information, but the exact manner of binning of time-series information will require judicious attention to the types of queries anticipated of such datasets.

4. *Central versus Distributed.* A business metaphor is appropriate here. Fledgling industries rarely do well when trying to establish standards, means of communication, and interoperability methods that are designed to result in a reliable and durable outcome for a given community. Examples abound including: telecommunications, aviation, electronics, meteorology, and others. In most of these cases, a well-designed, centralized approach established both the problems and the solutions that later led to deregulated, decentralized systems that were linked by regulatory groups, industrial standards, and meta-databases. Similarly, in the burgeoning field of neuroinformatics, an initial centralized approach appears both desirable and manageable. It allows for straightforward and easy monitoring of distributing datasets on a continuous basis. A centralized approach can also monitor the required submission of attributes derived from the datasets back into the atlas as a measure of successful, reciprocal sharing of data and results (Bloom, 1996; Pennisi, 1999). Finally, in order to even attempt such a project, there must exist a critical mass of data analysis tools, organization, and reputation to make participation attractive psychologically and sociologically. The common goal and ultimate result must also be sufficiently valuable to the contributing sites to make participation compelling. The results of participation must be worth more than the sum of the individual parts.

23.3.2 Three Types of Approaches

The methods associated with the development, maintenance, and ongoing evolution of three basic strategies in human brain atlas development are discussed in this section. Each has its own set of methodological problems, advantages, and limitations. Some strategies are more mature than others. Three approaches will be discussed.

The first employs the pooling of raw functional MRI (fMRI) data for future sharing, analysis, and interpretation. This approach is best exemplified by the National fMRI Data Center based at Dartmouth College in the United States (Van Horn and Gazzaniga, 2002).

The second is the development of a probabilistic atlas system and database (International Consortium for Brain Mapping (ICBM)) that captures data on variance in a large population of subjects (7000) in the normal state for both brain function and structure. A postmortem component of this approach allows microscopic data on cyto- and chemoarchitecture to be incorporated into the evolving atlas (Mazziotta et al., 1995, 2001a,b). This strategy employs both the original acquisition of data from a large number of subjects along with demographic information and DNA samples but also provides derived results that are a product of elaborate data processing strategies and database methodologies.

The third approach utilizes only derived data, in the form of coordinate points, from published articles on functional imaging, utilizing fMRI and PET. Best exemplified by BrainMap® (Fox and Lancaster, 1994, 1996, 2002), this project has the longest history of any of the three examples described in this chapter. In many ways this was the first attempt to pool information from individual experiments and make such datasets, in the form of reported coordinates, available for additional analyses different from those originally intended when the data were collected.

The ICBM probabilistic atlas focuses on new data collections obtained prospectively and expressly for the intent of developing a core atlas. The other two strategies take data acquired for other purposes and retrospectively pool these datasets for future use and community access.

23.3.3 Databases of Raw Data

The rapid proliferation of noninvasive functional imaging techniques using MRI strategies has resulted in a very rapid growth of the number of datasets of this type that have been, are being, and will be collected. Typically, such datasets are acquired at considerable cost and published by the group who obtained the source data, retaining proprietary control of these datasets, in conjunction with the policies of their respective universities and funding agencies. As such, the raw data and other uses for it are lost to the general neuroscience community. Prompted by this motivation, the fMRI Data Center was initiated at Dartmouth College in the United States in order to pool raw fMRI data for the purposes of sharing, subsequent data analysis, and interpretation. Such a strategy has numerous methodological and sociological challenges and opportunities.

Recent initiatives have been launched to begin the process of cataloguing datasets from functional magnetic resonance imaging (fMRI) studies of cognitive function with the hopes of organizing, searching, and mining these data. The fMRI Data Center (fMRIDC; http://www.fmridc.org) at Dartmouth College is one such example, seeking to advance understanding into cognitive processes through the archiving and open sharing of these large neuroimaging datasets (Van Horn and Gazzaniga, 2002). Presently, the fMRIDC receives complete neuroimaging datasets contributed from the authors of published fMRI studies. The current size of the fMRIDC archive is nearing 2.5 terabytes of data from nearly 40 different fMRI studies of cognitive function. These include investigations of human memory encoding and retrieval, the processing of visual information, attention, and language. In cataloguing these studies and their accompanying raw data, the fMRIDC seeks to provide the scientific community with access to complete study information at levels not before possible. Thus, researchers may not only validate fMRI results from the peer-reviewed literature, but may also examine the data in novel ways that promote new thinking about that data and encourage new experiments.

The fMRIDC is presently employing high-performance computing (HPC) hardware and software with a view toward large-scale modeling and "mega-analyses" of studies in the fMRI data archive. The HPC configuration includes an array of multiprocessor compute engines, each having an accompanying high-end server to handle user access. Multi-terabyte disk storage is in place with potential to grow to upwards of 10 TB of spinning disk space. The system also employs hierarchal storage management facilities to interface with a high-volume, near-line storage robotic library. Though seemingly small in contrast to the systems employed in the physical sciences, this particular effort represents a significant initiative for bringing computing power to the issues of modeling and visualization of complex cognitively induced patterns of functional activity.

Cognitive neuroscientists are now working closely with computer scientists to begin applying sophisticated computer search algorithms for looking through large amounts of data, extracting unique and potentially interesting features. To efficiently share neuroimaging data between investigators around the world, novel methods for loss-less wavelet compression of fMRI study data are being devised (Midtvik and Hovig, 1999). Published study articles and the data described in them are now being clustered so that unseen relationships among studies might be revealed and generate new hypotheses about fundamental brain functions (Yee and Gao, 2002; Goutte et al., 2001; Salli et al., 2001). For instance, methods for applying search queries across multiple levels of neuroimaging data are being investigated to identify similar studies from the literature based upon the published text of the article, the accompanying study metadata, and the functional imaging data itself. The combination of multiple "levels" of study data using information retrieval (IR) and machine learning (ML) algorithms permit computer programs to identify interesting patterns within and between these levels of data (Mitchell, 1999). Such methods, drawn from computer science, will permit neuroimaging data self-description, allowing the data to tell cognitive neuroscientists about themselves rather than through the fitting of statistical models as is currently practiced.

The enthusiasm accompanying the growth of cognitive neuroscience research is now encompassing the areas of functional brain imaging, high-performance computing systems, and leading-edge computer science (see Section 7.1, Chapter 7, this volume). A more thorough understanding of the brain and cognitive processes will necessitate advances in computer technology to accelerate data analysis and to mine vastly larger amounts of data. What is more, these data must be understood on a level that permits the representation of the dynamic interaction of brain regions and brain systems required for cognitive processes such as memory function, visual abilities, and motor skill. Efforts such as the fMRIDC are actively developing the methods needed to archive and manipulate large brain mapping studies with high-performance computing, paving the way toward modeling functional and structural brain data at the level at which earth ocean currents are currently modeled. Bringing high-end computational power to bear on the management, processing, and examination of these large fMRI datasets seeks to bring an unprecedented level of understanding of human cognitive function.

23.3.4 Probablistic Database—Raw and Processed Data

Overall Concept

The goal of the International Consortium for Brain Mapping (ICBM) is to develop a voxel-based, probabilistic atlas of the human brain from a large sample of normal individuals, aged 18 to 90, with a wide ethnic and racial distribution (Mazziotta et al., 1995, 2001a,b; www.loni.ucla.edu/ICBM). The dataset is designed to contain a substantial amount of demographic information describing the subjects' background, family history, habits, diet, and many other features. In addition, clinical and behavioral evaluations include neurological examinations, psychiatric screening, handedness scores, and neuropsychological tasks. One-cubic-millimeter multispectral MRI studies including T1-, T2-, and proton-density-weighted pulse sequences are obtained consistently. A subset of subjects also have functional imaging using a standardized battery of tasks and employing functional MRI, positron emission tomography, and event-related potentials. DNA samples have been acquired from 5800 of the subjects and are available for genotyping.

From an organizational point of view, eight laboratories in seven countries on four continents have participated in the core data collection and analysis. These sites were selected because of their expertise in brain imaging, capacity to perform a large number of studies in a consistent fashion, and the fact that most sites had different imaging devices and computer platforms, thereby requiring the consortium to solve problems of interoperability and data differences from different acquisition devices.

It was decided early in the planning for the program that in situations where the optimal solution to a given problem (e.g., data analysis pathway, visualization scheme, etc.) was not known, each laboratory would independently try to solve these problems. Once a laboratory-specific solution was obtained, appropriate algorithms would be distributed to consortium participants and evaluated. Ultimately, these algorithms were sent to outside laboratories for independent evaluation and comparison with methods developed by nonconsortium groups. In each case, the optimal strategy was then incorporated into the final approach used by the consortium. This was a "real-world" situation designed to produce the optimal result through competition. As each successful component of these competitions emerged, it was incorporated into the overall ICBM strategy for data analysis, visualization, and distribution. Thus, while each laboratory developed an independent strategy for processing data, the consortium as a whole made the commitment to a unified, centralized strategy for the pooled results, thereby resulting in a single atlas rather than a federation of atlases. The latter would result in inconsistencies in data analysis and confounding factors for users of the atlas in the long run.

The principles, practices, and tools developed through the ICBM consortium have also spawned a series of other atlas projects on different populations (see Section 25, Chapter 25, this volume). Probabilistic atlases for children (i.e., birth to age 18 years) and disease states (e.g., Alzheimer's disease, traumatic brain injury, multiple sclerosis, autism, schizophrenia, stuttering, cerebral infarction) are under development. These population- and disease-specific atlases have been developed for different reasons but employ similar principles and many of the same tools used for the normal adult brain atlas described here.

We also consider a part of this project to be the development of a reference system. The atlas describes brain structure and function in three spatial domains and a temporal one referenced to the age of the subjects. Attributes (e.g., blood flow, receptor density, behaviors inducing blood flow changes at specific sites, signs and symptoms associated with lesions at specific sites, literature references) are then superimposed on the basic atlas. As such, the atlas becomes the architectural framework for the reference system, the former being grounded in the four physical dimensions and the latter being extensible, based on the interests and datasets available by consortium participants and future users.

Probabilistic Approach

Since there is no single, unique structure for the human brain that is representative of the entire species, its variance must be captured in an appropriate framework. The framework that was chosen for the ICBM atlas was a probabilistic one in which the intersubject variability is captured as a multidimensional distribution. These probabilities can change if subpopulations are sampled because of the shifting distributions. The probabilistic approach was relatively new to neuroanatomical thinking when the ICBM group first proposed it in 1992. The only previous related strategies had to do with postmortem analyses that reported distributions for structure sizes and dimensions for certain select regions of the brain (Filimonoff, 1932). In recent years, the probabilistic strategy has been more widely used (Roland and Zilles, 1994, 1996, 1998; Mazziotta et al., 1995, 2001,a,b) and many probabilistic atlases are now being developed for such species as the monkey and the mouse (see Section 22.2, Chapter 22, this volume).

Use of a Large Population

In the literature of smaller-sample-size projects, it was clear that there was such a wide range of methodologies used and that a large population was required for the development of an atlas that is intended to capture the variance in structure and function of the human brain. Such a large population can be newly acquired, as was done in the ICBM project, or could be the result of pooling smaller studies to produce a metadatabase. The latter approach was rejected by ICBM because strategic differences among these studies would make their pooling difficult, if not impossible. Technical issues such as voxel size, slice thickness, scanning parameters, and many others would cause difficulties in any attempt to produce a homogeneous final product. The same can be said of subject selection and description. As might be expected, a wide range of criteria were used in selecting subject populations, including the definition of normality. Screening tests, demographic, and background information as well as neurological and psychiatric examinations vary from study to study, adding to the incompatibility of the pooled results. If one also includes functional information, the situation is far worse. Since brain function is obtained by having subjects perform tasks, any slight variation in the task presentation, psychophysics, or the strategy employed by the subject in performing the task will cause unpredictable differences among experiments, thereby adding methodological variance in the pooled data and confounding the final product. Thus, it was decided to prospectively collect a sample of a large number of subjects for which these confounding factors could be controlled if one were trying to describe variance in a population.

The ICBM atlas includes 7000 normal subjects obtained from geographical locations as disparate as Japan and Scandinavia and spanning the age range from 18 to 90 years. Special efforts were made to obtain a wide range of racial and ethnic diversity. In addition, 342 twin pairs (half mono- and half dizygotic) are also part of this sample. The dataset for each subject includes a detailed historical description of medical, developmental, psychological, educational, and other demographic features. In addition, behavioral data including neurological, neuropsychological, and neuropsychiatric examinations are part of the dataset. In 5800 subjects DNA samples are being collected, stored, and made available for genotyping. This large sample size allows the opportunity to provide realistic estimates about the variance of structure and function for brain regions, the relationships between structure and function at macro- and microscopic levels, and true phenotype–genotype–behavioral comparisons. The large sample size also increases statistical power in making such inferences about the population or when the atlas is used as a comparison sample for investigations involving other groups, be they normal or pathologic. Lastly, as the sample size increases, the opportunity to select subpopulations of meaningful sizes also increases.

Target and Reference Brains

A fundamental concept of the ICBM atlas was to distinguish between target and reference brains (Figure 23.3). The target brain was defined to be the dataset, derived from one or, at best, a few individuals, that has the richest collection of data available. Theoretically, this would be the brain of a normal individual studied with in vivo, high-resolution, structural and functional imaging and then, after death, having detailed postmortem analysis including cyto- and chemoarchitecture. If a series of such brains could be studied, then a probabilistic target brain would emerge. Given the high resolution of the postmortem data, target brains would be the most informative with regard to anatomical and chemical localizations.

In contradistinction to the target brain, reference brains are derived from large populations of subjects typically through in vivo imaging of structure and function. These datasets provide information about variance in the population for both structure and function, but at a three-dimensional spatial resolution that is three orders of magnitude lower than the target brains.

Target and reference brains are used for different purposes. Target brains are, as the name implies, the target to which an unlabeled dataset can be warped. The unlabeled dataset then picks up the anatomical, functional, or other attributes of each voxel. Once it is back-transformed to its original shape, the new dataset will have the appropriate anatomical and functional labels for all brain regions. A certain percentage of these labels will be erroneous based on imperfections of the warping system, an incomplete understanding of the anatomy of homologous brain regions between subjects, and errors in the primary labeling of the target brain. Reference brains provide data about distributions of brain regions and can be divided into subpopulations for specific purposes. Reference brains give estimates of anatomical and functional regions in a population of individuals and, as such, can be used to determine confidence limits when a new dataset falls outside the range of normality or expected variance for a given population. Taken together, these two tools provide important but very different vehicles for analyzing existing or new datasets with regard to brain structure and function.

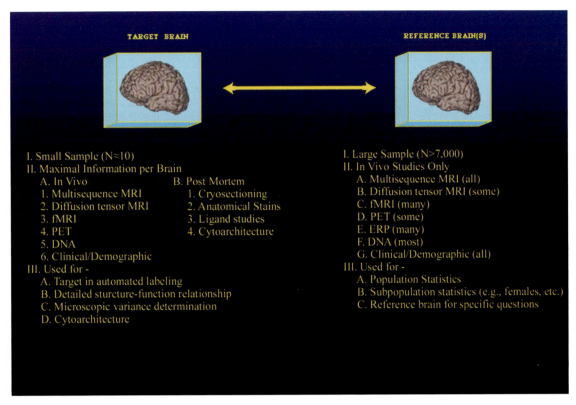

Figure 23.3   Target versus reference brains in the ICBM atlas. Fundamental to the development of the probabilistic atlas is the concept that there are two types of datasets that are required for a comprehensive system. The first is a target brain or brains. The criteria for this resource are listed in the figure. Ideally, single individuals who are normal and studied during life will have the complete complement of in vivo imaging studies performed. After death, these same individuals would be studied again using appropriate imaging studies such as MRI, and then detailed cytoarchitectural and chemoarchitectural analyses of their brains would result in a very information-rich ultimate dataset. Because it is unlikely that such a situation will occur (although a few individuals have been studied ante- and postmortem), typically, the postmortem and in vivo studies are obtained on separate individuals. These target brains are used to better understand microscopic and macroscopic structure–function relationships but also to take a newly acquired dataset and warp it to match the target thereby picking up the anatomical labels for each structure on a voxel-by-voxel basis. When the new study is back-transformed to its original shape and configuration, all structures in the brain will have the appropriate labels with an error rate defined by the confidence limits of the warping algorithm and other factors discussed in the text. Reference brains are also described in the figure. These represent large populations of in vivo studies where population statistics about variance for structure and function can be obtained. Reference brains can also be sampled to provide subpopulations appropriate to a unique set of descriptors. The size of the sample set for the reference brains will dictate whether subpopulations of sufficient size will be available for specific projects.

Function

The inclusion of functional landmarks in the ICBM atlas was completely analogous to the motivation for studying structural anatomic landmarks. Specifically, the ICBM atlas uses functional landmarks to augment atlasing methods that are currently based primarily on macroscopic structural anatomy in the same way that these anatomic methods now augment atlasing methods that were previously based on stereotaxis with simple proportional scaling. An important distinction must be made between functional imaging to answer neuroscience questions, which was not the intent of the ICBM project, and functional imaging to serve as a neuroinformatics tool, which was the intent. Whereas neuroscience functional imaging studies currently use individual macroscopic anatomic landmarks to define a common neuroinformatics framework for comparing and combining data from different subjects (see Section 22.2, Chapter 22, this volume), it was anticipated that future neuroscience functional imaging studies would complement this macroscopic anatomy with functional anatomic landmarks, identified in each individual through a selected battery of neuroinformatics tasks. In general, the neuroinformatics tasks and the functional landmarks that they produce may be completely unrelated to the tasks constituting the primary focus of the neuroscientific investigation. It was an objective to develop and validate tasks that are well-suited to producing functional landmarks. These tasks were the first of what is expected to be an ever-growing library of tasks that will constitute a Functional Reference Battery (FRB). The FRB was used to develop a new generation of brain atlases through novel warping techniques that move beyond macroscopic anatomy and into the realm of functional and cytoarchitectonic similarities as the fundamental basis for homologous mapping of one brain to another.

The major theoretical and practical issue in identifying homologous brain structures, as well as in warping strategies designed to compare brains within a population of subjects, is a critical issue with regard to both three-dimensional and surface geometries and representations. The ICBM project based all aspects of the atlas development on three-dimensional, voxel-based strategies. Nevertheless, this neither obviates nor limits one's capacity to address special issues related to cortical surface topology. It is important to understand the appropriate constraints that must be imposed to preserve cortical surface topology for both the cerebrum and the cerebellum (Felleman and Van Essen, 1991; Van Essen and Drury, 1997; Van Essen et al., 1998; Fischl et al., 1999).

To understand the motivation of identifying functional landmarks, it is important to understand the differences and similarities between functional landmarks and anatomic landmarks with respect to meeting the objectives of neuroinformatics research. Neuroanatomy, and specifically, neuroanatomic landmarks, have been the basis that formed the framework for indexing neuroscience information from a number of specific sources collected across spatial scales. Explicit in the plan was the notion that an atlas system would need to continue to adapt in an iterative fashion to accommodate improvements in spatial scale and in the models used to map data into a single neuroanatomic framework. A self-critical evaluation of the methodologies used for structural atlasing alone reveal areas in need of extension:

Macroscopic Landmarks from Structural MRI Studies. Macroscopic landmarks from structural MRI studies provide a suboptimal basis for appropriate mapping of individual anatomy into a unified neuroinformatics framework. Three independent lines of research serve to demonstrate the difficulties of relying exclusively on macroscopic anatomic landmarks as a neuroinformatics framework. The first evidence comes from the significant progress made in warping three-dimensional anatomic data to match a target template. In the absence of brain pathology, it is computationally feasible to use high-order, nonlinear warps to generate a one-to-one correspondence between brains, even while requiring perfect alignment of unambiguous cortical anatomic features such as the crests of gyri and the depths of sulci. However, even with the inclusion of such anatomic constraints, these mappings are not unique. Mechanical properties such as viscosity or elasticity must be ascribed to the brain tissues to find a solution that is optimal from the standpoint of those presumed mechanical properties (Christensen et al., 1993; Davatzikos, 1997; Schormann et al., 1996). In the absence of more restrictive constraints or independent external standards, different solutions can all lead to equally good (from the standpoint of visual inspection or image similarity criteria) but mutually inconsistent answers, indicating that with macroscopic anatomic data alone, the mapping problem is substantially underconstrained. While it might be a valid computer science goal to identify the transformation that perfectly maps one brain onto another, while minimizing some intuitively appealing quantity, this is not necessarily the best neuroinformatics goal. A more appropriate goal from a neuroinformatics standpoint is *to maximize the genuine homology of points that are brought into correspondence by the transformation.* Functional landmarks will provide additional constraints on intersubject warping that will help to meet this important neuroinformatics goal.

Various criteria can be used to define homology, and conflicts between macroscopic homologies and microscopic cytoarchitectonic homologies (Rademacher et al., 1993) constitute the second line of research demonstrating the problems of relying solely on macroscopic structural anatomy. Recent postmortem cyto- and chemoarchitectonic studies have shown that even some sulcal and gyral features that were once thought to be almost perfectly

correlated with nearby cytoarchitectonic boundaries are in fact only approximately correlated (Zilles et al., 1997; Geyer et al., 1997, 1999; Amunts et al., 1999, 2000). It is our explicit bias that homologies based on function and cytoarchitectonics are more fundamental to neuroscience, and hence to its informatics, than homologies based on sulcal and gyral anatomy.

The third line of research that highlights the difficulties of an informatics framework that is based solely on structural anatomy comes from fMRI studies. A decade ago, functional imaging with PET was of sufficiently low resolution that atlases based on the simple proportionality of the original Talairach system were adequate to ensure that homologous activation sites would overlap from subject to subject and that the resulting group results would be interpreted as consistent across laboratories. Subsequent improvements in PET image resolution have justified the adoption of the more sophisticated techniques based on structural MRI scanning and MRI–PET co-registration that are in widespread use today (Woods et al., 1993). The high resolution possible with fMRI, along with the fact that statistically significant responses are readily identified in fMRI data from single subjects, demand much more accurate mapping of homologous landmarks from every individual subject and threaten to make methods that rely solely on macroscopic anatomy obsolete. Ideally, a neuroinformatics framework should seek to stay a step ahead of such developments. Functional links may also be of particular value for patient populations where normal function may persist even in the presence of substantial anatomic distortions. Providing the necessary link between global and local anatomy is a problem that will require new tools, new approaches, and new population data acquired specifically for that purpose.

Cytoarchitectonic Studies. Cytoarchitectonic studies provide an insufficient basis for quantifying relevant intersubject variability in the population. As mentioned above, cytoarchitectonic studies in a small number of subjects can be extremely powerful in demonstrating the potential range of intersubject variability. This was the motivation to begin to incorporate such data into the ICBM atlas. However, the collection of cytoarchitectonic data is extremely demanding in terms of time and resources. These realities make it unlikely that reliable population estimates of the variability between structural and functional anatomy will be quantified for many brain regions any time soon using these techniques. Since cytoarchitectonics cannot be identified in vivo, such data may help to define general rules (e.g., a given cytoarchitectonic field is most likely to be located at position X in women and at position Y in men), but will not help to identify the individualized exceptions to such rules. In contrast, functional imaging is well-suited to population-based studies, and functional imaging can be applied routinely to living individuals. To a first approximation, functional landmarks can be viewed as an in vivo proxy for cytoarchitectonic landmarks. It should be explicitly stated that it is not our primary intent to equate a given functional landmark with a given cytoarchitectonic region. Indeed, it is clear that a one-to-one relationship will sometimes not exist, since functional subdivisions are present as maps within some cytoarchitectonic areas (e.g., M1 and V1) and since adjacent, functionally correlated areas can be distinguished cytoarchitectonically. Rather, cytoarchitectonic anatomy and functional anatomy were viewed as intrinsically intertwined features that reveal an underlying pattern of brain organization that provides an optimal framework for neuroscience research and neuroinformatics challenges. By warping brains in a way that brings homologous functional landmarks into concordance, it should be possible to simultaneously bring nearby cytoarchitectonic regions into better superimposition, even if one does not explicitly know the identities of the cytoarchitectonic regions or even the locations of their boundaries. Capturing the unique spatial information represented by functional landmarks is an important front for neuroinformatics research, one that will provide routine, direct access to this fundamentally important level of brain organization.

Properties of Good and Informative Functional Landmarks. The minimal attributes of a good functional landmark are that it be unambiguously detectable in individuals and that the variability in its location within individuals be small. "Small" is a relative term, and contexts for making this judgment will be explicitly defined below, along with specific consideration of how "landmarks" can be extracted from functional images. One must make an important conceptual distinction here between a "good" functional landmark and an "informative" functional landmark. In order to also be considered informative, a good functional landmark should provide unique information that could not have been determined purely on macroscopic anatomic grounds. For example, a functional task for identifying primary visual cortex might not prove to be particularly informative since human cytoarchitectonic data indicate that striate cortex consistently maps to the calcarine fissure (Polyak, 1957) (though some variability is present, as reviewed by Aine et al., 1996). In contrast, a good functional landmark in a frontal region where gyral anatomy is quite variable might be highly informative. Caution is generally indicated in trying to predict in advance which functional landmarks will ultimately prove informative since detailed studies of the relationship between cytoarchitectonic and macroscopic anatomy are still relatively rare, as are data comparing functional and structural anatomy. Indeed, with careful study, some traditional assignments of functional areas to specific sulcal or gyral locations are proving to be less reliable than previously expected (Aine et al., 1996; Zilles et al., 1997). Likewise, data from some functional studies have identified previously unsuspected function–structure correlations. A good example of this latter situation comes from studies of putative human area V5, where substantial variability in Talairach coordinate location across subjects turned out to be largely explained by a highly consistent relationship between V5 (defined functionally) and the intersection of the ascending limb of the inferior temporal sulcus and the lateral occipital sulcus (Figure 23.2A) (Watson et al., 1993). Because of the difficulty in predicting which landmarks will be informative, we have primarily focused our attention on identifying tasks that produce good functional landmarks and identifying these landmarks in a representative population. These data are being evaluated to determine how informative the landmarks actually are and to look for currently unrecognized structure–function correlations.

Those who are primarily involved in functional imaging neuroscience (as opposed to neuroinformatics) research may be surprised that the ICBM criteria for a good functional landmark do not include that the landmark should have a consistent location across subjects. When trying to answer neuroscience questions, there are situations where variability across subjects is undesirable; one hopes that a functional task will produce responses at a highly consistent anatomically standardized location across subjects so that overlapping regions of response will increase statistical significance and so that the consistency of location will increase confidence that the areas seen in each individual are truly homologous. It is, therefore, perhaps counterintuitive that the exact opposite situation applies to functional landmarks to be used for neuroinformatics. A functional landmark that is always present in the exact same location in every subject, when using current methods of anatomic standardization, is assured to be uninformative, providing only redundant information that could have been derived from the anatomic data alone.

The neuroinformatics goal is to use functional landmarks to provide a new source of valid, independent anatomic information that cannot be detected using macroscopic anatomy and to use this information to improve the homologous mapping of different subjects to one another or to an atlas. The result should be better mapping from one subject to another that will serve to improve local homology, a goal that should prove advantageous when subsequently analyzing neuroscience functional imaging data in these same subjects. Two major and one minor assumption are implicit in this line of reasoning and need to be explicitly stated: (1) Despite the variation in location, it is critical that the functional landmarks that are identified in each subject are truly homologous; (2) methodological variability in establishing the location of the functional landmark *within* each subject must be small when compared to the true anatomical variability in the standardized location of the landmark *across* subjects; and notable, though less important, (3) some preservation of local topology is assumed, so that establishing the location of a functional landmark will indeed improve the homologous mapping of nearby brain regions.

An important implication of the last two assumptions is that the value of a functional landmark will vary, depending on (1) the amount of within-subject variability (more variability decreases its value), (2) the amount of local intersubject variability (more variability increases its value), and (3) the functional landmark's proximity to the nearby regions where better mapping is desired (greater proximity increases its value). If the goal is to improve mapping throughout the brain, numerous functional landmarks may be needed, whereas local mapping may be improved with just one strategically placed functional landmark. The value of proximity raises an important consideration: In functional neuroscientific imaging experiments, why bother to use the locations of established functional landmarks that may be unrelated to the task of interest rather than simply using the locations produced by the primary task itself? There are at least two good answers to this question: (1) Unless landmarks produced by the primary task have been determined to be good landmarks (implying considerable prior investigation), the resulting mapping may actually lead to less reliable homologous mappings than anatomic data alone, and (2) statistical models for evaluating group significance would be invalidated by such a procedure unless separate trials were used for mapping and for addressing the primary neuroscience question. Consequently, appropriate use of the landmarks produced by the primary task being investigated as functional landmarks would require that these landmarks be validated and used in exactly the same way as any other nearby functional landmark. The use of landmarks will also depend, in part, on the brain region(s) of interest for a given experiment and the interests of the investigator.

Analysis Strategy

At the outset of the ICBM project, it was unclear what the optimal analysis strategy would be for both the structural and functional aspect of the program. Given the large number of subjects, each with multispectral MRI datasets and many with functional imaging studies as well, it was clear that the tools to be developed would have to function in an automated, or at least semi-automated, fashion to be feasible. Furthermore, reliable automaticity would be a general benefit to the brain imaging field, given the labor-intensive aspects of manual image editing. It was also clear that certain steps would be required to process data in what we have called an ICBM "analysis pipeline." These steps include:

- Screening data for obviously incomplete or artifact-laden studies and rejecting them
- Intensity normalization in three dimensions for each pulse sequence
- Alignment and registration across pulse sequences and studies within a given subject
- Tissue classification (i.e., gray and white matter, cerebrospinal fluid, other)
- "Scalping" whereby extracranial structures are removed
- Spatial normalization of each subject to a target where anatomical labels can be obtained automatically
- Surface feature extraction
- Visualization

Given this sequence of tasks, it was unclear, in most cases, what the optimal solution for each would be. Rather than making an a priori decision and have all ICBM consortium members work to achieve it, an alternate approach was chosen. It was decided that each of the primary laboratories in the consortium would work to solve each step in the analysis pipeline independently and in parallel. These laboratory-specific algorithms would then be locally optimized. Once a given laboratory was satisfied with the performance and documentation of their approach, it would be distributed to the other participating laboratories for alpha testing. If an algorithm failed to perform adequately or was awkward to use because of hardware platform incompatibilities or other factors, it was rejected. Those algorithms that performed well across consortium laboratories were ultimately sent to an independent group (David Rottenberg, M.D., Stephen Strother, Ph.D., and colleagues at the University of Minnesota) for beta testing. This independent testing included not only the ICBM algorithms for a given module in the analysis pipeline but also any other algorithms that could be identified worldwide that purported to perform the same functions. During beta testing, algorithms were evaluated with simulated as well as real datasets selected by the beta test laboratory and evaluated for documentation, ease of installation, computation time, accuracy, and precision. The results of these evaluations were then published (Strothers et al., 1994; Arnold et al., 2001). The winners of this competition were then selected for the ICBM analysis pipeline (Figure 23.4) and were then the basis for the mass data analysis of all datasets. While it was decided that it was important to analyze all 7000 studies in a consistent manner so that users would know the methodology, algorithms, and versions of the algorithms from which the results were derived, this in no way precluded individual laboratories in

the ICBM consortium or elsewhere from using their own strategies for data analysis on the original datasets which are provided through digital libraries. This strategy has been successful in that it established an internal competition whereby the best solution emerged rather than an a priori and hypothetical prediction that might have fallen far short of the optimal outcome.

Real World

The ICBM consortium always maintained a "real-world" environment such that the participating sites use different equipment, software, and protocols reflecting a microcosm of the larger neuroscience, neuroimaging, and neuroinformatics communities and forcing us to develop solutions to problems through flexible, compatible systems rather than rigid standards, protocols, and equipment requirements. The significance of this feature is that the products were not platform-, institution-, or protocol-specific.

Interoperability. Interoperability was an important concern early in the development of the ICBM atlas. So important was the requirement to develop interoperable tools and datasets that a conscious decision was made to deliberately utilize imaging instruments, computing hardware, and file formats that differed among the participating sites. This forced certain principles and rules to be utilized in the development of software and the exchange of data, the goal being accessibility of any ultimate end-user to all of these products (see Section 4.1, Chapter 4, this volume). The psychology and sociology of any advanced research field is to develop homemade tools and to maintain intralaboratory file structures. The experience in the ICBM consortium was no different. As such, translators were developed that allowed datasets to be transferred among sites with an agreed-upon file format (MINC; Neelin et al., 1998) but that was translated into the "home" file format upon receipt at any of the participating sites. A similar strategy was used for algorithms. This simplistic approach worked quite well, allowing a relatively seamless exchange of information.

Quality Control. Since the ICBM atlas is to be a growing resource, tools that have been developed, thus far, will ultimately be open to the entire neuroscientific community for the future additions of datasets. How then can one ensure the quality of data from investigators? Having pondered and debated this question for many years and having examined the approaches used by other fields, the simple answer was that no one can assure a certain level of quality control in a completely open data exchange program. Not only is this impractical, but it may also lead to the erroneous exclusion of data that might someday be deemed valuable. If there were some filter on the input of data, what would the review process be? How can one predict how tomorrow's observations will be judged by today's standards? One cannot. Furthermore, in a practical sense, such an approach would immediately become backlogged with datasets awaiting "review" by some "panel of experts" whose opinions might change as time and experience progresses. What was provided, however, was a system by which users of such datasets can select their own level of confidence about the populations or results that they sample. For example, a user might request all information about a certain region of the brain for a given demographic population of subjects. Most of these data would be of high quality and

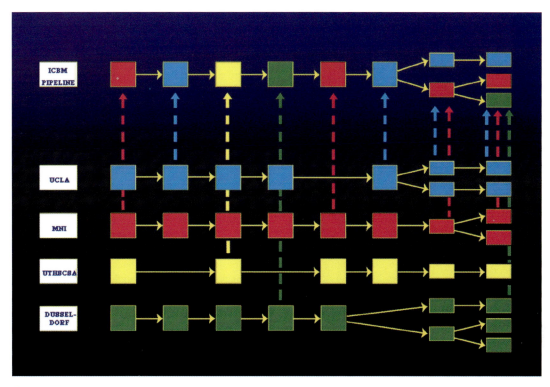

Figure 23.4 ICBM analysis pipeline. Since it was impossible to predict what specific algorithm or mathematical strategy would be optimal for each step in the analysis of structural and functional data collected for subjects in this consortium project, the original ICBM core laboratories elected to each develop independent strategies for each step. Once complete and tested within a given laboratory, they were distributed among consortium participants for alpha testing. After the consortium members were satisfied with the performance at this phase, all consortium-developed algorithms were delivered to an independent laboratory, not part of the consortium, for beta testing. The beta testing included not only the ICBM-developed algorithms but also any other algorithms identifiable worldwide that purported to perform the same function. The best (see text) algorithm was then selected for incorporation in the ICBM pipeline. All data in the final atlas were processed through this unified, single pathway. The bottom four boxes (white) in the left column represent consortium sites contributing algorithms to the pipeline.

reliably collected, but some of them would undoubtedly include experimental, methodological, and other errors. Nevertheless, it would give the user a complete picture of all of the information available about their query. At the other end of the spectrum, consider a user who is interested in only the most accurate information about a given site in the brain for a certain population. That user could request data that were only obtained from the results of peer-reviewed, published, and independently reproduced data collections. Thus, just as the datasets can be filtered using demographic, anatomical, or clinical criteria, they can also be filtered and queried by confidence level. "Let the user beware" is the only rational approach to developing such a system.

Postmortem Cryosectioned Material

Data Acquisition. Mapping the human brain and its functions requires a comprehensive anatomic framework. This reasoning dictated the need to obtain high-resolution, digital, whole-brain, postmortem datasets. The fact that recent advances in anatomic digital imaging techniques now permit unrestricted visualization in multiple cut planes and three-dimensional regional or subregional analyses when appropriate primary datasets are available (Spitzer and Whitlock, 1992; Wertheim, 1989) made this approach feasible. Digital representations also offer the opportunity for morphometric comparisons and sophisticated mapping between anatomic and metabolic imaging modalities (Payne and

Toga, 1990; Toga and Arnicar-Sulze, 1987). The primary source data for human brain atlasing must include not only very fine spatial detail but also image color and texture to convey the subtle characteristics that make it possible to distinguish subnuclear and laminar differences. Furthermore, the incorporation of an appropriate spatial coordinate system is critical as a framework for intersubject morphometrics. High-resolution anatomic datasets serve as references for the accurate interpretation of clinical data from the PET, CT, and MRI modalities as well as the mapping of transmitters, their receptors (Figure 23.5), and other regional biological characteristics.

Thus, a system was designed of histologic and digital processing protocols for the acquisition of high-resolution digital imagery from postmortem cryosectioned whole human brain and head for computer-based, three-dimensional representation and visualization (Cannestra et al., 1997; Toga et al., 1997). High-resolution (1024^2 pixel) serial images were captured directly from a cryoplaned blockface using an integrated color digital camera and fiber-optic illumination system mounted over a modified cryomacrotome. The system can process tissue treated in a variety of ways, including fixed, fresh, frozen, or otherwise prepared for sectioning at micron increments. Sometimes it is desirable to section the tissue while still in situ. Specimens frozen and sectioned with the cranium intact preserve brain spatial relationships and anatomic bony landmarks. Color preservation is

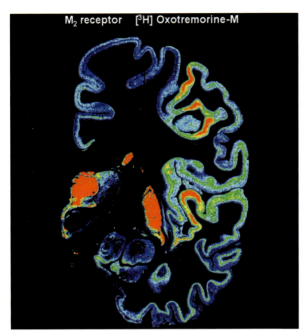

Figure 23.5 Coronal image demonstrating tritiated muscarinic receptors from one hemisphere of a cryosectioned brain and demonstrating the anatomical detail that such chemoarchitectural maps can provide. When serial sections are obtained and stained for a wide range of receptors, anatomical features, and gene expression maps, a tremendous wealth of information is available for comparison with sites of functional activation obtained using in vivo techniques and macroscopic brain structure (gyri, sulci, deep nuclei, white matter tracts). Having a probabilistic strategy for relating these different types of anatomies will provide previously unavailable insights about the relationship of structure and function on both microscopic and macroscopic levels for the human brain and, by analogy, for the brains of other species (see Figure 23.6). The analysis of the regional and laminar distribution patterns of transmitter receptors is a powerful tool for revealing the architectonic organization of the human cerebral cortex. We succeeded in preparing extra-large serial cryostat sections through an unfixed and deep-frozen human hemisphere. Neighboring sections were incubated with tritiated ligands for the demonstration of 15 different receptors of all classical transmitter systems. The distribution of [^{3}H]oxotremorine-M binding to cholinergic muscarinic M2 receptors is shown here as an example. Even a cursory inspection of a color coded receptor autoradiograph permits the distinction of numerous borders of cortical areas and subcortical nuclei by localized changes in receptor density and regional/laminar patterns. For example, the M2 receptor subtype clearly labels the primary sensory cortices (at the level of the section shown in the figure—for example, the primary somatosensory area BA3b and the primary auditory area BA41) by very high receptor densities sharply restricted to both areas. The different receptors allow the multimodal molecular characterization of each area or nucleus (K. Zilles, A. Toga, N. Palomero-Gallagher, and J. Mazziotta, unpublished observation).

superior in unfixed tissue, but unfixed heads were incompatible with decalcification and cryoprotection procedures. Thus, section collection from such specimens was complicated by bone fragmentation. Collection of 1024^2 images from whole brains results in a spatial resolution of 200 μm/pixel in a 1- to 3-gigabyte data space. Even higher three-dimensional spatial resolution is possible by primary image capture of selected regions such as hippocampus or brainstem or by using higher-resolution

cameras. Discrete registration errors can be corrected using image processing strategies such as cross-correlative and other algorithmic approaches. Datasets are amenable to resampling in multiple planes as well as scaling and transpositioning into standard coordinate systems. These methods enable quantitative measurements for comparison between subjects or to atlas data. These techniques allow visualization and measurement at resolutions far higher than those available through other in vivo imaging technologies and provide greatly enhanced contrast for delineation of neuroanatomic structures, pathways, and subregions.

The use of cryosectioned anatomic images as a gold standard for mapping the human brain requires a complete understanding of the assumptions and errors introduced by this method. While there are several obvious advantages to using these data as a reference for other tomographic and in vivo mappings, their collection requires sophisticated instrumentation and representative postmortem material. Spatial resolution, the inclusion of bony anatomy, full-color, blockface reference for histologically stained sections and the resulting registered three-dimensional volumetric datasets are important aspects of this method. Nevertheless, cryosectioning approaches, like all others, introduce distortion during acquisition and processing. Sources of errors include postmortem brain changes and artifacts associated with tissue handling. A major source of error is related to specimen preparation prior to sectioning. Removal of the cranium and subsequent brain deformation, perfusion protocols, or freezing altered the spatial configuration of the dataset.

While three-dimensional data at this resolution is difficult to acquire, it is necessary for careful studies of morphometric variability and the generation of digital comprehensive neuroanatomic atlases. Ultimately, what is needed is the combined use of cryosectioned data as the source of higher-resolution raw and stained anatomy spatially referenced to an in vivo electronically acquired dataset such as MRI.

Cyto- and Chemoarchitecture. A major effort of the ICBM project was to obtain cyto- and chemoarchitectural data from postmortem brains to enter into the probabilistic database for comparison with in vivo studies. An example of this approach can be described for Broca's area. The putative anatomical correlates of Broca's speech region—that is, Brodmann's areas 44 and 45 (Brodmann, 1909)—are of considerable interest in functional imaging studies of language. It is a long-standing matter of discussion whether or not anatomical features are associated with the functional lateralization of speech (Galaburda, 1980; Hayes and Lewis, 1995, 1996; Jacobs et al., 1993; Simonds and Scheibel, 1989; Scheibel et al., 1985). Furthermore, the precise position and extent of both areas in stereotaxic space and their intersubject variability still remain to be analyzed, since Brodmann's delineation is highly schematic, is not documented in sufficient detail, and does not contain any statement about intersubject variability.

The cytoarchitecture of Brodmann's areas 44 and 45 were studied in 10 human postmortem brains using cell-body-stained (Merker, 1983) 20-μm-thick serial sections through complete brains (Amunts et al., 1999). Cytoarchitectonic borders of both areas were defined using an observer-independent approach, based on the automated high-resolution analysis of the packing density of cell bodies (Gray Level Index = GLI) from

Probability map: BROCA's region

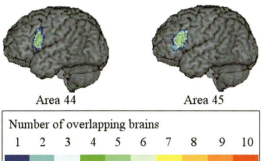

Area 44   Area 45

Number of overlapping brains

1   2   3   4   5   6   7   8   9   10

Figure 23.6 Location and extent of Broca's region (Brodmann areas 44 and 45). Areas 44 and 45 as defined in serial coronal sections of an individual brain after three-dimensional reconstruction; lateral views of the left hemisphere. Probability maps of Broca's region, based on microscopic analysis of 10 human brains, can be referenced, also in a probabilistic fashion, to functional activation sites associated with the functions of Broca's area using the multimodality, probabilistic atlas strategy. The overlap of individual postmortem brains is color-coded for each voxel of the reference brain (color bar); for example, seven out of 10 brains overlapped in the yellow-marked voxels. Left in the image is left in the brain. Source: Karl Zilles, M.D., Heinrich-Heinze University, Dusseldorf, Germany.

the border between layers I/II to the cortex/white matter border (Schleicher et al., 1999). These profiles are perpendicular to the cortical surface and define the laminar pattern of cell bodies. Thus, the profiles are a quantitative expression of the most important cytoarchitectonic feature. Multivariate statistical analysis was used for locating significant differences between the shapes of adjacent GLI profiles along the cortical extent. Those locations represent cytoarchitectonic borders. GLI profiles were also used for investigating interhemispheric differences in cytoarchitecture. Significant interhemispheric differences in cytoarchitecture (i.e., differences in GLI profiles between right and left areas) were found in both areas 44 and 45. Profiles obtained as internal controls from the neighboring ventral premotor cortex did not show any lateralization.

The position of the borders of areas 44 and 45 with respect to sulci and gyri showed a high degree of intersubject variability (Figure 23.6). This concerned the sulcal pattern—that is, the presence, course, and depths of sulci, as well as the spatial relation of areal borders with these sulci. The position of a cytoarchitectonic border could vary up to 1.5 cm with respect to the bottom of one and the same sulcus in different brains. Thus, sulci and gyri are not reliable and precise markers of cytoarchitectonic borders. Although there was a considerable intersubject variability in volume of areas 44 and 45 ($N = 10$), area 44 was larger on the left than on the right side in all cases of our sample. There was no significant left–right differences in the volume of area 45.

The extent and position of areas 44 and 45 were analyzed in the three-dimensional space of the standard reference brain of the European Computerized Human Brain Database (Roland and Zilles, 1996) after the above-described microstructural definition of the areal borders. MR imaging (3D FLASH-scan, Siemens 1.5-T Magnet) was performed on postmortem brains prior to histology. Corrections of deformations inevitably caused

by the histological technique were performed by matching MRI and corresponding histological volumes (Schormann and Zilles, 1997; Schormann et al., 1995). Brain volumes were finally transformed to the spatial format of the reference brain. For both steps, a movement model for large deformations was applied (Schormann et al., 1996, 1997; Schormann and Zilles, 1998). The superimposition of individual cytoarchitectonic areas in the standard reference format resulted in probability maps (Figure 23.6). These maps quantitatively describe the degree of intersubject variability in extent and position of both areas. They serve as a basis for topographical interpretations of functional imaging data obtained in PET and fMRI experiments (Amunts et al., 1998).

The observed intersubject variability in the cytoarchitectural extent of Broca's region has to be considered when correlating data of functional imaging studies with the underlying cortical structures. Interhemispheric differences in the volume of area 44 and in the cytoarchitecture of both areas may contribute to functional lateralization that is associated with Broca's region.

23.3.5 Metadatabases of Processed Data (BrainMap®)

BrainMap®: The Original Human Functional Imaging Database

BrainMap® is a community database of the human functional brain mapping literature structured as an electronic environment for retrieval, visualization, manipulation, and metanalysis (Fox and Lancaster, 1994, 1996, 2002). BrainMap® has many levels of use. It has been widely used as a tool for rapidly retrieving functional imaging studies using intelligent search criteria based on the BrainMap®'s Metadata Coding Scheme and brain anatomy. The highest level of use is for quantitative metanalysis. Functional brain mapping is fertile ground for the development of quantitative metanalysis methods because of its large size, rapid growth, highly quantitative nature, and methodological homogeneity. In particular, the field has made extensive use of spatial coordinates and statistical parametric imaging as analysis and reporting standards for more than 10 years. In anticipation of this need, Fox and Lancaster (2002) have developed an electronic environment to support Internet-based sharing and quantitative metanalysis of functional imaging data. This environment is the BrainMap® database. In addition to citational information, BrainMap® emphasizes two types of data: (1) highly structured metadata that document experimental conditions in great detail and (2) reduced results in the form of coordinate addresses of brain-activation sites. Detailed metadata are critical for any informed reinterpretation of published data. Reduced data are emphasized because they are easier to use and face less objections to sharing than do raw data. Simply stated, BrainMap® allows a user to relate brain locations with behavioral functions through retrieval and visualization of metadata and tabular format data (activated locations, behavioral conditions, etc.). For any brain region, the behavioral conditions (tasks) that have been associated with activation of that region can be returned. Conversely, for any task type, the brain regions that have been activated by such tasks can be retrieved, together with details regarding the specific tasks, imaging methods, analysis methods, citational information, and so on.

BrainMap® is a configured as a centralized database accessed over the Internet using dedicated software tools (rather than via a Web browser) for data contributions and data retrieval and manipulation. At the outset, a relatively small set of data (225 papers; 771 experiments, 7863 activation sites) were coded and entered by the developers (rather than by the originators of the data). The purpose was to test the conceptual structure of the BrainMap® database and gauge interest. Despite the acknowledged limitations of the first implementation of the BrainMap® database and the relatively small sample of literature provided, BrainMap® has attracted more than 1600 registered users. BrainMap® and its data have been used in several published metanalyses and has been the basis of metanalysis methods development (Fox, 1995; Fox et al., 1997, 2001; Nielsen and Hansen, 2002; Turkletaub et al., 2002). Based on this interest, development of the Brain Map® database has continued for more than a decade funded by the James S. McDonnell Foundation (1989–1990), the Office of Naval Research (1991–1992), the EJLB Foundation (1992–1996), the Human Brain Project (1992–present), and the National Library of Medicine (1999–present).

The BrainMap® concept was first presented to the brain imaging community in December 1992 at the first BrainMap® Workshop in San Antonio, Texas. Interest in BrainMap® and in neuroinformatics, in general, motivated Fox and Lancaster to host a series of seven annual "BrainMap® Workshops" each December from 1992 to 1998. Funding for the BrainMap® Workshop series came from a variety of sources, including the Human Brain Project. These workshop were well attended, most notably by persons who subsequently developed the ICBM Database, the fMRI Data Center, and the European Computerized Human Brain Database. Interest in this topic continues to grow, with symposia on databases and metanalyses being annual events at the Society for Neuroscience and the Organization for Human Brain Mapping.

The current implementation of BrainMap® is a journal-like system in which authors code their work for electronic submission and peer review (Figure 23.7A). A journal-like model was chosen for several reasons, including: familiarity to all investigators/authors; familiarity to BrainMap®'s developers; proven ability to scale to accommodate high submission volumes; ease of coordination with the process of scientific review of manuscripts; multiple levels of quality control (i.e., author, peer reviewer, and editorial staff); and the prospect of financial self-sufficiency, based on charging subscription fees for provided content. A fully electronic model maximizes submitter and reviewer convenience and minimizes costs. Because BrainMap® has become a database (DB) that emulates a journal (J), the name has been updated to BrainMap® DBJ. The software tools for submitting studies (BrainMap® Submit) and for retreiving and analyzing studies (BrainMap® Search and View; Figure 23.7B) are free-standing Java applications that can be downloaded from the BrainMap® website (www.brainmapdbj.org). The website also supports data submission and peer review. Plans are underway to link BrainMap®'s coding review with journal scientific review.

Metanalysis, as indicated above, is the highest level of use for BrainMap®. The first metanalysis of stored data was published in 1991 by Frith et al., computing a simple mean of previously reported activations of a language-related region to argue that a newly developed and seemingly different task activated the same location. At last count, there are 35 peer-reviewed human functional brain mapping (HFBM) meta-analyses based on standardized coordinates. Interest in meta-analysis in functional brain imaging is so high that novel, imaging-specific methods for performing meta-analyses are being published. Aware that traditional forms of meta-analysis had come under criticism of misuse, Fox and colleagues (1998) reviewed both the HFBM meta-analysis literature and the traditional meta-analysis literature to develop guidelines for performing HFBM meta-analysis. Functional volumes modeling (FVM) is a strategy introduced by Fox (Fox et al., 1997, 1999, 2001) to estimate functional-area location and location variance from the published literature. Input data are location coordinates judged by the investigator as performing the meta-analysis to be from a specific brain area, based on the tasks used and tightness of clustering. The output is a "probabilistic volume," setting confidence limits for the spatial distribution of a specific functional region. Activation likelihood estimation (ALE) is a new and very promising form of spatial probability estimation for (Turkletaub et al., 2002). As with FVM, input data are coordinates from the published literature. The output of an ALE meta-analysis is a "pseudo-image" expressing the likelihood of activation in a randomly chosen group performing the same task. Plans for extending BrainMap®'s capabilities include extending the supported data to include published meta-analyses and their output data (FVM's and ALE pseudoimages) and to provide statistical parametric images.

23.4 CURRENT APPLICATIONS

The majority of human brain atlases and databases thus far developed have been focused on the normal brain as an appropriate starting point. A number of the applications, however, have used the same tools, software, algorithms, and organizational principles as those applied to the normal brain to evaluate brain structure and function in the course of disease. Such applications have been utilized in degenerative diseases such as Alzheimer's disease, as well as for the purposes of monitoring therapeutic interventions in disorders where there is a well-accepted surrogate biomarker of disease burden. An example in the latter category is multiple sclerosis, where gadolinium-enhancing lesions provide a measure of disease activity by location and magnitude. What follows are a few examples where human brain atlases and databases, along with their associated software and developmental features, have been applied in disease states.

23.4.1 Surface Modeling

Vast numbers of anatomical models can be stored in a population-based atlas (Thompson and Toga, 1997, 2000). These models provide detailed information on the three-dimensional geometry of the brain and how it varies in a population. By averaging models across multiple subjects, subtle features of brain structure emerge that are obscured in an individual due to wide cross-subject differences in anatomy (Thompson et al., 2000b,c). These modeling approaches have recently

(A)

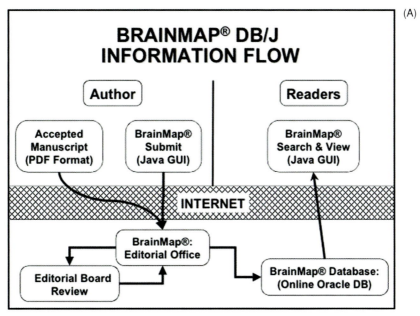

(B)

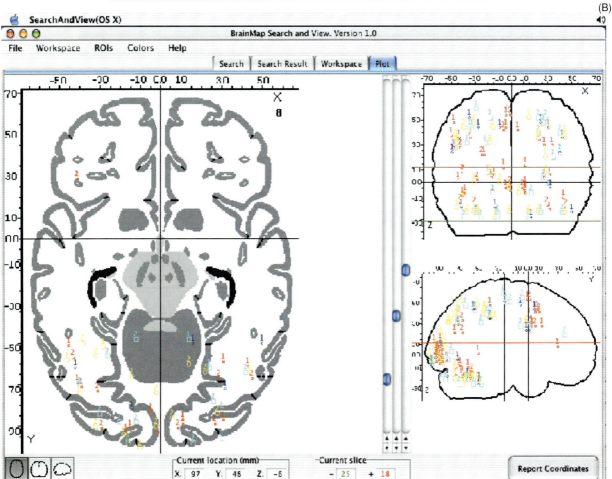

Figure 23.7 (**A**) Submission routing for BrainMap® DBJ. (**B**) BrainMap® Search and View. The locations illustrated are the basis of a metaanalysis of Braille reading. Note the extensive visual-system activations, indicating use of visual regions for tactile information processing. Source: Peter Fox, M.D., University of Texas, San Antonio.

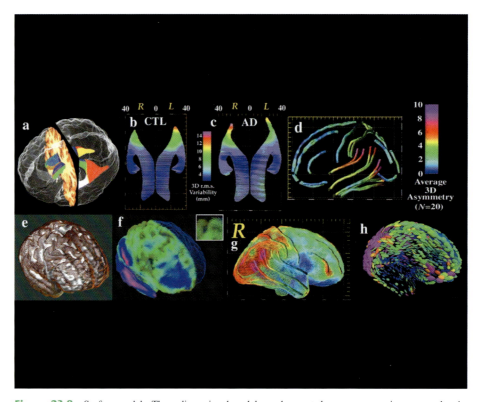

Figure 23.8 Surface models. Three-dimensional models can be created to represent major structural and functional interfaces in the brain. **(a)** Model of the lateral ventricles, in which each element is a three-dimensional parametric surface mesh. **(b, c)** Average ventricular models from a group of patients with Alzheimer's disease ($N = 10$) and matched elderly controls ($N = 10$). Note the larger ventricles in the patients, along with a prominent ventricular asymmetry (left larger than right). These features only emerge after averaging models for groups of subjects. **(d)** Average population maps of cortical anatomy reveal a clear asymmetry of the perisylvian cortex. **(e)** Individual's cortex (brown mesh) overlaid on an average cortical model for a group. **(f)** Differences in cortical patterns are encoded by computing a three-dimensional elastic deformation (pink colors: large deformation) that reconfigures the average cortex into the shape of the individual, matching elements of the gyral pattern exactly. These deformation fields (panel f) provide detailed information on individual deviations and can be averaged across subjects to create three-dimensional variability maps, demonstrating fundamental patterns of anatomical variability in the brain in **(g)**. [Thompson and Toga, 2000]. **(h)** Tensor maps using color ellipsoids reveal the directions in which anatomical variation is greatest. The ellipsoids are more elongated in the directions in which structures tend to vary the most. Pink colors denote the largest variation, while blue colors show the least. These statistical data can be used to detect patterns of abnormal anatomy in new subjects. Severe abnormality is detected (red colors) while corresponding regions in a matched elderly control subject are signaled as normal. From Thompson and Toga (2000).

uncovered striking patterns of disease-specific structural differences in Alzheimer's disease (Thompson et al., 1997, 1998, 2000b), schizophrenia (Narr et al., 2000), and fetal alcohol syndrome (Sowell et al., 2001), as well as strong linkages between patterns of cortical organization and age (Thompson et al., 2000a), gender (Thompson et al., 2000b), cognitive scores (Mega et al., 1997), and genotype (Le Goualher et al., 2000). To illustrate the approach, Figure 23.8 shows a model of the lateral ventricles, in which each element is represented by a three-dimensional surface mesh. These surface models can often be extracted automatically from image data, using recently developed algorithms based on deformable parametric surfaces (Thompson et al., 1996, 1996a,b; Thompson and Toga, 1996; MacDonald, 1998) or voxel-coding (Zhou and Toga, 1999). Once an identical computational grid (or surface mesh) is imposed on the same structure in different subjects, an average anatomical model can be created for a group. This is done by averaging the

three-dimensional coordinate locations of boundary points that correspond across subjects.

Figures 23.8b and 23.8c show average ventricular models from a group of patients with Alzheimer's disease ($N = 10$) and from matched elderly controls ($N = 10$). Not only are the ventricles larger in the patients, but a prominent ventricular asymmetry (left larger than right) is found in both groups, a feature that only emerges after surface averaging. Specialized approaches for averaging cortical anatomy can also be used to generate population-based maps of brain asymmetry (Figure 23.8d) and investigate its alteration in disease (Thompson et al., 2000b; Narr et al., 2000). Cortical anatomy can also be compared across subjects and its variability encoded to guide the detection of abnormal anatomy (Thompson et al., 1997). Figure 23.8e shows an individual's cortex (brown mesh) overlaid on an average cortical model for a group. Differences in cortical patterns can be encoded by computing a three-dimensional elastic deformation that reconfigures the average cortex into the shape of the individual, matching elements of the gyral pattern exactly (Figure 23.8f). These deformation fields store detailed information on individual deviations, and can be averaged across subjects to create three-dimensional variability maps, revealing fundamental patterns of anatomical variability in the brain (Figure 23.8g). The resulting confidence limits on the locations of cortical structures can be used in Bayesian approaches to guide the automated labeling of gyri and sulci (Pitiot et al., 2002) and to map profiles of abnormal anatomy in an individual patient or group of subjects (Thompson et al., 1997, 2000a,b,c; Cao and Worsley, 1999).

This strategy has been used to develop atlases and analysis methods for disease states. Specifically, it is both practical and desirable to build disease-specific atlases in order to observe the natural history of a disorder, compare it to normal, age-matched subjects, and use the disease-specific atlas as a comparison to populations of subjects undergoing conventional or experimental therapies. In this fashion, it is possible to have quantifiable, objective, and automated means by which to examine brain structure and function in the normal state, under pathologic conditions as well as during interventions designed to ameliorate or reduce the impact of the disorder. Such an approach using imaging as a surrogate marker of disease burden may greatly facilitate clinical therapeutic trials by providing objectivity and a quantifiable surrogate endpoint, both of which should increase the cost-effectiveness of these expensive undertakings.

Alzheimer's Disease

Probabilistic atlases based on populations of patients with specific disorders (Thompson and Toga, 2000; Thompson et al., 2000b,c) show enormous promise in advancing our understanding of the disease process. As imaging studies expand into ever-larger patient populations, population-based brain atlases offer a powerful framework to synthesize the results of disparate imaging studies. Disease-specific atlases, for example, are a type of probabilistic atlas specialized to represent a particular clinical group (for reviews see Thompson and Toga, 2000; and Section 24.3, Chapter 24, this volume). A disease-specific atlas of brain in Alzheimer's disease has recently been generated to reflect the unique anatomy and physiology of this subpopulation (Thompson et al., 1997, 1998; Thompson and Toga, 2000; Mega et al., 1997, 1998, 1999). Based on well-characterized patient groups, this atlas contains thousands of structure models, as well as composite maps, average templates, and visualizations of structural variability, asymmetry, and group-specific differences. It also correlates the structural, metabolic, molecular, and histologic hallmarks of the disease (Mega et al., 1997, 1999, 2000). Additional algorithms use information stored in the atlas to recognize anomalies and label structures in new patients. Because they retain information on group anatomical variability, the resulting atlases can identify patterns of altered structure or function and can guide algorithms for knowledge-based image analysis, automated image labeling (Collins et al., 1994; Pitiot et al., 2002), tissue classification (Zijdenbos and Dawant, 1994), and functional image analysis (Dinov et al., 2000). At the core of the atlas is an average MRI dataset based on a population of subjects with early dementia (Figure 23.9). Using specialized mathematical approaches for averaging cortical anatomy, the resulting average MRI template has a well-resolved cortical pattern (Figure 23.9a), with the mean geometry of the patient group (Thompson

and Toga, 2000). Surfaces for the cortex can include the external hull, the gray–white matter interface, or an average of the full cortical thickness. Figure 23.9b represents the external hull.

An example application of this type of atlas is in resolving the average profile of early gray matter loss in an Alzheimer's disease population. It would be ideal, for example, to calibrate the profile of gray matter loss in an individual patient against a normative reference population, for early diagnosis or for clinical trials. Since individual variations in cortical patterning complicate the comparison of gray matter profiles across subjects, an elastic matching technique can be used (driven by 84 structures per brain) that elastically deforms each brain into the group mean geometric configuration (Figure 23.9b). By averaging a measure of gray matter across corresponding regions of cortex, these shape differences are factored out. The net reduction in gray matter, in a large patient population relative to controls ($N = 46$), can then be plotted as a statistical map in the atlas (Figure 23.9b); (Thompson and Toga, 2000). This type of analysis uncovers important systematic trends, with an early profile of severe gray matter loss detected in temporo-parietal cortex, consistent with the early distribution of neuronal loss, metabolic change, and perfusion deficits at this stage of Alzheimer's disease. Finally, this local encoding of information on cortical variation can also be exploited to map abnormal atrophy in an individual patient (Thompson et al., 1997). Figures 23.8f and 23.9b illustrate the use of a probabilistic atlas to identify a region of abnormal atrophy in the frontal cortex of a dementia patient. Severe abnormality is detected, with a color code used to indicate the significance of the abnormality (*red colors*). As expected, corresponding regions in a matched elderly control subject are signaled as normal (Figure 23.9b).

23.4.2 Tissue Classification

Tissue classification strategies are equally applicable for population analysis of patients with brain disorders and for tracking structural change over time, such as the progressive tissue atrophy that occurs in some degenerative diseases. One illustrative example of this approach draws from a clinical trial of a new treatment for multiple sclerosis (MS) which used MRI observations as a surrogate marker for disease activity (Zijdenbos et al., 1996). Multi-spectral MRI data were collected at 14 sites in North America from 460 patients with relapsing-remitting MS. A total of 1850 datasets were available, each consisting of T1-, T2-, and PD-weighted volumes. After correction for MRI intensity inhomogeneity, interslice and intervolume intensity normalization and stereotaxic transformation the multi-spectral data were tissue classified to identify MS lesion voxels for each patient timepoint. Figure 23.10 shows a three-dimensional rendering of the probabilities for lesion distribution obtained from all datasets. This shows the most likely locations for MS lesions within a population and is a convenient way to distill a large amount of population data into a single entity. Tests of drug effect is reduced to testing for a significant group difference in the overall volume of this distribution above a given threshold when partitioned into drug and placebo groups. Tests for regional drug effects (i.e., changes in regional lesion probability) between groups become equivalent to the familiar test for CBF change between two stimulus conditions, using the same

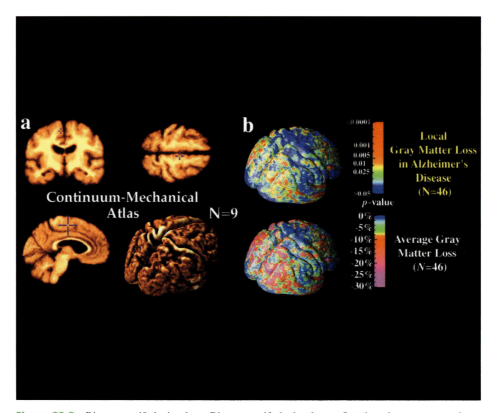

Figure 23.9   Disease-specific brain atlases. Disease-specific brain atlases reflect the unique anatomy and physiology of a clinical population, in this case an Alzheimer's disease population. Using mathematical strategies to average cortical anatomy across subjects [Thompson et al., 2000], an average MRI template can be generated for a specific patient group, in this case nine patients with mild to moderate Alzheimer's disease. The cortical pattern indicates clear sulcal widening and atrophic change, especially in temporo-parietal cortices. By averaging a measure of gray-matter across corresponding cortical regions, the average profiles of gray-matter loss can also be mapped. The net reduction in gray matter, in a large patient population relative to controls (Thompson and Toga, 2000), can then be plotted as a statistical map in the atlas. This type of analysis uncovers profiles of early anatomical change in disease. By encoding variations in gyral patterns and gray matter distribution, algorithms can detect a region of abnormal atrophy in the frontal cortex of a dementia patient. From Thompson et al., 2000.

statistical models as are used for detection of functional changes in activation experiments with PET and fMRI (Evans et al., 1997). Importantly, this approach allows for rapid re-analysis of clinical trials data in response to (i) new hypotheses, (ii) modification of the input data for discriminating lesion and non-lesion (types of image, noise filtering, image blurring), or (iii) modification of criteria for identifying lesions (threshold values for lesion probability, spatial constraints on lesion location, minimum voxels per lesion, etc.). This approach makes it feasible to extract the considerable and usually untapped information available in large clinical trial image databases.

23.5 LIMITATIONS

All projects have limitations. Those that are focused on a problem of the complexity of the human brain have limitations in scope and degree of success. Typically, projects of this magnitude have to be achieved in an iterative way where progressive breakthroughs and insights lead to further evolution, both in hypotheses and methodology. This is clearly the case with the burgeoning field of human brain atlases. A few well-defined limitations have already emerged and are discussed below.

23.5.1 Isolated Brain Regions

Developing atlas strategies that utilize datasets for the whole brain is a challenging problem. An even more difficult problem than working with whole brain, three-dimensional datasets is that of entering microscopic postmortem or surgical specimens or incomplete in vivo imaging data from brain sites that are analyzed on a regional basis (e.g., the study of the isolated hippocampus). Nevertheless, such data can still be incorporated into brain atlases. Such a problem requires landmarks to appropriately localize regional data in the global atlas brain.

Consider a series of postmortem cryomacrotome human brains that are stained with a series of conventional and commonly used neuroanatomical "landmark" stains (e.g., Nissl, acetylcholinesterase). Using state-of-the-art imaging devices, these sections would be digitized and sampled at a 20-μm resolution. The resultant datasets would be warped and entered into the brain atlas as an additional feature. Then consider an investigator who studies GABA receptors in the human hippocampus. This investigator would like to see where the receptors from the hippocampi of a given epileptic patient population fall with regard to other data in the probabilistic reference system. In preparing the tissue, this investigator would process every Nth section

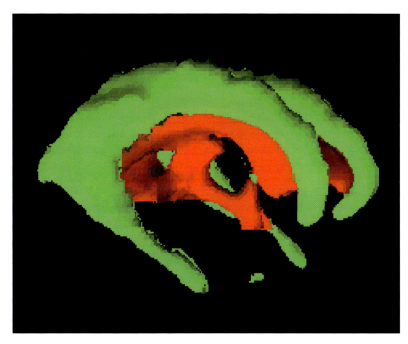

Figure 23.10 Probabilistic atlas of multiple sclerosis. This image is produced from 460 patients with multiple sclerosis derived from 5800 individual pulse sequences. The green areas demonstrate the probabilistic location of the actual MS plaques in the population referenced to the human ventricular system (red). This strategy gives a composite view of overall disease burden across a large population. Consider its use in a clinical trial where these patients were randomly assigned to treatment groups that included placebo, conventional therapy, and experimental therapy. By following the group for some period of time and serially imaging them, an automated, quantifiable, and objective measure of the relative effect of experimental therapy versus conventional therapy and the natural history of the disease could be obtained using MRI lesions as a surrogate marker of disease burden. Such strategies should lead to more efficient and cost-effective clinical trials (Zijdenbos et al., 1996; Evans et al., 1997).

using one of the "landmark" stains that are part of the probabilistic atlas. The investigator would then digitize the information from both the GABA receptor sections and the "landmark" stained sections. Using alignment, registration, and warping tools that are part of the atlas system, the investigator would register the "landmark" stained sections with the atlas and then use the same mathematical transformations to enter the GABA receptor information into the hippocampal region of the atlas. Once referenced, database queries and visualization of this new data could be performed in the atlas system. A similar approach allows referencing between newly acquired in vivo data and stored postmortem specimens that should aid in relating functional localization with macroscopic and microscopic anatomy (Rademacher et al., 1992; Larsson et al., 1999; Naito et al., 1999, 2000; Bodegård et al., 2000a,b).

23.5.2 EEG/MEG

The three atlas strategies discussed in this chapter are based on neuroanatomy. This is the most fundamental language of communication in neuroscience. As such, it allows appropriate reference and localization to any structure in the brain from any signal source. In the development of the brain atlas systems, cross-sectional and tomographic data were the initial datasets. Once established, however, appropriate vehicles for entering nontomographic data can be developed. For EEG data, for example,

systems already exist to localize scalp electrode placement three-dimensionally, either through the use of a paired tomographic image set or by nontomographic localization methods (Gevins et al., 1994).

23.5.3 Sociology

Any endeavor to organize information across laboratories, or especially across an entire field, requires attention to the sociology involved (Koslow, 2000). Frustration with existing methods must be high enough and the solutions good enough (in terms of practicality, economics and implementation) to be adopted (see Section 1.1, Chapter 1, this volume). Such a transition is made easier if rigid new standards are not imposed on the structure or organization of data generated in a given laboratory but rather the tools are available to translate such data into the framework and form required for interaction with the database and atlas. Perhaps the most important, if not critical, step is the willingness on the part of the community to share data in all its forms (including raw data) to allow for the full implementation of such a system. Such strategies require participation of traditional final end products of research (e.g., journals) as well as academic recognition for data provided to such systems. Lastly, it is always important to have a consensus from the community before embarking on the construction of such complex systems. Wide participation, frequent requests for input and distributed testing of products are all helpful in establishing a successful system that is accepted by the community for which it was intended.

23.5.4 Scope

Every project has limitations. The ones described here are no different. When faced with the opportunity to include data from large numbers of subjects, there is a tendency to be all-inclusive and attempt to collect every potential type of information available. At the onset of such projects the contributing investigators meet and discuss all of the possible datasets that could be collected from a human subject. The list is typically long. One must opt to start with those datasets that would provide structural imaging of the highest resolution in the largest number of subjects for the best price. In later years, and in subsequent iterations, it might be possible to add other datasets. In fact, this was done for the ICBM atlas with the addition of functional imaging using fMRI, PET, and event-related potentials. Nevertheless, it was not possible to add information about the vasculature from MR angiography, neurotransmitter systems through PET or SPECT ligand studies, cerebral perfusion through perfusion MRI or PET, chemical information about the brain from MR spectroscopy or datasets that describe major white-matter tracts in the brain using diffusion tensor imaging or connectivity using combinations of transcranial magnetic stimulation and PET or fMRI. These are all issues that would be extremely important and valuable to add in the future. Those that have been selected and implemented reflect the basic criteria list noted above as well as the realistic constraints associated with finances, subject risk, time burdens, and institutional review board criteria. In fact, the reason that only 5800 of the 7000 subjects have DNA samples in the ICBM atlas relates to IRB rules in certain countries with regard to the collection and distribution of genetic materials and information about subjects.

23.6 FUTURE IN NEUROSCIENCE

There is no question that the development of systems and tools such as human brain atlases will have a specific and not insignificant cost associated with them. Also true is the fact that increments in neuroscientific research funding have not kept pace with the growth of the field in terms of numbers of investigators or the magnitude of their projects.

The thoughtful response to this statement requires, however, an honest appraisal of the ultimate goals of neuroscientific research. If such research is designed to produce the most accurate understanding of normal brain function and diseases that affect it, then tools that will enhance the accuracy of results, enable us to compare results between subjects and laboratories, make more rigorous the confirmation or refutation of data, and guard against its loss should have a high priority. Brain atlases for a given species, or potentially across species, provide a means by which to rigorously store, compare, and analyze data over time and between laboratories. Furthermore, by virtue of data exchange and comparison, integration within the broad field of neuroscience will begin.

One could rephrase the above statement into a question and ask, "What would it cost not to develop such integrated systems?"

The costs would be the progressive and result in the continued reduction in the value of every dollar spent on future neuroscientific research because of the progressively unmanageable amounts and types of data that are generated by neuroscientists. Lacking the tools to manage, compare, and analyze these datasets will make funds spent for their acquisition of lesser impact than if such data could be preserved and referenced in an ever-evolving and integrated approach.

Clearly, it would be optimal if funding for systems and approaches to integrate data across not only neuroscience but also computer science, informatics, and potentially other related fields could come from new sources. In fact, this is already happening. Contributions to the funding of the initial round of the Human Brain Project (Huerta et al., 1993) in the United States came from sources that are both traditional and novel for funding neuroscientific research. By having small contributions from many countries and agencies, the burden on any one country or agency is small but the impact for the neuroscience community is significant. An expanded participation by agencies and contributors outside of traditional pathways as well as the potential for generating new appropriations based on interest in this international effort will hopefully result in the creation of systems such as the one described in this report as well as others contemplated or funded through the auspices of the Human Brain Project in the United States without detracting from traditional neuroscientific funding.

The development of atlases for the human brain is a formidable goal and one that involves participation from many sites around the world and investigators committed to the end product. The creation of a such atlases of the human brain is not an exercise in library science. It is a series of fundamental, hypothesis-driven experiments in merging mathematical and statistical approaches with morphological and physiological problems posed with regard to the nervous system. It will create new data and insights into the organization of the human nervous system in health and disease, its development, and its evolution. When successful, it will provide previously unprecedented tools for organizing, storing, and communicating information about the human brain throughout development, maturation, adult life, and old age. It will be a natural prelude to studies of patients with cerebral disorders and provide the first mechanism by which phenotype–genotype–behavioral comparisons can be made on a macroscopic and microscopic level. These results will provide the first insights into the structure–function organization of the human brain across all structures and a wide range of ages. Its design anticipates the continuing evolution in the quality, resolution, and magnitude of data generated by existing technologies that are used to map the human brain and even anticipates that many future technologies, unknown today, will be applicable because such systems are organized using the architecture of the brain as their guiding principle. The result will allow electronic experimentation and hypothesis generation, facilitated communication among investigators, and an objective way of assessing new information gleaned either at scientific meetings or through publications. Developing such systems are open-ended projects with constant evolution, improvement, and expansion both in the numbers of subjects included and the range of attributes associated with each. The results should be far more than a data structure and organizational system. Rather, such atlases should provide new insights and new opportunities for neuroscientists to utilize data in their own laboratories as well as others to more rapidly, effectively, and efficiently make progress in understanding human brain function in health and disease.

ACKNOWLEDGMENTS

This work was supported by a grant from the Human Brain Project (P20-MHDA52176), funded by the National Institute of Mental Health, National Institute for Drug Abuse, National Cancer Institute, and the National Institute for Neurological Disease and Stroke. For generous support, the author also wishes to thank the Brain Mapping Medical Research Organization, the Pierson–Lovelace Foundation, The Ahmanson Foundation, the Tamkin Foundation, the Jennifer Jones Simon Foundation, and the Robson Family. For materials about their respective projects, the author thanks Michael Gazzaniga, Ph.D., Jack Van Horn, Ph.D., and Peter Fox, M.D. Special thanks go to Kami Yaden for the preparation of the manuscript.

References

Ahlfors, S.P., Simpson, G.V., Dale, A.M., Belliveau, J.W., Liu, A., Korvenoja, A., Virtanen, J., Huotilanen, M., Tootell, R.B.H., Aronen, H.J., and Ilmoniemi, R.J. (1999). Spatiotemporal activity of a cortical network for processing visual motion revealed by MEG and fMRI. *J. Neurophysiol.* **82**, 2545–2555.

Aine, C.J., Supek, S., George, J.S., Ranken, D., Lewine, J., Sanders, J., Best, E., Tiee, W., Flynn, E.R., and Wood, C.C. (1996). Retinotopic organization of human visual cortex: Departures from the classical model. *Cereb. Cortex* **6**(3), 354–361.

Amunts, K., Klingberg, T., Binkofski, F., Schormann, T., Seitz, R.J., Roland, P.E., and Zilles, K. (1998). Cytoarchitectonic definition of Broca's region and its role in functions different from speech. *NeuroImage* **7**, 8.

Amunts, K., Schleicher, A., Bürgel, U., Mohlberg, H., Uylings, H.B.M., and Zilles, K. (1999). Broca's region revisited: Cytoarchitecture and intersubject variability. *J. Comp. Neurol.* **412**, 319–341.

Amunts, K., Malikovic, A., Mohlberg, H., Schormann, T., and Zilles, K. (2000). Brodmann's areas 17 and 18 brought into stereotaxic space— where and how variable? *NeuroImage* **11**, 66–84.

Arnold, J.B., Liow, J.-S., Schaper, K.A., Stern, J.J., Sled, J.G., Shattuck, D.W., Worth, A.J., Cohen, M.S., Leahy, R.M., Mazziotta, J.C., and Rottenberg, D.A. (2001). Quantitative and qualitative evaluation of six algorithms for correcting intensity non-uniformity effects. *NeuroImage* **13**(5), 931–943.

Bailey, P., and von Bonin, G. (1951). *The Isocortex of Man.* University Press, Urbana, IL.

Bartley, A.J., Jones, D.W., and Weinberger, D.R. (1997). Genetic variability of human brain size and cortical gyral patterns. *Brain* **120**, 257–269.

Bloom, F.E. (1996). The multidimensional database and neuroinformatics requirements for molecular and cellular neuroscience. *NeuroImage* **4**, S12–S13.

Bodegård, A., Geyer, S., Naito, E., Zilles, K., and Roland, P.E. (2000a). Somatosensory areas in man activated by moving stimuli: Cytoarchitectonic mapping and PET. *NeuroReport* **11**, 187–191.

Bodegård, A., Ledberg, A., Geyer, S., Naito, E., Larsson, J., Zilles, K., and Roland, P. (2000b). Object shape differences reflected by somatosensory cortical activation in human. *J. Neurosci.* **20**(RC51), 1–5.

Brodmann, K. (1909). *Vergleichende Lokalisationslehre der Grosshirnrinde in ihren Prinzipien dargestellt auf Grund des Zellenbaues.* Barth, Leipzig.

Cannestra, A.F., Santori, E.M., Holmes, C.J., and Toga, A.W. (1997). A three-dimensional multimodality map of the nemistrina monkey. *Brain Res. Bull.* **5**, 147–153.

Cao, J., and Worsley, K.J. (1999). The geometry of the Hotelling's T^2 random field with applications to the detection of shape changes. *Ann. Stat.* **27**, 925–942.

Christensen, G., Rabbitt, R.D., and Miller, M.I. (1993). A deformable neuroanatomy textbook based on viscous fluid mechanics, (invited paper). *Proc. 1993 Conf. Inf. Sci. Syst.* pp. 211–216.

Clark, S., and Miklossy, J. (1990). Occipital cortex in man: Organization of callosal connections, related myelo- and cytoarchitecture, and putative boundaries of functional visual areas. *J. Comp. Neurol.* **298**, 188–214.

Collins, D.L., Neelin, P., Peters, T.M., and Evans, A.C. (1994). Automatic 3D registration of MR volumetric data in standardized Talairach space. *J. Comput.-Assisted Tomogr.* **18**(2), 192–205.

Dale, A.M., Fischl, B., and Sereno, M.I. (1999). Cortical surface-based analysis. I. Segmentation and surface reconstruction. *NeuroImage* **9**(2), 179–194.

Davatzikos, C. (1997). Spatial transformation and registration of brain images using elastically deformable models. *Comput. Vis. Image Und.* **66**(2), 207–222.

Dinov, I.D., Mega, M.S., Thompson, P.M., Lee, L., Woods, R.P., Holmes, C.J., Sumners, D.W., and Toga, A.W. (2000). Analyzing functional brain images in a probabilistic atlas: A validation of subvolume thresholding. *J. Comput.-Assisted Tomogr.* **24**(1), 128–138.

Dumoulin, S.O., Bittar, R.G., Kabani, N.J., Baker, C.L., Le Goualher, G., Pike, G.B., and Evans, A.C. (2000). A new neuroanatomical landmark for the reliable identification of human area V5/MT: A quantitative analysis of sulcal patterning. *Cereb. Cortex* **10**, 454–463.

Evans, A.C., Frank, J.A., Antel, J., and Miller, D.H. (1997). The role of MRI in clinical trials of multiple sclerosis: Comparison of image processing techniques. *Ann. Neurol.* **41**, 125–132.

Felleman, D.J., and Van Essen, D.C. (1991). Distributed hierarchical processing in the primate cerebral cortex. *Cereb. Cortex* **1**(1), 1–47.

Filimonoff, I.N. (1932). Über die Variabilität der Großhirnrindenstruktur. Mitteilung II—Regio occipitalis beim erwachsenen Menschen. *J. Psychol. Neurol.* **44**, 2–96.

Fischl, B., Sereno, M.I., Tootell, R.B., and Dale, A.M. (1999). High-resolution intersubject averaging and a coordinate system for the cortical surface. *Hum. Brain Mapp.* **8**(4), 272–284.

Flechsig, P. (1920). *Anatomie des menschlichen Gehirns und Ruckenmarks.* Thieme, Leipzig.

Fox, P.T. (1995). Broca's area: Motor encoding in somatic space. *Behav. Brain Sci.* **18**, 344–345.

Fox, P.T., Mikiten, S., Davis, G., Lancaster, J.L (1994). Brainmap: A database of human functional brainmapping. In *Advances in Functional Neuroimaging: Technical Foundations.* Thatcher, R.W., Hallett, M., Zeffiro, T., John, E.R., and Huerta, M. (eds). Academic, Press, Orlando, FL pp. 95–105.

Fox, P.T., and Lancaster, J.L. (1996). Neuroscience on the net. *Science* **226**, 994–996.

Fox, P.T., and Lancaster, J.L. (2002). Mapping context and content: The BrainMap® model. *Nat. Rev. Neurosci.* **3**, 319–321.

Fox, P.T., and Mintun, M.A. (1989). Noninvasive functional brain mapping by change-distribution analysis of averaged PET images of $H_2^{15}O$ tissue activity. *J. Neurosci.* **7**, 913–922.

Fox, P.T., Lancaster, J.L., Parsons, L.M., Xiong, X.H., and Zamaripa, F. (1997). Functional volumes modeling: Theory and preliminary assessment. *Hum. Brain Mapp.* **5**, 306–311.

Fox, P.T., Parsons, L.M., and Lancaster, J.L. (1998). Beyond the single study: Function/location metaanalysis in cognitive neuroimaging. *Curr. Opin. Neurobiol.* **8**, 178–187.

Fox, P.T., Huang, A.Y., Parsons, L.M., Xiong, J.H., Rainey, L., and Lancaster, J.L. (1999). Functional volumes modeling: Scaling for group size in averaged images. *Hum. Brain Mapp.* **8**(2–3), 143–150.

Fox, P.T., Huang, A., Parsons, L.M., Xiong, J.H., Zamarippa, F., Rainey, L., and Lancaster, J.L. (2001). Location–probability profiles for the mouth region of human primary motor-sensory cortex: Model and validation. *NeuroImage* **13**, 196–209.

Friston, K.J., Frith, C.D., Liddle, P.F., and Frackowiak, R.S.J. (1991). Comparing functional (PET) images: The assessment of significant change. *J. Cereb. Blood Flow Metab.* **11**, 690–699.

Frith, C.D., Friston, K., Liddle, P.F., and Frackowiak, R.S.J. (1991). Willed action and the prefrontal cortex in man: A study with PET. *Proc. R. Soc. London, Biol. Ser.* **244**, 241–246.

Galaburda, A.M. (1980). La région de Broca: Observations anatomiques faites un siècle après la mort de son découvreur. *Rev. Neurol.* **136**, 609–616.

Gevins, A.S., Lee, J., Martin, N., Reutter, R., Desmond, J., and Brickett, P. (1994). High resolution EEG: 124-channel recording, spatial deblurring and MRI integration methods. *Electron. Clin. Neurol.* **90**, 337–358.

Geyer, S., Schleicher, A., and Zilles, K. (1997). The somatosensory cortex of human: Cytoarchitecture and regional distributions of receptor-binding sites. *NeuroImage* **6**, 27–45.

Geyer, S., Schleicher, A., and Zilles, K. (1999). Areas 3a, 3b, and 1 of human primary somatosensory cortex: 1. Microstructural organization and interindividual variability. *NeuroImage* **10**, 63–83.

Goutte, C., Hansen, L.K., Liptrot, M.G., and Rostrup, E. (2001). Feature-space clustering for fMRI meta-analysis. *Hum. Brain Mapp.* **13**, 165–183.

Hayes., T.L., and Lewis, D.A. (1995). Anatomical specialization of the anterior motor speech area: Hemispheric differences in magnopyramidal neurons. *Brain Lang.* **49**, 289–308.

Hayes, T.L., and Lewis, D.A. (1996). Magnopyramidal neurons in the anterior motor speech region. *Arch. Neurol.* **53**, 1277–1283.

Huerta, M., Koslow, S., and Leshner, A. (1993). The Human Brain Project: An international resource. *Trends Neurosci.* **16**, 436–438.

Jacobs, B., Batal, H.A., Lynch, B., Ojemann, G., Ojemann, L.M., and Scheibel, A.B. (1993). Quantitative dendritic and spine analysis of speech cortices: A case study. *Brain Lang.* **44**, 239–253.

Koslow, S.H. (2000). Should the neuroscience community make a paradigm shift to sharing primary data? *Nat. Neurosci.* **3**(9), 863–865.

Larsson, J., Amunts, K., Gulyás, B., Malikovic, A., Zillesm, K., and Roland, P.E. (1999). Neuronal correlates of real and illusory contour perception: Functional anatomy with PET. *Eur. J. Neurosci.* **11**, 4024–4036.

Le Goualher, G., Argenti, A., Duyme, M., Baare, W., Hulshoff Pol. H., Barillot, C., and Evans, A. (2000). Statistical sulcal shape comparisons: Application to the detection of genetic encoding of the central sulcus shape. *NeuroImage* **11**, 564–574.

MacDonald, D. (1998). A method for identifying geometrically simple surfaces from three-dimensional images. Ph.D. Dissertation, McGill University. Montreal, Quebec, Canada.

Mazziotta, J.C. (1984). Physiological neuroanatomy: Functional brain imaging presents a new problem to an old discipline. *J. Cereb. Blood Flow Metab.* **4**, 481–484.

Mazziotta, J.C., Huang, S.C., Phelps, M.E., Carson, R.E., MacDonald, N.S., and Mahoney, K. (1985). A non-invasive positron computed tomography technique using oxygen-15 labeled water for the evaluation of neurobehavioral task batteries. *J. Cereb. Blood Flow Metab.* **5**, 70–78.

Mazziotta, J.C., Toga, A.W., Evans, A.C., Fox, P., and Lancaster, J. (1995). A probabilistic atlas of the human brain: Theory and rationale for its development. *NeuroImage* **2**, 89–101.

Mazziotta, J., Toga, A., Evans, A., Fox, P., Lancaster, J., Zilles, K., Simpson, G., Woods, R., Paus, T., Pike, B., Holmes, C., Collins, L., Thompson, P., MacDonald, D., Schormann, T., Amunts, K., Palomero-Gallagher, N., Parsons, L., Narr, K., Kabani, N., Le Goualher, G., Boomsma, D., Cannon, T., Kawashima, R. and Mazoyer, B. (2001a).

A probabilistic atlas and reference system for the human brain. *Philos. Trans. R. Soc. London, Ser. B* **356**, 1293–1322.

Mazziotta, J., Toga, A., Evans, A., Fox, P., Lancaster, J., Zilles, K., Simpson, G., Woods, R., Paus, T., Pike, B., Holmes, C., Collins, L., Thompson, P., MacDonald, D., Schormann, T., Amunts, K., Palomero-Gallagher, N., Parsons, L., Narr, K., Kabani, N., Le Goualher, G., Boomsma, D., Cannon, T., Kawashima, R., and Mazoyer, B. (2001b). A four-dimensional atlas of the human brain. *J. Am. Med. Inf. Assoc.* **8**, 401–430.

Mega, M.S., Chen, S., Thompson, P.M., Woods, R.P., Karaca, T.J., Tiwari, A., Vinters, H., Small, G.W., and Toga, A.W. (1997). Mapping pathology to metabolism: Coregistration of stained whole brain sections to PET in Alzheimer's disease. *NeuroImage* **5**, 147–153.

Mega, M.S., Thompson, P.M., Cummings, J.L., Back, C.L., Xu, M.L., Zohoori, S., Goldkorn, A., Moussai, J., Fairbanks, L., Small, G.W., and Toga, A.W. (1998). Sulcal variability in the Alzheimer's brain: Correlations with cognition. *Neurology* **50**(1), 145–151.

Mega, M.S., Chu, T., Mazziotta, J.C., Trivedi, K.H., Thompson, P.M., Shah, A., Cole, G., Frautschy, S.A., and Toga, A.W. (1999). Mapping biochemistry to metabolism: FDG-PET and amyloid burden in Alzheimer's disease. *NeuroReport* **10**(14), 2911–2917.

Mega, M.S., Lee, L., Dinov, I.D., Mishkin, F., Toga, A.W., and Cummings, J.L. (2000). Cerebral correlates of psychotic symptoms in Alzheimer's disease. *J. Neurol. Neurosurg. Psychiatry* **69**(2), 167–171.

Merker, B. (1983). Silver staining of cell bodies by means of physical development. *J. Neurosci.* **9**, 235–241.

Midtvik, M., and Hovig, I. (1999). Reversible compression of MR images. *IEEE Trans. Med. Imaging* **18**, 795–800.

Mitchell, T. (1999). Machine learning and data mining. *Communi. ACM* **42**, 11.

Naito, E., Ehrsson, H.H., Geyer, S., Zilles, K., and Roland, P.E. (1999). Illusory arm movements activate cortical motor areas: A positron emission tomography study. *J. Neurosci.* **19**, 6134–6144.

Naito, E., Kinomura, S., Geyer, S., Kawashima, R., Roland, P.E., and Zilles, K. (2000). Fast reaction to different sensory modalities activates common fields in the motor areas, but the anterior cingulate cortex is involved in the speed of reaction. *J. Neurophysiol.* **83**, 1701–1709.

Narr, K.L., Thompson, P.M., Sharma, T., Moussai, J., Cannestra, A.F., and Toga, A.W. (2000). Mapping morphology of the corpus callosum in schizophrenia. *Cereb. Cortex* **10**(1), 40–49.

Neelin, P.D., MacDonald, D., Collins, D.L., and Evans, A.C. (1998). The MINC file format: From bytes to brains. *NeuroImage* **7**(4), S786.

Nielsen, F.A., and Hansen, L.K. (2002). Modeling of activation data in the BrainMap® database: Detection of outliers. *Hum. Brain Mapp.* **15**, 146–156.

Ono, M., Kubik, S., and Abernathy, C. (1990). *Atlas of the Cerebral Sulci.* Thieme Medical, Stuttgart.

Payne, B.A., and Toga, A.W. (1990). Surface mapping brain function on 3D models. *IEEE Comput. Graph.* **10**(5), 33–41.

Pennisi, E. (1999). Keeping genome databases clean and up to date. *Science* **286**, 447–450.

Pitiot, A., Toga, A.W., and Thompson, P.M. (2002). Adaptive elastic segmentation of brain MRI via shape-model-guided evolutionary programming. *IEEE Trans. Med. Imaging* **21**(8), 910–923.

Polyak, S.L. (1957). *The Vertebrate Visual System.* University of Chicago Press, Chicago, IL.

Rademacher, J., Galaburda, A., Kennedy, D., Filipek, P., and Caviness, V. (1992). Human cerebral cortex: Localization, parcellation, and morphometry with magnetic resonance imaging. *J. Cogn. Neurosci.* **4**, 352–374.

Rademacher, J., Caviness, V.S., Steinmetz, H., and Galaburda, A.M. (1993). Topographical variation of the human primary cortices: Implications for neuroimaging, brain mapping and neurobiology. *Cereb. Cortex* **3**(4), 313–329.

Roland, P.E., and Zilles, K. (1994). Brain atlases: A new research tool. *Trends Neurosci.* **17**(11), 458–467.

Roland, P.E., and Zilles, K. (1996). The developing European Computerized Human Brain Database for all imaging modalities. *NeuroImage* **4**, 39–47.

Roland, P.E., and Zilles, K. (1998). Structural divisions and functional fields in the human cerebral cortex. *Brain Res. Rev.* **26**, 87–105.

Salli, E., Aronen, H.J., Savolainen, S., Korvenoja, R., and Visa, A. (2001). Contextual clustering for analysis of functional MRI data. *IEEE Trans. Med. Imaging* **20**, 403–414.

Scheibel, A.B., Paul, L.A., Fried, I., Forsythe, A.B., Tomiyasu, U., Wechsler, A., Kao, A., and Slotnick, J. (1985). Dendritic organization of the anterior speech area. *Exp. Neurol.* **87**, 109–117.

Schleicher, A., Amunts, K., Geyer, S., Morosan, P., and Zilles, K. (1999). Observer-independent method for microstructural parcellation of cerebral cortex: A quantitative approach to cytoarchitectonics. *NeuroImage* **9**, 165–177.

Schmidt, D.M., George, J.S., and Wood, C.C. (1999). Bayesian inference applied to the electromagnetic inverse problem. *Hum. Brain Mapp.* **7**(3), 195–212.

Schormann, T., and Zilles, K. (1997). Limitations of the principal axes theory. *IEEE Trans. Med. Imaging* **16**, 942–947.

Schormann, T., and Zilles, K. (1998). Three-dimensional linear and non-linear transformations: An integration of light microscopical and MRI data. *Hum. Brain Mapp.* **6**, 339–347.

Schormann, T., Dabringhaus, A., and Zilles, K. (1995). Statistics of deformations in histology and improved alignment with MRI. *IEEE Trans. Med. Imaging* **14**, 25–35.

Schormann, T., Henn, S., and Zilles, K. (1996). A new approach to fast elastic alignment with application to human brains. *Proc. Visual. Biomed. Comput.* **4**, 337–342.

Schormann, T., Dabringhaus, A., and Zilles, K. (1997). Extension of the principal axis theory for the determination of affine transformations. In *Proc. of the DAGM: Informatik aktuell.* (Anonymous), pp. 384–391. Springer, Berlin.

Simonds, R.J., and Scheibel, A.B. (1989). The postnatal development of the motor speech area: A preliminary study. *Brain Lang.* **37**, 43–58.

Sowell, E.R., Mattson, S.N., Thompson, P.M., Jernigan, T.L., Riley, E.P., and Toga, A.W. (2001). Mapping callosal morphology and cognitive correlates: Effects of heavy prenatal alcohol exposure. *Neurology* **57**(2), 235–244.

Spitzer, V.M., and Whitlock, D.G. (1992). High resolution imaging of the human body. *J. Biol. Photogr. Assoc.* **60**(4), 167–172.

Strothers, S.C., Anderson, J.R., Xu, X.-L., Low, J.-S., Boar, D.C., and Rottenberg, D.A. (1994). Quantitative comparisons of image registration techniques based on high-resolution MRI of the brain. *J. Comput. -Assisted Tomogr.* **18**, 954–962.

Talairach, J., and Tournoux, P. (1988). Principe et technique des études anatomiques. In *Co-Planar Stereotaxic Atlas of the Human Brain—3-Dimensional Proportional System: An Approach to Cerebral Imaging.* Thieme Medical, New York.

Thompson, P.M., and Toga, A.W. (1996). A surface-based technique for warping 3-dimensional images of the brain. *IEEE Transa. Med. Imaging* **15**(4), 1–16.

Thompson, P.M., and Toga, A.W. (1997). Detection, visualization and animation of abnormal anatomic structure with a deformable probabilistic brain atlas based on random vector field transformations. *Med. Image Anal.* **1**(4), 271–294.

Thompson, P.M., and Toga, A.W. (2000). Warping strategies for intersubject registration. In *Handbook of Medical Image Processing* (I. Bankman, ed.). Academic Press, San Diego, CA.

Thompson, P.M., Schwartz, C., and Toga, A.W. (1996a). High-resolution random mesh algorithms for creating a probabilistic 3D surface atlas of the human brain. *NeuroImage* **3**, 19–34.

Thompson, P.M., Schwartz, C. Lin, R.T., Khan, A.A., and Toga, A.W. (1996b). 3D statistical analysis of sulcal variability in the human brain. *J. Neurosci.* **16**(13), 4261–4274.

Thompson, P.M., MacDonald, D., Mega, M.S., Holmes, C.J., Evans, A.C., and Toga, A.W. (1997). Detection and mapping of abnormal brain structure with a probabilistic atlas of cortical surfaces. *J. Comput.-Assisted Tomogr.* **21**(4), 567–581.

Thompson, P.M., Moussai, J., Khan, A.A., Zohoori, S., Goldkorn, A., Mega, M.S., Small, G.W., Cummings, J.L., and Toga, A.W. (1998). Cortical variability and asymmetry in normal aging and Alzheimer's disease. *Cereb. Cortex* **8**(6), 492–509.

Thompson, P.M., Giedd, J.N., Woods, R.P., MacDonald, D., Evans, A.C., and Toga, A.W. (2000a). Growth patterns in the developing brain detected by using continuum mechanical tensor maps. *Nature (London)* **404**, 190–193.

Thompson, P.M., Mega, M.S., and Toga, A.W. (2000b). Disease-specific brain atlases. In *Brain Mapping: The Disorders* (A.W. Toga, J.C. Mazziotta, and R.S.J. Frackowiak, eds.), pp. 131–177. Academic Press, San Diego, CA.

Thompson, P.M., Woods, R.P., Mega, M.S., and Toga, A.W. (2000c). Mathematical/computational challenges in creating deformable and probabilistic atlases of the human brain. *Hum. Brain Mapp.* **9**(2), 81–92.

Toga, A.W., and Arnicar-Sulze, T.L. (1987). Digital image reconstruction for the study of brain structure and function. *J. Neurosci. Methods* **20**, 7–21.

Toga, A.W., Goldkorn, A., Ambach, K., Chao, K., Quinn, B.C., and Yao, P. (1997). *Post mortem* cryosectioning as an anatomic reference for human brain mapping. *Comput. Med Image Graphics* **21**(2), 131–141.

Turkletaub, P.E., Eden, G.F., Jones, K.M., and Zeffiro, T.A. (2002). Meta-analysis of the functional neuroanatomy of single-word reading: Method and validation. *NeuroImage* **16**, 765–780.

Ungerleider, L.G., and Desimone, R. (1986). Cortical connections of visual area MT in the macaque. *J. Comp. Neurol.* **248**, 190–222.

Van Essen, D.C., and Drury, H.A. (1997). Structural and functional analyses of human cerebral cortex using a surface-based atlas. *J. Neurosci.* **17**(18), 7079–7102.

Van Essen, D.C., Drury, H.A., Joshi, S., and Miller, M.I. (1998). Functional and structural mapping of human cerebral cortex: Solutions are in the surfaces. *Proc. Natl. Acad. Sci. U.S.A.* **95**(3), 788–795.

Van Horn, J.D., and Gazzaniga, M.S. (2002). Opinion: Databasing fMRI studies towards a 'discovery science' of brain function. *Nat. Rev. Neurosci.* **3**, 314–318.

von Economo, C., and Koskinas, G.N. (1925). *Die Cytoarchitektonik der Hirnrinde des erwachsenen Menschen.* Springer, Wien/Berlin.

Wang, J.Z., Wiederhold, G., and Firschein, O. (1997). System for screening objectionable images using Daubechies' wavelets and color histograms, interactive distributed multimedia systems and telecommunication services. *Proc. 4th Eur. Workshop (IDMS'97)*, LNCS 1309.

Watson, J.D., Myers, R., Frackowiak, R.S., Hajnal, J.V., Mazziotta, J.C., Shipp, S., and Zeki, S. (1993). Area V5 if the human brain from a combined study using positron emission tomography and magnetic resonance imaging. *Cereb. Cortex* **3**, 79–94.

Wertheim, S.L. (1989). The brain database: A multimedia neuroscience database for research and teaching. *Proc. 13th Annu. Symp. Comput. Appl. Med. Care,* pp. 399–404.

Woods, R.P., Cherry, S.R., and Mazziotta, J.C. (1992). A rapid automated algorithm for accurately aligning and reslicing positron emission tomography images. *J. Comput.-Assisted Tomogr.* **16**, 620–633.

Woods, R.P., Mazziotta, J.C., and Cherry, S.R. (1993). MRI-PET registration with automated algorithm. *J. Comput.-Assisted Tomogr.* **17**(4), 536–546.

Yee, S.H., and Gao, J.H. (2002). Improved detection of time windows of brain responses in fMRI using modified temporal clustering analysis. *Magn. Reson. Imaging* **20**, 17–26.

Zeki, S., Watson, J.D.G., Lueck, C.J., Friston, K.J., Kennard, C., and Frackowiak, R.S.J. (1991). A direct demonstration of functional specialization in human visual cortex. *J. Neurosci.* **11**, 641–649.

Zhou, Y., and Toga, A.W. (1999). Efficient skeletonization of volumetric objects. *IEEE Trans. Vision Comput. Graphics* **5**(3), 196–209.

Zihl, J., von Cramon, D., and Mai, D. (1983). Selective disturbance of movement vision after bilateral brain damage. *Brain* **106**(2), 313–340.

Zihl, J., von Cramon, D., Mai, D., and Schmid, C. (1991). Disturbance of movement vision after bilateral posterior brain damage. Further evidence and follow up observations. *Brain* **114**, 2235–2252.

Zijdenbos, A.P., and Dawant, B.M. (1994). Brain segmentation and white matter lesion detection in MR images. *Crit. Rev. Biomed. Eng.* **22**(5–6), 401–465.

Zijdenbos, A.P., Evans, A.C., Riahi, F., Sled, J.G., Chui, H.-C., and Kollokian, V. (1996). Automatic quantification of multiple sclerosis lesion volume using stereotaxic space. *Proc. 4th Int. Conf Visual. Biomed. Comput VBC '96.*, Hamburg, pp. 439–448.

Zilles, K., Schleicher, A., Langemann, C., Amunts, K., Morosan, P., Palomero-Gallagher, N., Schormann, T., Mohlberg, H., Bürgel, U., Steinmetz, H., Schlaug, G., and Roland, P.E. (1997). Quantitative analysis of sulci in the human cerebral cortex: Development, regional heterogeneity, gender difference, asymmetry, intersubject variability and cortical architecture. *Hum. Brain Mapp.* **5**(4), 218–221.

Overcoming Challenges to Sharing Neuroimagery

Kenneth Smith

CONTENTS

Databasing the Brain. Edited by Stephen H. Koslow and Shankar Subramaniam
ISBN 0-471-30921-4 © 2005 John Wiley & Sons, Inc.

24.1 INTRODUCTION

The great promise of scientific data sharing is being met with growing interest. Empowered by the unprecedented Internet-driven opportunity to make datasets widely available, online sky surveys (Malik et al., 2003; Gray and Szalay, 2002) and genetic databases (NCBI, 2003; UCSC, 2003) have enabled researchers worldwide to analyze datasets collected by others, to produce new derived datasets, and to perform novel and massive collaborative analyses. Recognizing this potential, funding agencies specifically encourage data sharing especially for data which is unique and "cannot readily be replicated" (NIH, 2003), such as human neuroimagery.

The promises of shared human neuroimagery are especially appealing, including overcoming high intersubject variability through aggregated atlas representations (Van Essen and Drury, 1997; Toga and Mazziotta, 2000; Mazziotta et al., 2001), extending the utility of meticulously acquired datasets, making matched normal images widely available, and providing broader access to notable datasets. Additionally, large shared databases would enable novel means of *exploring* data: Given thousands of annotated structural MRI, a researcher could dynamically pose volumetric or morphometric queries to determine the mean and variance of anatomic features over demographic subpopulations.

However, despite successes in other fields and its clear potential, a "big science" approach currently remains elusive in neuroscience: massive amounts of data remain exclusively in the laboratory of origin, and heavily populated high-traffic public repositories do not yet exist on the same scale as in these other fields.

An extensive discussion in the literature has addressed this dilemma (Gardner et al., 2003; Kennedy, 2003; Koslow, 2002; Toga, 2002; Fox and Lancaster, 2002; Van Horn and Gazzaniga, 2002; Editorial, 2003; Koslow, 2000). Although perspectives are diverse, agreement is emerging on two significant data sharing challenges faced by the neuroimagery community:

1. Neuroimagery incurs complex and costly *metadata requirements*;

2. Significant *custodial responsibilities* must be fulfilled by data owners who share neuroimages.

Our focus in this chapter is on these two major challenges, and how they may be addressed. Both are being addressed by ongoing research efforts which should provide useful solutions for researchers.

Note that other important challenges to the sharing of neuroimagery have been described as well, such as the lack of a system rewarding the "publishing" of data into a shared public database. We do not diminish the importance of these challenges by our focus in this chapter; indeed, these are but two facets of larger interrelated whole. We believe this chapter's focus provides a useful context and foundation for the understanding and discussion of *all* facets of the topic of sharing neuroimagery.

24.1.1 Overview of Chapter

We begin with an example data sharing scenario. This is followed by an overview of neuroimagery's challenging metadata requirements and their impact on data sharing. These requirements have been the focus of significant research attention (Neu et al., 2003; Martin et al., 2001; Gardner et al., 2001; Gupta et al., 2000; Nadkarni et al., 1999) which has lead to many new tools and frameworks to enable the transfer of data from one community to another so it is well-understood by both. We describe the problem involved and give a brief overview of these efforts. The second challenge, the custodial responsibilities of data owners, has begun to be addressed by very recent research (Smith et al., 2003), which we present in some detail in the form of the Structured Sharing Communities (SSC) data sharing model and its implications for implementation architectures and assurance.

24.1.2 Example Data Sharing Scenario

Consider an Alzheimers study, conducted at Laboratory *A* by principal investigator PI_A. The protocol involves 25 Alzheimers subjects; each receiving 4 structural MRI scans per year for 4 years (400 scans total). In addition, each subject is given a standard physical exam and a battery of neuropsychological tests to measure cognitive function prior to each scan. Scanner parameters are similarly recorded for each scan. The goal of the study is to build a fine-grained, time-based anatomic atlas of disease progression. All neuroimages are shared freely within Lab *A* for processing and analysis.

Data sharing progresses as illustrated in Figure 24.1. After 3 years, it becomes clear that the data being collected is well-suited to an analysis performed at Lab *B*, run by PI_B, a colleague of PI_A. PI_B is contacted, and terms of collaboration are agreed upon: all current and future data from this study (i.e., images and their metadata, including updates such as new clinical notes) will be shared exclusively with members of Lab *B*, and the results will be jointly published.

After 5 years the dataset is completely collected and important publications have resulted. PI_A decides to entertain further collaborations, even with unfamiliar labs. He therefore sends an announcement to ALZ, an Alzheimers research community to which he belongs, inviting collaborative inquiries. This announcement includes: a succinct description of the dataset (e.g.,

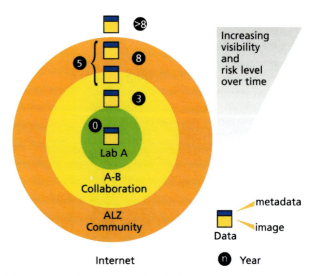

Figure 24.1 Progression of data sharing.

age range, scan protocol), a list of key publications, and his contact information. PI_A discusses each potential sharing arrangement with the corresponding PI to gain confidence that the new use is consistent with the original intent, that the data will be correctly interpreted, that fair attribution will be given, and that it will be handled appropriately (i.e., shared no further) for the duration of the study. When a new agreement is reached, the entire dataset is shared with the new collaborator as before.

In the 8th year, these further collaborations and publications are complete and the dataset is well known to the ALZ community. To further Alzheimer's research in general, PI_A decides to share the entire (de-identified) dataset with members of the ALZ community, but with nobody else.

PI_A also teaches pathology, and decides to develop a online tutorial illustrating disease progression in Alzheimers as a public resource to pathology students worldwide. Assume 10 specific images in the dataset crisply capture specific stages of Alzheimers, and that these subjects gave informed consent to broadly release their de-identified images in this case. Accordingly, these 10 images and a very limited subset of their metadata are incorporated into the tutorial and published on the Internet.

This example illustrates several aspects of data sharing. First, sharing can range in scope from local sharing with a specific partner to unrestricted public sharing. Second, the scope of sharing often increases with time. Third, metadata accompanies images to make them interpretable in their new context, but this information must be filtered in many situations (e.g., to preserve subject privacy). This example also raises some difficult questions illustrative of characteristic challenges associated with data sharing:

- Choice of metadata. *Which* tests should be administered, and which scan parameters should be used, given that this data will ultimately be shared? For example, in a Year 5 sharing agreement with previously unknown collaborator PI_X, did PI_A collect the metadata required for PI_X's proposed study?

- Interpretation of metadata. Will shared metadata terms be understood and correctly interpreted in their new shared environment?

- Expressing sharing policy. How can PI_A unambiguously express, communicate, and evolve the data sharing policy embodied in this scenario? For example, how does he state that all members of the ALZ community, but nobody else, can see data in Year 8?

- Enacting sharing policy. Once expressed, how is this sharing policy enacted without a heroic effort by PI_A? For example, how can he keep track of all the new Year 5 collaborations, and exactly what data is being shared with each?

Questions such as these currently impact data owners, and can limit data sharing. However, significant efforts are addressing these challenges as well.

24.2 BACKGROUND: METADATA AND DATA SHARING

Large quantities of metadata are typically associated with neuroimagery to enable its scientific interpretation. In this section, we give an overview of the metadata requirements of neuroim-

agery, how they interact with data sharing, and survey some technical solutions.

24.2.1 Metadata Requirements for Neuroimagery

For our purposes, we define three major categories of neuroimagery metadata: metadata describing the subject, the image acquisition event, and (for functional studies) the experimental design. High variability in the neuroanatomy and functional activation of human subjects necessitates a detailed characterization of individuals, including descriptors such as age, gender, handedness, race, medical history, and neuropsychiatric test results. family background, habits (e.g., addictives taken and their frequency), and genotype. Similarly, the diverse and growing array of imaging techniques necessitates a description of the image acquisition event, including descriptors for modality, resolution, scan date and time, software version, imaging and behavioral protocols, and file format. This metadata is recorded in a file format (e.g., DICOM, MINC, ANALYZE) providing standard fields names describing both the subject and the image acquisition.

For functional imagery, the experimental design must also be described so others can interpret the regions of activation. Important details include: a description of the stimulus, the time signature of that stimulus (i.e., when a switch is made from an *off* state to an *on* state, or the sequence of stimuli in more complex cases), and the correlated frequency of image acquisition. Important details such as these are frequently recorded in published papers [which are conveniently cited as metadata in functional neuroimagery databases, as in (Fox and Lancaster, 2002; Van Horn, 2001)]. Note that minute details can be quite important for interpreting the results. For example, letters on the screen in a Stroop test occupy an angle of the visual field (e.g., 5 degrees, 20 degrees), depending on their size and the subject's distance from the screen, which affects the degree to which the visual cortex is activated.

When considered together, a large amount of information has been assembled. Repositories have been reported with as many as 8700 descriptive metadata attributes (personal communication: Alex Zijdenbos of the Montreal Neurological Institute), although most are not this large. By contrast, data in more heavily shared fields has less comprehensive metadata requirements. For example, genetic sequence data can have a large scientific impact *without* individuating subject metadata (e.g., the *normal allele*) due to relatively low intersubject variability, and specific details of how the sequence data was aquired are not typically considered vital for its subsequent interpretation.

24.2.2 Sharing Interactions and Solutions

The sharing of neuroimagery is strongly influenced by the quantity of necessary metadata. First, capturing, entering, formatting, labeling, archiving, and finally transferring each field (and the associated image) are labor-intensive tasks. Second, metadata heterogeneity between the donor and receiver must be overcome, including reconciling the units, format, label or identifier, and semantics (Seligman and Rosenthal, 2001; Smith and Obrst, 1999) of each item, and relationships among items. The more metadata required, the larger this problem. Third, studies vary in their goals and thus in their specific metadata requirements, requiring reconciliation of mismatches between the fields collected by the *donor* and the fields needed by the *receiver* in the sharing relationship. The more metadata required, the more extensive such a mismatch is likely to be. This mismatch can be magnified by the loss of context which occurs when data is shared. When data leaves its laboratory of origin, it often leaves behind much implicit knowledge, such as local scanner artifacts and how they were addressed. Unless such information is made explicit, it is not conveyed and can be lost (Baringa, 2003).

Numerous tools and techniques exist which address these challenges to data sharing. Software tools minimize the labor involved in collecting and formatting large volumes of metadata. Software provided by scanner manufacturers will fill in image header fields. Information about the scanner and date/time are automatically filled in, while metadata about both the subject and the protocol are acquired through an interview. Similarly, software versions of neurocognitive tests (e.g., Stroop, handedness) administer tests and record the results. Such tools will continue to be refined and improve over time, significantly easing the acquisition of metdata. Important areas to be addressed include: the collection of metadata relevant to research (as opposed to clinical) users, customizing software for local requirements (e.g., Stroop in Canadian French), and the ability to automatically insert collected information into a database.

One common strategy for overcoming heterogeneity is to establish community standards. As illustrated in Figure 24.2, if *N* partners mutually share data, they can choose pairwise reconciliations, or they can agree on a community standard. The latter case only results in a number of reconciliations linearly proportional to the size of the community (each partner reconciles their local data with the standard), instead of proportional to the square of the community size, in the case of pairwise reconciliations.

A popular standarization strategy is to publish a reference for metadata terms, or an *ontology*, defining a set of labels, units, values, relationships, semantics, etc. Sharing partners then agree to map their local data representation (i.e., file format, database schema) to the common ontology to facilitate translation among their data sets. Software applications can improve interoperability by using terms defined in a common ontology. For example, if

a set of databases use a common definition for the spatial boundary of the anatomic term "amygdala," a query for the average volume of the amygdala can be expected to return comparable results from these databases. Ideally, the ontology would also provide guidance to new efforts seeking to define local data representations, and to efforts seeking to integrate disparate existing databases.

A number of current research efforts are actively developing ontologies, associated tools, and larger ontological frameworks relevant to neuroimagery. One example is the NeuroNames ontology (Martin et al., 2001) which addresses the need to standardize anatomic terminology. First published in 1995, by its third version it contained over 550 "primary" brain structures, hierarchical containment relationships to super- and substructures, unique abbreviations for each structure, more than 4000 synonyms, links to illustrations for 500 structures, and indications of the similarities and differences among primate species. NeuroNames itself has also been incorporated into the larger UMLS (unified medical language system) of the National Library for Medicine. Thus, beyond serving as a standard reference for terms and their definitions, ontologies can become a public (i.e., shared) repository for a rich variety of knowledge. Another ambitious neuroanatomic ontology project is being undertaken (Gupta et al., 2000), currently focusing on cellular neuroanatomy. Beyond containment relationships, many other types of relationships such as control and regulation are being recorded to capture detailed knowledge about both structure and function. The goal is to enable automated inference [e.g., If we know a Purkinje cell is a neuron, and neurons have axons, the system can infer that Purkinje cells have axons (Ludäscher et al., 2001)] by mediation tools.

Another, less obvious, source of standardization is an individual database schema. Database schemas are a template, defining the structure, labels, and format of the information within a particular database. As a database schema matures through years of use, user feedback, and practical refinement, it becomes a valuable artifact because of the domain knowledge it encodes. For example, consider an effort to database functional MRI data, which has spent considerable time and effort representing experimental design information. The associated schema encodes many lessons learned about good choices for specific metdata attributes, their labels and formats, useful groupings of attributes, and relationships among these groupings. It thus can serve as a valuable reference to others wishing to standardize aspects of experimental design (e.g., they can see which metadata attributes were chosen and what names they were given). Although schemas do not encode as much knowledge as an ontology, they have the advantage of directly enabling the definition of a database.

Some research focuses on neuroscience database schema design. A common data model is offered (Gardner et al., 2001) which provides high-level data structures to help guide, and standardize, neuroscience schema definition. Another project (Nadkarni et al., 1999) has applied and specialized the EAV (entity-attribute-value) modelling strategy to neuroscience schema design. The EAV model captures very heterogeneous and complex data within a relatively simple database schema. Note also that a growing body of research in the computer science community is addressing the problem of automatically reconciling database schemas (Rahm and Bernstein, 2001;

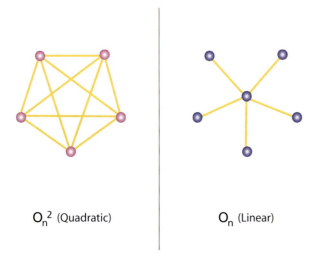

$O_n{}^2$ (Quadratic) O_n (Linear)

Figure 24.2 Reconciling heterogeneous data.

Madhavan et al., 2001; Doan et al., 2001; Doan et al., 2003), and that much of this research is applicable to neuroimagery schemas.

Despite these advantages, standardization can incur a high overhead in certain situations. In a dynamic field like neuroscience, new and evolving knowledge is fact of life. Thus even standards are subject to changes, affecting potentially every sharing partner. In addition, when only two partners are involved, it is cheaper to perform the single reconciliation than to define (or locate) an appropriate standard and reconcile both partners to that standard. Thus, pairwise reconciliation can be fairly lightweight and often makes good sense.

The pairwise strategy is greatly enhanced by automating the process of reconciling heterogeneity between donor and receiver. An example of this strategy is the Neurosys project (Pittendrigh and Jacobs, 2003), which is building semistructured database software to simplify and speed the construction of compatible neuroimagery database systems to support peer-to-peer file exchanges. Another example addresses the conversion of image headers. Note that each standard format differs from the others, making it challenging to simply "swap files." In fact, the *same* header format is frequently used differently by different scanners. This problem becomes quite evident when automated image analysis tools are chained together into *pipelines* (Toga et al., 2001), in which each tool expects a particular input format and produces a particular output format. When formats differ, the chain breaks. A tool called the *Debabeler* [Neu et al., 2003] has been developed which helps automate conversion among formats through a *conversion engine*, and an associated graphical mapping interface. Once a mapping has been defined, it can be embedded in automated conversion tools and used repeatedly. For an excellent treatment of the general issue of image header heterogeneity, see (Neu et al., 2003).

Finally the problem of mismatches between the fields collected by the *donor* and needed by the *receiver* could be addressed by simply attempting to aquire missing fields. Subjects can sometimes be reinterviewed, but frequently such information is simply lost. A more promising solution is to synchronize (standardize) image acquisition protocols. Standardized protocols are used by consortiums, such as the International Consortium for Brain Mapping (ICBM) and the Biomedical Informatics Research Network (BIRN). Each member agrees to acquire the same basic set of terms when doing local acquisitions (even if they are not directly relevant to local research) to facilitate reuse and combination within the consortium. To further facilitate reuse, common values for scanner settings, pulse sequences, and other acquisition paramenters are used to the greatest extent possible. Sources of variance persist (e.g., lack of funding to aquire certain metadata, differences in scanner capabilities) so the realistic goal is to reduce, instead of eliminate, variance in the set of metadata collected. Nonetheless, standard image acquisition protocols show great promise in making datasets comparable and interchangable.

24.3 BACKGROUND: THE CUSTODIAL RESPONSIBILITIES OF DATA OWNERS

Data owners face numerous *custodial responsibilities* when they share their data, including ensuring their data correctly inter-

preted in its new shared context, protecting the identity and privacy of human research participants, and safeguarding the understood order of use so researchers may "benefit from first use of their data" (NIH, 2003). The inability, or perceived inability, to perform any of the above responsibilities in the context of data sharing forms another type of barrier to data sharing which must be addressed.

These responsibilities affect how widely data may be shared at a given time. As the data sharing scenario illustrates, this appropriate scope of sharing changes with time. Note that data sharing does not typically begin only with the first publication, various types of limited-scope sharing occur long before this (e.g., collaborations in which one partner provides data and another provides analysis in order to produce publishable results). In addition, sharing is not always "total" (i.e., all image data and metadata fully released to anyone in the world) upon publication, due, for example, to legal or institutional review board (IRB) restrictions on subject privacy.

These responsibilites are strongly influenced by the aforementioned metadata requirements of neuroimagery. Incorrect understanding of the metadata can lead to a misinterpretation of the scientific relevance of the associated image. Data owners must therefore ensure metadata terms are sufficiently accurate, clear, and complete. For example, if PI_A saves effort by recording only certain aspects of the image acquisition in a local shorthand, the data may be fully understood locally, but future studies may not find it useful or may misinterpret it.

In addition, the need for significant amounts of individuating metadata increases concerns for the privacy of human subjects, and therefore the increases the responsibility of data owners to attend to issues of subject privacy when sharing data. This responsibility has been amplified by recent legislation, such as the Health Information Portability and Accountability Act (HIPAA) of 1996, and its subsequent Privacy Rule (effective April 14th, 2001). Finally, the large amount of skilled labor invested to produce exquisitely detailed metadata increases a dataset's intangible "worth" to its owner, making it more important to benefit from its first use and making the dataset more desirable as a target of theft. These responsibilities cause the owners of neuroimagery to take an intentional, and at times circumspect, approach to data sharing.

The problem of enabling data owners to fulfill their custodial responsibilities in the context of data sharing is only beginning to be addressed by research. At the core of this problem is the lack of a means for data owners to control exposure to the risks of sharing data while obtaining its benefits. We define a data owner's *sharing intent* to be a statement of the current intended scope of visibility for a particular dataset. Currently, a data owner has trouble expressing his or her sharing intent unless it is somewhat binary: share publicly and with little restriction, or not at all. In fact a data owner's sharing intent is likely to be much more complex, involving increasing degrees of visibility to specific communities over time and with the occurrence of events (such as publication). Where finer gradations of sharing visibility are currently possible, (e.g., peer to peer sharing via email or custom tools) there is currently little assurance the sharing intent will be respected once data leaves its laboratory of origin.

Stated succinctly, custodial responsibilities can be supported by two new capabilities. First, a greater *range* of sharing options is needed. Specifically, a widely-understood means for a data owner to express an appropriate range of sharing intent. Second, tools able to recognize and enforce that intent are needed. In lieu of such support, data owners who find themselves with little "say" about the visibility of their shared data may choose not to share, rather than accepting the risks of unrestricted sharing.

24.3.1 Expression of Sharing Intent

How should data sharing intent be expressed? What "language" should be used? The following are four requirements. First, it is critical that sharing intent be clearly recorded and interpretted by humans; ambiguity reduces its utility. Second, the language of sharing intent must be so easy to use as to be nearly invisible, or it is likely to fall into disuse because it affects daily work in a laboratory. Third, for sharing intent to be enforced by software systems, it must be in some unambiguous machine-readable format. Finally, the language must be able to easily reference arbitrary *communities* of individuals. Recall that PI_A's sharing intent is expressed in exactly this manner: data access provided to various communities at various times. We use the term *community of interest* (COI) (Renner, 2001), to describe a well-defined group of individuals sharing an interest in a specific task and in associated data. The language of sharing intent must provide the COI as a fundamental concept; without it we would lack a means to quantify the abstract notion of "data visibility."

Since sharing intent arises conceptually with the data owner, it could be recorded directly in a human language, such as English. However, this does not qualify as unambiguous and machine-readable; in fact natural language is frequently ambiguous for humans!

On the other extreme, computer scientists have recently developed numerous highly expressive software languages for expressing data access policy; these are capable of defining roles, permissions, prohibitions, duties, and the resolution of conflicts among these (Sandhu et al., 1996; Jajodia and Wijesekera, 2002). With their rich set of features, these languages could be adapted to express sharing intent. However, many of these languages are logic-based, requiring significant mathematical training to understand and correctly use. Thus, they do not qualify as easy to use for humans. Nor do they include the COI concept.

A promising approach which meets all the above criteria is a simple, yet powerful, language of sharing *labels*, as shown in Bell and LaPadula (1974), Denning (1976), and The Oracle Corporation (2003). In this methodology, COIs are represented by labels. Both individuals and data are assigned to COIs, and thus given a label. A data access request by an individual for a data item is arbitrated by simple rules, based on the relationship of their labels. Importantly, labels are readily understood by both humans *and* by software tools, enabling the automation of much of the labor in data sharing and the enforcement of sharing intent. This approach is used in the Secure Sharing Communities model described in Section 24.4.

24.3.2 Risk Management

A data owner's perception of risk is critical to data sharing, thus it is useful to briefly consider risk management strategies in the context of data sharing. Information security research describes three general strategies: *risk assumption*, *risk avoidance*, and *risk management* (Robinette and Marshall, 2001). Risk assumption treats risks as relatively inconsequential and shares information openly and without protective measures, while risk avoidance assumes risks are unassumable and the data owner expends great effort to remove all possibility of risk before sharing information. In contrast to these, risk management sees a tradeoff between the danger of realized risks and the benefits of realized sharing as in (Smith et al., 1996). PI_A exemplifies risk management by intentionally evolving his sharing intent with time so as to continually reap the benefits of sharing while simultaneously controlling exposure to risk. Because both the benefits and risks of sharing are undeniable, we believe risk management is the appropriate strategy for sharing neuroimagery.

Fundamental to the success of risk management is cataloging known risks. Through informal interviews with neuroimagery data owners, we have developed a list of the risks in sharing neuroimagery, summarized in Figure 24.3. These risks may be grouped into four broad categories: legal repercussions, time and effort, data misuse, and data theft. This list forms the core of an online survey at *neuroinformatics.mitre.org* whose intent is to clearly determine the data sharing threats perceived by neuroimagery researchers, and thereby serve as an online resource for software designers and granting agencies. (The results of this survey will be displayed once a statistically significant number of people respond. Readers are encouraged to participate).

As this list illustrates, data sharing risk is correlated with the community (people) with which data is shared. Some communities entail more risk than others; some communities provide a safe haven, insulating data from risk due to shared goals. Thus, risk can be managed by intentionally controlling the exposure of data to specific communities, as described in the following text.

24.4 A NOVEL DATA SHARING MODEL FOR NEUROIMAGERY

In the following, we describe a model for sharing neuroimagery, *Structured Sharing Communities* (SSC) (Smith et al., 2004), which provides well-defined intermediate levels of data visibility and supports the development of related data sharing tools. These intermediate visibility levels represent labeled *communities of interest* (COIs) defined by laboratories, collaborative agreements, consortium memberships, research organizations and other affliliations involving shared data. Labeled COIs are structured relative to each other into a *policy space*, which makes explicit the permissable paths of information flow (data sharing),

Within a policy space, a data owner expresses sharing intent by simply assigning their data a label. This can be done easily with a drag-and-drop graphical user interface (GUI), as illustrated in Figure 24.4. The operations of the SSC model, hidden from users by such GUIs, enable researchers and informatics applications to manage the visibility of their data and enforce access permissions and restrictions.

This model, through its Fairness property (described in the following text), directly addresses the concerns that "sharing should not imply relinquishing" and that "a data sharing

Legal Repercussions

1) Shared data is found to violate privacy laws, resulting in a financial penalty or incarceration.

Time and Effort

2) After sharing data, you are beseiged with questions and challenges.

3) You attempt to share data, but the effort to acquire, configure, and use sharing tools is too high.

4) Attempts to share are bogged down ensuring compliance with detailed sharing regulations.

Misuse

5) Data you share is altered into an incorrect form (e.g. through error or malicious hacking).

6) Data you share is misunderstood and improperly reanalyzed.

7) A study subject is denied coverage after an insurance company accesses his brain image.

8) Imagery is used for purposes opposed by the subject and PI (bioterror research, racist websites).

9) Imagery is used for purposes not originally envisioned by the subject (another study).

Theft

10) A rival researcher acquires your pre-publication data and "scoops" you.

11) Data you share is reused and published without proper citation.

12) Your entire carefully acquired database is downloaded without permission.

13) A commercial venture profits from your shared neuroimagery without compensating you.

14) A discgruntled lab worker de-anonymizes subject scans and posts them on the internet.

Figure 24.3   Risk examples.

policy should include safeguards against reuse of data without recognition of the original investigator" (Gardner et al., 2003). This model also supports proposals for rewarding data sharing (Koslow, 2002; Gardner et al., 2003) by quantifying the scope of sharing (e.g., the current location of data within a policy space), and distinguishing types of sharing. In addition, this model addresses the issue of sharing reciprocity reported in Dalton, 2000: "Scientists who reported refusing to share data were more than twice as likely to be the victims of data withholding as those who had not." by providing a neutral framework in which data sharing relationships can grow over time.

The SSC model has been adopted as a standard for the International Consortium for Brain Mapping (ICBM), and is being adapted for general use with other types of data (i.e., non-neuroimagery). The MITRE corporation is developing open source SSC-compliant "sharing-aware" neuroinformatics tools which can be deployed in a laboratory to manage and share neuroimagery.

24.4.1 Data Structures

The fundamental structure of the SSC model is the COI, Reasonable COI types for neuroimagery include: a data owner, a laboratory, a collaboration, and a research organization. In the example, lab A, lab B, the AB collaboration, and the Alzheimers research community ALZ are all COIs. Membership in one COI (e.g., lab A) may entail membership in another (e.g., the AB collaboration), however, we assume each individual has a unique "home" COI (e.g., their lab). We also assume each data object has a unique "home" COI. The COI is defined as follows:

Definition 1 (COI). Let $\mathcal{U}$ be the set of all users and $\mathcal{O}$ be the set of all data objects. A community of interest C consists of $\{U_C, O_C\}$ where

1. $U_C \subseteq \mathcal{U}$, *such that all users in U_C have C as their home COI;*
2. $O_C \subseteq \mathcal{O}$, *such that all objects in O_C have C as their home COI.*

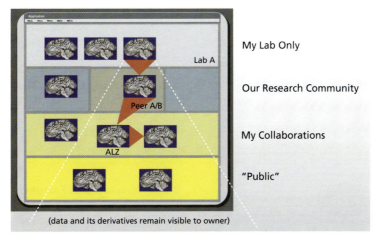

Figure 24.4   Administrative GUI for data sharing.

Note that $\mathcal{U}$ includes both human users and tools or applications which perform data access operations. Although the term *subject* is used in the security literature; we resort to *user* instead, to avoid a conflict with the *subject* studied in a scientific protocol. A user has access to all objects in their home COI. Within a COI, we assume data sharing furthers the goals of the community, and thus membership in a COI implies legitimate rights to access its data. Therefore, COI membership must be determined with certainty; we assume each member of a COI can prove that membership (e.g., by providing a password or membership number) when challenged.

A set of related COI's can be arranged into a *policy space*, as illustrated in Figure 24.5, in which explicit relationships between COIs denote paths of permissible data flow (sharing). When a data owner assigns a particular set of COI labels to data objects, this maps them onto a policy space, expressing the data owner's sharing intent. We use the term "policy space" because it defines a possibly large set of options (i.e., any assignment of data items to COIs) for expressing data sharing policy.

Definition 2 (Policy Space). A policy space P consists of a complete partial order (CPO) $(\mathcal{C}, \rightarrow, PUBLIC)$, where

1. $\mathcal{C}$ *is a finite set of COIs (i.e., labels);*
2. $\rightarrow \subseteq \mathcal{C} \times \mathcal{C}$ *forms a partial order over $\mathcal{C}$, the "shares to" relationship;*
3. $PUBLIC \in \mathcal{C}$, *and is its lower bound.*

The partial order $\rightarrow$ structures the COIs in a policy space so shared data *only* flows along explicit paths, from one trusted community (e.g., a lab) to another; ensuring shared data does not flow unaccounted-for along unpredictable paths. In a partial order, an element *dominates* the set of elements which includes itself and every other community reachable below it. Thus, in Figure 24.5, Lab A dominates itself, the AB collaboration, research group 1, and $PUBLIC$. A partial order ensures that, for any two distinct elements X and Y, either X dominates Y, Y dominates X, or X and Y are incomparable. As required for a partial order, $\rightarrow$ is reflexive: $C_1 \rightarrow C_1$, transitive: $(C_1 \rightarrow C_2) \wedge (C_2 \rightarrow C_3) \Rightarrow (C_1 \rightarrow C_3)$, and antisymmetric (acyclic): $(C_1 \rightarrow C_2) \wedge (C_2 \rightarrow C_1) \Rightarrow C_1 = C_2$. In graphic representation, such as Figure 24.5,

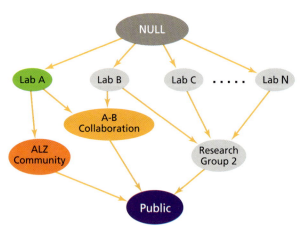

Figure 24.5   An example SSC policy space.

we follow the convention of omitting reflexive and transitive relationships, since they are readily inferred and paths of information flow are evident without them.

Every policy space contains two special elements of $\mathcal{C}$: *NULL* and *PUBLIC*, such that for any other COI $C_i \in \mathcal{C}$, $NULL \rightarrow C_i \rightarrow PUBLIC$. *NULL* is the least visible (most private) COI; it has zero users and zero data. *PUBLIC* is the most visible COI; it is accessible by all users, and corresponds well to our intuition about the visibility of an unrestricted website. The amount of data in *PUBLIC* at a given time is a function of previously executed sharing actions. All COIs in $\mathcal{C}$ represent graded intermediate degrees of visibility between *NULL* and *PUBLIC*. The existence of *PUBLIC* in $\mathcal{C}$ ensures $\rightarrow$ is a *complete* partial order, or CPO.

Note that the structure of labels in label-based methodologies has taken the form of a mathematical lattice (Denning, 1976); for data sharing purposes, we generalize this into a complete partial order (Winskel, 1993) which permits the realistic scenario of two organizations having multiple distinct concurrent data sharing agreements.

Finally, we associate with each data object O three *sharing* attributes:

- $O.coi$ is O's COI, its (current) position in the policy space.
- $O.owner$ is the data owner, with custodial responsibilities for O.
- $O.released$ is a boolean flag declaring O exempt from Fairness constraints (defined below).

In addition, each user U has an attribute $U.coi$, denoting its home COI.

24.4.2 Constructing a Neuroimagery Policy Space

A policy space, as just defined, could be constructed for *many* information domains (not just neuroimagery) by choosing the COIs and their relationships appropriately. To define a *neuroimagery* policy space, as in Figure 24.5, we provide the following guidelines.

In a neuroimagery policy space, we assign all COIs (other than *NULL* and *PUBLIC*) a type belonging to the set: {*Data owner, Lab, Collaboration, Research org*}. COIs of type *Data owner* are always immediate children of *NULL*, and contain only one person whose data is generated there. Data migrates from *Data owner* COIs down into the other types of COIs. COIs of type *Lab, Collaboration*, and *Research org* have the obvious semantics. Since they may share into COIs of their own type, hierarchical structures are permitted (e.g., labs and sublabs, research organizations and their parent organizations). Collaborations are always dominated by individual labs and other collaborations. Finally, any COI can share into *PUBLIC*, which models an unrestricted website.

24.4.3 Privacy and Fairness

Two principles, Privacy and Fairness, constrain data access within a policy space $P = (\mathcal{C}, \rightarrow, \text{PUBLIC})$.

Definition 3 (Privacy). Let $C_1, C_2 \in \mathcal{C}$, and let U_i be a user such that $U_i.coi = C_1$. U_i may read objects in C_2 if and only if $(C_1 \rightarrow C_2)$.

Privacy states that users may only read data belonging to dominated COIs, but all other read access is prohibited. As illustrated in Figure 24.5, data in laboratories A and B is private to those labs; however, data shared into the AB collaboration is mutually visible to users in both A and B, and remains inaccessible to people in laboratory C.

Definition 4 (Fairness). Let O_i be an object such that $O_i.coi = C_1$. O_i itself, and any derivative of O_i, may be written into C_2 if and only if $(C_1 \rightarrow C_2)$.

Fairness states that objects, and their derivatives, may only be written into dominated COIs, but all other write access is prohibited. This has several important effects:

- It prevents users from "unfairly" coopting data shared with them by copying it into a private COI.
- It ensures any new COI to which a data object is assigned will be more public than the last, corresponding to the intuition that data is shared with larger and larger communties over time.
- It ensures data owners can always see what has happened to their data, and to its derivatives.

We permit Fairness to be overridden to support the sharing of special reference data, for example, a commonly used anatomic atlas of the human brain (Holmes et al., 1998; Talairach and Tournoux, 1988). Even if such an atlas were placed in the *PUBLIC* COI, making it universally visible, other researchers could not use it for private research in their local COI. Under Fairness, all research results derived using this atlas must remain in *PUBLIC*, making them universally visible as well! We therefore permit data owners to relinquish their Fairness rights and unrestrictedly share certain data by marking it "released" (i.e., $O.released = \texttt{true}$), as explained in more detail as follows.

24.4.4 Operations

The following six operations provide a useful interface to data objects which is consistent with the principles of Privacy and Fairness.

Testing Dominance
The most fundamental operation in the SSC model is `dominates()`.

- *yesno* $\leftarrow$ `dominates`$(U.coi, O.coi)$.

For user U invoking any data access operation `A()`, both Privacy and Fairness require `dominates`$(U.coi, O.coi)$ be `true` for all objects O input to `A()` or generated by `A()`'s execution. For example, an image viewer must test for the dominance of the user invoking it over the object being viewed. (We also assume a tool acts with the COI of the user invoking it.)

Data Generation Operations
The two operations `new()` and `derive()` generate a new data object, initializing its sharing attributes.

- $O \leftarrow$ `new(data)`
- $O_N \leftarrow$ `derive`$(O_1, O_2, \ldots O_{N-1})$

`New()` generates new object O, based on the input data. and intializes: $O.coi \leftarrow U.coi$, $O.owner \leftarrow U$, and $O.released \leftarrow$ `false`. `New()` could be called by a data entry application adding new neuroimages to a data store.

`Derive()` returns a new object O_N based on one or more input objects $\{O_1, O_2, \ldots O_{N-1}\}$, initializes $O_N.owner \leftarrow U$, $O_N.released \leftarrow$ `false`, and $O_N.coi \leftarrow C_i$, where C_i is some COI dominated by $U.coi$. `Derive()` could be called by data analysis algorithms (e.g., normalization, probabilistic atlas generation) producing new images from one or more input images.

Note that `derive()` makes the simplifying assumption, more stringent than Fairness, that all input and output objects must reside in the *same* COI, C_i. If necessary data exists in a more private COI, it must first be shared downward into C_i; if necessary data exists in a more public COI, it must first be privatized (explained below) upward into C_i. This requirement corresponds to the intuition that a COI is the locus of shared work and data. To illustrate, let collaboration AB exist to produce an atlas from 30 input images. Let 10 images reside in lab A, 10 in lab B, and 10 be unreleased public data. The following steps produce the atlas; note that operations `share()` and `release()` are defined in Section 24.4.4.3.

1. The data owners in labs A and B execute `share()` (with *lowerCOI* $= AB$) for their 10 images.
2. The owner of the images in *PUBLIC* executes `release()` on those 10 images.
3. An investigator in AB executes `privatize()` (with *higherCOI* $= AB$) for the 10 images in *PUBLIC*.
4. Finally, the investigator executes `derive()` on the 30 images now in AB.

The resultant atlas resides in AB, owned by the investigator.

Data Sharing Operations
The following three operations administer the sharing status a data object, updating its sharing attributes.

- `share`$(O, lowerCOI)$.
- `release`(O).
- $O' \leftarrow$ `privatize`$(O, higherCOI)$

`Share()` moves O into a lower (more visible) COI, setting $O.coi \leftarrow lowerCOI$. `Share()` can only be executed by $O.owner$, and $O.coi$ must dominate $lowerCOI$. We restrict the execution of `share()` and `release()` to data owners. In the case of a neuroimage with HIPAA restrictions, clearly, only the data owner should be able to make sharing decisions.

`Release()` sets $O.released \leftarrow$ `true`, enabling a much broader mode of sharing. Specifically: nonowners may subsequently execute the `privatize()` operation below if O is released, with the result that O may be copied into a COI invisible to (i.e., not dominated by) its owner.

`Privatize()` migrates the information in *released* object O up the policy space, violating Fairness, by making a new private copy O', initializing: $O'.coi \leftarrow higherCOI$, $O'.owner \leftarrow O.owner$, and $O'.released \leftarrow$ `true`. $U.coi$ must dominate *higherCOI*, and *higherCOI* must dominate $O.coi$.

HigherCOI is "private" because Privacy may render it invisible to *O.owner*. *O′* can be used locally and further privatized, however, it retains original ownership. Retention of orginal ownership ensures "credit" is correctly assigned to a released neuroimage in its new privatized context (even though that context may be invisible to the original owner). Ownership retention also prevents others from resharing *O′* as if they had originated the image.

24.4.5 An Implementation

For a set of collaborating organizations in which users employ informatics tools (e.g., image viewers, analysis algorithms, databases) to access data, Privacy and Fairness are enforced in an *implementation* of the SSC model. More specifically: let *P* be a policy space, and let $\mathcal{U}$ be the set of all users whose home COI is in *P*, and let $\mathcal{O}$ be the set of all data objects in some COI of *P*.

Definition 5 (Implementation). A implementation I is a set of informatics tools T mediating access by $\mathcal{U}$ to $\mathcal{O}$, such that the following hold:

1. *User $U_i \in \mathcal{U}$ may only access object $O_j \in \mathcal{O}$ if* dominates $(U_i.coi, O_j.coi)$ *is* true.

2. *The data sharing attributes of O_j may only be initialized by* new(), derive(), *or* privatize() *and modified by* share() *or* release().

The second condition ensures changes made to $\mathcal{O}$ are consistent with Privacy and Fairness.

24.5 DISCUSSION OF IMPACT ON USERS

This model has, in general, a minimal impact on users. Specifically, $\mathcal{U}$ may be partitioned into three groups, each of which is impacted differently by an implemented SSC model.

- $\mathcal{U}_1$: Users who only interact with their home COI.
- $\mathcal{U}_2$: Users who use data from other COIs, but do not share their own data beyond their COI.
- $\mathcal{U}_3$: Users who share their data into other COIs, and therefore have custodial responsibilities.

Members of $\mathcal{U}_1$ experience little impact from the SSC model. COI labels on data, and calls to dominates() by their tools can be hidden. They require no knowledge of the policy space structure, or even of its existance. Members of $\mathcal{U}_2$ must understand Privacy and the local structure of the policy space in order to recognize which COIs they may access, and they must understand the use of privatize() to acquire released data. This does not seem to be a large impact, given the corresponding benefit of increased data access.

Members of $\mathcal{U}_3$, and their custodial responsibilities, are the focus the SSC model, and correspondingly require a more comprehensive understanding. They administer the sharing status of their data through share() and release(), positioning it appropriately for other users. This *data administration* task may be performed intuitively, however, as illustrated by the drag-and-drop graphical user interface (GUI) in Figure 24.4. The display

shows PI_A from our example the current sharing state of all his images. The levels depict, respectively: data in Lab *A*'s COI, data shared into the collaborations of Lab *A*, data shared with the research organizations Lab *A* participates in, and *PUBLIC* data. By dragging and dropping along the arrows, PI_A shares (i.e., the share() operation is executed by the GUI) an image into the *AB* collaboration, and then later into research Group *ALZ* where another image is derived from it. The cone illustrates how Fairness ensures PI_A's data, and its derivatives, remain visible to their owner.

We finally note that the policy space itself must also be administered. As sharing agreements begin and end, the policy space must be updated to reflect these changes through the addition, deletion, or reorganization of COIs and their relationships. This task is beyond the scope of this chapter.

24.6 DISCUSSION OF DATA SHARING ARCHITECTURES

Can an implementation of the SSC model be realized in actual laboratory computing systems? And if so, how? A data sharing architecture consists of a mapping of the components of an implementation (e.g., COIs, data) onto physical localities and systems (e.g., lab databases, webservers, neuroimagery analysis tools). We assume any such architecture includes a "policy space server" component, containing an internal representation of the policy space, and to which all dominates() queries are sent. For illustration, we compare two architectures representing spectral extremes: a centralized architecture relying on a single central database, and a distributed architecture relying on a peer-to-peer network of laboratory databases at separate sites. We also describe a key architectural component under development at the MITRE Corporation for open source distribution.

A centralized architecture is illustrated on the left side of Figure 24.6, consisting of a single central database, which also acts as the policy space server. A provider [e.g., (Van Horn, 2001)] may administer the central database on behalf of data owners. In our example, PI_A uploads image *O* to such a central database where new() is performed, intializing: $O.coi = A$,

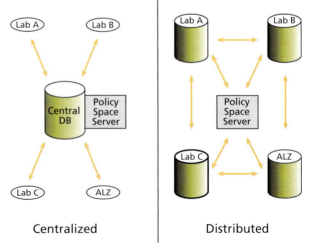

Figure 24.6   Example data sharing architectures.

O.owner = *PI_A*, and *O.released* = `false`. Users requesting access to *O* at the central database are subjected to a `dominates()` test, which will only succeed for members of *A*. If *PI_A* subsequently executes `share(O, AB)`, members of *B* (though still not *C*) become able to access *O*. Note that *O* has not moved physically, it has been reassigned to a more visible COI at the same physical location. This feature of the SSC provides a central database with significantly more utility, because it can manage both public and private data simultaneously.

A distributed architecture is illustrated on the right side of Figure 24.6, consisting of a local database *DB_X* for each community *X* and a policy server. We assume each *DB_X* supports remote login from other sites, and data transfers among sites. *PI_A* stores image *O* in his *local* database *DB_A*, instead of a centralized one. However, `new()` initializes *O*'s sharing attributes to the same values as above, enabling access for *A*, but not (remote) access for members of *B* or *C*. Once *PI_A* executes `share(O, AB)` locally in *DB_A*, members of *B* (but not *C*) can log remotely in and access *O*.

Although access patterns are similar, important differences exist. A distributed architecture offers greater autonomy to each site, since each administers their own database and local tools may be used on local data. But this autonomy also brings the price of increased operating costs. Multiple inconsistent copies of data may also exist in a distributed architecture. Let a member of *B* copt *O* (now belonging to the *AB* COI) from *DB_A* to *DB_B* for local use. If *PI_A* subsequently shares *O* with the *ALZ* community, the `share()` operation must be executed on *all* copies of *O* (not simply the local one) to maintain consistency. The problem is equivalent to the *mobile policy update* problem, addressed in Smith et al., 2002.

We anticipate many laboratories will participate in a distributed fashion: managing data in their local database, and sharing data with other labs doing the same. This will require the use of tools in each local laboratory which implement the SSC model by making calls to SSC operations (and the policy space server) at the appropriate points. However, it is unlikely that laboratories will want to develop such software themselves. To address

this, the MITRE Corporation is developing informatics tools for open source distribution to manage neuroimages locally in an SSC-compliant manner. Labeled *Sharing-Aware MRI Management Platform* in Figure 24.7, this software will perform standard data management functions for structural MRI, and later for other modalities, including: store, search, report, and analyze (through an integrated pipeline). It will also provide interfaces to administer the sharing status of owned data, and to mediate all accesses to local data consistent with its owner's sharing intent, whether that owner is local or remote.

24.7 DISCUSSION OF ASSURANCE

Assurance is confidence that the implemented model will perform as expected in actual practice. Assurance risks can be classified into three categories: risks derived from the model itself, risks due to an incorrect implementation of the model, and risks due to security vulnerabilities in a (correct) implementation of the model. Assurance is a detailed study, and we do not treat these categories exhaustively in the following. Instead, we raise central and illustrative issues for each category, and discuss reasonable counter-measures by which these, and other similar risks, may be addressed.

24.7.1 Model-Derived Risks

Can the SSC model itself be exploited, even when correctly and securely implemented, to circumvent secure sharing? Restated, are there exploitable flaws in SSC? In our example, let *PI_A* share *O* downward from COI *A* into the COI *AB* collaboration. Then let a unscrupulous member of *B* perform (within the *AB* COI): $O_{copy} \leftarrow$ `derive(O)`, where `derive()` simply copies *O* into O_{copy}, unchanged; O_{copy} now resides in *AB* and contains the same data as *O*. Note that O_{copy} is owned by the unscrupulous *B* lab member, because `derive()` assigns ownership of the result to the user executing it, who could share the data contained in

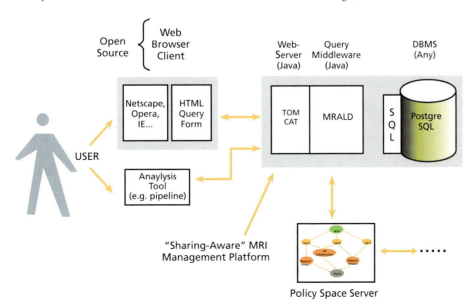

USER

Open Source { Web Browser Client

Web-Server (Java)   Query Middleware (Java)   DBMS (Any)

Netscape, Opera, IE...   HTML Query Form

TOM CAT   MRALD

SQL   Postgre SQL

Anaylysis Tool (e.g. pipeline)

"Sharing-Aware" MRI Management Platform

Policy Space Server

Figure 24.7   Open source sharing-aware software.

O more widely than PI_A would wish, and even release it! This does not violate Privacy or Fairness, but it *does* circumvent a data owner's control of the visibility of their data.

The SSC model could have been designed to prevent this, but in ways which introduce new problems. For example, if `derive()` preserved original ownership, a researcher legitimately deriving new data would be unable to share research results! *Compound* ownership would have unwieldy side-effects, such as the need to vote on sharing decisions among all (possibly many in the case of a large-N atlas) owners. We therefore choose to accept this risk as part of the SSC model and address its mitigation in two ways. First, each COI can securely audit every use of the `derive()` operation, including the time invoked, object(s) invoked on, the associated derivation algorithm, and the user invoking `derive()`. Thus, any abuse would be recorded and traceable. The risk of being caught attempting to steal a collaborator's shared data (and thereby losing the opportunity to participate in this and possibly many other collaborations) is likely sufficient to significantly diminish the threat. Second, the model itself provides a solution by enabling incremental exposure to communities of interest and their associated risks: if concern exists that members of a particular COI will exploit `derive()` in this way, one may simply choose to defer sharing their data with that COI until such concerns diminish.

24.7.2 Risks Due to An Incorrect Implementation of the Model

Developers may, through error or omission, incorrectly implement the model. This concern motivated the relatively simple design of the SSC model. An important subcase of omission is a site (e.g., lab) which does not implement the SSC model *at all*, yet wishes to collaborate and share in a distributed architecture. Certainly, a data owner at a *nonconformant* site can send data to *conformant* site implementing the model, where the data can be labeled and relabeled according to the wishes of the owner. However, tools at a nonconformant site cannot be trusted to respect the labels on any data sent to that site. To achieve full sharing reciprocity, each site must fully and correctly implement the SSC model. Enabling such widespread conformance to the SSC model is one goal of the sharing-aware software development project described in Section 24.6.

24.7.3 Risks Due To An Insecure Implementation of the Model

If the model were risk-free and its implementation correct, what assurance concerns would still exist? The remaining risks are due to attacks which exploit vulnerabilities to *circumvent* the model, as do traditional hacker attacks, for which the standard solution is to "harden" the implementation against its vulnerabilities. As an example, what if a vulnerability permits someone to alter the label of images stored at a site? Since sharing intent is embedded in labels, the entire model can be circumvented in this way. Such a need for stronger label integrity could be addressed by the use of cryptographic hashing (Menezes et al., 1997). Whenever object O is created or shared, we compute value V as the hash of $O.coi$ and a secret string S, and then store V and $O.coi$

securely (e.g., locally, or at the policy space server). Whenever O is accessed, V' is computed based on O's *current* label. If $V' = V$, O has not been tampered with. If $V' \neq V$, the original label can be recovered and reapplied. A similar technique can be used to protect O as it is being transported between sites.

24.8 FUTURE DIRECTIONS OF THE SSC MODEL

A number of issues remain to be explored, such as managing the evolution of a policy space. Clearly, COIs will appear and disappear: many IRBs approve collaborative agreements on a year-by-year basis. Rules are required to define valid updates to the policy space and to reassign or remove data belonging to a deleted COI. Rules are also needed to integrate two independently derived policy spaces.

Another important issue is the notion of a composite data object, in which components of the larger object (e.g., subject metadata, the actual image, an obscured version of the image) may be visible to different communities, as described in Toga, 2002. An extension of the SSC model which manages the integrity of such a multi-COI object will be valuable.

This research clarifies the importance of the concept of data ownership. The SSC model of ownership favors simplicity, however, as illustrated in Section 24.7.1, there are alternatives worth considering. Instead of having a single owner through its entire lifetime, an image might be owned by multiple persons at a given time or by different persons at different times. Ownership could also be held by individuals or by organizations (e.g., a journal, a corporation, or a data center). The concept of origination (who should be cited when this image is used) could also be separated from ownership (the right to share the image). Such variations merit further exploration.

Finally, from the computer science perspective, this model raises interesting questions concerning its interaction with existing mechanisms for ensuring security and privacy, such as traditional discretionary access controls.

24.9 CONCLUSIONS

The sharing of neuroimagery holds great promise, and has numerous facets. In this chapter we have addressed two of these: neuroimagery's substantial metadata requirements, and the need for data owners to fulfill certain custodial responsibilites. By focusing on qualities of the data itself, and on the needs of the key players (data owners), we believe this chapter provides a useful foundation for discussing aspects of data sharing we have not covered in any depth here, such as legal and financial issues, and the need for simplicity and ease-of-use in sharing tools.

Methodologies being developed to address these two challenges have been described. These include:

- Much-needed tools supporting automated metadata capture.
- Research efforts providing a diverse set of solutions for the problem of data heterogeneity, via standardization and rapid pairwise reconciliation.
- Tools automating data conversions.

- Standard data acquisition protocols which help reduce the variance in datasets, even where research goals differ.

- New research enabling data to be shared at intermediate levels of visibility over time.

- Sharing-aware tools, and data sharing architectures.

- An analysis of assurance in data sharing.

These efforts evidence an active and growing interest in neuroinformatic data sharing, and are providing enablers for its important goals.

ACKNOWLEDGMENTS

This work was supported by a grant from the Human Brain Project (R01-MH64417-01), which is funded by the National Institute for Mental Health and the National Science Foundation, and by funds from the MITRE Technology Program (project 51MSR203). We are indebted to *numerous* individuals who shared insights about the current practice, benefits, challenges, and risks of sharing neuroimagery. Annual meetings of investigators from the Human Brain Project and the International Consortium on Brain Mapping have been valuable forums for discussions as well. Sushil Jajodia, Vipin Swarup, and Donald Faatz helped develop the SSC model; Jeff Hoyt and Todd Cornett manage software development. Jolynn Smith provided helpful insights regarding sequence data, and Arnon Rosenthal provided valuable feedback on the label-based strategy.

References

Baringa, M. (2003). Still debated, brain image archives are catching on. *Science*, **300**:43–45.

Bell, D. E. and LaPadula, L. J. (1974). Secure Computer Systems: Mathematical Foundations and Model. Technical report, MITRE Corporation.

Dalton, R. (2000). Young, worldly and unhelpful all miss out on data sharing. *Nature*, 404:6.

Denning, D. (1976). A lattice model of secure information flow. *Communications of the ACM (CACM)*, **19**(5);236–243.

Doan, A., Domingos, P., and Halevy, A. (2001). Reconciling Schemas of Disparate Data Sources: A Machine Learning Approach. In *Proceedings: ACM SIGMOD*, pages 509–520, Santa Barbara.

Doan, A., Domingos, P., and Halevy, A. (2003). Learning to match schemas of databases: A multistrategy approach. *Machine Learning Journal*, **50**:279–301.

Editorial (2003). Neuroscience in the post-genome era. *Nature Neuroscience*, **6**(1):1.

Fox, P. T. and Lancaster, J. L. (2002). Mapping context and content: the brainmap model. *Nature Reviews Neuroscience*, **3**(4):319–321.

Gardner, D. et al. (2001). Common data model for neuroscience data and data model exchange. *Journal of the American Medical Informatics Association (JAMIA)*, **8**(1):17–33.

Gardner, D., Toga, A., et al. (2003). Towards effective and rewarding data sharing. *Neuroinformatics*, **1**(3):289–296.

Gray, J. and Szalay, A. (2002). The world-wide telescope. *Communications of the ACM*, **45**(11):50–55.

Gupta, A., Ludäscher, B., and Martone, M. (2000). Knowledge-based integration of neuroscience data sources. In *Proceedings of the 12th International IEEE Conference on Scientific and Statistical Database Management*, Berlin.

Holmes, C., Hoge, R., Collins, L., Woods, R., Toga, A., and Evans, A. (1998). Enhancement of MR images using registration for signal averaging. *Journal of Computer Assisted Tomography (JCAT)*, **22**(2):324–333.

Jajodia, S. and Wijesekera, D. (2002). Recent advances in access control models. In Olivier, M. and Spooner, D., editors, *Database and Application Security XV*, pages 3–15. Kluwer Academic, Boston.

Kennedy, D. N. (2003). Share and share alike. *Neuroinformatics*, **1**(3):211–213.

Koslow, S. H. (2000). Should the neuroscience community make a paradigm shift to sharing primary data? *Nature Neuroscience*, **3**(9):863–865.

Koslow, S. H. (2002). Sharing primary data: A threat or asset to discovery? *Nature Reviews Neuroscience*, **3**(4):311–313.

Ludäscher, B., Gupta, A., and Martone, M. (2001). Model-based mediatiohn with domain maps. In *Proceedings of the 17th International Conference on Data Engineering (ICDE)*, Heidelberg.

Madhavan, J., Bernstein, P., and Rahm, E. (2001). Generic Schema Matching Using Cupid. In *Proceedings: VLDB 2001*, pages 49–58, Rome.

Malik, T., Szalay, A., Budavari, T., and Thakar, A. (2003). Skyquery: A web service approach to federate databases. In *Proceedings of the 2003 Conference on Innovative Data Systems Research (CIDR)*, pages 188–196, Monterey, CA.

Martin, R., Mejino, J., Bowden, D., Brinkley, J., and Rosse, C. (2001). Foundational model of neuroanatomy: Implications for the human brain project. In *Proceedings of the AMIA Annual Fall Symposium*, pages 438–432, Washington, D.C.

Mazziotta, J., Toga, A., Evans, A., Fox, P., et al. (2001). A four-dimensional probabilistic atlas of the human brain. *Journal of the American Medical Informatics Association (JAMIA)*, **8**(5):401–430.

Menezes, A. J., van Oorschot, P. C., and Vanstone, S. A. (1997). *Handbook of Applied Cryptography*. CRC Press, Boca Raton.

Nadkarni, P., Marenco, L., Chen, R., Skoufos, E., Shepherd, G., and Miller, P. (1999). Organization of heterogeneous scientific data using the EAV/CR representation. *Journal of the American Medical Informatics Association (JAMIA)*, **6**:478–493.

NCBI (2003). Human genome resources. Online databases (OMIM, LocusLink, RefSeq); available at *www.ncbi.nim.nih.gov/genome/guide/human*.

Neu, S., Valentino, D., Oullette, K., and Toga, A. (2003). Managing multiple medical image file formats and conventions. In *Proceedings of the SPIE Medical Imaging Conference: PACS and Integrated Medical Information Systems*. In press.

NIH (2003). NIH policy on data sharing. Available at *grants2.nih.gov/grants/policy/data_sharing/*.

Pittendrigh, C. and Jacobs, G. (2003). Neurosys: A semi-structured database. *Neuroinformatics*, **1**(2):167–176.

Rahm, E. and Bernstein, P. (2001). On matching schemas automatically. *VLDB Journal*, **10**(4).

Renner, S. (2001). A community of interest approach to data interoperability. In *Proceedings of the Federal Database Colloquium*, San Diego.

Robinette, J. and Marshall, J. (2001). An integrated approach to risk management and risk assessment. *INCOSE Insight*, **4**(1):23–30.

Sandhu, R., Coyne, E., Feinstein, H., and Youman, C. (1996). Role-based access control models. *IEEE Computer*, **29**(2):38–47.

Seligman, L. and Rosenthal, A. (2001). XML's impact on database and data sharing. *IEEE Computer*, **34**(6):59–67.

Smith, K., Blaustein, B., Jajodia, S., and Notaragiacomo, L. (1996). Correctness criteria for multilevel transactions. *IEEE Transactions on Knowledge and Data Engineering (TKDE)*, **8**(1):32–45.

Smith, K., Fayad, A., Faatz, D., and Jajodia, S. (2002). Propagating modifications to mobile policies. In *Proceedings of the 17th International Coinference on Information Security (IFIP/SEC2002)*, Cairo.

Smith, K., Jajodia, S., Swarup, V., Hoyt, J., Hamilton, G., Faatz, D., and Cornett, T. (2004). Enabling the sharing of Neuroimaging Data Through Well-defined Intermediate Levels of Visibility. *NeuroImage* **22**, 1646–1656.

Smith, K. and Obrst, L. (1999). Unpacking the semantics of source and usage to perform semantic reconciliation in large-scale information systems. *SIGMOD Record*, **28**(1):26–31.

Talairach, J. and Tournoux, P. (1988). *Co-planar Stereotaxic Atlas of the Human Brain*. Thieme, New York.

The Oracle Corporation (2003). Oracle9i label security fact sheet. Available at *otn.oracle.com/products/oracle9i/datasheets/OLS9iR2_ds.html*.

Toga, A. (2002). Neuroimage databases: The good, the bad, and the ugly. *Nature Reviews Neuroscience*, **3**(4):302–309.

Toga, A. and Mazziotta, J. (2000). *Brain Mapping: The Systems*. Academic Press, San Diego.

Toga, A., Rex, D., and Ma, J. (2001). A graphical interoperable processing pipeline. *NeuroImage Abstract*, **13**(6:S266).

UCSC (2003). Genome bioinformatics. Reference sequences (human and *C. elegans*) and browsers, available at *genome.cse.ucsc.edu*.

Van Essen, D. and Drury, H. (1997). Structural and functional analyses of the human cerebral cortex using a surface-based atlas. *Journal of Neuroscience*, **17**:7079–7102.

Van Horn, J. (2001). The functional magnetic resonance imaging data center (fMRIDC): The challenges and rewards of large-scale databasing of neuroimaging studies. *Philosophical Transactions of the Royal Society of London: Biological Sciences*, **356**(1412):1323–1339.

Van Horn, J. and Gazzaniga, M. S. (2002). Databasing fMRI studies – towards a "discovery science" of brain function. *Nature Reviews Neuroscience*, **3**(4):314–318.

Winskel, G. (1993). *The Formal Semantics of Programming Languages*. MIT Press, Cambridge.

Probabilistic Brain Atlases of Normal and Diseased Populations

Arthur W. Toga, Ph.D., Paul M. Thompson, Ph.D.,
Katherine L. Narr, Ph.D., and Elizabeth R. Sowell, Ph.D.

CONTENTS

Databasing the Brain. Edited by Stephen H. Koslow and Shankar Subramaniam
ISBN 0-471-30921-4 © 2005 John Wiley & Sons, Inc.

25.1 INTRODUCTION

Comprehensive maps of brain structure have been derived, at a variety of spatial scales, from three-dimensional (3D) tomographic images (Damasio, 1995), anatomic specimens (Talairach and Szikla, 1967; Talairach and Tournoux, 1988; Ono et al., 1990; Duvernoy, 1991) and a variety of histologic preparations that reveal regional cytoarchitecture (Brodmann, 1909), myelination patterns (Smith, 1907), protein densities, and mRNA distributions. Most early atlases of the human brain were derived from one, or at best a few, individual *post mortem* specimens (Brodmann, 1909; Schaltenbrand and Bailey, 1959; Schaltenbrand and Wahren, 1977; Talairach and Szikla, 1967; Matsui and Hirano, 1978; Talairach and Tournoux, 1988; Ono et al., 1990). Such atlases take the form of anatomical references or represent a particular feature of the brain (Van Buren and Maccubin, 1962; Van Buren and Borke, 1972), such as a specific neurochemical distribution (Mansour et al., 1995) or the cellular architecture of the cerebral cortex (Brodmann, 1909).

Beyond the anatomic atlases based on postmortem and histologic material mentioned above, the application of magnetic resonance to acquire detailed descriptions of anatomy in vivo is a driving force in brain mapping research. MRI data have the advantage of intrinsic three-axis registration and spatial coordinates (Damasio, 1995) and have the ability to image a variety of structural signatures including tissue type and white matter tracts. Unfortunately, even high-resolution MR atlases, with up to 100–150 slices, a section thickness of 1 mm, and 256^2 pixel imaging planes (Evans et al., 1991; Lehmann et al., 1991), still result in resolutions lower than the complexity of many neuroanatomic structures. However, advances in the technology continue to push improvements in spatial and contrast resolution. A recent innovation in the collection of atlas quality MRI involves the averaging of multiple co-registered scans ($N = 27$) from a single subject to overcome the lack of contrast and relatively poor signal to noise (Holmes et al., 1998).

Other brain atlases map function, quantified by positron emission tomography (PET; Minoshima et al., 1994), functional MRI (Le Bihan, 1996) or electrophysiology (Avoli et al., 1991; Palovcik et al., 1992). Maps also have been developed to represent neuronal connectivity and circuitry (Annese et al., 2003; Van Essen and Maunsell, 1983; also see Chapter 21, this volume) based on compilations of empirical evidence (Brodmann, 1909; Berger, 1929; Penfield and Boldrey, 1937).

Disappointingly, each of the brain maps contained in these atlases (postmortem and in vivo) has a different spatial scale and resolution, emphasizes different structural or functional characteristics, and is inherently incompatible with the others. While each mapping strategy clearly has its place within a collective effort to map the brain, unless certain integrative approaches are implemented (including spatial normalization), these brain maps will remain as individual and independent efforts, and the correlative potential of the many diverse mapping approaches will be unrealized.

Population-based brain atlases (the focus of this chapter) offer a powerful framework to synthesize the results of disparate imaging studies. These atlases use novel analytical tools to fuse data across subjects, modalities, and time. They detect group-specific features not apparent in individual scans. Population-based atlases (see Chapter 23, this volume) can be stratified into subpopulations to reflect a particular (clinical) subgroup (Toga and Mazziotta, 1996). Design of appropriate reference systems for brain data presents considerable challenges, since these systems must capture how brain structure and function vary in large populations, across age and gender, in different disease states, across imaging modalities, and even across species.

Imaging algorithms are also significantly improving the flexibility of digital brain atlases. *Deformable brain atlases* are adaptable in that they can be individualized to reflect the anatomy of new subjects, and *probabilistic atlases* retain information on cross-subject variations in brain structure and function. These atlases are powerful new tools with broad clinical and research applications (Roland and Zilles, 1994; Kikinis et al., 1996; Toga and Thompson, 1998). Despite the significant challenges in expanding the atlas concept to the time dimension, dynamic brain atlases include probabilistic information on growth rates that may assist research into pediatric disorders (Thompson and Toga, 1999a,b) or degenerative tissue loss rates in aging and dementing diseases (Thompson et al., 2003a,b; Sowell et al., 2003).

Anatomic variations severely hamper the integration and comparison of data across subjects and groups (Meltzer and Frost, 1994; Woods, 1996). Motivated by the need to standardize data across subjects, analytic methods were developed to remove size and shape differences that distinguish one brain from another (Talairach and Tournoux, 1988). Spatially transforming individual brain maps onto a 3D digital brain atlas, removes subject-specific shape variations, and allows subsequent comparison of brain structure or function between individuals (Christensen et al., 1993; Ashburner et al., 1997). Conversely, *deformable brain atlases* are based on the idea that a digital brain atlas can be elastically deformed to fit a new subject's anatomy (Evans et al., 1991; Gee et al., 1993; Christensen et al., 1993; Sandor and Leahy, 1995; Rizzo et al., 1995; Toga and Thompson, 1997; Haller et al., 1997). High-dimensional brain *warping* algorithms (Christensen et al., 1993, 1996; Collins et al., 1994a; Thirion, 1995; Rabbitt et al., 1995; Warfield et al., 1995; Davatzikos, 1996; Thompson and Toga, 1996; Bro-Nielsen and Gramkow, 1996; Gee and Bajscy, 1998; Grenander and Miller, 1998) affect the transfer of 3D maps of structure, function, and other descriptions such as information on cytoarchitecture, histologic, and neurochemical content (Mega et al., 1997) onto the scan of any subject.

The coordinate system used to equate brain topology with an index must include carefully selected features common to all brains. Furthermore, these features must be readily identifiable and sufficiently distributed anatomically to avoid bias. Once defined, rigorous systems for matching or spatially normalizing a brain to this coordinate system must be utilized. This allows individual data to be transformed to match the space occupied by the atlas. In the Talairach stereotaxic system (Talairach and Szikla, 1967; Talairach and Tournoux, 1988), piecewise affine transformations are applied to 12 rectangular regions of brain, defined by vectors from the anterior and posterior commissures to the extrema of the cortex. These transformations re-position the anterior commissure of the subject's scan at the origin of the 3D coordinate space, vertically align the interhemispheric plane, and horizontally orient the line connecting the two commissures. Each point in the incoming brain image, after it is registered into the atlas space, is labeled by an (x, y, z) address indexed to the atlas brain. Although originally developed to help interpret brain stem and ventricular studies acquired using pneumoencephalography (Talairach and Szikla, 1967), the Talairach stereotaxic system rapidly became an international standard for reporting functional activation sites in PET studies, allowing researchers to compare and contrast results from different laboratories (Fox et al., 1985, 1988; Friston et al., 1989, 1991).

Perhaps surprisingly, few atlases of neuropathology use a standardized 3D coordinate system to integrate data across patients, techniques, and acquisitions. Atlases with a well-defined coordinate space (Evans et al., 1992; Friston et al., 1995; Drury and Van Essen, 1997), together with algorithms to align data with them (Toga, 1998), have enabled the pooling of brain mapping data from multiple subjects and sources, including large patient populations. Automated algorithms can then capitalize on atlas descriptions of anatomical variance to guide image segmentation (Le Goualher et al., 1999; Pitiot et al., 2002), tissue classification (Zijdenbos and Dawant, 1994), functional image analysis (Dinov et al., 2000), and pathology detection (Thompson and Toga, 1997, 2000).

Without methods to overcome the problems of anatomic variability, the statistical power to resolve disease and treatment effects is seriously undermined. First, normal anatomical variation results in an overlapping of diseased and normal subjects on most anatomical measures. Second, these difficulties are exacerbated in disease-related change such as atrophy (Meltzer and Frost, 1994; Woods, 1996; Mega et al., 1998; Thompson et al., 1998) or other progressive and dynamic anatomical changes. In the case of the cortex, profiles of gray-matter loss are difficult to calibrate against a reference population, due to the lack of statistics on expected changes in these populations. To fully capitalize on neuroimaging data in disease, an appropriately complex mathematical framework is needed to address these challenges. Once resolved, brain maps can then be compared across patients and across time (Mazziotta et al., 1995; Thompson et al., 1997, 2000d; Grenander and Miller, 1998).

This chapter not only discusses the construction and application of normal population-based atlases but includes descriptions of the concept of disease-specific atlases, designed to reflect the unique anatomy and physiology of a particular clinical subpopulation (Thompson et al., 1997, 1998; Thompson and Toga, 1999a; Mega et al., 1997, 1998, 1999, Narr et al., 1999, 2000). Based on well-characterized patient groups, these atlases contain composite maps and visualizations of structural variability, asymmetry and group-specific differences. This quantitative framework can be used to recognize anomalies and label structures in new patients. Because they retain information on group anatomical variability, disease-specific atlases are a type of probabilistic atlas specialized to represent a particular clinical group. The resulting atlases can identify patterns of altered structure or function, and they can guide algorithms for knowledge-based image analysis (Collins et al., 1994b; Dinov et al., 2000; Pitiot et al., 2002).

We present data from several ongoing projects, whose goal is to create disease-specific atlases of the brain in Alzheimer's disease, schizophrenia, and several neurodevelopmental disorders. Pathological change can be tracked over time, and disease-specific features can be resolved. Rather than simply fusing information from multiple subjects and sources, we describe strategies used to resolve group-specific features not apparent in individual scans.

25.2 THE APPROACH

To create atlases that contain detailed representations of anatomy, rather than utilizing intensity criteria to define structures, we have developed model-driven algorithms that deform them to match the anatomy of new subjects (Thompson and Toga, 1996, 1997, 2000; Toga and Thompson, 1997). Anatomic models provide an explicit geometry for individual structures in each scan, such as landmark points, curves, or surfaces. Because the digital models reside in the same stereotaxic space as the atlas data, surface and volume models stored as lists of vector coordinates are amenable to digital transformation, as well as geometric and statistical measurement (Thompson et al., 1996, 1998, 2000a,b; Mega et al., 1998; Zhou et al., 1999; Narr et al., 2000). The underlying 3D coordinate system is central to all atlas systems, since it supports the linkage of structure models and associated image data with spatially indexed neuroanatomic labels, preserving spatial information and adding anatomical knowledge.

25.2.1 Registration and Tissue Mapping

Three-dimensional volumetric MRI scans are first rotated and scaled to match a standardized brain template in stereotaxic space. This template may be either an average intensity brain dataset constructed from a population of young normal subjects (Mazziotta et al., 2001) or one specially constructed to reflect the average anatomy of elderly subjects (e.g., Thompson et al., 2000e; Mega et al., 2000; see these articles for a discussion of *disease-specific* templates). Once aligned, a measure of the brain scaling imposed is retained as a covariate for statistical analysis. If a given subject has been scanned repeatedly, the same scaling is applied to both baseline and follow-up scans, to ensure that observed differences reflect true brain atrophy. A tissue classification algorithm then splits up the scan into regions representing gray matter, white matter, cerebrospinal fluid (CSF), and nonbrain tissues. Stereotaxic maps of gray matter are retained.

25.2.2 Cortical Pattern Matching

MRI scans have sufficient resolution and tissue contrast, in principle, to track cortical gray-matter loss in individual patients. Even so, extreme variability in gyral patterns confounds efforts to (1) compare this loss against a normative population and (2) determine the average profile of tissue loss in a group (Figure 25.1). *Cortical pattern matching* methods (CPM; detailed further in Figure 25.2) address these challenges. They encode both gyral patterning and gray-matter variation. This can substantially improve the statistical power to localize deficits. These cortical analyses tease apart the effects of gyral shape variation from gray-matter change, and they can also be used to measure cortical asymmetries (Thompson et al., 2001a; Narr et al., 2001; Sowell et al., 2001b).

Briefly, a 3D geometric model of the cortical surface is extracted from the MRI scan, and flattened to a 2D planar format (to avoid making cuts, a spherical topology can be retained; Fischl et al., 2001; Thompson et al., 1997a, 2002). A complex deformation, or warping transform, is then applied that aligns the sulcal anatomy of each subject with an average sulcal pattern derived for the group. To improve feature alignment across subjects,

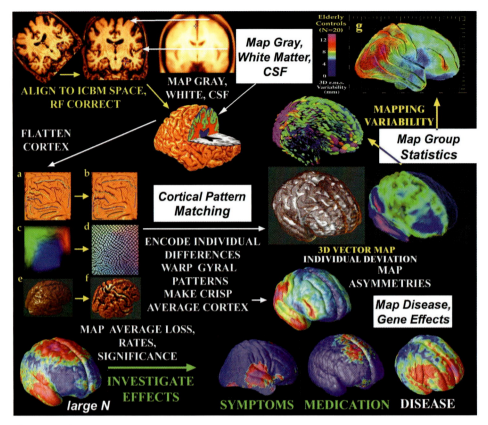

Figure 25.1 Analyzing cortical data. The schematic shows a sequence of image processing steps (Thompson et al., 2003a) that can be used to map how development and disease affect the cortex. The steps include aligning MRI data to a standard space, tissue classification, cortical pattern matching, as well as averaging and comparing local measures of cortical gray-matter volumes across subjects. (These procedures are detailed in the main text). To help compare cortical features from subjects whose anatomy differs, individual gyral patterns are flattened and aligned with a group average gyral pattern (**a–f**). Group variability (**g**) and cortical asymmetry can also be computed. Correlations can be mapped between disease-related gray-matter deficits and genetic risk factors. Maps may also be generated visualizing linkages between deficits and clinical symptoms, cognitive scores, and medication effects. The only steps here that are currently not automated are the tracing of sulci on the cortex. Some manual editing may also be required to assist algorithms that delete dura and scalp from images, especially if there is very little CSF in the subarachnoid space.

all sulci that occur consistently can be digitized (Thompson et al., 1997a; Sowell et al., 2001; Narr et al., 2001) and used to constrain this transformation. As far as possible, this procedure adjusts for differences in cortical patterning and shape, across subjects. Cortical measures can then be compared across subjects and groups. Sulcal landmarks are used as anchors, because homologous cortical regions are better aligned after matching sulci than by just averaging data at each point in stereotaxic space (see, e.g., functional MRI studies by Zeineh et al., 2001, 2003; Rex et al., 2001; Rasser et al., 2003). Given that the deformation maps associate cortical locations with the same relation to the primary folding pattern across subjects, a local measurement of gray-matter density is made in each subject and *averaged across equivalent cortical locations*. To quantify local gray matter, we use a measure termed "*gray-matter density*," used in many prior studies to compare the spatial distribution of gray matter across subjects. This measures the proportion of gray matter in a small region of fixed radius (15 mm) around each cortical point (Wright et al., 1995; Bullmore et al., 1999; Sowell et al., 1999, 2003; Ashburner and Friston, 2000; Rombouts et al., 2000; Mummery et al., 2000; Thompson et al., 2001; Good et al., 2001; Baron

et al., 2001). Given the large anatomic variability in some cortical regions, high-dimensional elastic matching of cortical patterns (Thompson et al., 2000b, 2001) is used to associate measures of gray-matter density from homologous cortical regions first across time and then also across subjects (as shown in Figure 25.2). One advantage of cortical matching is that it localizes deficits relative to gyral landmarks; it also averages data from corresponding gyri, which would be impossible if data were only linearly mapped into stereotaxic space. Annualized 4D maps of gray-matter loss rates within each subject are then elastically realigned for averaging and comparison across diagnostic groups (Figure 25.1). The effects of age, gender, medication, disease, and other measures on gray matter can be assessed at each cortical point (see Figure 25.3 for an example in Alzheimer's disease).

25.2.3 Statistical Maps

An algorithm then fits a statistical model, such as the *general linear model* (GLM; Friston et al., 1995) to the data at each cortical location. This results in a variety of parameters that characterize

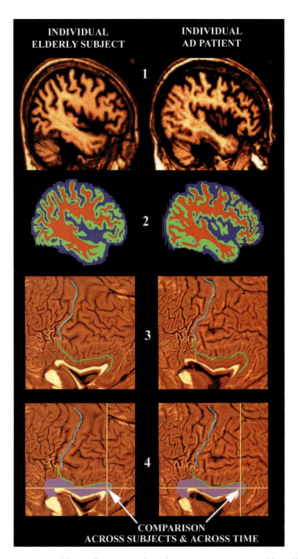

Figure 25.2   Comparing gray matter across subjects. Gray matter is easier to compare across subjects if adjustments are first made for the gyral patterning differences across subjects [data from Thompson et al., 2003a]. This adjustment can be made using cortical pattern matching (Thompson et al., 2000a), which is illustrated here on example brain MRI datasets from a healthy control subject (*left column*) and from a patient with Alzheimer's disease (*right column*). First, the MRI images (*stage 1*) have extracerebral tissues deleted from the scans and the individual pixels are classified as gray matter, white matter, or CSF (shown here in green, red, and blue colors; *stage 2*). After flattening a 3D geometric model of the cortex (*stage 3*), features such as the central sulcus (*light blue curve*) and cingulate sulcus (*green curve*) may be re-identified (http://www.loni.ucla.edu/~e.sowell/edevel/new_sulcvar.html).

An elastic warp is applied (*stage 4*) moving these features, and entire gyral regions (*pink colors*), into the same reference position in "flat space" (cf. van Essen et al., 1997). After aligning sulcal patterns from all individual subjects, group comparisons can be made at each 2D pixel (*yellow cross-hairs*) that effectively compare gray-matter measures across corresponding cortical regions. In this study, the cortical measure that is compared across groups and over time is the amount of gray matter (*stage 2*) lying within 15 mm of each cortical point. The results of these statistical comparisons can then be plotted back onto an average 3D cortical model made for the group, and significant findings can be visualized as color-coded maps. Such algorithms bring gray-matter maps from different subjects into a common anatomical reference space, overcoming individual differences in gyral patterns and shape by matching locations point-by-point throughout cortex. This enhances the precision of intersubject statistical procedures to detect localized changes in gray matter.

how gray-matter variation is linked with other variables. The significance of these links can be plotted as a significance map. A color code can highlight brain regions where linkages are found, allowing us to visualize the strength of these linkages (e.g., Thompson et al., 2001b). In addition, estimated parameters can be plotted, such as (1) the local rates of gray matter loss at each cortical location (e.g., as a percentage change per year), (2) regression parameters that identify disease effects, and even (3) nonlinearities in the rates of brain change over time (e.g.,

quadratic regression coefficients; Sowell et al., 2003). In principle, any statistical model can be fitted, including genetic models that estimate genetic or allelic influences on brain structure (Thompson et al., 2003b). Finally, permutation testing is typically used to ascribe an overall significance value for the observed map. This adjusts for the fact that multiple statistical tests are made when a whole map of statistics is visualized. Patients and controls are randomly assigned to groups, often many millions of times on a supercomputer. A null distribution is built

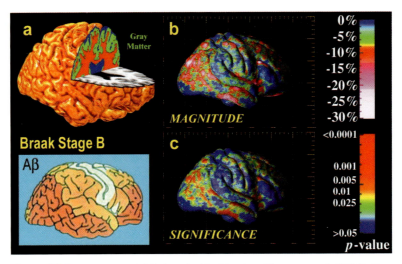

Figure 25.3   Gray-matter deficits in early AD. [Data from Thompson et al., 2003b]. Here the local amount of cortical gray matter (shown in *green colors* (**a**)) is compared across 26 patients with mild to moderate AD (age: 75.8 ± 1.7 years; MMSE score: 20.0 ± 0.9) and 20 matched elderly controls (72.4 ± 1.3 years). At this stage of AD, there is a reduction in gray matter, reaching 30% in the temporoparietal regions (**b**). (**c**) Map of the statistical significance of these deficits. Intriguingly, the pattern of temporal lobe gray-matter loss, seen with MRI, spatially matches the pattern of beta-amyloid (Aβ) deposition seen postmortem. The *inset panel* (Braak Stage B) is adapted from data reported by Braak and Braak (1997). It shows regions with minimal (*white*), moderate (*orange*), and severe (*red*) beta amyloid deposition. Because amyloid deposition and gray-matter loss may not be synchronized, these maps may represent different stages of AD; however, there is a clear spatial agreement in the severity of the deficits, between MRI and beta amyloid maps. Intriguingly, both maps indicated the relative sparing of primary sensorimotor regions (*white* in the amyloid map) and the superior temporal gyrus (*blue colors* in panel c) relative to other temporal lobe gyri. These overall MRI patterns have been replicated in independent studies by Baron et al. (2001), O'Brien et al. (2001), and Burton et al. (2002).

to estimate the probability that the observed effects could have occurred by chance, and this is reported as a significance value for the overall map.

When deforming an atlas to match a patient's anatomy, mesh-based models of anatomic systems help guide the mapping of one brain to another. Anatomically driven algorithms guarantee biological as well as computational validity, generating meaningful object-to-object correspondences, especially at the cortex. In this model-based approach (Thompson and Toga, 1996, 1998a, Thompson et al., 1997, 1998a), systems of surfaces are first extracted from each dataset, to guide the volumetric mapping. The model surfaces include many functional, cytoarchitectonic, and lobar boundaries in three dimensions. Both the surfaces and the landmark curves within them are reconfigured to match their counterparts in the target datasets exactly.

25.2.4 Anatomical Models

Since much of the functional territory of the human cortex is buried in *sulci*, a generic structure is built to model them (Thompson and Toga, 1996). The underlying data structure is a connected system of surface meshes, in which the individual meshes are parametric. These surfaces are 3D sheets that divide and join at curved junctions to form a connected network of models. With the help of these meshes, each patient's anatomy is modeled in sufficient detail to be sensitive to subtle differences in disease. Separate surfaces model the deep internal trajectories

of features such as the parieto-occipital sulcus, the anterior and posterior calcarine sulcus, the Sylvian fissure, and the cingulate, marginal, and supracallosal sulci in both hemispheres. Additional gyral boundaries are represented by parameterized curves lying in the cortical surface. The ventricular system is modeled as a closed system of 14 connected surface elements whose junctions reflect cytoarchitectonic boundaries of the adjacent tissue (Thompson and Toga, 1998b). Information on the meshes' spatial relations, including their surface topology (*closed* or *open*), anatomical names, mutual connections, directions of parameterization, and common 3D junctions and boundaries is stored in a hierarchical graph structure. This ensures the continuity of displacement vector fields defined at mesh junctions.

25.2.5 Surface Parameterization

After imposing an identical regular grid structure on anatomic surfaces from different subjects, the explicit geometry can be exploited to drive and constrain correspondence maps that associate anatomic points in different subjects. Structures that can be extracted automatically in parametric form include the external cortical surface, ventricular surfaces, and several deep sulcal surfaces. Recent success of sulcal extraction approaches based on deformable surfaces led us to combine a 3D skeletonization algorithm with deformable curve and surface governing equations to automatically produce parameterized models of cingulate, parieto-occipital, and calcarine sulci, without manual initialization (Zhou et al., 1999). Additional, manually segmented surfaces can also be given a uniform rectilinear parameterization using algorithms described in Thompson et al. (1996a,b) and can be used to drive the warping algorithm. Each resultant surface mesh is analogous in form to a uniform rectangular grid, drawn on a rubber sheet, which is subsequently stretched to match all data points. Association of points on each surface with the same mesh coordinate produces a dense correspondence vector field between surface points in different subjects. This procedure is carried out under stringent conditions to ensure that landmark curves and points known to the anatomist appear in corresponding locations in each parametric grid.

25.2.6 Maps of the Cortical Parameter Space

Detailed models of cortical anatomy are also created by driving a tiled, spherical mesh into the configuration of each subject's cortex (Thompson and Toga, 1996; MacDonald, 1998). Because these cortical models are obtained by deforming a spherical mesh, any point on the cortical surface must map to exactly one point on the sphere and vice versa. Each cortical surface is parameterized with an invertible mapping, so sulcal curves and landmarks in the folded brain surface can be re-identified in the spherical map [cf. Sereno et al. (1996) Fischl et al. (1999), for a similar approach]. To retain relevant 3D information, cortical surface point position vectors in 3D stereotaxic space are color-coded, so that a unique color is placed at each position on the spherical map indicating the 3D position of the cortical point that maps to it. In other words, the colors represent 3D locations, and the entire set of colors forms an image on the sphere in a color image format. To find good matches between cortical regions in different subjects, we first derive a spherical map for each respective cortical surface model and then perform a matching process in the spherical parametric space. A flow field is calculated on the sphere that brings corresponding gyral and sulcal regions into the same spherical locations across subjects (Davatzikos, 1996; Drury et al., 1996; Van Essen et al., 1997; Fischl et al., 1999). This warp can be set up in a variety of ways. Spherical harmonic functions are an orthonormal basis on the sphere, which means that any smooth flow on the sphere can be represented with arbitrarily high accuracy using a linear combination of these functions, so long as a sufficient number of functions is used. The resulting flow field can align structural or functional information across subjects, bringing curve and surface interfaces into exact register in the process (Thompson and Toga, 1996). Alternatively, an approach based on covariant partial differential equations can be used for matching cortical surfaces (Thompson et al., 2000a). This precisely matches cortical landmarks across subjects, and it creates maps that are independent of the surface metrics. The approach ensures that the way cortical structures are matched in 3D is independent of the way the surfaces are computationally represented (irrespective of their tile density and parameterizations).

25.2.7 Tensor Maps of Directional Variation

Structures do not vary to the same degree in every coordinate direction (Thompson et al., 1996), and even these directional biases vary by cortical system. The principal directions of anatomic variability in a group can be shown in a *tensor map* (Thompson et al., 2000a,b) (Figure 25.4). The maps have two uses. First, they make it easier to detect anomalies, which may be small in magnitude but in an unusual direction. Second, they significantly increase the information content of Bayesian priors used for automated structure extraction and identification (Gee et al., 1995; Mangin et al., 1994; Royackkers et al., 1996; Pitiot et al., 2002).

25.3 RESULTS

Using a variety of populations with imaging data collected in the same fashion, we have created a series of probabilistic atlases that retain information on anatomic and functional variability (Mazziotta et al., 1995; Thompson et al., 1997). Descriptions of several of these follow. As the subject database increases in size and content, the digital form of these atlases allows efficient statistical comparisons of individuals or groups. In addition, the population that an atlas represents can be stratified into subpopulations to represent specific disease types, and subsequently by age, gender, handedness, or genetic factors.

25.3.1 Pathology Detection

Normal anatomic complexity makes it difficult to design automated strategies that detect abnormal brain structure. At the same time, brain structure is so variable that group-specific patterns of anatomy and function are often obscured. Reports of structural differences in the brain linked to gender, IQ, and handedness are a topic of intense controversy, and it is even less clear how these factors affect disease-specific abnormalities (Thompson and Toga, 1999a). The importance of these linkages has propelled computational anatomy to the forefront of brain imaging investigations. To distinguish abnormalities from normal variants, a realistically complex mathematical framework is required to encode information on anatomic variability in homogeneous populations (Grenander and Miller, 1998). We employed elastic registration or *warping* algorithms to achieve distinct advantages for encoding patterns of anatomic variation and detecting pathology. Cortical patterns are altered in schizophrenia (Narr et al., 2000), Alzheimer's disease (Thompson et al., 1998, 2000a), and a wide variety of developmental disorders. By using specialized strategies for group averaging of anatomy, specific features of anatomy emerge that are not observed in individual representations due to their considerable variability. Group-specific patterns of cortical organization or asymmetry can then be mapped out and visualized (Thompson and Toga, 1999a; Narr et al., 1999).

25.3.2 Deformable Probabilistic Atlases

Warping algorithms create deformation maps that indicate 3D patterns of anatomic differences between any pair of subjects. By defining probability distributions on the space of deformation transformations that drive the anatomy of different subjects into correspondence (Thompson and Toga, 1997), statistical parameters of these distributions can be estimated from databased anatomic data to determine the magnitude and directional biases of anatomic variation. Encoding of local variation can then be used to assess the severity of structural variants outside of the

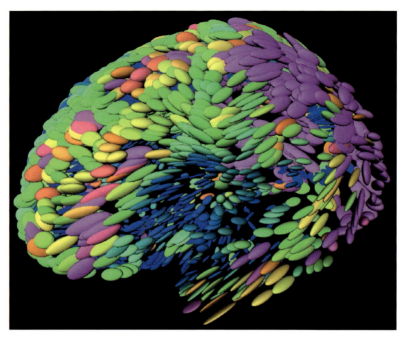

Figure 25.4   Mapping directional patterns of brain variation. This image shows a tensor map of variability for normal subjects, created after mapping 20 elderly subjects' data into Talairach space (all right-handed, 10 males, 10 females; Thompson et al., 2001a). Ellipsoidal glyphs indicate the principal directions of variation; they are most elongated along directions where anatomic variation is greatest across subjects. Each glyph represents the covariance tensor of the vector fields that map individual subjects onto their group average. Because gyral patterns constrain the mappings, the fields reflect variations in cortical organization at a more local level than can be achieved by only matching global cortical geometry. Note the elongated glyphs in anterior temporal cortex, as well as the very low variability (in any direction) in entorhinal and inferior frontal areas. By better defining the parameters of allowable normal variations, the resulting information can be invoked to distinguish normal from abnormal anatomical variants.

normal range, which, in brain data, may be a sign of disease (Thompson et al., 1997).

25.3.3 Encoding Brain Variation

To see if disease-specific features could be detected in individual patients, we developed a *random vector field* approach to construct a population-based brain atlas (Thompson and Toga, 1997). Briefly, given a 3D MR image of a new subject, a warping algorithm calculates a set of high-dimensional volumetric maps, elastically matching this image with other scans from an anatomic image database. Target scans are selected from subjects matched for age, handedness, gender, and other demographic factors (Thompson et al., 1997, 1998). The resulting family of volumetric warps provides empirical information on local variability patterns. A probability space of random transformations, based on the theory of anisotropic Gaussian random fields, with statistical flattening corrections (Worsley et al., 1999; Thompson et al., 2000a) is then used to encode the variations. For the cortex, specialized approaches are needed to represent variations in gyral patterns (Thompson and Toga, 1998a). Confidence limits in stereotaxic space are determined, for points in the new subject's brain, and probability maps can be created to highlight and quantify regional patterns of deformity (Thompson et al., 1997).

25.3.4 Brain Asymmetry

One feature observable from constructing average anatomical models is that consistent patterns of brain asymmetry can be mapped, despite wide variations in asymmetry in individual subjects (Figure 25.5). In dementia, the increased cortical asymmetry probably reflects asymmetric progression of the disease. Figure 25.6 shows average maps of the lateral ventricles, again from Alzheimer's disease and matched elderly normal populations. As expected, the ventricles are significantly enlarged in dementia. Notice, however, that a pronounced asymmetry is observed in both groups (left volume larger than right, $p < 0.05$). This is an example of an effect that becomes clear after group averaging of anatomy, and it is not universally apparent in individual subjects. It is, however, consistent with prior volumetric measurements (Shenton et al., 1992; Aso et al., 1995). Anatomical averaging can also be cross-validated with a traditional volumetric approach. Occipital horns are on average 17.1% larger on the left in the normal group (4070.1 ± 479.9 mm^3) than on the right (3475.3 ± 334.0 mm^3; $p < 0.05$), but no significant asymmetry is found for the superior horns (*left:* 8658.0 ± 976.7 mm^3; *right:* 8086.4 ± 1068.2 mm^3; $p > 0.19$) or for the inferior horns (*left:* 620.6 ± 102.6 mm^3; *right:* 573.7 ± 85.2 mm^3; $p > 0.37$). The asymmetry is clearly localized in the 3D group average anatomic representations. In particular, the occipital horn extends (on

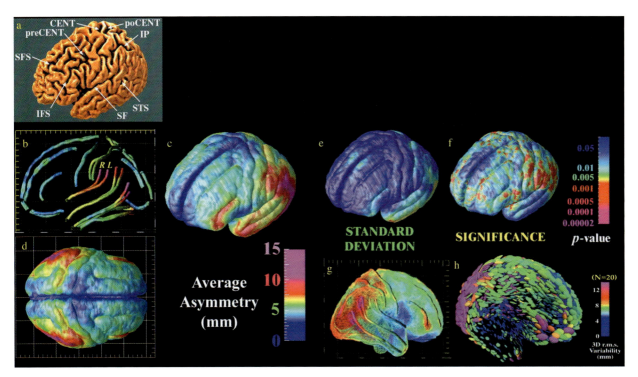

Figure 25.5 Multi-subject maps of brain asymmetry. Image analysis techniques make it possible to distinguish systematic asymmetries in a population, or a specific group of subjects, from random fluctuations in anatomy. After aligning and scaling individual MRI scans into a standard 3D space, 3D curves representing the primary sulcal pattern are digitized (**a**). [Sulci include central (CENT), precentral (preCENT), postcentral (poCENT), intraparietal (IP), superior frontal (SFS), inferior frontal (IFS), superior temporal and Sylvian fissures (SF)]. Averaging these curves across 20 normal subjects (**b**), the magnitude of asymmetry in the average anatomy is shown in color (red colors denote greater asymmetry). Extension of these methods to surfaces (**c, d**) reveals prominent asymmetries in Broca's anterior speech area and in language regions surrounding the Sylvian fissure. By comparing the average magnitude of these asymmetries to their standard error (**e**), regions of significant asymmetry are identified (**f**). Asymmetries are greatest in brain regions with greatest gyral pattern variability across subjects (**g, h**). The tensor map (**h**) shows that the preferred directions of inter-subject anatomical variability are also approximately aligned with the direction of interhemispheric asymmetry. Data from Thompson et al. (2000a) and Toga and Thompson (2003).

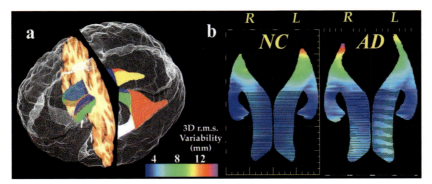

Figure 25.6 Population-based maps of average ventricular anatomy in normal aging and Alzheimer's disease. In patients and controls, 3D parametric surface meshes (Thompson et al., 1996) were used to model 14 ventricular elements, and meshes representing each surface element were averaged by hemisphere in each group. An average model for Alzheimer's patients (*red;* AD) is superimposed on an average model for matched normal controls (*blue;* NC). Mesh averaging reveals enlarged occipital horns in the Alzheimer's patients and shows high stereotaxic variability in both groups. Extreme variability at the occipital horn tips also contrasts sharply with the stability of septal and temporal ventricular regions. A top view of these averaged surface meshes reveals localized asymmetry, variability, and displacement within and between groups. These subcortical asymmetries emerge only after averaging of anatomical maps in large groups of subjects. Adapted from Thompson et al. (2000a).

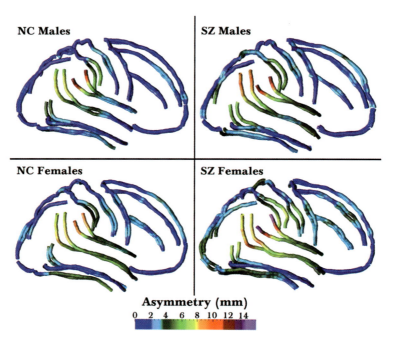

Figure 25.7   Brain asymmetry in schizophrenia. To map group variability and asymmetry in sulci and surrounding anatomy, 3D cortical extractions from MR scans were obtained and brought into register using surface warping algorithms that enable averaging of equivalent cortical regions delineated by tracing major hemispheric sulci across subjects. The atlas data can be stratified by gender and disease group (*SZ: schizophrenic patients; NC: normal controls*) to determine whether asymmetries are altered, on average in each group. Here the color code denotes the magnitude of the asymmetry in the average sulcal pattern, for each group, expressed in millimeters.

average) 5.1 mm more posteriorly on the left than on the right. The capacity to resolve asymmetries in a group atlas can assist in studies of disease-specific cortical organization (Thompson et al., 1997, 2000; Mega et al., 1998; Zoumalan et al., 1999; Narr et al., 1998a, 1999, 2000).

25.3.5 Asymmetry in Disease

To see if cortical asymmetries were disturbed in schizophrenia, we made average cortical representations for schizophrenic patients ($N = 25$; 15 males, 10 females; all right-handed) and matched controls ($N = 28$; 15 males, 13 females; right-handed). Thirty-six major sulcal curves were used to drive each subject's gyral pattern into a group mean configuration (Figure 25.7). The magnitude of anatomic variation in each brain region was also computed from the deformation vector fields and was shown in color as a variability map. Perhaps surprisingly, asymmetry was not attenuated in the patient group. Marked asymmetries were observed in the sagittal projections of average anatomy for each group. Significant asymmetries were confirmed by calculating curvature and extent measures from the parametric mesh models (Narr et al., 1999). In frontal cortex, the patients also displayed greater variability than did the controls.

25.3.6 Corpus Callosum Differences

We also attempted to identify regionally selective patterns of callosal change in patient groups with Alzheimer's disease and

schizophrenia (Thompson et al., 1998; Narr et al., 1999). The midsagittal cllosum was first partitioned into five sectors (Duara et al., 1991; Larsen et al., 1992). This roughly segregates callosal fibers from distinct cortical regions. In AD, focal fiber loss was expected at the callosal isthmus (sector 2) whose fibers selectively innervate the temporo-parietal regions with early neuronal loss and perfusion deficits (Brun and Englund, 1981). Consistent with this hypothesis, a significant area reduction at the isthmus was found, reflecting a dramatic 24.5% decrease from 98.0 ± 8.6 mm^2 in controls to 74.0 ± 5.3 mm^2 in AD ($p < 0.025$). Terminal sectors (1 and 5) were not significantly atrophied, and the central midbody sector showed only a trend toward significance (16.6% mean area loss; $p < 0.1$), due to substantial intergroup overlap. Average boundary representations, however, localized these findings directly (Figure 25.8).

25.3.7 Gender in Schizophrenia

Gender differences are observed in normal brain morphology (e.g., DeLacoste-Utamsing and Holloway, 1982; Harasty et al., 1997; Davatzikos and Resnick, 1998). Distinct gender differences are similarly observed in the clinical manifestation of schizophrenia and with respect to structural brain abnormalities (DeLisi et al., 1989; Gur et al., 1996; Leung and Chue, 2000). For example, in Figure 25.8, displacements representing and upward bowing of the corpus callosum are observed in patients with schizophrenia compared to controls and are more pronounced between male patients relative to controls (Narr et al., 2000,

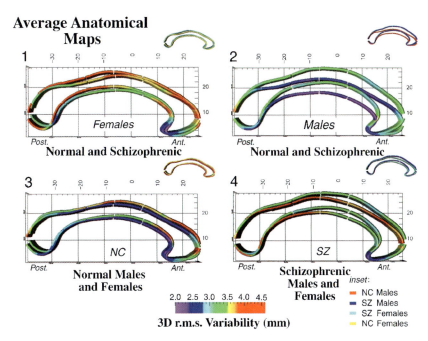

Figure 25.8 Average corpus callosum shapes in schizophrenia. Midsagittal *corpus callosum* boundaries were averaged from 25 patients with chronic schizophrenia (DSM-III-R criteria; 15 males, 10 females; age: 31.1 ± 5.6 years) and from 28 control subjects matched for age (30.5 ± 8.7 years), gender (15 males, 13 females) and handedness (1 left-handed subject per group). Profiles of anatomic variability around the group averages are also shown (*in color*) as an r.m.s. deviation from the mean. Anatomical averaging reveals a pronounced and significant bowing effect in the schizophrenic patients relative to normal controls. Male patients show a significant increase in curvature for superior and inferior callosal boundaries ($p < 0.001$), with a highly significant sex by diagnosis interaction ($p < 0.004$). The sample was stratified by sex and diagnosis and separate group averages show that the disease induces less bowing in females (**panel 1**) than in males (**panel 2**). While gender differences are not apparent in controls (**panel 3**), a clear gender difference is seen in the schizophrenic patients (**panel 4**). Abnormalities localized in a disease-specific atlas can therefore be analyzed to reveal interactions between disease and demographic parameters. Data from Narr et al. (2000).

2002). These gender differences in callosal displacements may relate to the sexual dimorphism in the etiology and course of schizophrenia where male patients appear to show increased negative symptoms, earlier age of onset, and a worse course of illness compared to female patients (e.g., DeLisi et al., 1989; Gur et al., 1996). Interestingly, genetic rather that shared environmental or disease-specific influences appear to contribute to displacements of the corpus callosum, where effects are related to both lateral and third ventricle enlargements (Narr et al., 2002). Stratification of probabilistic atlases by gender and other genetic factors provides a computationally fast way to visualize these effects and relate them to epidemiologic data (Mazziotta et al., 1995; Mega et al., 1998; Zoumalan et al., 1999; Blanton et al., 1999; Le Goualher et al., 1999).

25.3.8 Comparing a Subject with an Atlas

In one validation experiment (Thompson et al., 1997), probability maps were created to highlight abnormal deviations in the callosal and midline anatomy of a tumor patient. The two regions of metastatic tissue induced marked distortions in the normal architecture of the brain. After storing variations in deep surface anatomy as a spatially adaptive covariance tensor field, probability maps were generated for the tumor patient. In the tumor patient, the herniation effects apparent in the blockface imagery were detected in the probability maps of structures near the lesion sites.

In one experiment, mappings that deform one cortex into gyral correspondence with another were used to create an *average* cortex for patients with mild to moderate Alzheimer's disease (AD; Figure 25.11c). Thirty-six gyral curves for nine AD patients were transferred to the cortical parameter space uniformly re-parameterized, and a set of 36 average gyral curves for the group was created by vector averaging of point locations on each curve. Each individual cortical pattern was then aligned with the average curve set using a spherical flow field. These nine flow fields were then used to create an average cortex in 3D space, as follows. By carrying a code (that indexes 3D locations) along with the flow that aligns each individual with the average folding pattern, information can then be recovered at a particular location in the average folding pattern, specifying the 3D cortical points that map to it in each subject. By ruling a regular grid over the warped coded map, and reading off 3D position values for each subject, cortical positions in any subject's *original* 3D anatomy can be recovered. This produces a new coordinate grid on a given subject's cortex, in which particular grid points appear in the same location relative to the primary gyral pattern

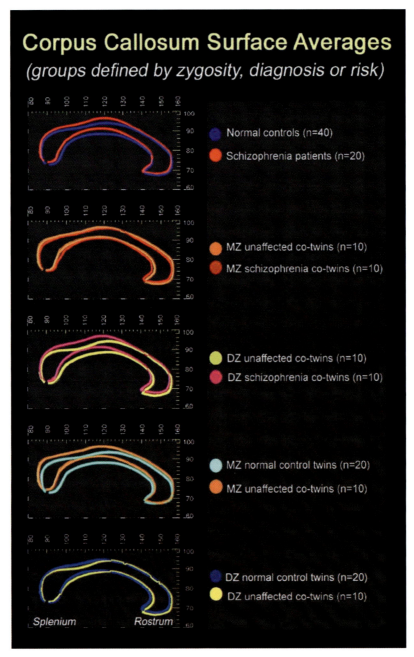

Figure 25.9   Callosal surface averages mapped in groups defined by biological risk for schizophrenia. Average anatomical mesh models of the corpus callosum are shown in different colors to illustrate differences between monozygotic and dizgotic co-twins discordant for schizophrenia and healthy monozygotic and dizygotic twin pairs. From the top, midsagittal callosal averages are mapped in (1) schizophernia patients and controls; (2) unaffected and affected monozygotic (MZ) co-twins; (3) unaffected and affected diaygotic (DZ) co-twins; (4) unaffected MZ co-twins of the schizophrenia prohands and MZ control twin pairs; and (5) unaffected DZ co-twins of the schizophrenia probands and DZ control twin pairs.

across all subjects (see Fischl et al., 1999, for a similar approach). By averaging these 3D positions across subjects, an average 3D cortical model was constructed for the group (Figure 25.9). The resulting mapping is guaranteed to average together all points falling on the same cortical locations across the set of brains, and it ensures that corresponding cortical features are averaged together.

25.4 DISCUSSION

The mathematical strategies employed in the construction of these atlases were needed to encode comprehensive information on structural variability in human populations. Particularly relevant is 3D statistical information on group-specific patterns of variation, and how these patterns are altered in

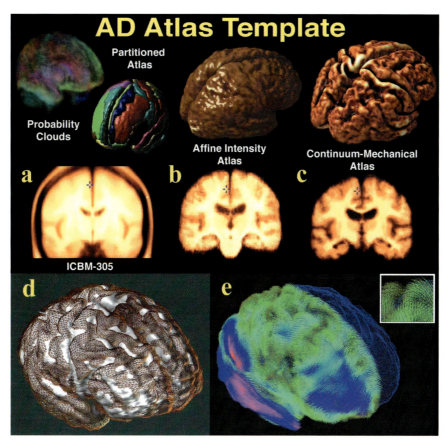

Figure 25.10 Methods for averaging brains. **(a)** In a widely used average brain image template (ICBM305) based on voxel-wise intensity averaging of 305 young normal subjects' scans (Evans et al., 1994), anatomical features are not well-resolved at the cortex. Cortical variability is represented using probability clouds (*top left*) that describe the frequency of incidence for each gyrus at each stereotaxic voxel, after linear registration and voxel-by-voxel comparison. In an affine brain template **(b)**, similarly constructed from Alzheimer's disease patients' scans, the cortical average is also poorly resolved. By contrast, anatomical features are highly resolved, even at the cortex, in the *Continuum-Mechanical Brain Template* **(c)**, which applies a continuum-mechanical transformation to each brain before intensity averaging. Scans are elastically reconfigured into a group mean configuration, using surface-based warping to match 84 surface models (including gyral pattern elements) across all subjects. Reconfigured scans are then averaged voxel-by-voxel, after intensity normalization, to produce a group image template with the average geometry *and* average image intensity for the group. Vector field transformations of extremely high spatial dimension **(d, e)** are required to resolve cortical features, in their mean configuration, after scans are averaged together **(c)**. Adapted from Thompson et al. (2000a).

disease. This information can be exploited by expert diagnostic systems, whose goal is to detect subtle or diffuse structural alterations in disease.

25.4.1 Pathology Detection in Image Databases

Pattern recognition algorithms for automated identification of brain structures can also benefit greatly from encoded information on anatomic variability. We recently developed a Bayesian approach to identify the *corpus callosum* in each image in an MRI database (Pitiot et al., 2002). The shape of a deformable curve is progressively tuned to optimize a mathematical criterion measuring how likely it is that it has found the corpus callosum. The measure includes terms that reward contours based on their agreement with a diffused edge map, their geometric regularity, and their statistical abnormality when compared

with a distribution of normal shapes. By averaging contours derived from an image database, structural abnormalities associated with Alzheimer's disease and schizophrenia were identified (Thompson et al., 1998; Narr et al., 1999). Automated parameterization of structures will accelerate the identification and analysis of disease-specific structural patterns.

25.4.2 Disease Progression

The atlases so far described, for the dementia and schizophrenia populations, have been based on homogeneous patient groups, matched for age, gender, handedness, and educational level. Since AD, in particular, is a progressive disease, the initial atlas was created to reflect a particular stage in the disease (MMSE score: 19.3 ± 2.0). At this stage, patients often present for initial evaluation, and MR, PET, and SPECT scans have maximal diagnostic value. Nonetheless, by expanding the underlying patient database, atlases are under construction to represent the more advanced stages of Alzheimer's disease (Thompson et al., 2003b,c). By stratifying the population according to different criteria, different atlases can be synthesized to represent other clinically defined groups.

25.4.3 4D Coordinate Systems

Atlasing of data from the developing or degenerating brain presents unique challenges. However, warping algorithms can be applied to serial scan data to track disease and growth processes in their full spatial and temporal complexity. Maps of anatomical change can be generated by warping scans acquired from the same subject over time (Thirion and Calmon, 1997; Thompson et al., 2000b,c). Serial scanning of human subjects (Fox et al., 1996; Subsol et al., 1997; Freeborough and Fox, 1998; Thompson et al., 1998) or experimental animals (Jacobs and Fraser, 1994) in a dynamic state of disease or development offers the potential to create 4D models of brain structure. These models incorporate dynamic descriptors of how the brain changes during maturation or disease. They are therefore of interest for investigating and staging brain development. In an atlas setting, these 4D maps can act as normative data to define aberrant growth rates and their modulation by therapy (Haney et al., 2000a,b,c).

In our initial human studies (Thompson et al., 2000b,c), we developed several algorithms to create 4D quantitative maps of growth patterns in the developing human brain. Time series of high-resolution pediatric MRI scans were analyzed. The resulting tensor maps of growth provided spatially detailed information on local growth patterns, quantifying rates of tissue maturation, atrophy, shearing, and dilation in the dynamically changing brain architecture. Pairs of scans were selected to determine patterns of structural change across the inter-scan interval. Deformation processes recovered by a high-dimensional warping algorithm were then analyzed using vector field operators to produce a variety of tensor maps. These maps were designed to reflect the magnitude and principal directions of dilation or contraction, the rate of strain, and the local curl, divergence, and gradient of flow fields representing the growth processes recovered by the transformation.

The growth maps obtained in these studies exhibit several striking characteristics. First, foci of rapid growth at the callosal isthmus appeared consistently across puberty. These rates appeared to attenuate as subjects progressed into adolescence (Thompson et al., 2000a). Rapid rates of tissue loss were also revealed at the head of the caudate, in an earlier phase of development.

25.4.4 Dynamic Maps of Brain Change

Statistical brain maps from large populations (Figure 25.10) are likely to help assess how different drug treatments affect the time course of aging and dementia. In developing dynamic atlases for clinical applications, there is a particular interest in modeling atrophic processes that speed up or slow down. Diseases may accelerate, or their rate of progression may be slowed down by therapy. If individuals are scanned more than twice over large time spans (e.g., Fox et al., 2001), brain change can be modeled more accurately. To compare atrophic processes in different groups of subjects, nonlinear mixed models can be used (Giedd et al., 1999; Thompson et al., 2003a) to analyze the registered degenerative profiles. For the ith individual's jth measure we have

$$Y_{ij} = f(\text{Age}_{ij}, \boldsymbol{\beta}) + \varepsilon_{ij}.$$

Here Y_{ij} signifies the outcome measure at a voxel or surface point, such as growth or tissue loss, $f()$ denotes a constant, linear, quadratic, cubic, or other function of the individual's age for that scan, and $\boldsymbol{\beta}$ denotes the regression/ANOVA coefficients to be estimated (Figure 25.11). In models whose fit is confirmed as significant (e.g. by permutation), loadings on nonlinear parameters may be visualized as attribute maps $\boldsymbol{\beta}(\mathbf{x})$. This reveals the topography of accelerated or decelerated brain change (Thompson et al., 2003a). The result is a formal approach to assess whether, and where, brain change is speeding up or slowing down, a key feature in medication studies.

In this statistical model, Age (Age_{ij}) may be replaced by time from the onset of disease or medication. This flexibility in parameterizing the time axis allows one to temporally register dynamic patterns using criteria that are expected to bring into line temporal features of interest that appear systematically in a group (Janke et al., 2001). For example, the independent variable could be a cognitive score such as Mini-Mental status, which declines over time in AD. Parameterization of dynamic effects using measures other than time (e.g., clinical status) also provides a mechanism to align new patients' time series with a dynamic atlas, potentially still further increasing the power to reveal systematic effects.

In the near future, 4D atlases will be able to map growth and degeneration in their full spatial and temporal complexity. Despite the logistic and technical challenges, these mapping approaches hold tremendous promise in analyzing the dynamics of degenerative or neoplastic diseases (Haney et al., 2000a,b,c). They will ultimately play a role in detecting how different therapeutic approaches modulate the course of disease.

25.4.5 Multimodality Atlases

Combining data derived from multiple subjects with data from multiple modalities can result in comprehensive representations of structure–function relationships and help elucidate subtle results difficult to appreciate in isolation. In the construction of anatomic atlases, the in vivo resolution available from MR is incapable of characterizing the cytoarchitectural detail available from postmortem material. Because of the superior anatomic resolution, several digital atlases have been created using cryosection imaging. This technique allows the serial collection of photographic images from a cryoplaned specimen blockface (Bohm et al., 1983; Greitz et al., 1991; Toga et al., 1994). Using 1024^2, 24-bits/pixel digital color cameras, cryosection imaging offers a spatial resolution as high as 100 μm/voxel for whole human head cadaver preparations, or higher for isolated brain regions (Toga et al., 1997). In the Visible Human Project (Spitzer et al., 1996), two (male and female) cadavers were cryoplaned and imaged at 1.0-mm intervals (0.33 mm for the female data), and the entire bodies were also reconstructed via 5000 postmortem CT and MRI images. The resulting digital datasets consist of over 15 gigabytes of image data. While not an atlas per se, the

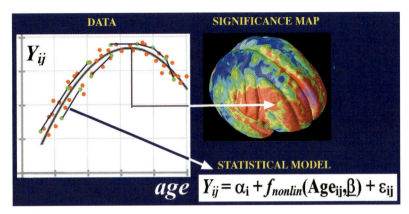

Figure 25.11 Data, statistical models, and maps. This schematic (Thompson et al., 2003a) shows some of the steps used in mapping cortical change. First, measures (Y_{ij}) are defined that can be obtained longitudinally (*green dots*) or once only (*red dots*) in a group of subjects at different ages. Fitting of statistical models to these data (*Statistical Model, lower right*) produces estimates of parameters that can be plotted onto the cortex, using a color code. These parameters can include age at peak (see arrow at peak of the curve), significance values, or estimated statistical parameters such as rates of change, and effects of drug treatment or risk genes. See Thompson et al. (2003)

Visible Human Project data have served as the foundation for developing related atlases of regions of the cerebral cortex (Drury and Van Essen, 1997) and high-quality brain models and visualizations (Schiemann et al., 1996; Stewart et al., 1996). Using multimodality data from a patient with a localized pathology, and more recently the Visible Human Project data, Höhne and co-workers developed a commercially available brain atlas designed for teaching neuroanatomy (VOXEL-MAN; Höhne et al., 1990, 1992; Tiede et al., 1993; Pommert et al., 1994).

Using 1024^2, 24-bits/pixel digital color cameras, spatial resolution can be as high as 50 μm/voxel for whole human head cadaver preparations, or higher for isolated brain regions (Toga et al., 1994). Cryosectioning in micron increments permits data collection with high spatial resolution in the axis orthogonal to the sectioning plane.

Integration of metabolic and functional images acquired in vivo with postmortem biochemical maps provides a unique view of the relationship between brain function and pathology. Mega et al. (1997) scanned Alzheimer's patients in the terminal stages of their disease using both MRI and PET. Using elastic registration techniques (Thompson et al., 1996), these data were combined with postmortem histologic images showing the gross anatomy (Toga et al., 1994), a Gallyas stain of neurofibrillary tangles, and a variety of spatially indexed biochemical assays. The resulting multimodality maps of the Alzheimer's disease brain relate the anatomic and histopathologic underpinnings of the disease in a standardized coordinate space. These data are further correlated with in vivo metabolic and perfusion maps of this disease. The resulting maps are key components of a growing disease-specific atlas (Mega et al., 2000).

25.5 CONCLUSION

The uses of brain atlases are as varied as their construction. They provide the ability to measure, visualize, compare, and summarize brain images. They encompass descriptions of structure or function of the whole brain to maps of groups or populations.

Individual systems of the brain can be mapped as can changes over time, as in development or degeneration. An atlas enables comparison across individuals, modalities, or states. But in most cases, the value added by brain atlases is the unique and critical ability to integrate information from multiple sources. The utility of an atlas is dependent upon appropriate coordinate systems, registration, and deformation methods along with useful visualization strategies. The probabilistic systems described here show promise for encoding patterns of anatomic variation in large image databases, for pathology detection in individuals and groups, and for determining effects in space and time on brain structure of age, gender, handedness, and other demographic or genetic factors.

ACKNOWLEDGMENTS

This work was supported by research grants from the National Center for Research Resources (P41 RR13642 and M01 RR00865), the National Institute of Mental Health, and the National Institute of Neurological Disorders and Stroke (P20 MH/NS65166), as well as by a Human Brain Project grant to the International Consortium for Brain Mapping, funded jointly by NIMH and NIDA (P20 MH/DA52176). Additional support was also provided by the National Library of Medicine (LM/MH05639) by the National Center for Research Resources (R21 RR19771), the National Institute for Biomedical Imaging and Bioengineering (R21 EB01651), and the Biomedical Informatics Research Network (BIRN, http://www.nbirn.net) which is funded by the National Center for Research Resources at the National Institutes of Health (NIH). Portions of this manuscript, including several figures, are reprinted with permission from Toga et al. (2001), copyright Springer-Verlag.

References

Annese, J., Pitiot, A., Dinov, I.D., and Toga, A.W. (2004). A myeloarchitectonic method for the structural classification of cortical areas. Submitted for publication *Neuroimage* **21** 15–26.

Ashburner J., and Friston, K.J. (2000). Voxel-based morphometry—the methods. *NeuroImage* **11**(6), 805–821.

Ashburner, J., Neelin, P., Collins, D.L., Evans, A.C., and Friston, K.J. (1997). Incorporating prior knowledge into image registration, *NeuroImage* **6**(4), 344–352.

Aso, M., Kurachi, M., Suzuki, M., Yuasa, S., Matsui, M., and Saitoh, O. (1995). Asymmetry of the ventricle and age at the onset of schizophrenia. *Eur. Arch. Psychiatry Clin. Neurosci.* **245**(3), 142–144.

Avoli, M., Hwa, G.C., Kostopoulos, G., Oliver, A., and Villemure, J.G. (1991). Electrophysiological analysis of human neocortex in vitro: Experimental techniques and methodological approaches. *Can. J. Neurol. Sci.* **18**, 636–639.

Baron, J.C., Chetelat, G., Desgranges, B., Perchey, G., Landeau, B., de la Sayette, V., and Eustache, F. (2001). In vivo mapping of gray matter loss with voxel-based morphometry in mild Alzheimer's disease. *NeuroImage* **14**, 298–309.

Berger, H. (1929). Uber das Elektrenkephalogramm des Menschen. *Arch. Psychiatr. Nervenkr.* **87**, 527–580.

Blanton, R.E., Levitt, J., Thompson, P.M., Badrtalei, S., Capetillo-Cunliffe, L., and Toga, A.W. (1999). Average 3-dimensional caudate surface representations in a juvenile-onset schizophrenia and normal pediatric population. *5th Int. Conf. Func. Mapp. Hum. Brain*, Dusseldorf, Germany, *1999*, Presentation No. 621.

Bohm, C., Greitz, T., Kingsley, D., Berggren, B.M., and Olsson, L. (1983). Adjustable computerized brain atlas for transmission and emission tomography. *Am. J. Neuroradiol.*, **4**, 731–733.

Brodmann, K. (1909). *Vergleichende Lokalisationslehre der Grosshirnrinde in ihren Prinzipien dargestellt auf Grund des Zellenbaues* Barth, Leipzig. [In *Some Papers on the Cerebral Cortex* (translated as: *On the Comparative Localization of the Cortex*), pp. 201–230. Thomas, Springfield, IL, 1960.]

Bro-Nielsen, M., and Gramkow, C. (1996). Fast fluid registration of medical images. In *Visualization in Biomedical Computing* (K.H. Höhne and R. Kikinis, eds.), Lect. Notes Comput. Sci. No. 1131, pp. 267–276. Springer-Verlag, Berlin.

Brun, A., and Englund, E. (1981). Regional pattern of degeneration in Alzheimer's disease: Neuronal loss and histopathologic grading. *Histopathology* **5**, 549–564.

Bullmore, E.T., Suckling, J., Overmeyer, S., Rabe-Hesketh, S., Taylor, E., and Brammer, M.J. (1999). Global, voxel, and cluster tests, by theory and permutation, for a difference between two groups of structural MR images of the brain. *IEEE Trans. Med. Imaging* **18**, 32–42.

Christensen, G.E., Rabbitt, R.D., and Miller, M.I. (1993). A deformable neuroanatomy textbook based on viscous fluid mechanics *27th Annu. Conf. Inf. Sci. Syst.*, pp. 211–216.

Christensen, G.E., Rabbitt, R.D., and Miller, M.I. (1996). Deformable templates using large deformation kinematics. *IEEE Trans. Image Process.* **5**(10), 1435–1447.

Collins, D.L., Neelin, P., Peters, T.M., and Evans, A.C. (1994a). Automatic 3D intersubject registration of MR volumetric data into standardized Talairach space, *J. Comput.-Assist Tomogr.* **18**(2), 192–205.

Collins, D.L., Peters, T.M., and Evans, A.C. (1994b). An automated 3D non-linear image deformation procedure for determination of gross morphometric variability in the human brain. *Proc. Visual. Biomed. Comput.* (*SPIE*) **3**, 180–190.

Damasio, H. (1995). *Human Brain Anatomy in Computerized Images*, Oxford University Press, Oxford and New York.

Davatzikos, C. (1996). Spatial normalization of 3D brain images using deformable models. *J. Comput.-Assist Tomogr.* **20**(4), 656–665.

Davatzikos, C., and Resnick, S.M. (1998). Sex differences in anatomic measure of interhemispheric connectivity: Correlations with cognition in women but not men. *Cereb. Cortex* **8**, 635–640.

DeLacoste-Utamsing, C., and Holloway, R.L. (1982). Sexual dimorphism in the human corpus callosum. *Science* **216**, 1431–1432.

DeLisi, L.E., Dauphinais, I.D., and Hauser, P. (1989). Gender differences in the brain: Are they relevant to the pathogenesis of schizophrenia? *Comp. Psychiatry* **30**, 197–208.

Dinov, I.D., Mega, M.S., Thompson, P.M., Lee, L., Woods, R.P., Holmes, C.J., Sumners, D.L., and Toga, A.W. (2000). Analyzing functional brain images in a probabilistic atlas: A validation of sub-volume thresholding. *J. Comput.-Assisted Tomogr.* **24**(1), 128–138.

Drury, H.A., and Van Essen, D.C. (1997). Analysis of functional specialization in human cerebral cortex using the visible man surface based atlas. *Hum. Brain Mapp.* **5**, 233–237.

Drury, H.A., Van Essen, D.C., Joshi, S.C., and Miller, M.I. (1996). Analysis and comparison of areal partitioning schemes using two-dimensional fluid deformations, poster presentation. *NeuroImage* **3**, S130.

Duara, R., Kushch, A., Gross-Glenn, K., Barker, W.W., Jallad, B., Pascal, S., Loewenstein, D.A., Sheldon, J., Rabin, M., and Levin, B. (1991). Neuroanatomic differences between dyslexic and normal readers on magnetic resonance imaging scans. *Arch. Neurol.* **48**, 410–416.

Duvernoy, H.M. (1991). *The Human Brain*. Springer-Verlag, New York.

Evans, A.C., Dai, W., Collins, D.L., Neelin, P., and Marrett, S. (1991). Warping of a computerized 3D atlas to match brain image volumes for quantitative neuroanatomical and functional analysis. *SPIE Med. Imaging* **1445**, 236–247.

Evans, A.C., Collins, D.L., and Milner, B. (1992). An MRI-based stereotactic brain atlas from 300 young normal subjects. *Proc. 22nd Symp. Soc. Neurosci.*, Anaheim, CA. p. 408.

Fischl, B., Sereno, M.I., Tootell, R.B.H., and Dale, A.M. (1999). High-resolution inter-subject averaging and a coordinate system for the cortical surface. *Hum. Brain Mapp.* **8**(4), 271–284.

Fischl, B., Liu, A., and Dale, A.M. (2001). Manifold surgery: Constructing geometrically accurate and topologically correct models of the human cerebral cortex. *IEEE Trans Med Imaging* **20**(1), 70–80.

Fox, N.C., Freeborough, P.A., and Rossor, M.N. (1996). Visualization and quantification of rates of cerebral atrophy in Alzheimer's disease. *Lancet* **348**, 94–97.

Fox, N.C., Crum, W.R., Scahill, R.I., Stevens, J.M., Janssen, J.C., and Rossor, M.N. (2001). Imaging of onset and progression of Alzheimer's disease with voxel-compression mapping of serial magnetic resonance images. *Lancet* **358**, 201–205.

Fox, P.T., Perlmutter, J.S., and Raichle, M. (1985). A stereotactic method of localization for positron emission tomography. *J. Comput. -Assisted Tomogr.* **9**(1), 141–153.

Fox, P.T., Mintun, M.A., Reiman, E.M., and Raichle, M.E. (1988). Enhanced detection of focal brain responses using inter-subject averaging and change distribution analysis of subtracted PET images. *J. Cereb. Blood Flow Metab.* **8**, 642–653.

Freeborough, P.A., and Fox, N.C. (1998). Modeling brain deformations in Alzheimer's disease by fluid registration of serial 3D MR images. *J. Comput.-Assisted Tomogr.* **22**, 838–843.

Friston, K.J., Passingham, R.E., Nutt, J.G., Heather, J.D., Sawle, G.V., and Frackowiak, R.S.J. (1989). Localization in PET images: Direct fitting of the intercommissural (AC-PC) line. *J. Cereb. Blood Flow Metab.* **9**, 690–695.

Friston, K.J., Frith, C.D., Liddle, P.F., and Frackowiak, R.S.J. (1991). Plastic transformation of PET images. *J. Comput.-Assisted Tomogr.* **9**(1), 141–153.

Friston, K.J., Holmes, A.P., Worsley. K.J., Poline, J.P., Frith, C.D., and Frackowiak, R.S.J. (1995). Statistical parametric maps in functional imaging: A general linear approach. *Hum. Brain Mapp.* **2**, 189–210.

Gee, J.C., and Bajscy, R.K. (1998). Elastic matching: Continuum-mechanical and probabilistic analysis. In *Brain Warping* (A.W. Toga, ed.). Academic Press, San Diego, CA.

Gee, J.C., Reivich, M., and Bajcsy, R. (1993). Elastically deforming an atlas to match anatomical brain images. *J. Comput.-Assisted Tomogr.* **17**(2), 225–236.

Gee, J.C., Le Briquer, L., Barillot, C., Haynor, D.R., and Bajcsy, R. (1995). Bayesian approach to the brain image matching problem. *SPIE Med. Imaging: Image Process.* **2434**, 154–156.

Giedd, J.N., Blumenthal, J., Jeffries, N.O., Castellanos, F.X., Liu, H., Zijdenbos, A., Paus, T., Evans A,C., and Rapoport, J.L. (1999). Brain development during childhood and adolescence: A longitudinal MRI study. *Nat. Neurosci.* **2**(10), 861–863.

Good, C.D., Johnsrude, I.S., Ashburner, J., Henson, R.N., Friston, K.J., and Frackowiak, R.S.J. (2001). A voxel-based morphometric study of ageing in 465 normal adult human brains. *NeuroImage* **14**(1 Pt 1), 21–36.

Greitz, T., Bohm, C., Holte, S., and Eriksson, L. (1991). A computerized brain atlas: Construction, anatomical content and application. *J. Comput.-Assisted Tomogr.* **15**(1), 26–38.

Grenander, U., and Miller, M.I. (1998). *Computational Anatomy: An Emerging Discipline*, Technical Report. Dept. of Mathematics, Brown University, Providence, R.I.

Gur, R.E., Petty, R.G., Turetsky, B.I., and Gur, R.C. (1996). Schizophrenia throughout life: Sex differences in severity and profile of symptoms. *Schizophr. Res.* **21**, 1–12.

Haller, J.W., Banerjee, A., Christensen, G.E., Gado, M., Joshi, S., Miller, M.I., Sheline, Y., Vannier, M.W. and Csernansky, J.G. (1997). Three-dimensional hippocampal MR morphometry with high-dimensional transformation of a neuroanatomic atlas. *Radiology* **202**(2), 504–510.

Haney, S., Thompson, P.M., Cloughesy, T.F., Alger, J.R., Frew, A. and Toga, A.W. (2000a). Prognostic value of growth rates and spectroscopic data in patients with malignant gliomas. *Proc. Soc. Neurosci.*

Haney, S., Thompson, P.M., Cloughesy, T.F., Alger, J.R., and Toga, A.W. (2000b). Tracking tumor growth rates in patients with malignant gliomas: A test of two algorithms. *Am. J. Neuroradiol.* **22**, 73–82.

Haney, S., Thompson, P.M., Cloughesy, T.F., Alger, J.R., Frew, A., and Toga, A.W. (2000c). Cross-validation of tissue classification and surface modeling algorithms for determining growth rates of malignant gliomas: Prognostic value of growth rates and MR spectroscopy. *Int. Conf. Math. Eng. Tech. Med. Biol. Sci.*, Las Vegas, NV, *2000*.

Harasty, J., Double, K.L., Halliday, G.M., Kril, J.J., and McRitchie, D.A. (1997). Language-associated cortical regions are proportionally larger in the female brain. *Arch. Neurol* **54**, 171–176.

Höhne, K.H., Bomans, M., Pommert, A., Riemer, M., Schiers, C., Tiede, U., and Wiebecke, G. (1990). 3D visualization of tomographic volume data using the generalized voxel model. *Visual Comput.* **6**, 28–36.

Höhne, K.H., Bomans, M., Riemer, M., Schubert, R., Tiede, U., and Lierse, W. (1992). A 3D anatomical atlas based on a volume model. *IEEE Comput. Graphics Appl.* **12**, 72–78.

Holmes, C.J., Hoge, R., Collins, L., Woods, R., Toga, A.W., and Evans, A.C. (1998). Enhancement of MR images using registration for signal averaging. *J. Comput.-Assisted Tomogr.* **22**(2), 324–333.

Jacobs, R.E., and Fraser, S.E. (1994). Magnetic resonance microscopy of embryonic cell lineages and movements. *Science.* **263**, 681–684.

Kikinis, R., Shenton, M.E., Iosifescu, D.V., McCarley, R.W., Saiviroonporn, P., Hokama, H.H., Robatino, A., Metcalf D., Wible, C.G., Portas, C.M., Donnino, R., and Jolesz, F. (1996). A digital brain atlas for surgical planning, model-driven segmentation, and teaching. *IEEE Trans. Visual. Comput. Graphics* **2**(3), 232–241.

Larsen, J.P., Høien, T., and Ödegaard, H. (1992). Magnetic resonance imaging of the corpus callosum in developmental dyslexia. *Cogn. Neuropsychol.* **9**, 123–134.

Le Bihan, D. (1996). Functional MRI of the brain: Principles, applications and limitations. *Neuroradiology* **23** (1), 1–5.

Le Goualher, G., Procyk, E., Collins, D.L., Venugopal, R., Barillot, C. and Evans, A.C. (1999). Automated extraction and variability analysis of sulcal neuroanatomy. *IEEE Trans. Med. Imaging* **18**(3), 206–217.

Lehmann, E.D., Hawkes, D., Hill, D., Bird, C., Robinson, G., Colchester, A., and Maisley, M. (1991). Computer-aided interpretation of SPECT images of the brain using an MRI-derived neuroanatomic atlas. *Med. Inf.* **16**, 151–166.

Leung, A., and Chue, P. (2000). Sex differences in schizophrenia, a review of the literature. *Acta Psychiatr. Scand., Suppl.* **401**, 3–38.

MacDonald, D. (1998). A method for identifying geometrically simple surfaces from three dimensional images. Ph.D. Thesis, McGill University, Montreal, Quebec, Canada.

Mangin, J.-F., Frouin, V., Bloch, I., Regis, J., and Lopez-Krahe, J. (1994). Automatic construction of an attributed relational graph representing the cortex topography using homotopic transformations. *SPIE* **2299**, 110–121.

Mansour, A., Fox, C.A., Burke, S., Akil, H., and Watson, S.J. (1995). Immunohistochemical localization of the cloned Mu opioid receptor in the rat CNS. *J. Chem. Neuroanat.* **8**(4), 283–305.

Matsui, T., and Hirano, A. (1978). *An Atlas of the Human Brain for Computerized Tomography*. Igako-Shoin, Tokyo.

Mazziotta, J.C., Toga, A.W., Evans, A.C., Fox, P., and Lancaster, J. (1995). A probabilistic atlas of the human brain: Theory and rationale for its development. *NeuroImage* **2**, 89–101.

Mazziotta, J.C., Toga, A.W., Evans, A.C., Fox, P.T., Lancaster, J., Zilles, K., Woods, R.P., Paus, T., Simpson, G., Pike, B., Holmes, C.J., Collins, D.L., Thompson, P.M., MacDonald, D., Iacoboni, M., Schormann, T., Amunts, K., Palomero-Gallagher, N., Geyer, S., Parsons, L., Narr, K.L., Kabani, N., Le Goualher, G., Boomsma, D., Cannon, T., Kawashima, R., and Mazoyer, B. (2001). A probabilistic atlas and reference system for the human brain: International Consortium for Brain Mapping (ICBM). *Philos. Trans. R. Soc. London*, Ser. B **356**, 1293–1322.

Mega, M.S., Chen, S., Thompson, P.M., Woods, R.P., Karaca, T.J., Tiwari, A., Vinters, H., Small, G.W., and Toga, A.W. (1997). Mapping pathology to metabolism: Coregistration of stained whole brain sections to PET in Alzheimer's disease. *NeuroImage* **5**, 147–153.

Mega, M.S., Thompson, P.M., Cummings, J.L., Back, C.L., Xu, L.Q., Zohoori, S., Goldkorn. A., Moussai, J., Fairbanks, L., Small, G.W., and Toga, A.W. (1998). Sulcal variability in the Alzheimer's brain: Correlations with cognition. *Neurology* **50**, 145–151.

Mega, M.S., Chu, T., Mazziotta, J.C., Trivedi, K.H., Thompson, P.M., Shah, A., Cole, G., Frautschy, S.A., and Toga, A.W. (1999). Mapping biochemistry to metabolism: FDG-PET and beta-amyloid burden in Alzheimer's disease. *NeuroReport* **10**(14), 2911–2917.

Mega, M.S., Thompson, P.M., Toga, A.W., and Cummings J.L. (2000). Brain mapping in dementia. In *Brain Mapping: The Disorders* (A.W. Toga and J.C. Mazziotta, eds.). Academic Press, San Diego, CA.

Meltzer, C.C., and Frost, J.J. (1994). Partial volume correction in emission-computed tomography: Focus on Alzheimer disease. In *Functional Neuroimaging: Technical Foundations* (R.W. Thatcher, M. Hallett, T. Zeffiro, E.R. John, and M. Huerta, eds.), pp. 163–170. Academic Press, San Diego, CA.

Minoshima, S., Koeppe, R.A., Frey, K.A., Ishihara, M., and Kuhl, D.E. (1994). Stereotactic PET atlas of the human brain: Aid for visual interpretation of functional brain images. *J. Nucl. Med.* **35**, 949–954.

Mummery, C.J., Patterson, K., Price, C.J., Ashburner, J., Frackowiak, R.S., and Hodges, J.R. (2000). A voxel-based morphometry study of semantic dementia: Relationship between temporal lobe atrophy and semantic memory. *Ann. Neurol.* **47**(1), 36–45.

Narr, K.L., Cannestra, A.F., Thompson, P.M., Sharma, T., and Toga, A.W. (1998). Morphological variability maps of the corpus callosum and fornix in schizophrenia. *NeuroImage* **7**(4), S506.

Narr, K.L., Thompson, P.M., Sharma, T., Moussai, J., Zoumalan, C.I., Wang, W., Rayman, J., and Toga, A.W. (1999). Cortical and subcortical asymmetries: Sex effects in schizophrenic and normal populations. *Proc. Soc. Neurosci.*, Miami, FL, *1999*, 5574.

Narr, K.L., Thompson, P.M., Sharma, T., Moussai, J., Cannestra, A.F., and Toga, A.W. (2000). Mapping morphology of the corpus callosum in schizophrenia. *Cereb. Cortex* **10**, 40–49.

Narr, K.L., Thompson, P.M., Sharma, T., Moussai, J., Blanton, R., Anvar, B., Edris, A., Krupp, R., Rayman, J., Khaledy, M., and Toga, A.W. (2001). 3D mapping of temporo-limbic regions and the lateral ventricles in schizophrenia: Sex effects. *Biol. Psychiatry* **50**, 84–97.

Narr, K.L., Cannon, T.D., Woods, R.P., Thompson, P.M., Kim, S., Asunction, D., van Erp, T.G., Poutanen, V.P., Huttunen, M., Lonnqvist, J., Standerksjold-Nordenstam, C.G., Kaprio, J., Mazziotta, J.C., and Toga, A.W. (2002). Genetic contributions to altered callosal morphology in schizophrenia. *J. Neurosci.* **22**, 3720–3729.

Ono, M., Kubik, S., and Abernathey, C.D. (1990). *Atlas of the Cerebral Sulci*. Thieme, Stuttgart.

Palovcik, R.A., Reid, S.A., Principe, J.C., and Albuquerque, A. (1992). 3D computer animation of electrophysiological responses. *J. Neurosci. Methods* **41**, 1–9.

Penfield, W., and Boldrey, E. (1937). Somatic motor and sensory representation in the cerebral cortex of man as studied by electrical stimulation. *Brain* **60**, 389–443.

Pitiot, A., Toga, A.W., and Thompson, P.M. (2002). Adaptive elastic segmentation of brain MRI via shape model guided evolutionary programming. *IEEE Trans. Med. Imaging* **21**(8), 910–923.

Pommert, A., Schubert, R., Riemer, M., Schiemann, T., Tiede, U., and Höhne, K.H. (1994). Symbolic modeling of human anatomy for visualization and simulation. *IEEE Visual. Biomed. Comput.* **2359**, 412–423.

Rabbitt, R.D., Weiss, J.A., Christensen, G.E., and Miller M.I. (1995). Mapping of hyperelastic deformable templates using the finite element method. *Proc. SPIE* **2573**, 252–265.

Rasser, P.E., Johnston, P., Lagopoulos, J., Ward, P.B., Schall, U., Thienel, R., Bender, S., and Thompson, P.M. (2003). Analysis of fMRI BOLD activation during the Tower of London Task using cortical pattern matching. *Int. Cong. Schizophr. Res. (ICSR),* Colorado Springs, CO, *2003.*

Rex, D.E., Pouratian, N., Sicotte, N.L., and Toga, A.W. (2001). Locational effect of brain masses on f MRI of tongue movement. *NeuroImage Abstr.* **13**(6), S232.

Rizzo, G., Gilardi, M.C., Prinster, A., Grassi, F., Scotti, G., Cerutti, S., and Fazio, F. (1995). An elastic computerized brain atlas for the analysis of clinical PET/SPET data. *Eur. J. Nucl. Med.* **22**(11), 1313–1318.

Roland, P.E., and Zilles, K. (1994). Brain atlases—A new research tool. *Trends Neurosci.* **17**(11), 458–467.

Rombouts, S.A., Barkhof, F., Witter, M.P., and Scheltens, P. (2000). Unbiased whole-brain analysis of gray matter loss in Alzheimer's disease. *Neurosci Lett.* **285**(3), 231–233.

Royackkers, N., Desvignes, M., and Revenu, M. (1996). *Construction automatique d'un atlas adaptatif des sillons corticaux, ORASIS 96,* pp. 187–192. Clermont-Ferrand.

Sandor, S.R., and Leahy, R.M. (1995). Towards automated labeling of the cerebral cortex using a deformable atlas. *Inf. Proc. Med. Imaging, 1995,* pp. 127–138.

Schaltenbrand, G., and Bailey, P. (1959). *Introduction to Stereotaxis with an Atlas of the Human Brain.* Thieme, New York and Stuttgart.

Schaltenbrand, G., and Wahren, W. (1977). *Atlas for Stereotaxy of the Human Brain,* 2nd ed. Thieme, Stuttgart.

Schiemann, T., Nuthmann, J., Tiede, U., and Höhne, K.H. (1996). Segmentation of the Visible Human for high-quality volume-based visualization. *Visual. Biomed. Comput.* **4**, 13–22.

Sereno, M.I,. Dale, A.M., Liu, A., and Tootell, R.B.H. (1996). A surface-based coordinate system for a canonical cortex. *NeuroImage* **3**(3), S252.

Shenton, M.E., Kikinis, R., Jolesz, F.A., Pollack, S.D., LeMay, M., Wible, C.G., Hokama, H., Martin, J., Metcalf, D., Coleman M. and McCarley, R. (1992). Abnormalities of the left temporal lobe and thought disorder in schizophrenia. *N. Engl. J. Med.* **327**(9), 604–612.

Smith, G.E. (1907). A new topographical survey of the human cerebral cortex, being an account of the distribution of the anatomically distinct cortical areas and their relationship to the cerebral sulci. *J. Anat.,* **41**, 237–254.

Sowell, E.R., Thompson, P.M., Holmes, C.J., Jernigan, T.L., and Toga, A.W. (1999). In vivo evidence for post adolescent brain maturation in frontal and striated regions. *Nat. Neurosci.* **2**(10), 859–861.

Sowell, E.R., Thompson, P.M, Tessner, K.D., and Toga A.W., (2001a). Mapping continued brain growth and gray matter density reduction in dorsal frontal cortex: Inverse relationships during post adolescent brain maturation. *J. Neurosci.* **21**(22), 8819–8829.

Sowell, E.R., Mattson, S.N., Thompson, P.M., Jernigan, T.L., Riley, E.P. and Toga, A.W. (2001b). Mapping callosal morphology and cognitive correlates: Effects of heavy prenatal alcohol exposure. *Neurology* **57**, 235–244.

Sowell, E.R., Thompson, P.M., Mattson, S.N., Tessner, K.D., Jernigan, T.L., Riley, E.P., and Toga, A.W. (2002). Regional brain shape abnormalities persist into adolescence after heavy prenatal alcohol exposure. *Cereb. Cortex* **12**(8), 856–865.

Sowell, E.R., Thompson, P.M., Peterson, B.S., Welcome, S.E., Henkenius, A.L. and Toga, A.W., (2003). Mapping cortical change across the human life span. *Nat. Neurosci.* **6**(3), 309–315.

Spitzer, V., Ackerman, M.J., Scherzinger A.L., and Whitlock, D. (1996). The visible human Male: a technical report. *J. Am. Med. Inf. Assoc.* **3**(2), 118–130.

Stewart, J.E., Broaddus, W.C., and Johnson, J.H. (1996). Rebuilding the visible man, *Visual. Biomed. Comput.* **4**, 81–86.

Subsol, G., Roberts, N., Doran, M., Thirion, J.P., and Whitehouse, G.H. (1997). Automatic analysis of cerebral atrophy. *Magn. Reson. Imaging* **15**(8), 917–927.

Talairach, J., and Szikla, G. (1967). *Atlas d'anatomie stereotaxique du telencephale: etudes anatomo-radiologiques.* Masson, Paris.

Talairach, J., and Tournoux, P. (1988). *Co-planar Stereotaxic Atlas of the Human Brain.* Thieme, New York.

Thirion, J.-P. (1995). *Fast Non-Rigid Matching of Medical Images,* INRIA Internal Rep. No. 2547. Projet Epidaure, INRIA, France.

Thirion, J.-P., and Calmon, G. (1997). *Deformation Analysis to Detect and Quantify Active Lesions in 3D Medical Image Sequences. INRIA Tech. Rep. No. 3101.* INRIA, France.

Thompson, P.M., and Toga, A.W. (1996). A surface-based technique for warping 3-dimensional images of the brain. *IEEE Trans. Med. Imaging* **15**(4), 1–16.

Thompson, P.M., and Toga, A.W. (1997). Detection, visualization and animation of abnormal anatomic structure with a deformable probabilistic brain atlas based on random vector field transformations. *Med. Image Anal.* **1**(4), 271–294.

Thompson, P.M., and Toga, A.W. (1998a). Anatomically-driven strategies for high-dimensional brain image warping and pathology detection. In *Brain Warping* (A.W. Toga, ed.), pp. 311–336. Academic press, San Diego, CA.

Thompson, P.M., and Toga, A.W. (1998b). *Mathematical/Computational Strategies for Creating a Probabilistic Atlas of the Human Brain,* Workshop on Statistics of Brain Mapping. Centre de Recherches Mathématiques, McGill University, Montreal, Quebec, Canada.

Thompson, P.M., and Toga, A.W. (2000). Elastic image registration and pathology detection. In *Handbook of Medical Image Processing* (I. Bankman, R. Rangayyan, A.C. Evans, R.P. Woods, E. Fishman, and H.K. Huang, eds.). Academic Press, San Diego, CA.

Thompson, P.M., and Toga, A.W. (2002). A framework for computational anatomy. *Comput. Visual. Sci.* **5**, 1–12.

Thompson P.M., and Toga, A.W. (2003). Cortical diseases and cortical localization. *Nat. Encycl. Life Sci.,* (in press).

Thompson, P.M., Schwartz, C., Lin, R.T., Khan A.A., and Toga, A.W. (1996a). 3D statistical analysis of sulcal variability in the human brain. *J. Neurosci.* **16**(13), 4261–4274.

Thompson, P.M., Schwartz, C., and Toga, A.W. (1996b). High-resolution random mesh algorithms for creating a probabilistic 3D surface atlas of the human brain. *NeuroImage* **3**, 19–34.

Thompson, P.M., MacDonald, D., Mega, M.S., Holmes, C.J., Evans, A.C., and Toga, A.W. (1997). Detection and mapping of abnormal brain structure with a probabilistic atlas of cortical surfaces. *J. Comput.-Assisted Tomog.* **21**(4), 567–581.

Thompson, P.M., Moussai, J., Khan, A.A., Zohoori, S., Goldkorn, A., Mega, M.S., Small, G.W., Cummings, J.L., and Toga, A.W. (1998). Cortical variability and asymmetry in normal aging and alzheimer's disease. *Cereb. Cortex* **8**(6), 492–509.

Thompson, P.M., Giedd, J.N., Woods, R.P., MacDonald, D., Evans, A.C., and Toga, A.W. (2000a). Growth patterns in the developing brain detected By using continuum-mechanical tensor maps. *Nature (London).* **404**, 190–193.

Thompson, P.M., Mega, M.S., Narr, K.L., Sowell, E.R., Blanton, R.E., and Toga, A.W. (2000b). Brain image analysis and atlas construction. In *SPIE Handbook on Medical Image Analysis* (M. Fitzpatrick, ed.). SPIE Press.

Thompson, P.M., Mega, M.S., and Toga, A.W. (2000c). Disease-specific brain atlases. In *Brain Mapping: The Disorders* (A.W. Toga and J.C. Mazziotta, eds.). Academic Press, San Diego, CA.

Thompson P.M. Narr, K.L., Blanton R.E., and Toga, A.W. (2000d). Mapping structural alterations of the corpus callosum during brain development and degeneration. In *The Corpus Callosum* (M. Iacoboni and E. Zaidel eds.). Kluwer Academic Press, Dordrecht,The Netherlands.

Thompson, P.M., Woods, R.P., Mega, M.S., and Toga, A.W. (2000e). Mathematical/computational challenges in creating population-based brain atlases. *Hum. Brain Mapp.* **9**(2), 81–92.

Thompson, P.M., and Cannon, T.D., Narr, K.L., van Erp, T., Poutanen, V.P., Hutteunen. M., Lonnqvist, J., Standertskjold-Nordenstam, C.G., Kaprio, J., Khaledy, M., Dail, R., Zoumalen, C., and Toga, A.W. (2001a). Genetic influences on brain structure. *Nat. Neurosci.* **4**(12), 1253–1258.

Thompson, P.M., de Zubicaray, G., Janke, A.L., Rose, S.E., Dittmer, S., Semple, J., Gravano, D., Han, S., Herman, D., Hong, M.S., Mega, M.S., Cummings, J.L., Doddrell, D.M., and Toga, A.W. (2001b). Detecting Dynamic (4D) Profiles of degenerative rates in Alzheimer's disease patients, using high-resolution tensor mapping and a brain atlas encoding atrophic rates in a population. *7th Annu. Meet. Org. Hum. Brain Mapp.* Brighton, England, *2001*.

Thompson, P.M., Mega, M.S., Vidal, C., Rapoport, J.L., and Toga, A.W. (2001c). Detecting disease-specific patterns of brain structure using cortical pattern matching and a population-based probabilistic brain atlas. *Lect. Notes Comput. Sci.* **2082**, 488–501.

Thompson, P.M., Mega, M.S., Woods, R.P., Blanton, R.E., Moussai, J., Zoumalan, C.I., Aron, J., Cummings, J.L., and Toga, A.W. (2001d). Early cortical change in Alzheimer's disease detected with a disease-specific population-based brain atlas. *Cereb. Cortex* **11**(1), 1–16.

Thompson, P.M., Vidal, C., Giedd, J.N., Gochman, P., Blumenthal, J., Nicolson, R., Toga, A.W., and Rapoport, J.L. (2001e). Mapping adolescent brain change reveals dynamic wave of accelerated gray matter loss in very early-onset schizophrenia. *Proc. Na. Acad. Sci. U.S.A.* **98**(20), 11650–11655.

Thompson, P.M., Cannon, T.D., and Toga, A.W. (2002a). Mapping genetic influences on human brain structure. *Ann. Med.* **34**(7-8), 523–536.

Thompson, P.M., Hayashi, K.M., de Zubicaray, G., Janke, A.L., Rose, S.E., Semple, J., Doddrell, D.M., Cannon, T.D., and Toga, A.W. (2002b). Detecting dynamic and genetic effects on brain structure using high-dimensional cortical pattern matching. *Proc. Int. Symp. Biomed. Imag. (ISBI2002)*, Washington, DC, *2002*.

Thompson, P.M., Hayashi, K.M., de Zubicaray, G., Janke, A.L., Rose, S.E., Semple, J., Hong, M.S., Herman, D., Gravano, D., Dittmer, S., Doddrell, D.M., and Toga, A.W. (2003a). Improved detection and mapping of dynamic hippocampal and ventricular change in Alzheimer's disease using 4D parametric mesh skeletonization. *9th Ann. Meet. Org. Hum. Brain Mapp.*, New York, *2003*.

Thompson, P.M., Hayashi, K.M., de Zubicaray, G., Janke, A.L., Rose, S.E., Semple, J., Herman, D., Hong, M.S., Dittmer, S.S., Doddrell,

D.M., and Toga, A.W. (2003b). Dynamics of gray matter loss in Alzheimer's disease. *J. Neurosci.* **23**, 994–1005.

Thompson, P.M., Rapoport, J.L., Cannon, T.D., and Toga, A.W. (2003c). Automated analysis of structural MRI data. In *Brain Imaging in Schizophrenia* (A.L. Lawrie, E.C. Johnstone, and D. Weinberger, eds.). Oxford University Press, London and New York.

Thompson, P.M., Rapoport, J.L., Cannon, T.D., and Toga, A.W. (2003d). Imaging the brain as schizophrenia develops: Dynamic and genetic brain maps. *Primary Psychiatry* (to appear).

Tiede, U., Bomans, M., Höhne, K.H., Pommert, A., Riemer, M., Schiemann, T., Schubert, R., and Lierse, W. (1993). A computerized 3D atlas of the human skull and brain. *Am. J. Neuroradiol.* **14**, 551–559.

Toga, A.W., ed. (1998). *Brain Warping*. Academic Press, San Diego, CA.

Toga, A.W., and Mazziotta, J.C. (1996). Introduction to cartography of the brain. In *Brain Mapping: The Methods* (A.W. Toga and J.C. Mazziotta, eds.), pp. 3–25. Academic Press, San Diego, CA.

Toga, A.W., and Thompson, P.M. (1997). Measuring, mapping, and modeling brain structure and function. *SPIE Med. Imag. Symp.*, Newport Beach, CA, *1997*, SPIE Lec. Notes, Vol. 3033.

Toga, A.W., and Thompson, P.M. (1998). Multimodal brain atlases. In *Medical Image Databases* (S.T.C. Wong, ed.), pp. 53–88. Kluwer Academic Press, Dordrecht, The Netherlands.

Toga, A.W., Ambach, K., Quinn, B., Hutchin, M., and Burton, J.S. (1994). Postmortem anatomy from cryosectioned whole human brain. *J. Neurosci. Methods* **54**(2), 239–252.

Toga, A.W., Goldkorn, A., Ambach, K., Chao, K., Quinn, B.C., and Yao, P. (1997). Postmortem cryosectioning as an anatomic reference for human brain mapping. *Comput. Med. Imaging Graphpics* **21**(2), 131–141.

Toga, A.W., Thompson, P.M., Mega, M.S., Narr, R.L., and Blanton, R.E. (2001). Probabilistic approaches for atlasing normal and disease-specific brain variability. *Anat. Embryol.* **204**, 267–282.

Van Buren, J.M., and Borke, R.C. (1972). *Variations and Connections of the Human Thalamus*, Vols. 1 and 2. Springer, New York.

Van Buren, J.M., and Maccubin, D. (1962). An outline atlas of human basal ganglia and estimation of anatomic variants. *J. Neurosurg.* **19**, 811–839.

Van Essen, D.C., and Maunsell, J.H.R. (1983). Hierarchical organization and functional streams in the visual cortex. *Trends Neurol. Sci.* **6**, 370–375.

Van Essen, D.C., Drury, H.A., Joshi, S.C., and Miller, M.I. (1997). Comparisons between human and macaque using shape-based deformation algorithms applied to cortical flat maps. *NeuroImage* **5**(4), S41.

Warfield, S., Dengler, J., Zaers, J., Guttmann, C.R.G., Wells, W.M., Ettinger, G.J., Hiller, J., and Kikinis, R. (1995). Automatic identification of gray matter structures form MRI to improve the segmentation of white matter lesions. *Proc. Med. Robotics & Comput. Assist. Surg. (MRCAS), 1995*, pp. 55–62.

Woods, R.P. (1996). Modeling for intergroup comparisons of imaging data. *NeuroImage* **4**(3), 84–94.

Worsley, K.J., Andermann, M., Koulis, T., MacDonald, D., and Evans, A.C. (1999). Detecting changes in non-isotropic images. *Hum. Brain Mapp.* **8**, 98–101.

Wright, I.C., McGuire, P.K., Poline, J.B., Travere, J.M., Murray, R.M., Frith, C.D., Frackowiak, R.S.J., and Friston, K.J. (1995). A voxel-based method for the statistical analysis of gray and white matter density applied to schizophrenia. *NeuroImage* **2**, 244–252.

Zeineh, M.M., Engel, S.A., Thompson, P.M., and Bookheimer, S. (2001). Unfolding the human hippocampus with high-resolution structural and functional MRI (invited paper). *New Anat. (Anat. Re.)* **265**(2), 111–120.

Zeineh, M.M., Engel, S.A., Thompson, P.M., and Bookheimer, S.Y. (2003a). Dynamic changes within the human hippocampus during memory consolidation. *Science* **299**(5606), 577–580.

Zeineh, M.M., Mazziotta, J.C., Thompson, P.M., Engel, S.A., and Bookheimer, S.Y. (2003b). Hippocampal flat maps of cortical thickness and power. *9th Annu. Meet. Org. Hum. Brain Mapp.* New York.

Zhou, Y., Thompson, P.M., and Toga, A.W. (1999). Automatic extraction and parametric representations of cortical sulci. *Comput. Graphics Appl.* **19**(3), 49–55.

Zijdenbos, A.P., and Dawant, B.M. (1994). Brain segmentation and white matter lesion detection in MR images. *Crit. Rev. Biomed. Eng.* **22**(5-6), 401–465.

Zoumalan, C.I., Mega, M.S., Thompson, P.M., Fuh, J.L., Lindshield, C., and Toga, A.W. (1999). Mapping 3D patterns of cortical variability in normal aging and Alzheimer's disease. *Annu. Conf. Am. Acad. of Neurol.*

Maximizing Information Content in Shared and Archived Neuroimaging Studies of Human Cognition

John Darrell Van Horn, Ph.D. and Michael S. Gazzaniga, Ph.D.

CONTENTS

26

Databasing the Brain. Edited by Stephen H. Koslow and Shankar Subramaniam
ISBN 0-471-30921-4 © 2005 John Wiley & Sons, Inc.

26.1 INTRODUCTION

The sharing of published functional imaging data between researchers and the contribution of data to centralized repositories has been the topic recent interest in the neuroimaging and cognitive neuroscience communities (Governing Council of the Organization for Human Brain Mapping, 2001; Kotter, 2001; Van Horn et al., 2001). Data-sharing initiatives for the neurosciences, in general, have been recently argued to be necessary, timely, and beneficial to the advancement of our understanding of fundamental brain processes (Koslow, 2002). Success stories from data-sharing initiatives in other fields, such as genomics (Gifford, 2001) and proteomics (Yarmush and Jayaraman, 2002), demonstrate the many scientific advantages to study data exchange and the unique educational opportunities available by providing access to biological data archives (Magee et al., 2001).

Several neuroimaging data exchange and database projects are now underway to archive brain imaging data for the purposes of developing brain atlases (Mazziotta et al., 2001), conducting detailed meta-analyses (e.g., Nielsen and Hansen, 2002), and mapping population-based activation patterns on flattened cortical representations (Van Essen et al., 2001). Available electronically, these efforts are exciting because they seek to provide the neuroscience community with immediate access to current findings in brain research that cannot often be conveyed in a printed form. By preserving the body of neuroimaging work, such neuroinformatics endeavors speak to a time when neuroimaging studies of brain function form a truly discovery-based science (Van Horn and Gazzaniga, 2002).

26.2 THE fMRI DATA CENTER

The fMRI Data Center (fMRIDC; http://www.fmridc.org) at Dartmouth College is one such example, seeking to advance progress in the understanding cognitive processes through the archiving and open distribution of neuroimaging datasets published in the peer-reviewed literature (Van Horn and Gazzaniga, 2002). The current size of the fMRIDC archive exceeds 2.6 terabytes of data from over 70 different fMRI studies of cognitive function. Study data are rapidly being added to the collection from neuroimaging laboratories around the world and made openly available to the community. These include a number of cognitive domains including investigations of human memory encoding and retrieval (Mechelli et al., 2003), the processing of visual information (Ishai et al., 2000), motor representation (Toni et al., 2002), attention (Kable et al., 2002), and working memory processes (Rypma et al., 2002). In cataloguing these studies, the fMRIDC has sought to provide the scientific community with access to complete study information at levels not before possible. By so doing, researchers can examine the statistical effects reported in the original peer-reviewed scientific article; they may explore alternative image preprocessing techniques; they can compare and contrast alternative statistical methods of analysis; and, they can explore homologies between the brain activation patterns from across different studies. Thus, they are in a position to not only validate fMRI results from the literature but to also examine the data in novel ways that promote new thinking about that data and encourage novel fMRI experiments. The potential of this neuroscientific resource is anticipated to attract researchers from within cognitive neuroscience, seeking data for the purposes of brain research and education, but also workers from fields such as computer science and mathematics, hoping to apply methods from their fields to brain imaging data (Van Horn et al., 2001). Indeed, over 1100 requests for datasets have been fulfilled to researchers worldwide for use in new research and education.

26.3 DATA SHARING AND NEUROIMAGING

A central question in the development of initiatives like the fMRIDC has been what data from neuroimaging studies ought to be shared (Koslow, 2000; Toga, 2002)? In a recent article, Fox and Lancaster (2002) argued that when constructing a publicly accessible archive of published neuroimaging studies, it is the test statistics and accompanying Talairach spatial coordinates from the processed statistical results that possess the greatest scientific worth. They argue that the open sharing of raw functional image data from published studies is problematic because the data have not been subjected to the processing used to arrive at the experimental conclusions and, therefore, are of little or limited value. With this in mind, Fox and Lancaster (2002) have proposed an admirable rule for the sharing of neuroimaging data; "share the most valuable data type first" (p. 320). They maintain that neuroimaging data increase in value with every step of mathematical and statistical processing. Implicit in their argument is the notion that the successive processing steps involved in the generation of statistical activation maps results in generally "improved" information content of the imaging data. That is to say that these data are somehow of increased purity and that greater inferences can be made from the data following such intelligent processing. Therefore, the processed statistical test results and accompanying spatial locations are intellectually and scientifically more valuable than the raw data from which they came.

Though theirs is an approach aimed at being practical and parsimonious, this component of the Fox and Lancaster model of data sharing may not be suitable to fully appreciating and characterizing the information contained in these routinely large, complex, and structured data sets. Study metadata are also important to consider, being essential to accurately describe the study protocols and experimental design matrix information. Moreover, the steps of processing themselves also require detailed description. In the present chapter, we discuss some key theoretical and practical challenges to the sharing and archiving of the data from fMRI studies of human cognitive function. We discuss the theoretical limitations of restricting data sharing to only a greatly reduced set of statistical results, the effective sharing of study metadata in the form of study ontologies, and some methods now being developed for accurately describing the pipeline image data processing, and we propose a more complete model for sharing neuroimaging data.

26.4 DATA PROCESSING AND STUDY INFORMATION CONTENT

Consider an event-related study performed to examine the differential hemodynamic response to certain types of visual stimuli. The investigator elects to collect data from eight subjects in four functional EPI runs where each run lasts 5 minutes with a TR of 2 seconds. Each 16-bit integer image volume has dimensions $64 \times 64 \times 27$. Each subject also undergoes a single high-resolution ($256 \times 256 \times 128$) SPGR anatomical image. This results in a raw dataset across all function and structural scans (including accompanying image header files) of all subjects in excess of a gigabyte (GB) of data. The raw functional time course data are then subjected to a chain of data processing involving the following: adjustment for slice acquisition order effects (Paradis et al., 2001); spatial realignment (e.g., Woods et al., 1998a,b); nonlinear spatial warping (Fox et al., 1985; Friston et al., 1991; Woods et al., 1999); spatial smoothing (Friston et al., 2000); and so on. This is by all means not an exhaustive list of the potential steps of processing that a dataset might have to pass through prior to being considered ready for statistical analysis or decomposition.

At each stage of this processing chain, however, any resulting output may experience round-off errors for parameter estimates, image interpolation effects, changes in image size, reslicing, filtering, and so on. Other researchers examining similar paradigms may make dramatically different choices in this data processing chain, substituting alternative methods, rearranging which steps are done in which order, and so on. With the outcome of each of these steps, there is the potential for some loss of information contained in the result about the original raw dataset. Finally, once significant local maxima have been obtained, a simple tabular representation of local maxima contains little information about underlying richness of the original dataset. Sharing only these data short-changes the research community by denying them the opportunity to reexamine potentially important datasets, examine ones for effects unreported by the original authors, and use the data to educate the next generation of neuroscientists.

With this in mind, a fundamental notion from information theory, termed the *data processing inequality* (Cover and Thomas 1991), may be illustrative for guiding which form of the image data may be most beneficial to share. This quantitative conjecture can be used to show that, in fact, no clever manipulation of a dataset can improve the inferences that can be made from that data. Additionally, this has important implications for gauging the value of shared fMRI data.

Suppose, for example, that we possess a neuroimaging dataset of voxel values that conform to some probability density function. Additionally, we have derived several more representations of this data through two stages of mathematical processing. Examples from functional neuroimaging may represent fMRI time course data at different stages of preprocessing before inferential statistical tests are applied—for example, raw, spatially realigned, spatiotemporally filtered, and so on. Let us call these representations X, Y, and Z, respectively. The random variables X, Y, and Z form a Markov chain, in that order (denoted by $X \rightarrow Y \rightarrow Z$), if the conditional distribution of Z depends only on Y and is conditionally independent of X. Specifically, X, Y, and Z form the Markov chain $X \rightarrow Y \rightarrow Z$ if the joint probability density function can be written as

$$p(x, y, z) = p(x)p(y|x)p(z|y).$$

The data processing streams often employed in functional neuroimaging often form such a Markov chain. Data are routinely transformed from one state of processed data to another that are then fed into still another transformation algorithm (Fig. 26.1).

These steps are often necessary to remove effect of physiological noise, de-trending of the data, and other general signal conditioning. However, with each step, one often incurs a cost due to potential reduction in the entropy (H) of the data. What is more, simple consequences follow from the above probabilistic relationship that has direct importance to the processing of fMRI voxel time course data.

First, $X \rightarrow Y \rightarrow Z$ if and only if (iff) X and Z are conditionally independent given Y. Markovity implies conditional independence because

$$p(x, z|y) = \frac{p(x, y, z)}{p(y)}$$
$$= \frac{p(x, y)p(z|y)}{p(y)}$$
$$= p(x|y)p(z|y).$$

This is the basic characterization of Markov chains and may be readily extended to define *Markov fields*, which are n-dimensional random processes in which the interior and exterior of the field are independent given the values on the field's boundary. In neuroimaging, the condition of Markovity is likely to be satisfied anytime an image time course is spatially resampled, interpolated, smoothed, or thresholded as well as when the data are temporally low-pass, high-pass, or otherwise filtered or adjusted for slice acquisition timing. For instance, data passed through a lossy compression scheme that then undergoes decompression again will not contain the full information content about the original data because a portion of the signal magnitude information will have been removed by the compression process itself.

Second, using the chain rule, the *mutual information* (denoted by I) of these variables can be represented in two different ways:

$$I(X; Y, Z) = I(X; Z) + I(X; Y|Z)$$
$$= I(X; Y) + I(X; Z|Y).$$

Since X and Z are conditionally independent given Y, we have $I(X; Z|Y) = 0$. Likewise, since $I(X; Y|Z) \geq 0$, we have $I(X; Y) = I(X; Z)$. We have equality iff $I(X; Y|Z) = 0$, wherein $X \rightarrow Y \rightarrow Z$ forms the Markov chain. Using this logic, one can show that

$$I(X; Y) \geq I(X; Z),$$

meaning that if $X \rightarrow Y \rightarrow Z$, then the mutual information between Y and X (i.e., $I(X; Y)$) is greater than or equal to the information shared between Z and X (i.e., $I(X; Z)$). In other words, the information in the raw image data X that is contained in the processed image Y is greater than or equal to the information in X contained in the further processed image Z.

Suppose now that Z is a function of the variable Y; for example, $Z = g(Y)$. Without loss of generality, we have $I(X; Y) \geq I(X; g(Y))$. Which simply indicates that $X \rightarrow Y \rightarrow g(Y)$ also forms a Markov chain and implies that mathematical functions or transformations of the data Y do not increase the information one has about X.

It is also possible to show that if $X \rightarrow Y \rightarrow Z$, then $I(X; Y|Z) \leq I(X; Y)$. From the fact that $I(X; Y, Z) = I(X; Z) + I(X; Y|Z) = I(X; Y) + I(X; Z|Y)$, and using the fact that $I(X; Z|Y) = 0$ by Markovity and $I(X; Z) = 0$, we have $I(X; Y|Z) \leq I(X; Y)$. This implies that even if we attempt to

Typical fMRI Data Processing Stream

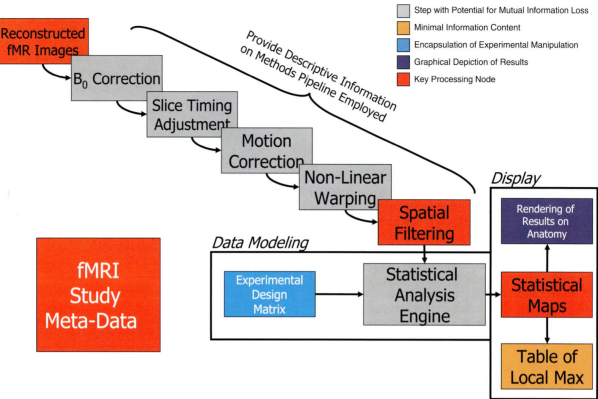

Figure 26.1   Similar to a biological tissue assay, a typical fMRI data processing chain involves a number of discrete steps in order to prepare the data for statistical analysis. This may involve several levels of preprocessing applied to the raw data such as B0 field inhomogeneity adjustment, slice timing phase modulation, image registration, spatial normalization, image smoothing, and so on. This figure depicts several such steps but is by no means exhaustive with respect to all possible such steps used by workers in the neuroimaging field. Each step is used to prepare the image data from the previous step for statistical modeling via the investigator's statistical engine of choice, potentially modifying the resultant image data in a way that reduced the amount of information it contains about the original version of the data. Some of these modifications are needed to correct the data for problems resulting from the measurement modality and are intended to make results obtained from the statistical analysis more robust. Many steps are not invertible without knowledge of the algorithms applied, without the parameters obtained, or without having access to the original data itself. However, in the absence of such accurate knowledge, these steps cannot amplify the amount of information contained in the data from any step about previous steps in the processing chain. By the end of processing and analysis, the statistical results represent a model summary of the processed information that remains at the last stage of processing before statistical analysis (i.e., t-maps from the General Linear Model). A further tabular summary of local maxima represents only a very small amount of the total information that was contained in the originally collected data. A thorough model for data sharing would include the data from the beginning (raw), the middle (after processing but before statistical modeling with a description of the processing chain), and the end (statistical results maps) of the processing chain. In this way, investigators may have access to the data at key stages of processing for the purposes of exploring a new analysis chain de novo, altering the statistical model applied to the processed data, or scrutinizing patterns of statistical results as obtained by the original authors.

take into account the influence of the processing Y given the processing Z, the amount of information is still less than between X and Y alone.

Lastly, it should be mentioned that one can show that $I(X;Y|Z)$ can be greater than $I(X;Y)$ when X, Y, and Z do not form a Markov chain. For instance, let datasets X and Y be independent, fair, binary random variables (perhaps obtained from two masks of the voxels contained within the boundary of the brain obtained from two subjects), and let $Z = X + Y$. Then $I(X;Y) = 0$, but $I(X;Y|Z) = H(X|Z) - H(X|Y,Z) = H(X|Z) = P(Z = 1)H(X|Z = 1) = 0.5$ bit. However, in such cases the process $X \rightarrow Y \rightarrow Z$ does not meet the definition of a Markov chain.

In the case of an image data processing chain that can be characterized as Markov, the information contained in Y about the original data X is decreased or remains unchanged through the observation of a "downstream" resultant variable Z. Specifically for the case of fMRI time course data, this notion of data processing concerns algebraic, trigonometric, and transcendental transformations, temporal and spatial filtering, and all forms of decomposition on the data time courses. In other words, irrespective of the amount or level of sophistication in the mathematical analysis, abstraction, and reduction of fMRI data, the amount of information concerning blood oxygenation level-dependent (BOLD) effects in fMRI can only be held constant or reduced through computational processing. Thus, a tabular representation of statistical results and accompanying Talaraich coordinates likely comprises only a very limited amount of information, in a mathematical sense, about the original raw fMRI time series. Though intended to reduce or remove artifactual influences on the effects of interest in the time course, these steps also have the effect of potentially reducing information content about the data collected from the scanner. When shared, such a reduced representation of study data is likely to greatly constrain the ability of other researchers to most broadly examine, scrutinize, and critique the data in a new meta- or mega-analysis.

26.5 STUDY METADATA AND THE DEVELOPMENT OF ONTOLOGICAL FRAMEWORKS

Critically important to the ability to work with the neuroimaging data itself is the accompanying study metadata. Metadata are the various parameters and measurements that describe the experimental data under manipulation. For functional neuroimaging, this includes subject demographic parameters, scanner protocol information, and the specifics of the experimental design (Van Horn et al., 2001). Basic metadata, such as scanner parameters, are often surprisingly difficult for investigators to manage and maintain, sometimes being recorded in laboratory notebooks, on the scanner, in computer files, and on a variety of pieces of scrap paper. fMRI experimental design information, in particular, may take a variety of forms that may make it difficult to effectively manage. For instance, a time course regressor for a block design experiment can be coded as a column of binary numbers pertaining to whether a particular stimulus was presented at each TR-interval. Alternatively, the same column vector can be run-length encoded to represent the stimulus type, the temporal onsets, and presentation duration. The software package preferred by an investigator for data analysis may represent this stimulus time course information in still another fashion (e.g., a Matlab-based structure file in SPM). Initiatives for the sharing of primary neuroscientific data, like the fMRI Data Center and other efforts, depend on the ability of researchers to coherently represent and organize these various forms of study metadata so that once it is contributed to the archive, it may seamlessly be incorporated with other studies in the study database.

Having such data available to researchers in an online form has been a goal in the brain imaging community for several years, and several methods of metadata organization have been suggested (Jakobovits et al., 1996, 2000, 2001). However, the conceptual relationships among study metadata elements may not preserved in flat data file formats or HTML-based systems. Nor are these always in the most suitable form for data exchange and large-scale archiving. This has necessitated the development of metadata *ontologies* for investigators to utilize in the management, visualization, and exploration of their study descriptive and metadata in addition to their fMRI results images (Van Horn et al., 2002). Ontological frameworks differ from other representations of data (text, binary, XML markup, commercial database schemas, etc.) in that they seek to capture the conceptual structure of a domain and support well-defined procedures for combining, searching, and extending sometimes widely varying views of data within the domain. Ontologies are a very practical way

of not only storing metadata but also storing the concepts and semantic content behind the metadata itself. Successful examples may be found in efforts to build biochemical (Rojas et al., 2002), pharmacogenetic (Oliver et al., 2002), and microarray databases (Fellenberg et al., 2002). Conceptually driven forms of data representation greatly simplify the entry, tracking, and searching of data relevant to a particular field of study, especially when the relevant data objects allow significant variability in the internal structure of its components and new variations evolve over time (Noy and Klein, 2002). Ontological metadata structures lend themselves naturally to sharing of study metadata between lab-mates, colleagues, and centralized data repositories (for discussion, see Sujansky, 2001). Finally, this new vision for data exchange is becoming a dominant theme in the notion of a "Semantic Web" architecture—that is, the emerging philosophy that the Internet may evolve from an ill-formed collection of unstructured, human-readable information sources into a global knowledge-base with additional layers providing deeper and more meaningful relationships between data sources (Jenssen and Hovig, 2002; Kamel Boulos et al., 2002; Stevens et al., 2002).*

Onotology development software may be used to easily construct frameworks for data representation via a graphical user interface and used to design and implement knowledge-based systems. One particular example of a software environment for ontology development is that of the Protégé project (http://Protégé.stranford.edu) from the Stanford Medical Informatics group (http://camis.stanford.edu/). Protégé utilizes the Resource Description Framework (RDF), an XML-based language for representing information about resources in the World Wide Web. This framework is specifically intended for representing metadata about Internet resources, such as the authors, title, and latest modification date of a Web page or copyright and licensing information pertaining to Web-based content. The broader goal of RDF is to design a mechanism for describing resources that makes no assumptions about a particular application domain, while not constraining the semantics of any application domain. The definition of the mechanism should be domain neutral, yet the mechanism should be suitable for describing information about any domain (see http://www.w3.org/TR/rdf-primer/ for discussion). The concept of RDF can also be extended to represent information about things that can be *identified* on the Web, even when they cannot be directly *retrieved* on the Web. What is more, RDF provides a common means for expressing this information so it can be exchanged between software applications without loss of its inherent meaning. Schemes developed for the management of biomedical data using RDF to preserve the semantic content have been shown to be very successful (Boulos et al., 2002). Sophisticated knowledge acquisition tools may be constructed around these formats to acquire instances of an ontology and thereby encompass even broader domain information. Software applications also may be created to formalize data acquisition and the use of the knowledge base in solving an end-user problem employing appropriate problem-solving, expert-system, or decision-support methodologies.

---

*For additional information on the development of the Semantic Web, see http://www.w3.org/2001/sw and, in particular, the 1998 Road Map article by Tim Berners-Lee that outlines its basic design philosophy, http://www.w3.org/DesignIssues/Semantic.

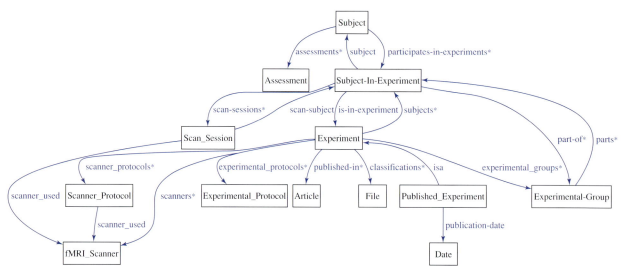

Figure 26.2 Study metadata are those measurements, values, and parameters that describe the various parts that comprise an fMRI experiment. These data often have complex interrelationships among each other which can be represented graphically. The representation of these relationships and their encapsulation in a structured data framework is referred to as an ontology. For a complete fMRI experiment, the ontology graph is very large (the subject of a manuscript in preparation) and contains portions pertaining to scanner, subject, and experimental protocol metadata. This figure presents a specific example of the ontological structure between *subjects* and *experiment*. The blue arrows indicate "has a" relationships between data nodes (i.e., "An experiment *has a* scanner") whereas the black arrows indicate "is a" relationships (i.e., "A published_experiment *is a* experiment"). Each node in the graph may have multiple synonyms associated with it, provided that each term is related to the other possible terms by a suitable transformation of the units associated with the values it contains. Ontologies are often constructed to be transparent to differences in the units of a particular field but accurately represent the relationships between the fields. For instance, units of magnetic field strength may be given in tesla or equivalently, by a scalar transform, in units of gauss. Coupled with raw, processed, and results level image data, these form a thorough representation of a neuroimaging study, permits others to reproduce published effects, and enables effective data sharing.

Using Protégé, The fMRI Data Center has begun actively working to bring the notion of ontological structures to bear on issues of encapsulating experimental metadata information from fMRI studies. An appropriately thorough ontology contains sections pertaining to scanner, subject, and experimental protocol metadata. These metadata often have intricate interrelationships among each other that can be drawn schematically using directed graphs. Figure 26.2 shows a graphical representation of how an ontology may be structured to contain the measurements, values, and parameters of an fMRI experiment. The complete, publicly available fMRI study ontology (http://www.fmridc.org/dmt/) being developed by the fMRIDC is, in fact, considerable in size; therefore, in this figure only a select portion of the ontological structure between *subjects* and *experiment* is presented. Java-based fMRI experiment management tools are being developed to concisely but more generally encode fMRI image data, such as header file information, designed to ensure that various units of measure for study metadata are used appropriately. In this way, it is easy to visualize the structured degree of connectivity among metadata objects. Accompanied by the image data from the critical image data processing nodes (see below), sharing contextual and conceptual neuroimaging metadata information in the form of ontological frameworks represents an effective approach for the archiving of the knowledge behind the fMRI experiment as well as the very structured datasets themselves. The fMRIDC invites input from the neuroimaging community on how to extend the ontology and its accompanying RDF framework to include more precise details about study protocols as well as to include modules for other imaging modalities (e.g., EEG, MEG, PET). Neuroimaging study management software projects, sim-

ilar to this one, are also underway at other centers (Jakobovits et al., 2000; Lober et al., 2001).

26.6 DESCRIPTION OF THE DATA PROCESSING CHAIN

The successful recreation and scrutiny of published activation results also require a complete description of the algorithms and methods that make up the chain of image data processing steps. It is often not enough to simply say that the data are mapped to Talairach space, for example, since many of the currently available software tools (e.g., SPM, AFNI, MNI-Minc) provide differing means for spatially warping brain data to a standardized template. Additionally, the programs used in volumetric registration, tissue classification, magnetic field inhomogeneity corrections, methods for statistical comparison, and so on, are continually undergoing development and enhancement, the results from which may change noticeably from revision to revision. Many such steps are often scripted or otherwise automated to maintain consistency in processing. However, the approaches chosen, algorithms applied, and the order in which they are performed can vary arbitrarily both within and between laboratories and can affect the obtained results. Indeed, some approaches are performed routinely and rarely questioned even though they may not be entirely appropriate. For instance, though initially proposed for sensible neurobiological reasons in PET imaging (Friston et al., 1990), global normalization by proportional scaling may, in fact, reduce sensitivity of results in studies using fMRI (Gavrilescu et al., 2002). On the other hand, simply because an analysis methodology happened to produce a particular area of activation commensurate with one's notion of brain function or dysfunction is not itself legitimate justification in favor of that methodology [see also Strother et al. (1995) for discussion].

Recent examination of processing methodologies has begun to measure the effects of differences in processing with an eye toward identifying unbiased criteria for evaluating data processing strategies (Lukic et al., 2002). In an examination of several processing strategies, Skudlarski and colleagues (1999) noted that, in general, the removal of intensity drift and the application of high-pass filtering applied to the voxel time-course level were largely beneficial to the outcome of an analysis, whereas temporally based normalization of the global image intensity, smoothing in the temporal domain, and low-pass spatial filtering did little to improve the statistical power of the results. This is consistent with conclusions from other examinations of the effects of spatial smoothing on statistical reproducibility in neuroimaging data (Van Horn et al., 1998). Most importantly, Skudlarski et al. observed that a careful design of task protocols was essential for maximizing statistical outcome. Nonparametric methods have also been applied to classifying the outcomes of data processing by considering an processing chain model in which all parameters of the preprocessing steps along with the final statistical model are treated as estimated model parameters (LaConte et al., 2003). The properties of the data-processing-dependent noise distribution, given a complete description of the steps in two processing chains, have also been considered (Strother et al., 2002). A further approach to examining data processing choices, not inconsistent with the quantitative argument offered above, suggests the evaluation of mutual information learning curves as a measure of model prediction error given a number of independent variables to examine bias/variance considerations in model estimation (Kjems et al., 2002). These novel assessment methods will undoubtedly achieve their greatest benefit for identifying the most efficient and effective chains of data processing in the brain imaging community when they can be applied to the broadest possible sample set of published data.

In order to examine the benefits and drawbacks in processing strategies from the peer-reviewed literature, there exists a strong need to accurately and comprehensively communicate analysis procedures to other investigators in the field, not only conceptually but also from a practical standpoint, allowing others to capitalize on the advances that often are difficult to glean from descriptions in journals [see Koslow (2000) for discussion]. One means for achieving consistent and comprehensively describing the steps performed in image data processing is through the use of graphical data pipelining software. These tools have been successfully applied to neuroimaging data (Matsumoto et al., 1999; Wei et al., 2002) and may have still broader scope for the coherent description of brain data processing. For instance, environments such as the LONI Pipeline (Toga et al., 2001) and the Rumba Tools (Bly et al., 2001) are specifically designed for constructing graphical descriptions of intricate brain image data processing chains.* They provide visually intuitive interfaces to data analysis while also allowing for interoperability between

*More information on the LONI Pipeline Project can be found at http://www.loni.ucla.edu/MAP/Software/Pipeline.html. The Rumba toolset is freely available for download from http://www.rumba.rutgers.edu.

diverse programs. Tools such as these allow researchers to share their methods of analysis with each other easily and provide a simple platform for distributing new programs, as well as program updates, to the desired community. Methods for evaluating and ranking the most efficient processing strategies such as those described above could be readily applied. Though, at present, data pipelining environments for neuroimaging are in their formative stages, they hold considerable promise for providing a means of thorough description of the algorithms applied to the data along the processing chain, permit software interoperability, and facilitate the fidelity of independent data reanalysis.

26.7 SHARING OF IMAGE DATA FROM CRITICAL NODES OF IMAGE PROCESSING

Different researchers, however, will have different needs from data at various stages of processing, depending on what their intentions with the data might involve. Study data archived after various stages of processing would be helpful for later researchers because they do not incur the cost of having to perform the (often computationally intensive) preprocessing operations themselves. Storing the data after every stage of the processing chain would likely not be necessary for enabling the preservation of the majority of study data information content. For instance, the amount of data from the single study of Buckner and colleagues (2000), available through the fMRI Data Center, exceeds 20 GB in size. In time, even typical fMRI studies may rival the size of the digital datasets from large biological datasets such as the Visible Human Project (Ackerman, 1999; Ackerman et al., 2001). In cases such as this, archiving brain image data from the outcome of every step of the processing chain of a study would surely be excessive from the point of view of storage and portability, greatly increasing the size of the disk storage required for just a single study.

One approach to the sharing and archiving of data from functional neuroimaging experiments is to preserve data at several key points in the processing chain: the raw fMRI data itself; the image data after all steps of processing (accompanied by descriptive information about that processing), just prior to its entry into the chosen statistical analysis engine; and the resultant statistical results maps. By archiving data from these nodes, researchers might choose to begin their secondary analysis de novo, at the level of the raw data; to start at the level of the processed data; or to use the statistical maps themselves in a meta-analytic framework. For example, if a researcher were to follow the same processing steps as the original authors on the raw data, then they should, in principle, obtain the same processed image data obtained by the original authors. Thus, the preprocessed data also provide added value to a raw dataset by serving as an importance cross-check that the steps of processing have been accurately followed. This opens the door for a contrasting the processing chain of one study with that of another (Strother et al., 2002). Additionally, differing statistical models can be explored and model averaging methods such as those recently proposed by Hansen and colleagues (2001) could be employed to identify a consensus of findings from across different analytic approaches. Software tools available for the surface mapping of statistical results (Van Essen et al., 2001) can be employed to examine the patterns of significant BOLD activity from across shared several datasets simultaneously. Therefore, the added importance of also including the preprocessed image data in an archive cannot be discounted because it clearly fills a central niche when coupled with the unprocessed dataset.

26.8 CONCLUSIONS

The archiving and indexing of data from studies of human brain function using fMRI has been a long-standing goal in the field of neuroimaging with a particular view toward the development of brain atlases (Roland and Zilles, 1994, 1996; Toga and Thompson, 2002). Neuroinformatics, computational neurobiological approaches, and efforts to design representative measurement systems for the brain will benefit from such databases. To be successful, however, these efforts require comprehensive image datasets as well as meta-information concerning the details of highly complex experiments as well as the widely variable methods used in the field for collecting functional and structural neuroimaging data (Beltrame and Koslow, 1999; Koslow, 2000).

Recent heuristic rules for the sharing of brain imaging data have been put forward in the neuroscientific literature. The advocates of these rules have suggested that the more the data have been processed, the greater value those data have for inclusion in study data archives or for use in meta-analytic assessment. Accordingly, greater study value is supposedly gained from the processing and statistical interpretation provided by the authors of the study than is present in the raw data itself. However, if the information content of the image time course data remains the same or is reduced by a series of processing steps, it is unclear as to the source of any added value for the sharing of only tabular representations of the results. Information-theoretic approaches have been previously explored in the context of neuroimaging data and shown to be valuable for evaluating the adverse effects of spatio-temporal filtering of data and any resulting loss of localization power (Descombes et al., 1998a,b). These studies give emphasis to the point that useful information contained within the unprocessed image data themselves might be implied from a tabular representation of the local statistical extrema. The present chapter has specifically avoided consideration of the monetary cost of fMRI data collection and processing that is incurred by the investigator. Monetary costs of scanning will vary across research centers and be dependent upon factors such as the research demands on a particular scanner or the additional costs associated with performing studies of patient populations.

It may be argued that the potential worth of a shared dataset needs to be measured by the potential information it contains about the data in its original raw form. Contrary to some proposed guidelines, this worth must therefore be greatest when those data are in its rawest form. Indeed, since the information content that a dataset contains about the original data is potentially reduced greatly through processing, it stands to reason that the scientific value of the data must influenced similarly. Even if only a portion of the data processing chain can be argued to form a Markov process (e.g., operations involving image volume resizing, re-slicing, spatial-temporal filtering, etc.), there remains the potential for, perhaps dramatic, information loss between the two datasets. In the case of a small set of heavily processed fMRI test statistics and Talairach coordinates, the amount of information contained in those results about the original time-series data will likely be considerably reduced. Therefore, as the information content of a shared data set is reduced, so, too, may be the potential worth of an archive of that data as a resource for promoting unique processing methodologies. This basic notion applies equally to the neuroimaging data as it does any accompanying behavioral or diagnostic data collected as part of the imaging study. Summary descriptive or inferential statistics are, in many instances, no substitute for the raw data when they can be made available to others for additional analyses.

In contradistinction to these heuristic approaches, greater scientific value of an fMRI dataset is maintained through the sharing of functional imaging data from across the processing chain including the raw, original data. The notion of the mutual information shared between levels of processing provides a strong argument in favor of sharing/archiving data in their most unprocessed form. Coupled with the data at later key stages of processing as well as metadata description provides the largest, reasonable amount of information that permits others to gain access to the processing chain dependent upon their purposes. Accurate metadata description is essential, and ontologies may prove particularly useful in this context as they go beyond being a simple list of study parameters and measurements to describing the conceptual relationships among the various data objects. In this way, the essence of the neuroimaging study may be more thoroughly encapsulated in a form that specifically permits data exchange. More broadly, maximizing the overall information content of shared and archived neuroimaging data impacts directly the quality and generalizabilty of large-scale re-analyses of brain function (e.g., Lloyd, 2002), may aid in making probabilistic brain atlases more representative of the natural anatomical variation among differing demographic and patient groups (Thompson et al., 2000; Toga and Thompson, 2001), and can affect a generalized improvement in how study data should be managed in theory and is communicated between investigators.

26.9 DATA SHARING MODELS AND THE POTENTIAL BENEFITS FOR BRAIN SCIENCE

Simply sharing and storing complete datasets are not enough, however, to embody the value contained in any particular study. That is to say, sharing a greater quantity of image or metadata does not necessarily mean that this study is of more value than another study. Methods exist for empirically assessing study quality on the basis of the effects of processing (Muley et al., 2001; Strother et al., 2002), although worth will most assuredly be afforded to a neuroimaging dataset when researchers have been permitted to independently examine, evaluate, and critique the information it contains. For instance, cross-institutional investigations into the reproducibility of fMRI findings (Casey et al., 1998) could further explore the characteristics of the BOLD response during working memory tasks beyond the comparison of activation results maps alone. Moreover, if novel ideas concerning brain function result and are contributed back to the literature, then scientific discourse has taken place and that should be strongly encouraged. In this way, the data may be shown to stand out dramatically relative to other such studies as being of seminal importance to the characterization and understanding of neurocognitive function. Finally, the true value of a study is realized when the raw data have been deemed to have so much worth as to have found a place in the education and training of new brain researchers.

This chapter has been concerned with discussing the scientific costs of processing in relation to open data sharing of complete study information. The monetary cost of fMRI data collection and processing that is incurred by the investigator has been specifically avoided. On the other hand, neuroimaging data have a scientific worth beyond the financial and personal expense that went into their collection. This scientific worth and potential is maximized when the study data are shared/archived as completely as possible and made available to the widest possible audience. True value may be achieved when the complete study dataset is utilized and examined by other researchers permitted to apply their own preprocessing techniques, statistical methodologies, or meta-analytic approaches, thereby arriving at their own, possibly divergent, conclusions. This serves to benefit the scientific community and drive forward scientific inquiry into fundamental neural processes in ways that small sets of heavily processed data and statistical results cannot.

ACKNOWLEDGMENTS

In particular, the authors wish to acknowledge the efforts of Jeffrey Woodward, fMRIDC Manager of Systems and Systems Integration. They also wish to thank Professor Karl Friston and Professor Scott T. Grafton for early comments, as well as two anonymous reviewers for helpful reviews on an earlier version of this chapter. Ms. Wendy Starr receives thanks for administrative assistance. The fMRIDC is supported by the National Science Foundation, The William M. Keck Foundation, and the National Institute of Mental Health and is a Sun Microsystems Center of Excellence for Neuroscience.

References

Ackerman, M. J. (1999). The Visible Human Project: A resource for education. *Acad. Med.* **74**(6), 667–670.

Ackerman, M. J., Yoo, T., and Jenkins, D. (2001). From data to knowledge—the Visible Human Project continues. *Medinfo* **10**(Pt 2), 887–890.

Beltrame, F., and Koslow, S. H. (1999). Neuroinformatics as a megascience issue. *IEEE Trans. Inf. Technol. Biomed.* **3**(3), 239–240.

Bly, B. M., Rebbechi, D., Grasso, G., and Hanson, S.J. (2001). *A Framework for Software Interoperability in Brain Imaging Data Analysis. Organization for Human Brain Mapping Annual Meeting, Brighton, England.* Academic Press, San Diego, CA.

Boulos, M. N., Roudsari, A. V., and Carson, E. R. (2002). Towards a semantic medical Web: HealthCyberMap's tool for building an RDF metadata base of health information resources based on the Qualified Dublin Core Metadata Set. *Med. Sci. Monit.* **8**(7), MT124–MT136.

Buckner, R. L., Snyder, A. Z., Sanders, A. L., Raichle, M. E., and Morris, J. C. (2000). Functional brain imaging of young, nondemented, and demented older adults. *J. Cogn. Neurosci.* **12**(Suppl. 2), 24–34.

Casey, B. J., Cohen, J. D., O'Craven, K., Davidson, R. J., Irwin, W., Nelson, C. A., Noll, D. C., Hu, X., Lowe, M. J., Rosen, B. R., Truwitt, C. L., and Turski, P. A. (1998). Reproducibility of fMRI results across four institutions using a spatial working memory task. *NeuroImage* **8**(3), 249–261.

Cover, T. M., and Thomas, J. A. (1991). *Elements of Information Theory.* Wiley, New York.

Descombes, X., Kruggel, F., and von Cramon, D. Y. (1998a). fMRI signal restoration using a spatio-temporal Markov Random Field preserving transitions. *NeuroImage* **8**(4), 340–349.

Descombes, X., Kruggel, F., and von Cramon, D. Y. (1998b). Spatio-temporal fMRI analysis using Markov random fields. *IEEE Trans. Med. Imaging* **17**(6), 1028–1039.

Fellenberg, K., Hauser, N. C., Brors, B., Hoheisel, J. D., and Vingron, M. (2002). Microarray data warehouse allowing for inclusion of experiment annotations in statistical analysis. *Bioinformatics* **18**(3), 423–433.

Fox, P. T., and Lancaster, J. (2002). Mapping context and content: the BrainMap model. *Nat. Rev. Neurosci.* **3**(April), 319–321.

Fox, P. T., Perlmutter, J. S., and Raichle, M. E. (1985). A stereotactic method of anatomical localization for positron emission tomography. *J. Comput.-Assisted Tomogr.* **9**(1), 141–153.

Friston, K. J., Frith, C. D., Liddle, P. F., Dolan, R. J., Lammertsma, A. A., and Frackowiak, R. S. J. (1990). The relationship between global and local changes in PET scans. *J. Cereb. Blood Flow Metab.* **10**, 458–466.

Friston, K. J., Frith, C. D., Liddle, P. F., and Frackowiak, R. S. J. (1991). Plastic transformations of PET images. *J. Comput.-Assisted Tomogr.* **15**, 634–639.

Friston, K. J., Josephs, O., Zarahn, E., Holmes, A. P., Rouquette, S., and Poline, J. (2000). To smooth or not to smooth? Bias and efficiency in fMRI time-series analysis. *NeuroImage* **12**(2), 196–208.

Gavrilescu, M., Shaw, M. E., Stuart, G. W., Eckersley, P., Svalbe, I. D., and Egan, G. F. (2002). Simulation of the effects of global normalization procedures in functional MRI. *NeuroImage* **17**(2), 532–542.

Gifford, D. K. (2001). Blazing pathways through genetic mountains. *Science* **293**, 2049–2051.

Governing Council of the Organization for Human Brain Mapping (2001). Neuroimaging databases. *Science* **292**, 1673–1676.

Hansen, L. K., Nielsen, F. A., Strother, S. C., and Lange, N. (2001). Consensus inference in neuroimaging. *NeuroImage* **13**(6 Pt 1), 1212–1218.

Ishai, A., Ungerleider, L. G., Martin, A., and Haxby, J. V. (2000). The representation of objects in the human occipital and temporal cortex. *J. Cogn. Neurosci.* **12**(Suppl. 2), 35–51.

Jakobovits, R. M., Modayur, B., and Brinkley, J. F. (1996). A web-based repository manager for brain mapping data. *Proc. AMIA Annu. Fall Symp.*, pp. 309–313.

Jakobovits, R. M., Soderland, S. G., Taira, R. K., and Brinkley, J. F. (2000). Requirements of a Web-based experiment management system. *Proc. AMIA Symp.*, pp. 374–378.

Jakobovits, R. M., Brinkley, J. F., Rosse, C., and Weinberger, E. (2001). Enabling clinicians, researchers, and educators to build custom web-based biomedical information systems. *Proc. AMIA Symp.*, pp. 279–283.

Jenssen, T. K., and Hovig, E. (2002). The semantic web and biology. *Drug Discovery Today* **7**(19), 992.

Kable, J. W., Lease-Spellmeyer, J., and Chatterjee, A. (2002). Neural substrates of action event knowledge. *J. Cogn. Neurosci.* **14**(5), 795–805.

Kamel Boulos, M. N., Roudsari, A. V., and Carson, E. R. (2002). A dynamic problem to knowledge linking Semantic Web service based on clinical codes. *Med. Inf. Internet Med.* **27**(3), 127–137.

Kjems, U., Hansen, L. K., Anderson, J., Frutiger, S., Muley, S., Sidtis, J., Rottenberg, D., and Strother, S. C. (2002). The quantitative evaluation of functional neuroimaging experiments: Mutual information learning curves. *NeuroImage* **15**(4), 772–786.

Koslow, S. H. (2000). Should the neuroscience community make a paradigm shift to sharing primary data? *Nat. Neurosci.* **2**, 863–865.

Koslow, S. H. (2002). Opinion: Sharing primary data: A threat or asset to discovery? *Nat. Rev. Neurosci.* **3**(4), 311–313.

Kotter, R. (2001). Neuroscience databases: Tools for exploring brain structure–function relationships. *Philos. Trans. R. Soc. London, Ser. B* **356**, 1111–1120.

LaConte, S., Anderson, J., Muley, S., Ashe, J., Frutiger, S., Rehm, K., Hansen, L. K., Yacoub, E., Hu, X., Rottenberg, D., and Strother, S. (2003). The evaluation of preprocessing choices in single-subject

BOLD fMRI using NPAIRS performance metrics. *NeuroImage* **18**(1), 10–27.

Lloyd, D. (2002). Functional MRI and the study of human consciousness. *J. Cogn. Neurosci.* **14**(6), 818–831.

Lober, W. B., Trigg, L. J., Bliss, D., and Brinkley, J. M. (2001). IML: An image markup language. *Proc. AMIA Symp.*, pp. 403–407.

Lukic, A. S., Wernick, M. N., and Strother, S. C. (2002). An evaluation of methods for detecting brain activations from functional neuroimages. *Artif. Intell. Med.* **25**(1), 69–88.

Magee, J., Gordon, J. I., and Whelan, A. (2001). Bringing the human genome and the revolution in bioinformatics to the medical school classroom: A case report from Washington University School of Medicine. *Acad. Med.* **76**(8), 852–855.

Matsumoto, S., Asato, R., and Konishi, J. (1999). A fast way to visualize the brain surface with volume rendering of MRI data. *J. Digit. Imaging* **12**(4), 185–190.

Mazziotta, J., Toga, A. W., Evans, A., Fox, P., Lancaster, J., Ziles, K., Woods, R. P., Paus, T., Simpson, G., Pike, B., Holmes, C., Collins, L., Thompson, P. M., MacDonald, D., Iacoboni, M., Schormann, T., Amunts, K., Palomero-Gallagher, N., Geyer, S., Parsons, L. M., Narr, K., Kabani, N., Le Goualher, G., Boomsma, D., Cannon, T., Kawashima, R., and Mazoyer, B. (2001). A probabilistic atlas and reference system for the human brain: International Consortium for Brain Mapping (ICBM). *Philos. Trans. R. Soc. London, Ser. B* **356**, 1293–1322.

Mechelli, A., Gorno-Tempini, M. L., and Price, C. J. (2003). Neuroimaging studies of word and pseudoword reading: Consistencies, inconsistencies, and limitations. *J. Cogn. Neurosci.* **15**(2), 260–271.

Muley, S. A., Strother, S. C., Ashe, J., Frutiger, S. A., Anderson, J. R., Sidtis, J. J., and Rottenberg, D. A. (2001). Effects of changes in experimental design on PET studies of isometric force. *NeuroImage* **13**(1), 185–195.

Nielsen, F. A., and Hansen, L. K. (2002). Modeling of activation data in the BrainMap database: Detection of outliers. *Hum. Brain Mapp.* **15**(3), 146–156.

Noy, N.F. and Klein, M. (2002). Ontology evolution: Not the same as schema evolution. Stanford Medical Informatics Technical Report SMI-2002-0926.

Oliver, D. E., Rubin, D. L., Stuart, J. M., Hewett, M., Klein, T. E., and Altman, R. B. (2002). Ontology development for a pharmacogenetics knowledge base. *Pac. Symp. Biocomput.*, pp. 65–76.

Paradis, A. L., Van de Moortele, P. F., Le Bihan, D., and Poline, J. B. (2001). Slice acquisition order and blood oxygenation level dependent frequency content: An event-related functional magnetic resonance imaging study. *Magma* **13**(2), 91–100.

Rojas, I., Bernardi, L., Ratsch, E., Kania, R., Wittig, U., and Saric, J. (2002). A database system for the analysis of biochemical pathways. *In Silico Biol.* **2**(2), 75–86.

Roland, P. E., and Zilles, K. (1994). Brain atlases–A new research tool. *Trends Neurosci.* **17**(11), 458-467.

Roland, P. E., and Zilles, K. (1996). The developing European computerized human brain database for all imaging modalities. *NeuroImage* **4**(3 Pt 2), S39–S47.

Rypma, B., Berger, J. S., and D'Esposito, M. (2002). The influence of working-memory demand and subject performance on prefrontal cortical activity. *J. Cogn. Neurosci.* **14**(5), 721–731.

Skudlarski, P., Constable, R. T., and Gore, J. C. (1999). ROC analysis of statistical methods used in functional MRI: Individual subjects. *NeuroImage* **9**(3), 311–329.

Stevens, R., Goble, C., Horrocks, I., and Bechhofer, S. (2002). OILing the way to machine understandable bioinformatics resources. *IEEE Trans. Inf. Technol. Biomed.* **6**(2), 129–134.

Strother, S. C., Kanno, I., and Rottenberg, D. A. (1995). Principal component analysis, variance partitioning, and "Functional Connectivity." *J. Cereb. Blood Flow Metab.* **15**, 353–360.

Strother, S. C., Anderson, J., Hansen, L. K., Kjems, U., Kustra, R., Sidtis, J., Frutiger, S., Muley, S., LaConte, S., and Rottenberg, D. (2002). The quantitative evaluation of functional neuroimaging experiments: The NPAIRS data analysis framework. *NeuroImage* **15**(4), 747–771.

Sujansky, W. (2001). Heterogeneous database integration in biomedicine. *J. Biomed. Inf.* **34**(4), 285–298.

Thompson, P. M., Woods, R. P., Mega, M. S., and Toga, A. W. (2000). Mathematical/computational challenges in creating deformable and probabilistic atlases of the human brain. *Hum. Brain Mapp.* **9**(2), 81–92.

Toga, A. W. (2002). Neuroimaging databases: The good, the bad, and the ugly. *Nat. Rev. Neurosci.* **3**(April), 302–309.

Toga, A. W., and Thompson, P. M. (2001). Maps of the brain. *Anat. Rec.* **265**(2), 37–53.

Toga, A. W., and Thompson, P. M. (2002). New approaches in brain morphometry. *Am. J. Geriatr. Psychiatry* **10**(1), 13–23.

Toga, A.W., Rex, D. E., and Ma, J. (2001). A Graphical Interoperable Processing Pipeline. *Organization for Human Brain Mapping Annual Meeting, Brighton, England.* Academic Press, San Diego, CA.

Toni, I., Shah, N. J., Fink, G. R., Thoenissen, D., Passingham, R. E., and Zilles, K. (2002). Multiple movement representations in the human brain: An event-related fMRI study. *J. Cogn. Neurosci.* **14**(5), 769–784.

Van Essen, D. C., Drury, H. A., Dickson, J., Harwell, J., Hanlon, D., and Anderson, C. H. (2001). An integrated software suite for surface-based analyses of cerebral cortex. *J. Am. Med. Inf. Assoc.* **8**(5), 443–459.

Van Horn, J. D., and Gazzaniga, M. S. (2002). Databasing fMRI studies —Toward a 'Discovery Science' of brain function. *Nat. Rev. Neurosci.* **3**(4), 314–318.

Van Horn, J. D., Ellmore, T. M., Esposito, G., and Berman, K. F. (1998). Mapping voxel-based statistical power on parametric images. *NeuroImage* **7**(2), 97–107.

Van Horn, J. D., Grethe, J. S., Kostelec, P., Woodward, J. B., Aslam, J. A., Rus, D., Rockmore, D., and Gazzaniga, M. S. (2001). The functional magnetic resonance imaging data center (fMRIDC): The challenges and rewards of large-scale databasing of neuroimaging studies. *Philos. Trans. R. Soc. London, Ser. B* **356**, 1323–1339.

Van Horn, J. D., Woodward, J. B., Simonds, G., Vance, B., Grethe, J. S., Montague, M., Aslam, J. A., Rus, D., Rockmore, D., and Gazzaniga, M. S. (2002). The fMRI Data Center: Software tools for neuroimaging data management, inspection, and sharing. In A *Practical Guide to Neuroscience Databases and Associated Tools* (R. Kotter, ed.), Kluwer Academic Publishers, Dordrecht, The Netherlands pp. 221–235.

Wei, X., Warfield, S. K., Zou, K. H., Wu, Y., Li, X., Guimond, A., Mugler, J. P., 3rd, Benson, R. R., Wolfson, L., Weiner, H. L., and Guttmann, C. R. (2002). Quantitative analysis of MRI signal abnormalities of brain white matter with high reproducibility and accuracy. *J. Magn. Reson. Imaging* **15**(2), 203–209.

Woods, R. P., Grafton, S. T., Holmes, C. J., Cherry, S. R., and Mazziotta, J. C. (1998a). Automated image registration: I. General methods and intrasubject, intramodality validation. *J. Comput.-Assisted Tomogr.* **22**(1), 139–152.

Woods, R. P., Grafton, S. T., Watson, J. D., Sicotte, N. L., and Mazziotta, J. C. (1998b). Automated image registration: II. Intersubject validation of linear and nonlinear models. *J. Comput.-Assisted Tomogr.* **22**(1), 153–165.

Woods, R. P., Dapretto, M., Sicotte, N. L., Toga, A. W., and Mazziotta, J. C. (1999). Creation and use of a Talairach-compatible atlas for accurate, automated, nonlinear intersubject registration, and analysis of functional imaging data. *Hum. Brain Mapp.* **8**(2–3), 73–79.

Yarmush, M. L., and Jayaraman, A. (2002). Advances in proteomic technologies. *Annu. Rev. Biomed. Eng.* **4**, 349–373.

Index

Databasing the Brain. Edited by Stephen H. Koslow and Shankar Subramaniam
ISBN 0-471-30921-4 © 2005 John Wiley & Sons, Inc.